Fault-finding in Computers and Digital Circuits

Measurement and Testing

Wolfgang Matthes

Fault-finding in Computers and Digital Circuits

Measurement and Testing

Elektor Electronics (Publishing)
Dorchester, England

Elektor Electronics (Publishing)
P.O. Box 1414
Dorchester
England DT2 8YH

British Library Cataloguing in Publication Data
A catalogue record for this book is available from the British Library.

ISBN 0-905705-60-2

Translation and interior layout: Kenneth Cox Technical Translations

First published in the United Kingdom 1998

Printed in the Netherlands by Wilco, Amersfoort.

Contents

1 Introduction

This book is concerned with the craft of fault-finding, measurement and testing in modern electronic systems and devices, which are primary digital. The typical 'IBM-compatible' personal computer (PC) can serve as an example of these devices. (Millions of such PCs can be found throughout the world, and they are much more inclined to develop faults than many other technical products. You are almost forced to fend for yourself if you want to assemble, maintain and modernize a PC system at an acceptable cost.) Of course, the basic techniques of fault-finding, measurement and testing can be applied to practically all electronic systems, devices and assemblies.

The problem area: fault-finding, measurement and testing. From a practical point of view, there are three situations in which these factors are important:

1. *Product development:* in the product development stage, the relevant issues are (1) test- and service-oriented design of the device or assembly, (2) getting the prototype device to work in the first place and (3) demonstrating that the hardware that has been developed is ready for production (formal handover and acceptance testing, certification and so on).
2. *Product assembly:* the product, once developed, must be assembled and delivered at a high level of quality. To assure this, control procedures and measurements are required at the various assembly stages.
3. *Product servicing:* if an item no longer functions as expected, it must be serviced and restored to proper operation. This requires the cause of the fault to first be determined.

Numerous procedures, presentation methods and intellectual approaches are available to deal with these problems. They range in level from basic research (which in turn is based on several specialized areas of mathematics), through the models and techniques used by professional engineers and technicians, and finally to the innumerable self-help tips that can be found in the popular periodicals and electronics books as well as in the 'bestsellers' in the realm of PC literature.

The personal computer as a preferred example. Modern personal computers are complex hardware/software systems, regardless of whether they are used in isolation or as part of a network. The technician is faced with the task of configuring such systems and maintaining them in good working order. However, the functional complexity which is represented by these innocent-looking devices is comparable to that of an earlier generation of mainframe computers. Previously, the system under investigation (the computer) was much more expensive than even the highest-quality test equipment (such as wide-bandwidth oscilloscopes and logic analyzers). Nowadays, high-performance test equipment is usually significantly more expensive than the devices and assemblies found in the PC market. Nonetheless, the critical objective remains the same: to satisfy the boss, the user or the customer in an acceptable period of time and at an acceptable cost, or – in the case of personal use – to keep your by-no-means inexpensive acquisitions 'alive and well'.

The otherwise obvious expedient of simply discarding defective (or merely suspect) hardware and replacing it with new, is thus not a real option.

Development trends. The technology that has been introduced into the market by the modern PC has also made inroads in many other areas. The trend can be expected to continue and to increase. This relates both to industrial products (with regard to microcontrollers, microprocessor systems and industrial PCs) and to the consumer market (in particular, multimedia). The main features of this technology are presented in Table 1.1.

The problems described in Table 1.1 are by no means insoluble. However, the typical 'tried-and-true' skills and knowledge of the hobbyist are hardly adequate to deal with them. Instead, it is necessary to appropriate the tricks and techniques that are used in industrial laboratories and test rooms, and which are also used by the true elite in the service industry.

There is an abundance of literature on the subject of basic measurement technology. By contrast, the whole area of tricks, workarounds, and exceedingly subtle relationships – 'gotchas' in the jargon – is not at all systematically documented. Sources for such information include (among others) device manufacturers' application notes and courses given by firms for their employees or customers. However, there is no doubt that the most important means for a beginner to gain such knowledge is to gradually pick it up from more experienced colleagues. Of course, you can also always learn things the hard way!

In this context, this book is primarily intended for the 'committed' PC user, for the person who is typically responsible for looking after the firm's PCs, for the beginner in the electronics trade and the electronics hobbyist. It is also suitable for persons who are taking job-related training in electronics and for students of electronics technology, for whom it provides a practically-oriented 'working' book to supplement the regular course material relating to electronics and measurement technology. It should prove helpful in the following areas:

- learning the practical aspects of servicing personal computers and other modern, primarily digital devices and assemblies;
- developing a working familiarity with the trade;
- developing and testing circuits and system configurations that can still be considered to be 'comprehensible'. (In this case, 'comprehensible' means (1) not all too complex in structure and not too big, and (2) constructed as one-offs or in a small quantity. This relates, among other things, to DIY projects, practical training projects and examination work, and the large majority of prototype development projects in industry.)

The structure of this book

Chapter 2 addresses the fundamentals of fault-finding. Here we deal with basic concepts and an introduction to the technique of logical deduction, which is so important for truly professional fault-finding.

Characteristic features	Description	Testing and measurement problems
functional complexity	Mutual interaction of hardware and software; complex operating principles, instruction sets, interface protocols and so on; networking.	It is difficult to understand how the devices and systems under investigation actually function; interactions are difficult to understand and you must master a collection of skills and tricks in a variety of different areas (software, digital technology, analogue technology, power supply technology, opto-electronics, EMC and so on).
high degree of integration	Many complicated functions are implemented in the internal structures of highly integrated circuits.	It is almost impossible to observe processes with simple means ("we can't tell what's happening"); checking or testing individual components is either impossible or highly laborious; attaching test leads to a component or replacing a component is often physically difficult (miniaturized components with many leads).
high working frequencies, unconventional signal levels, energy conservation measures	Clock frequencies of several hundred MHz; a variety of operating voltages (in part component-specific, such as with processor chips); special requirements placed on power supplies, such as rapid switching between normal and low-energy operating modes; allowing devices to be connected or disconnected with power applied (Hot Insertion).	Conventional (low-cost) service equipment is often no longer usable, simple rules of thumb and work-arounds are no longer applicable, and you need a profound knowledge of electronics technology to be able to find your way around (digital circuits no longer use 5 V alone, pulses are not just chopped DC levels, etc.).
mass market with a rapid succession of new products		Detailed technical documents and special replacement parts are difficult to obtain; service support is often inadequate or totally unavailable.

Table 1.1 *The current level of technology and its consequences for fault-finding, measurement and testing.*

Chapter 3 deals briefly with the testing and error-detection provisions that are found in modern computer hardware.

Chapter 4 is an introduction to test and measurement techniques, which in effect provides basic instructions (plus a certain amount of theoretical understanding) for working with various types of test and measurement equipment (and related aids) that have proven their worth in practical use.

Chapter 5 deals in more detail with the techniques of measurement and fault-finding.

The **Appendices** address the following:

- summary information regarding measures for working safety,
- selected problems in electronics and digital technology,
- sources of additional information (including a bibliography).

Related literature

This book is Volume 1 of a series of books which is planned to consist of three volumes. These are all separately available and can be used independently.

Volume 2 deals with test and verification software, and Volume 3 deals with home-made test equipment. You are thus referred to Volume 2 for issues which are characterized by the close interdependence of hardware and software, and to Volume 3 for DIY projects, to the extent that they go beyond very simple constructions.

In addition, refer to Appendices 7 and 8 for other sources of information.

2 Fault-finding fundamentals

2.1 Basic terms and concepts

Verification. Verification means proving whether or not a particular equipment item, functional assembly or the like is in good working order. The result of verification is a yes/no ('go/no-go') decision.

Localization. Localization means more-or-less precisely determining the source of a fault. Usually this involves figuring out which hardware items (assemblies or components) can be swapped out to try to remedy the fault.

Field-Replaceable Unit (FRU). A FRU is the smallest item (subsystem, device, assembly or component) that can be exchanged by means of simple disassembly and reassembly procedures (for example, removing/replacing screws and unplugging/reattaching connectors). 'On-site' repairs are normally performed by replacing devices or assemblies, with the occasional replacement of a component that can be removed without unsoldering (for example, a memory module (SIMM)). 'Shop' service, by contrast, also involves the unsoldering and replacement of individual electronic components and the replacement of individual mechanical parts.

Checking and measurement. Checking and measurement consist of the comparison of observed values or behaviour to expected (specified) values or behaviour. Measurement is a quantitative form of checking which yields numeric results.

Test object (Device Under Test, DUT). The test object is the system, equipment, assembly or component that is actually being checked or tested.

Testing. Testing means verifying whether a suspect item (test object) functions correctly, or proving whether a particular fault hypothesis is true or false.

Verification test. A verification test should yield a yes/no decision with regard to the correct functioning of the test object.

Localization test. A localization test should yield results that are useful for determining the source of the fault or identifying which FRU should be replaced.

Endurance test. An endurance test is a verification or localization test that is performed for an extended period of time. The objectives of endurance testing are (1) to detect intermittent faults, (2) to prove that a given fault has successfully been remedied or that the test object is fault-free and (3) to detect early failures (burn-in testing).

Self-test. The term 'self-test' is used for hardware testing by means of built-in test provisions. This may involve the use of resident programs or microprograms, as well as special hardware that can independently control and monitor test processes.

Diagnosis. 'Diagnosis' is a general term that is used (just as in the medical context) to refer to the determination of the exact cause of a fault.

Note: the terms *diagnosis, testing, checking* and *fault-finding* are often used interchangeably in the literature.

Permanent or 'hard' fault. A permanent or hard fault is always present and always occurs under a particular set of operating conditions.

Intermittent fault (transient fault, time-dependent error or malfunction). An intermittent fault only occurs at some particular time and is not always present or repeatable; that is, with a given set of operating conditions it sometimes occurs and sometimes does not.

Suspicion. The degree to which the term 'fault suspicion' is used in a fault-finding guide or diagnostic software is an indication of the extent of its author's 'professionalism'. Since no fault-localization procedure is foolproof, one avoids making absolute pronouncements regarding the cause or location of a fault. Something that in a given situation may well be defective is thus by no means to be referred to as such, but rather at most to be referred to as 'suspect'.

Complexity theory. Complexity theory is a field of mathematics. It is concerned with (among other things) the investigation, in a quite general and abstract sense, of how much processing capability or processing time, and/or how much storage capacity, is required just to execute particular algorithms (independent of considerations such as an actual program, an actual processor and so on). A particularly interesting question is how the demands on processing time or storage capacity increase with the dimension of the problem. Algorithms that are typically studied deal with solving systems of equations, searching data bases and the like. It is evident that in such cases there are 'smaller' and 'larger' problems (a set of equations can have a few or many unknowns, a data base can have a few or many fields).

However, the conclusions that can be drawn from complexity theory are also of decisive importance with regard to checkout procedures and test algorithms. For this reason, it is desirable for us to briefly outline the basic concepts of the theory. Naturally, we cannot in this context provide mathematically exact definitions; instead, we limit ourselves to reasonable explanations of the terms that are commonly encountered.

- Linear complexity: $O(n)$. This means that (for example) if the dimension n of the problem is doubled, then the required processing time is also doubled.
- Polynomial complexity: $O(n^p)$, where p is a fixed value. Typical relationships are $O(n^2)$ (quadratic complexity), $O(n^3)$ and so on. Quadratic complexity means (for example) that doubling the problem dimension n causes the processing time to quadruple ($2^2 = 4$) and tripling the problem dimension multiplies the processing time by nine ($2^3 = 9$). If the relationship is $O(n^3)$, then the required processing time increases as the cube of the problem dimension (doubling the dimension means nine times the processing time, tripling the dimension means 27 times the processing time and so on).

- Exponential complexity: $O(p^n)$. With exponential complexity, doubling the problem dimension n (for example) increases the processing time by a factor of p^2 and tripling the problem dimension increases the processing time by a factor of p^3. Suppose, for example, that a problem of a particular dimension requires 10 minutes of processing time. Doubling the dimension causes the processing time to increase to 100 minutes, tripling the dimension results in a processing time of 1000 minutes (roughly 17 hours) and so on.

Note: the theory does not give an exact expression for the rate of growth, but only an order of magnitude relationship or a general principle that describes how the dimension of the problem affects some factor, such as the processing time.

NP-completeness. This term must be taken as given. It means that in the worst case the required processing time increases in an exponential manner, and in the best case it has a polynomial relationship to the problem dimension.

The important factor is that in many cases the tasks associated with checkout and testing procedures are fundamentally NP-complete. The problem dimension is determined by (among other things) the number of inputs and outputs of the circuit being tested, the number of logic gates or memory cells and so on.

Notes:

1. There is ultimately nothing that can be done to counter the consequences of exponential complexity. Of course, the required processing time can be reduced by increasing the processing power. Nonetheless, the time required tends to increase faster than any possible increase in processing capacity, so that after a certain problem dimension is reached a purely computational approach is no longer practical. (*For example:* in the previously-mentioned exponential case, the processing capacity must increase by a factor of 1000 if the processing time is to remain the same when the problem becomes three times as large.) Possible remedies are (1) to divide the problem into several smaller problems (which is not always possible) and (2) to forgo an exact solution to the problem (which means that you must be satisfied with an approximate solution).
2. Be careful not to confuse the concepts 'complex' and 'complicated'. There are highly complex problems that at first glance do not appear to be at all complicated (for example, proving that a particular sequence of variables consisting of ones and zeros satisfies a given boolean equation).

2.2 Fault origins, failure mechanisms and fault models

2.2.1 Design faults and faults that occur in normal operation

Faults arising from the design and development process

Let's consider a typical situation in the product development process: the first model of a device (laboratory model or engineering prototype) has just been assembled and is ready to be put 'into action' (initial operation). At this stage the device has basically never before functioned, so that literally anything is possible. The numerous types of faults that can occur in the initial operational stage can be categorized as described below.

Specification faults. The design specification, or in other words the understanding of the problem which is to be addressed, is incorrect. Such faults, to the extent that they cannot be relatively easily corrected, are difficult to find at the beginning of the development process.

Common causes:

- Not all requirements have actually been taken into account.
- Complex (or even dangerous) side effects and interrelationships have not been recognized, or the specification is overly simple due to an insufficient understanding of the problem.
- The necessity of certain limitations has not been recognized. (Depending on one's point of view, either the specifications are ambiguous and thus allow for multiple interpretations of what was actually intended, or the designer has been allowed too much freedom.)
- The specification is too complicated or too sophisticated.
- Normal human errors and oversights.

Preventive measures:

- Never start a development without first having a carefully worked-out specification ("what should actually happen here?").
- Critically review the specification, as follows:
  a) Shortly after the specification has been fully worked out, read completely through it once again.
  b) Experience shows that attempting to explain your objectives to others (colleagues and acquaintances) will almost always bring to light specification errors (the 'aha!' effect)
  c) Arrange for a formal, systematic review of the specification (acceptance review, structured walk-through) performed by peers or acquaintances (plus external consultants, in the commercial realm).

d) Do a self-critical review of the specification based on the state of the technology: (1) do not dispense with proven ideas or disregard what you have learned from experience without a truly compelling reason; (2) beware of 'over-ambitious' specifications. (*For example:* if a high-quality piece of test equipment costs roughly £10,000 and you develop a specification for a DIY device with similar technical capabilities and a materials cost of £100 to £200, you must recognize that this will never work.)

e) Although formal correctness verification and simulation are better than nothing, in principle there is nothing that can help in the face of an erroneous conception of the problem. You must develop for yourself a clear and unambiguous idea of the problem that is to be solved, and you should select the most suitable method of presenting your ideas in each case, without regard to prevailing academic wisdom. (The selection is partly subjective and partly determined by the nature of the problem. For instance, you would right away describe a certain problem using a circuit diagram, a second problem with state diagrams and yet another problem using a high-level programming language.)

f) Employ a 'natural-language' description. Even with the most 'scientific' formal presentation methods, it is possible to express a lot of nonsense. Therefore, never attempt to do without an orderly description (which means one with understandable terms, a logical structure and a clear, concise presentation). Even when it seems natural to right away sketch a circuit diagram or type a C program into the computer – you will be amazed at how many difficulties you will encounter when you try to formulate what you have in mind in understandable, everyday language. It is a specific alarm signal when at some point you cannot clearly express the objective of the design: either you have not correctly understood how to perform the assigned task, you have overlooked something, or (all too often) you cannot see the well-known forest for the trees. Circuit designs, programs and the like that are developed on such a basis often contain an above-average number of faults. There is a good reason why highly-developed quality-assurance systems start the development process with a specification formulated in natural language.

'True' design faults: the design is incorrect.

Common causes:

- misunderstanding a portion of the specification,
- neglecting certain requirements of the specification,
- overlooking 'side effects',
- malfunctions related to the mutual interaction of multiple functions or assemblies (most design faults are found in the interfaces),
- failure to respect the specifications or characteristics of the components used,
- language problems: most documentation (standards, data sheets, application notes etc.) is written in English. All-too-literal translations often lead to false interpretations.
- ordinary errors and omissions.

Preventive measures:

- Perform a critical review of the design (see 'critically review the specification', above).
- Carefully read data sheets and application notes, and observe their restrictions and recommendations.
- Never use components outside of their specified operating conditions.
- Use components only in the manner in which they are intended to be used. If for some reason you must use a 'trick' circuit, first test it thoroughly for yourself (even if it has been published in thousands of books).
- Perform an initial evaluation of any new type of component (in a test jig or simple application circuit; an evaluation board or kit can sometimes be obtained from the manufacturer).
- Work in stages, and split the overall problem into several smaller problems that can largely be designed, assembled, checked out and tested independently of each other.
- Use (proven) software to perform whatever tasks can be performed by software (Electronic Design Automation, EDA).
- Employ formal correctness testing and simulation in order to detect design errors before the prototyping stage.

Component failure in the prototype

Common causes:

- early failures ('infant mortality'),
- using outdated components that have been lying around in the laboratory for years,
- overloading components during checkout and measurement (due to an inadvertent short circuit, for example),
- using components outside their allowed range of operating conditions (due to a design or assembly error) – this is in many cases the most common cause of component failures in the prototype.

Preventive measures:

- Screen (pre-test) components.
- Do not use outdated components. (Permanently dispose of old components, so that they will not at some later time be used by mistake.)
- Check the prototype for trivial faults (short circuits, reversed power supply leads, incorrect component insertion and the like) before putting it into operation.
- Don't be too keen to save money on your test equipment and test aids, and as a beginner, develop your skills on simple, low-cost projects.

Prototype assembly errors

Common causes:

- defective printed circuit board,
- defective cabling,
- incorrect mounting or insertion of components.

Preventive measures:

- Have the prototype assembled as much as is possible using automated equipment (this is of course a cost issue) and use 'true' printed circuit boards (not 'breadboards').
- Do formal plausibility checking (conformance to the design rules, correspondence between the schematic diagram and the board layout and so on).
- Be sure that the device is assembled in a tidy and workmanlike manner.
- Use good-quality sockets, connectors and so on (no second-hand junk).
- Individually check the circuit board, cables and so on before assembling the prototype.

Systematic faults (deficiencies) in the prototype. Such faults are especially to be expected when (1) you are dealing with a 'true' laboratory prototype (such as a handmade, improvised construction on a prototyping board) that (2) is supposed to provide a high level of performance (*for example:* running at a clock rate of umpteen hundred MHz).

Common causes:

- inadequate earth plane (insufficient cross-sectional area, excessive impedance, inadequate screening, inappropriate layout, excessively long traces, failure to separate analogue and digital earth and so on),
- inadequate power supply connection (wire size too small, inappropriate lead dress),
- inadequate decoupling capacitors (too few, incorrect values, inappropriate type, inappropriate placement),
- inadequate signal leads (too long, incorrect cable type, incorrect impedance, inadequate screening and so on),
- inadequate cooling,
- interference problems (due to poor construction or the absence of overall screening).

Preventive measures:

- Sometimes there's simply no alternative to finding the money to pay for an electrically 'clean' construction (using a double-sided or even multi-layer board).

- If the risk of problems with the prototype is too high, regard it as a 'provisional' model which is to be used only to investigate the functional capability of the design, and relax the specification in the high-risk areas for this first prototype. (*For example:* suppose that a pattern generator for a monitor tester is expected to run at video rates of 100 ... 200 MHz. Is it possible for a hand-wired breadboard construction to work at 200 MHz? Rather than attempt this, content yourself with an initial prototype that meets the usual VGA standard and runs at a 25 MHz video rate – while this is still not trivial, it is at least achievable with ordinary means.

Software faults. Modern electronics devices are often associated in some way or the other with programmable devices, ranging from the simplest microcontroller to a PC or an even more powerful processor. Thus we are dependent on software from the moment that the equipment is switched on – and software can also be faulty.

Preventive measures:

- Rely on tried and true solutions and/or build on proven software (such as familiar development environments, standardized interfaces and the like).
- Take care that basic functions are put into operation step-by step and in the simplest possible manner. (A counter-example, exaggerated for illustration, would be to try to check out several megabytes of software running under Windows 95 in combination with newly-developed hardware having its own high-performance processor and system software, interconnected via a network or a high-speed bus. What do you think are the chances that all of this will work properly, right off the bat?)

Operational faults

In contrast to when a prototype is first put into operation, in normal operation you can in principle assume that the equipment in question was originally fault-free, so that the fault must have been caused by some internal event or by an external influence. In other words, you can assume that some specific *failure mechanism* has been at work. Of course, it is not possible to definitely exclude 'all possible' other causes of the fault, but they are significantly less likely.

Correctly recognizing the cause of a fault. Faults, regardless of their causes, often show up either as hard failures or as malfunctions (intermittent faults). Some design and assembly faults show up as actual malfunctions (for instance, due to excessive susceptibility to external electromagnetic interference). Faults due to an improper operating environment or operator errors also mainly result in malfunctions. Apart from this, you usually have to deal with systematic (hard) faults, which always recur in certain circumstances. The necessary conditions sometimes occur relatively infrequently, so that the occurrence of a fault at an interval of several days (or even weeks) is not unusual, even during endurance testing. You are thus faced with the problem of deciding whether you should look for an actual fault in the system (a failure), whether some other action is indicated (such as registering a complaint with the manufacturer) or if something needs to done with the 'infrastructure' (such as modifying the mains installation, acquiring a non-interruptable mains supply, relocating the system, providing better training for the users and so on).

Taken all together, the causes of faults that have to be dealt with can be categorized as described below.

Latent ('hidden') design faults. Modern systems and devices are often so complicated that you can expect design faults even in equipment that has already entered mass production. Such faults can be suspected when all other possible causes have been eliminated. Typical fault symptoms are (among others):

- the fault always occurs under the same set of conditions, but a sufficiently thorough investigation uncovers no sign of a failure,
- the fault occurs with all systems of the same type or model which you are able to use for comparison,
- the software works absolutely correctly on compatible hardware of a different type or model,
- other users have similar difficulties with the same type of hardware.

Preventive measures:

- Do business only with vendors that have a good reputation for quality (and for redressing problems on the basis of 'goodwill' if it comes to that).
- Do not always blindly acquire the newest of everything, but take care that the hardware or software in question has proven itself for a certain length of time in practice. (Based on experience, the majority of the 'teething problems' will have been overcome after a product has proven itself on the market for a year's time, at least with reputable manufacturers who have a name to maintain.)
- Never introduce multiple changes at the same time:

  a) 'break in' new software initially on proven hardware, even if this means living with reduced performance for a certain period of time,

  b) initially check out new, unproven hardware using proven software.

Assembly faults. The manufacturer's quality control should actually prevent the shipping of faulty products. However, you cannot place absolute trust in this. An assembly fault (such as an incorrectly stuffed board) can affect a whole production run of products. Assembly faults have symptoms that are similar to those of design faults, and can they be suspected when other possible sources of the fault have been excluded. The most important sign of an assembly fault is that it is present from the very beginning.

Prevention: thoroughly check out the item in question immediately after it has been delivered; do not just set it aside or handle it with kid gloves. Another way of looking at this is the following: you should purchase an expensive hardware or software item only when you are in the position to be able to thoroughly exploit its capabilities – and keep in mind the duration of the warranty or the interval within which the item may be returned or exchanged.

Failures. The general characteristic of a failure (regardless of whether it shows up as a hard fault or as an occasional, intermittent fault) is that up until the time that it occurs everything functions correctly, and then the fault symptoms suddenly appear. Determining the cause of a failure and then remedying it is the 'classical' field of activity of the maintenance technician. We can distinguish the following types of failures:

- 'infant mortality'. These failures occur in the first few hours of operation. *Prevention:* an initial endurance test (burn-in).
- 'statistical' failures. These are the so-called 'normal' hardware failures, which must be expected from time to time.
- old-age failures. The older the equipment is, the more likely is the chance of a failure. Put another way, the occurrence of an unusually large number of failures in a relatively short period of time is a sign that the equipment is approaching the end of its useful life. *Prevention:* (1) maintain the equipment in good working order by replacing parts which wear out and subassemblies which tend to degrade when used over a long period of time, and (2) employ appropriate diagnostics (such as the SMART technology used in hard-disk drives) to detect 'gradual' aging effects.

Note: 'SMART' stands for 'Self Monitoring, Analysis and Reporting Technology'. In order to detect signs of aging, certain factors are regularly recorded or measured (the error rates during head positioning and reading, various parameters of the read and write channels and (technically the most difficult) the flying heights of the magnetic heads). Based on this information, a warning of impending disk failure can be given somewhere between 24 hours and 8 days in advance of an actual failure. This allows a replacement disk to be acquired and the data to be copied from the old disk to the new disk, without recourse to an intermediate tape backup.

Equipment failure is the subject of *reliability theory*. See Appendix 8 for references to additional information on this subject.

Software faults. Software faults are fundamentally latent design faults. With proven hardware that shows no evidence of failure, you will often suspect the software. With software that you have developed yourself, it is possible to trace its execution and identify specific faults (debugging). With purchased (commercial) software you can only appeal to the manufacturer. *Prevention:* see 'latent design faults' above.

Operator errors. The possibility of an operator error must always be taken into account ('to err is human'). Effective protection against operator errors can only be provided by software and – in the case of industrial products – by specific monitoring circuitry. The latter measure offers additional security (including the case of a software fault). The actual difficulty in fault-finding is to recognize operator errors as such.

Preventive measures:

- If you are responsible for systems that are subject to operator errors, familiarize yourself with the typical operator actions (or at least keep the relevant manuals accessible so that you can refer to them if necessary).

- Investigate what possibilities the system provides for logging operator actions and canceling operator errors (such as an UNDELETE function), and utilize them accordingly.

Environmentally-related faults. System malfunctions are sometimes caused by environmental factors. *Common causes:* disturbances arising from switching on and off nearby electrical equipment, vibrations (hard-disk drive malfunction or failure!), unacceptably high relatively humidity and so on. *Prevention:* check the environmental conditions using appropriate measuring equipment, and take prompt corrective actions if necessary (such as installing passive mains-interference suppression devices, moving the equipment to a different location or supplying the system from a non-interruptable mains power source).

2.2.2 Failure mechanisms and fault hypotheses

A *failure mechanism* is any process that results in a failure, in the sense of a hard fault or a malfunction. In the case of hardware failures, such processes are ultimately physical or chemical. We can distinguish the following types of hardware failure mechanisms:

- those occurring in semiconductors,
- those occurring in equipment or subassemblies,
- those due to environmental factors.

A *fault hypothesis* is a preliminary assumption regarding the cause and/or type of a given fault. Fault hypotheses are either confirmed or refuted by means of tests and measurements, and also by logical deduction. *For example:* suppose that we initially assume that the power supply is responsible for a particular fault. If subsequent measurements of the power supply's performance do not turn up any evidence of incorrect operation, then our fault hypothesis has been refuted.

Whenever you get stuck in the course of fault-finding, a proven technique is to systematically work through a variety of fault hypotheses, one after the other (the 'shotgun' approach; see Section 2.5.5). The common causes of faults that are identified in section 2.2.1, for example, can be used to form fault hypotheses for this approach. Additional fault hypotheses can be found in the various sections of Chapter 5.

Failure mechanisms in semiconductors

Experience shows that integrated circuits, transistors, diodes and the like tend to fail in the course of time, even though to all appearances these components do not wear out. A variety of mechanisms is responsible for such failures. We can briefly describe of some of the causes of failures and malfunctions in modern semiconductor devices.

Puncturing of an insulating layer (oxide failure). The insulating layers in the semiconductor structure have become increasingly thin with the increased level of integration. These extremely thin oxide layers can break down in the course of time, even if the normal failure voltage is not reached or exceeded (time-dependent dielectric breakdown, TDDB).

Injection of accelerated charge carriers (hot carrier injection). The charge carriers in semiconductor devices are accelerated by electric fields. The energy that is thus imparted to the carriers can be enough to allow them to penetrate an insulating layer in the MOS structure. This mechanism does not always lead to a complete failure, but it often results in intermittent malfunctions.

Note: a high electric field strength results from the combination of a sufficiently high voltage and a sufficiently thin insulating layer. It can be responsible for a variety of failure mechanisms. This is one of the reasons why modern hardware may be designed to use supply voltages significantly lower than the conventional 5 V. For 'normal' digital logic and memory components, 3.3 V represents a sort of 'industry standard'. (*For example:* all 16-Mbit DRAMs work with an internal operating voltage of 3.3 V, even if the external supply voltage is 5 V; the supply voltage is reduced to the internal operating level by circuitry inside the chip.) Modern high-performance processor chips, on the other hand, do not adhere to any industry standards. The design of each one is optimized without imposing any restrictions on the operating voltage, and a suitable operating voltage is determined once the design is complete. The result is a variety of nominal operating voltages and associated tolerances. Such processor chips are powered by separate voltage regulators. In practice, this means: be careful when 'swapping' processor chips!

Electromigration. In very thin conductors, metallic ions can 'migrate', resulting in a break in the conductor (see Figure 2.1a).

Stress migration. Thermal stresses in very thin conductors can cause the conductors to break (see Figure 2.1b).

Note: electromigration and thermal-stress effects are especially likely if excessive currents flow through a conductor. These mechanisms usually do not cause an immediate failure; instead, they first cause gradual changes in the component's characteristics. *Prevention:* take precautions to avoid situations in which excessive currents can flow. The actual cause of excessive currents is often an incorrect supply voltage or a reverse-polarity connection to the power source. In situations where either of these can occur, provide for overvoltage and/or reverse-polarity protection. Be especially careful when working with circuits where these measures are not present. Refer also to the description of the latch-up mechanism and the associated practical tips (see below).

Ruptures in the surface-passivation layer and metallization. These are especially prone to occur in integrated circuits with plastic packages, since the plastic material has a larger coefficient of thermal expansion than the silicon die. The result can be cracks in the protective passivation layer (see Figure 2.1c), and the conductive aluminium traces can be become broken due to crystallization.

Bonding-wire failure. The silicon die of an integrated circuit is normally connected to the external pins of its package by thin aluminium or gold wires, which are called 'bonding wires' after the manner in which they are attached. The bonds, or the wires themselves, can break loose or crack due to a variety of failure mechanisms (see Figures 2.1d and 2.2).

Moisture-induced corrosion. Plastic packages are very slightly porous to water vapour. When they are used under conditions of high relative humidity, the internal aluminium

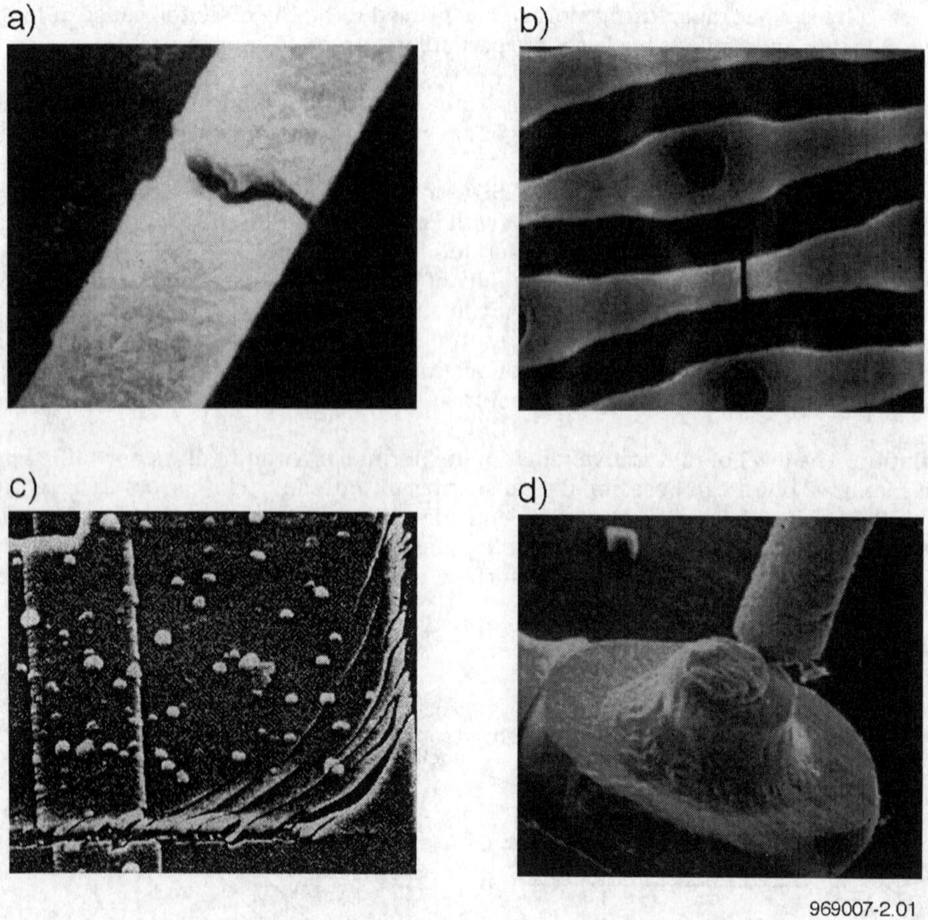

Figure 2.1 Examples of the effects of failure mechanisms in semiconductor circuits.
a) electromigration, b) stress migration,
c) cracks in the passivation layer, d) broken bonding wire.
(source: Quality/Reliability Handbook, OKI Europe GmbH, Neuss)

metallization can become corroded, which leads to a deterioration of the electrical characteristics of the integrated circuit or to its complete failure.

Cracks in the packaging. This effect is common with SMD packages. With conventional (dual in-line) packages only the pins are heated during soldering, but with a SMD component the entire package is heated to over 200 °C during the assembly of a surface-mounted board. This can result in cracks in the plastic package (Figure 2.3 illustrates the failure mechanism). It has been shown that components that have been stored for a year or more before assembly exhibit an especially high failure rate, whereby the moisture which is absorbed by the package during storage has proved to be the essential cause of failure. Especially thin and/or large packages are particularly prone to cracking.

Practical tip: always employ moisture-resistant packaging for long-term storage of integrated circuits. Use watertight containers or bags and include sacks of silica-gel desiccant.

Damage due to electrostatic discharge (ESD). An electrostatic charge is built up whenever unlike materials are brought into contact with each other and then separated. Charges of many thousand volts can readily be generated in this manner. Its discharge results in a short-lived current that has a high probability of damaging an integrated circuit. The heat that is generated by the discharge current can cause evaporation of the silicon and melting of the aluminium (since the structures in question have dimensions on the order of 1/1000 mm). Even if a discharge does not occur, the electrostatic charge can produce an electric field that is strong enough to puncture insulating oxide layers.

Radiation. The level of radioactive radiation in 'normal' environments is not sufficient to cause failures, but it can occasionally cause intermittent faults (soft errors). This applies especially to bit-loss errors in dynamic RAMs. Even a single alpha particle can result in a soft error in a RAM. Nowadays this problem has been mastered by means of improved packaging and the passivation of the surface of the silicon die. (Memories of several

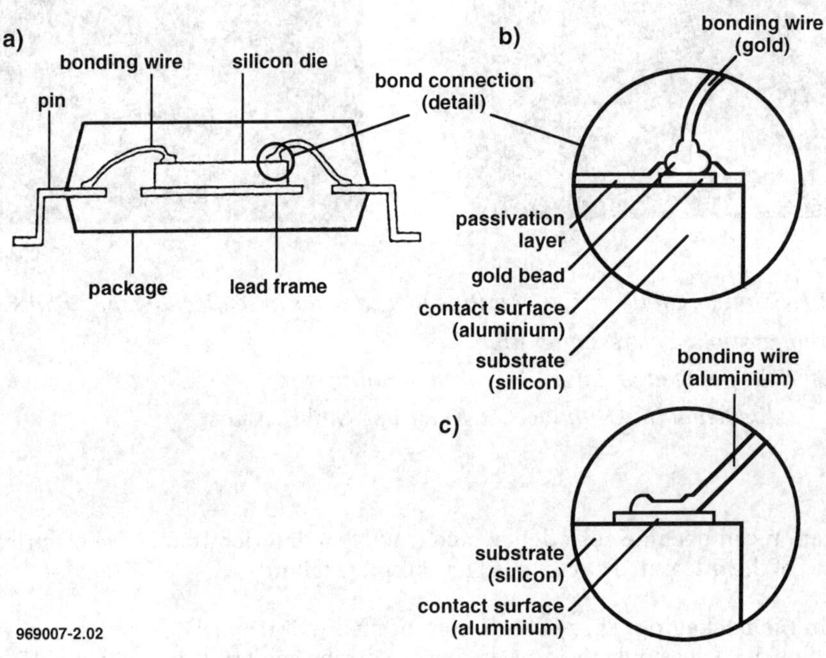

Figure 2.2 *Structure of a bonded connection.*

a) cross-section of a packaged IC, b) detail of a bonded connection with gold wire,

c) view of an ultrasonic bond with aluminium wire.

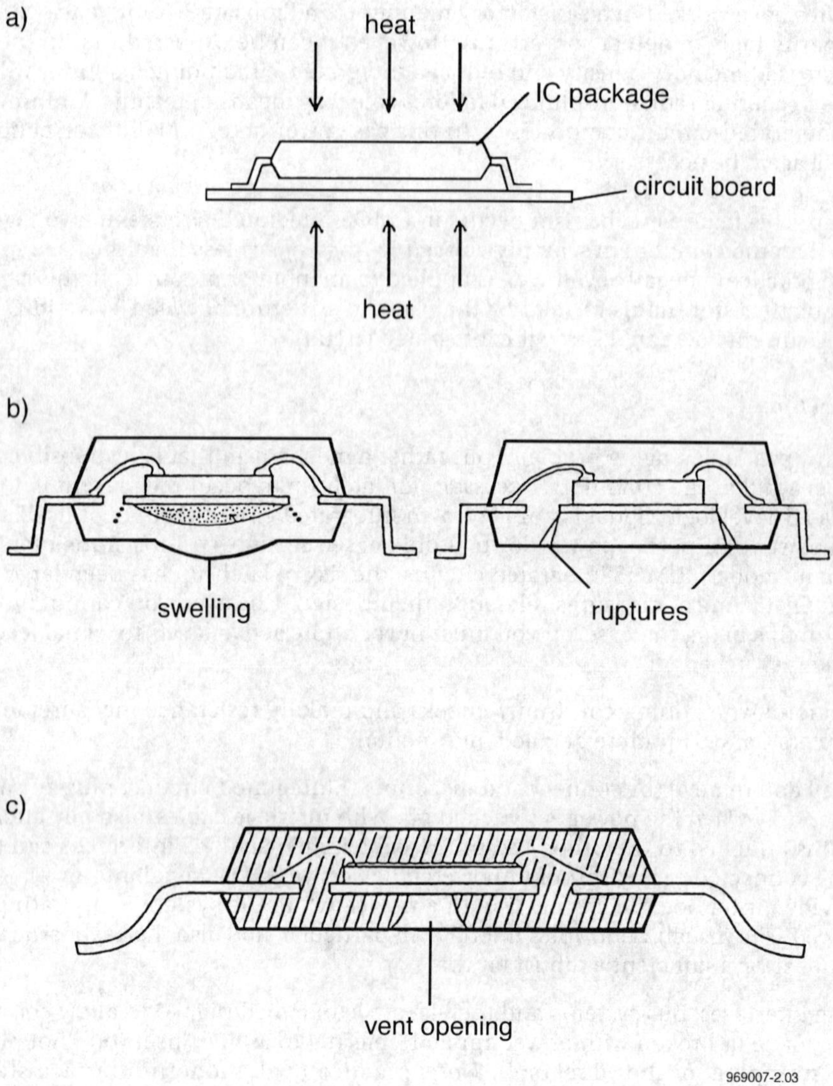

Figure 2.3 Cracks in the package.

a) SMD package heated during soldering,

b) fault mechanism (swelling and cracking),

c) improved package with relief vent (to prevent pressure buildup from steam inside the heated package)

megabytes have worked in a satisfactory manner in millions of PCs for years without the use of parity logic to detect soft errors.) However, it can be expected that in the future, with increased memory density and reduced integrated-circuit dimensions, it will not be possible to counter radiation-induced failures solely by means of technical improvements in the integrated-circuit components. In this case error detection and correction measures will again be necessary.

Latch-up. This failure mechanism occurs in CMOS integrated circuits. If a voltage that is greater than the component's supply voltage V_{CC} or V_{DD}, or less than the earth potential *GND* (in practice, a negative voltage), is applied to an input or output lead, this can trigger a parasitic thyristor that is formed by the p-n-p-n structure between V_{CC} and *GND*. The resulting current flow can destroy the integrated circuit.

Practical tips:

1 Latch-up is not an aging problem, but rather a mechanism that is responsible for subsequent faults. The conditions necessary for latch-up to occur are present if, for example, a HI-level logic signal is applied to an integrated circuit that is not itself supplied with power (the 'partial power-down' problem; see Section 4.4.1 for more information). In many modern CMOS integrated circuits, the risk of latch-up has been largely eliminated by technical measures related to their design. However, you cannot not always rely on this being the case, so you must never apply active levels to signal leads if the power is switched off.

2 Be careful when doing continuity checks and making resistance measurements (see Chapters 4 and 5 for more detailed information).

3 Install and remove (or connect and disconnect) integrated circuits, plug-in cards and cables only when the power is switched off. With interface cables, also pay attention to what is attached to the other end of the cable! The usual PC interfaces can tolerate being connected to any sort of unpowered device; with V24 (serial interface), SCSI and the video interface the connection of a device while the system is operating is also allowed. The (usual) Centronics interface in particular, and also the keyboard interface, are not always sufficiently robust for this.

4 Some interfaces, bus systems and the like are explicitly designed to allow connections to be made or broken with power applied (referred to as 'live insertion', 'hot plugging', 'hot swapping' or 'hot docking'). *Note:* practically all modern interface standards include specifications relating to this capability *(for example:* parallel interfaces compatible with IEEE 1284, the PCMIA PC-card interface and the USB serial bus system).

5 The Windows 95 hardware requirements also specify that all of the PC's external interfaces (and of course the peripheral devices that are connected to the interfaces) must tolerate having connections made while the equipment is powered on. This does *not* apply to the bus systems internal to the PC (such as ISA, PCI or IDE). Assertions to the effect that 'Plug and Play' means that you may exchange expansion cards at will while the PC is running without risking damage to the cards and/or the PC (and that the operating system will react appropriately to any change in the hardware configuration), are quite simply false.

How often do failures occur? Extensive tests and statistical surveys are needed in order to arrive at any more-or-less correct numbers. This takes time and occasionally leads to the situation in which a given technology is obsolete by the time that enough experience with it has been accumulated to allow its weaknesses to be recognized. For this reason, the data (Table 2.1 can serve as an example) must be revised from time to time. Tables 2.2 and 2.3 present an overview of methods that are used by integrated circuit manufacturers to check their products. The practical utility of this information is the following:

1 From the tables, you can see which conditions promote the occurrence of particular types of failures (or put differently, which environmental factors you must pay attention to in order to achieve increased operational reliability).

2 The tables can serve as a source of inspiration for devising your own endurance tests (*Caution:* first consult Section 2.4.4 and Volume 2).

Percentage	Fault mechanism
39 %	metallization-to-metallization shorts
2 %	metallization-to-substrate shorts
15 %	gross physical faults (scratches and cracks)
6 %	breaks in diffusion layers
14 %	shorts between diffusion layers
14 %	interruptions in metallization
10 %	undetermined (not visible)

Table 2.1 *Frequency of occurrence of various causes of failure in MOS integrated circuits, in percentage. (source: IEEE TC, Vol. C-29, No 6, June 1980)*

Failure mechanisms in equipment and subassemblies

Table 2.4 summarizes some additional mechanisms that can cause failures in equipment and subassemblies.

Practical tips:

1 Notice that Table 2.4 expressly notes the well-known fact that frequently switching equipment off and on, which is viewed by the manufacturers of so-called energy-saving hardware and certain members of the general public as an economical mode of operation, actually tends to reduce the useful life of the equipment.

Test	Test conditions	Examples of demonstrable failure mechanisms
high-temperature operation	ambient temperature 125 °C	time-dependent dielectric breakdown (TDDB)
low-temperature operation	ambient temperature −55 to 25 °C	hot carrier injection
operation at elevated temperature and high relative humidity	ambient temperature 85 °C, 85% relative humidity	aluminium corrosion
high-temperature storage	storage temperature 150 °C	thermal stress migration and other failure mechanisms of the intermetallic connections
storage with cyclically varying temperature	−65 °C, room temperature and 150 °C for 30 min, 5 min and 30 min respectively	cracks in the package and passivation layers, lifting of aluminium tracks
storage at elevated temperature and pressure	ambient temperature 121 °C at 2 atm pressure and 100% relative humidity	aluminium corrosion
accelerated stress test (operational)	ambient temperature 130 °C, 85% relative humidity	aluminium corrosion

Table 2.2 Endurance tests. (source: see Table 2.1)

2 Regarding the failure behaviour of hard-disk drives: the damage resulting from long-term use is less than you might initially think. Naturally, mechanical wear must be expected if the drive has been used for a long time (so that you should always replace the hard-disk drive when modernizing an older-model PC), but the experience of manufacturers of modern hard-disk drives (which have a lot of built-in electronics) suggests that the electronics is likely to fail before the mechanical assembly is worn out.

Environmentally-related failure mechanisms

Unfavourable ambient conditions increase the probability of failure. Failure mechanisms arising from environmental factors often show up as repeatedly-occurring intermittent faults, but they can also occasionally lead directly to hard faults. Typical causes of faults due to environmental conditions are summarized in Table 2.5.

Test	Test conditions
resistance to soldering temperatures	260 °C for 10 seconds
temperature cycling	cyclic variation between –65 °C, room temperature and 150 °C for periods of 30 min, 5 min and 30 min respectively
thermal shock	10 cycles between 0 and 100 °C
vibration (used for ceramic packages)	20 test cycles in the range of 100 to 2000 Hz, acceleration up to 20 g in three axes
constant acceleration (used for ceramic packages)	2000 g, 3 axes, 1 minute
solderability	230 °C for 5 seconds (with application of flux)
mechanical pin loading	depends on package type

Table 2.3 Environmental and handling tests. (source: see Table 2.1)

Cause of failure	Affected devices	Countermeasures
excessive mechanical wear	drives of all types, in particular hard-disk drives; keyboards, mice and the like	Perform regular checkouts, avoid frequently switching off and on, maintain uniform environmental conditions.
reduced electron emission (cathode activity)	monitors	Avoid frequently switching off and on, avoid setting the brightness unnecessarily high (proper setup, use a screen saver if appropriate).
demagnetization processes	magnetic data-storage media	regular use, periodic maintenance with suitable programs

Table 2.4 Important failure mechanisms in devices and assemblies.

Cause of failure	Countermeasures
mains power disturbances	Check out the mains installation and modify it if necessary; supply the computer via a mains filter or uninterruptable power supply.
lightning strike	Check the lightning-protection provisions of the mains installation; install surge suppressors (overvoltage protectors) ahead of the equipment's mains cable; use countermeasures such as optocouplers or fibre-optics links for critical interfaces (such as network cables or terminal connections that lead to other parts of the building).
electrostatic discharge	Check the earthing system; correct conditions such as excessively low relative humidity or inappropriate floor coverings.
earth loops	Check the earthing and mains connections of devices that are connected to each other; modify problem interfaces (for example, change to a differential-signal standard or use electrically isolated adapters using optoelectronics or fiber-optics cable).
screening problems	Check enclosures, earthing and cabling; if appropriate employ a better-quality cable with adequate screening.
interference on data lines	Check lead dress and cable routing and modify as necessary; use a better-quality cable with adequate screening; modify problem interfaces (for example, change to a differential-signal standard or employ electrically isolated adapters using optoelectronics or fiber-optics cable).
interference from the surroundings	This actually should not occur with equipment that meets EMC standards (a given piece of equipment should neither generate interference nor be susceptible to incorrect operation from external interference). In practice: check enclosures, earthing and cabling; check equipment and installations in the immediate vicinity of the affected item; set up the affected item(s) in a different location. If all else fails, seek help from a specialized firm or the local authority to locate the source of the disturbance.
temperature	Improve the ventilation (clean the insides of enclosures, clean ventilation openings, check that ventilation grills are not blocked, install an additional fan); move the equipment to a different location.
humidity, dust, vibration etc.	Use a suitable protective enclosure or move the equipment to a different location.

Table 2.5 Environmentally-determined causes of failures.

2.2.3 Fault models

Murhpy's Law ("whatever can go wrong will go wrong") is in practice valid, and sometimes even Callahan's Corollary ("Murphy was an optimist") is also applicable. Based on this, you can hardly expect to be able to locate faults by logical deduction alone, and the question "what else has to be checked?" always arises when one is writing test software or developing test hardware. For this reason as well, all fault-finding methods that are based on logic, and all attempts to automate testing and fault-finding (or to at least provide automated support), must be based on certain assumptions about what kinds of faults are to be looked for. Such assumptions are called *fault models* (and in this sense Murphy's Law is also a fault model, but since it is of the most extreme sort it is hardly helpful for fault-finding).

Single-fault hypothesis. The assumption that *there is only a single fault at any given time in the system under investigation* lies behind almost all test programs, built-in test provisions, checkout aids, trouble-shooting guides and the like. In other words: there are not different faults in different parts of the system at the same time. Experience and (in particular) statistical data confirm that this assumption is justified in the vast majority of cases. There are however certain details and some departures from this rule that must be taken into account.

Multiple symptoms of a single fault. Don't let yourself be misled: even if you observe fault symptoms at a variety of different locations, they usually have a single common cause. In such cases the source of the fault is frequently to be found in a part of the system that must be in good working order before other parts of the system can function properly. *For example:* with a PC, if it's not possible to read a floppy diskette and at the same time there is no image on the monitor, it is highly unlikely that the two associated subsystems (drive plus controller, monitor plus video adapter) are both defective. Instead, there is probably a fault in the motherboard or the power supply.

Consequential faults. 'True' multiple faults are most likely to occur as a result of a single initial fault which causes additional, subsequent faults to occur. *For example:* a defective extension card overloads the system bus and causes the failure of a bus driver on the motherboard.

Independent faults. The concurrent occurrence of multiple, truly independent faults can naturally not be entirely ruled out, but it is highly unlikely. (The probability of this is equal to the product of the probabilities of the individual faults.)

Consequential faults from a single latent fault. Some faults, in particular those that occur in parts of the system that are themselves intended to respond to faults and other infrequently occurring conditions, will not be immediately observed because (1) the specific situation in which the associated system element is supposed to be active has not yet arisen and (2) preventive-maintenance checks have not been performed. *For example:* suppose a power supply incorporates short-circuit protection that is intended to switch off the entire power supply in the event of a short circuit on one of its outputs, and that in the course of time the short-circuit protection hardware has developed a fault. If an output short circuit now occurs, it is likely to cause permanent damage to the power supply.

Practical tips:

1 The single-fault hypothesis is the basis for almost all fault-finding charts and guides. Only after an investigation based on this hypothesis has proved to be unsuccessful should you consider the possibility of multiple faults, and even then first in the sense of consequential faults and only afterwards as independent faults.

2 Although consequential faults are relatively rare in purely digital hardware, they are quite frequently encountered in analogue circuits and power circuits (especially power supplies).

'Stuck-at' fault. This is the simplest model of a hard fault: a single line (or a single pin of an integrated circuit) is held constantly at either a LO level ('stuck at LO') or a HI level ('stuck at HI'). It is the basis for most checkout procedures, test-data generation algorithms, test programs and self-test provisions.

How valid is the 'stuck-at' fault model? Only a minority of faults that occur in modern hardware are isolated stuck-at faults. However, practical experience has shown that good checkout and test procedures which are based on this model are also effective in a large number of other fault situations. It is extremely unlikely that a test that (for example) recognizes more than 90% of all stuck-at faults will deliver a result of 'OK' or 'passed' in the presence of an arbitrary fault of some other type. In such cases, you are usually presented with a fault indication that is either incorrect or so general that it is useless, or (most often) the test simply hangs.

2.3 Principles of testing and diagnosis

2.3.1 Test and checkout procedures

Every test or checkout is a comparison of the actual behaviour of the item being investigated (the test object or device under test, DUT) to the behaviour expected in the absence of faults. The behaviour to be observed is provoked by a particular test stimulus, either by operating the device in its normal manner or by the use of special diagnostic stimuli (test signals or test vectors).

In this context, the device being investigated may be regarded as a sort of 'black box' with a certain set of inputs and outputs (see Figure 2.4).

Test steps. A test step is a sequence of events consisting of the application of a certain stimulus to the inputs, observing or measuring the outputs, and comparing the actual result to the expected result. Usually, based on this comparison, a decision is made either to proceed with the test or (occasionally in connection with an error message) to end the test. A variation on this scenario is to simply record the result of the test step and postpone the evaluation to a later time (certain types of test equipment work this way, and it is the typical manner in which a logic analyzer is used).

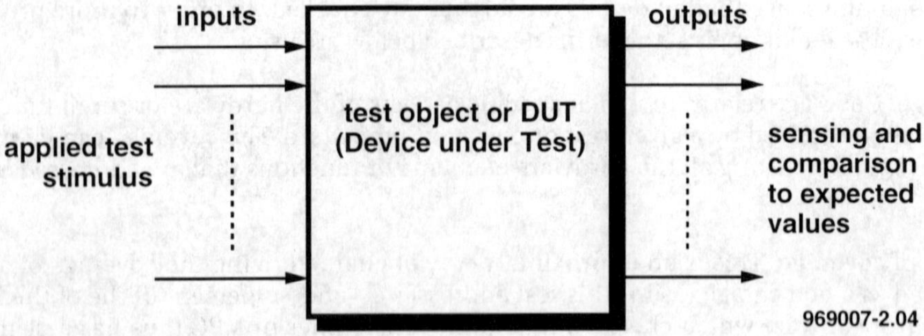

Figure 2.4 The test object as a black box.

Accessibility, reachability and observability. These terms reflect the fact that certain parts of the test object can be directly 'approached' by a test or checkout procedure while others cannot. In this context, something can be 'approached' if it is possible to apply input stimuli and to observe or record output values.

Examples:

1 The display memory of the video adapter in a computer is accessible to test software (it is no problem to execute read and write accesses). However, no test program is able to determine whether the video signal actually reaches the monitor, since the video output is not software-accessible and thus not observable.

2 The floppy-disk controller in a computer is accessible to the processor, but the interface to the drive and the drive itself are not. Operations relating to some of the controller's internal registers are directly observable (by reading back their contents), but the processes used in the drive for writing the data are not. However, they are indirectly observable by means of writing test data and then checking it.

2.3.2 Quality criteria for test procedures

The following situation should actually never occur in a fault-finding situation (although it does sometimes happen!):

1 the test object is quite obviously faulty, and

2 the test software, test equipment or the like does not recognize any fault (the test runs 'without a hitch', or the result is 'OK' or 'passed').

The following requirement can thus be placed on test procedures and equipment: *the result of a test shall never indicate that the test object is fault-free when a fault is actually present.* This means that checking and testing must be (1) comprehensive, in the sense that all aspects of the test object and all of its available functions are addressed, and (2) rigorous, in the sense that all details are adequately checked. In order to more precisely describe these requirements, the terms described below are used.

Coverage. Coverage relates to which specific aspects of the hardware or set of functions are actually addressed by a given test or checkout procedure. The coverage can be practically described by listing all the hardware elements or functions that are addressed by the test.

Degree of coverage. This is an approximate way of characterizing the coverage of a test, expressed as a percentage (as in "this test addresses so-and-so percent of the of the hardware"). Test software which checks out all of the assemblies of a PC thus has a degree of coverage of (practically) 100%. If certain assemblies are not checked (such as the floppy-disk drive or part of the working memory) then the degree of coverage is accordingly lower.

Coverage matrix. Consider a system that consists of a number of elements (devices, subassemblies, components) which can be labeled *A, B, C* and so on. Various tests and other procedures are intended to be used to check the system. For convenience, we number these tests and procedures sequentially, starting with '1'. A matrix chart can then be used to show, in a simple and elegant manner, which test or procedure checks which element(s), as illustrated in Figure 2.5.

The coverage matrix has one row for each test and one column for each element of the system. A '1' at a particular intersection indicates that the associated test checks the indicated element (otherwise a '0' is present). Once we have produced such a diagram, we can immediately see whether all of the system elements have actually been accounted for in our test and checkout scheme; it is only necessary to see that there is at least one '1' in every column. (In boolean-logic terms, this means computing the logical OR column by column; the result should be all '1's.) If this is not the case, then not all elements have been taken into account. *Note:* the coverage matrix does not say anything about how rigourous the tests or checkout procedures may be.

Test severity. Test severity describes how rigourous a test is, that is, how thoroughly a given test or checkout does its work or how many of the possible faults it can actually detect. A numerical value for test severity is only possible with reference to a specific fault model. Thus, you will see statements such as "this test will detect up to 90% of all stuck-at faults". The test severity depends primarily on the fault model, the extent of the test and the hardware level at which the test takes place.

Practical tip: when 'spot checks' or sample measurements do not indicate a fault, this by itself is no guarantee that the hardware in question is actually fault-free. When you are shopping for test software (for example), you should look for an adequate degree of coverage and sufficient severity. In the absence of contrary evidence, you should assume that all test measures and test equipment work on the basis of spot checks alone. Whenever you are trying to locate a fault by measurement, you should at least keep this possibility in mind. (A typical scenario: you have checked out every device or subassembly in turn and not found anything that allows you to regard any of them as suspect, yet the behaviour of the system clearly indicates the presence of a fault.)

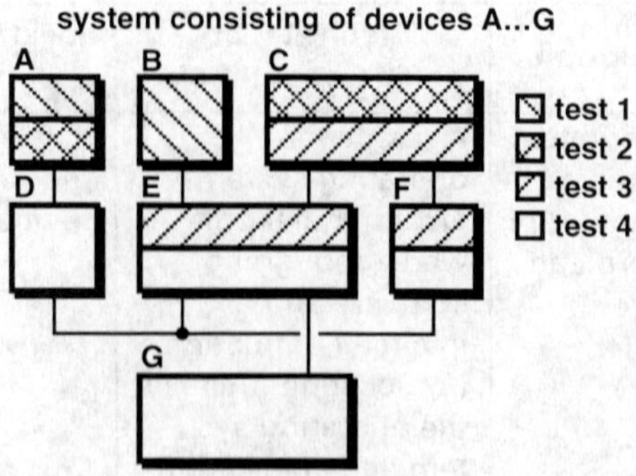

coverage matrix

test	device tested						
	A	B	C	D	E	F	G
1	1	1	0	0	0	0	0
2	1	0	1	0	0	0	0
3	0	0	1	0	1	1	0
4	0	0	0	1	1	1	1

969007-2.05

***Figure 2.5** Sample coverage matrix.*

2.3.3 Level, coverage and severity

These concepts of test level, test coverage and test severity form the basis for investigating the following questions:

1 Where can you come to grips with the hardware (that is, where can you apply a stimulus or initiate a test process, and where can you observe or measure the results)?

2 How 'large' is the test (how much effort is required, how long does it take to execute)?

3 How many faults, and which faults, can the test identify?

Test level. It is clear that there are various levels at which you can investigate a system that is to be tested. Table 2.6 presents a scheme of 'stacked' levels that directly follows from the technologies used in the systems under consideration.

The costs of fault-finding at different levels. There are two rules of thumb: (1) the lower the level at which a fault can be found and remedied, the better, and (2) each time you go up one level, the cost increases by roughly a factor of ten. These rules apply at any rate to manufacturing and assembly (including the construction of one-offs). Obviously, it is

Level	Description	Typical test equipment and aids	Locating and remedying faults in use and in service
System	the complete system as a combination of hardware and software	test software, operating system error handler software, debugging software and in-circuit emulation (for locating faults in the operating system and application software)	checking the overall system and individual devices, device substitution
Device	PCs, monitors, printers, modems etc.	test software, built-in error detection provisions, testers	individual testing, device substitution, localization of suspect subassemblies
Subassembly	expansion cards, power supplies and other replaceable assemblies ('circuit board level')	test software, built-in error detection provisions, testers, measuring equipment and test aids	substitution of subassemblies, localization of suspect components
Component (integrated circuits and 'discrete' components)	processors, RAMs, resistors, transistors etc.	test software, built-in provisions for fault detection, testers, measuring equipment and test aids	Identify suspect components using measurements and signal tracing in the affected subassembly or individual testing. Faults in processors, memory components and so on can often be localized at the component level with test software or built-in error detection provisions (self-testing, scan techniques etc.).
Gate	logic gates, flip-flops, driver stages etc.	test software, built-in fault-finding provisions, testers	Fault localization at the gate level is sometimes possible (with individual integrated circuit tests or based on the scan technique), but this is only useful for reconfirming the suspected fault (an IC is the smallest replaceable part).

Table 2.6 *PCs and comparable systems can be broken down into various levels, based on technology and type of testing. (part 1 of 2)*

Level	Description	Typical test equipment and aids	Locating and remedying faults in use and in service
Transistor (elementary devices in integrated circuits)	bipolar or MOS transistor structures	process control, (very rigorous) testers	not possible
Electrical design and IC layout	physical semiconductor structures and metallization layers	process control, (very rigorous) testers	not possible
Physical effects		basic research, process control, (very rigorous) testers	not possible

Table 2.6 PCs and comparable systems can be broken down into various levels, based on technology and type of testing. (part 2 of 2)

much easier to test an integrated circuit before it has been built into a circuit than to laboriously search for a fault in a completed piece of equipment. In the first case, the faulty IC can simply be rejected (and the manufacturer's warranty will be honoured without question). In the second case, a time-consuming investigation of the finished product is necessary to find the faulty item, and then it must be replaced (which means rework). Shipment delays and a possible reduction in the value of the product may also result, and it may be difficult to make a warranty claim for the faulty component.

With servicing, the situation is less clear. The problem is that low-level testing requires an investment in suitable test equipment, while a PC (by contrast) is in a certain sense a natural test environment for a processor or SIMM, for example – just plug it in and see whether it works! Why then should you use test equipment in the service environment?

- If you do not want to limit yourself to simply replacing equipment or subassemblies that you suspect to be faulty, then testing costs a certain amount of time. The system is not available for normal use during this time.
- Items that are at all reasonably replaceable (FRUs) are typically rather complicated and also expensive. Anyone who values his professional reputation can hardly afford to frequently resort to fruitlessly swapping out motherboards, disk drives and the like.
- In modern systems, simply 'looking for a fault' without using any equipment is actually not all that simple. In addition, quite economical test equipment can be developed using current technology. Finally, you can build your own, which also has the advantage that you can learn something in the process (see Volume 3 for more information).

Complete (exhaustive) testing

The highest degree of test severity is obviously achieved when the test object is checked using all available inputs and all possible combinations of input values. The simple black-box model (Figure 2.4) is not sufficient for determining what the extent of such a test will be. Instead, a certain amount of knowledge of the internal structure of the test object is needed (see Figure 2.6), and a specific fault model must be assumed.

Test object without memory or feedback (Figure 2.6a). A given set of input values always produces the same output, regardless of any previous input values.

Test object with memory (Figure 2.6b). This object includes some sort of memory, so that its output depends not only on the input values but also on the content of the memory.

Test object with feedback (Figure 2.6c). Outputs are coupled back to inputs, so that the object's behaviour cannot be described by simple signal flows from the inputs to the outputs. Its behaviour is fundamentally time-dependent, and a given set of input values can result in different outputs, depending on when they are applied.

Test object with memory and feedback (Figure 2.6d). At least one of the memory elements is coupled back to at least one of the inputs. The content of the memory at a specific time is called the *state* of the object at that time (or *machine state*). The object's behaviour depends on both the current machine state and the input values.

Combinational digital circuits. A combinational digital circuit with n inputs can have 2^n different input states. In principle, such a circuit can be completely tested in 2^n steps.

Sequential digital circuits. The behaviour of a sequential digital circuit depends on both the inputs and the current machine state. A complete test must investigate, for each possible machine state, whether the correct subsequent state is reached for each possible set of input values. A more precise statement requires additional assumptions about the structure of the circuit. Suppose we consider an easily-understood example: a circuit with n inputs which has (1) an unambiguous initial state and (2) well-defined final states, and with (3) a maximum number z of state transitions, where z characterizes the longest path from the initial state to one of the end states. In addition, (4) it is possible to reset the machine state to the initial state at any desired time and (5) all states and state transitions are observable (which in practice means that we can measure the status of every flip-flop and, if necessary for logistical reasons, halt the circuit in any state (step-by-step operation or quasi-static testing). The first test steps are very simple:

1 Reset the circuit and check whether the initial state is actually achieved.

2 Provoke the first state transition with the first set of input values. Check whether the correct subsequent state is achieved. Then reset the circuit again.

3 Repeat step 2 for each of the other possible sets of input values.

In this way, we can check whether each possible state that can be reached from the initial state is correctly achieved. We could obviously continue the test by bringing the circuit to

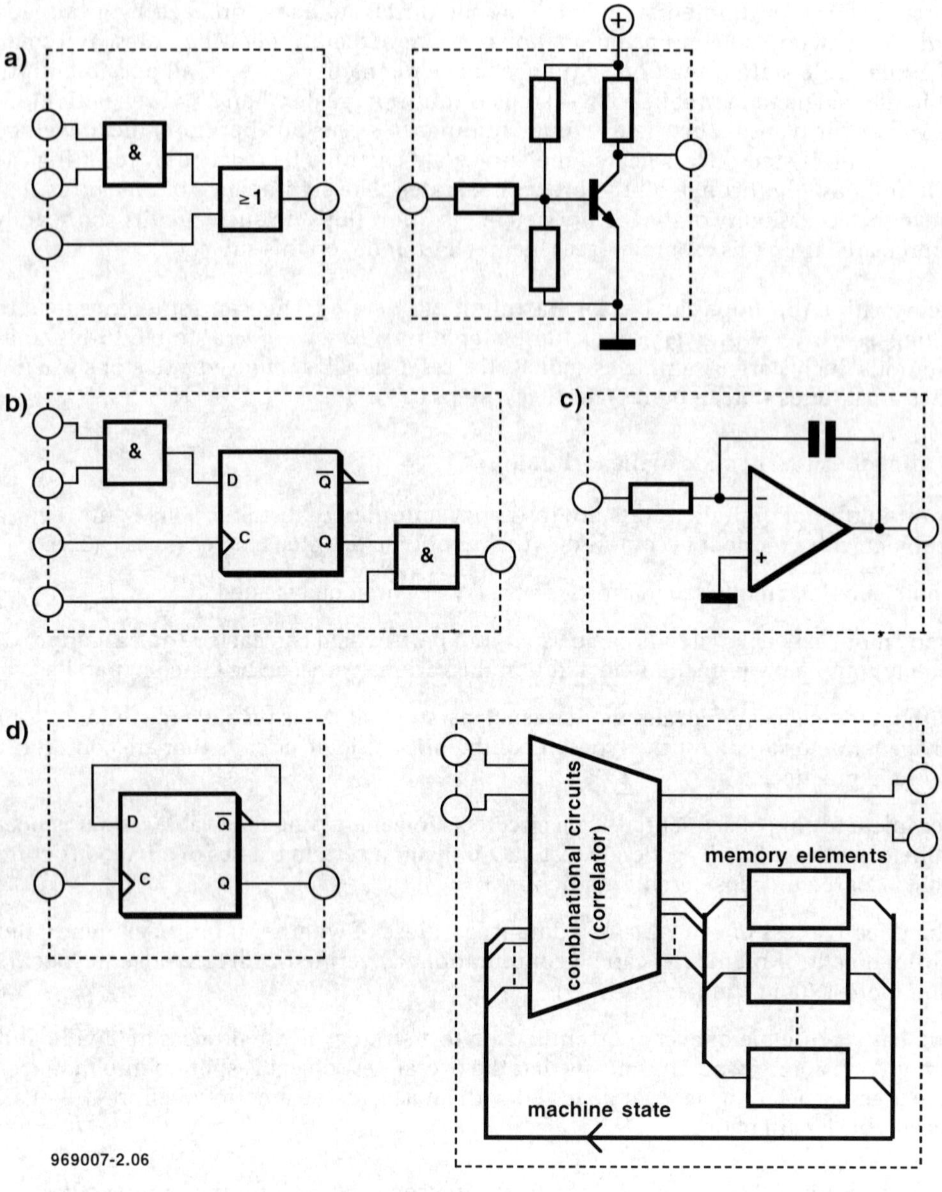

Figure 2.6 Examples of basic test object structures.

a) combinational logic circuit and transistor amplifier

b) digital circuit with memory element (flip-flop)

c) operational amplifier with feedback , wired as an integrator

d) 2:1 divider with D-type flip-flop, and block diagram of a synchronous automaton (state machine).

the first of the states immediately following the initial state and proceeding as already described for all possible input combinations and immediately following states, and so on until all possible state transition sequences have been checked with all possible input combinations. The total number of test steps is thus not greater than $2^{n \cdot z}$. Even with moderate values for n and z this yields a huge number of steps! Furthermore, it can well be that for each test step the associated machine state can only be reached by resetting the circuit and passing through all the preceding states. The significance of measures that improve the accessibility of the memory elements (flop-flops) in such circuits can readily be appreciated from this example – see Chapter 3 for more on this subject.

A circuit with n flip-flops can have a maximum of 2^n states. This can sometimes lead to very long tests even with a very small number of inputs. *For example:* a simple 16-bit binary counter, which starts counting as soon as the reset signal is removed and stops when it reaches an all-ones state, requires 65,536 test steps (2^{16}).

Regarding the current state of the technology

- Combinational digital circuits are in theory completely testable. This is however a problem with exponential complexity (the number of test steps increases as $O(2^n)$).
- Small (simple) circuits can in practice actually be completely tested.
- Many more extensive circuits can be split into parts which can each be at least approximately completely tested (and in addition the various parts can be tested in parallel).
- With reference to simple fault models (such as 'stuck-at'), it is possible to computationally generate test data for the types of combinational-logic circuits that are commonly found in practice.
- Complete testing of sequential-logic circuits is in general not achievable, and it is possible to computationally generate test data only for a certain classes of circuit structure (that is, based on considerable restrictions).
- The effectiveness of available test data (regardless of whether it has been generated automatically or manually) can be computationally verified with reference to specific fault models (fault simulation).
- The basic principle of every currently used test strategy is the process of 'divide and conquer' in some form. This means that the overall test object is split up into manageable pieces, each of which can be tested with an acceptable amount of effort and in an acceptable length of time.

Analogue circuits. With analogue circuits, it is not possible to limit the inputs to certain well-differentiated regions, such as the HI and LO inputs of digital circuits. A complete test would thus take infinitely long. But even here we can realize manageable tests if we make additional assumptions about the nature of the circuit and/or the restrict the range of the possible input stimuli, which means limiting the minimum and maximum values of the amplitude and frequency.

Examples:

- using sine wave or square wave signals as test stimuli (linear analogue circuits can even be tested exactly in this way, based on the frequency response and transfer function),
- checking the DC behaviour (such as recording voltage/current characteristic curves),
- testing in certain operating regions (small-signal behaviour, step response and so on),
- testing with the feedback path opened.

Complete testing in practice. Whether or not complete testing can be performed is in practice determined almost solely by the amount time that is either allowed or considered acceptable for conducting the testing:

- With automatic testing (using test software or suitable test equipment), even combinational-logic circuits with (for example) 20 inputs are currently completely testable. (2^{20} = 1,048,576. If each test step takes 1 µs then the entire circuit can be checked in about 1 second. This is without question acceptable in the service environment.)
- In the manufacturing (mass production) of integrated circuits the time allowed for testing is typically only a few milliseconds, so complete testing is out of the question.
- With a very large number of inputs, complete testing is fundamentally not practical. *For example:* a 32-bit adder has 65 inputs (two 32-bit operands and an input carry bit). This would require 2^{65} test steps. Even if each test step takes only 100 ns, a complete test would take 10,000 years. ($2^{65} \approx 32 \times 10^{18}$, so the test would last approximately 32×10^{11} seconds. Since 1 year is roughly 32×10^{6} seconds, this is around 10^{5} years.)

Testing with reference to a specific fault model. According to Murphy's Law, testing even the simplest circuit would tend to be a never-ending task. For this reason, all test procedures are based on considerably simplified fault models. The simplest of these (the 'stuck-at' model) is by far the most often used, and a more complex model is only used when it is required by the nature of the technology in question. Figure 2.7 shows a simple example that illustrates how a specific fault model (the 'stuck-at' model) works at various levels.

Test vectors. Since complete testing is often practically impossible, testing must be limited to *selected* operand values. Such values are called *test vectors* or test samples.

Test vectors and fault models. The relationship between test vectors and the fault model can be explained by means of a simple example (Figure 2.8 and Table 2.7).

The fault table. We start by assuming the 'stuck-at' fault model and the single-fault hypothesis, and on this basis we work through the circuit to determine every possible fault. Every signal line can be affected by a fault, and a fault can result in either a fixed '0' level or a fixed '1' level. If we observe the output *G*, we can only detect a fault if it causes the actual output value to differ from the value expected in the absence of a fault. Each observable fault is marked in the fault table with an asterisk '∗'. The fault table can be read as follows: "a given fault (such as S-A-0) can only be observed on output G with the marked set(s) of input values". Table 2.8 shows in detail the conditions under which each of the fourteen possible 'stuck-at' faults can be observed at the output *G*.

In order to completely test the circuit with respect to the 'stuck-at' fault model (that is, with 100% coverage), it is sufficient to apply one suitable set of input values (a test vector) for each possible 'stuck-at' fault – and given the single-fault hypotheses, it does not matter if a particular test vector relates to more than one fault. A possible set of test vectors (a *test suite*) for this example is listed in Table 2.9. (These four test vectors suffice to test the circuit in Figure 2.8 for input and output 'stuck-at' faults with 100% coverage.)

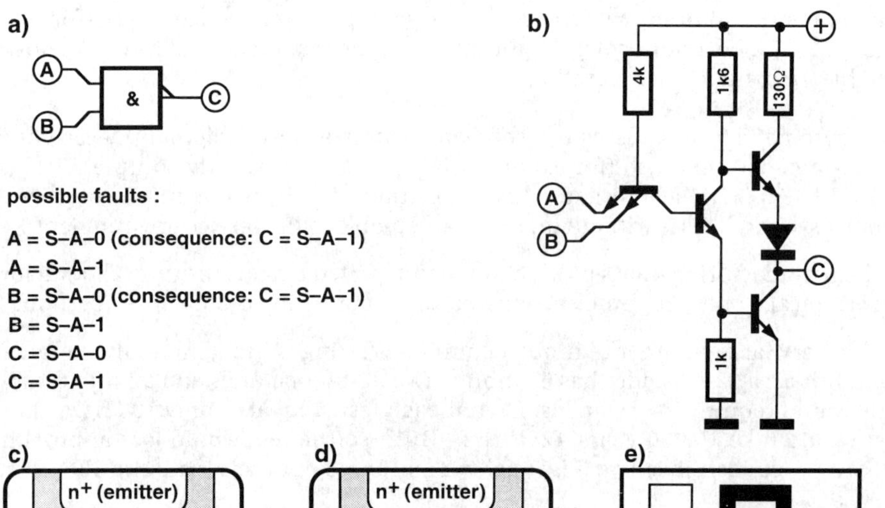

Figure 2.7 Stuck-at faults with a 2-input NAND gate.

a) gate level, b) basic TTL circuit, c … h) silicon structure

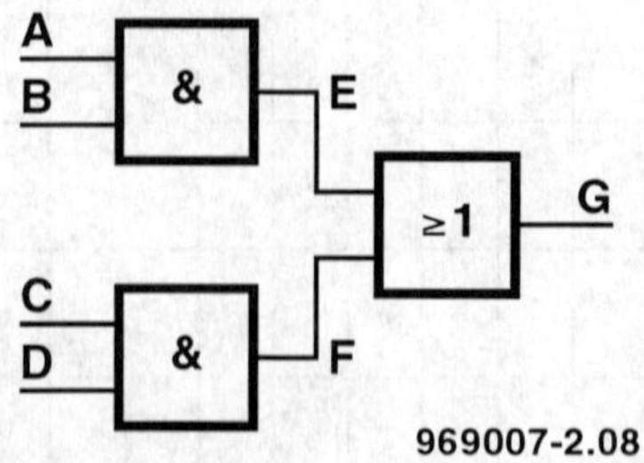

Figure 2.8 An example of a combinational circuit.

Partially complete (pseudo-exhaustive) testing. A large circuit can often be partitioned into sections that are completely testable (either by applying all 2^n input values or by assuming a particular fault model, as in the above example). Such tests can often be performed in parallel (concurrently). Partitioning is especially likely to be possible with circuits that have multiple outputs, and in particular if some of the the outputs do not depend on all of the inputs (see Figure 2.9). Each output can then be individually tested by subjecting the related portion of the circuit to a complete test.

Testing individual signals: the 'sensibilized path' approach. In this approach the circuit is tested by checking each of the inputs one after the other. For each input to be checked, all other inputs are set to values such that (in a fault-free test object) a change in the value applied to the input currently being checked must cause a change in the observed output

---

Explanation of Figure 2.7:

a) The logic gate as a functional element. Based on the single-fault hypothesis, there are six possible faults, whereby an S-A-0 fault at an input results in a S-A-1 fault at the output ('S-A-0' = 'stuck-at-0' or LO, 'S-A-1' = 'stuck-at-1' or HI).

b) The logic gate as a TTL transistor circuit. At this level, several types of faults are possible (shorts, opens, incorrect resistance values and so on). It can be shown that many of these faults will show up as 'stuck-at' faults at the inputs or output of the gate. Even 'trivial' faults, such as a broken bonding wire, will typically show up as 'stuck-at' faults (a broken bonding wire typically shows up as S-A-1).

c ... e) Details of an integrated-circuit transistor: (c) shows the cross section of a fault-free transistor, (e) shows the plan view and (d) illustrates the effect of a process fault: the collector and emitter are practically shorted together, so that the transistor acts more or less like an ohmic connection. If this is the upper transistor of an output pair, the gate will have an S-A-1 fault on the output.

f ... h) Example of part of an integrated circuit. Complete layout diagrams take up a lot of space, and it requires a certain amount of experience to understand them. For this reason, we limit ourselves to a simple multiple-transistor structure. (f) shows the schematic diagram and a fault-free metallization; (g) and (h) illustrate faults that can occur: the metallization can be broken or completely missing as in (g), there can be extra metal which causes shorts as in (h), and so on.

Number (hex)	Truth table							Fault table													
	A	B	C	D	E	F	G	A – 0	A –1	B – 0	B –1	C – 0	C – 1	D – 0	D – 1	E – 0	E – 1	F – 0	F – 1	G – 0	G – 1
0	0	0	0	0	0	0	0										*		*		*
1	0	0	0	1	0	0	0						*				*		*		*
2	0	0	1	0	0	0	0								*		*		*		*
3	0	0	1	1	0	1	1					*		*				*		*	
4	0	1	0	0	0	0	0		*								*		*		*
5	0	1	0	1	0	0	0		*				*				*		*		*
6	0	1	1	0	0	0	0		*						*		*		*		*
7	0	1	1	1	0	1	1					*		*				*		*	
8	1	0	0	0	0	0	0				*						*		*		*
9	1	0	0	1	0	0	0				*						*		*		*
A	1	0	1	0	0	0	0				*		*		*		*		*		*
B	1	0	1	1	0	1	1					*		*				*		*	
C	1	1	0	0	1	0	1	*		*						*				*	
D	1	1	0	1	1	0	1	*		*						*				*	
E	1	1	1	0	1	0	1	*		*						*				*	
F	1	1	1	1	1	1	1											*		*	

Table 2.7 *Truth table and fault table for the circuit of FIgure 2.8. 'A-0' stands for 'A = S-A-0' (A stuck at 0), 'B-1' for 'B = S-A-1' etc.*

The fault	can be observed under the following conditions ...	because ...
A – 0	A = 1 & B = 1 & (C, D are not both 1)	output G should be 1 but stays at 0
A – 1	A = 0 & B = 1 & (C, D are not both 1)	output G should be 0 but goes to 1
B – 0	A = 1 & B = 1 & (C, D are not both 1)	output G should be 1 but stays at 0
B – 1	A = 1 & B = 0 & (C, D are not both 1)	output G should be 0 but goes to 1
C – 0	C = 1 & D = 1 & (A, B are not both 1)	output G should be 1 but stays at 0
C – 1	C = 0 & D = 1 & (A, B are not both 1)	output G should be 0 but goes to 1
D – 0	C = 1 & D = 1 & (A, B are not both 1)	output G should be 1 but stays at 0
D – 1	C = 1 & D = 0 & (A, B are not both 1)	output G should be 0 but goes to 1
E – 0	E = 1 & F = 0	output G depends only on E and should be 1, but stays at 0
E – 1	E = 1 & F = 0	output G depends only on E and should be 0, but goes to 1
F – 0	E = 1 & F = 0	output G depends only on E and should be 1, but stays at 0
F – 1	E = 1 & F = 0	output G depends only on E and should be 0, but goes to 1
G – 0	G = 1	output G should be 1 but is held at 0
G – 1	G = 0	output G should be 0 but is held at 1

Table 2.8 Explanation of the fault table shown in Table 2.7.

Number (hex)	A	B	C	D	A–0	A–1	B–0	B–1	C–0	C–1	D–0	D–1	E–0	E–1	F–0	F–1	G–0	G–1
5	0	1	0	1		*				*				*		*		
7	0	1	1	1					*		*				*		*	
A	1	0	1	1				*				*		*		*		*
C	1	1	0	1	*		*						*					
all test vectors:					*	*	*	*	*	*	*	*	*	*	*	*	*	*

Table 2.9 A minimal set of test vectors for the circuit shown in Figure 2.8.

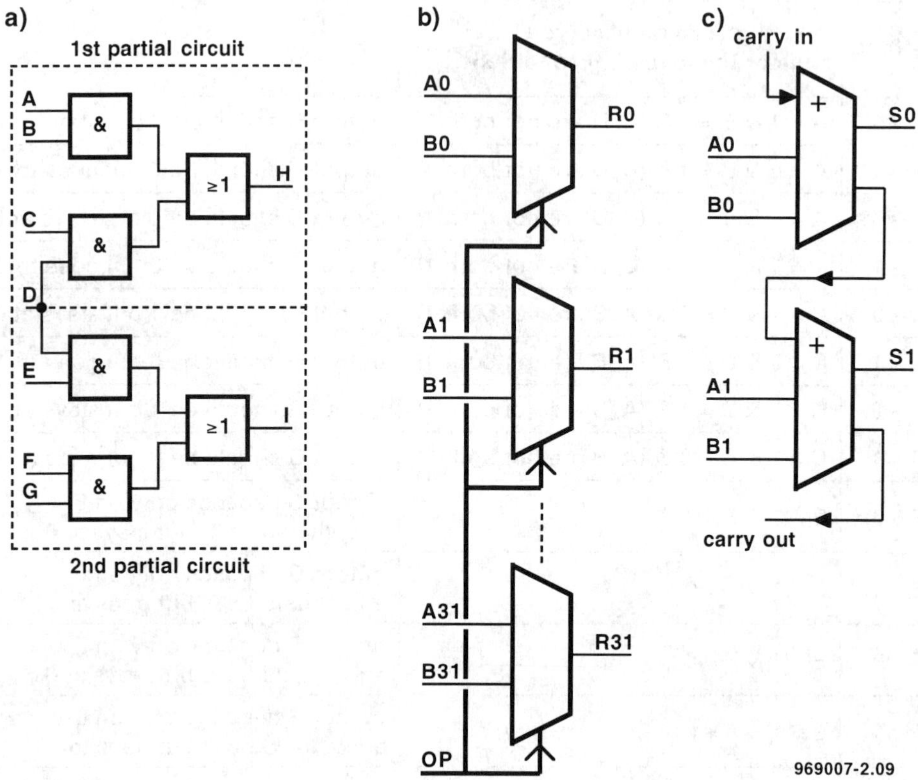

Figure 2.9 *Examples of circuits that are piecewise completely testable.*

a) This easily-understood example can obviously be split into two circuits corresponding to Figure 2.8. Just as obviously, each of these partial circuits can be tested using the test vectors listed in Table 2.9. The test vectors can also be applied simultaneously to the two partial circuits, as long as they do not have different values for the input D.

b) A simplified representation of the arithmetic logic unit (ALU) of a processor chip. The logical operations (AND, OR and so on) are individually implemented for each bit position. The line marked 'OP' determines the type of operation to be executed. For an OR operation, $R_0 = A_0 \vee B_0$, $R_1 = A_1 \vee B_1$ *etc.; for an AND operation* $R_0 = A_0 \bullet B_0$, $R_1 = A_1 \bullet B_1$ *and so forth. Clearly, all bit positions can independently be exhaustively tested, and every test step can be performed for each bit position simultaneously.*

c) A simplified representation of an adder (which could for example be part of an ALU). It consists of bitwise-independent addition circuits that are only interconnected by the carry-out and carry-in lines. The test thus can be divided into (1) an independent test of the addition logic in each bit position and (2) testing the carry-logic circuitry.

value. The set of selected input values thus causes the output response to depend only on the value applied to the input being checked. If the expected change in the output does not occur then the circuit has been shown to be faulty. A sort of 'fault-sensitive' signal path ('sensibilized path') has thus been configured between the input and the output. (This is also referred to as 'responsible switching' of the signal to be checked: a signal which is introduced into the hardware to be tested is said to switch 'responsibly' when a change (switch) in this input signal alone causes a change in related observable output value(s).) Figure 2.10 illustrates this principle, based on the circuit shown in Figure 2.8.

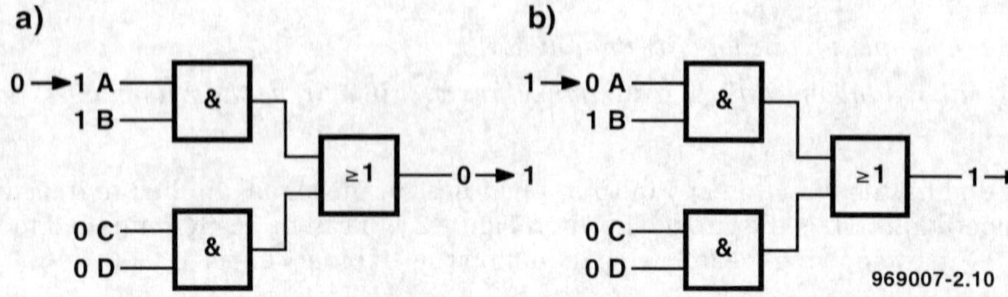

Figure 2.10 The principle of the 'sensibilized path' or 'responsible' switching.

a) Input A is switched from 0 to 1. Since B is held at 1, the output should also go to 1. The second AND gate does not affect the output, since C and D are both 0.

b) Input A is switched from 1 to 0. The second AND gate does not affect the output, because C and D are both 0. Since B is held at 1, switching A to 0 should cause the output to also go to 0.

Practical tip: it is only rarely possible to check whether all of the tested signals have actually switched responsibly, and in addition the specification of the test severity (for example, with purchased test software) is not always accurate. However, it is frequently sufficient to verify that all signals which must switch actually do switch (regardless of whether or not they switch responsibly), at least as long as you also check the signals with an oscilloscope at the same time to see whether the switching appears to be 'reasonable' with regard to signal levels and edges, pulse lengths and possible disturbances.

Incomplete testing

Testing is obviously incomplete whenever there is not an adequate set of test vectors available or when the available test vectors cannot all successfully be applied to the inputs. The latter situation can occur when the circuit to be tested is embedded in another circuit, so that the test vectors cannot be directly applied (the key factor is *reachability*).

However, there are faults that in principle cannot be detected by applying test vectors to the inputs and observing the outputs (*redundant* faults). This can be illustrated by a quite simple example (Figure 2.11a). A 'stuck-at-1' fault at the input *E* is not observable in this example because the inputs *C* and *D* must be active for such a fault to be recognizable, but then *A* and *B* are also active and the output *H* goes to 1 regardless of the state of *E*.

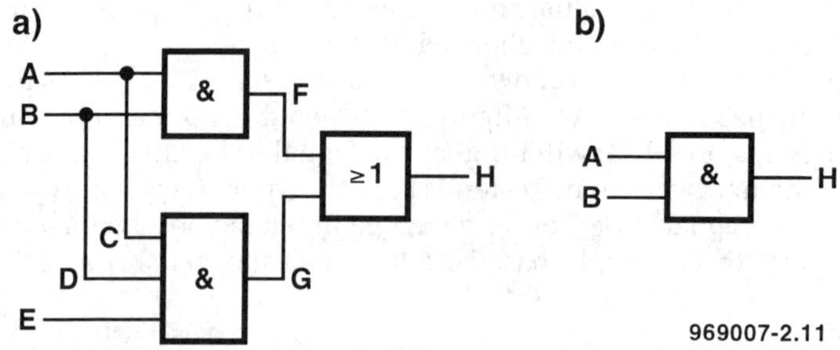

Figure 2.11 Example circuits for a redundant fault.
a) Circuit with redundant fault, b) redundancy-free circuit with the same function.

An input that has no effect on any output, regardless of the signal applied to it, is quite simply superfluous. Thus the circuit shown in Figure 2.11a can be easily simplified to that of Figure 2.11b by applying the following reduction rule of binary logic:

$f_1 f_2 \vee f_2 = f_2$ (where f_1 and f_2 may be any desired logic functions); thus $(A\,B\,E \vee A\,B) = A\,B$.

It is obvious that a redundant fault is equivalent to redundant logic. Put another way, the testability of a circuit that has been reduced to a minimum (in the sense of boolean algebra) cannot be affected by redundant faults as long as the circuit is accessible and observable, which means that test vectors can be applied to its inputs and its outputs can be sensed. (For a counter-example, see below.) In practice it is not always possible to completely avoid redundancy. The reasons for this are (among others):

- only a partial minimalization can be achieved (because the circuit is so complicated),
- there are other design objectives or constraints that prevent a logically 'perfect' minimalization (such as being required to partition the circuitry into two or more integrated circuits or circuit cards, and having to manage this with a limited number of pins or interconnections),
- the circuit is not implemented at the gate level, but is instead put together from already-available 'building blocks', and in addition the design must fit into a certain technological framework (such as using circuit cards with prescribed dimensions, implementing the design using PLDs or FPGAs with pre-defined cell and interconnection structures, using ASICs with a particular technology and so on).

Note: redundant faults must not be confused with insufficient test severity or inadequate coverage. The coverage can be improved by means of additional test vectors (assuming that the circuit is accessible) without intervention in the hardware to be tested or the test 'philosophy'. Redundant faults, by contrast, fundamentally cannot be recognized without significant interventions. An example of such an intervention would be the direct observation of input E in Figure 2.11a (to see whether or not it switches). But for this to be possible the input E must in the first place be accessible (which means, for example, that you can attach a test probe to it or sense its state using test software).

Fault simulation. In fault simulation the hardware of the computer (or other device) is simulated by software and the individual test routines are used as simulation examples. In the initial absence of faults, each test should run in an orderly manner. Thereafter, faults are simulated one after another (fault provocation: for example, the program can simulate 'stuck-at' faults by setting each signal of the simulation model in turn permanently LO or HI) and the test is repeated for each such fault. It is thus possible to check whether or not the fault in question is recognized by the test, and thereby to exactly determine the test coverage. (Usually, you must then decide whether to accept this degree of coverage or to improve the test). In addition, by logging the simulation results you can obtain an exact listing of the response of the test to each individual fault (the *error dictionary* or *fault dictionary*). When the test is run in actual practice, you need only look up the error messages in this 'dictionary' in order to identify the suspect circuit, connection and so on. From the yes/no results of a set of individual test routines, you can determine which parts of the system are the most suspect by seeing which ones are identified by all tests in common (differential diagnosis; see Section 2.8).

Test severity with respect to verification and localization. An ideal verification test should meet the requirement that it yields a correct yes/no response for any arbitrary fault. The test severity in this case does not necessarily depend on the number of available test vectors. It is sufficient that each possible fault simply yields a 'no' result. It is not specifically necessary to individually check out each unit, subassembly, IC and so on in turn. Instead, 'complex' tests, using a relatively small number of test vectors that have been selected for their effectiveness, can without question be completely adequate. An ideal localization test, on the other hand, should exactly identify the suspect unit, subassembly or component. This is only possible if the available test vectors, test steps and so on have been specifically selected for the individual hardware items in question.

Causes of insufficient test severity and/or coverage

Inadequate reachability or observability. A combinational circuit can be much more complicated than the one shown in Figure 2.8, but it can still be readily testable as long it is actually possible to apply all of the test vectors to the inputs and to sense the output values. This is true if (for example) its inputs and outputs are wired to the pins of an IC or to a circuit-board connector. Frequently, however, the circuits that we wish to test are embedded in other circuits and thus are not easily accessible. Sometimes this difficulty can be overcome, but often it cannot (see Figure 2.12).

Note: sometimes it is still possible to completely test such circuits. You simply have to examine the fault table to see if there are other test vectors that can be used together with the original vectors to allow all 'stuck-at' faults to be checked, in spite of the given limitations. Figure 2.13 shows examples of how to do this for the circuits shown in Figure 2.12. First, strike out all test vectors in the table which cannot be applied because of the given limitations. Then compute the columnwise logical OR of all remaining entries, and enter the results in an additional row. (In other words, examine each column in turn to see if it has at least one '×' that is not stricken out, and if so then enter a '×' in the associated cell of the new row. If the resulting row has a '×' in every cell, then the circuit is still completely testable.)

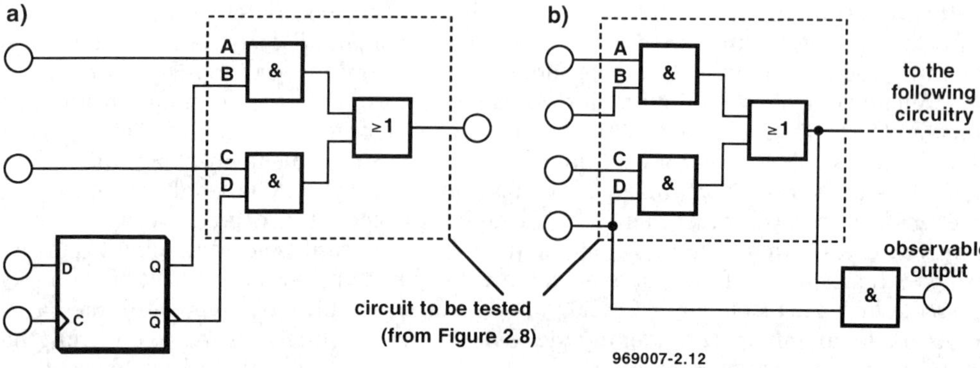

Figure 2.12 *Insufficient reachability and observability (circuits as shown in Figure 2.8 are imbedded in a larger circuit.).*

a) *Inputs B and C of the circuit under test are controlled (here, by a flip-flop) such that they cannot be both HI or both LO at the same time. Even if the state of the flip-flop can be set as desired during the test, it is not possible to apply test vectors 5_H and A_H of Table 2.9 to the circuit to be tested. It would thus not be possible to detect stuck-at faults on either the B or C inputs using the test vectors of Table 2.9.*

b) *The output of the circuit under test cannot be observed directly. Instead, it is only accessible via an additional AND gate (located at an IC pin or circuit-board connector, for example), and then only when input D is HI. In this case the behaviour of the circuit cannot be checked when D is LO, as with test vectors A_H and C_H of Table 2.9.*

Examples (see Figure 2.13):

a) For the circuit of Figure 2.12a: after all cells where *B* = *D* have been struck out, there are still marked cells in all columns (100% test coverage is still possible).

b) For the circuit of Figure 2.12a: after all cells where *D* = *0* have been struck out, there is no longer any marked cell in the column '*D-1*', so that this fault is not testable (the degree of coverage drops to 13 of the 14 faults, or roughly 93%).

The fundamental way to avoid the problem of incomplete testability due to insufficient accessibility or observability is to use technical means to enforce their presence. These range from test points and simple measures for connecting adapters, to test provisions built into the circuit (the scanning technique). See Section 3.3 for more information.

Incomplete test vector suites. Test vectors can be generated manually, with the help of a computer program or (as it were) by the 'lottery' principle (manual and automatic test data generation, (pseudo)random test patterns). With manually or randomly-generated test vectors it is clear that you cannot expect absolute perfection (in the form of 100% coverage). But 100% coverage is not always possible even with automatic (computerized) test vector generation, even if you can be satisfied with the simple 'stuck-at' fault model. This is because the computation time increases enormously with increasing circuit size; com-

a

Number (hex)	A	B	C	D	E	F	G	A·0	A·1	B·0	B·1	C·0	C·1	D·0	D·1	E·0	E·1	F·0	F·1	G·0	G·1
	Truth table							Error table													
0	0	0	0	0	0	0	0										x		x		x
1	0	0	0	1	0	0	0						x				x		x		x
2	0	0	1	0	0	0	0								x		x		x		x
3	0	0	1	1	0	1	1					x		x				x		x	
4	0	1	0	0	0	0	0		x								x		x		x
5	0	1	0	1	0	0	0		x				x				x		x		x
6	0	1	1	0	0	0	0		x						x		x		x		x
7	0	1	1	1	0	1	1					x		x				x		x	
8	1	0	0	0	0	0	0				x						x		x		x
9	1	0	0	1	0	0	0				x		x				x		x		x
A	1	0	1	0	0	0	0				x				x		x		x		x
B	1	0	1	1	0	1	1					x		x				x		x	
C	1	1	0	0	1	0	1	x		x						x				x	
D	1	1	0	1	1	0	1	x		x						x				x	
E	1	1	1	0	1	0	1	x		x						x				x	
F	1	1	1	1	1	1	1											x		x	
								x	x	x	x	x	x	x	x	x	x	x	x	x	x

b

Number (hex)	A	B	C	D	E	F	G	A·0	A·1	B·0	B·1	C·0	C·1	D·0	D·1	E·0	E·1	F·0	F·1	G·0	G·1
	Truth table							Error table													
0	0	0	0	0	0	0	0										x		x		x
1	0	0	0	1	0	0	0						x				x		x		x
2	0	0	1	0	0	0	0								x		x		x		x
3	0	0	1	1	0	1	1					x		x				x		x	
4	0	1	0	0	0	0	0		x								x		x		x
5	0	1	0	1	0	0	0		x				x				x		x		x
6	0	1	1	0	0	0	0		x						x		x		x		x
7	0	1	1	1	0	1	1					x		x				x		x	
8	1	0	0	0	0	0	0				x						x		x		x
9	1	0	0	1	0	0	0				x		x				x		x		x
A	1	0	1	0	0	0	0				x				x		x		x		x
B	1	0	1	1	0	1	1					x		x				x		x	
C	1	1	0	0	1	0	1	x		x						x				x	
D	1	1	0	1	1	0	1	x		x						x				x	
E	1	1	1	0	1	0	1	x		x						x				x	
F	1	1	1	1	1	1	1											x		x	
								x	x	x	x	x	x	x	!	x	x	x	x	x	x

969007-2.13

Figure 2.13 Investigating the testability of the circuit of Figure 2.12.

putational processes for test data generation belong to the NP-complete class of problems (see Section 2.1). The size of the problem is determined by factors such as the number of inputs and outputs, the number of gates or the number of storage elements (flip-flops).

Expedients:

1 'Divide and conquer': partition the circuit into manageable pieces (which usually requires that each piece must itself be accessible and observable)

2 Accept less than 100% coverage.

2.3.4 System design for testing and maintenance

It is not possible to successfully produce complex electronic systems, or to operate them for any extended period of time, if the importance of testing and fault-finding is totally neglected. Since we are only interested in issues relating to equipment servicing and relatively easily-understood design projects (based on standard integrated circuits), we can limit our attention to the technological levels that cover the range from the system level to the component level.

Test and maintenance provisions that are 'acceptable' to the user (that is, that the user or maintenance technician can actually employ) are commonly provided (in line with the current state of the technology) for verifying functional assemblies, units and systems and localizing any fault that may occur to an FRU (such as a plug-in card). These provisions include test programs, built-in test hardware, test and measurement equipment, test adapters and specialized tools, plus – not to be forgotten – fault-finding guides and service documentation.

When the system is being designed, two fundamental decisions must be taken:

1 Will the normal operational processes also be used for testing and fault-finding, or will normal operational processes be kept completely separate from test and fault-finding processes? The first option is called *functional* and the second one *structural.*

2 Will suitable provisions be built directly into the equipment, subassemblies and so on, or does the plan call for supplementary test equipment and accessories? The first option is called *self-testing* and the second one *external testing.*

Test- and service-oriented system design ultimately comes down to a question of optimizing a cost/benefit ratio, as is usual with technological issues, and the solution naturally involves compromises. In practical system designs, we thus find primarily 'mixed' forms of the above options.

Functional diagnosis, functional testing. This type of testing consists of verifying whether the test object (DUT) correctly executes the functions for which it is designed. *For example:* functional testing of an arithmetic logic unit (ALU) consists of executing all of the logical operations which the unit can perform (AND, OR, addition and so on) for a selected set of inputs and comparing the actual results to the correct results.

Structural analysis, structural testing. A device or system obviously must function correctly if (1) it consists of fault-free parts that are (2) correctly interconnected. You can thus replace the verification of the actual functions of the overall item by the combination of (1) individual verification of each its parts and (2) verification of the interconnections (interfaces).

For example: suppose that a processor's ALU is constructed using integrated-circuit gates on a printed circuit board. We verify each gate individually by measuring its truth-table behaviour and comparing the results to the correct values. Then we check all interconnections (continuity and short-circuit tests). The advantages of this approach are obvious: we have (1) reduced the complexity (by testing connections and individual gates each having only a few inputs, instead of the much more complicated arithmetic operations), and (2) our testing is independent of the actual function of the circuit. For each individual gate we can – with regard to the required number of test steps – certainly afford to do a complete test. Both types of test (gates and interconnections) are fully independent of the function and/or the intended use of the circuit in question. It practically does not matter whether the overall circuit is quite simple or very complicated.

However, there is no advantage without some sort of disadvantage: since we no longer can make use of the normal functions of the test object, we need additional measures (1) to ensure the accessibility and observability of the test processes and (2) to control the tests (generating and applying test data, executing the test steps, collecting and evaluating results).

Functional testing and diagnosis of computer systems. The basic idea is to use the normal operating functions for the purposes of testing and diagnosis. In the most extreme case, all that is needed to allow program execution to be traced step by step is some sort of control and display unit. The following features have proven themselves to be suitable for such testing (see Figure 2.14):

- a choice of normal operating mode, single-instruction mode (instruction step), single-cycle mode and breakpoint mode (comparison stop),
- the ability to enter a starting address for the program, and to run the program starting at a user-entered address,
- at least one type of breakpoint mode (the most important is 'stop on instruction match', but 'stop on data match' is also useful, and in the latter case it is advantageous to be able to distinguish between data reads and data writes),
- some way to display the contents of the instruction counter and the most important processor registers,
- some way to display and alter the contents of registers and user-specified memory locations,
- a 'start' button which either triggers the execution of the next clock cycle or the next instruction, or causes a return to the normal operating mode.

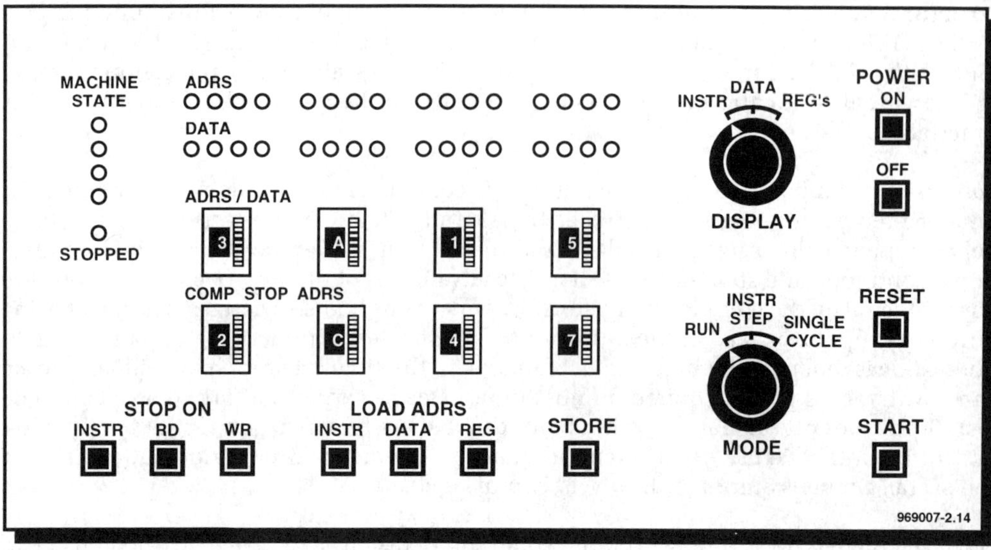

Figure 2.14 The control panel of a (hypothetical) computer.

With the present state of technology, we would naturally use a monitor and a keyboard (an operator terminal or service processor) in place of arrays of LED displays, switches and buttons. Most microprocessor systems (and PCs) do not have built-in provisions for controlling their own operation and displaying data. In such cases, the usual techniques are: (1) to use a 'debugger' program that runs in the background and can be activated by the interrupt system, and which exploits the available control and display hardware (keyboard and monitor), or (2) to connect a special development-system computer to the processor, or to replace the normally-installed processor with a special version which is itself connected to a control computer (in-circuit emulation; see Volume 2).

With such facilities, it is possible to follow the execution of any sort of program instruction by instruction (and sometimes even to observe individual instructions clock cycle by clock cycle), and thus to attempt to track down faults (with the help of hardware measurements if necessary). For many years, this was common practice in the service environment, and it is still indispensable for the designer.

The impossibility of pure functional diagnosis. The more complicated the relationship between the hardware and the software, the less that diagnosis can be based on normal application functions alone. Clearly, it is totally impractical to use a program such as Windows, Word, Excel or the like in single-instruction mode to try to locate hardware faults. Consequently, test-mode operation must be separated from normal operation.

Elementary functional checking in modern service practice. In spite of this, in some cases functional testing is practically the only way to make progress. This applies to fault situations in which 'nothing works' after the computer (or microprocessor) is switched on or reset. (When a microprocessor system is operated for the fist time, this is more often the rule than the exception, but 'true' failures can also cause such fault symptoms.) Obviously, either the processor itself or its environment is suspect, but where should you start looking for the fault when not even a single bit of software can be executed? An self-evident solution is to individually check the processor and its environment. Quite simple test equipment is adequate for 'manually' checking the processor's functions; see Volume 2 for more information.

Self-testing and self-diagnosis. Here, the test provisions are built into the system, unit or assembly. In the ideal case, no additional equipment or aids are required for verification testing or tracking down whatever FRUs may be present. There are several types of self-testing and various design approaches, as follows:

- Provisions built into the hardware (built-in test equipment, or BITE). The hardware itself takes care of applying the test data to the inputs and comparing the output values to the correct values (see Figure 2.15). Naturally, this requires a certain amount of extra expense, but frequently it is possible to utilize portions of the existing circuitry to generate the test stimuli and control the progress of the individual tests. Such tests are commonly referred to as Built-in Self Tests (BISTs) (see Chapter 3).
- Checking with test software. In practically all software-controlled devices (controllers and processors) it is possible to arrange for self-testing using test software. The most significant difference between this and 'built-in' self-testing (which is purely hardware-controlled) is that the testing is based on the execution of machine instructions that are read from addressable memory. Normally, at least a ROM-resident test program is provided which is executed after switch-on or after a hardware reset (common designations for this are 'Power-on Self Test' or POST, 'Basic Assurance Test' or BAT and 'Power-on Diagnostics' or POD). This can be supplemented by plug-in diagnostic ROMs and test software that can be loaded from storage media. *Note:* do not confuse BIST with POST! A BIST is exclusively a processor function that does not involve external items such as the bus system, memory devices and the like when it is run (or put differently, a BIST will usually execute successfully even when something in the processor's environment is defective). With the exception of microcontrollers with built-in (on-chip) ROM, a POST, BAT or the like is executed from memory external to the processor, which means that for it to run at all the bus system and memory must be able to function at a certain minimum level.
- Self-testing using redundancy. You can check a hardware device for any arbitrary failure or malfunction by constructing two such devices and continuously comparing their outputs while operating them in parallel. This type of checking offers virtually unsurpassed severity, but it is quite expensive. For more information see Section 3.1.
- Plausibility checking. Here you limit yourself either to making spot checks (such as checking data paths or memory contents for individual errors) or to checking general operating conditions (supply voltages, ambient temperature and so on). See Chapter 3 for more information.
- Measures to ensure accessibility, reachability and observability. The applicable measures range from providing test points to the incorporation of additional hardware which makes it possible to selectively affect the circuits to be checked and/or more exactly sense their internal status. See Section 3.3 for more information.

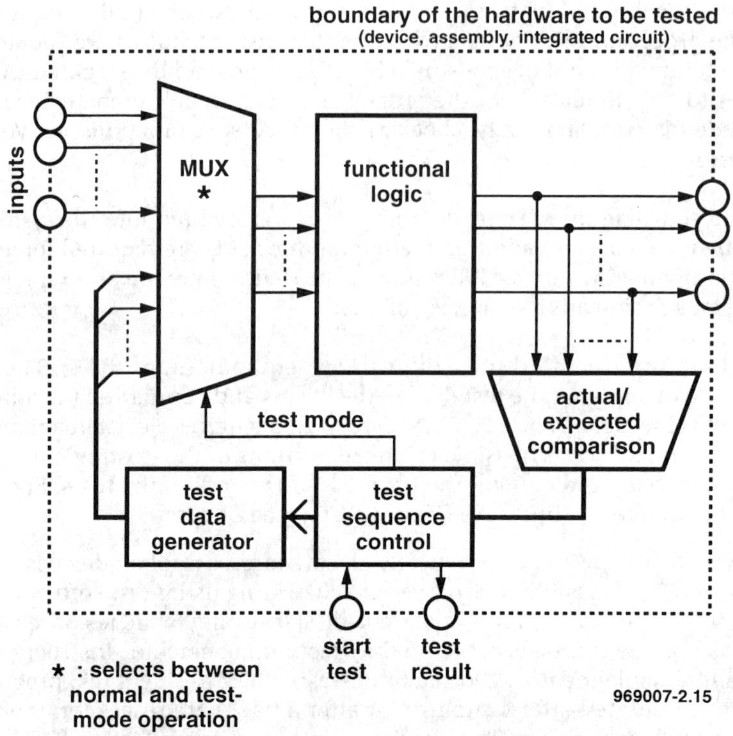

Figure 2.15 Hardware with built-in test provisions.

- Loopback tests. Some circuits can be checked by feeding their outputs back to their inputs (loopback connection). This technique is used in particular with interfaces. In the simplest case suitable 'loopback connectors' are used. However, there are also interface ICs which allow loopback connections to be switched on or off under software control. See Volume 2 for more information.

- Built-in test equipment. In more complex systems or devices, it can make sense not only to provide self-testing provisions in the 'functional' hardware, but also to provide special equipment or assemblies for recording faults, controlling test processes and the like. Given the state of the technology, it is an obvious choice to implement such equipment on the basis of programmable devices, so that you have a sort of 'computer inside a computer' (service processor). See Volume 2 for more information.

The core system. A test can only succeed if the test equipment is itself fault-free. In particular, every sort of self-testing is only possible if certain parts of the hardware actually work properly (for example, no software can run without a functioning power supply or with a completely defective processor). The circuitry that provides these essential elementary functions is called the *core system.*

The bootstrap diagnostics principle. Usually it is not necessary that the entire system be functional before testing can be performed. Test software can be run with only a portion of the total available memory, not all of the peripheral functions are necessary, and even the processor only needs to be capable of executing a few machine instructions. Its most important tasks are (1) controlling the progress of each test, (2) executing elementary test programs (with software comparison of actual and expected results) and test stimulus loops and (3) providing operation and display functions. (The basis of self-testing in a nutshell is that the core system is only a part of the overall system to be tested). Obviously it makes sense to keep the core system as small as possible. However, a disadvantage of relying on only a few basic functions is that it takes quite a long time to check out a complicated system. For this reason, we start with the smallest possible core system and gradually adds functions to it as they are checked out. (*For example:* a memory module that has been tested and found to be fault-free can be used to store additional test programs, an output device that has been proved to be in good working order can be used to output more extensive test results, and so on). The core system thus gradually becomes larger and larger until at last the entire system has been verified. If a particular step in this process fails, then the initial suspicion falls on the hardware that was added to the core system in that step. This manner of systematically increasing the extent of the core system is called *bootstrapping* or *bootstrap diagnostics*.

External testing. Here the test measures are not built in, but instead take the form of additional equipment or software. These items, which must themselves function correctly, then practically represent an 'external' core system. The choice of such an approach is sometimes dictated by the cost/benefit calculation. The reasons for choosing external testing are usually the following:

- The test object is mass-produced in large quantities (emphasis on minimum cost).
- Faults are relatively rare.
- Specialists must always be called in to do fault-finding (and they can bring the necessary equipment with them).

It is thus understandable that extensive self-test provisions are not provided for items such as washing machines, television sets and the like. Typical compromise solutions are:

- providing elementary self tests (ROM-resident test programs),
- in addition, monitoring certain ambient conditions and logging detected faults, to the extent that this is practical (using a non-volatile RAM, for example),
- providing connectors in the hardware for external testers (diagnostic connectors).

(The electronics of almost all modern autos, for example, is designed in this manner.)

On the other hand, external testing can be compulsory under certain conditions: if the bootstrap diagnostic approach cannot be readily used (which is the case for most of the commonly-available modern PCs – see Volume 2) then one is forced to use additional, external test measures. Strictly speaking, this means that it must be possible to check out every exchangeable item or assembly (FRU) at its interfaces using known-good test equipment or software, so that instead of actually searching for the fault we can break down the faulty system into its FRUs and check each of them individually. This can certainly take less time than meticulous testing and measurement (in most cases it is in fact the fastest approach), but it requires effective (and affordable) test equipment (testers, emulators, master machines, test PCs, test jigs, testbeds and the like). See also Section 2.7 and (for practical examples) Volume 3.

Test equipment. If you have decided for external testing, then you must next decide what sort of test equipment to use: general-purpose, commonly-available equipment (such a multimeter or oscilloscope) or special diagnostic and test equipment? As a rule, the larger the quantity of a given kind of test object, the more worthwhile is specialized test equipment. (With the present state of the technology, such equipment can largely be assembled from ready-made subassemblies or can even be based on a standard personal computer.) The advantages of specialized test equipment (compared to relying on general-purpose equipment) are: (1) less time is spent on fault-finding and (2) lower demands are placed on the qualifications of the technician doing the testing ("connect the tester, press the button and follow the instructions displayed the screen"). The other side of the coin is that in the course of time the 'button-pusher' loses touch with the technology (and sometimes just pushing a button is not enough!). In addition, such testers are only as good as the person(s) who design and build them. (With current technology, it is easy to produce a quite convincing user interface, but the actual test functions behind the interface are often much more modest than you might assume from the 'professional' user interface.)

Practical tips:

1 Keep testability in mind even with relatively simple, easily-understood (DIY) projects. Naturally, in such cases you will preferably resort to inexpensive test equipment and aids. Still, a few test points, a test connector or a DIP switch can work wonders (these don't cost a lot, but you must plan for them from the beginning and know what is actually needed). In addition, the already-present PC can often support the testing activities. We hope that the suggestions in Chapter 4 (as well as in Volume 3) provide a certain amount of help.

2 As a technician, to avoid becoming degraded to being a mere button pusher: keep in practice! (but naturally not at the expense of your regular work or the paying customer). Without 'true' ongoing training you won't achieve very much. You can get 'hands-on' experience via DIY projects (among other things).

2.4 Fault detection (verification)

2.4.1 Observing and monitoring normal operation

The observation of normal operating processes is the simplest form of verification. If departures from 'normal' operation are noticeable, then you can assume that a fault has occurred. In order to localize it you must individually investigate each such departure. Numerous measurements, experiments, and thinking processes are often necessary.

This manner of working has been common for years in this field and has proven itself with many traditional technical devices. Here 'traditional' means that the structure is directly related to the function ('single-purpose' hardware, as opposed to programmable hardware which must have software before it can be used for anything at all).

2.4.2 Monitoring diagnostic processes

It is not always effective to base fault-finding on normal operational processes, and sometimes it is not even possible. The following scenarios can illustrate this.

Note: the first two scenarios are valid regardless of working principles and technical implementations. Even in the case of a function that works directly (such as a mechanical function), special provisions for testing are required in these two scenarios. The last three scenarios, by contrast, relate specifically to all types of programmable devices and equipment.

1 *Testing measures or functions related to security and safety.* Dangerous situations are exceptional cases which normally do not occur. Measures designed to avoid or counter such situations must however work properly under all circumstances. In order to assure this, they must be checked from time to time. Functional testing in this case means reproducing the associated conditions as a test case. Sometimes this is literally done, as in the crash-testing of autos. Obviously you can only afford to perform a limited number of such experiments, and they are absolutely not feasible in everyday operation.

2 *Checking functions which relate to rarely-occurring events.* In this case the problem is not the risk but rather the amount of time needed. *For example:* the date display of a clock normally changes only once every 24 hours (and you may have to stay awake until midnight to observe it.)

3 *The interrelationships are very complicated.* In such cases the time required for testing becomes unacceptably long. Anyone who wishes to do detailed fault-finding using normal operational functions must be familiar with all features of the test object.

4 *Individual functions can be observed only with great difficulty or not at all.* For example, they may be implemented inside a highly-integrated chip. In addition, it is perfectly typical of systems based on computers that their functions can only be observed at their external interfaces. On the inside it is, practically speaking, only possible to monitor memory accesses, bus transactions and the like; these are elementary processes which appear the same for all functions. To be able to associate the bus cycles of a microprocessor with concrete functional processes (for example), you will need an in-circuit emulation system as well as exact program documentation.

5 *The documentation is insufficient.* Let's assume that we have the necessary test equipment and the technical knowledge to be able to use it. Even then, exact and complete documentation for the application in question is required, which means that at least the source code must be available. This is however almost never available, at least with commercial software.

From the above scenarios, we can conclude that it is just about impossible to rely on normal functions or applications for fault-finding in the service environment. Even when a fault has been detected and the suspected source of the fault has been indicated by an application, you must employ test software, self-test provisions, test equipment and the like to more closely check out the device or assembly in question.

This situation is by no means always disadvantageous. The more that complicated functions are moved from hardware to software, the simpler the testing and fault-finding measures can be. In fact, a special advantage of items designed in this way is that they can be tested using independent test functions, regardless of the type or degree of complexity of the application functions.

A simple example can illustrate this: suppose that parts that are transported by a conveyer belt are to be counted by an optical-barrier device (interrupted light-beam detector). The counting range shall be 0 to 99,999, and an output signal is to be produced for counts of

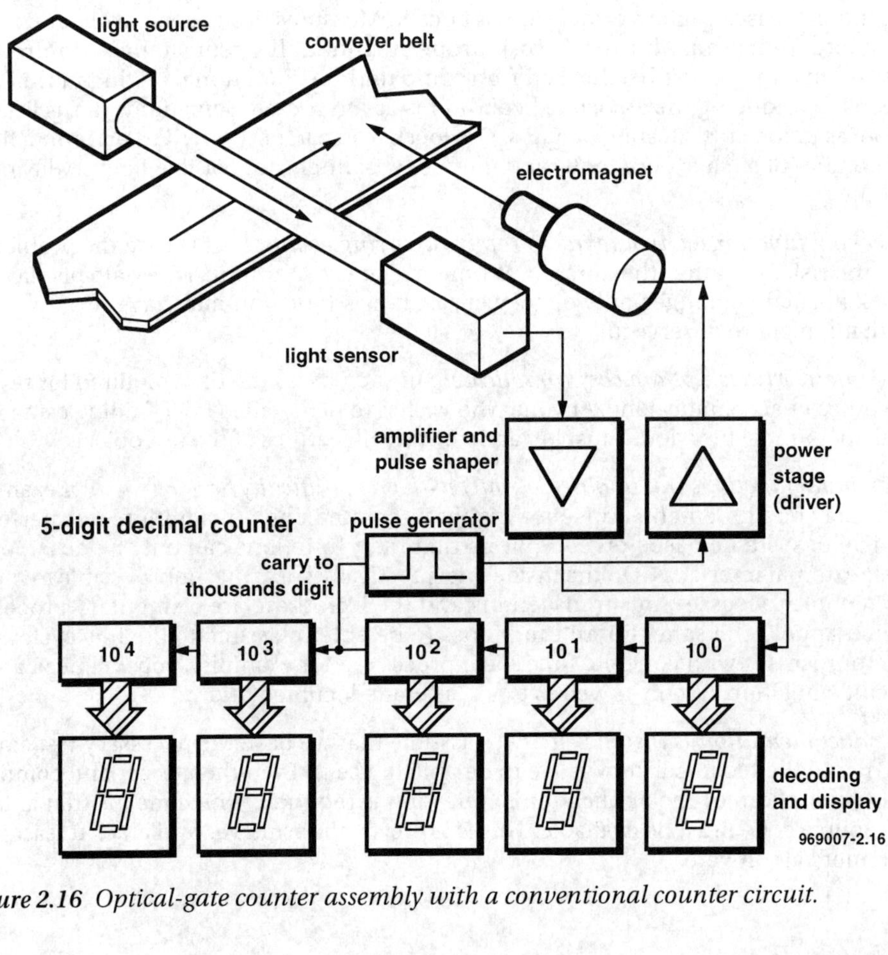

Figure 2.16 Optical-gate counter assembly with a conventional counter circuit.

1000, 2000 and so on, to actuate an electromagnet that pushes a part from the belt for sample testing. Conventionally, a special counter circuit would be used (see Figure 2.16). However, if a microprocessor is used (or in general, some sort of computer) you need only provide hardware to allow the status of the optical barrier to be sensed and the electromagnet to be actuated (Figure 2.17).

In the first case, you would have to actually transport 100,000 parts over the conveyer belt in order to completely test the hardware, or you would have to use special test equipment (such as a motor-driven disk to interrupt the light beam, a pulse generator connected in place of the optical gate or the like). In addition, purpose-built hardware such as that of Figure 2.16 becomes increasing complicated when it is required to perform more and more functions. If (for example) you wish to allow the count interval for selecting the sam-

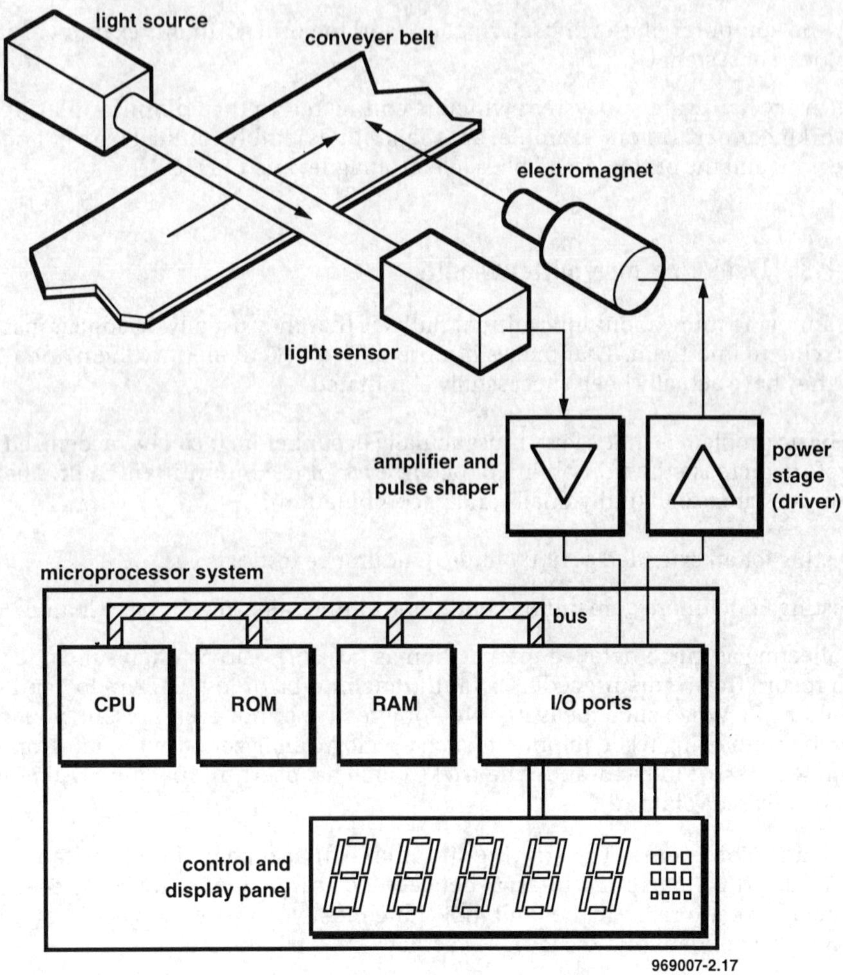

Figure 2.17 Optical-gate counter assembly with a microprocessor.

ple part to be configurable, to be able to count normal parts and sample parts separately and so on, then each such additional function requires additional hardware, all of which has to be tested.

In the second case it is sufficient to test the computer platform using test software. This is completely independent of the application in question, regardless of how complicated it may be. In addition, you merely have to check the operation of the optical barrier and the electromagnet subassembly. This requires only a simple test program which displays the status of the optical barrier (perhaps via an LED) and operates the electromagnet in response to pressing a button.

A 'computer-based' design thus makes it possible to reduce the complexity of testing. It is not necessary to check all functional relationships in order to locate hardware faults. Instead, it is sufficient to check two things:

1 Is the computer platform itself in good working order? (In our example: is the microprocessor system OK?)

2 Is every hardware subsystem which is connected to the computer platform in good working order? (In our example: are the input assembly (optical barrier and detection region) and the output assemblies (electromagnet and LED) OK?)

2.4.3 Detecting intermittent faults

The annoying thing about intermittent faults is that they usually disappear just when you are trying to find them. That makes it more difficult (1) to identify them and (2) to prove that they have actually been successfully eliminated.

The basic problem is that a fault must actually occur before it can be recognized. You must therefore increase the probability of occurrence of the intermittent fault. The following techniques are useful (individually and in combination):

1 testing for an extended period of time (endurance testing),

2 testing under more demanding conditions ('limit-value' testing; see Chapter 5),

3 collecting as much detailed information as possible. Large systems include provisions to record the events preceding a fault (machine-error log, backup log and so on; see Volume 2). When such measures are not effective or not even present, you must make do by connecting test equipment (such as a logic analyzer or pulse detector; see Chapter 4) or resort to tried-and-true tricks (such as inserting specific breakpoints in the software; see Volume 2).

Verification of repairs. The starting situation is that repeated intermittent faults have occurred, with a maximum interval between occurrences of n hours. After some action has been taken to remedy the fault, how long must you test the object in question to be reasonably confident that the fault has actually been eliminated?

Rule of thumb for test duration: the test must last at least twice as long as the interval between occurrences of the fault (thus $2n$ hours). *Practical tip:* extending the test duration to $3n$ hours usually does not yield any additional insights.

For the design of endurance tests, refer to Volume 2.

2.4.4 Burn-in tests

A burn-in test is an endurance test applied to a device before its actual use. The objective of burn-in testing is to detect early failures (reducing the 'running-in' time, or 'artificial aging'). The typical test duration is 24, 48 or 96 hours.

Practical tips:

1 All self-respecting manufacturers subject every deliverable piece of hardware to a burn-in test (and advertise the fact). Nonetheless, burn-in tests performed by the user are also worthwhile: (1) for newly-acquired hardware before it is put into service and (2) after remodelling or updating the hardware.

2 Test duration: systems of the size of a typical personal computer should be tested for at least 24 hours. No additional early failures are likely to occur after 200 hours of testing.

3 If you test for early failures, you must expect to find them. Your planning should thus take into account the time required to repair the faults. Repairing faults that occur during burn-in testing not only takes time itself, it also increases the overall testing time.

4 When 'running in' new hardware pay attention to its specified operating conditions. 'Strenuous' burn-in testing at elevated temperatures, high relative humidity and so on (see Tables 2.2 and 2.3) should be left to the manufacturer. The user risks voiding the warranty by employing such techniques.

For planning burn-in testing, see Volume 2.

2.5 Fault localization

2.5.1 Fault-finding strategies

The actual process of fault-finding (that is, fault localization) is fundamentally based on *comparing* and *deciding*. In contrast to verification, a simple yes/no result is almost never adequate. Instead, it is often necessary to evaluate a number of measured values, test results and so on. In this regard, a unique mixture of intuition, experience and precise logical thinking is required. Naturally, the preparation of test procedures, service manuals, test systems and the like is primarily based on exact logic, and it would seem obvious to consider the use of software to support the fault-finding process (expert systems; see Volume 2). However, in practice we see that intuition and experience often play an important role, and not infrequently a truly decisive role.

Fault-finding strategies describe basic techniques for systematic fault-finding (see Figure 2.18).

Note: see Section 3.3 regarding structural testing (the scanning technique) and Section 5.1.2 regarding in-circuit testing. The remaining strategies are individually described in the present chapter, starting with Section 2.5.3.

Localization precision. In any given case, the desired degree of fault-localization precision is the deciding factor in choosing a particular fault-finding strategy. Localization precision is often specified in a practical manner in the form of a 'down-to' specification (such as 'down to the unit level/card level/component level'). An exact specification consists of a list of all items that can identified for exchange or repair in order to remedy the fault. *Note:* in the practical service environment, such replaceable items are FRUs. To determine what can be considered to be an FRU, see Section 2.6.

Localization precision and verification. In order to remedy any given hardware fault by replacing an FRU, it must be possible to actually prove that each FRU present in the system is either suspect or fault-free. This means that whenever a fault is present in a given system, it must be possible to *verify* each FRU in the system in order to *localize* the suspected source of the fault to one of the FRUs.

2.5.2 Logical deduction in fault-finding

The decision process in fault-finding is based on simple principles (see Figure 2.19).

The simplest case. In figure 2.19a we see a single device, which has a fault that can be recognized at its outputs (by means of simple observation or measurement; the exact means does not matter). We can be relatively confident that this device is suspect if its inputs are correctly driven (fault-free). There are basically two ways to check this:

1 Examine the actual input signals.

2 Provide good input signals, using a known test stimulus. If no fault is seen, then presumably the device itself is OK but it is being incorrectly driven in normal operation.

The inverse conclusion (a device that produces a correct output is basically not suspect) is not always valid! Many faults do not show up under all possible operating conditions (latent faults); they are only evident when suitable input values are applied. See Section 2.5.4 for a more extensive discussion.

Two connected devices. The situation illustrated in Figure 2.19b can be reduced to testing each of the two devices separately, as in Figure 2.19a. Certain special situations can occur:

1 A test stimulus can be applied to device *1* but not to device *2* (device *2* receives only the input values that can be provided by device *1*), so the possibility of latent faults in device *2* must be taken into account.

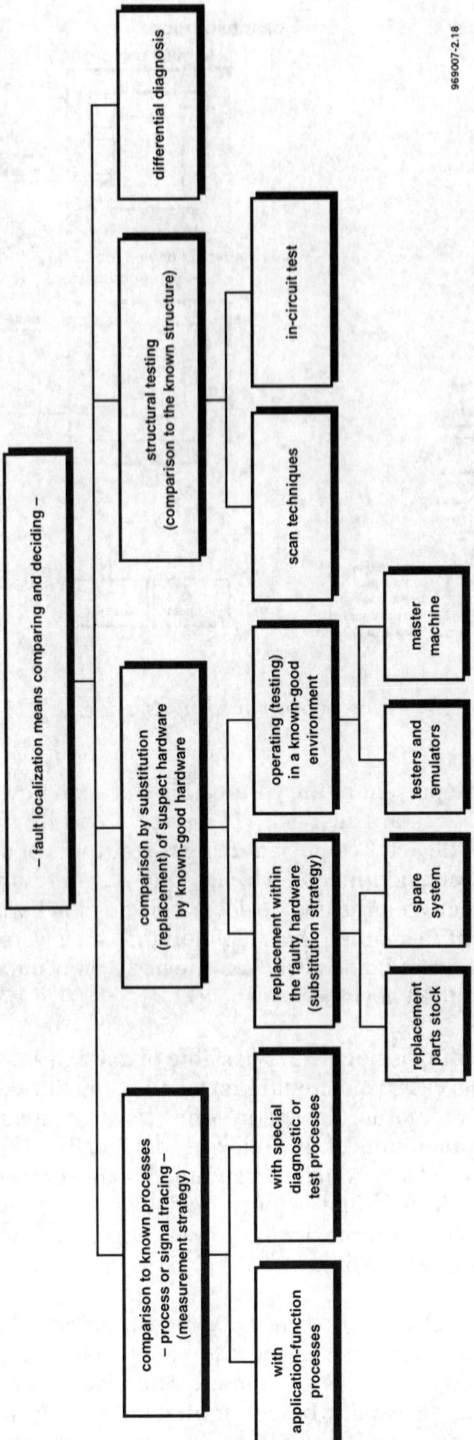

Figure 2.18 *Fault-finding strategies.*

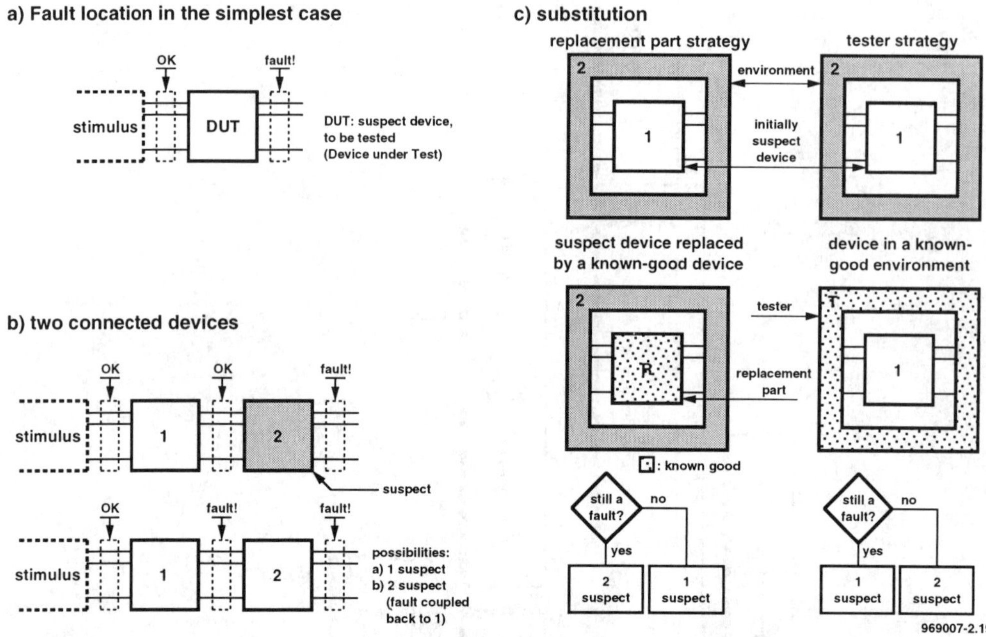

Figure 2.19 Elementary logical decisions in fault-finding (see text).

2 If a fault can be shown to be present at the connection between device *1* and device *2*, then in principle both devices are suspect: (1) device *1* can be incorrectly driving device *2* or (2) the inputs of device *2* can be affecting the outputs of device *1* (for example, due to a 'stuck-at' fault on an input of device *2*). One way to resolve the uncertainty is to break the connection between the two devices. You can check whether the outputs of device *1* are then different (if so, then the suspicion that they were being affected by the inputs of device *2* is confirmed) and also check device *2* with suitable test stimuli (if no fault is found in device *2* then device *1* is suspect).

Practical tip: the second procedure is not always possible (if suitable test equipment is too expensive, for example). In this case, you should first suspect the device that (apparently) is supplying incorrect signals to another device (in other words, first suspect the outputs of the source device and then the inputs of the following device). If this does not yield any result (or if it is not possible to collect enough measurements and/or test results to be able to draw a conclusion), you should then first suspect the device that is fundamentally more fault-prone (based on the level of complexity, number of components, number of contacts, use of special components and so on).

Substitution. This fault-finding strategy does not use measurements, but relies solely on exchanging parts of the system or unit (see Figure 2.19c). Suppose there is an initially suspect device that operates in a particular environment. The distinction between the suspect device and its environment is usually based on direct observation (a rough distinction can be made from the obvious fault symptoms) and/or systematic checking. This is the usual situation when (for example) one or more functional assemblies of a PC (such as the power supply, various plug-in cards, the motherboard and so on) appear to be directly suspect based on the observed fault symptoms. (In principle, if a fault is observed in a particular function then the initial suspicion should fall on the device or assembly that is responsible for this function.) There are then two possible ways to perform substitution:

1 The replacement-part approach: the suspect item is replaced by an equivalent item that is known to be fault-free (the replacement part). If the fault then disappears, the part that was replaced is suspect; otherwise its environment is suspect.

2 The tester approach: the suspect item is checked out in an environment that is known to be fault-free (a tester, emulator, master machine or the like). If the fault still occurs in the new environment, then the item being tested is suspect; otherwise its original environment is suspect.

Note: the decision processes used in substitution are usable in a similar manner for distinguishing between hardware and software faults (for example, by trying out suspect software on known-good hardware).

2.5.3 Process monitoring and signal tracing

Process monitoring and signal tracing mean checking whether or not the actual process values (signals) and/or their behaviour over time agree with the correct values and/or time progression of these signals.

Static, quasi-static and dynamic testing

Depending on the time-related conditions under which testing is performed, testing can said to be static, quasi-static or dynamic.

Static testing. In static testing the signals do not change during the test (or at least they *should not* change; measurable changes are thus signs of a fault). Static testing can be used to check power supplies, interfaces that can be operated statically and so on, but it can also be used for testing combinational logic circuits if it is possible to stop the clock. This is not always possible (for instance, dynamic RAMs and various types of processors must be continuously (dynamically) driven, or they may require a certain minimum clock frequency). An important facility for static testing is the ability to reset the hardware.

Practical tip: simple test and measurement equipment (a digital multimeter or logic tester) is usually adequate for static testing. However, a check with an oscilloscope is always a good idea whenever you suspect oscillations, open leads or the like.

Quasi-static testing. The device under test is checked out using a sequence of test steps executed directly one after the other. The duration of each individual test step is determined by the test equipment. From the point of view of the test equipment this is a series of static tests. However, they are executed more quickly than fully static tests in which the test stimuli are manually generated or applied and the outputs manually checked or measured. Quasi-static testing is the usual operating mode of many types of automated test equipment. Automatic testing using diagnostic connectors (test interfaces; see Section 3.3) is also quasi-static in nature.

Dynamic testing. The device is tested while it is running at its full working rate. This is the case with the usual test software. If you wish to examine operating processes in detail, then test and measurement equipment with suitable capabilities is required, such as oscilloscopes, logic analyzers and automated testers.

Procedures for signal tracing

Signal tracing is a stepwise process, in which all signals in question are checked in a particular sequence. Each signal has a source, which is the directly preceding device (assembly, component) that supplies the signal. The directly following device is the signal destination or 'consumer'. You can follow the signal from its original source through each of its 'consumers' in turn to its ultimate destination (*forward tracing*) or from its ultimate destination back to its original source (*backward tracing*).

First consider the simplest case: signal flow with no feedback (Figures 2.20, 2.21 and 2.22).

Fundamentals of the decision-making process. Which components are suspect when a signal value or waveform is found to be incorrect? A quite important technique in fault-finding consists of systematically eliminating various possibilities one after the other. This process is made a lot easier if we base our work on the single-fault hypothesis, since we then do not need to take combinations of faults into account. It is also useful to distinguish between *functional* and *electrical* faults:

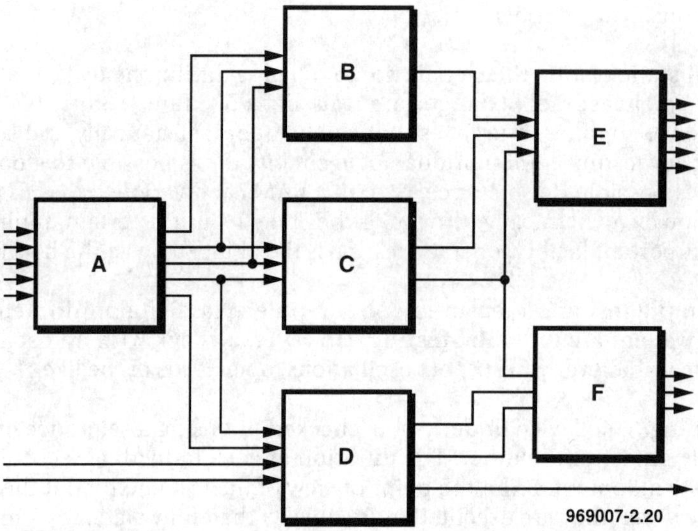

Figure 2.20 Signal flows without feedback.

- functional fault: a particular item (IC, assembly or the like) yields incorrect output values when correct input values are applied, but the electrical characteristics of the signals lie within the boundaries of the applicable specifications (in other words, the input and output levels and the supply voltage are apparently OK, so that the fault lies somewhere 'inside' the item). Functional faults in digital circuits can usually be localized using 'digital' test measures (test software, logic probes, logic analyzers and so on).
- electrical faults: these include not only rather trivial faults (shorts, intermittent contacts, open leads) but also incorrect input and output levels or supply voltages. Such faults can be recognized by the fact that the signal values do not meet the relevant specifications (levels out of bounds, incorrect switching times (pulse edges), large over- or undershoots, oscillations, induced disturbances and so on). Occasionally it is a question of 'trite' stuck-at faults, which can also occur as multiple faults (*for example:* all of the outputs of an integrated circuit are inactive due to a fault in its supply voltage connection). Other symptoms are an excessively high or suspiciously low operating temperature, abnormally high or very low current consumption and the like. (See Chapter 5 for related plausibility testing.) Electrical faults in digital circuits cannot always be localized with 'digital' test equipment; instead, analogue measurements (with a multimeter, oscilloscope or the like) are necessary (plus tricks such as breaking connections to check for possible coupling effects).

Consider now a single component or assembly with inputs and outputs, such as item *C* in Figure 2.20, that is embedded in a larger assembly. We check both the input and the output signals. Depending on the results of the individual checks, only certain items can possibly be suspect (see Table 2.11).

Figure 2.21 (overleaf)) and Table 2.10 further illustrate this. Figure 2.21 shows the transistor output stage of IC *A* connected via a signal line to the input transistor stages of ICs *B* and *C*. With functional faults, these stages will all still operate correctly, while electrical faults could affect precisely these stages. Table 2.10 lists the symptoms of the various types of faults. *Note:* 'opening' a suspect lead is a proven fault-finding trick. See Section 5.5 for more details and other tricks.

Note: in a bus system it is not always easy to distinguish between faults due to sources (drivers) and destinations (receivers). The effect of a fault can be: (1) no signal appears on the bus (the actual source, which is the device that should drive the bus, is not functioning) and (2) unallowed signals appear on the bus (bus conflicts). Bus conflicts may appear to be electrical faults when the signals are measured, but they are always due to functional faults (incorrect control of driver stages). See also Section 5.3.

Forward signal tracing. With forward signal tracing (Figure 2.22a) we follow signals from the input to the output. The first deviation from the expected values indicates a suspect component or assembly.

Backward signal tracing. With backward signal tracing (Figure 2.22b), we start with the output at the 'outside edge' (the right-hand edge in Figure 2.22b). If there is a deviation from the expected values, then we must follow the signal back until we encounter an item which is correctly driven but produces incorrect output values.

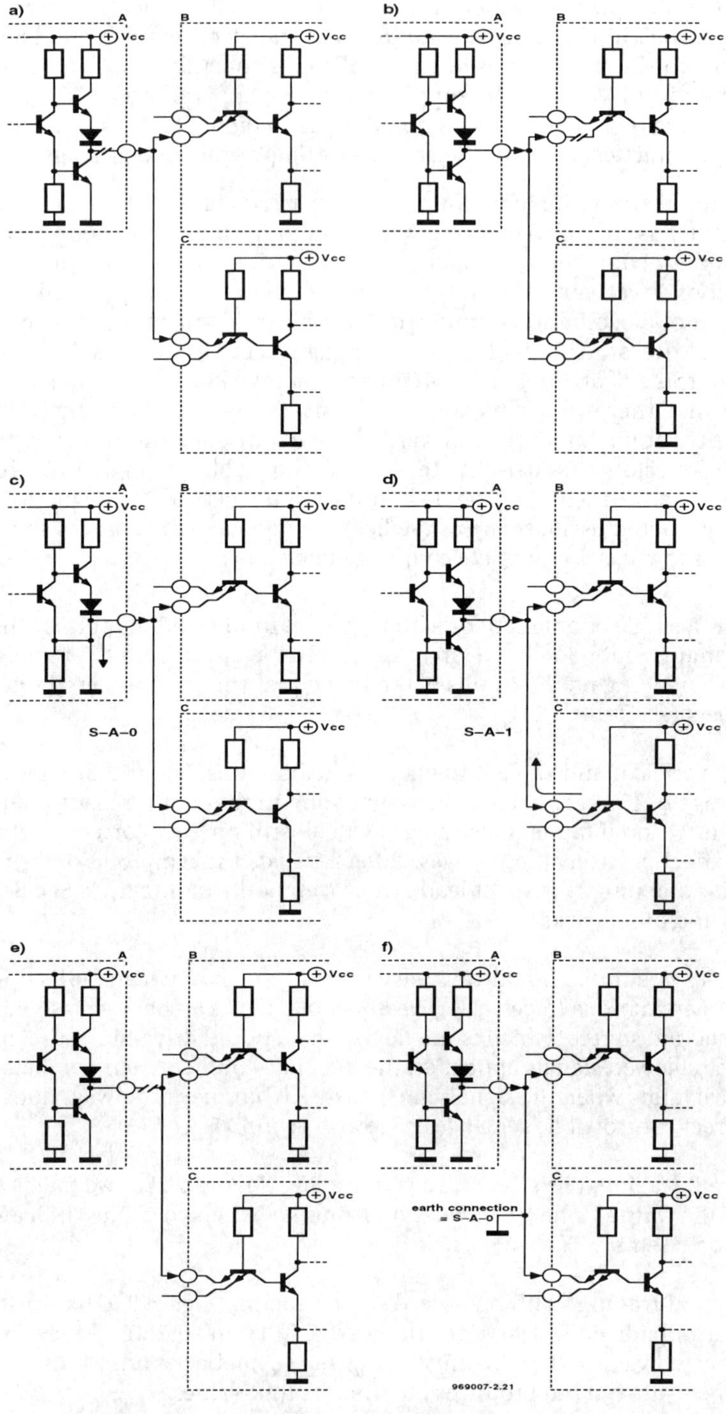

Figure 2.21 Examples of faults relating to a simple connection between digital logic ICs.

Fault	Figure 2.21	Observable behavior	Notes
functional fault in source *A*	not shown	Output A and the following inputs (*B* & *C*) switch identically and incorrectly.	The signals are electrically within specification.
functional fault in one of the receivers	not shown	Output *A* and the following inputs (*B* & *C*) switch identically and correctly.	Faulty behavior is first observable at the outputs of *B* & *C*; the signals are electrically within specification.
open output	a)	Output *A* and the following inputs (*B* & *C*) do not switch at all.	Fault shows up as 'stuck-at' or an electrically incorrect signal. No change when the lead is opened.
open input	b)	Output *A* and the following inputs (*B* & *C*) switch identically and correctly.	Faulty behavior is first observable at the output of *B*.
'stuck-at' at output	c)	Output *A* and the following inputs (*B* & *C*) do not switch at all.	After the lead is opened, the incorrect behavior at the output of A remains and the inputs of *B* & *C* change to 'open' inputs.
'stuck-at' at input	d)	Output *A* and the following inputs (*B* & *C*) do not switch at all.	A typical example of a reverse coupling effect. After the lead is opened, the output of *A* again switches (correctly), but the faulty behavior at the inputs of *B* & *C* remains.
connection fault: open	e)	Output *A* switches (correctly); the following inputs (*B* & *C*) do not switch at all.	The fault shows up as a 'stuck-at' fault or as electrically incorrect signals at the unconnected inputs. Simple proof: continuity test.
connection fault: short or 'stuck-at'	f)	Output A and the following inputs (*B* & *C*) do not switch at all.	The fault shows up as 'stuck-at'. After the lead is opened, the output of *A* again switches (correctly). Shorts to earth or the supply voltage can usually be detected by continuity testing.

Table 2.10 *Explanation of Figure 2.21.*

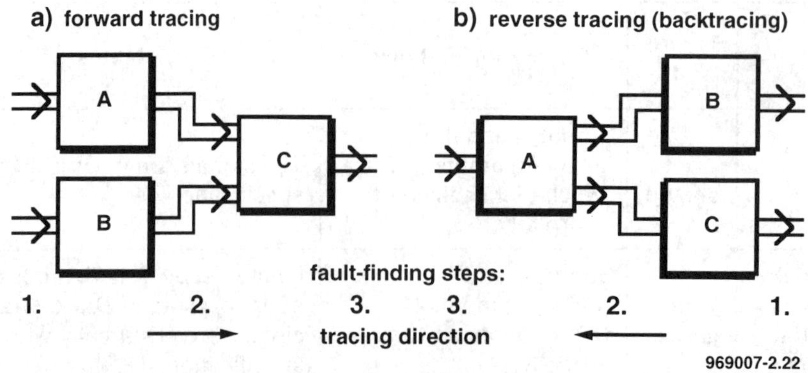

Figure 2.22 Forward and backward signal tracing in simple circuit structures.

Choosing the direction. This is ultimately a question of experience and depends on both the structure of the circuit in question and the specific circumstances. With easily-understood circuit structures (such as those shown in Figure 2.21,) it is reasonable to select the direction based on the following considerations:

1 If the structure is an 'inverted' tree (many input elements affect a few output elements, as in Figure 2.22a) then forward tracing is often better. We know that the overall assembly is faulty and we want to better localize the fault by signal tracing. The element *C* will doubtless have an incorrect output and the only problem is to find out which element is responsible (*A, B* or *C* itself). If we start with *A* or *B* then we save ourselves the trouble of checking *C* if it is not itself the source of the fault.

2 If the structure resembles a true tree (a few input elements affect many output elements, as in Figure 2.22b) then backward tracing is often better. If we start with *B* or *C* then we may find the faulty element right away. In any case we can avoid doing unnecessary tests (*For example:* if the output of *B* is correct then you need not check the path between *A* and *B*).

Expected behaviour

How do we know how the device in question should actually behave? This question reflects what is certainly the greatest single difficulty in fault-finding based on measurements. You can fundamentally resort to the following sources of information (but for some devices either there are no sources of information, or the information is incomplete or hard to come by):

- specific fault-finding documentation. At least in the case of high-quality assemblies, units and systems, this should actually exist: it should be possible to acquire service manuals and fault-finding guides ('service', 'field engineering', 'hardware' or 'troubleshooting' manuals), schematic diagrams and the like. (A typical example of a fault-finding document is the service schematic for a television set. It depicts the structure of

Suspect item	Symptoms	Functional fault	Electrical fault
the DUT itself	1) output faults 2) input faults	possible (but only for output faults)	possible
a prior device (signal source)	1) faults on the associated inputs (incorrect drive) 2) the output values are OK for the actual input values	possible	possible
a subsequent device (signal destination)	faults at the associated outputs	no	yes (A subsequent device can only influence a signal electrically, for example by excessive loading or some other type of coupling effect.)
interconnections	1) incorrect outputs from a prior device or inputs of a subsequent device 2) general input and output faults	no	yes (possible faults: open leads, shorts to other leads (including supply power and earth), coupled-in interference)
'infrastructure' and the general surroundings	1) faults on inputs and outputs, power-supply leads etc. 2) multiple faults	no	yes (typical fault sources: power supply, earth connection, decoupling capacitors, coupled-in interference)

Table 2.11 Suspect items, based on the results of signal tracing.

the equipment and shows expected signal values at salient test points in the form of voltage readings or oscillograms.) But in the PC world in particular, such documentation is not commonly available – only a few suppliers even have such documentation and they often only make it available to authorized service agencies.

- data sheets and standards. We can use these to learn how components and standardized interfaces function.
- general technical knowledge. This is the truly decisive factor. You must understand the technical fundamentals before you can work with data books, standards and specialist literature. Even the best fault-finding documentation is written for the specialist and not for persons who cannot tell an error correction circuit apart from an ammeter.

- 'learning' from hardware known to be in good working order. When doing fault-finding on the basis of technical knowledge, we actually do the following: we measure the signals on the processor bus (for example) and, based on a data book or a fault-finding guide, consider whether there might be something wrong with them. This sort of brain-racking would be totally unnecessary if we had an equivalent hardware configuration available which we knew to be fault-free. We could then simply operate both systems in the same way (provide the same stimulus) and compare the measured output data from the two systems. Automated testers used in manufacturing in fact work in exactly this way (with the correct behaviour partly 'learned' from a properly-functioning item and partly provided by computer simulation). Such comparison data cannot be purchased (it would anyway only be useful in the service environment if you could reproduce the tests used, without having access to the rather expensive automated testers). Nonetheless, the principle of 'learning' from properly-functioning hardware can be applied in various ways to practical servicing. For more information see Chapters 4 and 5, and Volume 2.

Simple examples

Example 1: a combinational digital circuit. Figure 2.23 shows a simple combinational-logic circuit, along with a set of measured input and output values.

Is the circuit behaving correctly or is it not? In order to answer this question, we must examine the schematic diagram and figure out what the output values should be for each of the six possible sets of input values (based on the truth tables of the NAND gates), and then compare these values to the measured values. We can conclude the following:

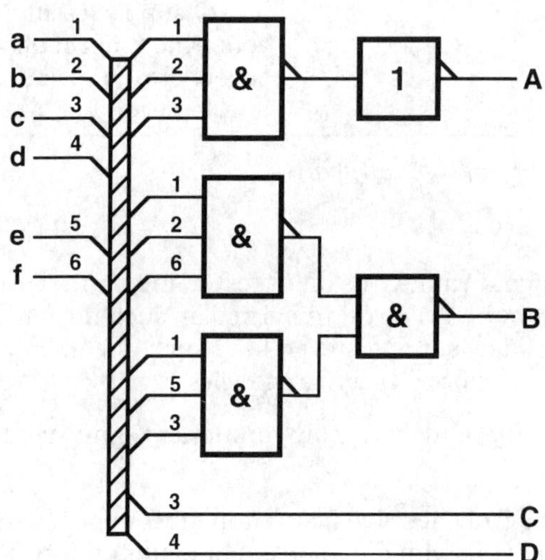

	measurement					
	1	2	3	4	5	6
a	H	L	L	H	L	L
b	H	H	H	L	H	L
c	H	H	L	H	L	L
d	H	H	H	L	H	H
e	L	L	L	H	L	H
f	L	L	H	L	H	H
A	H	H	L	L	L	L
B	L	L	L	H	L	L
C	H	H	H	H	H	H
D	H	H	H	L	H	H

969007-2.23

Figure 2.23 A combinational-logic circuit.

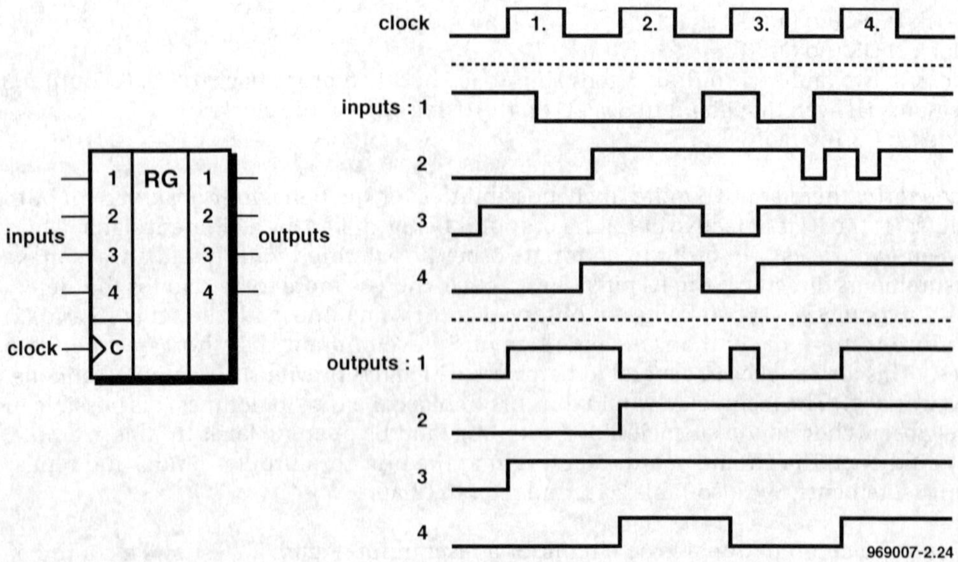

Figure 2.24 A D-type flip-flop register with timing diagram.

- measurement 1: OK (no fault)
- measurement 2: output *A* is not LO although *a* is LO – fault 1!
- measurement 3: output *C* stays HI although *c* is LO – fault 2!
- measurement 4: OK (no fault)
- measurement 5: the same fault as with measurement 3
- measurement 6: the same fault as with measurement 3

Now we examine the two faults more closely:

- Fault 1: with the exception of measurement 2, the signal at output *A* switches correctly between HI and LO. This means that both the inverter and output of the preceding NAND gate are OK. Only the input *a* to the NAND remains suspect. There could be a defect in the logic circuit or the connection to input *a* could be open (this question can be answered by a direct measurement at the gate input).

- Fault 2: there is no gate at all between input *c* and output *C*. Such a fault indicates that there is most likely an open connection (check this by means of a continuity test with the power off).

Example 2: a D-type flip-flop register. Figure 2.24 shows a 4-bit D-type flip-flop register, whose timing diagram has been measured using an oscilloscope or logic analyzer.

In order to evaluate the timing diagram, we must know how a D-type flip-flop should behave: (1) the current input value is stored on the rising edge of the clock and (2) the outputs must not change state in the interval between positive clock edges. If we investigate the timing diagram shown in Figure 2.24 based on this, we can conclude the following:

- clock 1: OK (no fault)
- clock 2: OK (no fault)
- clock 3: two faults! (1) output *1* goes HI initially but then drops back to LO, (2) output *3* remains HI even though input *3* is LO on the rising edge of the clock.
- clock 4: OK (no fault)

Practical tip: there are two quite likely possibilities for the behaviour observed on both outputs: (1) the IC is defective or (2) the output is being 'pulled' by a different signal (short between two traces). (In order to eliminate other possibilities, you should always make measurements directly at the IC pins.) In practice, there is more to be said for the 'defective IC' hypothesis: (1) two faults are observed at the same time with the same IC and (2) IC failure is more likely than the 'spontaneous' development of a short between two traces. (This also applies in general to hardware which has previously functioned and just stops working.) The technician would thus first replace the IC. (Sometimes it is possible to more closely check out a suspicion of a coupling fault by opening leads. In this example, you could open the circuit-board traces right at the pins for outputs 1 and 3 and repeat the measurements. See also Table 2.11, and refer to Chapter 5.)

Example 3: a combinational logic circuit for a reset input. Figure 2.25 shows a combinational-logic circuit connected to the reset (CLR) input of an IC (such as a processor). The associated timing diagram has been determined by measurement.

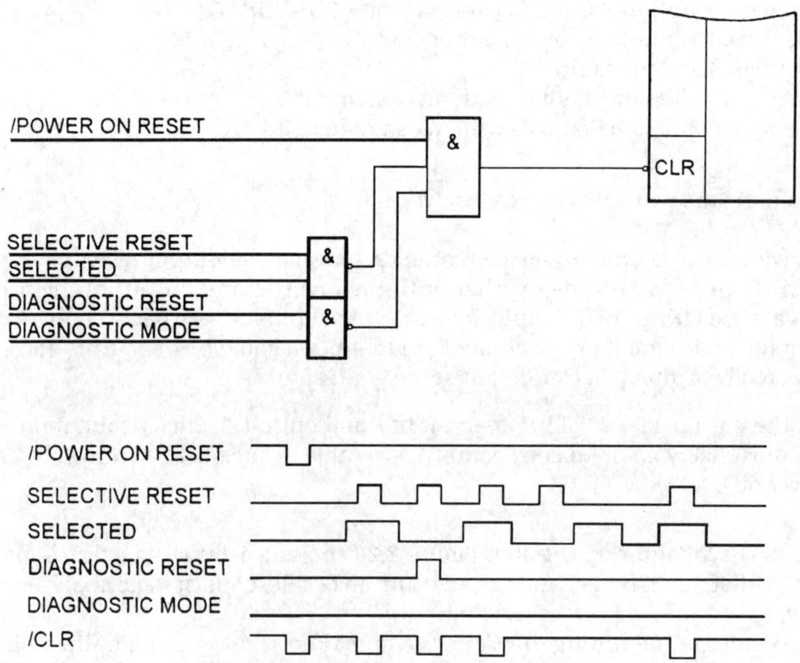

Figure 2.25 Reset circuit with timing diagram.

The signal names are so suggestive that the 'correct' behaviour is almost self-evident. There are three possible sources for a reset: (1) switch-on reset (POWER ON RESET), (2) a 'selective' reset (SELECTIVE RESET & SELECTED) and (3) a reset for testing purposes (DIAGNOSTIC RESET & DIAGNOSTIC MODE). We only need to examine the individual CLR pulses to see if they are valid or if any one is missing. The result is that the third CLR pulse should not be present: it is generated by DIAGNOSTIC RESET even though DIAGNOSTIC MODE is not active. The fault suspicion falls on the associated NAND gate, since it has an active output even though this should be prevented by the input values. (A quite likely fault hypothesis is an internal 'stuck-at 1' fault. Such a fault can result from a broken bonding wire on the DIAGNOSTIC MODE input; see also Figure 2.21b).

Note: a pulse is also present on the SELECTIVE RESET line at the same time as the faulty CLR pulse. Why then is the associated gate less suspect? The answer is that it otherwise functions properly. An unambiguous indication of this can be seen with the fourth pulse on the SELECTIVE RESET line: at this time the SELECTED line is inactive and no pulse appears on the CLR line, so the gate must be working properly.

Latent faults and signal tracing

A latent (hidden) fault is actually present but is not noticed. This is always the case if the operating conditions, system state, signal values and so on that are necessary for the fault to become effective simply do not occur. In other words, the faulty hardware has not been properly 'stressed', or at least not all of its functions have been used or tested. *Examples:* (1) on a graphics card, the hardware that supports CGA emulation is faulty, but the card is used exclusively in SVGA mode; (2) on a graphics card, there is a fault in the memory addressing hardware for the video RAM, but the selected operating mode does not use the affected region of memory; (3) a 486 processor has a fault in the debug register hardware, but no program is ever run which addresses the debug register. *Note:* a symptom that occurs quite frequently with PCs which run under Windows 3.x is related to latent faults: the PC appears to be OK and in fact functions properly under DOS and under Windows in Standard mode. However, errors occur under Windows in the 386-Enhanced mode (frequently as undefined 'system hangs'). A possible cause is a latent fault in the processor's '32-bit' hardware. (Under DOS, and under Windows in the Standard mode, the processor works in real mode with a 16-bit-wide data path and a simple memory segmentation technique, so that a significant part of the hardware is not utilized. However, in the 32-bit mode everything must actually function: the full-width data path, the complicated memory segmentation technique based on segment descriptors, the page management logic, the TLB hardware and so on.)

At the gate level the difficulties associated with latent faults can be illustrated in detail. Let's again look at Figure 2.23: output *A* always behaves correctly with the exception of measurement 2, in spite of the fact that the a input of the associated NAND gate is defective. In all other cases the signal values applied to the other gates serve to prevent any fault from being observed at the output *A*. If we had not performed measurement 2 (or not used the associated set of input values in our testing) then we would not have detected the fault at all.

In practice you must always take into account the possibility of latent faults. In particular, when you must improvise (without truly severe test software and without exact documentation) it is usually simply not possible to check out all conceivable combinations of inputs, functional situations and so on. You should thus treat 'OK' results arising from signal-tracing tests with a certain amount of caution. (The statement "we've checked it out and didn't find anything wrong" can always be true, but the statement "our testing has proven that the device is fault-free" need not always be correct.)

Feedback effects in signal paths

Feedback is present when a signal that is output by an item affects the same item as an input (either directly or via one or more intermediate stages). Even the simplest combination of a microprocessor plus memory represents a typical example of a system that includes feedback (see Figure 2.26).

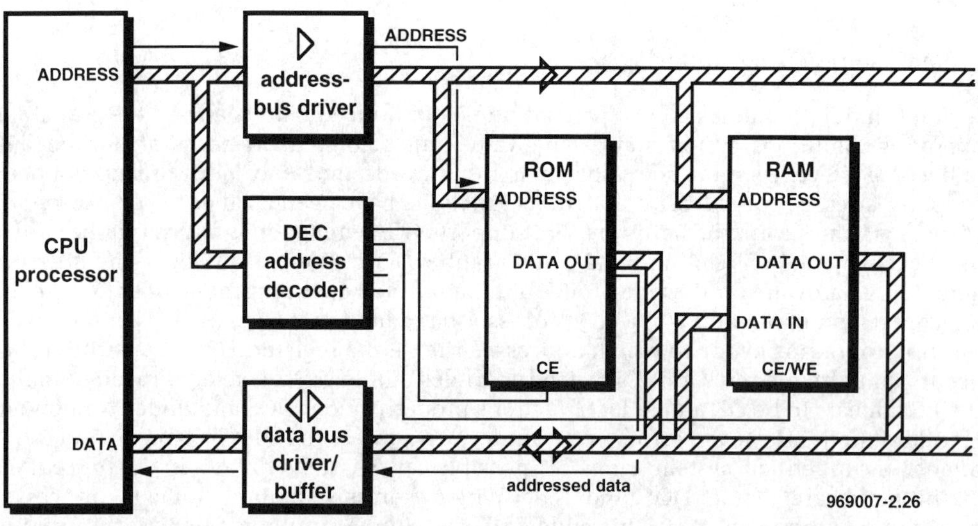

Figure 2.26 Microprocessor with memory as an example of a system with feedback.

After the power is switched on, the processor supplies an address to the memory and reads the first executable machine instruction from the memory. This in turn causes a subsequent address to be issued, which fetches the next instruction, and so on. Clearly, an incorrectly addressed memory cannot deliver the correct (intended) data or instructions, just as a processor which receives incorrect data or instructions cannot properly supply addresses to the memory. (This can happen on any PC motherboard. In such a situation, nothing at all works properly after the power is switched on). How should we locate the suspect component in this situation?

Feedback versus reverse coupling. You must distinguish between a fault in a logical *feedback* process and an electrical *reverse coupling* fault. (In the latter case, an input fault causes incorrect behaviour at an output connected to the faulty item; see Figure 2.21d.) The 'foolproof' fault-finding trick is to open the feedback path. However, this is (1) not always easy to do and (2) it often means that it is no longer possible to maintain the signal flow (test stimulus) that is essential for actually performing the test. (See also Chapter 5 and Volume 2.)

Fault-finding with closed-loop logical feedback. You should split up a complex problem of this sort into smaller, more easily understood problems (the 'divide and conquer' technique) – but this is ultimately a question of practice and experience, so that we can only address some basic principles here.

Well-defined starting conditions. A 'service-friendly' digital circuit should have a well-defined, unambiguous initial state, which can be reached from any arbitrary operating state (for example, by activating a reset line). This initial state should be checked out first (for example, the motherboard reset). In the above example of a microprocessor system, the logical feedback process is initiated at this point in time: the processor must provide the correct initial address after it has been reset, and the memory must in turn provide the correct data. (For practical testing see Volume 2).

Process recording. Assume that we can record the complete set of bus signals at the microprocessor, clock cycle by clock cycle from the moment of the reset (using a logic analyzer, for example; see also Section 4.4.2). We then need only compare the recorded (actual) data to the expected data and look for discrepancies. The source of the first discrepancy (in time) is then suspect.

Process comparison. In principle we can start the same process in a second, identical device at the same time and also compare the signals clock by clock. Here again, the first deviation indicates the suspect device. See also Section 3.2.1.

Incremental testing (bootstrap diagnostics). Start by testing subcircuits that are largely free of feedback, and then add the subcircuits that incorporate feedback. A microprocessor-based circuit can also be run without feedback: simply replace the memory with a device connected to the data bus (a test data generator) that is forced to supply the correct data. The simplest situation is to let the processor 'free run', for example by continuously supplying it with a single instruction (for practical information see Volume 2). If the processor provides correct addresses under these conditions, the fault should be sought in the memory, the bus interface ICs and so on. Further checking can then be performed on each functional group in turn (*for example:* first the ROM, then writing to the RAM, followed by reading from the RAM and so on).

Rule of thumb: whenever something that used to work no longer works immediately after some new item has been added to the overall system being tested, then the last-incorporated item is usually (but not always!) suspect of being faulty.

Partial substitution. You can check whether a given item functions correctly by replacing it briefly by an identical, known-good item (a 'reference device' or 'golden device'). In the system of Figure 2.26, for example, you could replace either the processor or the RAM.

Practical tips:

1 Select reference devices that are found in as many as possible of the units or systems to be tested, and used in the same manner. Thus, with PCs the processor is usually a better choice than a BIOS ROM. A processor can be used with any motherboard designed for that processor type, while a BIOS ROM is much more restricted.

2 If the substitution causes the fault to disappear, then you're in luck. Otherwise, if there are more than two devices involved in the signal path, then an 'unsuccessful' substitution does not help with the localization and also does not guarantee that a more useful stimulus is now present for further fault-finding. (The system continues to behave just as it did before the substitution).

External test equipment versus self-testing. Any self-test (with test software, for example) can provide useful information for fault localization only if certain basic functions work faultlessly (the *core system* concept). However, when 'nothing at all' works from the very start, measurement-based testing unconditionally requires external test equipment, at least to observe and record test results (oscilloscope, logic analyzer and the like) and sometimes also to generate a suitable test stimulus (test data generator, various test aids).

2.5.4 Signal tracing with incomplete documentation

Often (and in the PC world usually) there is not any usable service documentation available. In spite of this it is possible to do fault-finding based on measurements. There are two circumstances that make this feasible:

1 When doing fault-finding, you need only identify the suspect item from among a relatively small number of components, subassemblies or the like. (With a mainframe computer of the 80's you would have to find the faulty card among perhaps 50 cards, whereas nowadays with a similar fault only the processor IC need be shown to be suspect). In other words, the demands placed on test severity have become much smaller.

2 By far the majority of the hardware is built using mass-produced, practically standardized integrated circuits (processors, motherboard ICs, bus components, controllers), and only a relatively small number of components are found between any two standardized or well-known interfaces. These interfaces are also adequately documented even in the absence of any device-specific material. This applies in particular to processors, memory devices and bus systems as well as to the interfaces commonly used in the PC world (Centronics, RS232, keyboard, video, IDE, SCSI and so on).

A proven technique that promises success even without extensive documentation is to use a cyclic test routine that serves only to stimulate the hardware to be checked (test-stimulus generation) but not to actually test it. Check the relevant signals using an oscilloscope, namely by working with generally known (and adequately documented) interfaces (such as DRAM connections, the ISA bus and so on). In modern hardware there are rarely more than two or three items between any two such interfaces – for example, between the processor and the RAM there is at most a controller IC plus an easily-checked driver stage. With a little practice, the suspect item can be quickly and quite accurately identified.

For example: suppose that the RAM complex is suspect. First start a cyclic test routine that only executes write accesses to a number of memory addresses, with the addresses and data chosen so that at least one '0' and one '1' occur successively on each address and data line. The signals at the DRAM ICs are precisely defined in the DRAM data sheets. You must check whether the actual signals look 'reasonable' in this regard (do the control signals CAS, RAS and OE switch, do all of the address signals reach the DRAMs?). The next test step consists of cyclically performing write accesses immediately followed by read accesses. With this test routine you can observe data read operations from the memory back to the processor. If these functions all work OK and the signals themselves appear to be in order (levels, edges, under- and overshoots) then the hardware in question is a lot less suspect. Some examples of logical deduction here are:

- If it is not even possible to start the test-loop routine: the fault lies in the core hardware system. In certain circumstances the processor itself is suspect. (For fault-finding in the core system see Volume 2.)
- If the processor apparently works (the test loop routine runs), but RAM accesses do not work at all: the responsible motherboard controller IC is suspect.
- If the processor apparently works and the RAM works 'partially' (symptom: the test software has indicated that certain bit positions are faulty): by means of signal tracing based on the test loop technique you can decide whether the memory ICs in question or the bus ICs are suspect.

2.5.5 The 'shotgun' approach

This term, in its typically American directness, accurately suggests what is involved: when we do not know exactly where to look, we tackle the problem with a large number of tests in the hope of thereby scoring a hit. In other words, we construct the most manifold fault hypothesis possible, which means that every possible component, assembly, unit and so on is suspect, and then we investigate whether our suspicion is confirmed. You should however not succumb to the temptation to randomly 'poke around' in such a situation. There are fundamentally two systematic strategies, each of which has its own justification:

1 Work from the less likely to the more likely faults. (This idea is that if the usual techniques have not located the fault, you must consider quite unlikely causes.)

2 Start with the suspicion of a trivial fault (such as a switched-off monitor, a loose cable, intermittent contacts of all sorts, a defective power supply and so on) and work in the direction of increasingly unlikely faults. (This is based on the idea that by far the most faults are so trite that you don't even consider them at first.)

Practical tips:

1 Only use the first strategy when you simply cannot get anywhere in a particular fault situation (assuming that you have adequate practical experience). Otherwise this approach only represents a waste of time, and not infrequently results in considerable consequential faults. The second strategy, on the other hand, is often the preferred choice when you want to quickly restore normal system operation.

2 Occasionally, the differential diagnosis technique (see Section 2.8) will yield some starting points. To proceed, (1) use all of the available test possibilities one after the other, (2) record the results of each test and then (3) based on the collective test results. decide which devices are more suspect and which less suspect. *Note:* when working through a series of tests, always start from the beginning each time. (The idea is to have the same initial conditions, which can for example be assured by resetting the system before each test. However, sometimes this can only be achieved by switching the power off and back on. Since making a regular practice of this tends to reduce the reliability of the equipment, it should be reserved for use only when it is absolutely necessary.)

2.6 Substitution instead of measurement?

"What is the point of making measurements when you anyhow have to replace the defective item?" In the professional service sector such a point of view would be considered absurd, at least if taken to its logical conclusion (if all that is involved is just plucking some defective item out of the system so that it can be discarded and replaced by a new item, what then is the meaning of 'service'?), but in the case of self-help it is not all that unreasonable. Here you must compare the cost of stocking replacement parts against the cost of the equipment that is required to localize a fault to the component level and then replace the component so identified.

The pure replacement strategy. This is in itself an absolutely reliable means of remedying faults, although at times a rather expensive one: start somewhere and replace parts until the fault has disappeared. (If you want to go a bit further you can then reinstall the FRUs that were unnecessarily replaced in the process.)

The limits of the replacement method. Can we really manage completely without making measurements? Many of the 'tips and tricks' in the computer literature are basically rather adventuresome replacement strategies. They amount to using expensive spare parts to demonstrate the presence of trivial faults (the effects of which may also damage the spare parts!). It is not particularly workmanlike to ruin a spare hard-disk drive just to be able to conclude that there is a fault in the power supply. Thus, even in the self-help environment, you should only replace a part when you are sure that the replacement part will survive the exchange undamaged. In practice, ICs are quite robust. They can usually withstand overloaded outputs and incorrect inputs for a while, at least as long as the maximum allowed operating conditions are not exceeded. We can thus formulate the following rules for exchanging parts:

1 At the very least, thoroughly verify the power supply voltages before making any substitutions. (Strictly speaking, you should also make at least spot checks of the logic levels, signal edges, pulse durations and periods at the relevant interfaces, and also the ambient temperature. In regard to the latter, also pay attention to soiling, ventilation and the condition of ventilation grilles, and the routing of flat cables.)

2 If the replacement does not immediately result in success, re-install the original part.

Determining which parts are exchangeable (FRUs). Components, subassemblies and so on that meet the following criteria are all suitable candidates for being regarded as FRUs:

1 Only 'nondestructive' procedures are required to remove and reinstall the item (removing and replacing screws, unplugging and reinstalling connectors and so on).

2 The item can actually be acquired in single quantity at an acceptable cost and with a reasonable amount of effort.

The following items are 'natural' FRUs:

- all parts of the system that are connected via interfaces (in a PC system, these include the keyboard, mouse, monitor, printer and so on),
- replaceable assemblies within individual units (plug-in cards, power supplies etc.),
- removable components (especially ICs and plug-in modules) within individual assemblies (processors, RAM and the like).

You can add the following systematic considerations to these initial selection criteria:

Do several devices or assemblies have identical replaceable (sub)assemblies or components? If so, then it usually is sufficient to stock only one such item as a spare part.

How likely is it that individual components will fail, and how expensive are they, compared to the cost of the complete assembly? The price of a component increases as the quantity delivered decreases. PC manufacturers thus buy DRAM modules and processors at a low cost per item, since they purchase thousands of pieces. Before purchasing an expensive component that is not likely to fail, look at the cost of the next-higher assembly (*for example:* compare the price of a Pentium processor to that of a complete motherboard).

Will the components be available at all, especially after several years? If a component is too difficult or too expensive to acquire, consider sparing the next-higher assembly instead (up to the complete unit). Alternatively, you can provide yourself with complete spares coverage from the beginning (which however means 'dead' capital; see below).

Will you actually be able to properly replace the part? You should give it a pass if disproportionately expensive equipment is required (as for some SMD components) or if the process can be dangerous (such as replacing the picture tube in a monitor).

How long will the FRUs be stored unused? You should seriously consider this question if you contemplate holding spares of expensive assemblies (such as a graphics card) or complete units (such as a laser printer). Keep in mind the course of technical progress. If immediate availability is not an overriding consideration, it is often sufficient to acquire commonly-available hardware only when it is actually needed (the outage time should not exceed a few days). If you can manage to bridge over such gaps, consider building up *financial reserves* for acquiring equipment as needed, instead of holding spare parts (and such reserves pay interest, in contrast to the 'dead' capital invested in spare parts).

A limited selection of spare parts. Suppose that the system contains n different kinds of assemblies or components, which are candidates for holding as spare parts. If system availability does not have absolute priority, then it is sufficient to keep $n - 1$ spare parts on hand, on the basis that if all of these parts have been replaced without eliminating the problem then the nth part is suspect. It is reasonable to choose the most expensive part for the nth part. In addition, spares are not needed for equipment that you can do without for a certain length of time.

Minimal spare parts selection. If minimum cost is more important than ready availability, then the spares stock can be even further reduced on the basis of the single-fault hypothesis. The principle consists of holding a spare for only one of any pair of coupled devices or assemblies (selecting either the less expensive of the two or the one most likely to fail, depending on the relative importance of cost and availability). In case of a fault you can replace the one item with its spare; if this does not eliminate the fault then the second item is suspect and you can then acquire a suitable replacement (and reinstall the original example of the other item).

Caution: beware of obsolescence due to 'responsible' use. This has two aspects: (1) the hardware being maintained becomes obsolete, so that you can no longer obtain spare parts, and (2) the spare parts become obsolescent, so that they cannot be used in equipment acquired at a later date.

Actual obsolescence. Technical devices that are simply stored unused last longer than those that are used. This is an obvious rule, which however has its exceptions. Some devices and assemblies actually become less reliable when stored: exposed metal surfaces corrode, lubricants dry out, IC packages absorb moisture (see Section 2.2.2), electrolytic capacitors transform themselves into something resembling a resistor, batteries become unusable and so on.

The solution. Carefully plan how long to hold spare parts, and intentionally 'use up' stored parts. For instance, put a spare PC or monitor into normal use after a year of storage and acquire a new spare part. (It goes without saying that spare parts should be stored in an appropriate manner: properly packed, correctly labelled, protected against moisture and damage, and locatable even after years of storage.)

Rule of thumb for repair time: just replacing a part takes about half an hour when everything is handy, regardless of what is being replaced. The specification "MTTR = 0.5 h" found in many product brochures ('MTTR' means 'Mean Time to Repair') is based on this well-know fact. The process is: (1) switch off the equipment, (2) open the enclosure, (3) dismount the FRU, (4) mount the replacement part, (5) close the enclosure, (6) switch on and (7) do a global functional test (such as a cold start followed by running an application). Actual fault-finding and obtaining the replacement part are not included in this calculation! The time needed to restore the system to a usable state is also not included. With a PC, this essentially involves restoring data (loading software and backup data). You can thus figure on a MTTR of at least one hour plus the backup time (which depends on the amount of data to be restored and the performance of your backup hardware). Any fault-finding activities (more exact fault localization, repair) in addition to the basic activities of replacement, restoring data and running a primitive functional test cost additional time!

2.7 'Inverted' substitution: testers and master machines

Substitution basically amounts to the verification of an FRU by replacing it with a known good item. If the fault no longer occurs after the substitution, then the FRU that has been replaced is suspect; if the fault remains then the working environment of the FRU is suspect. Installing a replacement part practically amounts to checking the faulty system with a known-good FRU. This principle can also be applied in reverse: we can employ the FRU to be checked in a known-good environment. If the fault then occurs, the FRU is suspect; otherwise the suspicion is refuted and we must check other parts of the system being investigated. There are in principle two types of known-good environments: (1) testers, emulators and test aids and (2) master machines.

2.7.1 Testers, emulators and test aids

A *tester* is a device for checking the operation of another device, which can apply input signals to the device under test and evaluate its outputs (ranging from a simple display, readout and the like up to fully automated comparison to expected values). All mass-produced computer hardware is checked using automated testers (Automatic Test Equipment, ATE). Such equipment is much too expensive for the service environment, even if repairs are made on a grand scale.

For this reason we limit our attention to test equipment that replaces the devices or assemblies that are normally connected to the interfaces of an FRU. For purposes of clarity we can divide these into three groups (but also see the following note):

1 *Testers.* A tester is a piece of equipment that works 'stand-alone'. Testers are usually named after the type of device that they test: a drive tester, a monitor tester and so on.

2 *Emulators.* In contrast to a tester, an emulator is only a sort of 'substitute circuit' that can be connected to a given interface in order replace the device or assembly that is normally connected to the interface, or to imitate its behaviour. In this case the DUT remains in its usual environment (for example, in the PC). An emulator is named after the item that it replaces: thus (for instance) a floppy-disk emulator can be connected to a floppy-disk controller in place of a floppy-disk drive. Suitable test software is normally required to use an emulator. Compared to using test aids, testing with an emulator is largely automatic (the emulator provides a known-good environment that allows the test program to run). The simplest form of emulator is the loopback connector, which feeds the output of a given interface back to its input so that a test program can check the functionality of both the output and input signal paths (loopback test).

3 *Test aids.* Test aids are items that makes 'manual' testing easier. These include relatively simple assemblies with control and display functions (such as a set of switches and LEDs that allows a given interface to be operated and checked 'manually'), adapters and special connectors with test points for attaching test and measurement equipment, and devices that allow signal lines to be opened and (inter)connected (breakout boxes).

Note: many items that are referred to as 'testers' in the catalogues are in fact only test aids. This does not necessarily mean that such items are not useful (many test aids are indispensable), but you should at least know what you are dealing with. Many fully-fledged testers can be used as 'push-button' devices (you need only connect the DUT, press the 'start' button and proceed as directed), but a certain amount of technical expertise and patience is necessary for the successful use of test aids.

The testing principle. We connect the item to be tested to the test equipment and start the predefined test processes (or do 'manual' testing ourselves). From the test results we can see whether or not the item in question is actually suspect.

How many different types of testers or emulators do we need? Strictly speaking, we need as many different types as are necessary to check all of the various types of FRUs and interfaces, or to emulate all types of devices connected to the 'other sides' of the interfaces. With the usual PC, we could for example consider the following:

- a floppy-disk tester,
- a hard-disk-drive emulator (for the IDE interface),
- a hard-disk tester,
- a monitor tester,
- a graphics-card tester,
- interface emulators/testers (V24, keyboard, Centronics and SCSi interfaces, etc.).

Here again certain savings can be realized, just as with spare parts. Thus, for a given pair of devices that are normally connected together, you could provide suitable test equipment for only one of the two. *For example:* if a monitor is checked with a monitor tester and found to be in good working order, then the associated video adapter is suspect.

For more information about testers, emulators and test aids see Volume 3.

2.7.2 Master machines

Instead of providing special test equipment, you can keep a complete system on hand which is known to be in good working order, and in which any item to be checked out can be installed and tested. We call such a system a 'reference system' or a *master machine*. *For example:* suppose that you have a known-good (properly-working) PC, and the video adapter of another PC appears to be suspect. Using the substitution strategy, you would replace the suspect video adapter with the known-good video adapter from the PC which is known to be in good working order. With the master-machine strategy you would check the suspect video adapter in a known-good environment (the master machine).

This strategy is even used in production, especially with expensive and high-performance hardware. The advantage of such an approach is that it provides functional testing under the stringent dynamic conditions in which the hardware will subsequently be used. Such conditions cannot be practically reproduced by universal testers in this environment.

For more information about master machines, see Volume 3.

2.8 Differential diagnosis

This term, borrowed from the medical practice, refers to an enlightening concept. Consider a device consisting of a number of assemblies or parts. In order to determine which of these are suspect we apply a set of tests, each of which checks just one of these parts or a set of them. Each test yields a yes/no result relating to the part or set of parts that it tests (verification test). If several tests give fault indications, then the suspect parts are those which are common to the set of all parts that are checked by this group of tests (that is, the subset of parts that is checked by each of the tests in question).

You need a certain stock of individual tests that cover different parts of the device under test to perform differential diagnosis. These can be tests that are performed automatically as well as tests that require manual intervention. You must run all of these tests and record their results in the form of 'OK/fault' or 'go/no-go'. If an individual test indicates a fault, then in principle all of the parts that it covers are suspect. By evaluating the results of the whole set of tests to determine which parts are commonly indicated as suspect, you can narrow down the number of suspect parts to a few items or even a single one.

Practical tip: the best basis for differential diagnosis is an appropriate coverage matrix table (see Figure 2.5). You can then mark the rows of the individual tests that have indicated a fault. The common parts can then be determined by a columnwise AND operation on these rows. The parts that are associated with a '1' in the resulting product term are the most suspect.

For example: suppose that there are six devices *A* ... *F* and four tests *T1* ... *T4*. Table 2.12 is the coverage matrix, which shows which test checks which device(s). All tests have been run, and tests *T2* and *T4* have indicated a fault. The associated rows are ANDed together column by column. The resulting row (*T2* & *T4*) contains '1's only for devices *B* and *C*, so these are the most likely suspects.

	A	B	C	D	E	F	
T1	1	0	0	0	0	0	
T2	1	1	1	0	0	0	Fault!
T3	0	0	1	1	0	1	
T4	0	1	1	0	1	0	Fault!
T2 & T4	0	1	1	0	0	0	
The common set of the tested devices {A, B, C} ∩ {B, C, E} = {B, C}							

***Table 2.12** Differential diagnosis, based on a sample coverage matrix.*

3 Built-in testing and fault-finding provisions

3.1 Overview

Almost all electronic systems and devices of any size include built-in provisions for self-testing and error detection. These range from simple microprocessor self-tests to highly-sophisticated, integrated hardware/software solutions (see Figure 3.1).

System design perspective. The requirements placed on 'ideal' hardware and software provisions for testing and fault-finding can be easily summarized:

1 Cost and effort: negligible.

2 Operation:

- Applications are not affected by of faults, regardless of type (fault tolerance).
- All faults are automatically recognized, and fault propagation is prevented.
- Faults are remedied as quickly and simply as possible.

Built-in test and fault-finding provisions

- **On-line error detection**
  - duplicated hardware with comparison
  - parity checking
  - CRC error-checking codes
  - time-related monitoring
  - additional (application-specific) plausibility tests
  - monitoring environmental and operating conditions
  - protective provisions (in the processor)
- **Provisions for observability and reachability**
  - test and measurement points
  - test interfaces and circuit-based solutions:
    - polling
    - addressing
    - shift registers (scan techniques)
- **Self-testing and other testing provisions**
  - built-in test circuits
  - microdiagnosis
  - testing using pseudo-random test patterns
- **System-level configuration (error-management systems)**
  - error signalling
  - error recording
  - error handling

969007-3.01

Figure 3.1 Overview of built-in test and measurement provisions

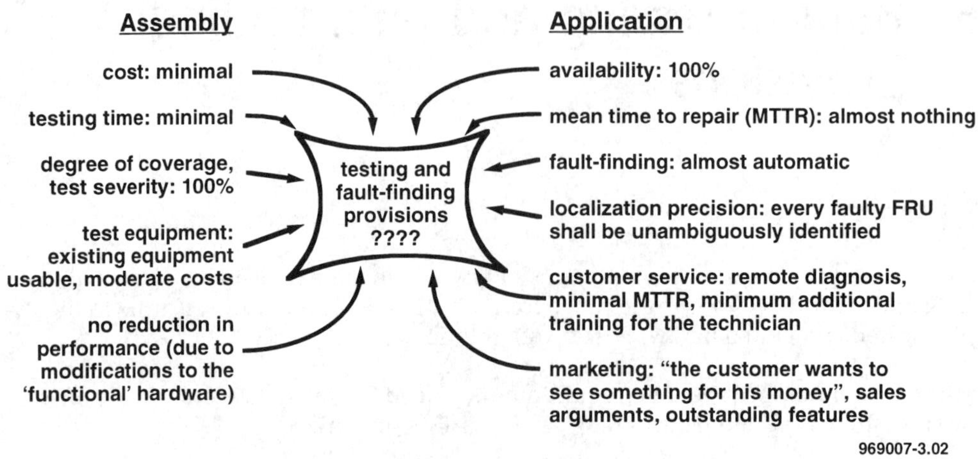

Figure 3.2 *What to do? The requirements for testing and fault-finding provisions.*

In practice, there are numerous conditions and desires that must be taken into account (see Figure 3.2).

On-line and off-line. In this context, these terms refer to the time at which testing and fault-finding measures are effective, and have nothing to do with data processing:

- on-line = effective during normal operation
- off-line = effective outside of normal operation

Protection objectives: what is to be assured or prevented? Table 3.1 lists both conditions that must be *protected against* and the *provisions* that can be employed.

Strictly speaking, suitable provisions have three tasks:

- *detecting* faults that occur during normal operation,
- *preventing* the propagation of faults that have arisen,
- *supporting* fault-finding (up to and including automating the fault-finding process).

How much is affordable? Testing and fault-finding provisions cost money, but do not contribute directly to productive capacity. In this regard, it is understandable that some manufacturers are in principle inclined to simply leave them out. (In the 70's, when practically all suppliers of mainframe systems had introduced suitable provisions and promoted their systems on that basis, the mordant question was raised of whether those systems that were constantly referred to as 'easily maintainable' were not precisely the ones which most needed maintenance. After all, Rolls-Royce does not advertise its motors as being 'easy to repair'). In fact, there is no substitute for traditional quality, and there is definitely a lot to be said for the design philosophy of employing as little hardware as is absolutely

Protection against ...	Provisions for ...
hardware failures	fault tolerance: continued operation even when faults are present
hardware malfunctions	error monitoring: recognizing, indicating and responding to faults
general design faults	testing and diagnosis: hardware verification, fault detection and localization
software faults	debugging facilities: support for searching for software faults and other design faults
operator errors, environmentally-caused faults, sabotage	protection against the effects of unintentional and intentional faults (operator or program errors, software sabotage)

Table 3.1 What must be protected against and what can be provided.

necessary (since what is not present cannot fail) and ensuring high quality down to the least component. When this approach is taken, special equipment is provided for testing and fault-finding, and it is only connected as needed. The other extreme is also possible: duplication (two-fold redundancy) allows for practically 100% monitoring, and triplication (three-fold redundancy) allows a system to survive the failure of any one of its parts. However, doubling or even tripling the hardware cost is hardly warranted in the vast majority of application situations. In between these two extremes, there are system configurations in which testing and fault-finding provisions are provided only at certain particularly critical locations and for especially important tasks. In a good system design, online fault detection measures, error-handling systems and testing and diagnostic provisions are matched to each other, specifically in the form of integrated hardware and software solutions. (The software includes programs for handling and where possible working around faults, fault-evaluation and fault-statistics programs, test programs, utility programs and so on.)

Software costs. The cost of software does not depend on the physical size of the system, but only on its functional extent. A functional extent that is comparable to that of a mainframe computer results in test, diagnostic and maintenance software that easily matches the size of an operating system or a demanding application (a few to many megabytes). For quickly checking out the usual PC, on the other hand, you can manage with a single diagnostic diskette (a few hundred kilobytes).

Hardware costs. Here there is a proven rule of thumb: 10% to 15% additional cost is bearable and also justifiable, based on the utility of the provided features.

The remainder of this chapter essentially relates to those provisions which either facilitate fault-finding or which can be utilized more or less artfully for fault-finding purposes. For questions relating to subjects such as the selection of a particular system, availability planning, preventive data security (backups), periodic maintenance and so on, you are referred to the literature (see Appendix 8). This applies equally well to the mutually-related issues of fault tolerance and error correction. The rest of this chapter basically deals with pure hardware issues. For issues relating to system design and the combination of hardware and software, see Volume 2.

3.2 On-line error detection

3.2.1 Duplication (two-fold redundancy)

Quite clearly, fool-proof error detection is possible if you duplicate the system in question, provide both systems with the same input data, and continuously compare their outputs (see Figure 3.3).

Such an arrangement has two obvious advantages:

1 Each error shows up immediately as a discrepancy in the compared output values, and does so in the simplest imaginable manner (without the need for any sort of special error-detection circuitry built into the 'functional' hardware).

2 The duplicate system is effectively a complete spares stock, so that in case of an error you can still continue to work with the hardware that is fault-free (error assurance or error tolerance).

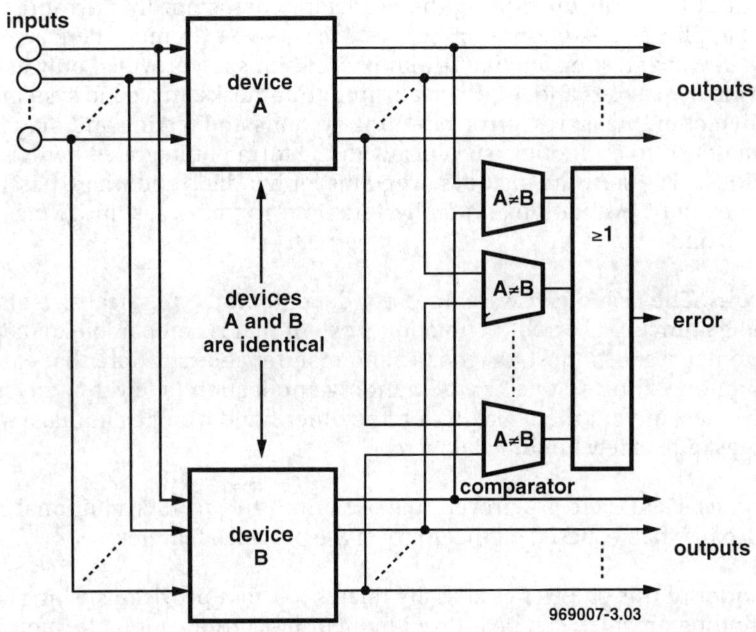

Figure 3.3 On-line fault detection by means of duplicate construction (two-fold redundancy).

The drawbacks can also be easily recognized:

1 Both systems must work exactly in tandem (in other words, you have to ensure that the outputs are synchronous – the data can only be compared when both results are actually available).

2 Faults in the common input data cannot be detected.

3 Although such an arrangement is outstandingly verifiable, it is not possible to localize an error by means of the comparison results. In other words, two-fold redundancy does not eliminate the need for fault-localization measures.

The master/checker technique

The master/checker technique is a rather elegant manner of monitoring the correct operation of digital logic circuits. Of course, the circuitry must be designed to support this.

A master/checker configuration consists of two identical circuits whose inputs and outputs are interconnected, except for one lead each (see Figure 3.4). This lead controls whether each of the two circuits functions as the master or as the checker. The master is the circuit that is functionally active; it evaluates the applied input signals and places its outputs on each of the output lines. The checker also receives the input signals and executes the same functions as the master, but it does not place its outputs on the associated output lines. Instead, it receives the outputs from the master via these lines and compares them to its own output signals. If there is a mismatch then it generates an error signal.

This technique is incorporated in various types of circuits that are intended to be used for constructing high-performance systems. The Pentium processors also include such provisions (Functional Redundancy Check or FRC).

Fault-finding with two-fold redundancy

The necessity of fault assignment. If a comparison error is detected in a duplicated system, then we know for sure that a fault has occurred but we do not know which of the two redundant systems is suspect. If we were to immediately start with detailed fault-finding then we would quite probably have twice as much work to do (since there is twice as much hardware present as in a non-redundant system). It is thus advisable to first verify each of the two systems independently. (If we rely on the single-fault hypothesis we can terminate the verification as soon as the first evidence of a fault is found.) Separately verifying the individual systems has the following advantages:

1 We can identify the unit which warrants more thorough investigation, and

2 we know which units are (most likely) fault-free and thus can thus be used for continued operation if necessary.

Verification can (for example) be performed by checking the hardware using test software.

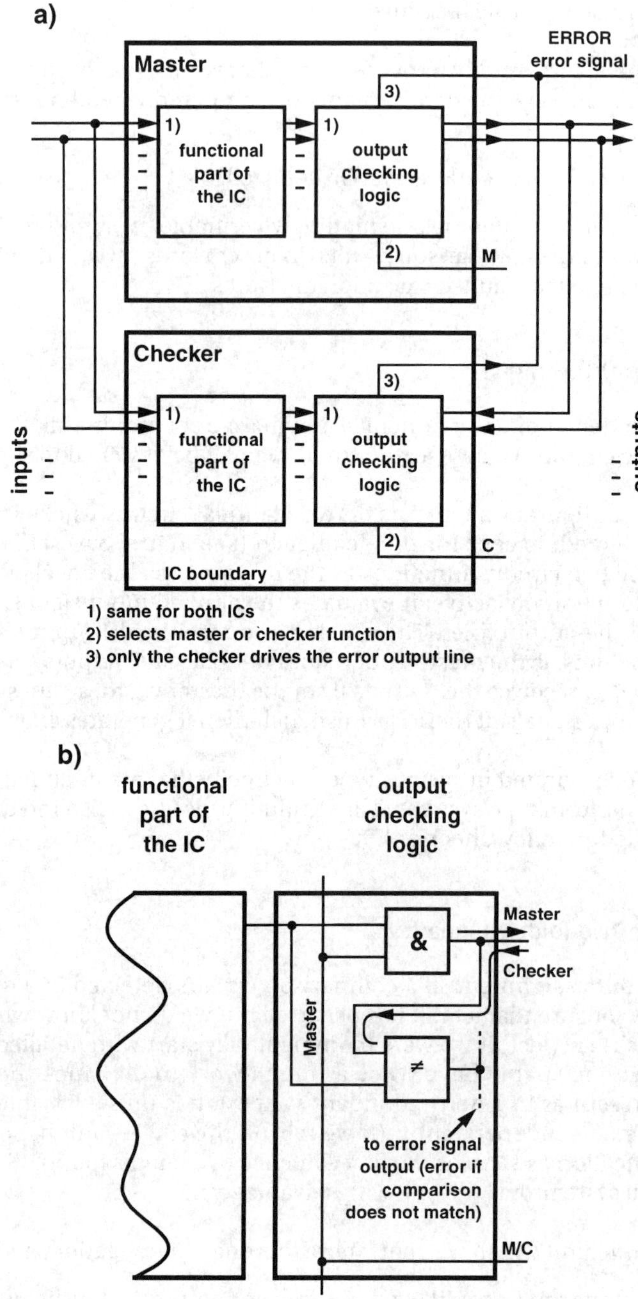

Figure 3.4 *The master/checker technique.*
a) overview, b) detail of a single bit position of the output control logic.

Verifying a master/checker configuration. Master/checker arrangements consisting of two devices A and B should allow the following operating configurations to be selected (for example, by means of jumpers or switches):

1 A is the master, B is the checker.

2 B is the master, A is the checker.

3 A operates alone (or: mode 1 with the error signal ignored)

4 B operates alone (or: mode 2 with the error signal ignored)

Drawing logical conclusions: suppose that an error signal is generated in configuration 1. We can try to check out the hardware with test software (or in the case of a PC, resume our normal work and see whether or not further faults occur). We can choose between the following manners of proceeding:

a) verify A and B independently (using configurations 3 and 4 respectively),

b) select configuration 2 (swap the functional roles). If no error is detected with this configuration, then either the outputs of A are suspect (which can be checked by measurement in configuration 3), or the comparison logic of B is suspect. (*Note:* if this is (for example) a Pentium processor, it can certainly still be used as long as only the comparison logic is defective.) If on the other hand the fault is still present, then we can draw no conclusion about its location.

Two-fold redundancy for fault-finding purposes. Does it make any sense to put together a redundant system yourself for purposes of fault-finding, even only provisionally? This amounts to connecting a known-good device to the device under test via a comparator, in order to be able to detect faults with near-absolute certainty. How can this help in fault localization?

1 If you keep an exact log of the number and types of faults that are detected, you can draw quite far-reaching conclusions by analyzing the log.

2 If the comparison results can be individually evaluated (and in the ideal case bit-by-bit), and if in addition comparisons have been made at a sufficient number of locations, the suspect item(s) can to a certain extent be identified on the basis of the comparison results alone (see Figure 3.5).

The following remarks should help to clarify the basic relationships:

- Every comparison to 'learned' reference data (or data generated by a simulation) basically works on this principle.
- Comparison is only truly helpful in fault localization if the comparisons are sufficiently detailed (not just a global comparison) and the results can be individually evaluated.
- The most rigorous test method uses an adapter that makes contact with the test object at as many locations as possible (a bed-of-nails adapter, for example). The comparison

results for all test points can be individually evaluated. The reference values ('correct' values), which are learned from a known-good object or derived from a simulation, are stored in the tester. This method, which is often used in manufacturing, does however require automated test equipment with prices that easily run to six figures.

- True duplication makes it unnecessary to learn reference values (or generate them by simulation) and to store them. It would certainly be possible to organize a comprehensive test on this basis, using relatively simple equipment (that is, without using a powerful computer to run a simulation model, or an automated tester). You could even do

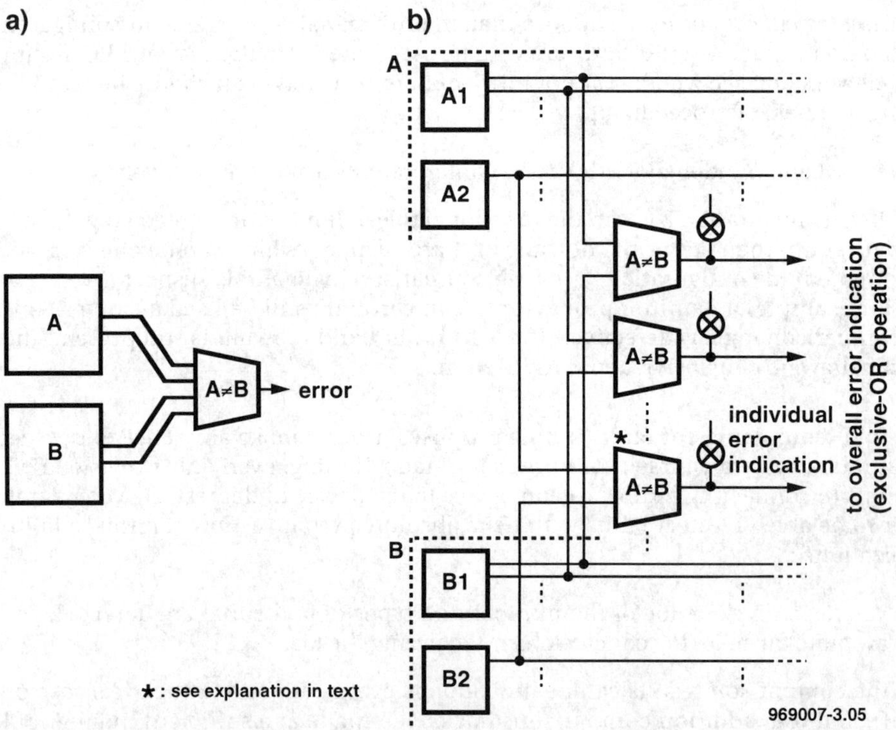

Figure 3.5 Fault localization by means of comparison with duplicated hardware.

a) Unfavourable: only a global comparison; no conclusions can be drawn regarding the fault location.

b) Favourable: (1) numerous outputs of individual subassemblies or components are compared to each other, and (2) the comparison results are recorded individually and bit-for-bit. In this way, suspect signals and subassemblies are directly recognizable (so that you can concentrate on them for more detailed investigation).
*For example: if the comparator marked with * indicates an error, then clearly either subassemblies A and B or the signals being compared are immediately suspect (there is either an internal 'logical' fault or an 'electrical' fault, in the I/O stages or signal lines respectively – see Section 2.5.3).*

without special test software; with continuous comparison you could certainly test the hardware using application functions alone.

- The comparison technique is suitable for testing at the integrated circuit level, assuming that a suitable tester is available (comparing the suspect IC to a known-good example by applying identical inputs and comparing the outputs; similar to the master/checker technique). See Section 5.1.2 for more information.
- *Scenario 1:* the hardware is simple, and it is possible (1) to devise a working method that is synchronized to the clock of the test object and (2) to access an adequate number of test points (*for example:* a (small) microprocessor-based assembly.) In this case, you could certainly consider building a sort of master machine that has comparison circuitry for the bus system and the outputs. The advantages are: (1) lower cost than a 'full-fledged' automated tester, (2) fully dynamic ('severe') functional testing is possible and (3) comprehensive testing is possible using application functions (avoiding the need for test software in the ROM, which is anyhow always too small).
- *Scenario 2:* the hardware is complicated, and synchronization can only be achieved at considerable effort, if at all (*for example:* the usual PC). We will need fully identical hardware with exactly the same complement of software. Assuming that this is even feasible, the next question is, "where shall we make the comparisons?" Let's look at two possibilities:

  a) The ISA bus alone: this is technically possible, but it doesn't help much (only ISA bus faults can be precisely localized, and other than this we don't have much more than a global error indication – but we already know that the PC being tested is faulty).

  b) The PCI bus or the processor: this is quite expensive and difficult (a Pentium test adapter alone costs several thousand pounds, and you have to master working at several hundred MHz) – and even then large parts of the hardware are excluded from comparison (motherboard, RAM, drives, interfaces and so on).

3.2.2 Parity checking

Parity checking is a form of plausibility testing for the detection of individual bit errors: a supplementary parity bit makes it possible to detect a single incorrect bit in an arbitrary number of associated bits. The fundamental fault mechanism is that the incorrect bit has taken on a value that is the opposite of what it should be. (For this reason we speak of bits rather than data lines, since parity checking can be applied equally well to a serial stream of bits, a group of stored bits or a group of bits on a parallel set of lines.)

Figure 3.6 illustrates both the principle and an important application: parity checking can be used in particular for checking 'bundles' of data lines that run via various assemblies over relatively long distances, such as a data bus running from an expansion card source via a motherboard, backplane or interface cable to a destination device. Faults in such data paths, which are interrupted at various points by connectors and components such as drivers and receivers, can be relatively easily detected by a parity test. The contents of memory components can also be checked by supplementary stored parity bits.

Parity circuits for parallel data lines

Parity checking basically requires two types of circuitry:

1. circuitry to produce the parity bit (parity generator), located in the data transmitter, which is assumed to output the bits correctly,
2. circuitry to check the incoming data plus parity bit for parity errors (parity checker), located in the data receiver, which must be prepared to deal with erroneous bits.

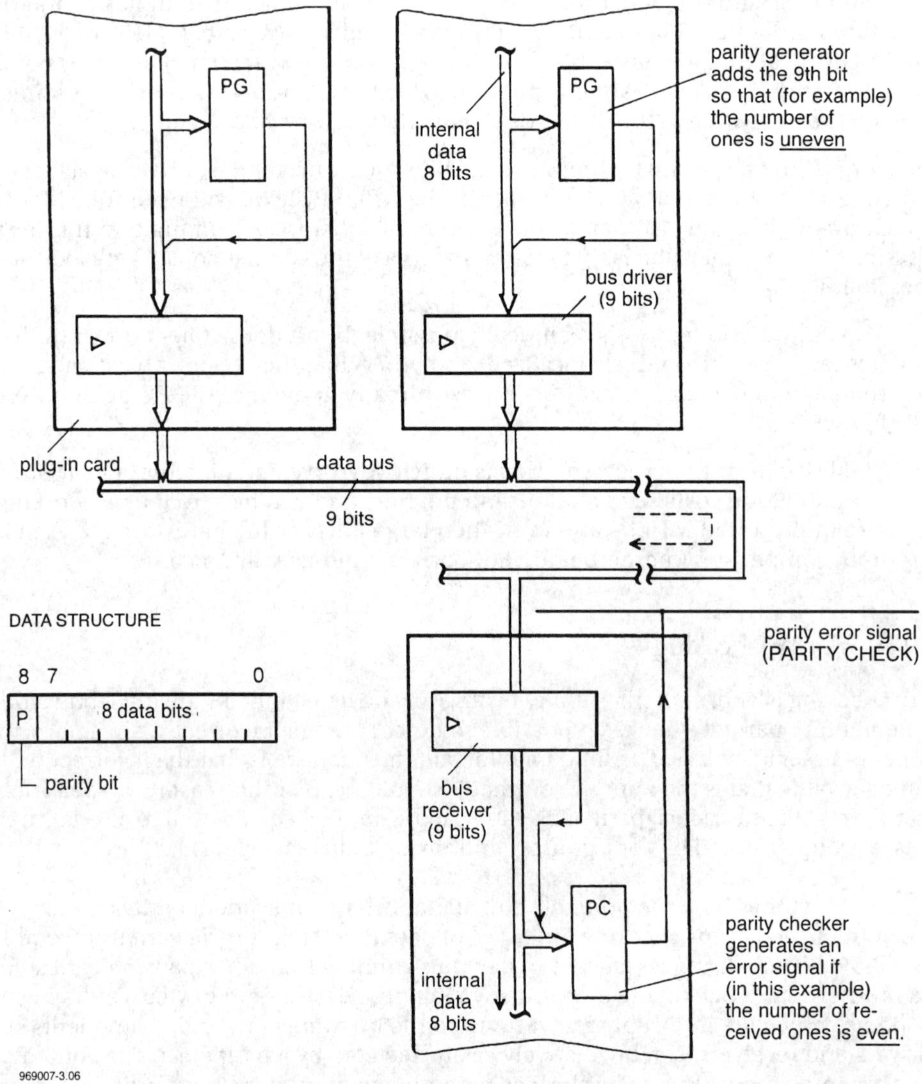

Figure 3.6 Parity generation and checking.

Both of these tasks (parity generation and parity checking) can be accomplished by a 'tree' structure of exclusive-OR (XOR) gates. Figure 3.7 illustrates how an odd-parity bit can be generated as the XOR of *n* bits (the output signal has a value of '1' when an even number of ones is present at the inputs). An even-parity bit can be generated by inverting the output signal.

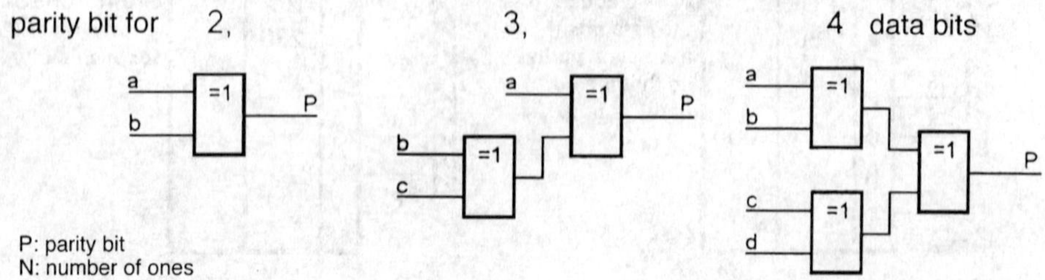

a	b	P	N
0	0	0	0
0	1	1	1
1	0	1	1
1	1	0	2

a	b	c	P	N.
0	0	0	0	0
0	0	1	1	1
0	1	0	1	1
0	1	1	0	2
1	0	0	1	1
1	0	1	0	2
1	1	0	0	2
1	1	1	1	3

a	b	c	d	P	N
0	0	0	0	0	0
0	0	0	1	1	1
0	0	1	0	1	1
0	0	1	1	0	2
0	1	0	0	1	1
0	1	0	1	0	2
0	1	1	0	0	2
0	1	1	1	1	3
1	0	0	0	1	1
1	0	0	1	0	2
1	0	1	0	0	2
1	0	1	1	1	3
1	1	0	0	0	2
1	1	0	1	1	3
1	1	1	0	1	3
1	1	1	1	0	4

969007-3.07

Figure 3.7 Parity bit generation.

Integrated-circuit parity generators and parity checkers. The most-commonly used IC is the xx280 (9-bit odd/even parity generator/checker). It has nine inputs, plus outputs for both even and odd parity bits. Figure 3.8 shows how the parity bit can be generated and parity can be checked for a data byte, using two such ICs (the illustrated circuit is set up for odd parity).

Even and odd parity. Parity generation and checking can be designed to work with either even or odd parity. This always refers to the total number of ones in the data plus the parity bit itself. The outputs always relate to the parity of the *entire* set of input bits (data bits to be checked plus the parity bit). Table 3.2 shows how the parity bit is generated and how the parity of a set of bits is checked, for both even and odd parity.

Applications. The parity-checking technique can only be used with a group of related bits. It is 'economical' only when the circuitry that is to be monitored cannot actively modify

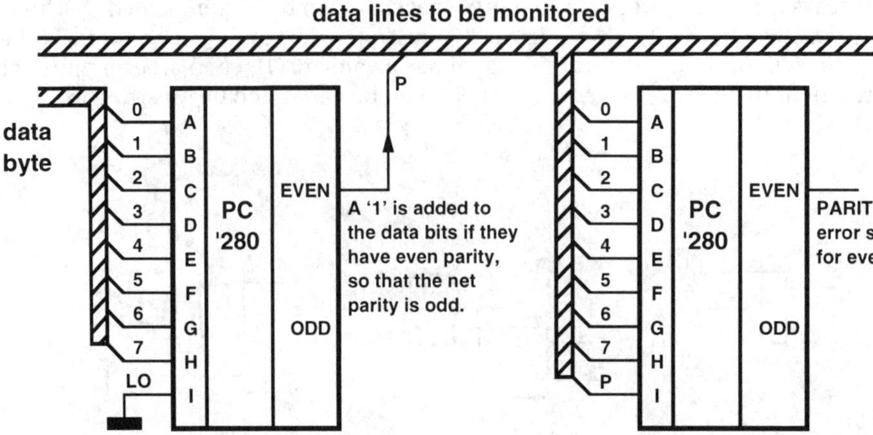

number of ones * in the 9 output bits	output EVEN	output ODD
even	HI	LO
odd	LO	HI

* ≙ HI value

969007-3.08

Figure 3.8 *Bytewise parity-bit generation and checking (odd parity). The operation of this circuit is explained in Table 3.2.*

Type of parity checking	Requirement	Parity generator circuit output	A parity check error occurs if …
even parity	the total number of 1s must be an even number	odd parity (a 0 if there is an even number of 1s on the signal lines, otherwise a 1)	an odd number of 1s is present on the input lines (data bits plus the parity bit): PARITY CHECK = ODD PARITY
odd parity	the total number of 1s must be an odd number	even parity (a 0 if there is an odd number of 1s on the signal lines, otherwise a 1)	an even number of 1s is present on the input lines (data bits plus the parity bit): PARITY CHECK = EVEN PARITY

Table 3.2 *Parity generation and parity checking.*

the values of the individual bits. Its utilization is thus normally limited to checking stored data (memory contents) or transmitted data (the parity bit is stored together with the data or transmitted along with the data).

Note: it is in principle possible to use parity checking for data-processing operations. This can be done by *parity prediction* (using additional circuitry that calculates the parity of the result of an operation for a given set of operands) and comparing the predicted parity to the actual parity of the result. Hardware based on this principle has in fact been developed for on-line monitoring of working hardware (parity adders, for example), and it is used in some modern systems (such as the IBM ES/9370). However, where true 100% checking is desired, two-fold redundancy is increasingly common, since (1) it is better (true comparison instead of plausibility checking, plus the possibility of continued operation in case of a fault by reconfiguring the hardware) and (2) it is also usually less expensive with highly-integrated circuitry (master/checker techniques with the same type of IC for each function; no extensive monitoring circuitry is needed apart from that which compares the output data).

How many bits should be included in the parity check? The following factors must be taken into account:

1 A parity bit can only be generated for a group of associated bits that 'belong together'. Suppose that we have a memory with a 32-bit-wide access path. A single parity bit is actually enough to check all 32 bits, but then all 32 bits must always be read or written at the same time. However, if the memory allows individual bytes to be accessed (byte-selective access), then writing a single byte requires first reading the associated 32-bit word, replacing the byte in question with the new data, generating a new parity bit for the resulting word and finally writing the entire word back to the memory (which requires a read-modify-write cycle). This is time-consuming and requires extra hardware. For this reason a byte-oriented memory structure is preferred for microprocessor systems (and PCs), which means that bytewise parity must be used. A counter-example is the Pentium address bus, which has a single parity bit for the entire bus.

2 Parity generation and checking takes time. A parity bit can only be generated when all of the actual data bits are available. Parity generation and checking is always based on the exclusive-OR operation. The number of inputs of a practical elementary XOR gate cannot exceed 3 or 4, so that several cascaded gates must be used for a larger number of data bits (compare to Figure 3.7). This results in a significant increase in the time required to generate or check the parity (due to gate propagation delays).

3 The smaller the number of bits to which parity checking is applied, the more exactly a fault can be localized. A 32-bit parity check, for example on an address bus, can only indicate that something is wrong, while bytewise parity checking allows the fault to be localized to a specific byte. In this case it is only necessary to more closely investigate 9 lines, and a suspect IC can usually be directly identified from the fault symptoms.

4 The parity checking hardware itself must also be tested. In this regard, exclusive-OR gates are somewhat nasty: the only practical way to test them is to check the complete truth table. (This means 2^n test steps for an exclusive-OR gate with n inputs, unless you can reduce the number of steps required based on knowledge of the internal structure

of the gate.) Obviously, it is not a problem to test all 512 (2^9) possibilities for 8 data bits plus 1 parity bit, but a test with 2^{33} (≈ 8 billion) steps would take a relatively long time.

Even or odd parity? In theory, both even and odd parity work equally well, but there is an important additional consideration: what happens with a fault that affects all bits in the same way? *Examples:* no data is placed on the bus, due to a faulty bus-interface IC (or faulty control circuitry); a memory IC is not correctly selected and thus does not supply any data; an interface connector has come loose and so on. In such cases, we must examine the electrical behaviour: is it more likely that all zeros or all ones will be 'seen' by the data receiver? Some circuits interpret 'open' (floating) inputs as logical zeros, which an odd-parity detector would recognize as an error. On the other hand, some circuits interpret open inputs as logical ones (this applies to TTL inputs and also to bus systems that have pull-up resistors); in this case odd parity is not suitable (a set of 9 ones has odd parity), so that even-parity checking is preferable (*examples:* xx486 and Pentium processors).

Integrated circuits with parity provisions. The extensive use of parity checking has encouraged IC manufacturers to add an additional bit to bus-related ICs and memory components. There are thus bus interface ICs, registers, memory chips and memory modules organized as '9 ×', '18 ×' and '36 ×', as well as 8- and 9-bit bus interface ICs with built-in parity generation or parity checking capability.

Parity checking for serial data. Data that are transmitted or stored serially (bit-by-bit) can also be checked by a parity bit 'added on' to the actual data bits. For instance, the RS 232 and V.24 serial interfaces have parity options; the interface ICs can be configured (1) to either check or ignore parity and (2) to generate or check either even or odd parity.

The effectiveness of parity checking

- Basically, only single-bit errors can be detected by parity checking.
- Total failure of a signal path (due for example to an IC not being driven, a loose connector or the like) is only detectable under the following conditions: (1) the inputs of the receiving device assume fixed and known values in the presence of the fault (such as 'all ones' due to the presence of pull-up resistors) and (2) parity checking is implemented such that the associated input values cause a parity error (even-parity checking in the case of an uneven number of input bits that are all ones (or vice versa), and odd-parity checking in the case of all zeros).
- A 'stuck-at' fault at a given bit position will only result in a parity error if the value applied to that position is the opposite of the 'stuck-at' value (thus a '1' for 'stuck-at-0' and a '0' for 'stuck-at-1').
- Errors in two bit positions at the same time are fundamentally not detectable.

Effectiveness of parity checking for serial data. Just as for parallel data, only single-bit errors can be detected. However, fault mechanisms that result in the generation of single-bit errors are quite rare in hardware that works serially. Thus parity checking for serial data is only effective for the following cases:

1 bit errors are due to intermittent disruptions of the data path (disturbances that are shorter than the duration of an individual bit cell),

2 bit errors occur in stored data (where the error is limited to a single bit cell),

3 bit errors result from synchronization problems (for example, baud-rate tolerances in serial interfaces), although here the probability of detecting an error is not 100%.

Note: parity checking works in an entirely satisfactory manner for single-byte data transfers (asynchronous or start/stop operation) as long as the data rate is not too high (typically 300–1200 bit/s). In practice it is quite rarely used. In most interface protocols, parity checking is disabled (PARITY OFF), and error checking is moved to a different protocol level or communications layer (such as software or a modem error-correction protocol).

Testing parity logic. On-line error detection by means of parity checking can only give the desired results if the following conditions are satisfied:

- The signal lines are free of 'stuck-at' faults.
- The parity-generation logic generates the correct parity bit for all possible input data.
- The parity-checking logic correctly recognizes all possible sets of error-free and erroneous data.

In order to ensure this, the parity-generation and checking logic must itself be tested. There are two proven ways to do so:

1 Parity injection: the parity bit can be directly set to '1' or '0' (typically under software control). In this way (1) all bit positions (including the parity bit) can be tested for 'stuck-at' faults, and (2) data with both correct and incorrect parity can be selectively applied to the inputs of the parity-checking logic (functional checking to determine whether parity errors are correctly detected).

2 Parity inversion: the parity generator can be switched over to the opposite parity sense by means of a jumper or switch (or more elegantly, under software control). This allows the following test process:

   a) With 'normal' parity selected, all possible combinations of input values are applied one after the other, and the output value is checked for each one. No parity error should be detected.

   b) With 'inverted' parity selected, all possible combinations of input values are again applied one after the other and the output values are again checked. A parity error should be detected in every case (error provocation).

Note: provisions for testing parity generation and checking cannot always be taken for granted. *For example:* the 'classic' IBM PC/AT allows only parity *checking* to be switched off at the motherboard RAM (via control port 61_H, bit 7). The parity bits are still generated and written to the RAM, so that it is not possible to provoke parity errors by disabling the parity logic, writing data to the RAM and then re-enabling the parity logic. (This would be desirable not only to check the operation of the parity logic, but also to check whether there is actually any effective parity logic in the RAM). Other motherboards, by contrast, allow the parity to be inverted and/or parity errors to be provoked under software control (Force Bad Parity). This is also true of SCSI host adapters. The practical problem (if such provisions are present) is that software that utilizes such provisions must also be available – and you cannot write suitable test programs without knowing the details of the hardware design.

Practical tip: when developing hardware that includes parity logic, keep in mind self-test provisions. Useful features are:

1 Being able to sense or evaluate parity-error signals using software (error signalling via an interrupt or reset).

2 Being able to force parity errors to occur (parity inversion function).

For the design of suitable test software, refer to Volume 2.

3.2.3 CRC error-detection codes

CRC (Cyclic Redundancy Checking) is an error-detection technique that is applied to an entire block of data. It is particularly important for serial data transmission and data storage (in mass-storage devices). It is based on the following:

1 Additional check bits (the CRC bits) are added to the block of data that is to be checked, at the time that it is transmitted or stored. These bits are generated according to the rules of a specific *cyclic code*.

2 When the data block is received or read, the entire data block plus the CRC bits is subjected to a checking process. (The simplest example of such a process is to generate the CRC bits anew and compare the newly-generated bits to the bits accompanying the data block.)

CRC and parity checking. Parity checking can only exactly detect a single bit error at any one time, in the set of bits being checked. However, single-bit errors are quite uncommon for data that are transmitted serially or stored in a serial structure. External disturbances, and defects in the storage medium, naturally do not always limit themselves to affecting only a single bit, and data synchronization is hardly likely to be 'off' by only a single bit. Instead, it is rather more likely that a *group* of consecutive bits will be affected. Such errors are called *burst errors*. Cyclic codes are particularly well suited to detecting such errors with a high probability of success.

Generating the CRC bits. The objective is to calculate a series of test bits (the CRC bit chain) that can be attached to the (unaltered) block of data bits, so that a structure results as shown in Figure 3.9.

The sequence of check bits is generated according to the formation rule of a particular CRC code. This is described by a *generator polynomial* $G(x)$. (In other words, every CRC code is unambiguously characterized by a particular generator polynomial $G(x)$; see Appendix 6 for more information.)

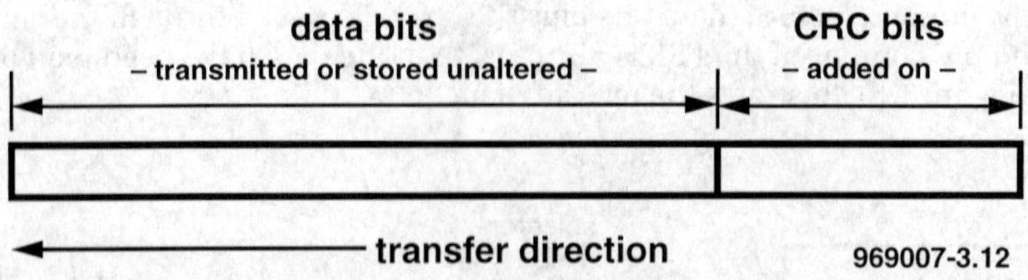

Figure 3.9 Data structure with CRC error-checking field.

CRC formation in polynomial notation:

1 We refer to the given series of data bits as $I(x)$ and the CRC bit chain as $C(x)$.

2 We first form the product $x^k \cdot I(x)$ (in binary terms, this means that we initially attach k zeros to the data in place of the k CRC bits – see Figure 3.9).

3 We divide $x^k \cdot I(x)$ by $G(x)$. The remainder of this division corresponds to the desired CRC bits. They are attached to the block of data bits as shown in Figure 3.9, in place of the k zeros that were attached in step 2.

In other words, $C(x)$ is the remainder of the division $(x^k \cdot I(x) / G(x))$. We can write this as $C(x) = \mathbf{rem}(x^k \cdot I(x) / G(x))$, where **rem** means 'the remainder of'.

CRC checking. The following procedure is an obvious way to do CRC checking:

1 Generate the CRC bits anew from the read or received series of data bits that make up the data block.

2 Compare the newly-generated CRC bits to the CRC bits that have been read or received. Any mismatch indicates an error.

In practice, a simpler procedure is used. If we refer to the stored or transmitted data structure (in the sense of Figure 3.9) as $T(x)$ and the read or received data structure as $R(x)$, then the CRC checking process consists of dividing $R(x)$ by $G(x)$. If the result is zero then no error is present (that is, $T(x)$ and $R(x)$ are identical.). We thus have:

$R(x) = T(x)$ if $\mathbf{rem}(R(x)/G(x)) = 0$; thus everything is OK;

$R(x) \neq T(x)$ if $\mathbf{rem}(R(x)/G(x)) \neq 0$; thus there is an error.

CRC logic. We have two bit series, one of which must be divided by the other. Since the division is performed 'modulo 2', the hardware is remarkably simple if the generator polynomial is represented by a feedback shift register (FSR) (see Appendix 6). The FSR is first loaded with an starting value. When the data are stored or transmitted they are also passed through the shift register (although the data bits themselves reach the storage medium or transmission channel unmodified). Once all the data bits have passed through the shift register, the data remaining in the shift register are the CRC bits, which are then 'pushed out' and attached to the block of data bits. When the data block is read or received, the entire structure (data bits plus CRC bits) is passed through the shift register. If at the end the content of the FSR is all zeros, then the data have been read or received error-free. Figure 3.10 illustrates the relevant data flows.

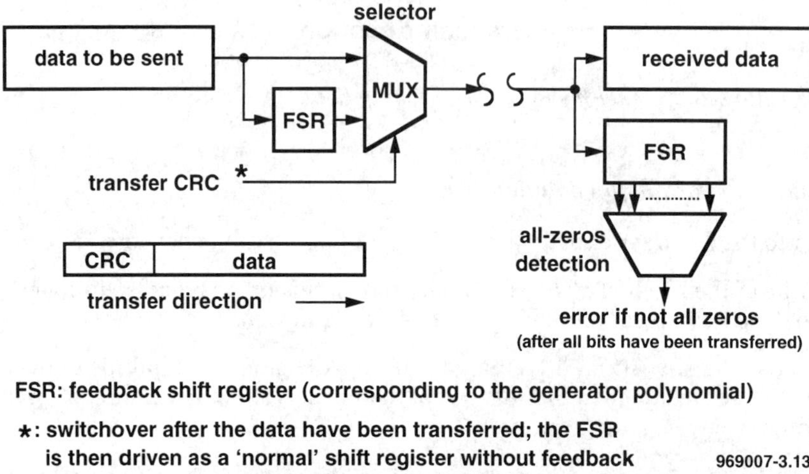

Figure 3.10 *CRC generation and checking.*

The effectiveness of CRC checking

Let us again consider the CRC error-detection process. We receive a bit sequence $R(x)$ consisting of the data and the CRC bits; we divide this by the generator polynomial $G(x)$ and interpret a zero result as 'no error'. Suppose that $R(x)$ has been changed from the original (correct) sequence $T(x)$ such that $\mathbf{rem}(R(x)/G(x)) = \mathbf{rem}(T(x)/G(x)) = 0$ even though $R(x) \neq T(x)$. This error cannot be detected (this is comparable to two-bit errors with parity checking). How probable is this? Table 3.3 shows the reliability of a particular, frequently-used CRC code with a 16-bit generator polynomial (the CCITT CRC code). Basically, a CRC with a generator polynomial of degree n (an n-bit generator polynomial) has an error detection probability of $(1 - 2^{-n})$. The probability that a 16-bit generator polynomial will detect an arbitrary series of incorrect bits $R(x) \neq T(x)$ is thus 99.9984%. A 32-bit CRC will detect practically all errors (the probability is $1 - 2^{-32} \approx (100\% - 0.25 \times 10^{-7})$.

Important CRC polynomials. The 'secret' of a good CRC polynomial (or in other words, the reason why certain polynomials are used) is that it guarantees that the most likely types of errors (such as isolated burst errors up to a certain length) will not result in a received bit sequence $R(x)$ that yields a zero-value remainder after being divided by $G(x)$. The most important CRC polynomials are summarized in Table 3.4.

Note: the 12-bit CRC and both 16-bit CRCs are designed to have a 100% probability of detecting all burst errors containing an odd number of bits; all non-detectable errors consist of an even number of bits.

Type of error	Probability of detection
single-bit error	100%
double-bit error	100%
triple-bit error	100%, if the total data stream is less than 32,767 bits
burst error with an odd number of incorrect bits	100%
burst error, up to 16 incorrect bits	100%
17-bit burst error	99.9969%
any other burst error	99.9984%

Table 3.3 Error-detection probability for the CCITT CRC code.

Code	Generator polynomial	Application
CRC–12	$x^{12} + x^{11} + x^3 + x^2 + x + 1$	Protocols with 6-bit character codes (eg BSC)
CRC–16	$x^{16} + x^{15} + x^2 + x + 1$	Protocols with 6-bit character codes (eg BSC)
CRC–CCITT	$x^{16} + x^{12} + x^5 + x + 1$	Floppy disks and hard-disk drives; modem protocols
CRC–32	$x^{32} + x^{26} + x^{23} + x^{22} + x^{16} + x^{12} + x^{11} + x^{10} + x^8 + x^7 + x^4 + x^2 + x + 1$	LANs (IEEE 802 standard); modem protocols

Table 3.4 Important CRC polynomials.

Testing CRC logic. Strictly speaking, the CRC hardware must itself be tested. The technical requirements are that either (1) error provocation is made possible (by allowing arbitrary bit sequences to be inserted, in particular those with CRC errors) or (2) the CRC hardware can be structurally tested due to the provision of suitable accessibility measures (see Section 3.3). An exhaustive test is however hardly possible. In practice, therefore, the following approaches are often used for approximate testing:

1 Special provisions are either not present at all or are not used. Data transfers are tested using numerous test vectors on the basis of actual/expected comparisons (the received or read data bits are compared to the transmitted or written data bits). If no errors are detected by the comparisons and the CRC logic does not detect any error, then presumably everything is in order. Otherwise you can draw the following conclusions: (1) comparison errors but no CRC errors detected, or (2) no comparison errors but CRC errors detected: the CRC hardware is suspect; (3) both comparison errors and CRC

errors: the CRC hardware is OK, and the fault must lie in the transmission medium or the storage medium.

2 It is possible to provoke errors. In this case the test of step 1 is augmented by several spot checks (for example, does the hardware detect maximum-length burst errors; does it react correctly to non-correctable errors and so on).

Notes:

1 With appropriate cyclic codes, it is possible to not only detect burst errors but also to correct them. This capability is utilized in particular in hard-disk drives.

2 The usual floppy-disk controller does not allow the CRC checking process to be externally influenced. The command sets of IDE, ATA and SCSI drives, on the other hand, do include commands that allow data including the CRC or ECC bits to be written and read. This makes it possible to include error-provocation tests in the test software.

3 Cyclic codes can be constructed algorithmically (by software). Refer to Volume 2.

Using cyclic codes to compress test data

Cyclic codes are also suitable for checking whether bit sequences (especially longer ones) are error-free, or for comparing actual values to expected values. It is costly to directly store expected values that take the form of long bit sequences. Thus it is natural to look for some way to compress the data. One idea is to apply a cyclic code to the reference data and to then store the resulting CRC bits instead of the actual data. In the test procedure, the data resulting from the test are subjected to the same CRC process, and then the CRC bits calculated from the test data are compared to the stored CRC bits from the reference data. The advantage of this is that only CRC strings of perhaps 16 bits have to be stored, instead of reference strings that can be several thousand bits long. The fault-detection probability is in practice usually adequate. (This is the principle of *signature analysis*; see Section 4.4.3.)

3.2.4 Time-based monitoring and time measurement

Time-based monitoring methods

Time-based monitoring of programs and hardware processes (for example, bus wait states) is a simple and effective means of detecting general malfunctions. In regard to fault localization, such techniques are effectively very rough plausibility tests. We can distinguish two forms of time-based monitoring Figure 3.11 (overleaf) shows state diagrams and circuits that illustrate the principles involved:

1 Elapsed-time monitoring (*timeout* checks). The basic idea is that if a particular routine or process does not complete within a specific, predefined time interval, then a fault must be present. A timer is started at the beginning of the routine or process. This timer is set to run for the maximum time allowed for the routine or process to complete. If

this time is exceeded then an error indication is generated; if the process being checked completes within the allowed time then the timer is reset. A variation is to check for too-early completion or for a specific duration (plausibility checks for certain real-time activities: early completion is just as suspect as late completion).

2 Activity monitoring (*watchdog timer*). The basic idea is that a properly-working system must be able to periodically give some indication that it is active. If no indication of activity occurs for an extended period of time, there must be a fault present (typical faults: the system is 'hung' or 'crashed', the software is stuck in an endless loop, and so on). The principle of operation is that a timer is periodically re-triggered. If the trigger signal does not recur within a predefined time interval (the watchdog interval) the timer expires and an error is signalled.

Triggering the timer. A timeout or watchdog monitor timer can be triggered (1) by hardware alone, (2) by software alone or (3) by a combination of hardware and software.

Examples:

1 Checking the timing of a bus system. Specific bus activities (phases) are checked to see whether they last longer than prescribed intervals. This can relate to complete bus cycles or to certain parts of the bus cycle, such as bus-master arbitration or the durations of wait states.

2 Checking interrupt-handling latency. A signal on any of the interrupt lines directly activates the timeout hardware. The interrupt handler program resets the timer (for example, via an I/O instruction). If the interrupt handler is not called before the timeout expires, an error indication is generated.

3 Checking a running program (watchdog monitoring). The program that is being monitored recurringly 'reports' to the timer (for example, via an I/O instruction) to show that it is still running. An error is indicated if no 'report' signal (trigger) is received from the program for the duration of the prescribed watchdog interval. When writing the program, you must therefore ensure that instructions that retrigger the watchdog timer will be executed sufficiently often, for all relevant branches of the program. (A typical example of such an instruction is an I/O instruction that generates a trigger signal).

Setting the time interval. Here there is a self-evident requirement – and often a problem.

- The requirement: a fault should be detected as quickly as possible, so you should not be too generous with the timeout or watchdog interval. For real-time applications, the interval should be chosen such that if a fault occurs there will still be enough time to take any required emergency measures.
- The problem: (1) the allowable timeout or watchdog interval cannot always be precisely specified (in other words, the time relationships cannot be predicted accurately enough) or (2) it would require too much effort to keep within a tightly-specified interval. For this reason, it is often necessary to settle for being able to detect a generally 'hung' system (by specifying a quite generous interval).

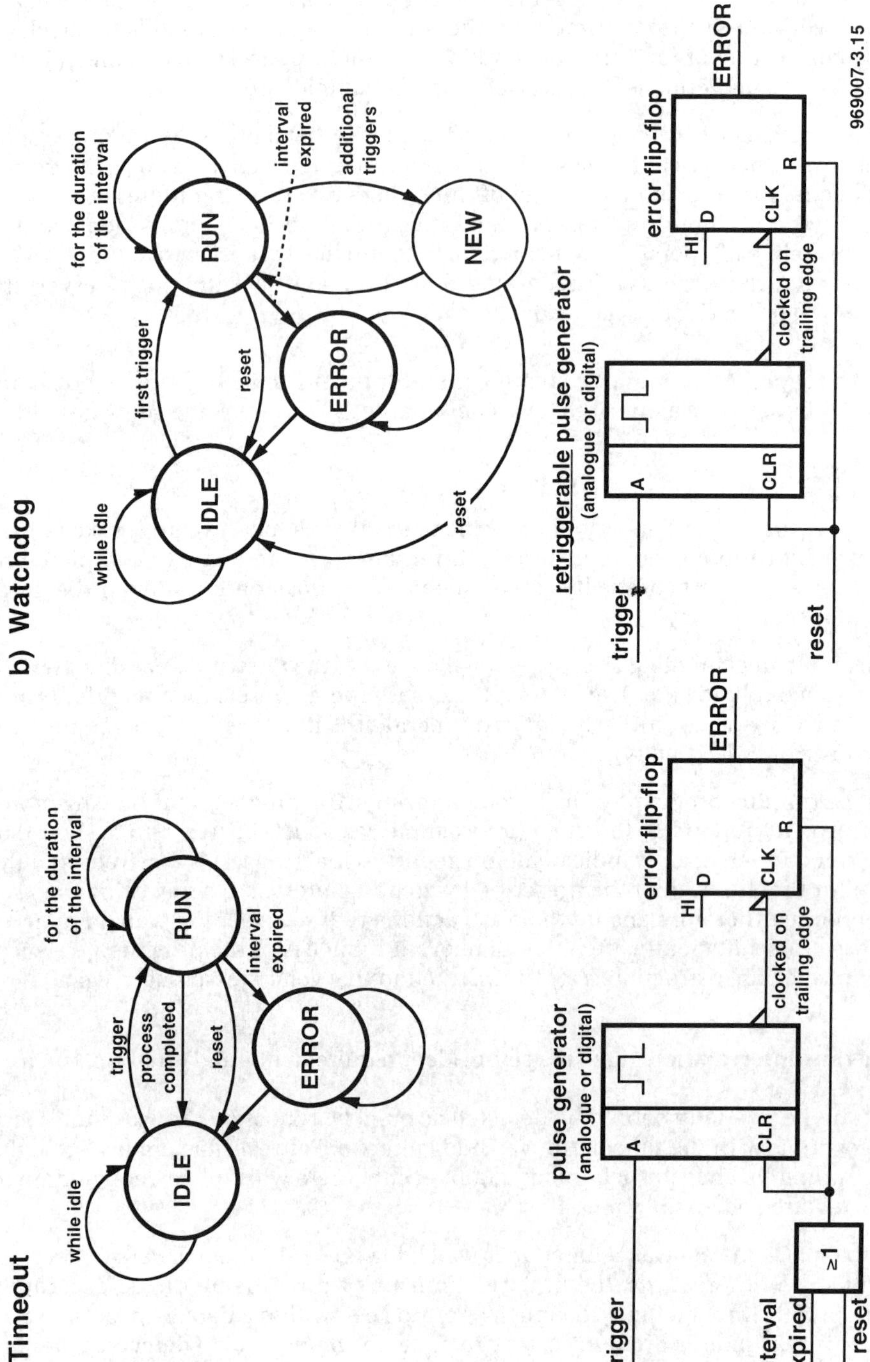

Figure 3.11 Timeout and watchdog timer processes. top: state diagrams, bottom: simplified schematics.

Examples of 'precise' time checking: two timeout checks are provided for the EISA bus:

1 8 µs Bus Timeout (64 bus clocks). If the central bus control logic deactivates the MACK line of the active master (in order to force it off the bus), the active master must deactivate its MREQ line and release the bus within the specified interval. (The bus control logic waits until the currently active cycle is completed and then starts the timeout.)

2 32 µs Bus Timeout (256 bus clocks). A timeout error occurs if the CMD line stays active for more than 32 µs in a given bus cycle, that is, if the cycle remains in the Wait state for that long. The timeout interval only starts when a new device requests the bus. (Under normal operating conditions a refresh request occurs every 15 µs, which guarantees general timeout monitoring.)

A timeout error in either of these cases results in an NMI at the processor plus a bus reset.

An example of 'loose' time checking: device selection on the SCSI interface. A selected device must reply within 200 µs by activating BUSY. (This interval has been chosen to allow the interface to work properly with relatively slow devices, without the use of complicated circuitry.) This means that only true 'bus hung' situations can be detected. In reality a device must react much faster to being selected (within a few microseconds).

Explanation of Figure 3.11:

a) *Timeout: as soon as the process that is being monitored starts, the pulse generator is triggered (it exits the idle state; IDLE→RUN). If the process completes within the prescribed length of time, the pulse generator is reset (it returns to the idle state; RUN→IDLE). However, if the entire interval passes before the process completes, then the circuit goes to the error state (RUN→ERROR), which it cannot exit without some external signal (such as a reset). The circuit shows a pulse generator that produces a pulse ab whose length corresponds to the timeout interval. The trailing edge of this pulse triggers the Error flip-flop. If the process completes normally, the pulse generator and the error flip-flop are simultaneously reset (the Error line will not become active because the reset input to the flip-flop overrides the clock input).*

b) *Watchdog timer: the circuit is initially in the idle state. When the activity that is to be monitored starts, its first trigger signal causes a transition to the active state (IDLE→RUN). If everything is working properly, the interval between successive triggers is always less than the prescribed maximum interval. Each trigger signal causes the circuit to start timing again from the beginning of the interval (this is provided by the transition sequence RUN→NEW→RUN). In normal operation (no error), the circuit thus remains in the RUN state, with brief excursions to the NEW state (every time that a new trigger occurs). However, if the trigger signal is absent for too long, the timeout interval expires. The circuit then goes to the error state (RUN→ERROR). The circuit shows a pulse generator that is repeatedly retriggered. In normal operation (no error) it thus never times out, so that a falling edge never occurs at the clock input of the Error flip-flop. A falling edge can only occur if the trigger does not arrive within the timeout interval, so that the pulse terminates.*

Watchdog example 1: IBM PS/2 computers. These systems have a built-in watchdog timer (timer channel 3). It checks the handling of interrupts that are periodically triggered by timer channel 0 (the general system clock). An NMI is generated in case of an error.

Watchdog example 2: PIC microcontrollers. These ICs have a built-in watchdog timer which, once it has been activated (this decision is made at the time that the IC is programmed and cannot be subsequently changed), expects to see recurring watchdog instructions. Refer also to Figure 3.18 and Volume 2.

Time monitoring in standard PCs. PCs as such are not conceived as real-time systems. For this reason, they do not have any time-monitoring hardware (with the occasional exception; see the examples of the EISA bus, SCSI bus and PS/2 machines). There are two situations in which effective time monitoring would be desirable:

1 Unattended operation (such as a server or a system that runs overnight to perform a computationally intensive task). It should at least be possible to recognize a hung system (in order to suspend operation or alert maintenance personnel, for example).

2 When the PC is used as a real-time system (industrial PCs).

Some monitoring tasks can be implemented in software (by utilizing the system clock, other timer channels or the real-time clock). But beware: (1) this approach only works if the software is able to run under all conditions, which means that the fault situations for which time monitoring is provided must not be able to affect the interrupt control logic or the timing mechanism that is used, and (2) it is not particularly easy to implement time-monitoring measures with modern-day operating systems (such as Windows). The solution is to use supplementary time-monitoring hardware. Suitable expansion cards are available from various suppliers. Additional hardware for time monitoring can also be constructed as not-too-difficult DIY projects (see Volume 3).

Time measurement

Absolute time (date and time of day). It is a significant advantage to always have the current date and time of day available in a computer system (and not only available, but – ideally – fully automatic, which means (1) without having to type in the current date and time every time the system is switched on and (2) with protection against all kinds of intentional corruption). The following are some of the common applications:

- time stamping of files (so that you can immediately recognize the most recent version and see when the file was last modified),
- automatic insertion of the current date and time in applications (such as inserting the current the date in a letter),
- keeping track of the 'expiry date' of demonstration software,
- logging error messages together with the date and time. (Such logs often allow reasonable conclusions to be drawn regarding the actual cause of a fault; see also Volume 2.)

Time intervals. In the computer realm, measurement of time intervals (that is, measuring the *duration* of a process) is especially important for making and comparing performance measurements. The fundamental technical basis for such measurements consists of fast, multi-digit, high-resolution timers which must be completely independent of the software as far as the actual timing function is concerned. Only on such a basis is it possible to accurately measure things such as data rates under specific operating conditions, the relative performance of alternative codings of time-critical algorithms, the performance of caches and the like. An ideal solution would be time measurements based on the processor clock of the system in question. This can however only be realized within the framework of the processor IC (this is for example included in the Pentium as a part of Performance Monitoring: see Volume 2). The only other alternative is to attempt to make the performance-related measurements as accurately as possible using external timer circuitry or standard measurement equipment (such as a universal counter/timer).

Circuits for time monitoring and time measurement

Pulse generators. We use the term 'pulse generator' to refer to a circuit that produces a single output pulse with a specific duration, in response to an input trigger signal. We can distinguish various types of pulse generator according to the following characteristics:

1 Timing behaviour: the important factor is how the pulse generator responds to a trigger pulse at its input while the output pulse is active. In this regard we can differentiate between *retriggerable* and *non-retriggerable* pulse generators (see Figure 3.12).

2 Implementation: there are digital and analogue pulse generators.

Digital pulse generators. Digital pulse generators are based on counters. The pulse duration is determined by the number of counter stages (bits) and the clock frequency. Figure 3.13 shows the construction of a retriggerable pulse generator that can (for example) be used as a digital watchdog timer.

Analogue pulse generators. With analogue pulse generators, the duration of the output pulse is determined by a resistor-capacitor time constant. A high degree of accuracy is not needed for time monitoring, so that high-precision implementations are rarely required. A monostable multivibrator ('monostable') is often adequate (these are available as ICs and need not be constructed from individual gates – see Figure 3.14). With microprocessors that have built-in watchdog timers in their internal monitoring logic (supervisory circuits), it's even simpler (refer to Figure 3.19). The advantage of a monostable is that the duration of the timing pulse is determined by external components, and thus can be set by the designer (or user) within a quite large range. Simple monitoring ICs, by contrast, usually have fixed intervals (typically on the order of a second).

Analogue or digital? It depends on the individual situation. The advantages and disadvantages of the two types are listed in Table 3.5.

Supervisory ICs. These types of ICs are essentially intended to be used for monitoring the supply voltage and generating reset signals, but some types also include a pulse generator. The manufacturers typically regard the pulse generator as an 'extra' that can be used as a

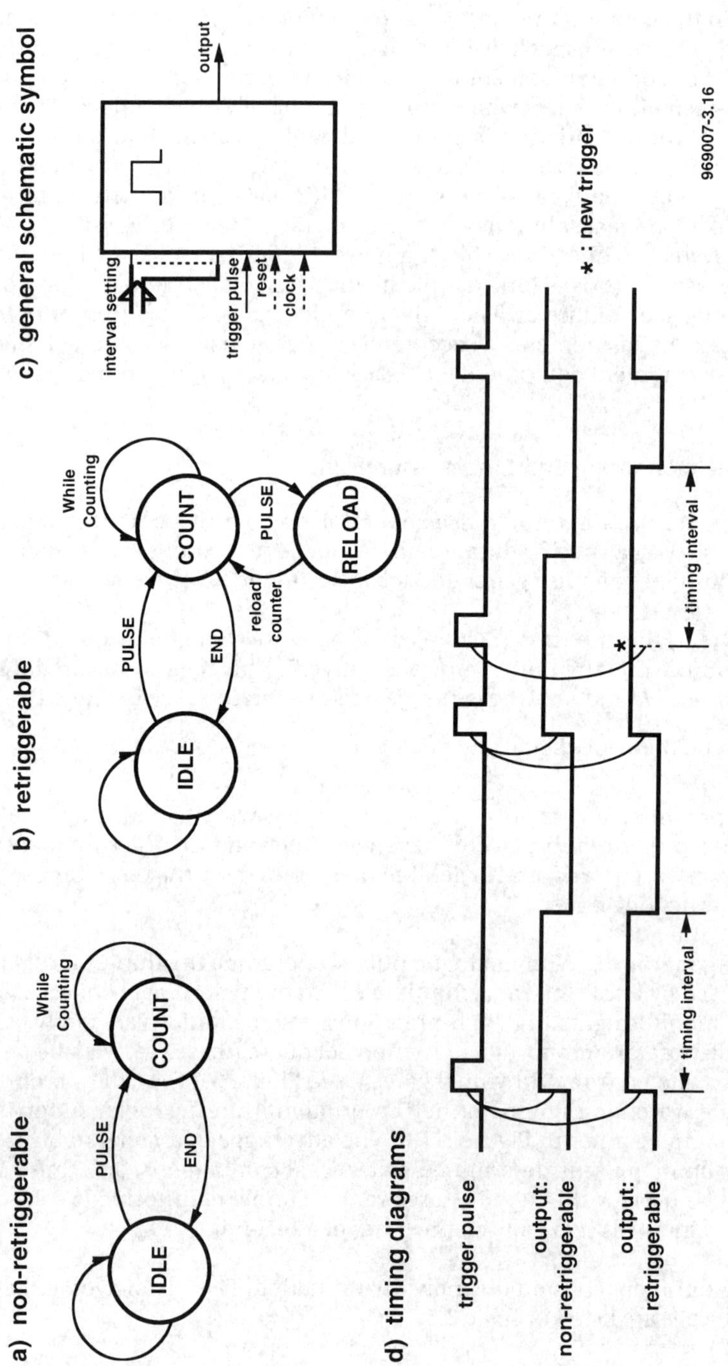

Figure 3.12 Operating principles of pulse generators.

watchdog timer (at least this is the term that is used in the data sheets, and the application notes mostly relate to this application). Naturally there are no limits to the uses to which such ICs can be put with a little imagination. There are supervisory ICs with fixed timing intervals and with adjustable intervals (see Figures 3.15 and 3.16).

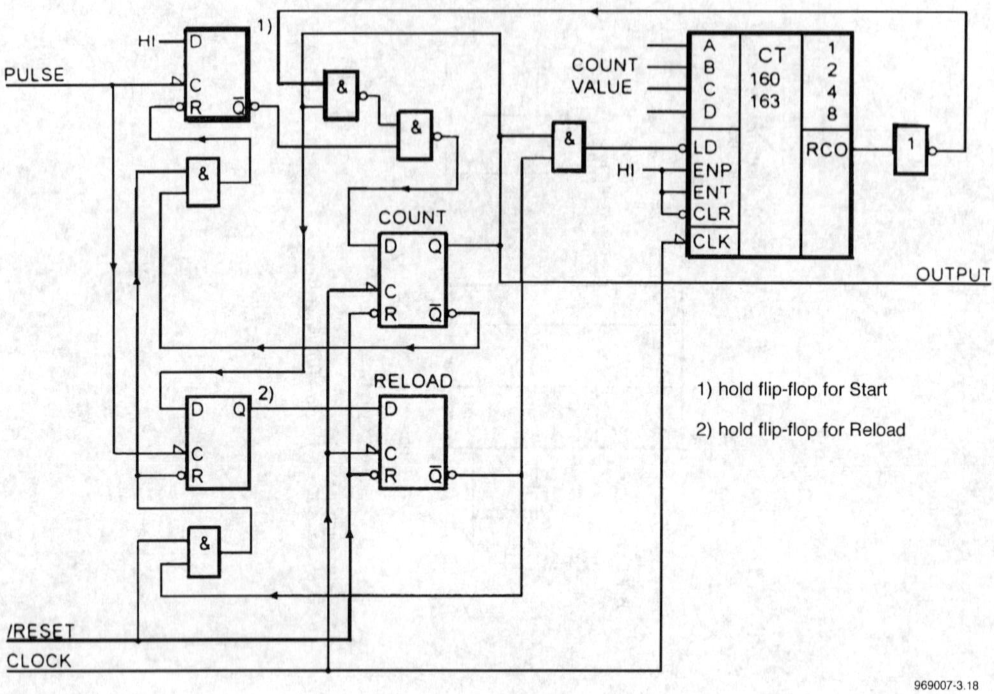

Figure 3.13 *Retriggerable digital pulse generator.*
This circuit implements the state diagram of Figure 3.12b. The counter can be extended as desired by cascading several counter ICs. For variations and further details, see Volume 3.

Explanation of Figure 3.12:

a) State diagram of a non-retriggerable digital pulse generator: trigger signals applied to the input while the output pulse is active do not have any effect on the duration of the output pulse.

b) State diagram of a retriggerable digital pulse generator: an input trigger signal applied while the output pulse is active causes the timing of the output pulse to start again from the beginning, which results in a corresponding lengthening of the output pulse.

Note: IDLE = rest state, COUNT = counting state (maintained for the duration of the interval – While Counting), PULSE = trigger pulse, END = internal end-of-count signal.

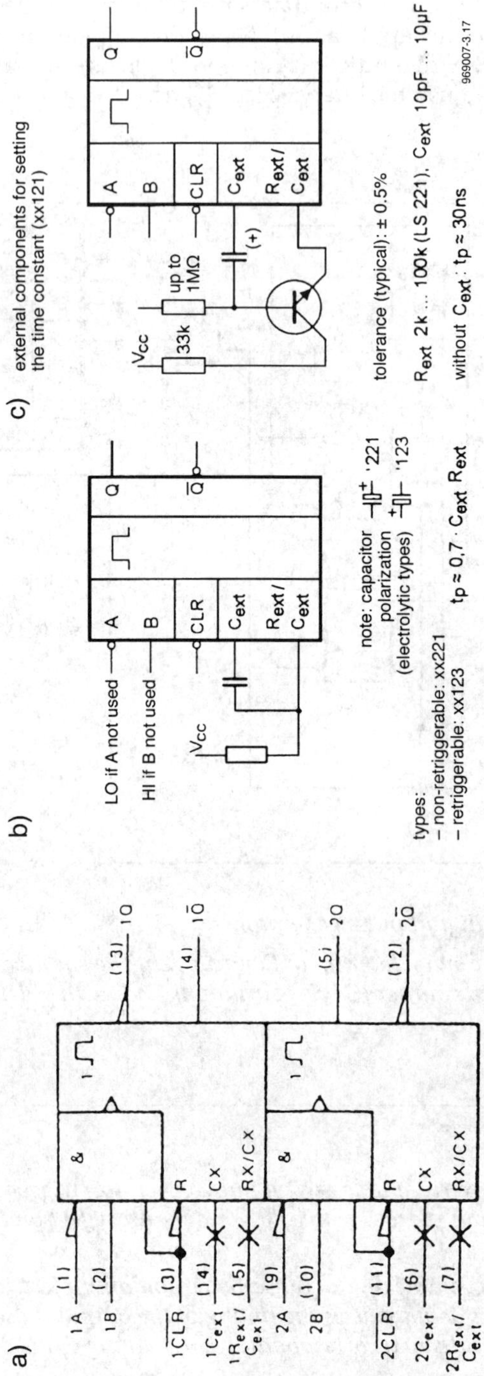

Figure 3.14 Monostable multivibrator ICs (source: Texas Instruments).

In Figure 3.15, it does not matter whether the pulse generator is digital or analogue (digital operation based on an internal clock generator is characteristic of modern ICs). The precision of the timing interval is not especially high. In the example IC, the watchdog period lies in the range of 1.0 to 2.25 s (typical value: 1.6 s). The WDI input must therefore be activated at least once per second for the WDO output to remain HI. Either the rising edge or the falling edge of the signal at the WDI input can trigger the pulse generator. If WDO goes LO due to a timeout, it remains LO until the next trigger is applied to WDI. If WDI is not connected then WDO remains permanently HI (this allows the watchdog function to be

Characteristics and special features	Analogue pulse generator	Digital pulse generator
prerequisites for operation	supply voltage alone (no clock necessary)	clock required
adjustable range of pulse duration	nanoseconds to minutes (reasonable upper limit: several seconds)	microseconds to 'forever', depending only on the number of counter stages
accuracy	no better than 1% to 2% (around 0.1% for special, precision circuits)	Depends on the clock generator; in principle 'as accurate as desired'. The accuracy of inexpensive quartz-crystal oscillators is 200 ppm (0.02%). Clock quantization error must be added to this (average value: 0.5 clock period).
testability	less favourable ('analogue' functional testing is necessary)	more favourable (can be functionally or structurally tested, just as other digital circuits, and can be divided into individually testable portions)

***Table 3.5** Comparison of analogue and digital pulse generators.*

Explanation of Figure 3.14:

a) Symbolic representation of one IC. Each IC contains two monostables. Both of the types shown here have the same pinout arrangement.

b) Simplified schematic, showing how the external components are connected.

c) Circuit for extremely long time constants. The transistor allows large time constants to be achieved with high-value resistors and relatively small-value capacitors (the basic time constant is multiplied by the current gain of the transistor).

Notes:

1 Monostable ICs with type numbers xx221 (non-retriggerable) and xx123 (retriggerable) are available in a variety of logic families (including 74LS and 74HC).

2 Triggering occurs on the rising edge at input B and on the falling edge at input A. An active input signal must be applied to the other input!

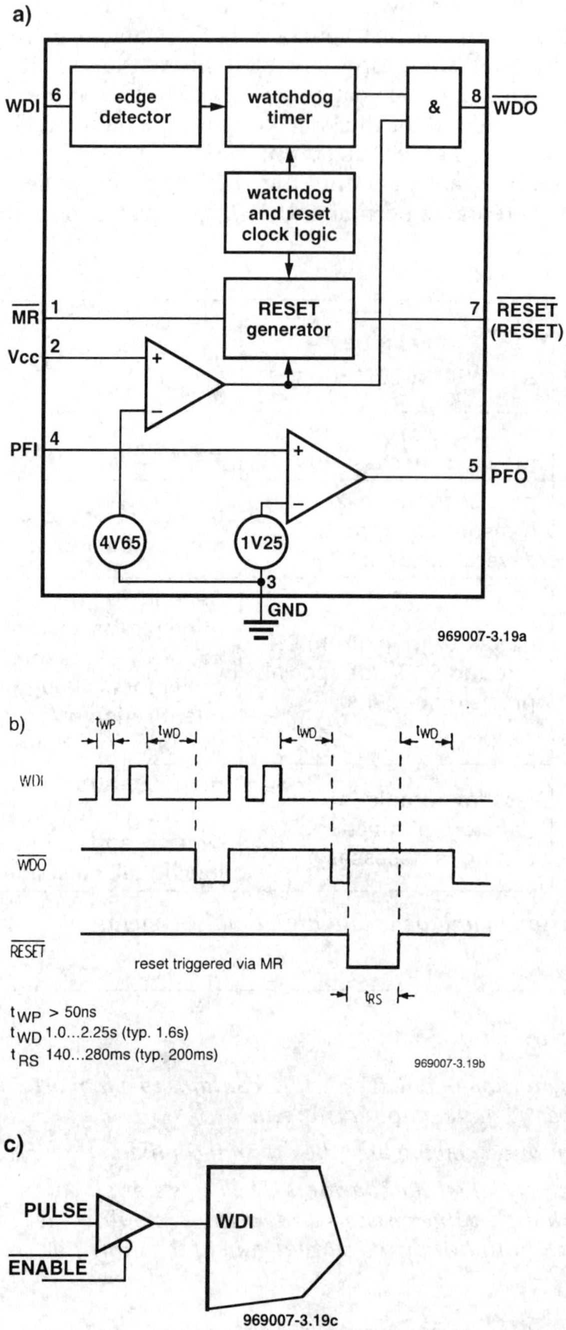

***Figure 3.15** Supervisory IC type MAX705 with a fixed timing interval (source: Maxim). a) block diagram, b) watchdog timer state diagram, c) watchdog control logic.*

enabled or disabled via a tri-state gate; see Figure 3.15c). When the IC is reset the watchdog timer is also reset (which causes WDO to go HI). If it is desired that a reset signal is generated on a watchdog timeout, then WDO can be connected back to the MR input.

In Figure 3.16, the length of the watchdog interval is determined by a counter. The frequency of the clock that drives the counter is determined by how the inputs OSC SEL and OSC IN are connected (to an external clock source, in which case 'quartz precision' is possible; to an external capacitor or to an internal oscillator). In certain operating modes the

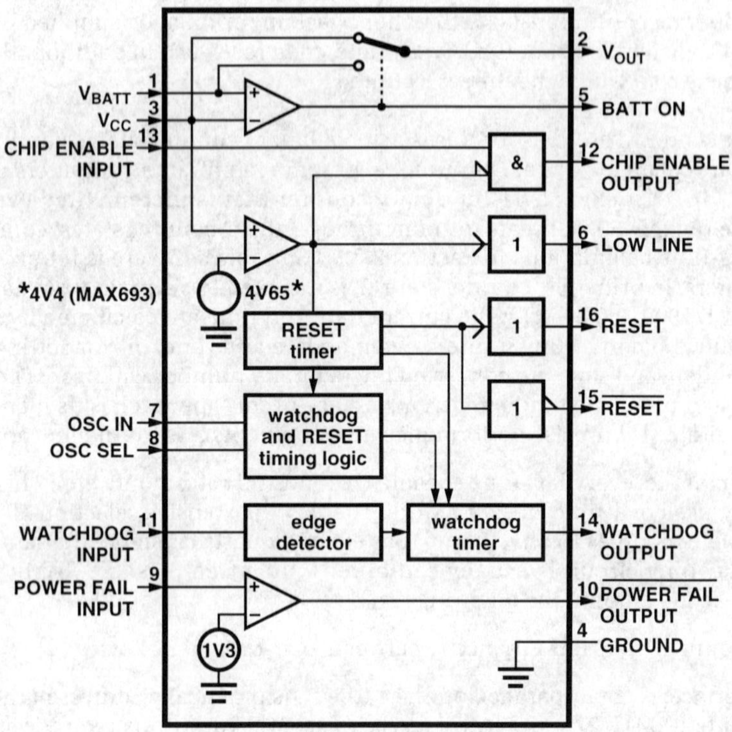

OSC SEL	OSC IN	Watchdog interval [2] Normal	Watchdog interval [2] after a reset	RESET [2] pulse duration
LOW	external clock[1]	1024 clocks	1096 clocks	512 clocks
LOW	capacitor C to earth	$\frac{400\text{ ms}}{47\text{ pF}} \times C$	$\frac{1{,}6\text{ s}}{47\text{pF}} \times C$	$\frac{200\text{ ms}}{47\text{ pF}} \times C$
open	LOW	100 ms	1.6 s	50 ms
open	open	1.6 s	1.6 s	50 ms

1) max. 250kHz
2) typical values

969007-3.20

Figure 3.16 Block diagram and timing characteristics of the type MAX691 supervisory IC with adjustable watchdog period (source: Maxim).

first watchdog interval following a reset is longer than normal. The watchdog hardware starts working as soon as a HI or LO level is applied to the WDI input (if WDI is not connected, the watchdog function is suppressed; this can be used for control purposes in the same way as is shown in Figure 3.15c). The watchdog pulse is triggered by either a rising edge or a falling edge at the WDI input. If WDO goes LO due to a timeout, it remains LO until the next trigger is applied to WDI. *Caution:* If the watchdog time is exceeded, a reset signal is also generated.

Time-of-day or real-time clocks. Electronic time-of-day or real-time clocks are timekeeping circuits that allow the current date and time to be read out at any time. Such hardware must either operate continuously (with battery backup) or must be supplied with the current date and time data from outside (via a time-code receiver). In addition, there are differences in the way in which the time is 'counted':

1 Pure binary count. The basis for this is a 'long' binary counter that is continuously driven at a particular clock rate. *For example:* in some mainframe computers the time-of-day clock (ToD Clock) is a 64-bit binary counter that is incremented every 1 µs. (It would take roughly 50,000 years to count through all 2^{64} counter states; compare this to the testing-time calculations in Section 2.3.3.) Suitable software is required for using such counters. In principle, counter state '0' is assigned to a certain fixed date and time (such as 1/1/1990, 0:00 GMT); the current date and time are calculated by reading the binary value and converting it into a decimal date and time, or (for setting the clock) the current date and time are converted into a binary number with respect to the reference date and time. The fundamental advantage of this approach is its high resolution; it is thus suitable for precise performance measurements and journalling applications.

2 Replication of the 'everyday' time scheme (seconds:minutes:hours etc). This is how the commonly-used real-time clock ICs work. There is an extensive selection of suitable ICs and complete modules (with a lithium battery and quartz crystal in the same package). The timekeeping circuit is usually combined with a freely-usable RAM (timekeeper RAM). These ICs differ in the following respects:

   - the amount of 'free' RAM (typically between 24 bytes and 32 kbytes),
   - the interface: (1) byte-parallel interface (the IC is practically addressed the same as a byte-wide RAM), (2) serial interface, with several variations (vendor-specific, I^2C bus, Microwire bus and so on),
   - time resolution (typically 1 s or 1/100 s),
   - additional functions provided (alarm, supplementary watchdog timer, mains power-on detector and so on).

Additional features:

- Data retention and timekeeping when operating under battery power are guaranteed for a period of between 3 and 10 years for modern real-time clock modules with built-in lithium batteries.
- Accuracy: ± 1 minute/month at 25 °C (according to the data sheet of a leading manufacturer).

Figure 3.17 shows how such an IC can be used with a conventional microcontroller. For more detailed information, refer to the data sheets for the individual components (also see Volume 2).

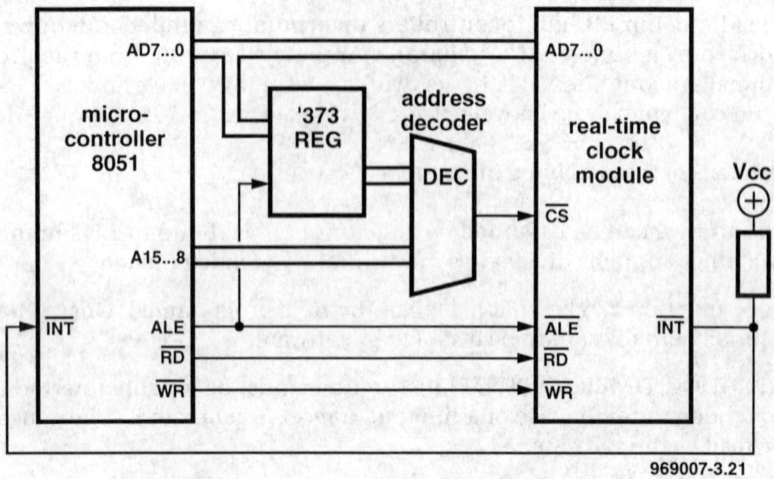

Figure 3.17 *Example usage of a real-time clock module (source: Dallas Semiconductor).*

Using real-time clock circuits for time measurement and monitoring. There are three possibilities:

a) reading and evaluating the date and time via software,

b) utilizing the alarm function. This relates to the comparison of two times: the current time and an alarm time that is stored in certain RAM locations. If these match then an a signal is output; this can for example be used to generate an interrupt. (It is also possible to implement a watchdog monitor using this function: the software must repeatedly set the alarm time forward by a certain increment, and if this fails to happen then the real time will 'catch up' with the alarm time.)

c) utilizing supplementary built-in watchdog timer functions.

Time code transmitters. The DCF77 transmitter (located in Mainflingen, Germany) provides time code signals that can be received up to 1500 km from the transmitter (see Volume 2 for more details). The advantages of these time code signals are:

- high stability and accuracy: ± 1 second in 1 million years,
- fully automatic setting of the date and time (even after changing the battery or IC), and automatic switchover between winter and summer time.

Note: devices that work with transmitted time code signals (such as expansion cards) should have their own quartz-crystal clocks to bridge over occasional losses of reception.

Time-monitoring circuits in processors. Time-monitoring techniques have proven themselves to be particularly effective in embedded systems. For this reason, some processor ICs that are especially intended to be used in such application environments incorporate built-in monitoring circuits.

Example 1: the Microchip PIC microcontrollers incorporate an independent watchdog timer that is driven by a built-in RC oscillator, completely separate from the processor clock. The nominal timeout interval is 18 ms (with a 5 V supply voltage and 25 °C ambient temperature) but can typically lie between 9 and 30 ms (see Figure 3.18).

Some further details of the watchdog timer are:

- The timeout interval can be extended by a factor of 1 ... 128 (up to 2.3 s nominal) by connecting a programmable divider in series with the watchdog timer.
- The watchdog function can be enabled when the IC is programmed. Once it has been enabled, it fundamentally cannot be disabled by software.
- CLEAR WATCHDOG TIMER (CLRWDT) instructions must be executed periodically to reset the watchdog timer. In case of a timeout, a reset is generated. (The cause of the reset can be read by the software.)

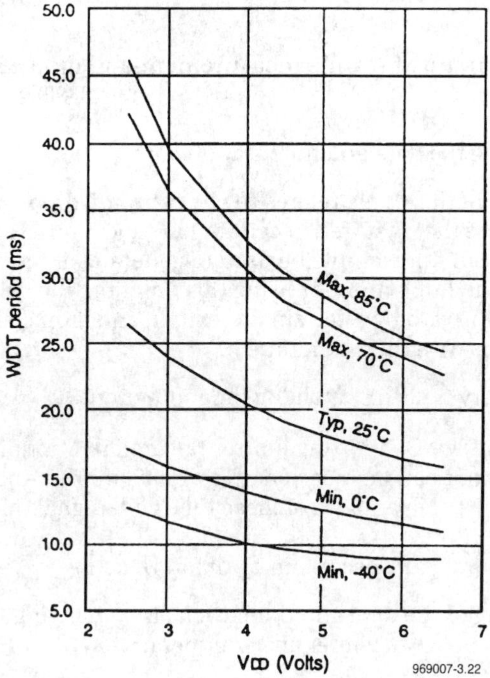

Figure 3.18 An example of 'global' time monitoring: the time vs voltage and temperature characteristics of the watchdog timer in PIC microcontrollers (source: Microchip).

Example 2: the Motorola MC68340 includes the following time-monitoring circuits:

- a supervisory circuit for monitoring program execution (watchdog timer),
- a supervisory circuit to check for bus transactions that exceed specified time limits (bus timeout monitor).

Practical tips:

1 If you employ a watchdog timer, then you must be sure that watchdog-reset instructions are in fact properly 'salted' into all program branches (just before every jump, ahead of every subroutine call and so on, for example).

2 Internal or external monitoring hardware? The PIC approach (an independent, if somewhat 'coarse' pulse generator built into the IC) provides a certain degree of protection, but it is not a perfect solution. It can detect 'hung' software, but it naturally cannot work if the whole IC is non-functional. If the requirements are demanding, then independent monitoring hardware is always required. (In principle it must be guaranteed that every fault mechanism against which the system is to be protected can affect either the system itself or the monitoring hardware, but never both at the same time.)

Software-based timekeeping and time measurement. There are two types of software-based timekeeping with regard to practical applications: (1) time measurement or time monitoring software in the 'actual' application processor and (2) simple microcontrollers that are converted into time measurement, timekeeping or time monitoring devices by means of software. The programming problems are the same in both cases:

- The minimum resolvable time interval is determined by the shortest required program loop in any given situation. It depends on (1) the processor clock frequency and (2) how many processor clock cycles are required to run through the loop.
- With simple processors or controllers having a fixed number of clock cycles for each instruction, the program execution time is always predictable. This applies equally well to processors with extremely simple instruction sets (RISC types) and those with complex instruction sets (although it is more of a mental challenge to write good timing loops in the latter case). As a rule, the more highly developed the processor and the system environment, the more difficult it is to exactly predict instruction execution times down to the clock cycle. (The time required to execute a particular instruction depends on the cache contents, the current bus loading, pending interrupts and so on.)

Practical tip: in highly-developed system environments (which include the common PC), *never* program explicit timing functions! If you do so you will only create problems for yourself, and – ultimately – incompatibilities. A better approach is to always rely on the system's timekeeping hardware. By contrast, it is not difficult to configure commonly-used microcontrollers as pulse generators or time-measurement devices (see Volume 2).

Fast, high-resolution counters. Such circuits are indispensable when times must be measured very accurately and with high resolution. However, there are certain problems with trying to construct such circuits using relatively simple means:

1. Maximum counting frequency: the usual TTL or CMOS ICs are specified for clock rates up to around 100 MHz (for example, the AC and F type TTL families), and some programmable ICs can be used up to roughly 150 MHz. But be careful: this specification relates to the *toggle rate*, which is the maximum rate at which a flip-flop of the family in question can switch states back and forth (a 2:1 divider; 100 MHz on the clock input and 50 MHz at the data output). This by no means says anything about the speed of other operations, such as resetting, preloading, reading out and so on, in such a counter circuit.

2. Evaluating or reading out the count: a fast, high-resolution counter circuit is ultimately only truly useful if it can be read with an accuracy equal to the clock interval. *For example:* if you have a 64-bit counter, it must be possible for the complete 64-bit counter state to be read in response to a single suitable instruction (thus, if the processor reads in 32 bits at time, it is not acceptable for the remaining bits to continue to accumulate counts while the processor reads the first 32 bits).

Naturally, the whole assembly should also not cost all too much. A few circuits that are useful in the PC realm for time monitoring and performance measurement are collected in Volume 3.

Testing time monitoring and time measurement functions

Checking time intervals. Time intervals can most easily be checked by comparison (the 'stopwatch' technique) or by measuring their duration (using a universal counter/timer, for example). See Section 4.4.5 for more information.

Practical tips:

1. *Checking 'long' analogue time constants:* reduce the time constant. The time constant of an RC combination can be reduced by connecting a second resistor in parallel with the existing resistor and/or a second capacitor in series with the existing capacitor. *Tip:* construct such timing circuits using two parallel-connected resistors and two series-connected capacitors. One of the resistors is the normal working resistor and has the value required for the desired time constant, while the second resistor can be connected by a jumper to reduce the time constant. Similarly, one capacitor has the correct value for the working time constant while the second one has a smaller value. Provide jumpers so that in normal operation the second resistor is not connected and the smaller capacitor is bridged over. For checking, you can reposition the jumper to connect the second resistor in parallel with the 'operating' resistor and insert the second capacitor in series with the 'operating' capacitor, so that the resistance and the capacitance are both reduced (see Figure 3.19). Since the components that determine the time constant in normal operation are also included in the checking, this test addresses a variety of fault mechanisms.

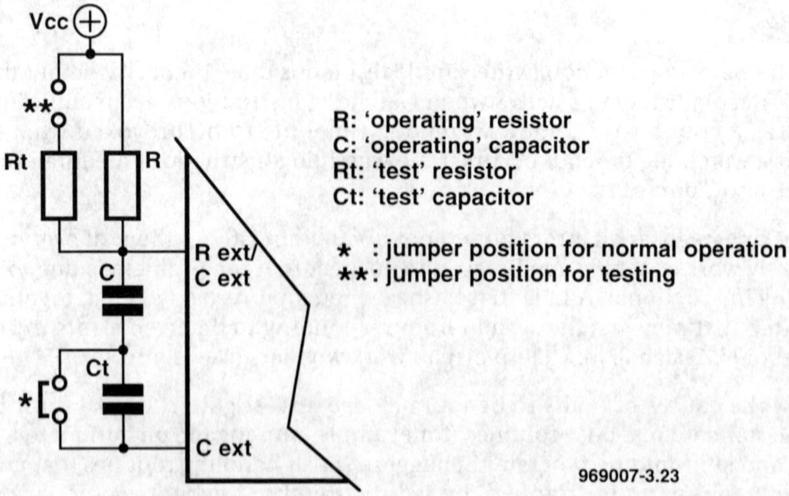

Figure 3.19 Test provisions for the time constant network of a monostable. (see Figure 3.17)

2 *Checking 'long' counter chains.* There are basically four possibilities: (1) apply a faster clock (only possible when the normal operating rate is much less than the counter's maximum operating rate), (2) split the counter into sections which can be individually functionally tested (that is, by letting each one count; this may also be done in parallel), (3) check selected steps in the counting process (for example, the transition from all ones to all zeros) and (4) do pure structural testing (under the condition that all counter stages (flip-flops) are accessible and observable for testing purposes.). See Volume 2 for designs of counter stages which lend themselves to testing.

3 *Checking software-generated intervals.* The interval in question must somehow be observable in the hardware. If this is not already the case, at the time that you are checking out the software you must insert suitable instructions that will cause some action that is 'externally' observable. The simplest possibilities are (1) bus accesses that can be used for triggering a logic analyzer or comparable test aid (for example, a 'blind' read at a specific address), (2) write instructions that address an I/O port. In the latter case, a good technique is to output short pulses (for example, set a given output to '1' and then reset it to '0') and connect a 2:1 divider (a flip-flop) to the output in question. This results in a symmetric square wave whose period can readily be measured – see Section 4.4.4.

Initial testing of software that uses a watchdog timer. The potential problems are (1) the watchdog reset interval has been misjudged (the time between successive reset instructions is too long), (2) watchdog-reset instructions have not been inserted *everywhere* in the software and (3) the error-handling routines are not working (so that you cannot tell whether the observed behaviour of the software is due to a watchdog timeout or some other fault).

Practical tips:

1 Check the watchdog reaction: write simple test loops that (1) check whether the watchdog timer actually remains active when watchdog instructions occur sufficiently often and (2) as a counter-test, allow watchdog timeouts to be provoked (that is, check whether a watchdog timeout occurs if the watchdog instructions are intentionally left out). See also Volume 2.

2 For checking, register all errors but do not do anything about them (if everything then apparently works, there is good reason to suspect that the problem is due to 'missing' watchdog instructions). A LED that flashes when an error is detected, together with a DIL switch that prevents the system from responding to detected errors (by disabling the reset or NMI signal, for example), can work wonders (see Figure 3.20).

3 To allow the causes of faults to be more closely investigated, 'divert' error handling from the standard response routines (for example, retrying the operation) to a 'manual' display and fault-finding program (debugger). (Error handling routines that are activated via interrupts can be 'trapped' by redirecting the interrupt vectors or by setting breakpoints; see Volume 2).

Checking real-time clocks. A real-time clock can be checked just like any other piece of hardware, with one special consideration: the current date and time will be lost in the process. It thus only makes sense to include a real-time clock in routine hardware testing (for example, every time the computer is switched on) if it is possible to also automatically restore the current time (for example, via a time-code receiver). Otherwise, you can only make a 'visual' test as part of routine maintenance (checking to see if it 'keeps good time'), supplemented by proper tests at a longer interval (such as every six months).

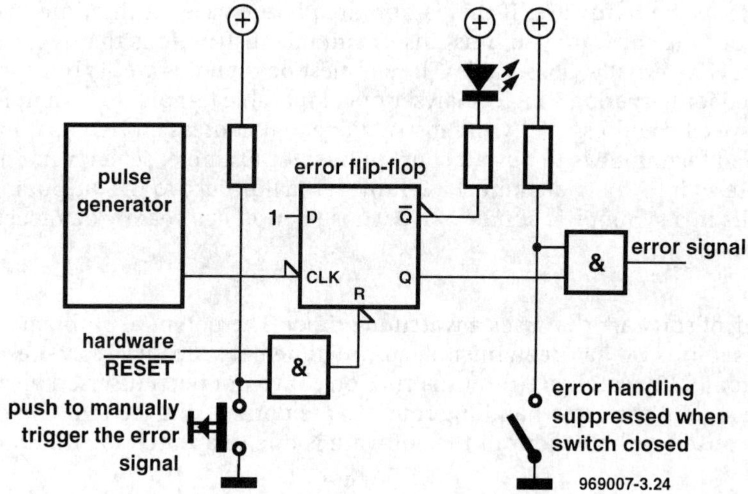

Figure 3.20 Timer-based monitoring logic with output disable switch.

3.2.5 Additional plausibility testing

It is worthwhile to monitor some other simple relationships. Some examples are:

- In a multi-master bus system, if a bus request is generated and the then arbitrated, some device must actually take over control of the bus.
- If a slave device is addressed on a bus system, only one device should respond or connect itself to the bus.
- For every bus system that employs handshaking signals, the signals which can be active for each bus phase are precisely defined. Missing signals can be detected by timeout monitoring, but unallowed signals can also be signs of a fault.
- In the selection and reselection phases of the SCSI bus, exactly two lines of the data bus must active at any given time.
- In block-associative cache memories and TLBs, at most one block may indicate a 'hit' resulting from the associative comparison of the address segments in question.
- A second error condition while the system is handling a previous error condition. *For example:* the supervisory logic of the Motorola MC68340 detects incorrectly-signaled interrupts and double bus errors (a single bus error can be handled by the software).

Suitable provisions should always be considered in such cases, in order to (1) monitor critical functional assemblies that (2) can otherwise only be checked at considerable effort or expense.

Figure 3.21 illustrates a typical example of such a monitoring circuit. It shows a portion of a plug-in card that has its own microcomputer and is designed to be connected to a universal bus system. The internal memory can be accessed from the bus via slave requests. There are three sources of RAM access requests: (1) the local microprocessor (LOCAL REQUEST), (2) the external bus (SLAVE REQUEST) and (3) the refresh logic (REFRESH REQUEST). Some sort of memory-access arbitration hardware is thus required. If this hardware is faulty and occasionally allows more than one memory cycle to take place at the same time, the resulting errors will show up as intermittent malfunctions. The cause of these malfunctions will be difficult to find. (By contrast, it would be quite simple to identify a total failure of the arbitration logic using test software, possibly together with a watchdog monitor, since in this case no memory cycles at all would take place). Since each source of memory requests has its own clock and they all basically work asynchronously, it would be very difficult to force all three types of requests to occur at the same time for testing. The alternative is a monitoring circuit that indicates an arbitration error (ARBITRATION CHECK) if more than one cycle is active at the same time (for example, REFRESH CYCLE together with SLAVE CYCLE).

3.2.6 Monitoring environmental and operational parameters

The probability of a fault is increased by unfavourable ambient or operating conditions. Occasional departures from proper conditions (such as short-term mains dropouts) frequently cause malfunctions. Long-term operation under unfavourable conditions (such

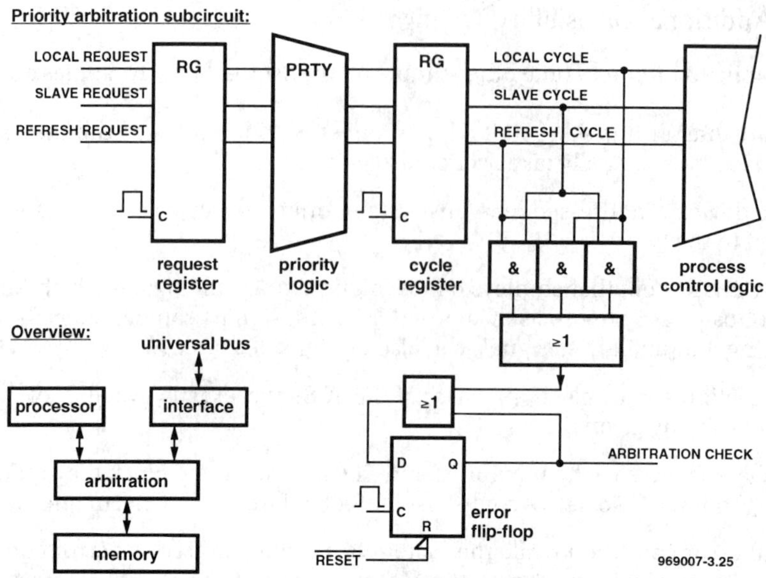

Figure 3.21 Plausibility checking in memory arbitration logic.

as an excessive operating temperature due to a defective fan) can cause premature component failures. It is a good idea to monitor the following factors, in particular:

- power sources: over- and undervoltage and dropouts, including short-duration disturbances (this applies to the mains power and internal supply voltages),
- temperatures inside the enclosure, for important assemblies that are sensitive to temperature variations (working memory, processor, mass-storage devices),
- relative humidity (inside the equipment),
- vibrations, especially those which due to their nature, orientation and intensity are prone to cause connectors to become loosened or, in particular, to affect the operation of mass-storage devices with moving mechanical parts,
- external magnetic fields,
- any signs of unauthorized activities(opening enclosures, theft and the like).

The fundamental feasibility of such measures essentially depends on the availability of suitable (and affordable) sensors (see Table 3.6). Their practical utility is determined by how their outputs are evaluated (see Table 3.7).

Type of sensor	Suitable for ...
comparator	voltage monitoring
current-sensing resistor, transformer current probe, Hall-effect sensor	current monitoring
thermistor, Pt100 resistive sensor, semiconductor sensor, thermocouple	temperature monitoring
humidity sensor	monitoring relative humidity
ultrasound transmitters and receivers, mercury-filled tilt detectors, light sensors (photodiode/phototransistor), mechanical switches	protection against theft and sabotage
optical-barrier detector (light-beam detector)	fan monitoring, protection against theft and sabotage
accelerometer	detecting vibrations, protection against theft
Hall-effect sensor	detecting magnetic fields that cause interference

Table 3.6 Sensors for monitoring ambient and operating conditions.

Type of evaluation	Technical implementation	Utilization
simple indication of a discrepancy (indicator light, acoustic signal)	simple hardware, including 'analogue' hardware (for example, a comparator driving a LED)	Can be better than nothing, but of little use in tricky fault-finding situations since short-duration deviations are practically not recognizable.
registering the discrepancy (so that you can later see if anything has happened)	simple hardware, including 'analogue' hardware (for example, a comparator driving a flip-flop)	Better than nothing, but of limited use for locating causes of crashes and malfunctions, since there is no time correlation.
recording discrepancy events together with the time and possibly other data	evaluation performed under control of a microcontroller (for example) — the simplest form of a service processor	This is the minimum capability that is necessary for a truly 'professional' arrangement; it can also be used to help find the causes of crashes and other malfunctions.
permanently logging error messages in the form of a 'historical log'	recording and evaluation performed by a combination of hardware and software (service processor)	This allows gradual changes to also be recognized.

Table 3.7 Evaluation of sensor indications.

Minimum technical requirements

- Sufficient accuracy. The accuracy is sufficient if it is possible to detect deviations from specified operating regions. A 'global' check is usually of little use. *For example:* a simple LED indicator connected to the 5 V supply line will be lit regardless of whether the actual voltage is 4.3 V, 5.0 V or 5.8 V (the difference in brightness is noticeable, but only if you look very carefully.) However, the allowed range is typically only 4.5 V to 5.5 V.
- Recording short-lived deviations. For example, short-term voltage dips or spikes could be the actual cause of a fault that appears to have some other cause.
- Recording the date and time and (if possible) other accompanying conditions. If this requirement can be met, it is a great help for fault-finding. (*For example:* suppose that the system crashed 'around four o'clock'. If it can be shown that at 3:58 the mains voltage sagged to less than 50% of the nominal value for 300 ms, then clearly there is no difficulty in determining the cause of the fault.)

What is 'short-term'? This depends on how the data are evaluated:

a) In the simplest case, the mere occurrence of a discrepancy is registered. The sensor circuit should not be so sensitive that it reacts to disturbances that are so short-lived that they do not affect the operation of the system.

b) If multiple disturbances can be registered and logged, then in principle you should in fact detect 'everything'. If an actual fault occurs, you can study the disturbance log and draw your own conclusions. *For example:* according to the CE regulations, a mains-operated device should be able to withstand dropouts lasting up to 80 ms (four 50-Hz cycles). Although this means that in principle it should be sufficient to only record dropouts that last longer than this, there could be a fault in precisely that part of the power supply that is supposed to protect it against these short dropouts.

Monitoring provisions in the PC realm. Monitoring *everything* is a rather costly proposition, so monitoring is usually restricted to voltage and temperature, whereby the latter can be replaced or supplemented by a thermostatically-controlled fan (the temperature is not measured as such; the fan is just switched on as necessary). In mainframe computers and other high-performance computers, such monitoring provisions have been a matter of course for years. There are also 'compatible' PCs that include monitoring provisions as part of their standard configuration (high-end servers).

Practical tips:

1 As an alternative to built-in monitoring hardware, you can use suitable equipment to measure the ambient and operating conditions (prior to the installation of the equipment, during its initial operation and in case of a relevant fault suspicion.) *Note:* measurement equipment (such as mains disturbance analyzers) can be leased as needed. The advantage of this is that it makes the use of high-performance measurement equipment affordable.

2 Monitoring circuits can easily be added as 'upgrades'. Such circuits are quite popular as DIY projects, and there is an ample selection on the market. *Caution:* more-or-less

jury-rigged solutions (which do not meet our minimum requirements) are not of much use. (For some concrete suggestions, see Volume 3.)

3 A compromise solution is to provide connection points inside the PC so that external measurement equipment can be conveniently attached when needed, instead of building provisions into the PC.

4 *Monitoring circuits in uninterruptable power sources.* Some uninterruptable power sources have quite extensive monitoring circuitry whose outputs can be evaluated by a PC via an extension card or a standard interface. Most network operating systems include suitable evaluation software (for example, UPS-Monitoring in Netware), but there is also suitable software that can be used under DOS or Windows. When the objective is to keep an extensive PC-based system (such as a network) 'alive and well', such a 'better-quality' UPS is a quite an effective piece of test equipment, since (1) it provides a thoroughly usable mains disturbance analyzer and (2) in case of a suspected fault, you can always check whether or not a mains disturbance is responsible for an 'unexplained' crash (by supplying the PC in question via the UPS).

Basic aspects of monitoring circuitry

Figure 3.22 shows the basic block diagram of a monitoring device.

Signal conditioning. Normally, sensor output signals are converted to analogue voltages. Since each type of sensor requires its own specific type of signal conditioning, it is not worthwhile to go into further details here. Instead, you are referred to the specific example circuits described in Volume 3.

Signal assessment. Signals can be assessed according to their amplitudes and their durations. Figure 3.23 shows some typical basic circuits.

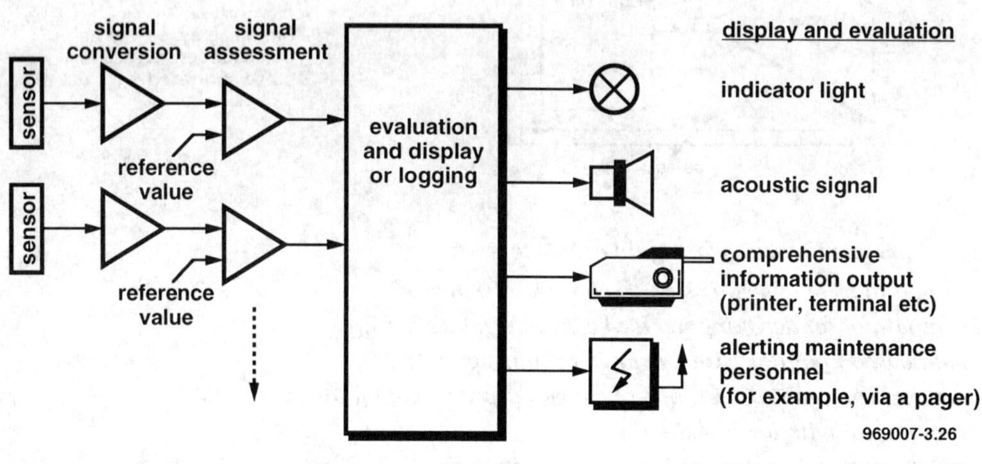

Figure 3.22 Monitoring environmental and operating conditions.

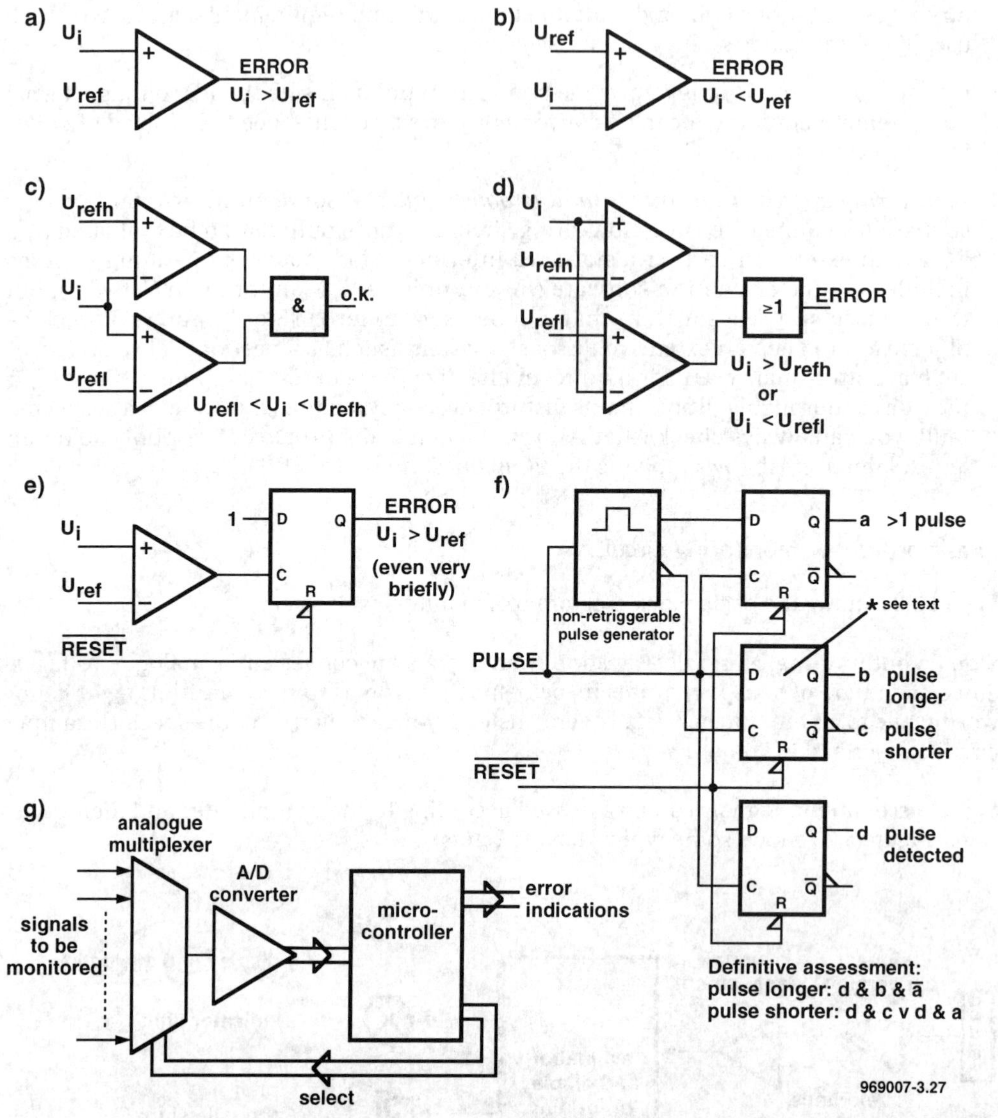

Figure 3.23 *Basic circuits for signal assessment.*

a) comparator for detecting signals above a reference level,

b) comparator for detecting signals below a reference level,

c) window comparator with 'in-range' output signal ('OK'),

d) window comparator with 'out-of-range' output signal ('fault'),

e) comparator with latch register,

f) pulse-duration assessment (pulse discriminator),

g) software-based signal assessment using a microprocessor.

Evaluation, recording, and signalling. In the simplest case, it is sufficient to display the registered error indications using (for example) a set of LEDs. More demanding tasks can readily be handled by software, based for example on a DIY microcontroller system (service processor). One specific difficulty is data retention when the supply voltage fails. Possible solutions are:

- battery backup for the service processor (current consumption may be a problem),
- bridging over short-term dropouts with a large buffer capacitor,
- storing the error messages in a battery-backed CMOS RAM (perhaps the RAM in a real-time clock, which anyhow should be present in the service processor). Here short-term dropouts are handled by a buffer capacitor, and the circuit goes into 'sleep' mode for longer-term outages.

Signalling (to the outside world) does not basically differ here from the signalling of other types of faults. Provisions for this can thus be consigned to the error-handling system (see Volume 2).

Notes for Figure 3.23:

1 *Notation: U_i = monitored input voltage, U_{ref} = reference voltage, U_{refh}, U_{refl} = high and low reference voltages (for a window comparison).*

2 *Comparator operation: the output is active when the '+' input is more positive than the '–' input. For convenience, the comparator outputs are all shown as active HI in the illustration, but many comparator ICs have active-LO outputs. The behaviour of the comparator is reversed if the inputs are reversed.*

3 *The pulse discriminator is a simple circuit with several classification flip-flops. The objective of this circuit is to determine whether a given input pulse (from a comparator, for example) is longer or shorter than a specific reference interval, which is provided by the pulse generator in the discriminator circuit. The flip-flop marked * is loaded with the value of the input line on the trailing edge of the reference pulse. If the input pulse is still active at this time, then it is longer than the reference pulse and this flip-flop is set HI. If the input pulse is shorter than the reference pulse, this flip-flop remains in its initial state (reset). You must keep the following subtleties in mind: (1) there can be multiple input pulses during the reference interval, with the last input pulse still active at the end of the reference pulse; (2) there may not be any pulse at all on the input. In order to detect such conditions, additional flip-flops are needed. The logical assessment of the outputs of these flip-flops is not described here, but it is shown in boolean form in the figure. Pulse discriminators are useful both for monitoring purposes and as test aids. For more information see Section 4.4.4.*

4 *Software signal assessment would at first glance appear to be the ideal solution for all monitoring tasks. However, extremely short-duration deviations (spikes) cannot be detected by such an arrangement. If precise monitoring is desired, then a combination approach is required: (1) special event-capturing and assessment circuits for detecting short-lived variations plus (2) software-driven sensing to detect 'slow' deviations and to record any short-term events that are registered.*

Voltage monitoring

Mains voltage monitoring

The simplest solution is a power-on indicator. Every item of equipment should have such an indicator (for practical circuit considerations see Section 5.2.1), since 'no mains voltage' (switch off, plug pulled out, fuse blown) is one of the most common trivial faults (and one of the most aggravating, if you have placed a service call because of it).

Checking for mains disturbances. From a circuit-design standpoint, this task is not trivial (and not for beginners!), but it is possible (see Volume 3). The only equipment that is of any real value is that which can actually detect all practically significant disturbances, classify them (at least roughly) and record or log them along with the time of occurrence.

Supply voltage monitoring

Voltage indicators: simple LED indicators (which may not even be externally visible) are of only limited value: to be sure, a total failure can be recognized at a glance, but otherwise you must make measurements. A solution that is usually better is to provide easily-accessible test points, so that a multimeter can be conveniently connected. *Note:* in the 50's and 60's it was fashionable to build a switchable voltmeter into the operator's control panel. Perhaps such a feature would also make a good impression in a modern PC (see also Section 4.2.2).

Voltage monitoring: the first requirement is to signal *undervoltage* conditions (voltages less than the minimum specified values). *Overvoltage* conditions are usually detected by the power supply, which handles them by turning itself off. (It is however a good idea to also record overvoltage events and the times when they occur, so that you can determine why the power supply switched itself off.) An important consideration is that supply-voltage fault indications should be 'saved' even after the power has (been) switched off: simple indicator lights that go off when the power itself goes off are of very limited value.

In modern systems, the voltage-monitoring circuitry also has two 'special' tasks: (1) resetting the hardware and (2) monitoring the processor supply voltage.

Hardware reset: this is often designated as 'Power On Reset'. Nonetheless, modern hardware should always be held in a 'hard' reset state if the supply voltage is not within specifications, since signal levels are not properly defined when the supply voltage is not correct. This can lead to irregular behaviour in the overall system. This applies to (1) switching on, (2) switching off (often overlooked!) and (3) voltage dropouts due to other causes.

An example of the effects of irregular operating conditions: in tri-state bus systems (and almost all bus systems in modern PCs are tri-state), irregular signal conditions may exist if the supply voltage is incorrect. In particular, it is possible for more than one driver to be active on the bus at the same time (bus contention). Although bus contention should actually not be allowed to occur, traditional bus-interface components (only a few signals per device, DIL packaging, moderate drive capacity) can withstand bus contention conditions for short times; switching times on the order of a few hundred milliseconds will not do any harm. Modern bus interface components (with many signals per device, very small SMD packages and high drive capacity) are by contrast extremely sensitive to bus contention. Therefore, a hardware reset that is a direct as possible must always be provided whenever the supply voltage is not within the specified operating range.

The purpose of the hardware reset is not just to provide the hardware (and in particular the processor) with a well-defined 'starting point'. Old-fashioned, simple reset circuits (see Figure 3.24a overleaf) are not adequate (in case of a voltage dropout they only generate a reset after the voltage has fallen to about 1 V – and at that point it no longer helps very much). The modern alternative is a microprocessor monitoring IC (µP supervisory circuit) as shown in Figure 3.24b; see Figures 3.15 and 3.16 for comparison.

Monitoring the processor supply voltage. Modern processors are often designed to have their 'own' supply voltages (that occasionally have quite narrow tolerances). They are usually powered via a separate voltage regulator or DC-DC converter. Since processors are not exactly cheap, a supply voltage monitor is certainly a good idea. Special ICs are available for this purpose.

Note: the number of readily-available monitoring ICs (voltage monitors and supervisory circuits) is so large that you should only resort to a design based on discrete comparators, voltage references and the like when faced with extremely demanding requirements. (And in passing, you do not always have to connect an IC output that is labelled 'Reset' to a hardware reset line. Such ICs can also be used for a variety of other purposes.)

Battery monitoring. Voltages provided by batteries (primary or secondary) are basically monitored just as carefully as any other type of supply voltage. With secondary (rechargeable) batteries, the *charge state* must also be monitored. This is based on measuring the current flowing into or out of the battery (charge or discharge current) together with the time, voltage and cell temperature.

Temperature monitoring

In contrast to voltage monitoring, temperature monitoring usually does not have to be all that precise. Simply checking whether the fan is running (via an optical-gate sensor, an inductive pickup or measuring the current flowing through the fan motor) is often used in place of actually measuring the temperature. Table 3.8 lists commonly-used sensors.

Sensor type	Temperature range	Accuracy	Cost
thermocouple	−200 to >1000 °C	1 %	£7–£15
platinum film resistor (Pt100)	−50 to >1000 °C	0.2 %	£7–£40
thermistor	−50 to 150 °C	10 %	£1–£3
semiconductor sensor	−50 to 150 °C	1 %	£1–£10

Table 3.8 *Characteristics of temperature sensors.*

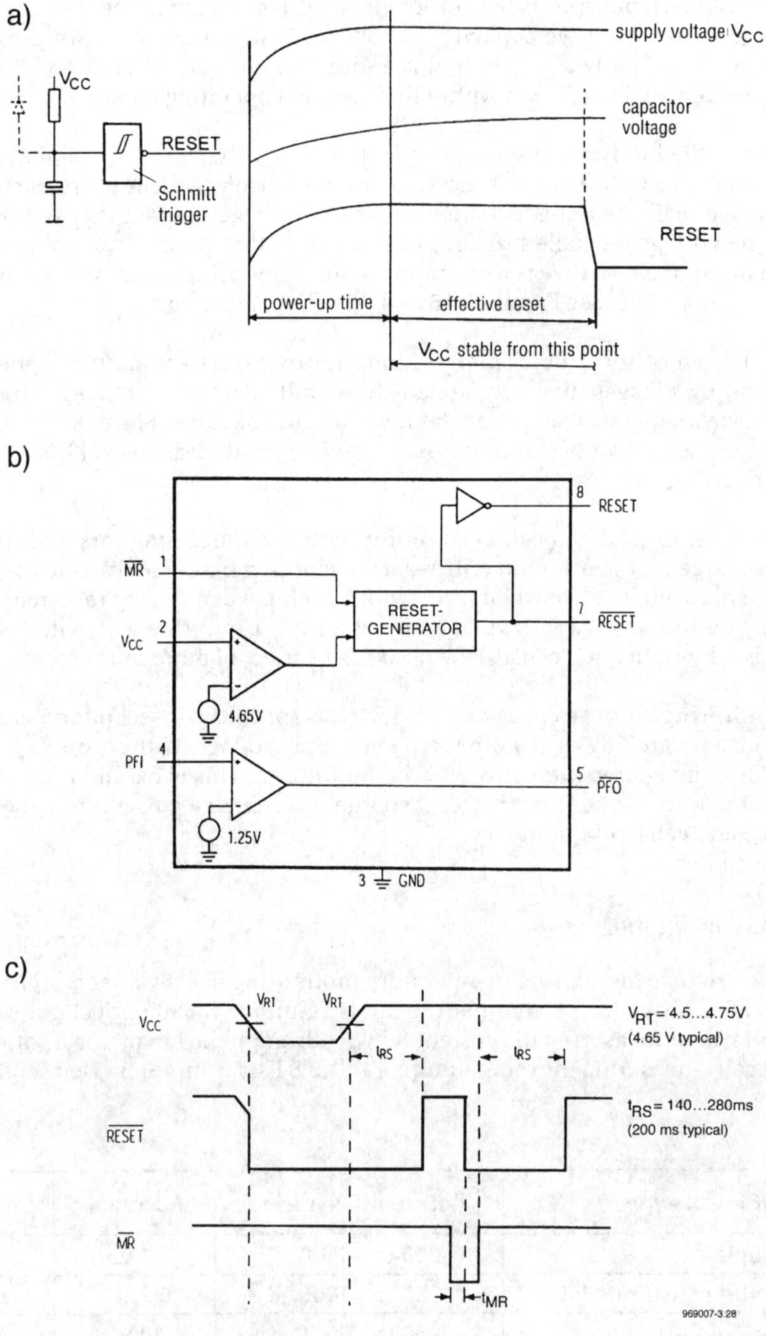

Figure 3.24 Reset circuits.
a) simple reset circuit with RC network and Schmitt trigger
b) and c) MAX707 supervisory IC (source: Maxim)

Notes regarding Table 3.8:

1 The values listed are very rough representative numbers. The specifications and prices of various types of sensors vary considerably.

2 The temperature ranges for thermocouples and platinum-film resistors refer to the sensing elements. The actual working temperature ranges (and prices) of such sensors depend mainly on the package type (metal, plastic and so on).

3 Thermocouples and platinum-film resistors are available in tape form for sticking onto the object to be measured (such as an IC). The listed prices refer to this form of sensor.

4 The preferred devices for temperature monitoring are thermistors and semiconductor sensors. Some models have mounting flanges or threaded studs, so that they can be mounted directly on the heat sinks of power components.

5 The location and means of attachment of the sensor are the critical factors for accurate temperature measurements (the issue is *thermal resistance*). The sensor must actually absorb heat from the object to be measured, and the transfer of heat from other objects to the sensor must be avoided.

Explanation of Figure 3.24:

a) When the supply voltage 'goes up', the capacitor initially holds the Schmitt trigger input at the earth potential. The Schmitt trigger only switches when the capacitor has charged to a sufficient voltage. The RC time constant must be chosen such that the reset pulse adequately covers the switch-on time interval.

b) This IC contains a reference voltage source and a comparator that triggers a RESET generator whenever the supply voltage crosses the reference level in either direction (the reference level is typically 4.65 V). The RESET pulse is active whenever the monitored voltage lies below the threshold level, and it typically persists for 200 ms after the voltage has risen above the threshold value. (The switch-on time of the power supply thus does not affect how long the operating hardware is held reset).

Notes:

1 The reverse-connected diode shown in dotted outline in (a), which is often omitted, comes into action when the power supply is switched off. It then discharges the capacitor through the 'parasitic' resistances of other components (the diode is forward biased as long as $V_{CC} = 0$ and the capacitor is not fully discharged). This is intended to ensure that the capacitor is discharged, so that a new reset signal will be generated if the power is switched on right after it has been switched off (in itself a bad practice that reduces the reliability of the equipment!).

2 Some ICs allow a simple RC combination to be connected directly to the Reset input.

3 In (b), there are two additional possibilities: (1) a switch can be connected from the MR input to earth (no pull-up resistor is necessary). Pressing the switch triggers a (bounce-free) reset pulse with a typical duration of 200 ms. (2) a second comparator with a reference voltage of 1.25 V is available. It can be used as desired for other monitoring functions (such as a power fail indicator).

Recommendations:

a) For surface temperature measurements (of components, heat sinks and the like), the sensor should make direct contact with the object. A certain amount of mechanical pressure should be applied and a heat-conductive paste should be used – but the paste alone is no substitute for good physical contact.

b) For some components (including processors), the manufacturer specifies suitable locations for sensing the package temperature (data sheet temperature specifications relate to these locations). The best place to measure the temperature of a processor is usually in the middle of the top surface of the package (and not near a corner of the heat sink). A sensor mounted 'somewhere near' a component will of course measure *something*, but not the temperature that you actually want to know.

c) The trend is to build the temperature sensor into the component (for example, the Pentium Pro processor).

d) Ambient (air) temperature measurements: metal has the least thermal resistance, so if the sensor is 'suspended' in the air the critical factor is where and how it is mounted. *For example:* a sensor soldered directly to the motherboard will measure a temperature, but certainly not the temperature of the air stream coming from the fan. (*Important:* the sensor should not be mounted such that it can absorb heat from another object – in this example the motherboard – via its leads. One way to achieve this is to mount the sensor on a small, separate piece of circuit board.

Protection against theft and sabotage

It is not all that difficult to detect attempts to carry off a piece of equipment or open an enclosure without permission. Table 3.9 lists examples of electronic monitoring measures.

Note: the actual problem is not the sensor, but rather what to do in the event that something actually happens. This is of course not the subject of this book. (We only address protection against theft and sabotage here because suitable preventive measures can reasonably be grouped together with 'general-purpose' monitoring provisions – a collection of unrelated 'single-purpose' DIY circuits would by contrast only have drawbacks in many aspects (cost, utility, reliability).

Example of a monitoring IC

Figure 3.25 shows the block diagram of a monitoring IC containing assessment and interface circuitry that can perform all of the previously-described monitoring functions. It has the following features:

- 7 analogue inputs for monitoring 5 positive voltages (*1*) and 2 negative voltages (*2*). The illustration shows two examples of how the voltage to be monitored is input to the IC via a voltage divider.
- a digital sensor input (Chassis Intrusion) (*3*) for protection against theft and sabotage,

Sensor	Detects ...
mercury-filled tilt switch	removing the equipment
accelerometer	removing the equipment
light sensor	opening the enclosure (the sensor detects the entry of external light)
ultrasonic barriers (sender/receiver)	opening the enclosure
mechanical contacts	opening the enclosure, removing the equipment

Table 3.9 Sensors for protection against theft and sabotage.

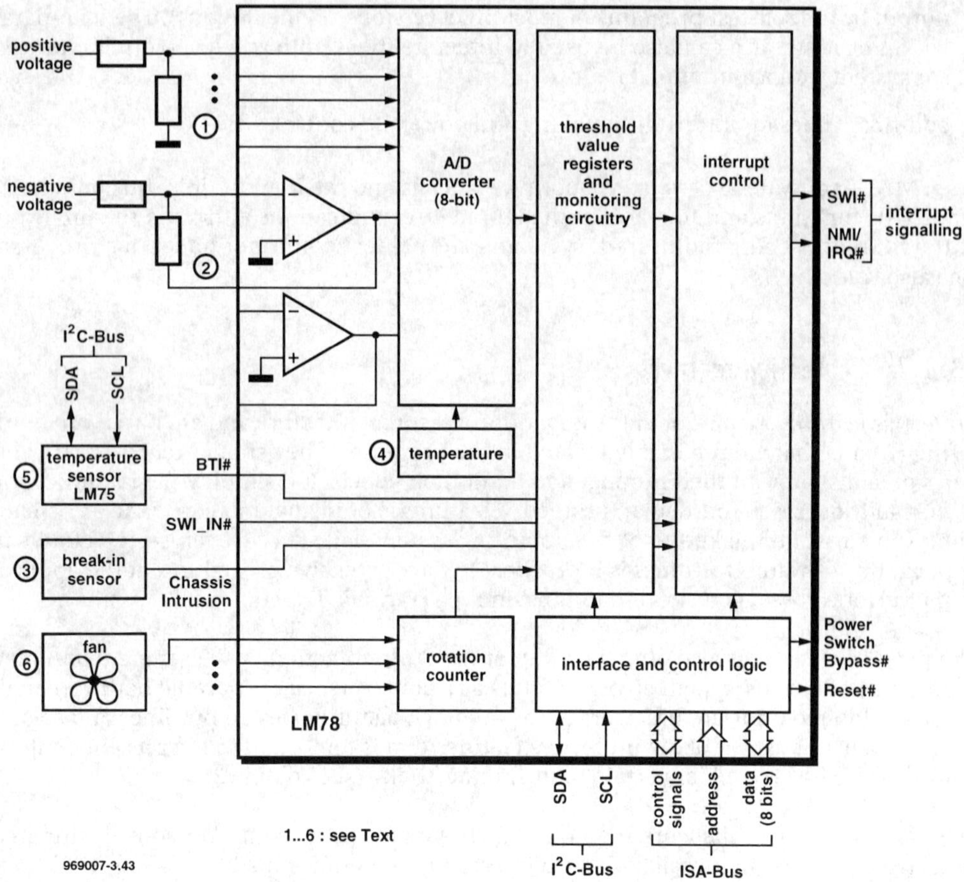

Figure 3.25 LM78 monitoring IC (Microprocessor System Hardware Monitor). (source: National Semiconductor)

- a built-in temperature sensor (*4*),
- a digital overtemperature-indication input (BTI = Board Temperature Interrupt) (*5*). The outputs of up to eight LM75 temperature sensors can be connected in a wired-OR configuration to the BTI input. The LM75 itself has an I^2S interface, so the temperature sensors can be individually interrogated to determine which one(s) generated an overtemperature indication.
- 3 digital inputs for fan-speed sensors (*6*).
- Threshold-value registers are provided for each of the analogue inputs as well as for the temperature sensor and fan speed inputs. Upper and lower threshold values can be set for each voltage to be monitored by writing the appropriate data to these registers.
- a synchronous serial interface (I^2S),
- address, data and control signals for direct connection to the ISA bus. The IC can store up to 32 POST codes via this interface (POST codes are single-byte error codes that are output by the self-test programs of most BIOS versions via the ISA bus). The byte-parallel ISA bus interface can also be used to integrate the IC into any desired microprocessor system with a minimum of effort.
- additional lines for interrupt signalling and power-up control.

The LM78 is software-accessible via both the I^2S bus and the ISA bus. It is thus possible to control its functions and to read out the stored error indications either via the processor of the PC that is being monitored or via an external test computer or service processor (see also Volume 2).

3.2.7 Protective features

Protective features are part of the system or processor architecture and are based on monitoring circuits that are preferably located in the processor. They should recognize various types of faults, prevent their propagation, limit their effects and simplify the identification of the fault sources. In particular, protective features should guarantee that the error-handling software and the kernel of the operating system will survive even serious faults in application software. Only then is it practical to have extensive procedures at the application level for error logging, error handling and retrying failed operations.

The protective features also offer a certain amount of protection against the uncontrolled occurrence of serious system errors or faults as a consequence of sporadic hardware malfunctions (due to transient bit errors in the memory, disturbances on bus lines and so on). It is precisely this aspect of the protective features that is important in the present context. They can also be utilized for certain fault-finding 'tricks' (see Volume 2).

Most monitoring circuits work in parallel with the data processing and control circuitry. Detected errors are mainly signalled via interrupts. Sometimes the error is so serious that it can no longer be handled by software. This is expressly indicated by certain types or processor (for the example, the Shutdown state of the xx86 family), but not by all.

What is the significance of protective features in practical fault-finding? Occasionally you must investigate error indications or error messages that are generated by the protective features. In such cases there are a few difficult questions that have to be answered:

- Is the problem purely due to software?
- Is the problem complex (resulting from the interplay of software and hardware)?
- Does the problem indicate that the hardware is defective (or on the verge of failure)?
- Are the protective features themselves suspect?
- Is the problem due to short-term disturbances (perhaps in the mains network)?

In the PC realm, we have to deal with the protective features of the xx86 series of processors. They are described in summary form in Table 3.10 (for the 486 and Pentium processors). Any errors that are detected are signalled via the interrupt system. For more detailed information, refer to the xx86 programming literature.

3.3 Observability and accessibility

3.3.1 The simplest forms of accessibility

Every device to be tested is accessible and/or observable via its inputs and outputs. However, if testing is limited to this level of access then only functional testing is practically possible (which means long test durations and expensive test equipment). The objective of measures that improve accessibility is to effectively divide a given device into relatively simple subcircuits for test purposes, without actually having to take the device apart.

Test and measurement points. These are provisions that allow external test and measurement equipment to be conveniently connected to pre-defined subcircuits of the device under test. We can distinguish test and measurement points according to (1) their technical implementation (Table 3.11) and (2) their intended use (Table 3.12 and Figure 3.26).

Normally, test and measurement points are primarily provided at the following locations:

- interfaces between functional groups,
- in feedback paths (for interruption and signal injection),
- at points where several input signals are combined to form a single output signal,
- on signal lines that affect multiple inputs,
- for signals that serve as typical reference or trigger signals for making measurements (such as clocks, validation signals, wait-state signals and handshaking signals),
- for analogue signals that are used as references (for example, reference potentials) or that must be observed for alignment or calibration purposes,
- for supply voltages.

Protective features		
Feature	**Effect**	**Disabled by**
basic segment checking	checks access authorization	permitting all accesses
privilege setting (4 privilege levels)	checks the authorization of program calls and data accesses and the executability of privileged instructions; protects system programs	setting the highest privilege level (0) for all segments
I/O instruction privilege levels (IOPL)	checks the executability of I/O instructions and other sensitive instructions	always setting the lowest IOPL (3)
separate task state segments (TSS) for individual tasks	protects tasks from each another	
I/O permission list in the TSS	selective access protection for each individual I/O port	always setting the lowest. IOPL (3)
multiple stacks	protects the stack regions of tasks with higher privilege levels	
two levels of page-management protection	checks write permission and general access authorization (system / user)	full access rights for all pages; permanent privilege level 0

Additional security measures	
Measure	**Effect**
checking operation codes	instructions with illegal operation codes are not executed
INTO instruction	detects overflows for calculations with binary integers
BOUND instruction	checks field boundaries
operand checking	error reported for division by zero, over- and underflows, non-standard operands and insufficient accuracy (the latter two only for the FPU)
special formats, such as non-standard numbers and NaNs	The FPU can continue operation even with critical values, and critical results are recognizable as such.

***Table 3.10** Security and protective measures in the xx86 architecture (example, i486)*

Implementation	Advantages, disadvantages and remarks
test pads on the IC die (silicon)	The test pads are contacted by special needle-probe adapters and are only accessible during IC manufacturing.
individual contacts distributed over the circuit board	Simple, inexpensive and requires almost no additional circuit board traces. Disadvantage: each test lead must be individually connected.
test connector for the quick attachment of a large number of leads	Suitable for routine connections to a number of test leads; economical for mass production. Disadvantage: additional circuit-board traces are needed for the test connector (the associated leads are thus lengthened, which causes higher capacitive loading and greater risk of radiated interference).
test pads on the circuit board	Contacts made by needle-probe adapters. Lower cost than a test connector; advantageous for mass production. Not as good for servicing (high cost of needle-probe adapters).
attaching a test clip to an IC, or (if the IC is socket-mounted) plugging in a test adapter between the IC and its socket	No special provisions are needed on the circuit board (although there must be enough free space around the IC!). Disadvantages: 1) Test clips or adapters are relatively expensive and unreliable. 2) Usually only one IC at a time can be so connected. 3) The test equipment must itself be somehow connected to the test clip or adapter.

Table 3.11 Technical implementations of test and measurement points.

Type of use	Explanation
passive use: sensing (observing) signals	Can be provided 'everywhere'. In practice, limited to 'critical' points.
active use: injecting signals (test stimuli)	Requires some means to switch between normal and test mode, or can only be used if normal and test stimuli can safely 'overlap' (for example, pulling a TTL output LO or coupling a test signal via an XOR gate).
combined active and passive (signal injection and sensing)	On bus leads, for example. Either it must be possible to safely superimpose normal and test stimuli, or they must be mutually exclusive (to avoid bus contention).
redirecting or interrupting signal paths (especially feedback paths), setting up diagnostic states	

Table 3.12 Utilization of test and measurement points.

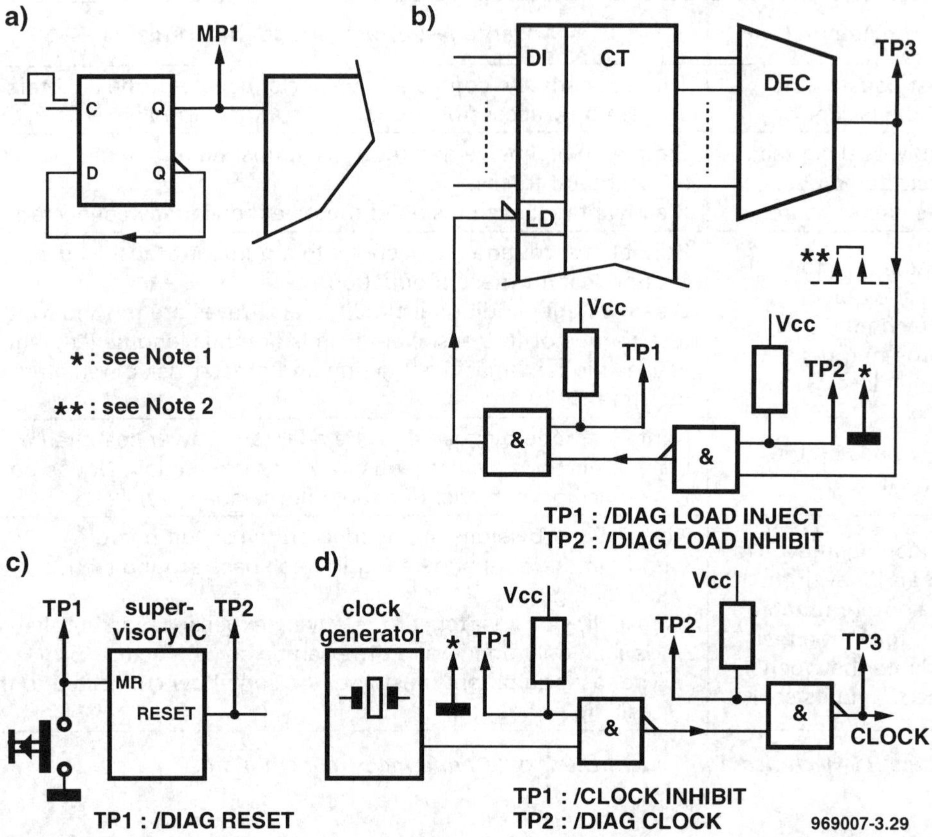

Figure 3.26 Typical uses for test and measurement points.

a) 2:1 divider with a passive measurement point MP1.

b) Counter with decoder and feedback to the Preload input. (the counter is reloaded when it reaches the decoded state.) There are two active test points (TP1, TP2) as well as a passive test point (TP3). A load pulse can be injected via TP1, to force the counter to be loaded with a specific value. TP2 suppresses counter loading (pulling TP2 LO interrupts the feedback path, so that the counter counts 'through'; you can observe at TP3 whether the decoder detects the correct counter state).

c) Monitoring IC (compare to Figure 3.24b) with an active test point (TP1) for triggering a reset signal and a passive test point (TP2) for observing the reset signal.

d) Clock source with a clock generator and provision for injecting an alternative test clock via TP2. When a test clock is injected, the output of the clock generator must be suppressed by pulling TP1 LO. TP3 is a passive test point for observing the clock signal.

Notes:

1) Pulling LO: when putting test points for this purpose in a circuit, remember that an earth connection is also required. Tip: put the earth connection point adjacent to the test point, so that the connection can be made with an ordinary jumper.

3.3.2 Test interfaces

A high degree of test severity, or precise fault localization, requires a correspondingly large number of test points. Since it is tedious to have to connect test equipment to a large number of test points, it is more convenient to have a regular or standardized interface – ideally with only a few leads – between each testable subcircuit and the test equipment. Such an interface is called a *test interface* (see Figure 3.27).

In general, a test interface supports the following:

- sensing signals ('logging out'),
- injecting signals (test stimulus) and/or setting state parameters ('logging in'),
- controlling operational modes in the device under test (for example, switching between normal operation and test mode).

We will look at three proven ways of implementing test interfaces: (1) the polling or serialization technique, (2) the addressing or de-serialization technique and (3) various types of shift-register or scanning techniques.

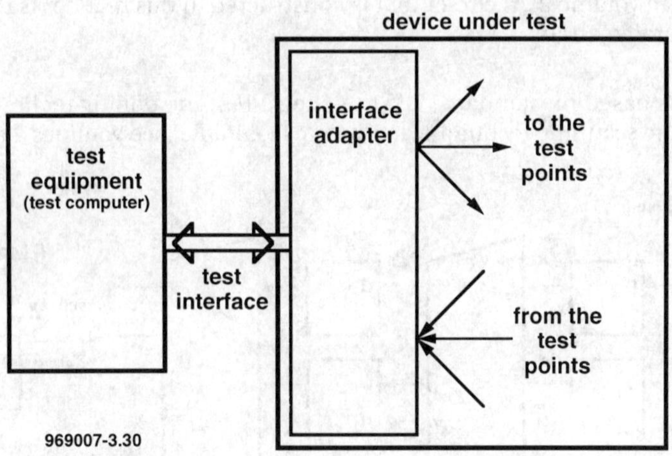

Figure 3.27 *Test interface (conceptual).*

---

2) *Probes for measurement equipment (oscilloscope probes, logic analyzer probe pods and the like) also need earth connections. If no earth points are provided (bad practice), the user is forced to make the earth connection by clipping onto a suitable IC pin. It is a good idea to provide specific earth points.*

3) *Opening a lead: of course you can simply interrupt the trace and provide a jumper to bridge the gap for normal operation. However, 'logical' interruption (via a gate, for example) is preferred for reasons of reliability.*

Accessibility and possible types of testing. Test interfaces are particularly important for *automated* testing. Table 3.13 summarizes the basic possibilities that are provided by various types of test interface in this regard.

Comments on Table 3.13:

- Adding test provisions to a functional design: for the designer, it is a clearly an advantage to only have to be concerned with how the circuit should ultimately function, without having to take testing provisions into account from the very beginning. Ideally, it should be possible to automatically merge in accessibility measures after the functional design is complete, without thereby impairing any of its functional properties.
- In a digital circuit, *complete accessibility* means that all flip-flops are available to the test interface (they can all be set/reset and their states sensed via the test interface). The objective is to be able to divide every (clock-driven) digital circuit into a combination of storage elements (flip-flops), storage structures (RAM and ROM arrays) and combinational networks (see Figure 3.28). Combinational networks can be completely tested (at least with reference to the 'stuck-at' model) using test vectors, and accessible flip-flops can be simply tested by alternatingly writing ones and zeros to them and checking that the proper values appear at their outputs. Naturally, storage structures whose data and address lines are accessible can be tested using the usual memory test procedures (see Volume 2). A circuit that is constructed in this manner is thus to a large degree structurally testable.

Note: for projects based on standard ('off-the-shelf') ICs, the polling technique and certain versions of the scanning technique are the most suitable (see Volumes 2 and 3). Provi-

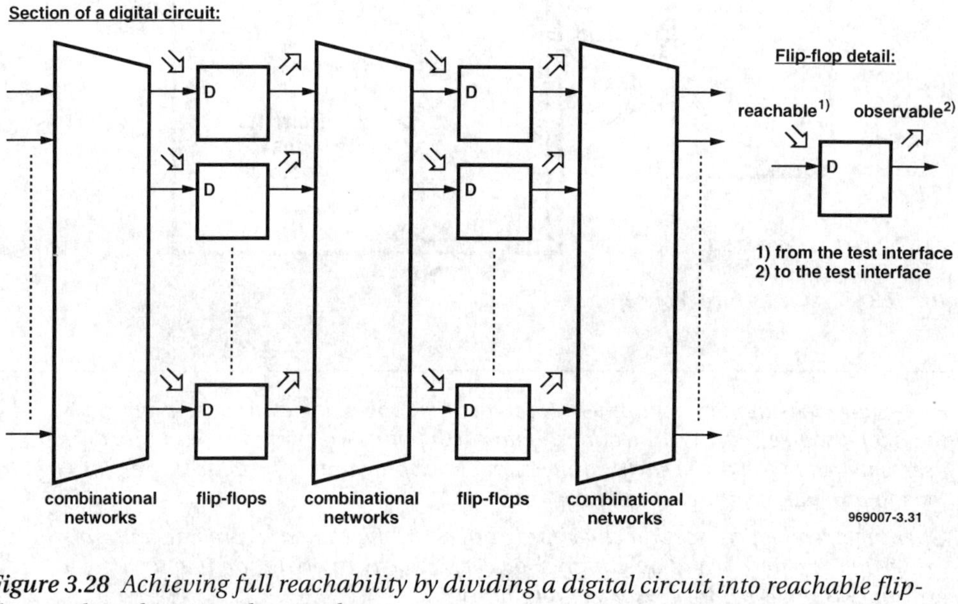

Figure 3.28 Achieving full reachability by dividing a digital circuit into reachable flip-flops and combinational networks.

Test interface	Accessibility	Usable test methods	Special advantages	Typical disadvantages
polling or serialization technique	sensing only (observation)	functional testing; high level of localization accuracy due to detailed observability	– Also usable in analogue circuits. – Simple. – Can be readily added to completed functional designs. – Has almost no adverse effect on the functional circuitry. – Selected signals can also be observed in dynamic operation.	– Self-testing of the selection hardware (multiplexer) is either incomplete or not possible.
addressing or de-serialization technique	signal injection and sensing	structural testing is possible	– Can be added to completed functional designs (although this is not always easy) with only a small adverse effect on the functional parts of the circuit. – Can be made to support dynamic testing. – Short test times are possible if the test data bus is wide enough.	– Self-testing is either incomplete or not possible. – Relatively expensive if the objective is complete reachability.
shift register or scanning technique	signal injection and sensing	structural testing is possible (and is the preferred technique)	– Excellent self-test characteristics. – The number of leads in the test interface is independent of the number of test points. – Complete reachability can be achieved at an acceptable cost.	– Strict design rules must be adhered to in order to allow scanning to be easily added to completed functional designs (scanning limits the designer's 'artistic freedom'). – Only conditionally suitable for dynamic testing. – Long test times.

Table 3.13 Characteristics of various types of test interface.

sions that go beyond this are only practical in the context of integrated-circuit design. We thus limit our discussion of these techniques to brief presentations of their structures and working principles.

3.3.3 The polling or serialization technique

This approach is advantageous when it is only necessary to be able to sense signals from the device under test (passive test points). The signals to be sensed are routed to supplementary selector circuits that are addressed by the test equipment (Figures 3.29 and 3.30). A suitable selector circuit is provided by a proven basic circuit element, the multiplexer, which can be used in both analogue and digital applications. (Briefly, a multiplexer has n analogue or digital inputs and one output, plus $\log_2(n)$ address (select) inputs. If a binary-coded number corresponding to a given signal input is applied to the select inputs, the signal applied to the selected input is passed through to the output.)

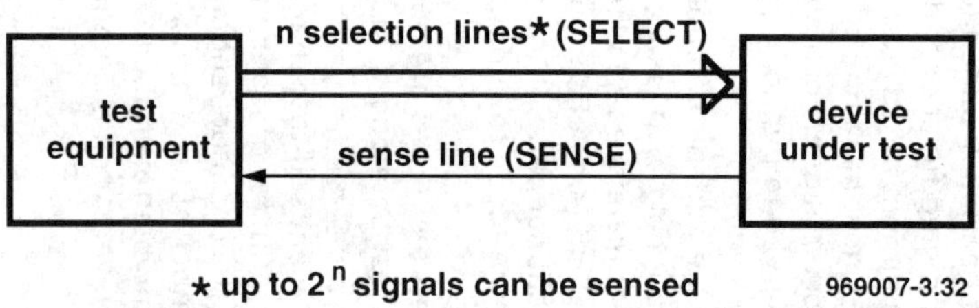

Figure 3.29 The polling or serialization technique: test interface.

Explanation of Figure 3.30:

a) Signal selection in a digital circuit using a multiplexer (left: standard symbol, right: simplified representation).

b) Signal selection in an analogue circuit using an analogue multiplexer. The voltages to be sensed are reduced if necessary (via voltage dividers) so that they lie within a predefined test range, before being applied to the multiplexer inputs (see the output of the second amplifier). A buffer is always recommended at the output of the multiplexer.

c) Selecting a large number of signals by means of cascaded multiplexers (digital or analogue). In the example shown, up to 64 signals can be sensed by means of nine cascaded 8:1 multiplexers.

*d) Cascaded multiplexers with partial self-testing: it is possible to check whether the selection signals actually switch the multiplexers, by activating the test inputs to the XOR gates at the multiplexer outputs (testing by signal inversion). Only the first-level multiplexers, which are directly connected to the signals to be sensed, cannot be checked in this way. The INVERT lines augment the test interface. Note: from the standpoint of functional testing (verification), only the XOR gates marked * are essential (and thus also only the control line INVERT1). The remaining XOR gates increase the degree of fault localization in the sensing hardware.*

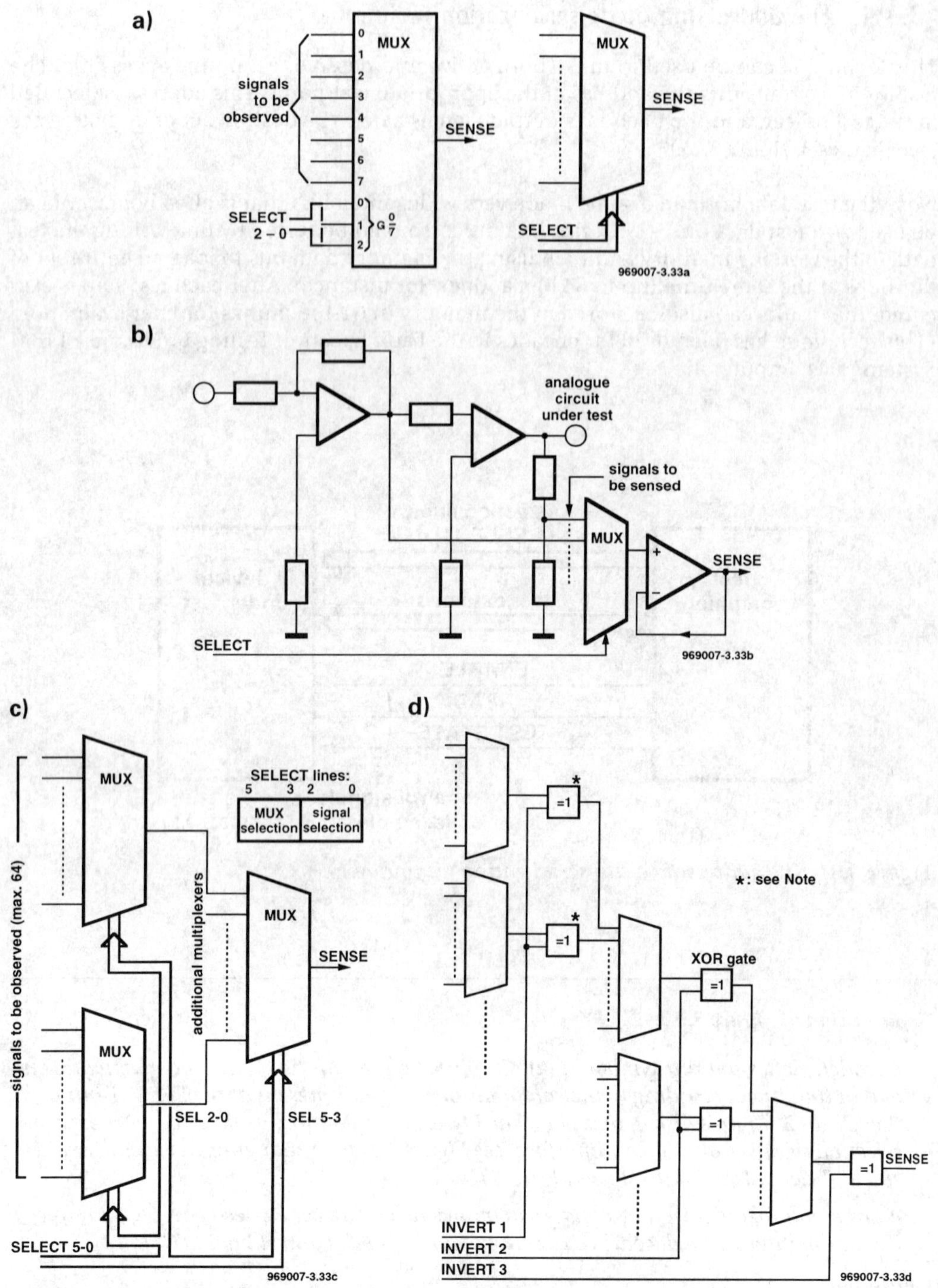

***Figure 3.30** The polling or serialization technique: circuit variations.*

3.3.4 The addressing or de-serialization technique

This technique can be used to make both active and passive test points accessible. The test interface transmits the address of the appropriate test point. This address is decoded in the test object, and the decoder's output signals cause the desired action to take place (see Figures 3.31 and 3.32).

Note: the test data bus can also be made very wide, and a bus that is already present can be used as a test data bus. *For example:* a processor can be tested by first writing the test data to the working memory and then using the memory data bus to load an entire set of flip-flops at the same time (up to 64 bits at once, for instance). After each test step is executed, the results can also be stored in the memory in 64-bit chunks for later evaluation. (This technique has been used for decades in the Fault Location Testing provisions of IBM System/360 computers).

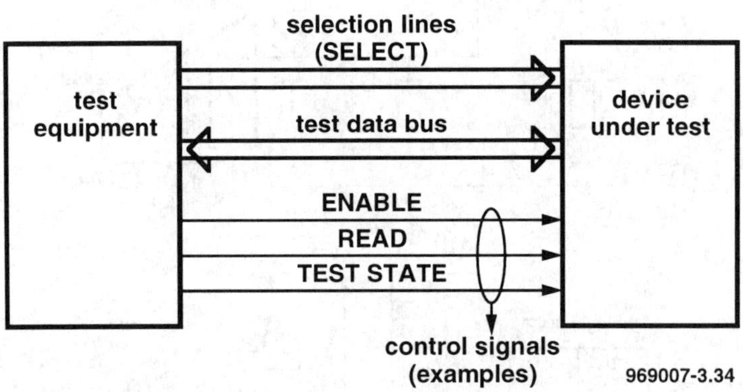

Figure 3.31 *The addressing or de-serialization technique.*

Explanation of Figure 3.32:

a) Decoders (left: standard symbol, right: simplified representation). A decoder activates one of it output lines according to the combination of signals present at the SELECT inputs, if the ENABLE input is also active. Note: the ENABLE signal must be inactive whenever the selection signals from the test interface (see Figure 3.31) are changing, in order to avoid spikes (brief pulses) on the output lines.

b) A larger number of selection signals can be decoded by cascaded decoders. For example, a cascade of nine 1-of-8 decoders can activate up to 64 test control lines (TCTL...).

continued ...

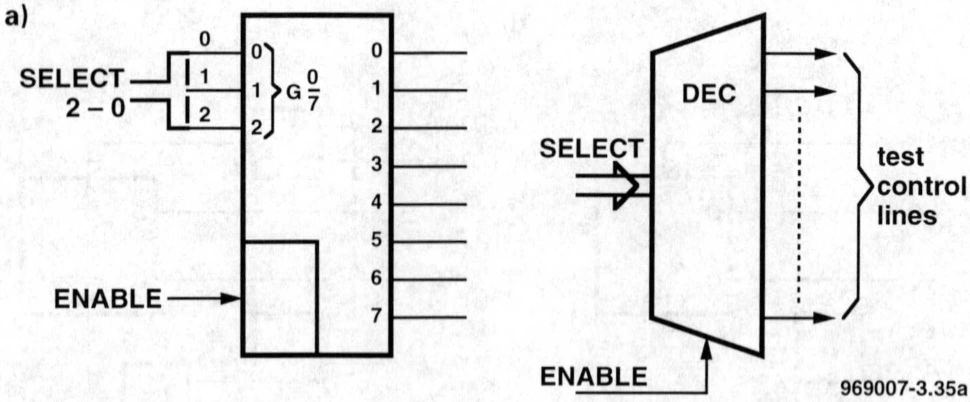

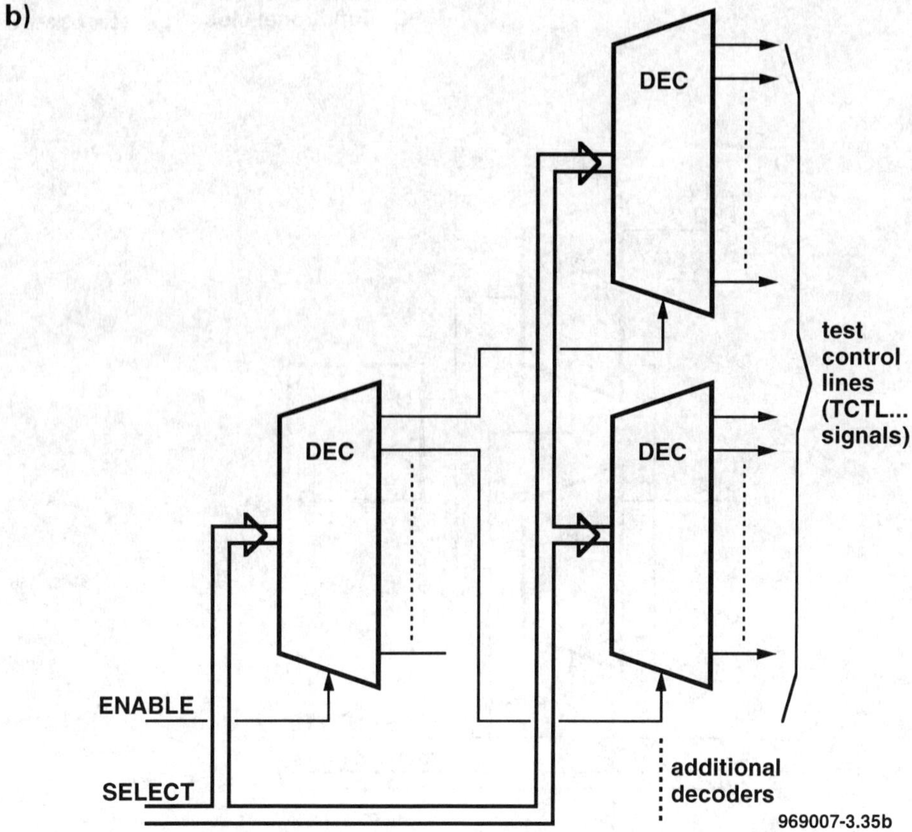

Figure 3.32 The addressing or de-serialization technique: circuit variations (1 of 4).

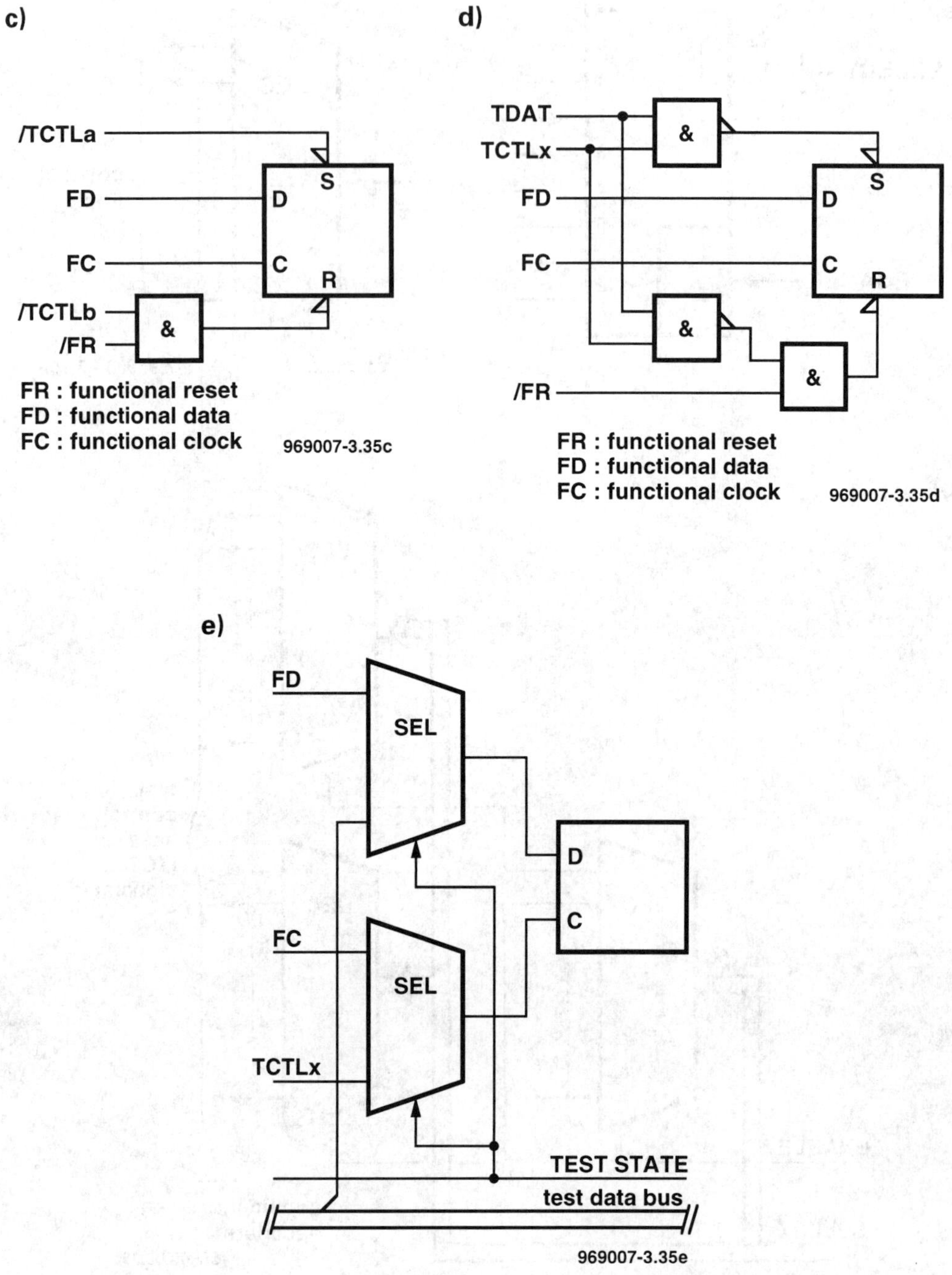

Figure 3.32 The addressing or de-serialization technique: circuit variations (2 of 4).

Explanation of Figure 3.32 (continued):

c) *Setting and clearing a flip-flop (1): two separate test control lines can be used, one connected to the D input (TCTLa) and the other to the RESET input (TCTLb).*

d) *Setting and clearing a flip-flop (2): the SET and RESET inputs of the flip-flop are connected to a test control line (TCTLx) and a test data line via intermediate logic, so that when the test control line is active the flip-flop is set or reset according to the value present on the test data line (TCTLx acts as a transfer clock).*

e) *Loading a flip-flop: selector logic is placed ahead of the data and clock inputs of the flip-flop so that its data and clock inputs come from the test interface (a test data bus signal and a test clock) in the TEST state, instead of the normal functional logic.*

f) *Sensing an output value by connecting it to a test data bus: the driver is selected by a test control signal TCTLx and enabled by the test interface READ signal.*

g) *Loading flip-flops from the test data bus: see Figure 3.32e for details.*

h) *Matrix decoding (RAM decoding) with row and column decoders. Each test control signal TCTLxy (which can be used as shown in (c) through (g)) is generated as the logical AND of a row decoder output and a column decoder output. The advantages are: (1) reduced decoder cost (up to n · m flip-flops can be selected via n columns and m rows) and (2) the arrangement can be addressed (and tested) as a RAM.*

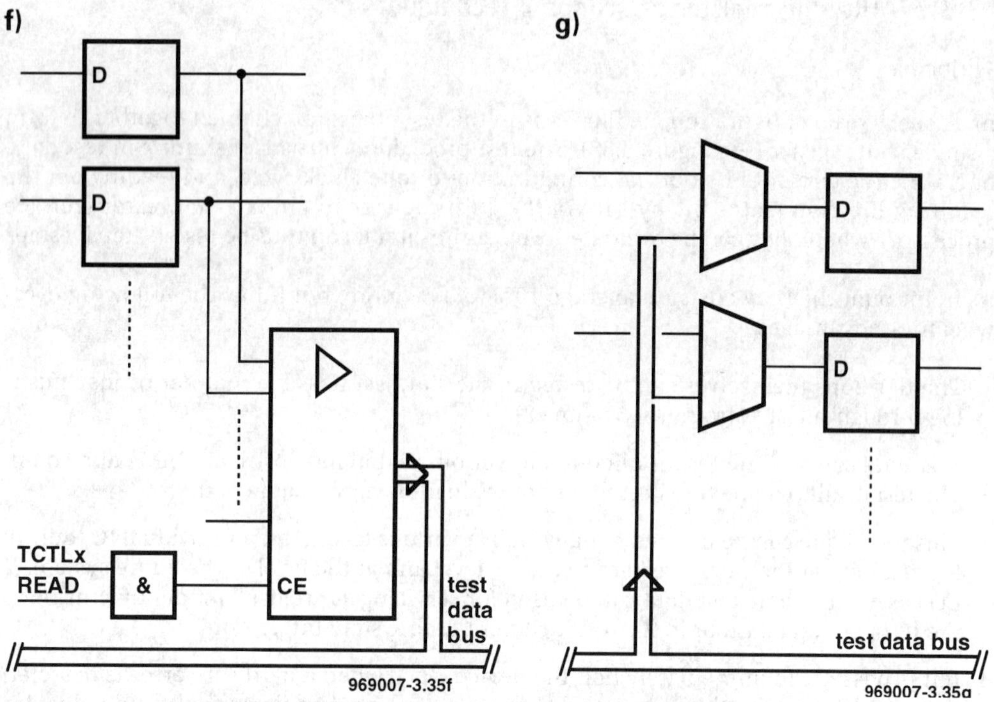

Figure 3.32 The addressing or de-serialization technique: circuit variations (3 of 4).

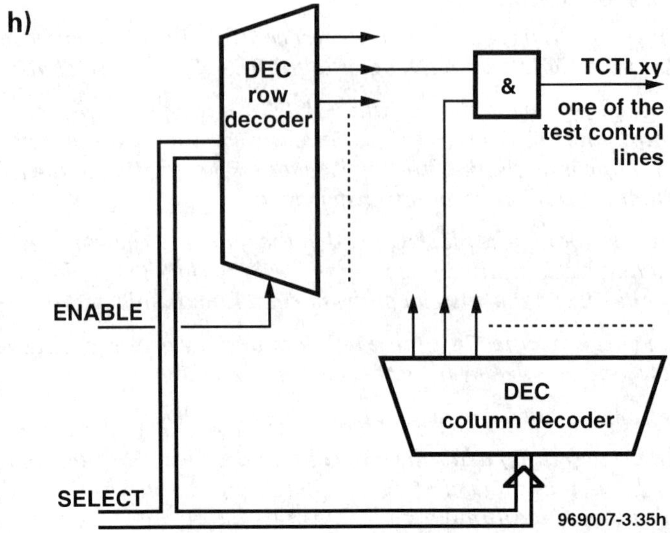

Figure 3.32 The addressing or de-serialization technique: circuit variations (4 of 4).

3.3.5 The shift-register or scanning technique

Principle

In a special test configuration, all flip-flops in the test object are chained together to form a single shift register (see Figure 3.33). The test procedure consists of shifting in test data, operating the circuit in its normal configuration for one clock cycle, and reading out the resulting flip-flop states bit-by-bit via the shift register to an external tester (service processor), where they can be evaluated. The entire circuit can thus be tested step by step.

A circuit built this way costs on average 10% to 20% more, but it has the following overwhelming advantages:

1 The flip-flops themselves can be tested in the simplest possible manner by just 'pushing through' a test pattern (see Volume 2).

2 Test data can be directly supplied to all combinational networks, all RAMs and so on. The test results can be transferred out for evaluation via the same path.

3 This technique can be applied equally well at various levels: in an individual IC (during fabrication), at the card level, at the device level and at the level of the entire system. If necessary, the same test data can be used for servicing as are used in manufacturing of the IC or circuit board or in the test facilities of the system integrator.

4 If it is possible to immediately halt the device clock when a hardware error is detected in normal operation, the complete device state can accessed down to the individual bit level via the shift register (Machine Error Logout).

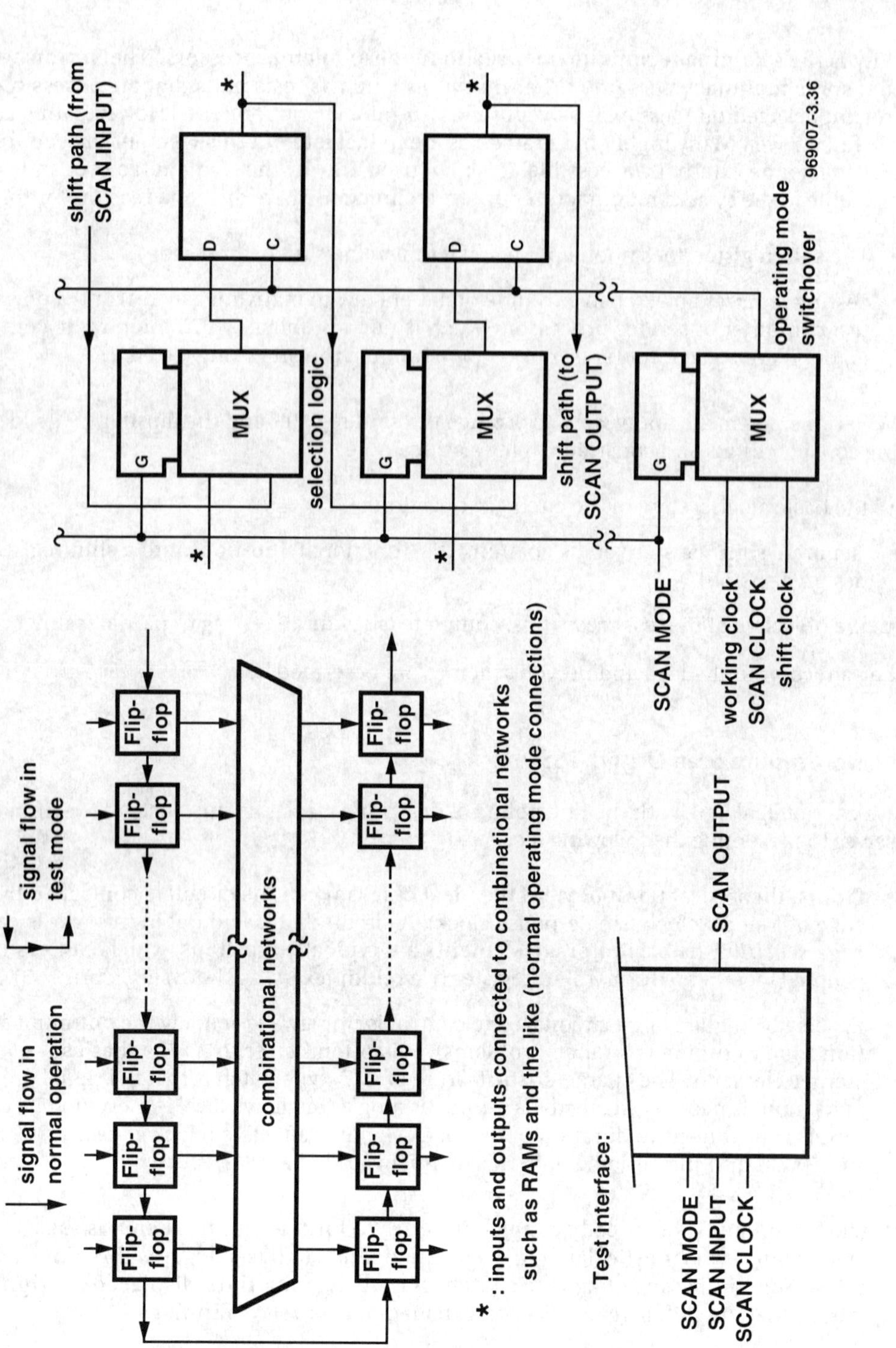

Figure 3.33 *The shift-register or scanning technique.*

Why is the scanning technique not used in modern microprocessors? The first answer is that some manufacturers find other measures (such as tests pads that are accessible on the unpackaged die) less expensive for mass-produced components. However, this is not the full answer. Many highly-integrated ICs are in fact laid out this way, but the shift register may not be externally accessible. It is then used only by the IC manufacturer and is not available to the system manufacturer or test technician. There are good reasons for this:

- The shift register does not help much if it only relates to the processor.
- Testing requires an extensive volume of data (a few to many megabytes) and a powerful test computer or service processor (which is not a problem with a mainframe computer, but if the test computer is more expensive than the one being tested ...?).

Variations. There are many ways to arrange the shift register and the flip-flops. The following considerations and constraints play a critical role:

- the basic clock system of the circuit in question,
- separate shift register versus chaining the 'functional' flip-flops into a shift register (a design decision),
- the objective of the test provisions: complete (structural) testing or partial testing?

Certain commonly-used and proven schemes are described below.

Level Sensitive Scan Design (LSSD)

We cannot deal here with the fine points of designing clock systems. Instead, we limit ourselves to presenting the following facts:

- From a theoretical point of view, the ideal case is when the circuit is controlled by the edge of a single clock (single-phase clock). A circuit that is laid out in this way is called *edge-sensitive*. Its basic storage elements are typically flip-flops, which can easily be coupled together to form a shift register (via multiplexers, as shown in Figure 3.33).
- Especially in quite large circuits, edge control is not always actually the optimum solution (due to timing tolerance problems). It is often better to use latches as the basic storage elements. The characteristic feature of a (D-type) latch is that the signal present at its input is passed directly to its output as long as an active level is present at its clock input. Thus the entire duration of the clock signal is effective (clock state or level control). A circuit which is laid out in this manner is called *level sensitive*.

Latches cannot just be directly coupled in series to form a shift register. It is thus necessary to couple a 'diagnostic' latch to each 'functional' latch (see Figure 3.34). For shifting data the functional and diagnostic latches are driven by time-displaced (antiphase) clocks, so that the shift register acts like a chain of master/slave flip-flops.

The additional latches need not always result in higher costs. Level-sensitive digital circuits are always based on multiple sets of latches driven by non-overlapping clock signals

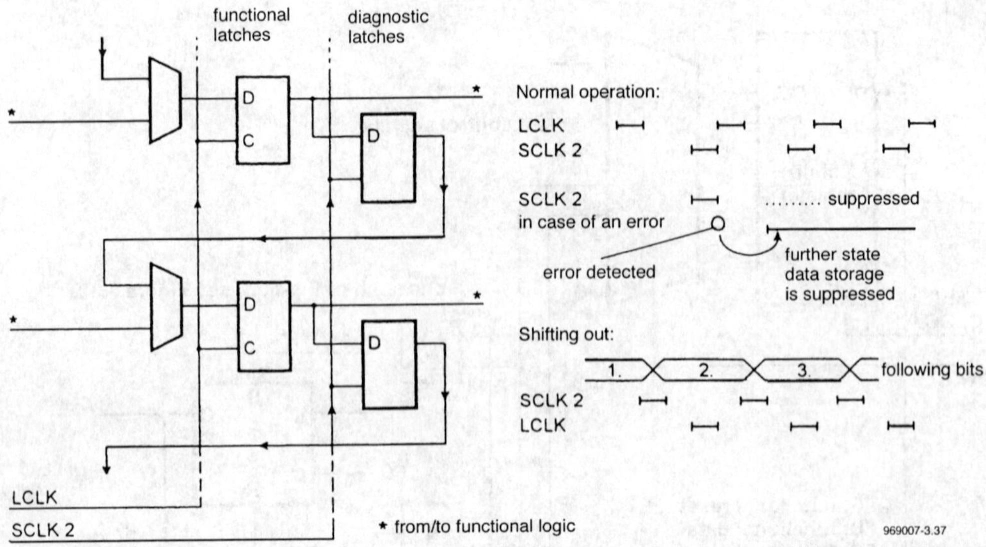

Figure 3.34 Latch arrangement for LSSD. The diagnostic latches are used as a backup register in normal operation.

(multi-phase clocks), since latch outputs cannot be fed directly back to latch inputs via combinational networks. It is sometimes possible to partition the circuit such that the functional latches of the different sets can be coupled together to form a shift register.

The set of diagnostic latches can also serve as a backup register for the the functional latches (see Figure 3.34, for example). In normal operation, LCLK is the working clock and SCLK2 the time-displaced backup clock, which will be masked out in case of an error. In the shift-register mode, LCLK and SCLK2 are used alternately to shift the data. SCLK2 thus copies the data from the functional latches to the diagnostic latches, and LCLK copies the data from the diagnostic latches to the next-following set of functional latches. In normal operation, the diagnostic (backup) latches are loaded with the output values of the functional latches, in parallel but time-displaced. This time delay gives the error-detection logic enough time to prevent further loading if an error is detected. In order to recover the backup data, it is necessary to switch to shift-register mode and then start with SCLK2 instead of LCLK, taking each data bit value *before* the SCLK2 pulse.

Random access scan

This method is essentially a version of the addressing or de-serialization technique. Instead of a broad test interface with a test data bus and test clock signals, random access scanning uses a 'narrow' serial interface. The selection signals are fed serially into a shift register located in the device under test. The flip-flops or latches are loaded and sensed individually (see Figure 3.35).

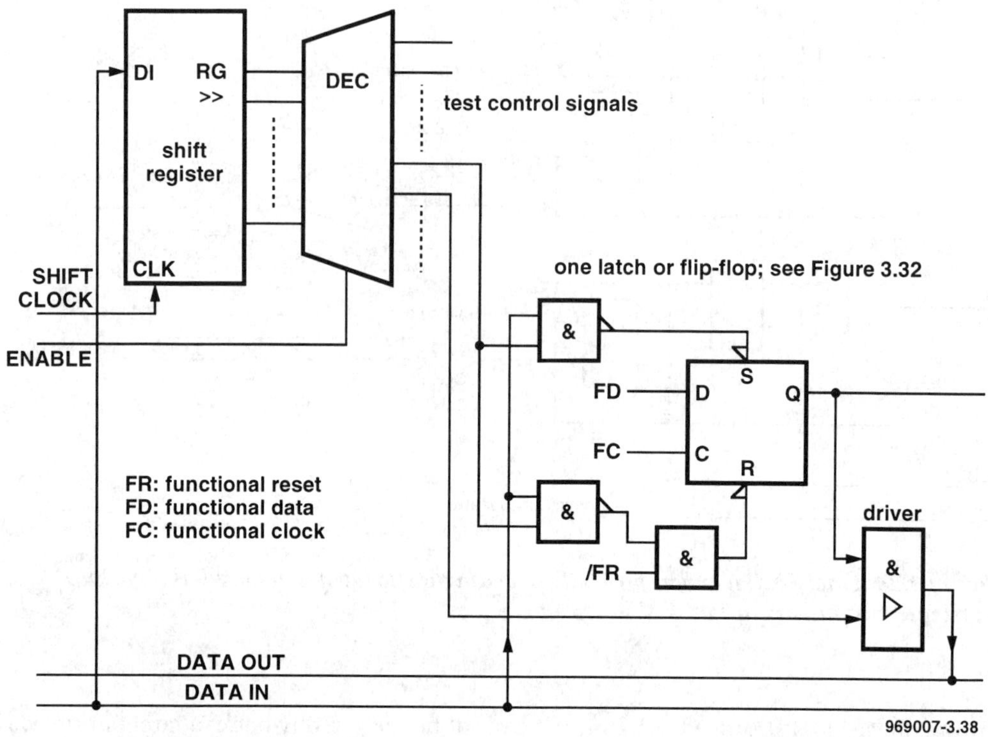

Figure 3.35 Random access scan (implementation example).

In the circuit shown in Figure 3.36, the test interface consists of a data input line (DATA IN), a data output line (DATA OUT), a shift clock input (SHIFT CLOCK) and the control line ENABLE. Practically the only difference between this arrangement and that of Figure 3.32 is the use of bit-serial data input and output. (The advantage of this is that only four lines are needed for the test interface; this is important when such provisions are built into complex ICs.)

The scan/set technique

The scan/set technique is based on independent shift paths, with a shift flip-flop associated with every functional storage element that should be accessible (see Figure 3.36). The obvious disadvantage of higher cost is balanced by the advantage that no specific limitations are placed on the design of the functional logic (basically, flip-flops, latches or a combination of the two can be used as desired).

Figure 3.36b shows a scan flip-flop with a storage element. The test interface consists of the control line SHIFT MODE (which selects between copying the flip-flop output and

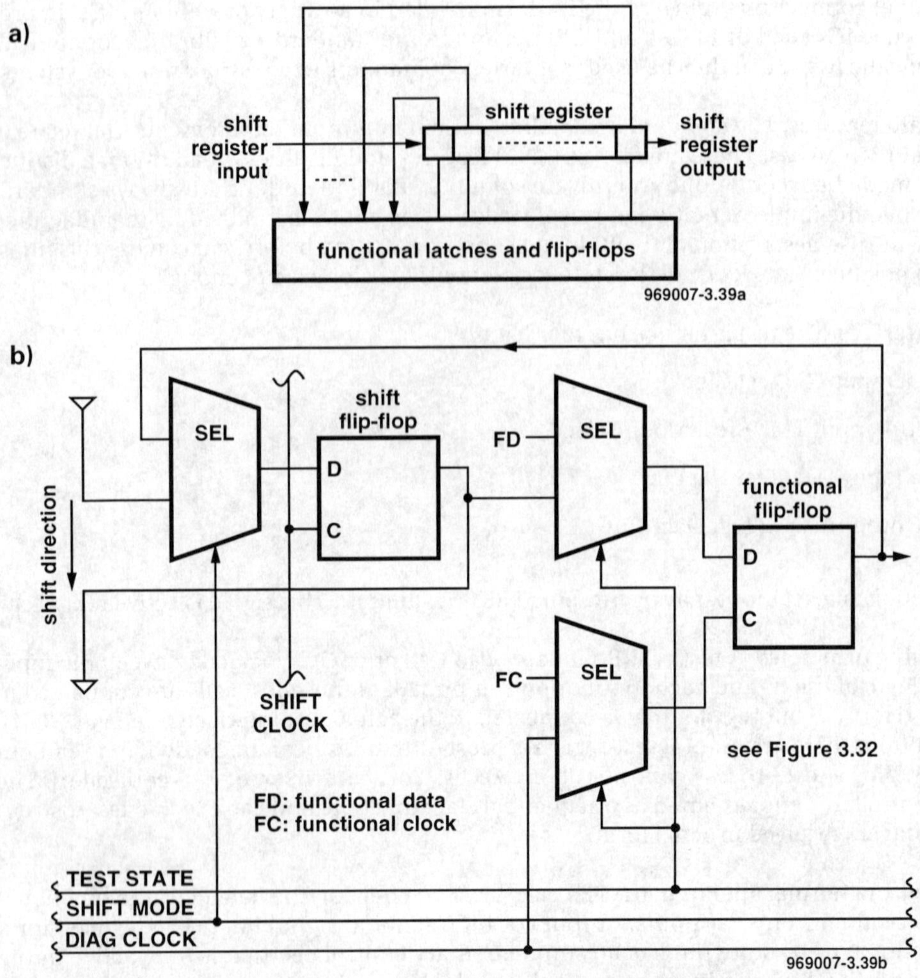

Figure 3.36 The scan/set technique. a) concept, b) sample implementation.

shifting data), the SHIFT CLOCK line, the copy clock line DIAG CLOCK, the control line TEST STATE and the shift register's data input and output lines (not shown in the figure).

Boundary scan

Boundary scan is a version of the scanning technique that is limited to the *external connections* (pins) of an integrated circuit. It is primarily intended to be used for checking interconnections on a printed circuit board. The test interface is described in the IEEE (or JTAG) 1149.1 standard ('IEEE' stands for *Institute of Electrical and Electronics Engineers* and 'JTAG' stands for *Joint Test Action Group.*) If the IC has been suitably laid out, this test interface allows data to be read out and test data to be fed in via serial paths, using the functional connections (pins) of the IC. It is provided in a variety of standard ICs, in programmable ICs such as FPGAs and CPLDs, and in the standard-cell libraries for application-specific ICs. It can thus be used in a variety of manners to construct testable systems.

Standard features. The standard first defines certain minimum requirements that relate to mandatory features. There are also optional features and facilities. In addition, a distinction is made between public and private solutions. Public solutions are always standardized, while the implementation of private solutions is left to the individual manufacturer. *Note:* a precise description of the IC in question is necessary before you can use this interface in practice; knowledge of the shift-register outputs alone is not enough.

Test interface. The test interface has four leads:

- clock input TC (Test Clock),
- control input TMS (Test Mode Select),
- data input TDI (Test Data In),
- data output TDO (Test Data Out).

In addition, a reset input may be present (TRST, Test Reset). This signal is active LO.

Operating principles. A test cell (Boundary Scan Cell or BSC) is located between the functional logic of the IC and each input or output pin. All of these test cells are connected in series to form a shift register that is connected to the actual test interface (see figure 3.37). In addition, there are several test registers present that also can be loaded and read out serially, plus a test process controller (Test Access Port controller, or TAP controller). The TAP controller is implemented as a sequential state machine and can execute test instructions that are received in serial form.

Test cells. In normal operation the test cells have no effect on the data flow; the functional logic uses the IC pins as inputs, outputs or bidirectional signal lines. The test cell for a bidirectional or tri-state pin has an supplementary control cell that is only active in the test mode, in which it generates control signals that replace those from the functional logic. Each test or control cell has a Halt flip-flop to store its current state as well as a Shift flip-flop for the serial link to its neighbouring cells. A cell can store the current value of its associated pin without affecting the normal function of the pin. The stored value can be transported via the serial path (shift register). For active testing, test data can be entered into the cells. All cells are active simultaneously at the IC pins; serial shifting of their data does not produce any noticeable effect on the functional inputs or outputs.

Test mode and normal mode. These operating modes relate to the functional logic of the IC. Some test instructions do not affect the normal operation of the functional logic (normal mode), while certain other instructions disable normal operational functions (test mode).

Test registers. Three test registers are mandatory. These are the instruction register and two data registers. Additional data registers may optionally be included (see Figure 3.38).

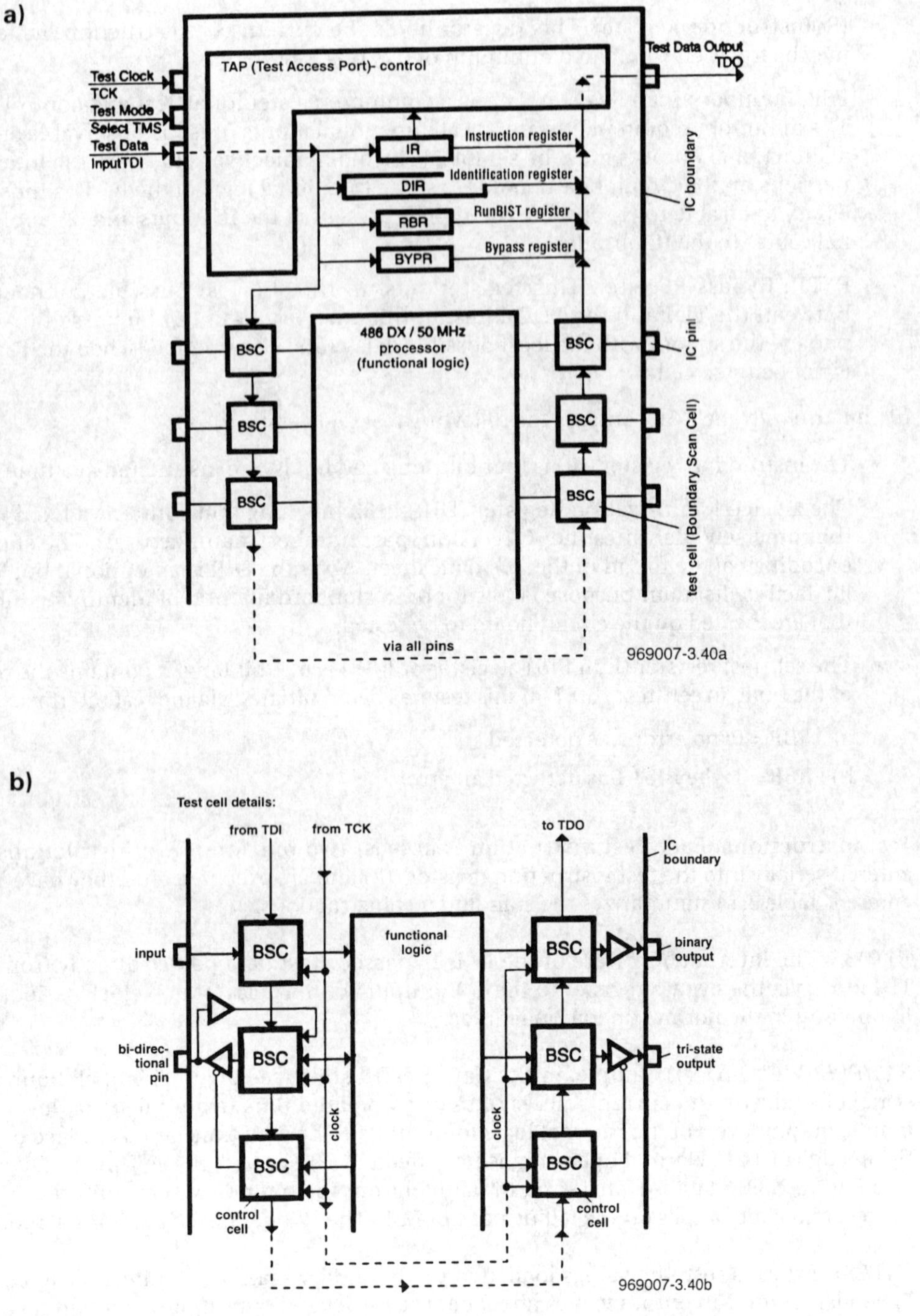

Figure 3.37 *Boundary scan: the JTAG/IEEE 1149.1 test interface (example from the i486).*

a) mandatory test registers:

- IR: Instruction Register. This register holds the current test instruction. Its length may be freely chosen, but a minimum of two bits is required.
- BSR: Boundary Scan Register. This is a common register for all test and control cells. The minimum requirements are: (1) all functional inputs must be observable and (2) all functional outputs must be setable. (The main objective is to check the interconnections on the circuit board, not to test the functional logic of the IC. It is thus necessary to be able to (1) observe the signals present at the IC inputs and (2) apply signal values to the IC outputs.)
- BYPR: Bypass Register. This register sets up the shortest possible connections between the TDI and the TDO lines. In this way, test data can be passed to other parts of the system with the least possible delay (one clock period) when the IC itself is not being tested.

b) an example: the i486 contains the following test registers:

- The instruction register (IR) is four bits long and holds one instruction at a time.
- The Device Identification Register (DIR) is 32 bits long and contains a fixed value that precisely identifies the IC (version, part number, manufacturer). The specific encoding can be found in the IC's data sheet. *Note:* the DIR is an optional but 'public' facility. Its main purpose is to support a standard method of identifying the ICs that are located on the circuit board to be tested.
- The self-test register (Run BIST Register or RBR) is one bit long. It contains the result of the built-in self-test (BIST) if this test has been initiated via the test interface:

  a) RBR = 0: no error was detected,

  b) RBR = 1: the BIST has detected an error.

Test instructions. Each test instruction is at least two bits long. Test instructions are entered serially into the test instruction register (IR) and are executed after they have been entered. Table 3.14 summarizes the standard test instructions.

BYPASS. The internal hardware of the IC is bypassed. Test data passes directly from the TDI input via the bypass register to the TDI output, with a delay of one clock period. The IC operates in the normal operational mode.

SAMPLE/PRELOAD. (1) Sample: in the Capture-DR state (see Figure 3.39), all boundary scan cells take on the current values of their associated pins (external signal levels for inputs, output levels of the internal logic for outputs). The data transfer takes place on the rising edge of TCK. Normal operation is not affected. (2) Preload: in the Update-DR state (see Figure 3.39), data is transferred in parallel from the boundary-scan shift register to the test and control cells on the falling edge of TCK. The IC operates in the normal mode.

EXTEST. External test: the values loaded in the boundary scan register become effective. Previously shifted-in output values appear at the associated output pins, and input values are stored in the test cells. Bidirectional and tri-state pins are configured as appropriate by their associated control cells. The IC works in the test mode. *Note:* many ICs must be reset

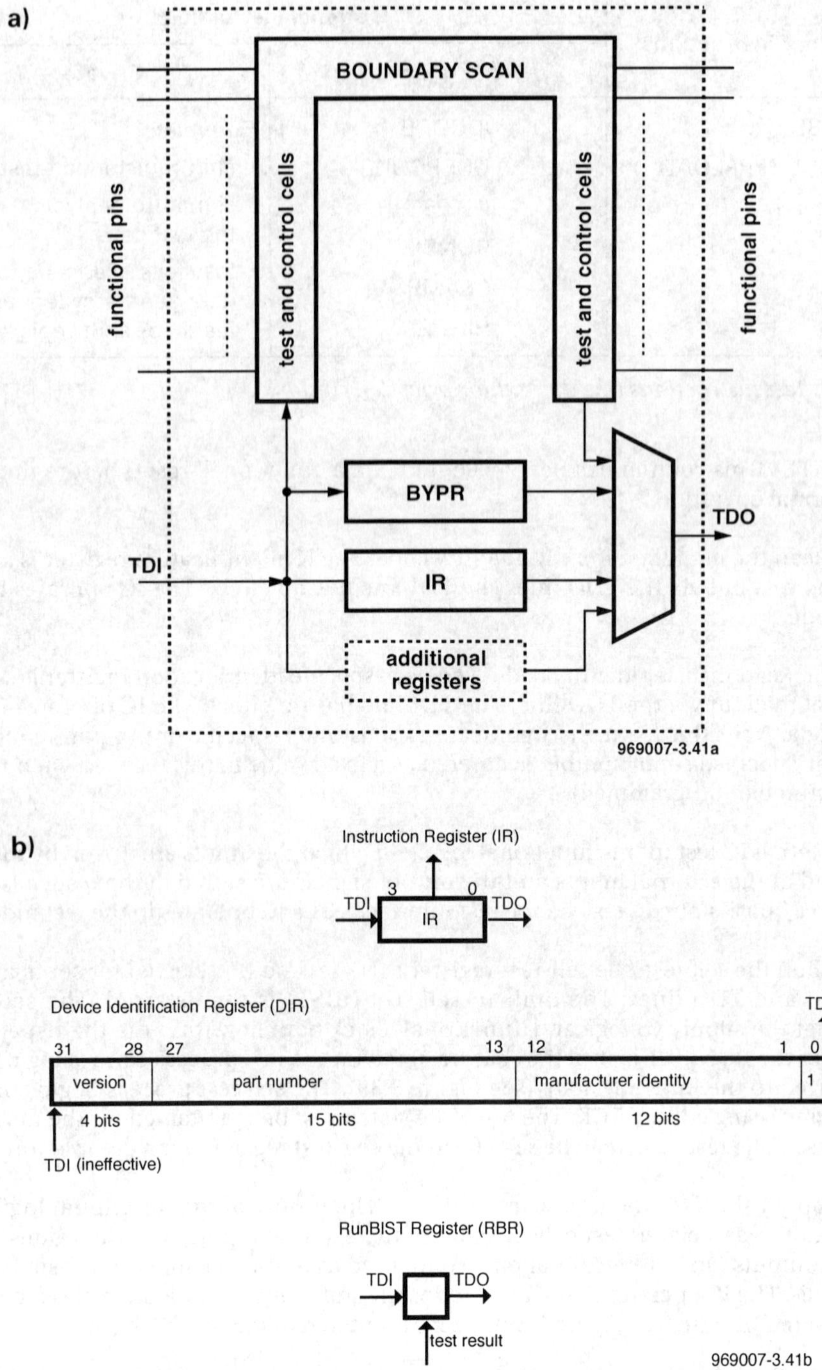

Figure 3.38 Test register. a) minimum configuration, b) example from the i486.

Mandatory instructions	Optional instructions	
	Public	Private
BYPASS SAMPLE / PRELOAD EXTEST	IDCODE USERCODE INTEST RUNBIST CLAMP HIGHZ	*Examples:* • shift-register self test • signature analysis using the test cells • test cells used as a test data generator (eg, as a feedback shift register)

Table 3.14 *Test instructions (see the text for brief descriptions).*

after an EXTEST instruction has been executed (via a 'functional' reset) before they will resume normal operation.

IDCODE. Read the manufacturer's IC identity code. The IC identification register is selected for serial readout via the TDO line. The TDI line has no effect. The IC operates in the normal mode.

USERCODE. Read the user identity code. The user-specific identification register is selected for serial readout via the TDO line. The TDI line has no effect. The IC operates in the normal mode. *Note:* this feature is significant for customer-specific or programmable ICs (a customer-specified serial number is entered as a fixed value in this register when the IC is manufactured or programmed).

INTEST. Internal IC test (of the functional logic), in which the inputs are driven by the test cells instead of the external inputs and the output signals are stored in the test cells (this amounts to a quasi-static IC test via the test interface). The IC operates in the test mode.

RUNBIST. Run the self test: the self-test register is (1) set and (2) selected for serial access via the TDI and TDO lines. The built-in self test (BIST) is then started. The self test requires that the supply voltage and 'functional' clock both lie within the their specified ranges. After the RUNBIST instruction has been fed in, the test process control logic must be transferred to the Run/Idle state (see Figure 3.39). The self test process begins on the next following rising edge of TCK. The Run/Idle state must be maintained for the duration of the self test. The test result can be read from the self-test register upon completion.

CLAMP. Apply values to the inputs and outputs. The inputs of the functional logic are taken from their associated test cells instead of the external inputs, and the values of all functional outputs (and bidirectional pins configured as outputs) come from their associated test cells. The IC operates in the test mode. The shift path runs via the bypass register. In the Capture-DR state (see Figure 3.39), a zero is entered in the bypass register.

HIGHZ. High impedance. All functional outputs are switched to a high-impedance state. The functional inputs and the functional logic remain operational. The IC operates in a modified test mode. The shift path runs via the bypass register. In the Capture-DR state (see Figure 3.39), a zero is entered in the bypass register.

Process control logic. The test process control logic operates as a sequential state machine. Figure 3.39 shows its state diagram. The state transitions are controlled via the Test Mode Select (TMS) input. Each transition occurs on the rising edge of TCK. You can picture the process control logic as a stepping switch that is advanced on the rising edge of TCK by applying the appropriate TMS inputs, and which thus selects the individual test functions in turn. Starting from the initial state, the first state that it reaches is the 'Run Test/Idle' state. The state sequences for accessing data and instruction registers, which have identical structures, can all be reached from this state.

Notes:

1 All data registers (except the instruction register) are test registers. The instruction that is being executed selects the appropriate data register for each test function.

2 The Pause states allow the testing device (automated tester, test computer and the like) to interrupt the shifting process in order supply new data or store data that have been read. The pause state is followed by an intermediate state (Exit2 DR/Exit2 IR) that allows the choice of resuming the shifting process or canceling the test (transition to Update DR/Update IR).

3 In the End state, it is possible to transfer to a new selection state or to branch to the Idle state.

System testing. Test steps can be built from the elementary control processes. Each test step can inject test vectors (stimulus bit chains) into the system, read out the results and compare actual to expected results. This corresponds to the way in which a conventional automated tester works, with the difference that in this case the data transfer is performed serially instead of via a large number of test-adapter probes. This means that we are dealing with a quasi-static test at a lower speed than that used for normal system operation. (*For example:* with a 50-MHz i486, the maximum rate of TCK is 25 MHz).

System design considerations. The boundary-scan technique can be implemented in an entire system, in order to ensure its ongoing testability. (The extent of boundary-scan testing could include all interconnections (including bus structures), all logic that can be driven quasi-statically and all static memory.) Basic bus interface components with built-in test interfaces are available, with which you can construct testable systems using processors that do not themselves have test interfaces.

The utilization of built-in scan paths not only in the manufacturing process but also in working systems seems like an obvious idea. Of course, some sort of programmable equipment with sufficient built-in memory capacity will be needed for controlling the test process. Possible alternatives are: (1) another processor (in multiprocessor systems), (2) a dedicated service processor and (3) an externally-connected test computer (such as a notebook PC).

Note: refer to Volumes 2 and 3 for more information, including descriptions of basic control software and sample circuits.

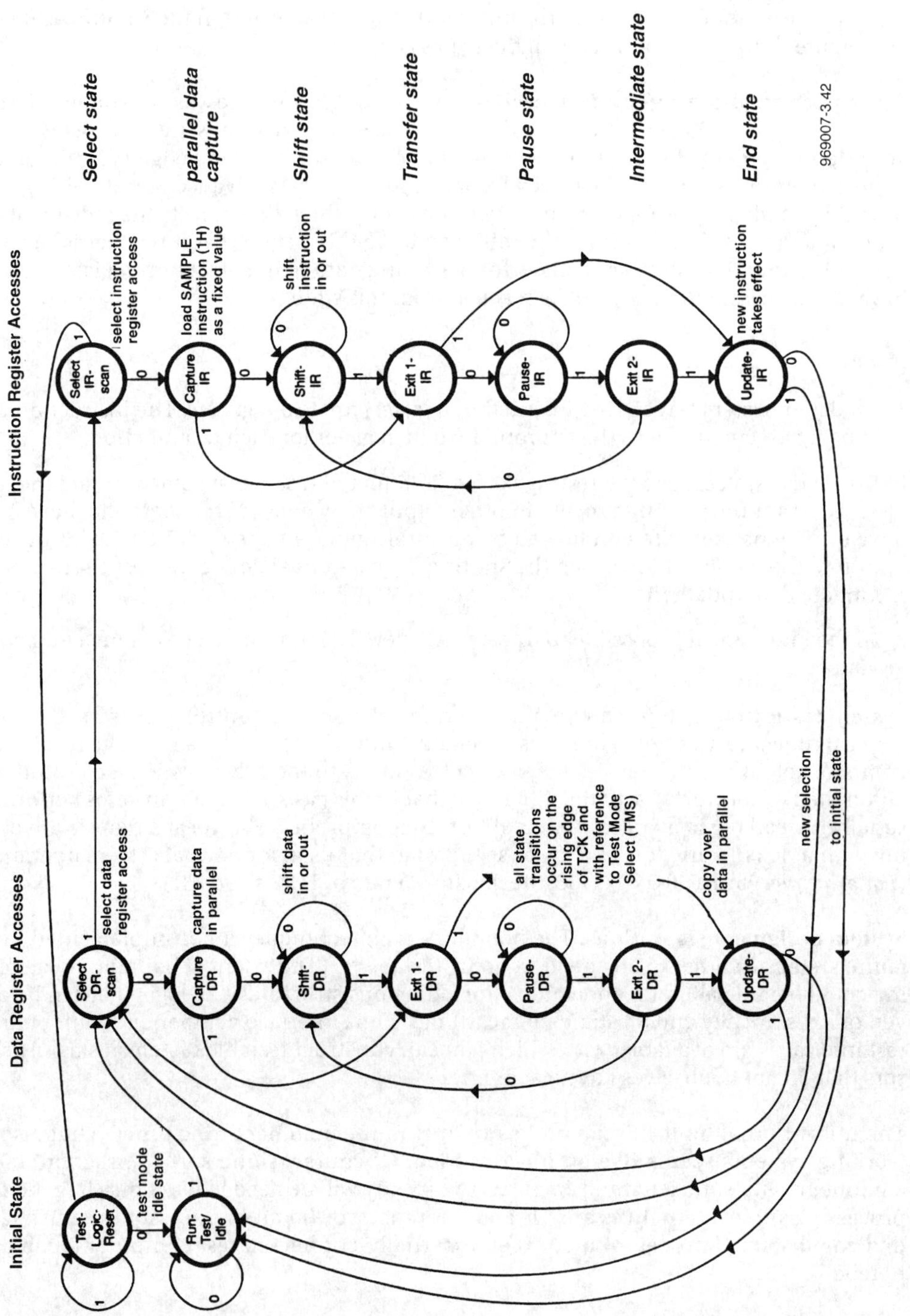

Figure 3.39 Test process control logic state diagram.

3.3.6 Observability and accessibility in your own projects

Even relatively small circuits can be mischievous. You should never blindly assume that everything will work properly; rather, from the very beginning you should keep in mind that the hardware not only has to be assembled, it also has to be put into service and – later – maintained in good working order. Strictly speaking, test- and service-oriented system design is a science in its own right. Nonetheless, it is possible to come up with quite acceptable solutions – at least for relatively simple projects – on the basis of good common sense (for examples see Volume 3). The following approaches in particular have proven themselves for projects of this size (that is, they are truly effective and their costs are acceptable):

- arrangements of test and measurement points,
- utilization of the serialization technique for sensing device status,
- utilization of the addressing technique for sensing device status and/or configuring the device for testing purposes,
- implementing the scanning technique in subcircuits where it is appropriate. (Basic ICs with boundary scan capability can be used for this. In addition, there are various types of shift registers that can be used to achieve partial accessibility using the scanning technique, without significantly increasing the overall cost.)

3.4 Built-in self-test provisions and other test provisions

3.4.1 Self-test provisions

A device can be called self-testing or self-checking if it contains circuitry that supplies the test stimulus, controls the test process and evaluates the test results. If we leave aside 'true' redundant hardware (such as the master/checker technique), we can distinguish the following types of arrangements:

1 The device under test is programmable, and the self-test functions are mainly implemented as special program routines. This is the case with most devices or systems that incorporate microcontrollers or microprocessors. Self-tests within a controller or processor IC are sometimes based on *test microprograms* (microdiagnostics). These are basically the same as 'ordinary' test programs, except that they are executed directly from the microprogram (or control) ROM and their test severity is usually significantly better than that of test routines based only on normal machine instructions, since they act directly on the functional logic and storage elements of the processor.

2 The device under test and the test equipment are in essence independent of each other, but they form a permanently-connected assembly. The test equipment may itself be programmable (a test or service processor), but it can also be a 'genuine' tester that (for example) contains ROM structures (for storing test stimuli and expected results data), comparators, address counters and the like. (For the construction of such testers see Chapters 4 and 5, and Volume 3.)

3 The test provisions are 'uniformly distributed' in the device under test: that is, parts of the functional hardware are used in the test mode to form the testing device. In the case of built-in test provisions (Built-in Test Equipment or BITE) for highly-integrated circuits, this arrangement is usually what is used. One technique that has proven to be particularly useful in this situation is to interconnect the input flip-flops to form a feedback shift register in the test mode. This allows any desired pulse sequence to be generated, depending on the specific feedback arrangement used. For checking, the output flip-flops are also interconnected to form a feedback shift register. The feedback can be arranged such that a type of CRC checking is realized: the shift register on the output side must contain a specific bit sequence (for example, all zeros) on completion of the test sequence if the hardware is fault-free. By far the majority of faults can be detected by the fact that they cause the data in the output shift register to differ from the expected data. This technique is sometimes referred to as Built-In Logic Block Observation (BILBO). See the following section for more details.

3.4.2 Pseudo-random tests

Truly 'severe' test processes require a considerable amount of test data (stimuli and reference data). However, particularly with built-in test provisions, it is hardly possible to generate comprehensive test stimuli and to store a large amount of reference data. Possible solutions are:

1 Test stimuli: you can content yourself with a rather 'global' test. In this case, random testing often has a better chance of detecting a fault than testing with a reduced set of deterministically generated test vectors (that is, generated by calculation or by examination). However, truly random mechanisms are quite difficult to produce. In addition, there is the problem of having to have all imaginable test vectors 'on hand'. For these reasons 'pseudo-random' test processes are usually used; these are processes that appear to be random but which are actually deterministic in nature.

2 Reference values: *data compression* techniques are used to reduce the volume of the data to be stored. The most important of these is storing the CRC bit sequence instead of the actual data (signature analysis; see Section 4.4.3).

What does 'random' mean? A truly random test pattern could for example be generated by tossing a die (or a coin, since only ones and zeros are needed). The following properties are characteristic of random bit patterns:

- ones and zeros occur (on average) practically equally often,
- the value of a particular bit does not depend on the value of the bit that precedes it,
- a series of two or more ones or zeros occurs occasionally, but the longer such a series is the less often it occurs.

Series that approximately meet these conditions can be easily generated, namely by linear feedback shift registers. Such shift registers differ from those used for CRC purposes in the following aspects:

1 there is no input,

2 the feedback is wired such that a maximum-length sequence is generated. (A shift register with n stages can generate a cyclic sequence with a length of $2^n - 1$ bits. A register with 12 flip-flops thus produces a cyclically repeating sequence of 4095 bits, while a 16-bit register produces a sequence that is 65,535 bits long.) See Appendix 6 for more information.

Figure 3.40 shows how a circuit can be made self-testing on this basis. The flip-flops of the device under test are arranged as three BILBO blocks *X*, *Y* and *Z*. The combinational circuitry forms three corresponding combinational networks *A*, *B* and *C*. Each BILBO block can operate in three different modes: (1) normal, (2) test stimulus generator and (3) signature analyzer. Mode selection can be explained with reference Figure 3.33: a multiplexer placed ahead of the inputs of each flip-flop allows the inputs to be switched between normal operation and shift-register operation. Furthermore, a BILBO block operating in the signature analysis mode produces an error signal if its register content does not match the expected 'signature' value. A sample self-test process consists of the following steps:

1 logic block *A*: BILBO block *Z* = stimulus generator, BILBO block *X* = signature analyzer;

2 logic block *B*: BILBO block *X* = stimulus generator, BILBO block *Y* = signature analyzer;

3 logic block *C*: BILBO block *Y* = stimulus generator, BILBO block *Z* = signature analyzer.

This process can be controlled by a relatively simple control circuit (test sequencer).

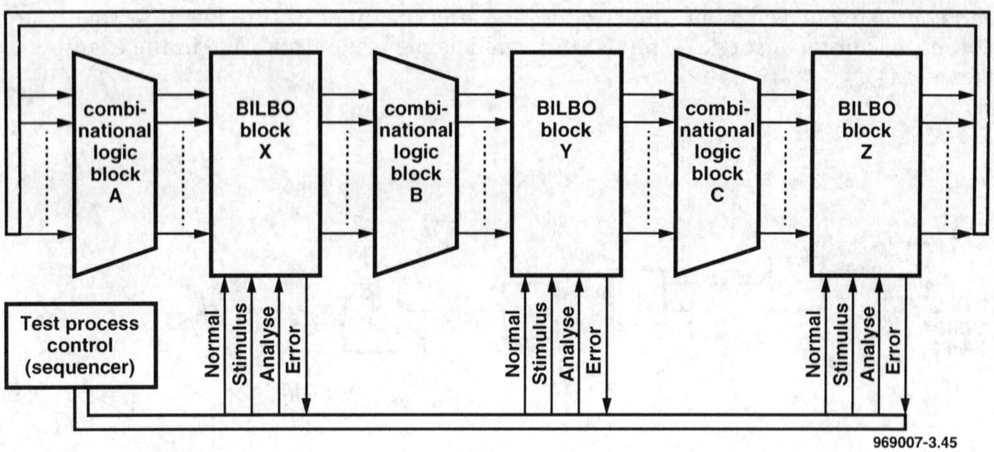

Figure 3.40 Self-testing based on feedback shift registers (BILBO technique).

Note: ICs with boundary-scan provisions are available, in which the scan hardware can be configured as a linear feedback shift register, either for generating test data or performing signature analysis as desired. See Section 4.4.3 for more information.

3.4.3 High-impedance busses

Some highly-integrated circuits can be put into a state in which they no longer drive their functional pins (tri-state test mode). For example, the i486 and Pentium processors, as well as various motherboard ICs, can be switched to a mode in which all outputs and bidirectional pins assume a high-impedance state. In this mode the lines connected to these pins are no longer driven by the IC logic and thus can be easily driven by external signals (from an automated tester, for example) to test memory and peripheral logic circuitry and the like. (This is effectively an elegant substitute for an IC socket which, particularly with modern ICs having many pins, is costly and represents a source of unreliability.)

It is usually necessary to apply a certain set of signal levels or a certain sequence of signals to specific pins of the IC in order to put it into this mode. (There is no general 'industry standard' for this; you must instead consult the data sheet for the device in question.) As an example, the i486 and Pentium processors enter the tri-state test mode if the FLUSH# pin is held LO while a HI-LO transition occurs on the RESET pin. ICs with boundary scan provisions can usually be put into the tri-state test mode via the HIGHZ instruction.

3.4.4 NAND trees

The NAND tree technique is an extension of the high-impedance bus test technique, which allows IC connections to be tested for 'continuity' (see Figures 3.41 and 3.42).

Certain IC pins are provided for enabling this test mode, just as for the 'high-impedance bus' test mode. In the test mode, the pins that are to be tested are configured as inputs and connected to a chain of NAND gates, one after the other (the designation 'tree' is not entirely accurate). A single IC pin is connected as the test output (*A30* in the example of Figure 3.41).

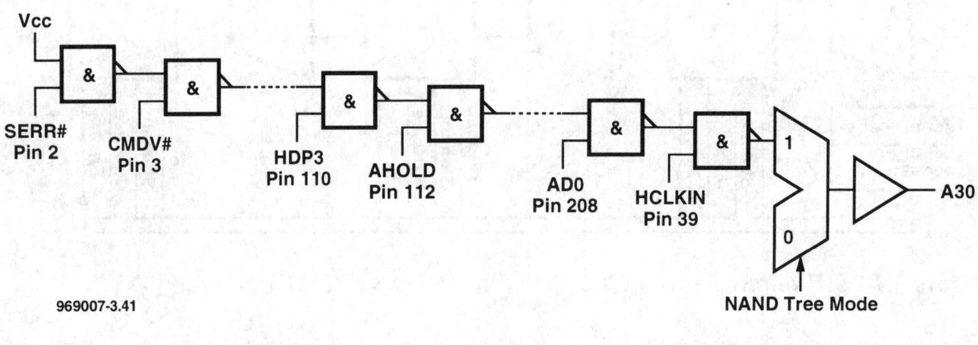

Figure 3.41 *Example NAND-tree circuit for a motherboard IC. (source: Intel)*

(Extracted from a data sheet. The exact arrangement of all pins to be tested is given in the data sheet in tabular form; it normally corresponds to order of the pins on the IC package.)

Test process:

- The tester initiates the test mode. At this time no signals are applied to any inputs.
- The tester applies HI signals to all pins that provide inputs to the NAND chain.
- The value at the test data output pin (A30 for example) is sensed. If the number of NAND gates is even, it should be HI; otherwise it should be LO.
- The tester then applies LO signals to each of the inputs of the NAND chain in turn. This should cause the output signal to change levels.

Note: a typical situation for using this technique is the testing of 'stuffed' motherboards on an automated tester. The input signals are then applied via a bed-of-nails adapter. (Alternatively, if each of the lines to be tested is connected to an IC with boundary scan capability, then all of the input values can be applied via the boundary-scan shift registers.)

In the circuit shown in Figure 3.42, there are four test inputs and one test output. The signals that are applied to the test inputs for the first four test steps are:

- Step 1: the IC is put into the NAND-tree test mode. A HI level is applied to each input. Since there is a even number of NAND gates, the output signal is also HI.
- Step 2: a LO level is applied to the first test input. The test output changes to LO.
- Step 3: a LO level is also applied to the second test input. The test output goes to HI.
- Step 4: a LO level is also applied to the third test input. The test output again goes LO.

In the fifth test step (not shown), a LO level is applied to the fourth test input and the test output again goes HI.

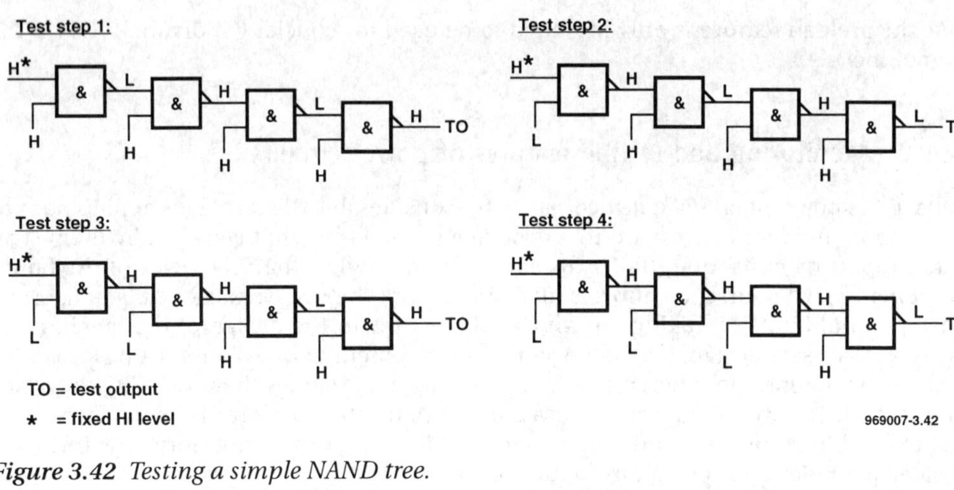

Figure 3.42 Testing a simple NAND tree.

The advantages of a NAND tree are: (1) each signal to be tested only needs a single NAND gate (which is much less costly than a boundary-scan cell) and (2) each signal to be tested only requires a single test probe connection (compared to a connection at each end for conventional continuity testing).

In addition, the NAND-tree technique provides a simple means to check each input for reliable switching with maximum LO and minimum HI input levels (V_{IL}/V_{IH} test). The respective levels are applied in sequence to each input in turn with all other inputs held HI, for example, and the output is checked for proper switching between LO and HI.

The device under test is OK if the test output changes level each time the level applied to a test input is changed from HI to LO. In addition, at a given test step the input level can be switched repeatedly between HI and LO, which should cause the output signal to switch in step. The NAND tree thus allows a certain degree of dynamic testing of the interconnections on the circuit board and the input and output stages of the ICs.

3.4.5 Test provisions in programmable logic

Programmable ICs that contain flip-flops (PALs, GALs) have provisions for directly loading the flip-flops (register preload). There are two typical versions:

1. The IC has its own preload pin (PL): the data are loaded in parallel via the functional pins: (1) the outputs are switched to a high-impedance state by deactivating the output-enable signal, (2) the bit pattern to be loaded is applied to the output pins, (3) a pulse is applied to the PL pin to copy the bits into the IC, (4) the signals are removed from the output pins and (5) the outputs are re-enabled.
2. The flip-flops are loaded serially via a shift register: the shift-register mode is selected by applying a special voltage to a particular input pin (for example, 15 V). Certain other pins then function as data and clock inputs for the shift register.

Note: the preload feature is only intended to be used for checking individual ICs (in the programmer).

3.4.6 Monitoring and testing features of power circuits

Faults in the power stages of a device can cause considerable damage. This applies both to the components themselves and to any device or assembly that is driven by them. The cost of repairing consequential faults can be much higher than the price of the failed component(s). Of course, an effort is made to prevent the occurrence of truly dangerous consequential faults by taking appropriate measures in the design of the mechanical assembly. (Fail-safe design: if a fault occurs the equipment in question must be forced (in the simplest manner possible) into a 'safe' state, which excludes the possibility of consequential faults.) It would naturally be even better to be able to avert the actual fault, or at least to be able to detect it and limit its effects. Various types of monitoring features are provided in modern power circuits for this purpose:

- current limiting, or at least short-circuit protection,
- overvoltage protection,
- overtemperature protection,
- detection of an open earth connection,
- detection of a no-load condition,
- reporting internal status.

Short-circuit protection. Shorting the output of a power stage to earth is a very common cause of faults. A simple protective measure is based on the fact that a short forces the output to earth potential ('0 volts'). You can exploit this to cut off the input to the power transistor.

Current limiting. Current limiting means restricting the load current that flows in a short-circuit situation to a non-damaging level. The simplest technique is to exploit the heat produced by the load current. The more current that flows, the warmer the component becomes and more its resistance increases, so that the maximum current is limited. This is only practical for relatively low-power components. A more effective solution is to monitor the component's temperature and cut off the load current if the temperature becomes too high. If it is important to monitor the actual current, then it must be measured. This requires connecting a low-resistance sensing resistor (shunt) in series with the load. The monitoring circuitry is usually designed to tolerate currents that briefly exceed the limiting value (for example, switch-on transients), and only to take action after a certain time has elapsed. Occasionally, special connection schemes are provided for conveying the measured voltage to the IC (Kelvin sensing or four-wire technique). *Note:* for a discussion of current measurement, see Section 4.2.3.

Under- and overvoltage protection. Zener diodes are preferred for simple and reliable detection of under- and overvoltage conditions. Modern power circuits contain zener diodes at various locations in order to ensure that 'critical' voltages are never applied to the power transistors. Sometimes this can be achieved by a simple clamping circuit, and sometimes it is necessary to cut off the power components in case of an overvoltage, or even to switch off the entire circuit. Undervoltage protection relates to monitoring the supply voltage(s). Its job is to make sure that the power stages are put into a safe operating state in case of a fault (usually, all power transistors are forced off).

Overtemperature protection. Temperature monitoring is a simple and effective way to protect against overloads. Many power ICs are equipped with such monitoring circuits. Normally, the power stages are cut off if the temperature of the PN0 junctions rises above approximately 150 °C.

Diagnostic reporting. Monitoring circuits prevent lasting damage, but they do not help in fault-finding. Sometimes fault-finding is even made more difficult by such measures (that is, you must determine whether the IC is defective or the protective circuitry has 'had a go'; in this case the possible sources of the fault must be systematically checked out). It is thus useful if the power IC can directly indicate an error condition. In the simplest case all of the monitoring circuits produce a common 'fault' signal that can be displayed (most simply) by an LED. In equipment that is controlled by a microcontroller or microprocessor, such error indications are usually sensed by the software.

Note: this type of monitoring is not always helpful for fault-finding. In most PCs, the error signals are sensed by the associated controller (such as the hard-disk controller) and processed to yield a general error indication. However, the power ICs are not accessible to 'normal' software (that is, via the main processor). The result is that: (1) operational security is improved and (2) detecting a faulty *assembly* is simplified (since the related controller provides an unambiguous indication of a fault), but (3) localizing a fault to an individual *component* is not made any easier. (*Tip:* check the error signals with a logic probe or pulse catcher. You will however need the data sheets for the components in question.)

4 Test and measurement techniques

4.1 Basics

4.1.1 Selecting test and measurement equipment

You should select your equipment according to the type of testing to be done and/or the particular devices to be tested. We limit our attention here to equipment that (1) is adequate for the vast majority of tests and measurements in the service environment and small development projects, and that (2) has a reasonable cost relationship to the test objects. With regard to testing, we essentially concentrate on interfaces, digital circuits and power supplies. There is certainly a place for 'classical' analogue circuits, but these form a small part of the total picture (although they can cause a lot of trouble).

Prices. Naturally, you must always consider whether the cost of any given piece of equipment is justified in a given situation, but it is reasonable to assume that the 'pain threshold' starts somewhere around £1000 and can range up to several tens of thousands of pounds for professional equipment.

Ease of use. Modern test and measurement equipment is remarkably easy to use, with features such as automatic range selection, signal processing, graphical displays and computerized signal evaluation. However, the 'basic' specifications (such as accuracy, number of channels and so on) are still the most important.

Multi-function equipment. Modern technology (microprocessor control, LCD screens, digital signal processing) allows manufacturers to offer quite powerful and truly economical multi-function equipment (such as a combined oscilloscope and digital multimeter). Such equipment is particularly attractive for 'on-site' servicing. On the other hand, although a collection of separate devices that provide the same functions is more cumbersome, it is often less costly than the equivalent multi-function device. *Practical tip:* don't just look at the the specifications and price; you should also check out the ease of use and the power-supply provisions of a potential purchase. ('Features' such as an awkwardly-written user's manual, a confusing display, an LCD screen with poor contrast, rechargeable batteries that go flat after an hour of use or a mains adapter that must be carried separately, can seriously spoil your pleasure with your brand-new purchase.)

DIY. Obviously, making some of your own test or service equipment provides a good opportunity to practice basic skills and to make something useful in the bargain. Certain suggestions can be found in this chapter and in Chapter 5, as well as in Volume 3.

Second-hand equipment. The used-equipment market is worth looking at, particularly for high-grade equipment. It is possible to acquire older-model equipment, manufactured by leading suppliers, at entirely affordable prices. The specifications of such equipment are usually significantly better than those of new equipment at the same price, although older equipment is often bigger (and heavier) and not always as easy to use.

4.1.2 Basic specifications

The specifications of a piece of test or measuring equipment are listed on its data sheet, which usually accompanies the user's manual. The main specifications can also be found in brochures and catalogues. Manufacturers of such equipment (and their representatives) will supply detailed information on request, and can also provide additional advice. We can thus limit our attention to the most basic features.

Measurement ranges. Any measuring instrument can only evaluate the quantity to be measured (whether it be current, voltage, resistance or whatever) if it exceeds a certain minimum value. At the other end of the scale, there is an upper limit or maximum value; input levels which exceed this value also cannot be measured and may result in damage to the instrument. The range between the minimum and maximum measurable values is called the *measurement range*. An instrument may have (1) only a single, fixed measurement range, (2) multiple ranges that are automatically selected (*autoranging*) or (3) multiple ranges that are manually selected.

Frequency limits and bandwidth. Whenever we make measurements on signals that change value during the course of the measurement (such as AC voltages or currents, pulse trains and the like) we can only expect satisfactory results if the instrument is able to follow the changes in the signal. This characteristic is described by the *cutoff frequency* specification. In some cases both upper and lower frequency limits are specified (which means that a *bandwidth* is specified), while in other cases only an upper frequency limit is specified. In the latter case the instrument is said to be *DC-coupled*, since the instrument can correctly process signals that are absolutely static.

Note: with equipment for measuring analogue quantities (such as multimeters and oscilloscopes), the upper cutoff frequency refers to the maximum acceptable frequency of a sinusoidal input signal, regardless of whether the internal operation of the equipment in question is analogue or digital. With purely 'digital' equipment (such as pulse generators and logic analyzers) the upper cutoff frequency refers instead to the maximum pulse repetition frequency that the equipment can handle.

Input resistance, and loading the device under test. Connecting test or measurement equipment to a test object should not affect its operation. This effectively means that:

- If a *voltage* is measured or checked, no current should flow through the instrument; in other words, the input impedance of the measuring equipment must be as high as possible (ideally infinite). If a *current* is measured or checked, no voltage drop should be caused by the instrument; in other words, the insertion resistance of the measuring equipment must be as low as possible (ideally zero).
- When signal levels are checked (for example, logic levels) the effect of the instrument must be negligibly small.
- When precise measurements are to be made, errors due to the finite input resistance or non-zero insertion resistance of the instrument may need to be taken into account.

In practice, the vast majority of checks and measurements are effectively voltage measurements (the instrument is connected 'in parallel' to the device under test and not 'in series' in a current path). Modern instruments have input resistances in the megohm region and thus usually do not cause significant errors in the measured values.

Note: sometimes an extremely high input resistance can lead to measurement errors, since the measurement may be corrupted by coupled-in interference. In such cases a supplementary load may be necessary.

Input impedance. The input impedance is often specified as a combination of a resistance and a capacitance (such as 1 MΩ || 5 pF). The capacitive load cannot be ignored at high frequencies. (*For example:* 5 pF represents an additional IC input load , and the effective impedance of a 5 pF capacitor is around 3 kΩ at 10 MHz.) This is especially significant for signal tracing with an oscilloscope or logic analyzer. In particular, problems can occur with time-critical signals that are connected to many inputs, as follows:

1 The circuit being tested may not work at all when the instrument is connected, due to the additional loading.

2 The circuit being tested may work all too well when the instrument is connected, due to the additional loading (*for example:* the fault disappears when a measurement is made, because the additional capacitance reduces the amplitude of noise pulses).

Accuracy. The accuracy specification indicates how much the measured value differs from the actual value of the measured quantity, or how much the indicated value is allowed to differ from the actual value. *For example:* if a digital voltmeter has a specified accuracy of ±1% and we read a value of 5.00 V, then the actual value can lie between 4.95 V and 5.05 V. Accuracy is characterized in terms of the applicable *measurement error* (see Section 4.1.3).

Ambient conditions. The specified performance of the equipment is only guaranteed if it is operated under the specified ambient conditions. Table 4.1 shows an example of typical ambient condition specifications.

Ambient temperature	Relative humidity
Operation	
standard: 15 to 35 °C [1),2)]	typical values: < 65 % at 30 °C [2),3)]
basic operational range: 0 to 50 °C [4)]	90% at 30 °C
	75% at 45 °C
	45% at 50 °C
Storage	
standard: −15 to +50 °C	< 95% (non-condensing)

Table 4.1 *Typical ambient condition specifications for test and measurement equipment (typical values taken from relevant data sheets). See the text for the notes.*

Notes:

1 Some equipment achieves its specified accuracy only within a limited range of operating conditions (such as an ambient temperature of 18–28 °C). High accuracy over an extended temperature range requires temperature-compensation circuitry, which adds to the cost of the equipment.

2 Ideal (measurement laboratory) operating conditions are 20 °C ± 0.5 °C at 58% relative humidity.

3 Some equipment can be both stored and operated at up to 90% relative humidity. Operation under conditions of high relatively humidity requires additional expenditures on insulation and encapsulation. *Note:* bear in mind that even if the measurement equipment can tolerate high relative humidity conditions, such conditions may damage the test object!

4 The equipment will not fail at temperatures below 0 °C or above 50 °C, but it may not maintain the specified accuracy.

Practical tips:

1 Allow equipment to acclimatize for 30 minutes (nominal) after it is brought into a room from outside (especially from colder conditions).

2 Allow equipment to warm up before attempting to make exact measurements (nominally 15–30 minutes; precision equipment may need even longer).

3 For long-term storage, pack the equipment appropriately (protected against dust and moisture), and include sacks of silica-gel desiccant. Remove batteries if necessary!

Requirements for practical testing. Table 4.2 lists the electrical quantities to be measured and summarizes the typical values that are encountered in modern hardware.

Notes:

1 Battery voltages are often multiples of the individual cell voltage (2.4 V, 4.8 V, 7.2 V etc.).

2 Supply currents may be only a few milliampères in the 'power-saving' operating mode.

3 The highest pulse rate for video signals is one half of the pixel clock rate (representing a sequence of alternating 'on' and 'off' pixels).

4 The specified values relate to typical logic signals internal to the equipment. The higher the pulse rate, the lower the allowable capacitive loading; the lower the drive capacity, the lower the allowable current load.

5 Some processor ICs require closely-regulated supply voltages (for example, ±2%); for some analogue assemblies the required accuracy is significantly higher.

Item	Characteristic values
supply voltages [1]	+5 V, +12 V, –12 V, –5 V, +3.3 V; +1.8 to +3.6 V (processors)
supply currents	less than 1 mA to roughly 20 mA [2]
special voltages	• LCD modules: ≈ 40 V • CCFL backlights: ≈ 1.5 kV • EL backlights: ≈ 150 V • CRTs: ≈ 25 kV
mains voltage	230 V~ (U_{pp} ≈ 650 V), < 10 A (switch-on currents may be higher)
logic levels	LO: 0 V to 0.8 V HI: 2 V to 5 V
signal edges	rise times ≥ 2 ns, slopes up to 4 V/ns
pulse repetition rates	≥ 66 MHz (processor bus clocks) > 100 MHz (video signals) [3]
allowable signal loading [4]	capacitive: 5–20 pF current: < 100 µA
allowable tolerances	±5 V to ±10% (supply voltages and logic levels) [5]

***Table 4.2** Summary of characteristic electrical values for modern PC hardware. See the text for the notes.*

4.1.3 Measurement error, sensitivity and resolution

Measurement error

The measurement error is determined by the difference between the measured (indicated) value and the actual (error-free) value of the quantity being measured. We use the following notation:

- X_m is the measured (indicated) value (*for example:* 4.8 V)
- X_a is the actual (real) value (*for example:* 5.1 V)
- X_{fs} is the maximum (full-scale) value of the measurement range (*for example:* 6.0 V)

The following error specifications are commonly used:

- the *absolute error* E_a is the difference between the measured and actual values:

  $E_a = X_m - X_a$ (*for example:* E_a = 4.8 V – 5.1 V = – 0.3 V)

- the *relative error* E_r is the ratio of the absolute error to the actual value:

$$E_r = \frac{X_m - X_a}{X_a} \qquad \left(\textit{for example: } E_r = \frac{4.8\,\text{V} - 5.1\,\text{V}}{5.1\,\text{V}} = -0.0589\right)$$

- the percentage relative error E_{rp} (also referred to as the 'reading error' or the 'measured-value error') is:

$$E_{rp} = E_r \cdot 100\% \qquad (\textit{for example: } E_{rp} = -0.0589 \cdot 100\% = -5.89\%)$$

- the relative full-scale *display error* E_{fr} is the ratio of the absolute error to the full-scale range value:

$$E_{fr} = \frac{X_m - X_a}{X_{fs}} \qquad \left(\textit{for example: } E_{fr} = \frac{4.8\,\text{V} - 5.1\,\text{V}}{6.0\,\text{V}} = -0.05\right)$$

- the percentage display error E_{frp} (also referred to as the 'full-scale error') is:

$$E_{frp} = E_{fr} \cdot 100\% \qquad (\textit{for example: } E_{frp} = -0.05 \cdot 100\% = -5\%)$$

Measurement error and instrument error. The measurement error is the actual difference between the indicated and true values when a measurement is made. It is a signed value; the measurement error is positive if the reading is greater than the true value and negative if the reading is less than the true value. The instrument error, by contrast, specifies the worst-case inaccuracy of the instrument itself. It can also be positive or negative. In many cases the positive and negative instrument errors have the same absolute value, so that the instrument error is stated without a sign or as a '±' value. If the positive and negative errors are not equal, then both values are always stated (for example, +0.1%/–0.05%). *Note:* the accuracy specification is usually valid only for certain specified ambient conditions (see for example Table 4.1).

Accuracy specifications of analogue meters. According to the VDE 0410 standard, meter accuracy is specified with respect to the maximum range value or the full-scale value and is stated as a percentage (the specified value is thus the percentage display error, E_{frp}). *Note:* analogue meters are grouped into accuracy grades according to the magnitude of the error (see Section 4.2.1).

Accuracy specifications of digital meters. Digital meters are fundamentally subject to an error of ±1 count in the least-significant digit position (digitization error, specified as '±1 digit'). Some devices have even larger digitization errors.

The causes of errors in digital instruments are the following:

- quantization or digitization errors in the acquisition of the analogue quantity (analogue-to-digital conversion errors),
- display rounding errors ('intermediate' values cannot be displayed, in contrast to an analogue meter which can be read 'between' the scale divisions, so rounding of the last

display digit must be taken into account). *For example:* a measured voltage between 100.0 V and 100.1 V will be displayed either as '100.0' or as '100.1',

- sampling and synchronization errors in acquiring digital signals (with logic analyzers, universal counters and the like – see Section 4.4.1).

An example of a complete accuracy specification is:

$E_{max} = \pm(0.05\%$ of reading + 0.1% of range + 1 digit$)$

Note: the accuracy of digital instruments is often characterized by a single percentage value. This usually refers to the reading (the measured value).

Practical tip: the full-scale range plays a role in the accuracy specification for every measuring instrument (including digital instruments: it makes a difference whether one least-significant digit represents 0.1 mV or 10 mV). The associated absolute error represents a larger relative error for readings near the lower end of the measurement range than for readings near the upper end of the range. You should always try to utilize the full measurement range! (This means selecting the range so that the reading lies as close as possible to the full-scale value.)

Sensitivity

A measuring instrument is said to be *sensitive* if a small change in the quantity being measured causes a relatively large change in the indicated value. The sensitivity is defined as the ratio of the change in the display indication to the change in the measured quantity:

$$S = \frac{\Delta I}{\Delta X} \quad \text{(analogue instrument)},$$

$$S = \frac{\Delta N}{\Delta X} \quad \text{(digital instrument)}.$$

where S is the sensitivity, ΔI is the change in the analogue display (deflection angle, scale divisions, bar length), ΔN is the numerical change in a digital display and ΔX is the change in the measured quantity.

Resolution

Resolution is defined as the smallest change in the measured quantity which results in an observable change in the indication of the measuring instrument:

- analogue display: the resolution is the amount by which the measured quantity must change to cause the display to change by one scale division,

- digital display: the resolution is the amount by which the measured quantity must change to cause the display to change by ±1 count in the least significant digit position.

Notes:

1 Measuring equipment with high sensitivity or high resolution does not necessarily also provide high accuracy, since the displayed value will change with every minute variation in the value of the measured quantity. However, a highly-sensitive or high-resolution instrument is the right choice for detecting very small changes in the measured quantity (for example, in comparison measurements).

2 The resolution is frequently greater than the accuracy. *For example:* consider a digital multimeter with a measurement range of 0–400 mV, a specified resolution of 10 μV and a specified accuracy of ±0.1% +1 digit. Since 0.1% of 400 mV is 400 μV, the resolution is clearly better than the accuracy. If we have to make an adjustment by (for example) turning a potentiometer shaft, the high resolution is useful since we can easily observe very small changes in the measured voltage. However, the accuracy of the voltage measurement is not as good as the resolution with which it is displayed. In this case the measured value appears to be more accurate than it actually is.

Selecting equipment

Accurate, high-resolution equipment is correspondingly expensive. A moderate level of accuracy is usually sufficient, since most of the measurements that we need to make are for checking purposes.

Accuracy guidelines:

1 'Control' measurements: the allowed measurement error (inaccuracy) is one half of the allowed deviation of the measured quantity from its nominal value.

2 Precise measurements: the allowed measurement error (inaccuracy) is one tenth of the allowed deviation of the measured quantity from its nominal value (the measurement error should be an order of magnitude less than the allowed tolerance).

Table 4.3 shows how the minimum required measurement accuracy and resolution can be computed, given a known tolerance for the value of the measured quantity (as an absolute value x or a percentage p).

Note: For the required resolution, the symbol '≤' should be read as 'as good as or better than'. If alignment or comparison work is contemplated, an even better resolution (such as $\leq 0.1 \cdot x$) is advantageous.

Accuracy guidelines for 'control' measurements:

1 1.5% of the measured value is more than adequate (inexpensive digital multimeters),

2 2.5% of the full-scale value is usually adequate (analogue meters in the 2.5% accuracy class and most inexpensive analogue instruments)

Equipment specifications	Quantity X with specified tolerance	
	absolute: $X \pm x$ *)	in percentage: $X \pm p$ % *)
Resolution [1]	$\leq x$	$\leq \frac{X \cdot p}{100}$
Control measurements		
Absolute measurement error	$\leq 0.5\,x$	$\leq \frac{X \cdot p}{200}$
Relative measurement error in percent	$\leq 50\,\frac{x}{X}$ %	$\leq 0.5\,p$ %
Exact measurements		
Absolute measurement error	$\leq 0.1\,x$	$\leq \frac{X \cdot p}{1000}$
Relative measurement error in percent	$\leq 10\,\frac{x}{X}$ %	$\leq 0.1\,p$ %
*: Conversions: $x = \frac{X \cdot p}{100}$; $p = \frac{x}{X} \cdot 100$ % Note 1: see text		

Table 4.3 Minimum resolution and accuracy requirements.

An example of a precise measurement: suppose that the specified value of the quantity to be measured is 4.8 V, with a tolerance of ±0.2 V. This means that the instrument should display the measured value with a resolution of at least 0.2 V and an accuracy of ±0.02 V. Since 0.02 V is around 0.4% of 4.8 V, an instrument with 0.25% or 0.1% accuracy would be fully satisfactory. But what about the resolution (how many digits in the display)? A 3½-digit multimeter has an additional display error of at least one least-significant digit, which corresponds to an error of 0.01 V or 0.2% in round numbers. (Refer to Figure 4.6 and Table 4.5 for an explanation of display specifications for digital meters.) You would thus be completely on the safe side with a basic accuracy of 0.1%. An instrument with 0.25% accuracy would be just barely adequate (0.25% + 0.2% = 0.45%, compared to the required 0.4%, but most instruments are in practice better than their specifications). For comparison, the least-significant-digit error of a 4½- digit display would be 0.001 V or 0.02%.

Errors due to the measurement setup

In practice, measurement accuracy depends not only on the inherent accuracy of the instrument but also on the way that the measurement is made. The following factors contribute to the total measurement error:

- an incorrectly-selected measurement range,
- insufficient input resistance (for voltage measurements) or excessive insertion resistance (for current measurements),
- excessively long probe leads,
- poor connections,
- interference,
- earth loops,
- thermocouple potentials,
- temperature effects.

Calibration versus certification

The accuracy that the manufacturer has 'built in' to a piece of measurement equipment does not last forever. Component values can change over time, factory alignments can drift and parts can simply fail. (It is particularly annoying when some part fails in a way that does not have any evident effect on the device's operation. You can still use the equipment in this case, but the measurements will be incorrect.) It is therefore necessary to pay attention to the following:

- All measurement equipment should be checked periodically and/or immediately prior to being used (even if this is not required as a matter of policy).
- High-accuracy measurement equipment should be calibrated from time to time (for example, every two years). (*Note:* do not confuse calibration with *certification*, which is sometimes used to mean the same thing. Properly speaking, certification is a form of supervised calibration that is performed with reference to official standards. This is not really necessary for everyday work.)

Practical tips:

1 Normally, a few control measurements whose expected results are well known are sufficient for checking ordinary utility and service equipment (*for example:* measuring the output voltage of a regulated power supply or the voltage of a fresh battery, or measuring the current through a known resistance with a known voltage).

2 A very simple check can be made by making the same measurement with a second instrument (which need not be identical to the one being checked).

3 If you want to accurately calibrate your equipment, suitable devices are commercially available (for example, voltage and current reference standards with accuracies of 0.01%), and calibration standards for 'household' use can be quite easily constructed using reference voltage sources, precision resistors and the like (see Section 4.3.3, Section 5.1.3 and Volume 3).

4 Manufacturers and vendors of high-grade measurement equipment usually offer a calibration service.

5 Always use good common sense when making measurements. Do not just blindly accept whatever the instrument displays! (This applies in particular to digital displays, which the inexperienced user is often inclined to regard as 100% accurate.)

4.2 Measuring fundamental electrical quantities

4.2.1 Measurement equipment for voltage and current

Analogue measuring instruments

Analogue measuring instruments are transducers that directly exploit physical effects to produce observable indications of electrical quantities. For our purposes, there are only three basic types that are important: moving-coil meters, moving-iron meters and cathode-ray tubes. The first two types belong to the class of moving-pointer instruments, while the last type – in the form of the oscilloscope – forms a class on its own, which is treated in Section 4.3. In terms of their physical working principles, moving-coil and moving-iron meters measure current, while cathode-ray tubes measure voltage.

The scale of a moving-pointer instrument. Figure 4.1 shows the scale of a typical moving-pointer instrument.

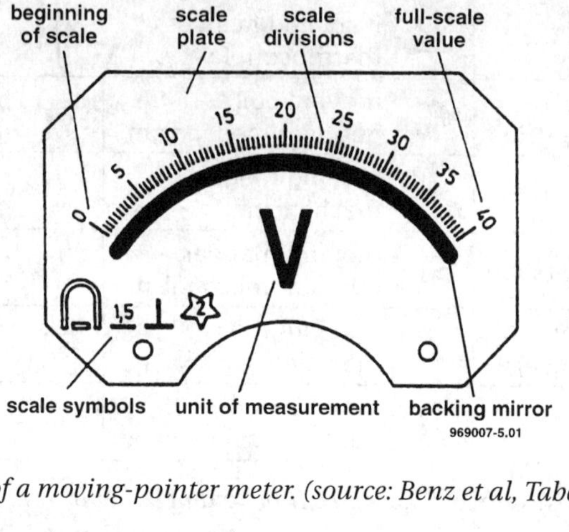

Figure 4.1 The scale of a moving-pointer meter. (source: Benz et al, Tabellenbuch Elektronik)

Notes:

1 The scale divisions allow the pointer position to be read. The scale may be linear (as in the illustration, where the separation of any two scale divisions is constant) or non-linear (for example, logarithmic).

2 A backing mirror is common in high-accuracy meters. It helps prevent reading errors.

3 The scale symbols provide information about the meter. They are standardized according to VDE 0410 (see Table 4.4).

—	for DC current		indicates a separate external shunt resistor		iron-free electrodynamic mechanism
≂	for AC and DC currents		indicates a separate external series resistor		iron-free electrodynamic quotient mechanism
∼	for AC current	○	magnetic screen (iron screen)		iron-screened electrodynamic mechanism
	for 3-phase current with 1 mechanism		electrostatic screen		iron-screened electrodynamic quotient mechanism
	for 3-phase current with 2 mechanisms	AST	antistatic mechanism		induction mechanism
	for 3-phase current with 3 mechanisms	⚠	*Caution!* (note the usage instructions!)		quotient induction mechanism
1,5	accuracy grade, referred to maximum range value		moving-coil mechanism		hot-wire mechanism
1,5	accuracy grade, referred to scale length or plotting width		as rectifier heated thermocouple insulated heated thermocouple	}	supplementary to the moving-coil symbol
1,5	accuracy grade, referred to true value		moving-coil quotient mechanism		bimetallic mechanism
⊥	vertical orientation		moving-magnet mechanism		electrostatic mechanism
⊓	horizontal orientation		moving-magnet quotient mechanism		vibrating-reed mechanism
∠60°	tilted orientation (at specified angle)		moving-iron mechanism		with internal amplifier
☆2	test voltage		moving-iron quotient mechanism		

***Table 4.4** Scale symbols (source: Benz et al, Tabellenbuch Elektronik)*

Accuracy grades. Moving-pointer instruments are graded according to their accuracy; the grades are defined in the VDE 0410 standard. The classification is based on the maximum error as a percentage of the full-scale value. The following grades are defined: 0.1, 0.2, 0.5, 1, 1.5, 2.5, 5. Meters in the 0.1, 0.2 and 0.5 grades are precision instruments, while those in the remaining grades are designated as 'utility' instruments. Meters with even lower accuracy are considered to be merely indicators and are not classified.

For example: a meter in the '2.5' class has a maximum allowed error of 2.5% of the full-scale value. If the full-scale value is 10 V, then every measurement reading can be in error by up to ±0.25 V. If we read exactly 3.3 V, the true voltage lies somewhere in the range between 3.05 V and 3.55 V (and in this case the 'actual' error can be up to 7.6% of the measured value!).

Note: the specified accuracy is only guaranteed when the instrument is used in the specified orientation (nominal position) - see the relevant symbols in Table 4.4.

Reading a meter scale. In order to read the pointer position correctly, you must view it straight on (perpendicular to the scale plate) and not from the side (in other words, viewing parallax must be avoided – see Figure 4.2). A backing mirror (if present) aids in avoiding parallax errors: if the pointer and its mirror image coincide, then there is no viewing parallax. If the pointer is not located exactly above a scale division, the intermediate value must be estimated.

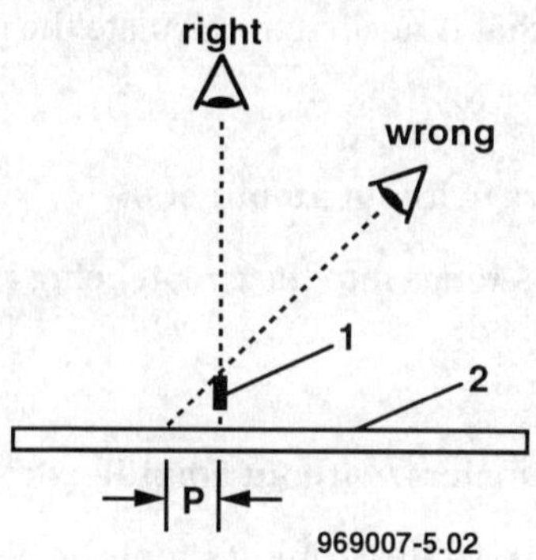

Figure 4.2 How to read a scale. 1 – pointer, 2 – scale plate, P – parallax.

Reading a multimeter. It's all a matter of practice:

1 Place the meter in the prescribed orientation (if it is specified for vertical operation, then tilting it or placing it horizontally reduces its accuracy). The pointer must point exactly at the zero mark before the measurement is made (adjust this if necessary).

2 Make sure you have the right polarity. If you are unsure of the magnitude of the quantity to be measured, always start with the highest available range.

3 Read the indicated value as previously described (avoiding viewing parallax).

4 The scale divisions and/or labeling do not always correspond directly to the measurement range (that is, several ranges may use the same scale). You thus typically have several values to deal with: (1) the reading and (2) the full-scale value (both according to the markings on the scale) plus (3) the maximum value of the measurement range (according to position of the range selector switch). You can calculate the measured value as follows:

$$\text{measured value} = \text{reading} \cdot \frac{\text{maximum range value}}{\text{full-scale value}}$$

For example: the full-scale value is 30, the measurement range is 6 V and the reading is 23. The measured value is:

$$\text{measured value} = 23 \cdot (6 \div 30) = 4.6\,\text{V}.$$

Notes:

1 The ratio of the maximum range value to the full-scale value is a constant *scale factor* for any given measurement range, so that it has to be calculated only once (in the above example, the scale factor is 6 ÷ 30 = 0.2).

2 The full-scale and maximum-range values of modern analogue multimeters are selected to simplify calculating the scale factors. *For example:* there may be two scales with full-scale values of 30 and 120, and the measurement ranges have maximum values of 300 mV, 1.2 V, 12 V, 30 V and so on. The scale factors in this case are multiples or submultiples of ten, so that all that is needed to calculate the measured value is to correctly place the decimal point.

Advantages of analogue meters (without amplifiers):

1 They do not inject any interference into the circuit being checked.

2 They do not require a power supply.

Problem areas with analogue meters (without amplifiers):

1 In many cases, the internal resistance cannot simply be neglected.

2 The accuracy near the beginning of the scale is relatively low.

3 The mechanical movement of the pointer mechanism is subject to friction, which also reduces the accuracy. *Practical tips:* (1) tap the face of the meter lightly before making a reading (especially for small pointer deflections), (2) meters with taut-band suspensions are better than pivot-type meters.

4 The movement of the pointer mechanism creates a reverse induction potential (counter-emf) that can upset sensitive circuits. (*Remedy:* connect a capacitor across the meter's terminals. The size of the capacitor must be determined empirically; try various values until the circuit behaves practically the same with the meter connected and not connected.)

Moving-coil meters. Moving-coil meters are among the most sensitive types of mechanisms (analogue multimeters also employ such meters). Figure 4.3 shows the construction of a moving-coil meter. The operating principle is as follows: the current to be measured flows through a coil of wire that is mounted in the field of a strong permanent magnet such that it can rotate. The magnetic field produced by the current flowing through the coil is either attracted to or repelled by the field of the permanent magnet, depending on the direction of current flow. The coil thus rotates and causes a deflection of the attached pointer. Moving-coil meters can be used only with direct currents. The direction of the pointer deflection depends on the direction in which the current flows (in other words, on the polarization of the mechanism). Some meters have the zero point in the middle of the scale, so that the pointer can deflect in both directions.

Sensitive, mass-produced meters have typical measurement ranges of 50 µA, 100 µA or 1 mA (full-scale). However, meters with full-scale deflections of 10 µA and 20 µA are also manufactured.

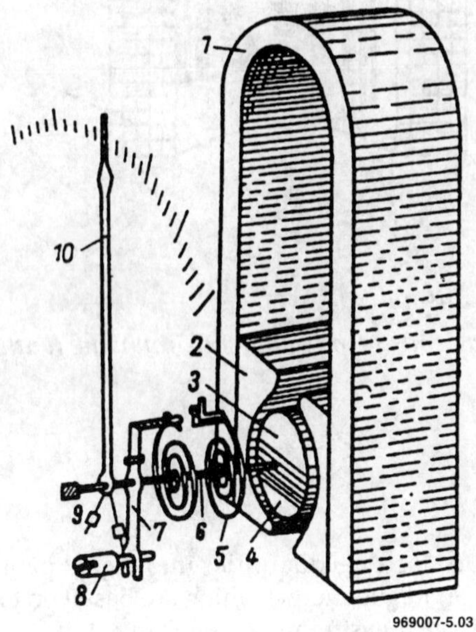

Figure 4.3 Moving-coil mechanism.

1 – permanent magnet, 2 – pole piece, 3 – fixed iron core, 4 – moving coil (wound on an aluminium frame), 5 – spiral springs (which also conduct the current to and from the coil), 6 – shaft, 7 – guide strut, 8 – eccentric screw for adjusting the zero point, 9 – pointer counterweights, 10 – pointer.

Moving-iron meters. A moving-iron meter does not have a permanent magnet. Instead, the magnetic field produced by a current-carrying coil acts on a soft-iron armature (the simplest conceptual model consists of a thin rod of soft iron, suspended from a spring, that is pulled into a cylindrical coil by the magnetic field). Figure 4.4 shows the basic construction. The pointer deflection of such a meter is independent of the direction of current flow, so that it can be used without a rectifier for measuring alternating currents. Moving-iron meters are more robust and less costly than moving-coil meters, but not as sensitive. They are mainly used as dashboard instruments.

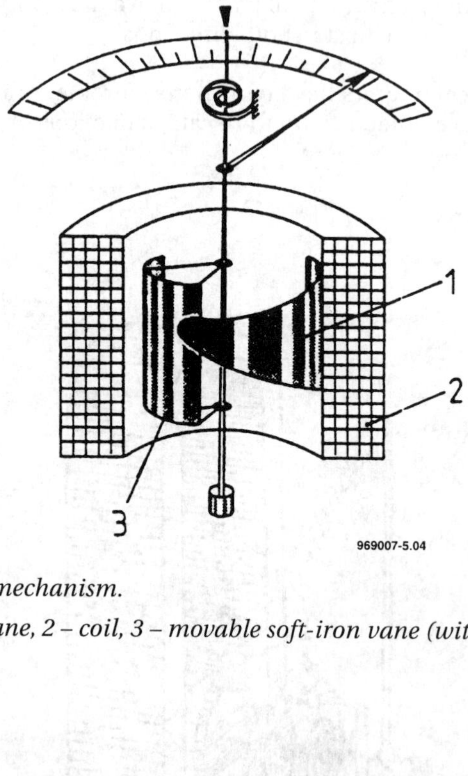

Figure 4.4 Moving-iron mechanism.

1 – stationary soft-iron vane, 2 – coil, 3 – movable soft-iron vane (with a pointer attached to its shaft)

Digital meters

Digital meters convert analogue electrical quantities into digital data that are displayed in numerical form. In contrast to analogue meters which are based on current measurement, digital meters are based on voltage measurement (see Figure 4.5).

A digital meter consists in principle of nothing more than an analogue-to-digital converter and a display. Figure 4.5 shows one of the simplest conversion methods. The voltage to be measured is directly compared to an internally-generated test voltage that is produced in discrete steps by the combination of a counter and a digital-to-analogue converter. As soon as the test voltage is equal to or greater than (more positive than) the voltage being measured, the comparator output switches and the counter is stopped. The content of the counter then represents the value of the measured voltage. After the content of the

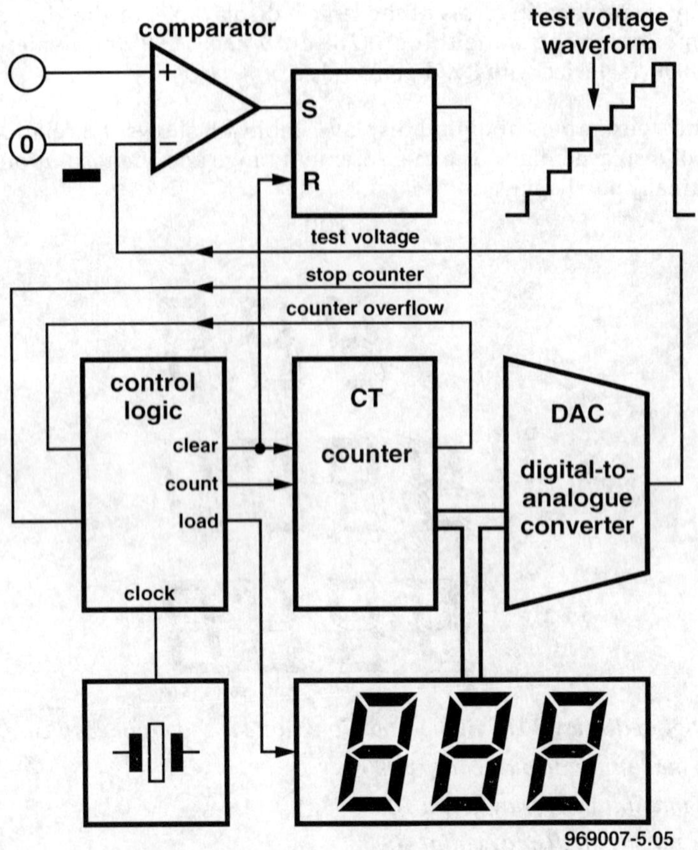

Figure 4.5 Block diagram of a digital voltmeter (greatly simplified).

counter has been transferred to the display, the control logic starts the conversion process again from the beginning. In practice, the dual-slope conversion technique is normally used instead, due to its superior accuracy. Digital measurements of DC voltages can be made with accuracies on the order of ± 0.005 to 0.02% using the dual-slope conversion technique (inexpensive service instruments have accuracies on the order of ± 0.1 to 0.5%). A display error of at least ±1 digit must of course be added to all of these values.

The number of displayed digits. There are two common ways in which the number of displayed digits is specified:

1 The display is specified as '3½ digits', '3¾ digits', '4½ digits' or the like. The expression '*N*½ digits' means that there are *N* digit positions that can display the numerals 0 through 9, plus one most-significant digit position that can only display 0 and 1. Some meters can display a larger range of numerals in the most significant position (for example, 0 through 4); the designation '*N*¾' is commonly applied to such devices.

2 The display is specified in terms of the largest displayable, or the 'display size in digits' (which amounts to the same thing). (*For example:* '1999' (equivalent to a 3½-digit device), '4000' (a device with 3¾ digits) and so on.).

Figure 4.6 shows examples of digital displays. Table 4.5 shows the relationship between the number of displayed digits, the measurement range and the resolution. *Note:* leading zeros are normally not displayed.

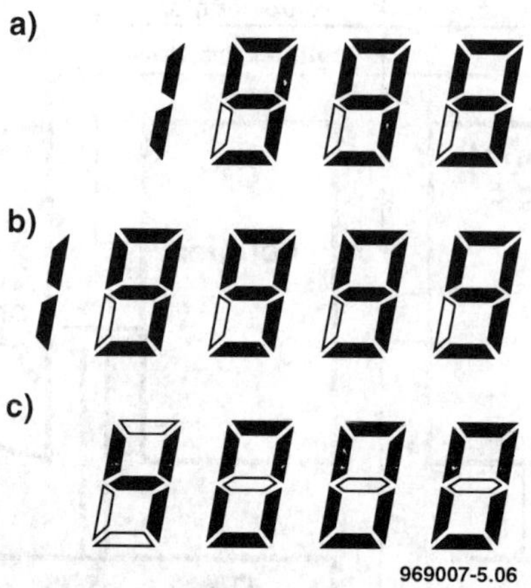

Figure 4.6 *Digital displays. The highest possible value is shown in each case.*
a) 3½ digits, maximum display count 1999
b) 4½ digits, maximum display count 19999
c) 3¾ digits, maximum display count 4000

For example: consider a 20-V measurement range. A 3½-digit instrument can display a maximum of 19.99 V, while a 4½-digit instrument can display 19.999 V. In this range, the resolution of the least-significant digit corresponds to 0.01 V (10 mV) for the 3½-digit instrument and to 0.001 V (1 mV) for the 4½-digit instrument. This amount is the *display error*, which must be added to the relative measurement error. The specified display error of some devices is as much as two or three times this amount (that is, 2 or 3 digits).

Note: a 3¾-digit meter does not have a higher display resolution than a 3½-digit meter. Instead, its higher inherent resolution is 'spread out' over a larger measurement range. *For example:* a voltage of 300 mV can be read with a resolution of only 1 mV using a 3½-digit meter (since the 2 V range must be used), while a resolution of 0.1 mV is possible with a 3¾-digit meter since the 400 mV range can be used.

Display range: 1999 – 3½ digits –		
Measurement range	**Maximum value**	**Resolution (1 digit)**
200 mV	199.9	0.1 mV
2 V	1.999	1 mV
20 V	19.99	10 mV
200 V	199.9	100 mV
Display range: 19999 – 4½ digits –		
Measurement range	**Maximum value**	**Resolution (1 digit)**
200 mV	199.99	0.01 mV
2 V	1.9999	0.1 mV
20 V	19.999	1 mV
200 V	199.99	10 mV
Display range: 4000 – 3¾ digits –		
Measurement range	**Maximum value**	**Resolution (1 digit)**
400 mV	400.0	0.1 mV
4 V	4.000	1 mV
40 V	40.00	10 mV
400 V	400.0	100 mV

Table 4.5 Digital displays and measurement ranges (examples).

4.2.2 Voltage measurements

A voltmeter is always connected in parallel to the voltage to be measured (see Figure 4.7a). Ideally, no current should flow through the voltmeter; in other words, the voltmeter should have an infinite internal resistance.

Analogue voltage measurements. Analogue meters are basically current-measuring devices. A voltage is thus determined indirectly by measuring the current that flows through a series resistor (see Figure 4.7b). Analogue meters that are supplied as 'voltmeters' have built-in series resistors.

Digital voltage measurements. Digital meters are by nature voltage-measuring devices.

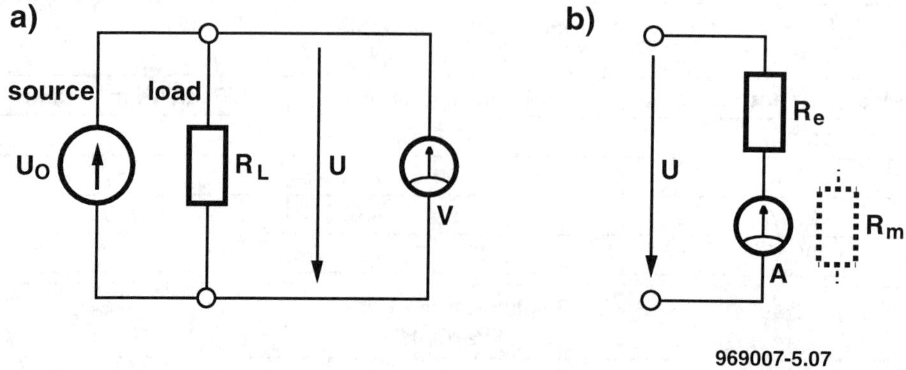

Figure 4.7 Voltage measurement.
a) basic circuit, b) using a current meter to measure voltage.

Relationships:

$$R_e = \frac{U_{Rm}}{I_{fs}} - R_m = U_{Rm} \cdot R_{iV} - R_m$$

R_m = internal (DC) resistance of the meter
U_{Rm} = desired maximum measured voltage at full-scale deflection
I_{fs} = full-scale current (rated meter current)
R_{iV} = instrument sensitivity (in kΩ /V)

Typical internal resistance values for moving-coil meters:

100 mA: R_m = 500 Ω to 3 kΩ; R_{iV} = 10 kΩ /V
1 mA: R_m = 75 Ω to 150 Ω; R_{iV} = 1 kΩ /V

Extending the measurement range. The measurement range of a voltmeter can be extended by adding an external series resistor (Figure 4.8a) or by using a voltage divider. A selector switch can be used to provide several ranges (see Figure 4.8b–d).

Internal resistance. The internal resistance of a practical instrument cannot be infinite. For analogue meters (without an external amplifier), the internal resistance is specified in the form of the 'specific' resistance R_{iV}, which is the ratio of the series resistance to the applied voltage that produces a full-scale indication. This is commonly referred to as the *sensitivity* of the instrument and is specified as 'kΩ /V' (kilo-ohms per volt). Good-quality meters for DC measurements have sensitivities ranging from 20 to 100 kΩ /V. If the meter specification is 20 kΩ / V then the internal resistance is 20 kΩ for a 1-V measurement range, 100 kΩ for a 5-V range and so on.

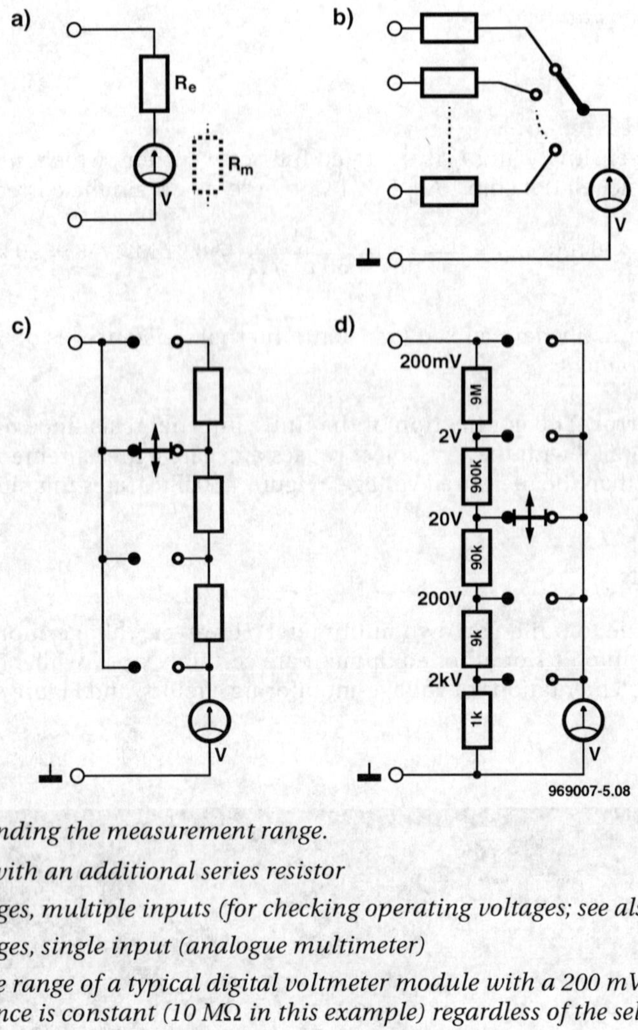

Figure 4.8 Extending the measurement range.

a) a voltmeter with an additional series resistor

b) multiple ranges, multiple inputs (for checking operating voltages; see also Figure 4.10)

c) multiple ranges, single input (analogue multimeter)

d) extending the range of a typical digital voltmeter module with a 200 mV range. The input resistance is constant (10 MΩ in this example) regardless of the selected range.

Calculating the value of the external range resistor:

$$R_e = R_m \cdot \left[\frac{U_{Rm}}{U_{fs}} - 1 \right] \qquad (R_e \approx U_{Rm} \cdot R_{is})$$

R_e = external (series) range resistor
R_m = internal resistance of the voltmeter
U_{Rm} = desired maximum range value
U_{fs} = rated full-scale voltage of the voltmeter

Calculating the sensitivity:

$$R_{iV} = \frac{1}{I_{fs}}$$

where R_{iV} is the sensitivity and I_{fs} is the rated full-scale current, which must be represented according to the relationship 1 A = 1 V / 1 Ω in order to obtain the correct result.

For example: $I_{fs} = 50\ \mu A;\ R_{iV} = \dfrac{1}{50\ \mu A} = \dfrac{1\ \Omega}{50\ \mu V / \Omega} = 0.02\ M\Omega / V = 20\ k\Omega / V.$

Digital meters normally have a fixed (constant) internal resistance, as do analogue meters with internal amplifiers.

Measurement error. The connection of the finite internal resistance of the measuring instrument in parallel with the test object causes a negative measurement error (the true voltage is higher than the measured voltage). Figure 4.10 illustrates this situation.

Practical circuits

It's not worthwhile to build you own multimeter. However, voltage-monitoring circuits that can be built into PCs or other equipment are certainly worthwhile. Figure 4.9 shows two practical implementations of voltage-monitoring circuits, and Figure 4.11 shows three possible designs.

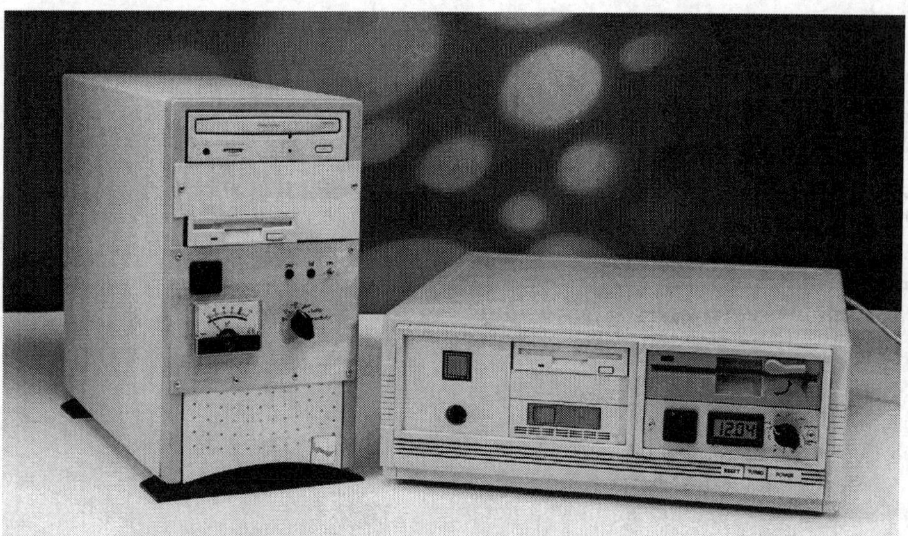

Figure 4.9 *PCs with retrofitted provisions for checking operating voltages. This illustration also shows an example of a decent mains switch (see Section 5.2.1).*

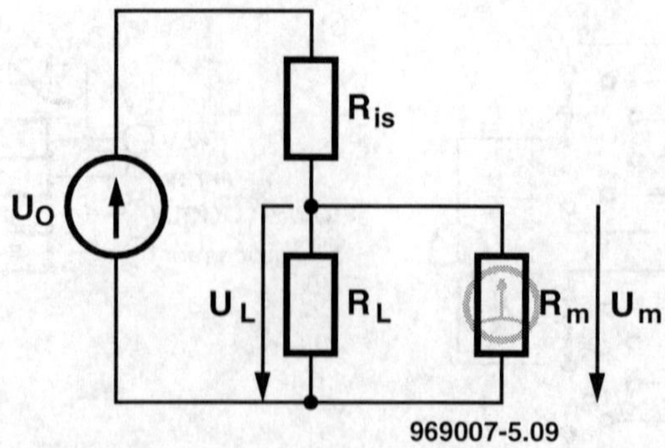

Figure 4.10 Voltage measurements under real conditions: the meter loads the voltage divider formed by the source resistance and the load resistance.

Voltage measurement error:

correct voltage:

$$U_L = U_O \cdot \frac{R_L}{R_{iS} + R_L}$$

measured voltage:

$$U_m = U_O \cdot \frac{R_L \parallel R_m}{R_{iS} + R_L \parallel R_m} = U_O \cdot \frac{R_L\, R_m}{R_{iS}\, R_L + R_{iS}\, R_m + R_L\, R_m}$$

relative measurement error: 1), 2)

$$E_r = \frac{U_m}{U_L} - 1 = \frac{R_{iS}\, R_L}{R_{iS}\, R_L + R_{iS}\, R_m + R_L\, R_m} \approx \frac{-\, R_{iS}\, R_m}{R_m\, (R_{iS} + R_L)}$$

1): The percentage error is: $E_{rp} = E_r \cdot 100\ \%$
2): The approximate formula can be used if $R_m \gg R_L$ (that is, if $R_m > 10^3 \cdot R_L$).

error correction:

$$U_L = \frac{U_m}{E_r + 1} = \frac{U_m}{1 - \dfrac{R_{iS}\, R_L}{R_{iS}\, R_L + R_{iS}\, R_m + R_L\, R_m}}$$

U_O = source voltage
R_{iS} = internal resistance of the voltage source
R_L = load resistance
U_L = (correct) voltage across the load resistor
R_m = internal resistance of the voltmeter
U_m = measured (indicated) voltage

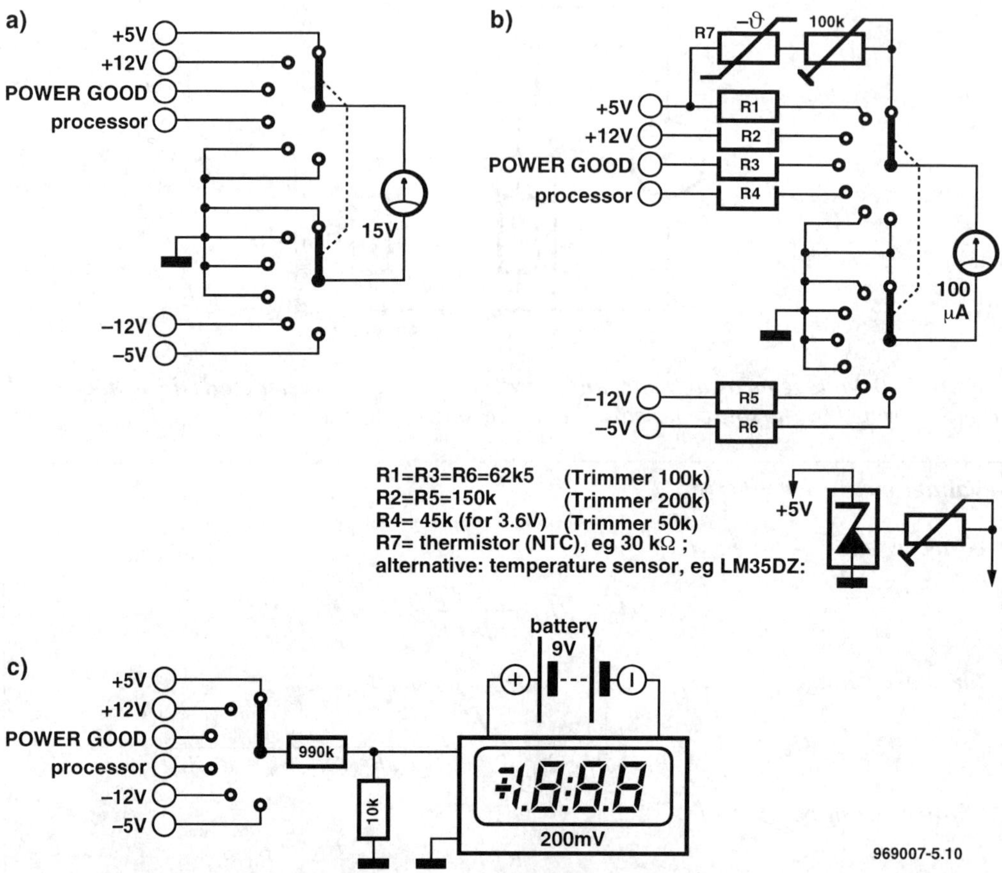

Figure 4.11 *Voltage-monitoring circuits for personal computers.*

a) This circuit is based on an analogue voltmeter (a moving-coil meter with a built-in series resistor).

The following voltages can be checked in the usual PC: +5 V, +12 V, –12 V and –5 V. If you are going to go to this much trouble, it makes sense to think about checking other signals as well. The sample circuit shows two possibilities: the POWER GOOD signal (from the power supply) and the separate processor supply voltage (which is very common these days). A two-pole selector switch is necessary, so that negative voltages can be connected to the voltmeter with the proper polarity. The meter circuit can be connected to the voltages to be checked via wires attached to the motherboard (for example). The disadvantage of this simple circuit is its relatively poor accuracy at low voltages. This applies in particular to the (sometimes tightly-specified) processor supply voltage, which ranges around 3 V and thus may not be measurable with sufficient accuracy on a 15-volt scale.

b) *This circuit is based on a 'pure' moving-coil ammeter which, in the interest of readability, should have a scale calibrated in percentage. The idea is that 80% of the meter's full-scale deflection should correspond to 100% of the voltage being measured (in other words, the full-scale current – in this case 100 µA – should correspond to 125% of the nominal value of the measured voltage). The series resistors can be calculated on this basis (the full scale value for 5 V is 6.25 V, for example) according to the sensitivity of the meter (in this case, 10 kΩ / V). A practical approach is to use variable resistors (high-precision cermet trimpots) that are set to the required values using a digital multimeter.*

The diagram also shows how this arrangement can be extended to monitor temperatures by means of a thermistor (NTC resistor) or a semiconductor temperature sensor. The sensor can be calibrated using a digital thermometer with an air-temperature or surface-temperature probe, depending on the location of the sensor. Allow the sensor to reach its normal operating temperature (or warm it up with a hair dryer) and adjust the series resistor so that the meter reading is 80% of full scale.

c) *This circuit is based on a digital panel meter module. Inexpensive digital panel meters are available with a measurement range of ±199.9 mV (3½-digit display) and an input resistance of 100 to 1000 MΩ. The measurement range must be extended to ±19.99 V via a 100:1 voltage divider in order to handle the voltages usually found in the PC. The advantage of this arrangement is that the input polarity does not have to be switched; the disadvantage is that the meter module must be supplied with power.*

Notes for Figure 4.10:

1 *Caution:* the voltage selector switch must always be a non-shorting type!

2 It goes without saying that these arrangements can be extended to include other monitoring functions (such as activity displays – refer to Figure 4.31).

3 As an alternative, the selector switch and a pair of test-probe jacks can be built into the equipment (that is, without the meter), so that an external multimeter can be connected to read the voltages. The selector switch need have only one pole, since digital multimeters (which are most often used nowadays) can display both positive and negative voltages.

4 The arrangement shown in Figure 4.10b can be made more handsome by colouring the region around the nominal value green (eg, from 75% to 85% of full scale). (There are meters available with exchangeable scale plates.)

5 Voltage dividers that are suitable for the arrangement shown in Figure 4.10c are commercially available (100:1 ratio, 1 MΩ input impedance and the necessary accuracy). Panel voltmeter modules are also available with a measurement range of ±19.99 V; these have built-in voltage dividers.

Power supplies for digital panel meters

It would seem obvious to dispense with the battery and utilize a supply voltage that is already available (such as +12 V). *Caution:* first check the specifications of the voltmeter module – some of them do not allow the supply earth to be connected to the measurement earth! Certain models, however, are explicitly designed to be powered from the same voltage as they (can) measure (the two earth terminals are either already connected together or may be connected together). The disadvantage of such an arrangement is that it is not possible to make any measurements at all if the supply voltage itself fails. Figure 4.12 shows an arrangement that can partially remedy this problem.

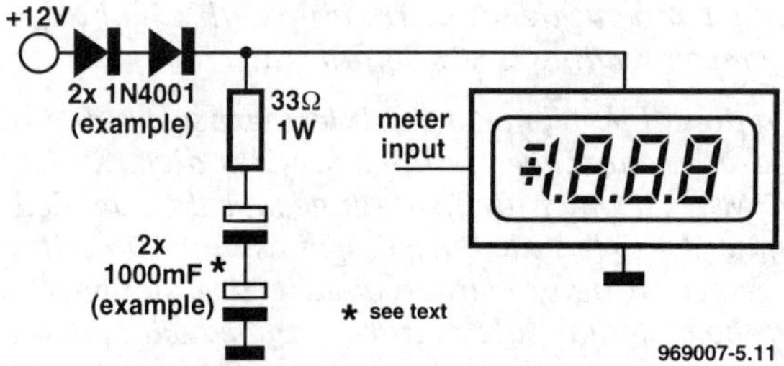

***Figure 4.12** Backup supply for a digital panel meter, using energy-storage capacitors.*

Relationships:

reserve operating time:

$$t = C \cdot \frac{U_i - U_f}{I_L}$$

U_i = initial voltage (≈ 11 V)
U_f = final voltage (minimum operating voltage; 7 V)
I_L = load current (3.5 mA)

The reserve operating time is around 570 s ≈ 10 min with the illustrated pair of series-connected capacitors.

charging time:

$$t_C \approx 4 \cdot R\,C$$

R = external series resistance plus the ESR of the capacitor bank

t_C is typically 3 to 6 minutes.

Notes:

1 The most important specification is the *breakdown voltage* of the storage capacitors. Typically, the working voltage is 5.5 V and the breakdown voltage is > 6.5 V. This means that the maximum voltage that should be applied to the capacitor is 5.5 to 6 V. With a 12-V supply voltage we are compelled to use two capacitors in series. To be on the safe side, we also use two series diodes in the power-supply connection, in order to reduce the voltage applied to the capacitors by 1.4 V (two diode drops).

2 The storage capacitors (which have trade names such as Supercaps™ or GoldCaps™) are short-circuit protected and cannot be overcharged. The series current-limiting resistor is intended to protect the connection to the power supply, not the capacitors.

3 Typical 3½-digit digital panel meter modules require a minimum supply voltage of 7 V and have a current consumption of 1 to 3.5 mA.

4 The storage capacity can be increased by connecting additional pairs of series-connected capacitors in parallel with the first pair. Two pairs of series-connected capacitors (4 × 1000 µF) can power the voltmeter module for 20 minutes from their stored charge.

5 Don't overlook the charging time. It takes several minutes for the capacitors to be fully recharged.

6 You must always be careful if the supply voltage (regardless of its source) falls below the minimum specified value. Some panel meters measure rubbish in this situation, instead of displaying a suitable warning (such as LO BATT).

Interference problems with DC voltage measurements

Excessive voltmeter internal resistance. The wires connected to the instrument act as antennas that pick up disturbances from their surroundings. These can cause problems in sensitive instruments with very high internal resistances. (Some meters have special measurement ranges with reduced internal resistance.) If a problem occurs, a suitable load must be provided by connecting an external resistor in parallel with the meter. An extremely high internal resistance (ranging from 100 MΩ to the GΩ region) can also cause oscillations in the measurement setup. This can be remedied by connecting a resistor (on the order of 30 to 100 kΩ) in series with the test lead, as close as possible to the probe tip. (With an internal resistance in the GΩ region, the voltage drop across a 100 kΩ resistor is negligibly small.)

Disturbances caused by digital multimeters. A digital multimeter must always be regarded as a noise source, due to the way that it works (even if it is battery-powered!). Figure 4.13 shows a remedy (see Volume 3 for its mechanical construction). The filter arrangement should prevent noise signals originating within the instrument from reaching the circuit being tested. The exact values of the filter components must be determined experimentally.

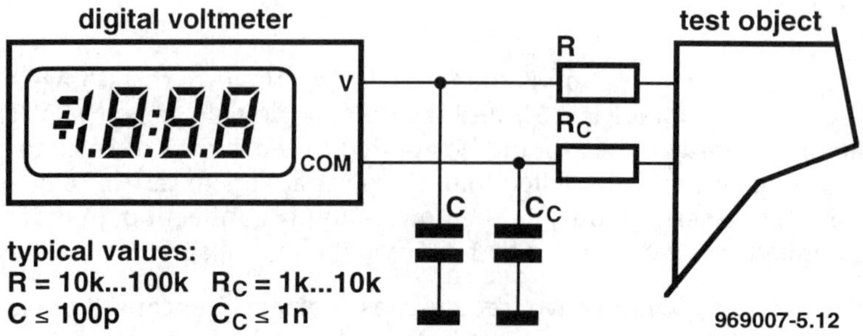

Figure 4.13 *Filter for the leads of a digital multimeter (source: EDN)*

Measuring voltages more accurately: floating measurements and in-circuit measurements

The smaller the measurement range, the smaller the absolute error. There are two 'tricks' that exploit this principle to allow more accurate measurements to be made in certain situations (see Figure 4.14):

1 floating measurements: the measurement is not made with respect to earth, but with respect to a fixed reference voltage (the instrument 'floats' at this potential),

2 in-circuit measurements: the measurement is made directly across a resistance where a voltage drop occurs.

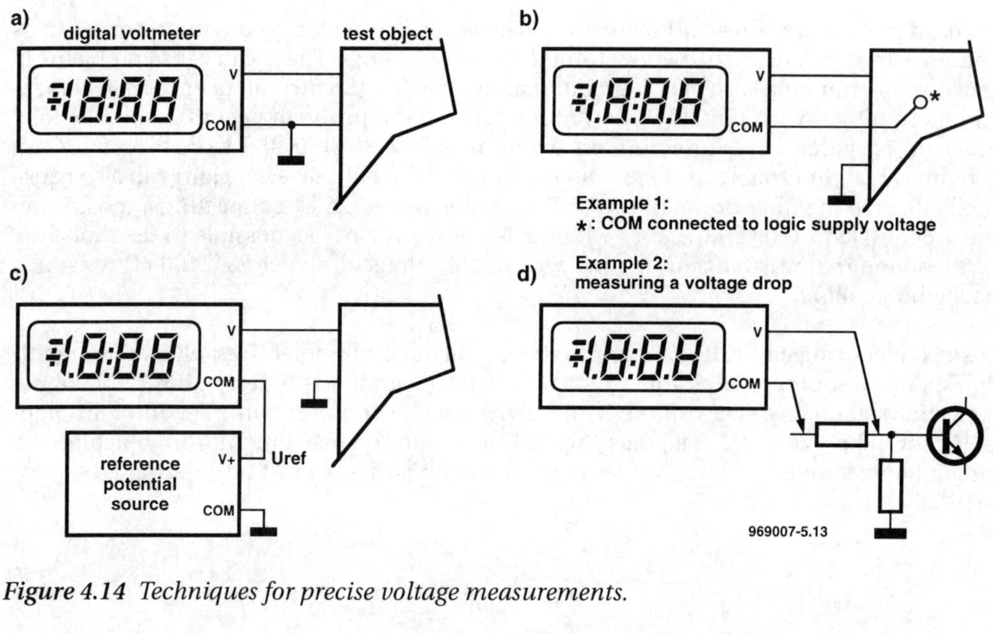

Figure 4.14 *Techniques for precise voltage measurements.*

Figure 4.14 illustrates the following techniques:

a) The conventional arrangement: the voltage is measured with reference to the common earth potential.

b) Floating measurement (1): the earth terminal of the instrument (the COM jack) is connected to a particular voltage inside the test object. This can be used for relative and differential measurements. *For example:* suppose you want to investigate the logic supply voltage (+5 V), and in particular to measure the exact value of the supply voltage at each IC. If you measure the voltage with reference to earth, then (in the case of a 3½ digit meter) you must work with a 20-V measurement range, in which one digit corresponds to 20 mV. If you instead connect the COM lead of the meter directly to the +5 V output terminal of the power supply, then you can directly read the voltage difference between this point and the V_{CC} pin of each IC (this will be a negative voltage). It will normally be possible to use the meter's most sensitive range for this (such as 200 mV, with a resolution of 0.1 mV), so that you can make measurements that are much more precise. In a similar manner, you can 'float' the instrument in order to more precisely measure voltage drops across specific components.

c) Floating measurement (2): the instrument's COM terminal is connected to a precise reference potential (see Section 4.2.11) that is in turn referred to the earth potential. *For example:* suppose that you want to accurately measure the +5 V logic supply voltage. A 3½-digit instrument with an accuracy of ±1% + 1 digit can measure 5 V with a net error of 1.2% = ±60 mV. If you connect the COM terminal to a reference potential source with an output voltage of 5.00 V ±0.1% (±5 mV absolute error) then you can normally measure the difference between the reference potential and the supply voltage using the most sensitive range (for the 200 mV range the full-range error is 2.1 mV; with the error term of the 5 V reference source this yields an overall error of ±7 mV).

d) In-circuit measurement: this example shows how the voltage drop across the base resistor can be read directly by connecting the test leads to either end of the resistor.

Note: refer to Appendix 3 for a discussion of the limitations and problem areas of floating and in-circuit measurements.

Measuring high voltages

Caution! Before starting, carefully check the specifications of the measuring equipment.

Notes:

1 See Appendix 2 regarding general safety precautions.

2 Many instruments have 1000 V DC and AC measurement ranges.

3 Use only instrument accessories (test leads, clamp terminals, plugs, probes and so on) with rated test voltages that are at least four times the highest voltage to be measured. For AC voltage measurements, this is the peak-to-peak voltage U_{pp} (230 V ~ ≙ 640 V_{pp}, 1000 V ~ ≙ 2800 V_{pp}).

4 Exercise extreme caution when floating the earth connection (the COM terminal of the measuring equipment). Some instruments (such as certain battery-operated oscilloscopes) allow the COM or reference input to be connected to potentials that lie several hundred volts above the earth potential. See Appendix 3 for more information.

5 Always use a special high-voltage probe (see Figure 4.15) to measure very high voltages (several kV, such as CRT anode voltages). Do not use attempt to use DIY arrangements based on ordinary shop material!

6 Keep probe assemblies, test leads and cables clean and dry. Do not measure high voltages under conditions of high relative humidity. Always keep one hand in your pocket when measuring high voltages!

High-voltage probes

Construction: suitable probe assemblies typically have divider ratios ranging from 1000:1 (40 kV ≙ 40 V) to 50,000:1. They contain series resistors with values on the order of 500 to 1000 MΩ. Some models have built-in meters, and some models are intended to be used with a digital multimeter or oscilloscope. Probes for use with multimeters typically have a smaller divider ratio than those for use with oscilloscopes (which makes for higher measurement accuracy).

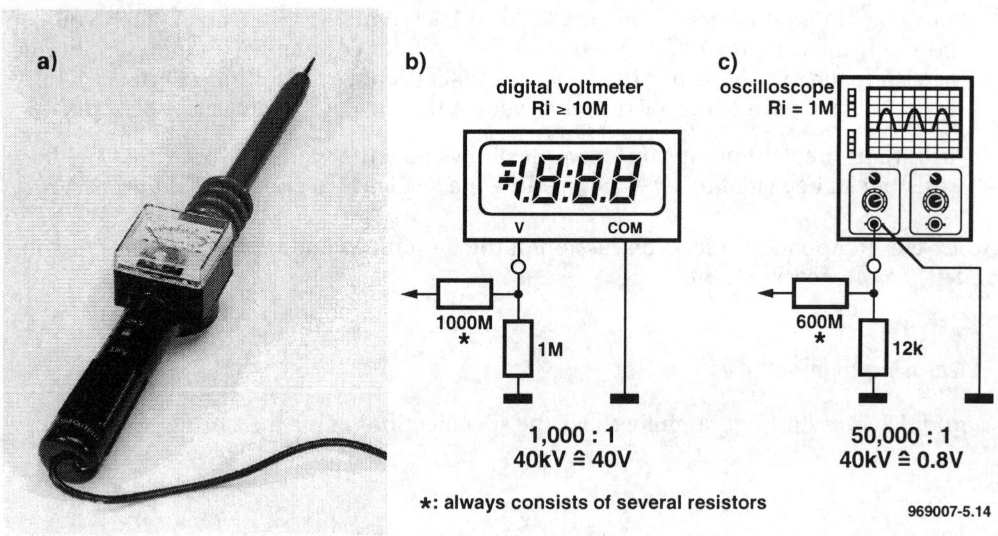

Figure 4.15 High-voltage probes. (Resistor values and divider ratios are typical.)

a) typical construction (switchable between 30 kV and 45 kV; measurements can be made using the built-in meter or via an oscilloscope; measurement error ±3 %).

b) for connection to a digital multimeter

c) for connection to an oscilloscope

Caution:

1 The input resistance of the connected instrument forms part of the voltage divider (it is in parallel with the 'lower' element). You must therefore check whether a given probe is intended to be used with a digital multimeter (typical input impedance 10 MΩ) or an oscilloscope (typical input impedance 1 MΩ). If a probe intended to be used with a multimeter (see Figure 4.15b) is used with an oscilloscope, the effective division ratio is 2000:1 (since the 1 MΩ input impedance of the oscilloscope is in parallel with the 1 MΩ of the probe divider). This means that 20 V is present at the oscilloscope input if the voltage being measured is 40 kV!

2 Pay attention to the input sensitivity when making measurements using an oscilloscope (see Section 4.3.3). Many instruments have a minimum sensitivity of 5 V/cm, some even higher (eg 2 V/cm). An input potential of 40 V (see Figure 4.15b) could be high enough to damage the oscilloscope.

3 *AC voltage measurements:* simple probe assemblies as shown in Figure 4.15a are in principle only suitable for DC measurements. If you want to accurately measure AC voltages, pulses, noise and the like (or observe them using the oscilloscope), then you will need a high-voltage probe assembly with a frequency-compensated voltage divider (see Section 4.3.1), which is not exactly inexpensive.

Practical tip: the divider ratio of a ready-made probe assembly can be modified by exchanging the 'lower' resistor (which does not carry a high voltage). *For example:* the 12 kΩ resistor in Figure 4.15c could be replaced by a 60 kΩ resistor to create a divider ratio of 10,000:1. *Caution:* always select the divider ratio such that the resulting output voltage does not exceed 40 to 50 V.

4.2.3 Current measurements

A current meter (ammeter) is inserted into the path of the current to be measured (series connection, see Figure 4.16). Ideally, there should be no voltage drop across the current meter; this means that it should have zero internal resistance.

Analogue current measurements. Analogue meters are by nature current-measuring devices.

Digital current measurements. Digital meters are by nature voltage-measuring devices. The current thus has to be determined indirectly, by measuring the voltage drop across a resistance through which the current flows. To determine the value of this resistance, see Figure 4.16.

Extending the measurement range. The range of a current meter can be extended by connecting a resistor in parallel with it (a current-divider resistor or *shunt*) as shown in Figure 4.17a. A selector switch can be used to provide several ranges (Figure 4.17b–d).

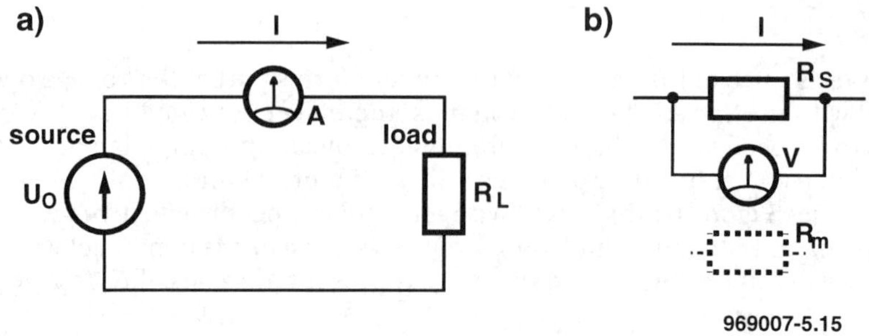

Figure 4.16 *Current measurement.*

a) basic circuit

b) using a voltage meter for current measurements

Calculating the value of the voltmeter shunt resistor:

$$R_S = \frac{U_{fs} \cdot R_m}{I_{Rm} \cdot R_m - U_{fs}} \approx \frac{U_{fs}}{I_{Rm}} \quad \text{(if } R_m \text{ is nearly infinite)}$$

R_S = shunt resistance
U_{fs} = full-scale meter voltage (rated meter voltage)
R_m = internal resistance of the voltmeter
I_{Rm} = desired full-scale current reading (full-range value)

The relationship can also be expressed as:

$$U_{fs} = I_{Rm} \cdot (R_S \parallel R_m)$$

Explanation of Figure 4.17:

a) An ammeter with a shunt resistor.

b) Multiple ranges, multiple current paths (applications: checking current consumption in operating equipment and power-supply testing – see Section 5.2 and Volume 3 for details). The current-carrying leads must remain operational (and thus intact) even when their current is not being measured. The simplest solution is to use permanently mounted resistors that are connected to a voltmeter via a two-gang selector switch.

c) Multiple ranges, single current path (application: analogue multimeters).

d) Ayrton circuit for use in analogue multimeters. The advantage over circuit (c) is that the contact resistance of the switch does not affect the current division between the shunt and the meter (since the entire current always flows through the switch).

e) Range extension circuit for a typical digital panel meter module with a 200 mV measurement range and 10 MΩ internal resistance. The voltage drop across the meter is 200 mV when the current is the maximum value for the selected range.

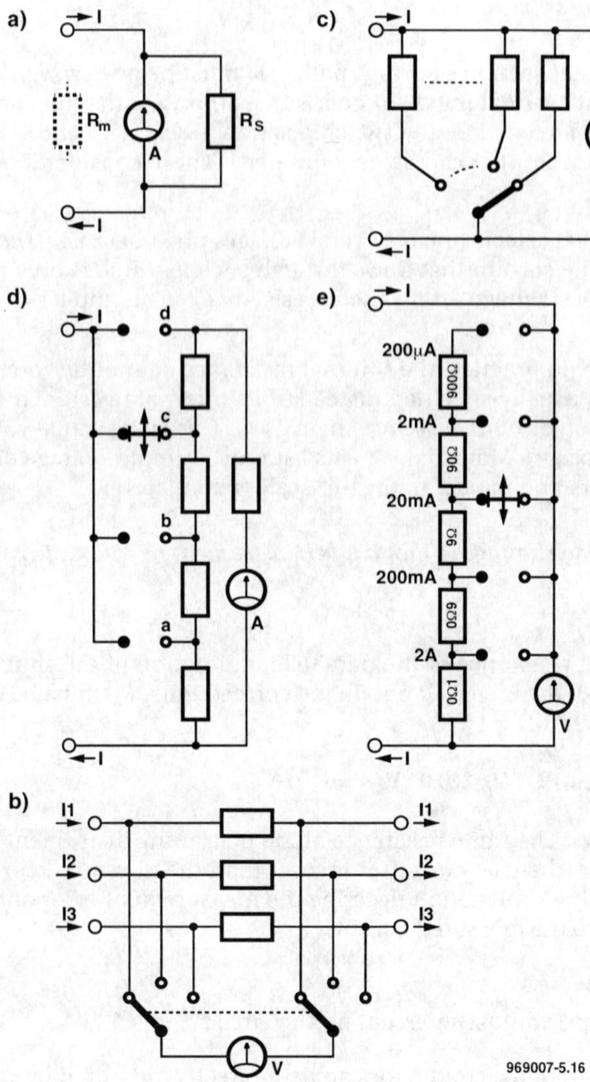

Figure 4.17 Extending the current-measurement range.

Calculating the value of the ammeter shunt resistor:

$$R_S = \frac{I_{fs} \cdot R_m}{I_{Rm} - I_{fs}}$$

R_S = shunt resistance
R_m = internal resistance of the voltmeter
I_{fs} = full-scale meter current (rated meter current)
I_{Rm} = desired full-scale current reading (full-range value)

Notes:

1 A range switch that switches current paths (1) must be a *shorting* type (make-before-break) to prevent overloading the meter (since otherwise the full current would briefly flow through the meter when the switch position is changed) and (2) must be rated for the maximum current that can be encountered. (These considerations apply to Figure 4.17c–e.)

2 A range switch that selects one of several voltages must be a *non-shorting* type (break-before-make). The current that flows through such a switch is only the meter current, which is negligible (voltage drop ÷ meter resistance). (This applies to Figure 4.17b.)

Internal resistance. In practice, the internal resistance cannot be zero. The voltage drop across the meter is usually specified instead of its internal resistance, in the form of the 'specific' voltage drop or *burden voltage* (in mV / A). This is the actual voltage drop divided by the measured current. Many data sheets list only a simple 'voltage drop' specification, which is then understood to refer to the full-scale (rated) current.

Calculating the burden voltage U_{iA} for a given instrument resistance R_M:

$$U_{iA} = R_M$$

where R_M is the net resistance of the parallel combination of the shunt resistor and the internal resistance of the meter. To obtain the correct units, R_M must be expressed according to the relationship $1\ \Omega = 1\ \text{V/A}$.

For example: $R_M = 0.01\ \Omega$; $U_{iA} = 0.01\ \text{V/A} = 10\ \text{mV/A}$.

Measurement errors. The finite resistance of the measuring instrument causes a negative measurement error (the true current is greater than the measured current). Figure 4.18 and its accompanying explanation describe the measurement error and the relationship of the true current to the measured current.

Measuring a current without interrupting the circuit

A current flowing through a conductor can be indirectly measured by sensing the associated magnetic field. This is the operating principle of clip-on current probes and current converters. Two physical phenomena can be utilized to detect and measure the magnetic field: (1) the Hall effect and (2) electromagnetic induction (the transformer principle). Hall-effect current probes can be used to measure both DC and AC currents, but transformer probes can only be used to measure AC currents. (The latter type is technically simpler and thus less expensive). Typical specifications are: resolution 10–100 mA, accuracy ±1–2%, measurement range 20–600 A. *Note:* the probe (or in general the magnetic circuit) must enclose a single current-carrying conductor; it is normally not possible to measure currents in cables that carry currents in both directions. (A non-zero current measurement in such cases means that the net currents in the two directions are not equal. This fact is utilized in earth-leakage current breakers, for example).

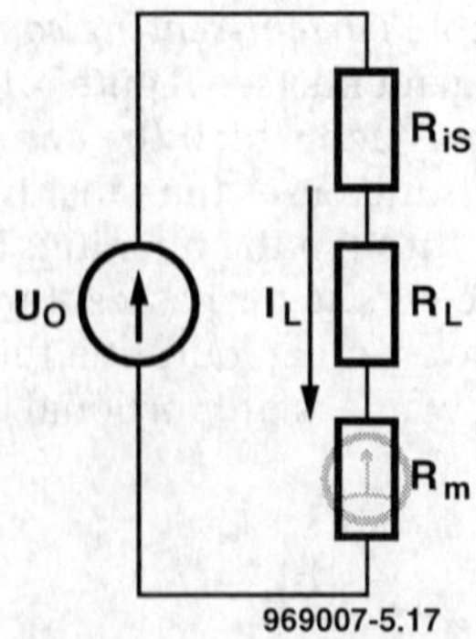

Figure 4.18 *Measurement under actual conditions: the internal resistance of the meter increases the total circuit resistance.*

Current measurement error:

correct current:

$$I_L = \frac{U_O}{R_{iS} + R_L}$$

measured current:

$$I_m = \frac{U_O}{R_{iS} + R_L + R_m}$$

relative measurement error: [1)]

$$E_r = \frac{I_m}{I_L} - 1 = \frac{-R_m}{R_{iS} + R_L + R_m}$$

1): The percentage error is: $E_{rp} = E_r \cdot 100\ \%$

error correction:

$$I_L = \frac{I_m}{1 - \frac{R_m}{R_{iS} + R_L + R_m}}$$

U_O = source voltage
R_{iS} = internal resistance of the voltage source
R_L = load resistance
I_L = (correct) current through the load resistor
R_m = meter resistance (ammeter internal resistance || shunt resistance) *
I_m = measured (indicated) current

* If the voltage across the shunt resistor is measured as shown in Figure 4.16b, the internal resistance of the voltmeter is usually negligible.

There are also commercially-available *current-sensing converters* that can be used to construct suitable measurement arrangements (see Figure 4.19). The illustrated converter has a magnetic core with an opening through which the conductor is fed. Several turns may be wound through this opening; this increases the sensitivity in proportion to the number of times that the conductor passes through the opening. The current rating is 50 ampère-turns (A-t), which means 50 A if the conductor passes through the opening only once, 5 A for 10 windings and so on. The converter can output either (a) a measurement current or (b) a measurement voltage, each of which is proportional to the primary (sensed) current.

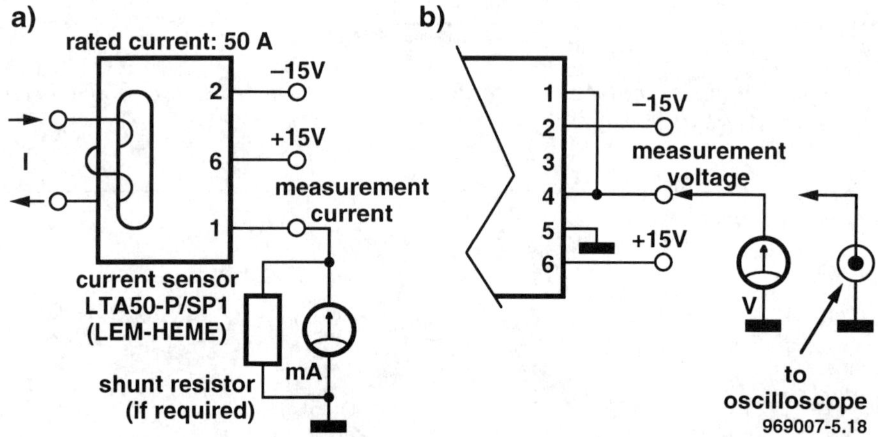

Figure 4.19 Hall-effect current sensor.

Specifications of the type LTA50-P/SP1 current sensor:	
calibration error:	*±0.5% of the rated current*
linearity error:	*±0.1% of the rated current*
bandwidth:	*DC to 100 KHz*
output current:	*50 mA ≙ 1 mA/primary A-t*
output voltage:	*5 V ≙ 0.1 V / primary A-t*

Notes:

1 Various types of converters are available, including ones with pivoting side arms ('openable core') and with pre-wound primary coils. There are also converters that sense the effective value (true-RMS value) of the current.

2 The illustrated converter represents the magnitude and polarity of the sensed current as either an output current or an output voltage. If the primary current flows in one direction the output current or voltage is positive, and if the primary current flows in the other direction the output is negative. Such a transducer is thus suitable for measuring both AC and DC currents.

3 One passage of the primary conductor through the core opening is one 'winding'; one loop around the core (twice through the opening) equals two windings, and so on.

4 DIY current-measuring aids based on the converter illustrated in Figure 4.19 are significantly less expensive than commercial current probes. For additional practical information see Volume 3.

5 The special advantage of a current-sensing converter is the complete galvanic separation of the current path and the measuring instrument.

6 A drawback of all current-sensing converters is that they rely on magnetic effects and are thus sensitive to external magnetic fields.

7 All current-sensing converters that have an openable magnetic core suffer from the fact that the opening acts as an air gap, which noticeably affects the magnetic circuit. You should thus take care that the mating surfaces are always clean and not damaged. (Good-quality current probes have a sliding mechanism instead of a pivoting mechanism, since the sliding action of one part of the core over the other part has a certain self-cleaning effect.)

8 For the subtleties of using current-sensing converters and current probes, see the available literature (refer to Appendix 8).

Practical issues

Measuring currents directly versus voltage measurements. With modern equipment, voltage measurements are in practice usually more accurate than current measurements. For this reason, currents are frequently measured indirectly by measuring the voltage drop across a sensing resistor with a known value (according to the formula $I = U / R$).

Notes:

1 The internal resistance of ordinary instruments (moving-coil meters and multimeters) is usually too large for measuring currents in the µA or mA range. For instance, the internal resistance of a standard 3½ digit digital multimeter is 1 kΩ for the most sensitive current range (200 mA; see Figure 4.17e) – and this is often too much for measuring currents in low-impedance circuits (such as measuring the current consumption of CMOS ICs in general, checking energy-saving measures and so on). All power-supply networks have a low impedance, even if only a few microampères are drawn from them, but you should watch out for switch-mode operation. Your only option in such situations is to measure the voltage drop across a suitable low-value resistor.

2 Sometimes there is already a resistor in the current path that can be used for an indirect current measurement (refer to Figure 4.14d).

Current-sensing resistors. Resistors for use as current-sensing shunts are available with a variety of current ratings and accuracy grades, as shown in Table 4.6.

Calculating the value of the current-sensing resistor:

$$R_S = \frac{U_{fs}}{I_{max}}$$

where R_S is the value of the shunt resistor, U_{fs} is the full-scale voltage of the voltmeter and I_{max} is the maximum value of the current to be measured ('full-scale' current)

Notes:

1 Power dissipation: $P_S = I^2 \cdot R_S$. Choose a resistor with a power rating of at least $2 \cdot P_S$, since heating the resistor degrades the accuracy of the measurement.

2 Some high-current sensing resistors must be mounted on heat sinks.

3 Various special types of packaging are available (for example, for incorporating the resistor in a bus bar).

4 Sensing resistors are available that can be directly attached to the terminals of certain types of moving-coil meters and digital panel meters.

5 High-current and high-accuracy types usually have separate connections for the current to be measured and the sensing leads (four-wire connection).

6 The greater the current, the more important it is to pay attention to minute 'parasitic' resistances in the current path (check for adequate conductor cross-sectional area, proper junctions and firmly-tightened binding screws). *Practical tip:* connect a known load (such as a power resistor) and check that the measured current value is roughly correct. It is astonishing how much the voltage drop is affected by small changes in the parasitic resistance (for instance, by how well the binding screws are tightened).

Resistance	Burden voltage	Current range	Typical values [2]
1 Ω	1 mV/mA	200 mA [1]	200 mA: 200 mV; 40 mW
0.1 Ω	100 mV/A	2 A [1]	2 A: 200 mV; 0.4 W
0.01 Ω	10 mV/A	20 A [1]	20 A: 200 mV; 4 W
0.001 Ω	1 mV/A	50 to > 100 A	100 A: 100 mV; 10 W
0.0001 Ω	100 mV/A	≥ 150 A	150 A: 15 mV; 2.25 W

Typical values:
Accuracy: ±0.1% (precision types) to 1%
Temperature coefficient: 30–50 ppm

1: Applies to digital voltmeters with a range of 200 mV (maximum displayed value 199.9 mV)

2: the values listed are:
- current: voltage drop;
- power dissipation

***Table 4.6** Characteristics of selected current-sensing resistors.*

Current measurement adapters. It is not particularly difficult to insert an ammeter into a current path of a lashed-together laboratory construction, but in practical service situations (such inside a PC) it is rarely all that easy. The remedy is provided by suitable test aids (measurement adapters). Here we only consider the basic options for current measurement adapters (see Figure 4.20). For specific implementations see Section 5.2 and Volume 3.

Shunt resistors versus current converters. The larger the current to be measured, the more compelling are the advantages of current converters: (1) no voltage loss in the current path and (2) galvanic isolation. The prices of high-current shunts and current converters (similar to that illustrated in Figure 4.19) are roughly the same. The disadvantages of current converters are: (1) they require operating power and (2) they are somewhat less accurate than sensing resistors.

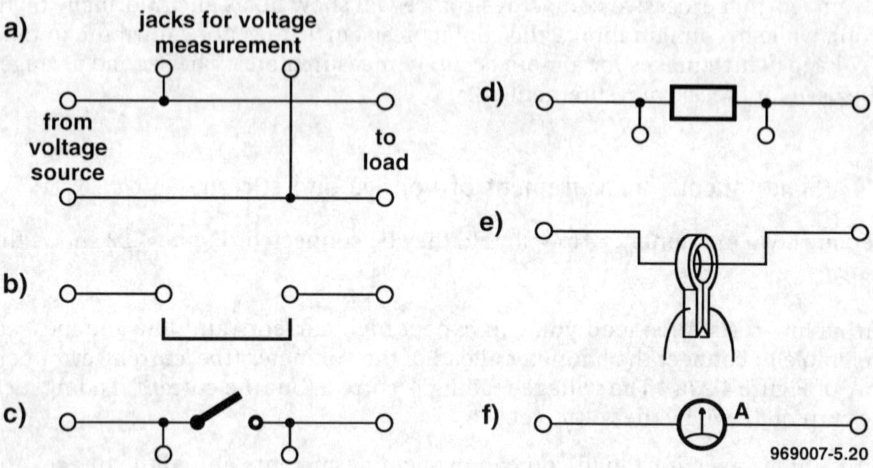

Figure 4.20 Current measurement using a measurement adapter.

a) Basic measurement adapter construction, with jacks for voltage measurements. The following items illustrate the how current measurements can be made.

b) Ammeter jacks that can be bridged by a shorting plug. This is simple but not very convenient: the ammeter can only be connected or disconnected when no current is flowing (commonly used for inexpensive mains-network measurement adapters.)

c) Ammeter jacks that can be bridged by a switch. This is more convenient than (b), but costly for high currents (greater than 1–2 A). An operator error is possible (accidentally opening the current path), especially when disconnecting the meter.

d) Permanent, built-in sensing resistor. This is more expensive than (b) or (c), but it allows dynamic measurements (with an oscilloscope). The current path cannot be interrupted. The sensing resistor may be overloaded in case of a fault.

e) Using a current-sensing converter or clamp-on current probe. This is relatively expensive; it has the same advantages as (d) with a reduced danger of overloading.

f) Permanent, built-in current meter (digital or analogue). This is convenient but costly. Inexpensive meters have low accuracy, and there is a risk of overloading the meter.

Measuring voltages across low-value resistors: the four-wire connection (Kelvin sensing)

If you want to measure the voltage drop across a low-value resistor that is carrying a heavy current, then you must connect the voltmeter directly to the resistor via a separate pair of wires, since otherwise you will obtain an incorrect reading (since the resistance of the sensing-resistor leads will be included in the measurement). See Figure 4.21.

Precision current-sensing resistors (shunts) have separate connections for the current path and the sensing voltage. Some current sensing resistors in application ICs (such as power ICs) are also connected in this manner.

Note: the diagram at the bottom of Figure 4.21 shows what you should pay attention to when making connections to a high-current sensing resistor: in configuration *1* the contact resistance of the binding posts contributes to the measured voltage drop (which has the 'advantage' that excessive contact resistance will show up as an abnormally high voltage drop), while in configuration *2* the contact resistance does not contribute to the measured voltage (which makes for a more accurate measurement). The second arrangement must be used for precise measurements.

4.2.4 Simultaneous measurement of voltage and current

This requires two instruments. How should they be connected? Figure 4.22 shows the two alternatives:

a) With a low-resistance load you can expect high currents and low voltages, so it is preferable to connect the ammeter ahead of the voltmeter (the 'current-error connection' of Figure 4.22a). The voltage reading is correct, but the current reading includes the current drawn by the voltmeter.

b) With a high-resistance load you can expect low currents and high voltages, so it is preferable to connect the voltmeter ahead of the ammeter (the 'voltage-error connection' of Figure 4.22b). The current reading is correct, but the voltage reading includes the voltage drop across the ammeter.

If you know the internal resistance of the instrument that causes the incorrect measurement, then you can calculate the correct value (see Table 4.7).

4.2.5 Resistance measurements

Resistance values can be measured using a bridge circuit (see Section 4.2.9) or by utilizing Ohm's law to calculate the resistance from the measured values of voltage and current ($R = U / I$).

Calculating the resistance from the voltage and current

To calculate the resistance, it is necessary to measure both the voltage and the current (see Figure 4.22 and Table 4.7 for comparison).

Basic principles:

- Use the current-error connection for low resistance values and the voltage-error connection for high resistance values.
- For accurate results, measure the voltage directly at the resistor terminals and keep the current-carrying conductors separate from the voltage-sensing leads (four-wire or Kelvin sensing; see Figure 4.21).

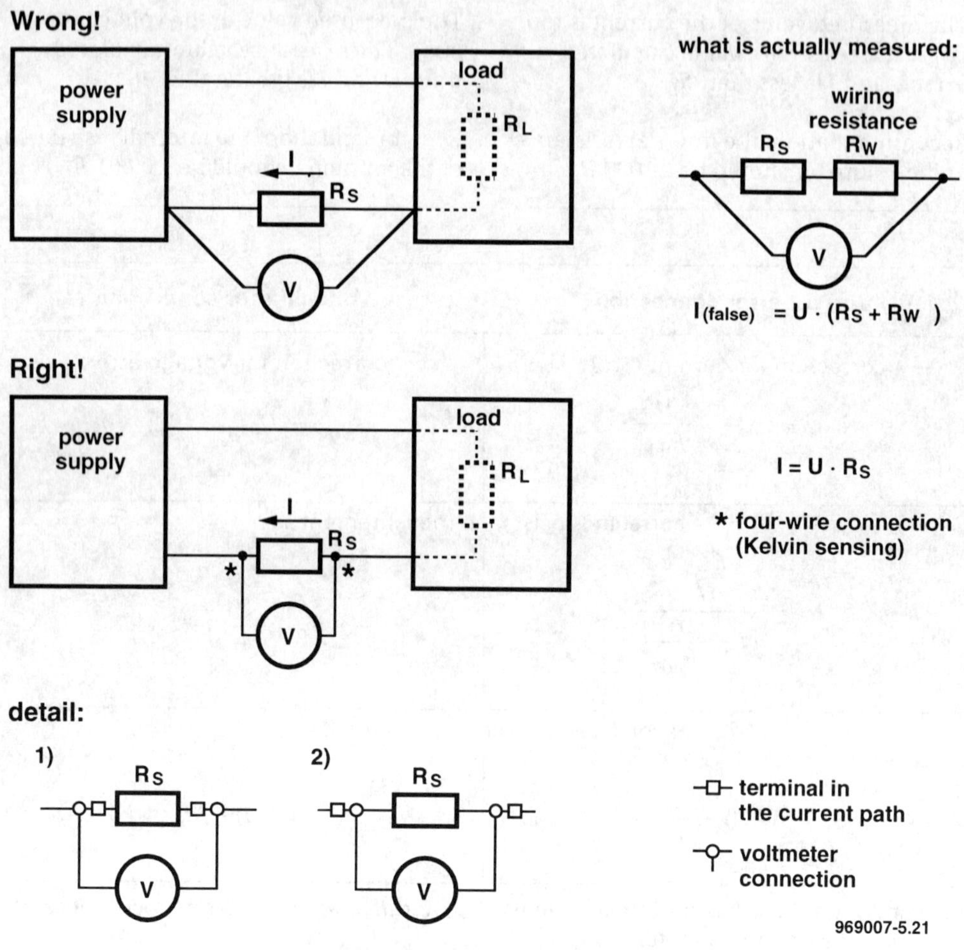

Figure 4.21 The four-wire technique (Kelvin sensing).

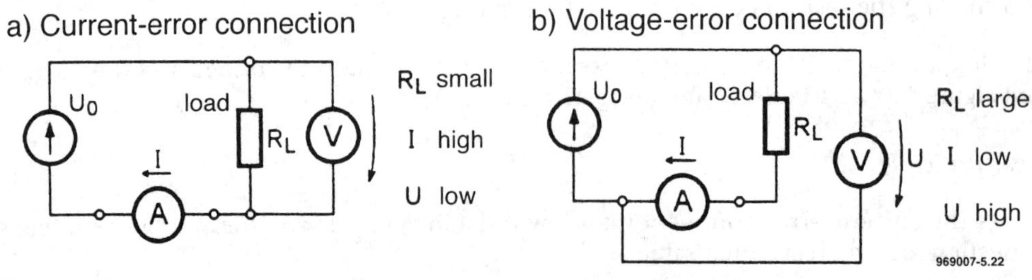

Figure 4.22 Simultaneous voltage and current measurements.

a) current-error connection:	*b) voltage-error connection:*
The measured value of the current is too large. The resistance value calculated from U and I is less than R_L.	The measured value of the voltage is too large. The resistance value calculated from U and I is greater than R_L.
Recommendation: the internal resistance of the voltmeter should be $\geq 1000\ R_L$.	Recommendation: the internal resistance of the ammeter should be $\leq 0.001\ R_L$.

Current-error connection	**Voltage-error connection**
correction for current error: $I_{true} = I_{meas} - \dfrac{U}{R_{mV}}$	correction for voltage error: $U_{true} = U_{meas} - (R_{mA} - I)$
corrected resistance measurement *)	
$R_x = \dfrac{U}{I - \dfrac{U}{R_{mV}}}$	$R_x = \dfrac{U}{I} - R_{mA}$
corrected power measurement *)	
$P_x = (U \cdot I) - \dfrac{U^2}{R_{mV}}$	$P_x = (U \cdot I) - (R_{mA} \cdot I^2)$
*: U and I are the measured (read) values; R_{mV} and R_{mA} are the internal resistances of the voltmeter and ammeter respectively.	

Table 4.7 Correcting errors arising from simultaneous voltage and current measurements.

Direct-indicating resistance meters (ohmmeters)

Ohmmeters measure only one quantity (either current or voltage), and the resulting pointer deflection or numerical display is interpreted in terms of resistance.

The accurate method. One of the two reference quantities (voltage or current) is held at a known constant value. Basically, high resistances should be measured by applying a known voltage and measuring the resulting current, and low resistances by applying a known current and measuring the resulting voltage (which is how commercial milliohm-meters work).

The simple method. The resistor to be measured forms part of a test circuit, so that it affects both the test current and the voltage drop. Since the other measurement parameters (source voltage and internal resistance) are known, the resistance can be determined from either the current or the voltage drop across the resistor. The disadvantage of this method is that the resistance scale is non-linear. It is used in most ohmmeters and for the resistance ranges of most analogue multimeters (see Figure 4.23).

Continuity checking. A continuity check is a sort of cursory test for a low-resistance connection. Frequently, all that is needed is a yes/no indication (connected/not connected), which can be provided by a simple visual or acoustic (beep) indication. This is the operating principle of commonly-available continuity testers. Many digital multimeters also have a continuity mode (usually with a piezo-electric transducer to provide a beep signal). The typical decision level is: <400 Ω = continuity ('beep'); >400 Ω = no continuity ('no beep'). For practical circuits see Section 5.3.1 and Volume 3.

Test voltage. When individual components are being checked, the test voltage must not exceed the rated voltage of the component in question. This is normally not a problem (the test voltage of most multimeters is at most a few volts). However, critical situations can arise when resistance measurements or continuity checks are made within a circuit (with the power off!). The problem is that the protective diodes within the ICs will conduct as soon as the test voltage exceeds their forward-bias voltage (0.45 V for Schottky diodes, 0.7 V for normal silicon diodes; see also Sections 4.4.1 and 5.1.2). There is nothing in the IC to limit the current that flows when a protective diode conducts. The resulting current can destroy the IC if the measuring device does not itself limit the current to a safe level. Strictly speaking, this means that the test voltage should be restricted to less than 0.45 V for making in-circuit measurements (so that it is less than the forward-bias voltage of a Schottky diode). *Note:* although limiting the test current (to ≪ 20 mA) will in fact prevent damage to the IC in case the test voltage causes the protective diodes to conduct (so that an ohmmeter constructed according to Figure 4.23c (test voltage = 1.5 V, test current ≤100 µA) can safely be used in semiconductor circuits), relatively high test voltages can yield invalid measurements if they cause the protective diodes to conduct (see Section 5.1.2).

Practical tip: the test voltage should be stated in the equipment specifications. Devices with test voltages ≤ 400 mV are commercially available. In case of doubt, measure the test voltage using a digital multimeter. For DIY cable testers and continuity testers, see Section 5.3 and Volume 3.

Accurate resistance measurements

If both the voltage and the current are measured, or if you can be sure that one of these quantities (voltage or current) is held constant, then the only errors that have to be taken into account are systematic errors in the measuring setup. These errors can be reduced by measuring only one quantity (voltage or current) and determining the resistance by comparing it to the measured value of the same quantity for a known resistance (the comparison method; see Figure 4.24).

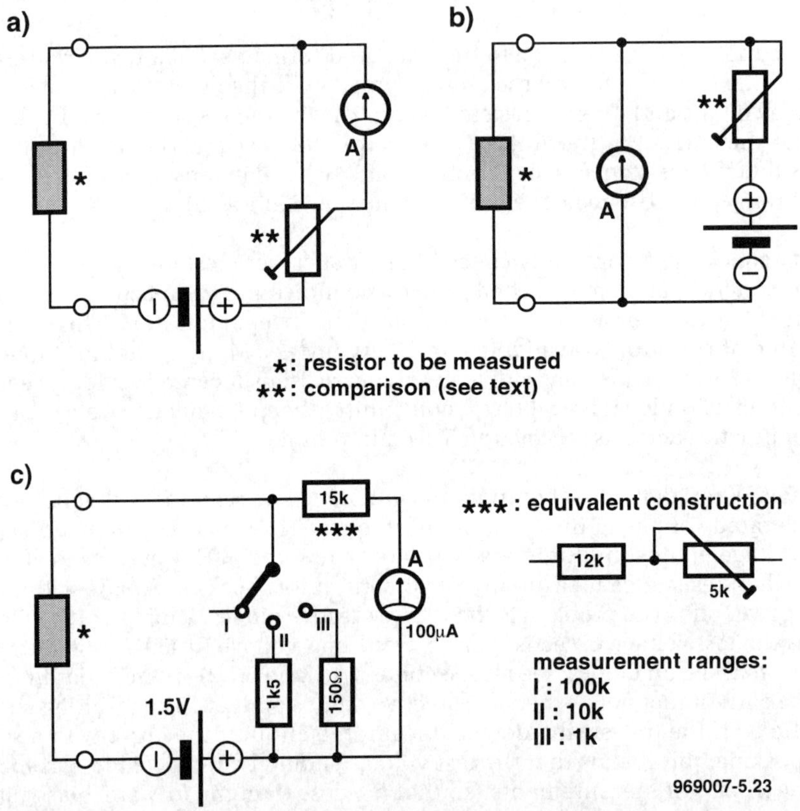

Figure 4.23 Basic circuits for direct-indicating resistance meters (ohmmeters).

a) *Measuring high-value resistors. The resistor to be measured completes the circuit. The smaller the resistance, the greater the deflection of the meter pointer. The meter is zeroed by adjusting it for a full-scale indication with zero resistance (test leads shorted together).*

b) *Measuring low-value resistors. The resistor to be measured forms part of a current divider. The greater the resistance, the greater the deflection of the meter pointer. The meter is zeroed by adjusting it for a full-scale indication with infinite resistance (test leads open).*

c) *An example of circuit (a), with representative component values.*

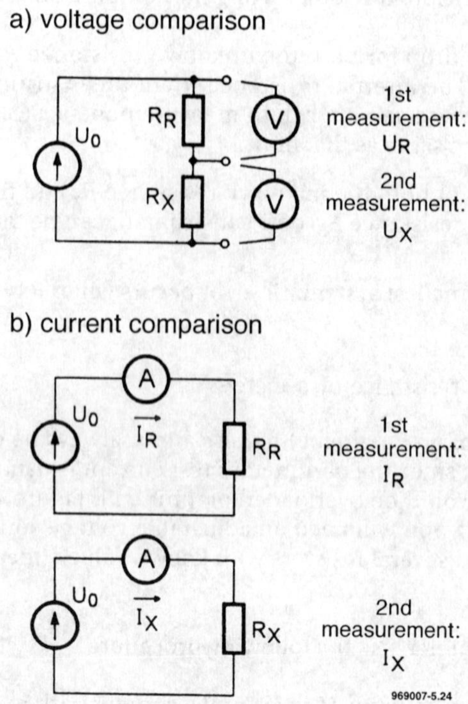

Figure 4.24 Measuring resistance by comparison.

a) Voltage comparison: *suitable for low-value resistors*

$$R_X = R_R \cdot \frac{U_X}{U_R}$$

b) Current comparison: *suitable for high-value resistors*

$$R_X = R_R \cdot \frac{I_R}{I_X} - R_m \cdot \frac{I_X - I_R}{I_R}$$

$$R_X \approx R_R \cdot \frac{I_R}{I_X} \qquad (\text{if } R_m \approx 0)$$

R_R = reference resistance (known value)
R_X = the resistance to be measured
R_m = the internal resistance of the meter

There are two ways to measure resistance using the comparison method:

a) Measure the voltage drop for both the unknown resistance R_x and the known (reference) resistance R_R. The internal resistance R_M of the instrument does not matter as long as either (1) it is the same for both measurements or (2) R_R is chosen to have a value that is nearly the same as that of R_x.

b) Measure the current for both the unknown resistance R_x and the known resistance R_R. The effect of internal resistance R_M of the instrument can be taken into account.

Note: in both cases, the smallest systematic error occurs when $R_R \approx R_x$.

Measuring the internal resistance of a meter

The internal resistance of an (analogue) meter cannot always be determined by measuring it with an ohmmeter, since the ohmmeter's test current is usually too high (so that the meter being measured would be overloaded or 'pinned'). Figure 4.25 illustrates a simple method that can be used. You will need an adjustable voltage source, a sufficiently accurate digital voltmeter and several resistors with known values (ideally, a resistance decade box).

For a (micro or milli)ammeter, use the following procedure:

1 Calculate the meter's sensitivity. (*For example:* a meter with a 250 μA full-scale deflection (rated current) has a sensitivity $R_{iV} = 1 \div 250\ \mu A = 4\ k\Omega\ /V$.)

2 Use the lowest possible test voltage (the lowest that the voltage source can deliver and that the digital voltmeter can still accurately measure). (*For example:* with a 400 mV measurement range, use a 300 mV test voltage.)

3 Calculate the value of the series resistor that must be used with the selected test voltage. (*For example:* $R_s = (300\ mV) \cdot (400\ k\Omega\ /V) = 1200\ \Omega$.)

4 Connect a resistor having the calculated value in series with the meter being measured. Increase the test voltage from the initial value (300 mV in our example) until the meter displays its exact full-scale value.

5 Use the appropriate formula in the box under Figure 4.25 to calculate the meter's internal resistance, based on the measured test voltage. (*For example:* the measured voltage is 355.8 mV, $R_s = 1200\ \Omega$ and $I_{fs} = 250\ \mu A$. R_m is then 223 Ω.)

For a voltmeter (with an internal series resistor), use the following procedure:

1 First connect an external resistor R_1 in series with the meter, and then adjust the test voltage until the meter displays its exact full-scale value. Measure the resulting test voltage U_1. (*For example:* $R_1 = 4\ k\Omega$, $U_1 = 18.18$ V.)

2 Replace the series resistor with a resistor R_2 having a different value, and again adjust the test voltage to obtain an exact full-scale indication. Measure the resulting test voltage U_2. (*For example:* $R_2 = 2\ \text{k}\Omega$, $U_2 = 14.06$ V.)

3 Use the appropriate formula in the box under Figure 4.25 to calculate the internal resistance of the meter from the measured test voltages and resistor values. (In our example, $R_m = 4.8\ \text{k}\Omega$.)

4.2.6 Power measurements

Power is the product of current and voltage, and can thus be calculated from individually measured current and voltage values (see Figure 4.22 and Table 4.7). One of these quantities can also be replaced by the corresponding resistance value, according to the following formulas:

$$P_r = U \cdot I = \frac{U^2}{R} = I^2 \cdot R$$

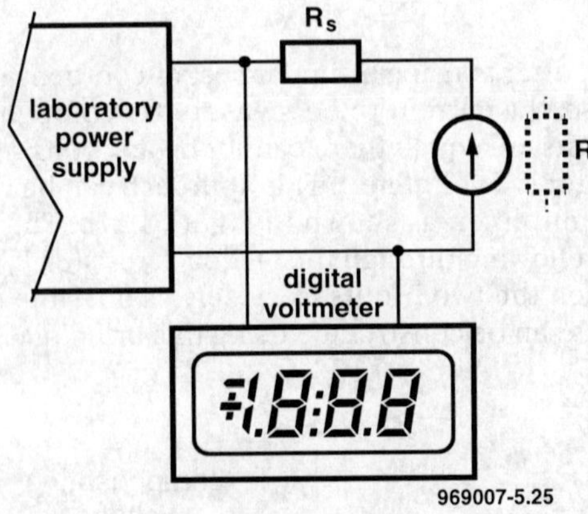

Figure 4.25 Determining the internal resistance of an (analogue) meter.

Ammeter:

$$R_m = \frac{U}{V} - R_s$$

Voltmeter:

$$R_m = \frac{U_2 R_1 - U_1 R_2}{U_1 - U_2}$$

R_1 and R_2 are two resistors with different values, that are used in turn as R_s.

4.2.7 Compensation measurement techniques

Compensation measurement techniques are based on comparing the quantity to be measured to a known quantity, instead of converting the quantity to some sort of indicated value. The principle is illustrated in Figure 4.26 for a voltage measurement. The voltage U_x that is to be measured is connected to a second voltage source, and the current that flows between the two devices is is shown by a null indicator (a sensitive moving-coil meter with a zero-centre scale). Potentiometer P is adjusted until the null indicator shows that no current is flowing, which means that the compensation voltage U_{comp} is equal to the unknown voltage U_x. The voltage can be directly read from the position of the potentiometer wiper arm (the potentiometer can be provided with a calibrated dial).

The advantages of the compensation method:

1 The measurement does not load the device under test (since no current flows when the compensation voltage is correctly adjusted).

2 The measurement accuracy does not depend on the scale accuracy of the meter, since all that is required is a single reference point (null reading). The compensation voltage source can however be implemented with very high precision (for example, by using a finely-graduated set of switchable precision resistors instead of a potentiometer).

It is possible to measure currents using a similar sort of compensation method, as shown in Figure 4.27. In this case, the current to be measured is compensated by an adjustable test current. If both currents are equal, there cannot be any voltage drop between the two terminals (the null indicator is a voltmeter). This approach can be very simply implemented using an operational amplifier, as shown in Figure 4.27b. The unknown current I_x is compensated by a current flowing through the feedback resistor R_f (the feedback serves to hold the potential between the two inputs as closely as possible to zero). Very sensitive current measuring devices can be constructed using this principle (see Volume 3).

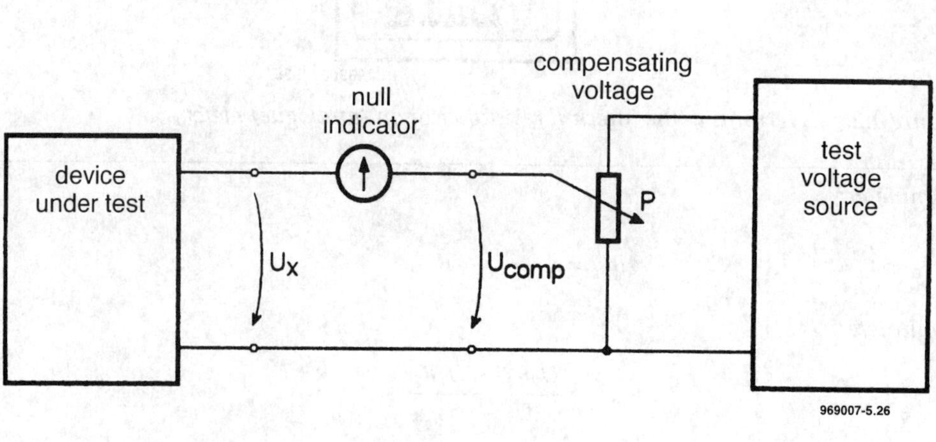

Figure 4.26 *The principle of the compensation measurement method.*

4.2.8 Measuring AC quantities

Voltage and current measurements

Meters are connected in the same way for measuring AC voltages and currents as for DC measurements. Depending on the particular type of meter and any accessory circuitry that is used, either the peak value, the DC value or the effective value of the quantity in question is measured. *Note:* see Appendix 5 for basic concepts and characteristic values.

Moving-iron meters. A moving-iron meter indicates the true effective value (true RMS value) of the current that flows through it. Its cutoff frequency is roughly 1 kHz. Moving-iron meters are not especially sensitive and are thus only suitable for relatively coarse measurements.

Moving-coil meters and digital meters. These types of meters can only directly measure DC currents or DC voltages, respectively. Rectifiers must therefore be connected ahead of such meters for AC measurements (see Figure 4.28).

Measuring the peak value. This requires a rectifier with a storage capacitor. The operating principle is that the capacitor is charged to the peak value of the input signal. The load current of the meter must be small enough that the amount that the capacitor is discharged between successive peaks is negligible. A half-wave rectifier (Figure 4.28a) passes the peak value of one of the two half-waves of the AC signal. A rectifier circuit that takes both half-waves into account, such as the voltage-doubler circuit of Figure 4.28b, must be used to measure the peak-to-peak value (U_{pp}).

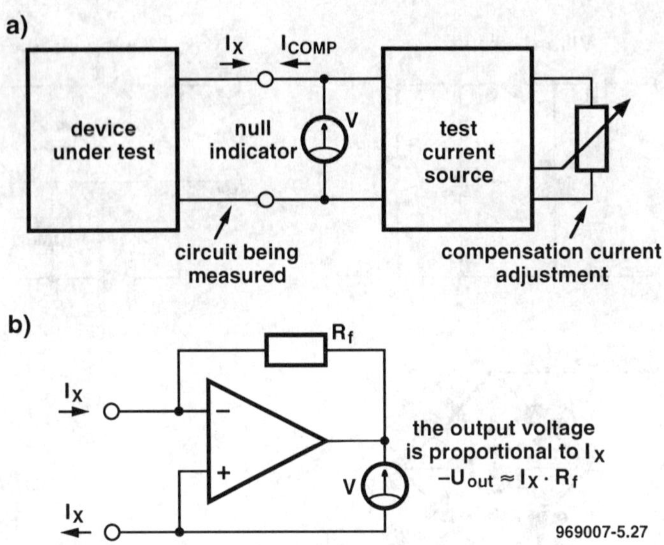

Figure 4.27 Current measurement using the compensation method.
a) working principle, b) current-measuring circuit based on an operational amplifier.

Measuring the average value. The average value can be measured using a moving-coil meter by (for example) connecting a full-wave rectifier ahead of the meter (Figure 4.28c). The necessary averaging (integration over time) is provided by the mechanical inertia of the meter movement.

Measuring the effective value. If the quantity to be measured has a sinusoidal waveform, the following simple relationships apply (refer to Appendix 5):

- effective value = (average value)·(form factor)
- effective value = peak value (0.5 U_{pp}) ÷ (crest factor)

Although effective values can be read from most meters, these are only valid for sinusoidal waveforms. The indicated value is thus incorrect if (for example) a 'truncated' sine wave is being measured. In order to correctly display the effective value of the measured quantity, the meter must be able to perform the necessary mathematical operations by means of analogue or digital circuitry (squaring the instantaneous value of the measured quantity,

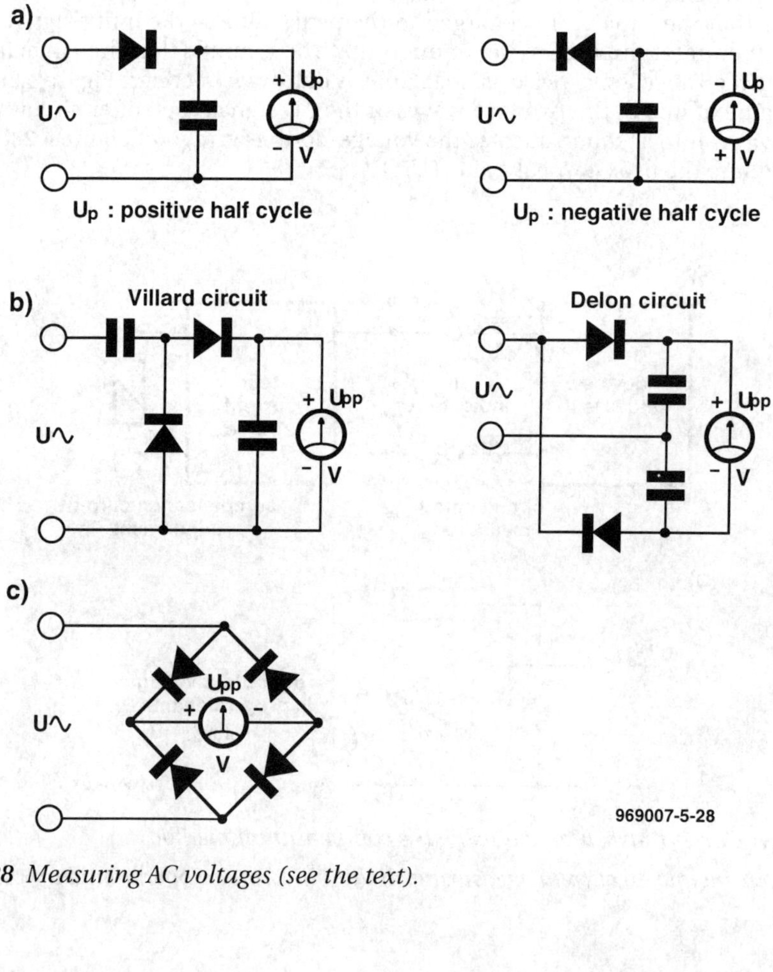

Figure 4.28 Measuring AC voltages (see the text).

averaging or integrating the result and calculating the square root; see Appendix 5). This means that the true effective value can only be measured using a moving-iron meter or an instrument that has been specifically designed to compute 'true RMS' values. With ordinary multimeters (both analogue and digital), the indicated effective value is only valid for sinusoidal current or voltages waveforms. This is however adequate for pure 'control' measurements of the mains voltage or mains-transformer secondary voltages.

Note: the bandwidth of the meter must be suitable for the waveform being measured. In practical terms, the bandwidth should include at least the fifth harmonic of the signal being measured for true-RMS measurements of non-sinusoidal waveforms. Put another way, if the specified cutoff frequency of a meter is f_u, then it can correctly indicate the true RMS value of a 'harmonic-rich' waveform only if the fundamental frequency of the waveform does not exceed 0.2 f_u. See also Section 5.2.1.

Practical tips:

1 *What does a given instrument display?* You can check whether a device really displays true-RMS values by making a set of measurements on different types of waveforms having the same fundamental frequency and peak value U_p (or U_{pp}). Table 4.8 shows the reciprocal conversion formulas for the most important waveforms. *Procedure:* measure U_p using a half-wave rectifier (if necessary with a jury-rigged arrangement corresponding to Figure 4.28a), or measure U_{pp} using an oscilloscope. Then, for each measurement, calculate the ratio $U_{measured} / U_p$ (or U_{pp}) for each waveform and locate the entry in Table 4.8 which most closely matches the calculated value.

2 *Measuring the effective value of an AC voltage with a DC component.* The AC ranges of most meters are usually AC-coupled. In such cases the indicated effective value is incorrect if the voltage being measured has a DC component, since the indicated value relates to the AC component alone. Two measurements are necessary to arrive at the actual effective value: (1) an AC measurement that yields the true-RMS voltage of the AC component (U_{ac}) and (2) a DC measurement that yields the DC voltage U_{dc}. The true effective value can then be calculated as:

$$U_{eff} = \sqrt{U_{ac}^2 + U_{dc}^2}$$

Extending the measurement range. AC measurement ranges can be extended using transformers (see Figure 4.29), as well as by means of series or parallel resistors. AC measurement ranges can also be 'reduced' by using step-up transformers. Suitable voltage and current conversion transformers are commercially available.

Note: a suitable load resistor (with a specified value and sufficiently narrow tolerance) must be connected across the secondary winding of a voltage or current conversion transformer. The secondary of a voltage conversion transformer must never be short-circuited, and the secondary of a current-conversion transformer must never be open-circuited, since otherwise the transformer or the attached instrument can be destroyed. (An excessive current will flow in a voltage transformer with a shorted secondary, and an excessive voltage will be induced into the secondary of a current transformer with an open secondary. In this regard, the relatively high input resistance of the instrument connected to

Type of AC voltage	Waveform	Effective (true RMS) value U_{eff}	DC value $\overline{U}$
Sinusoid	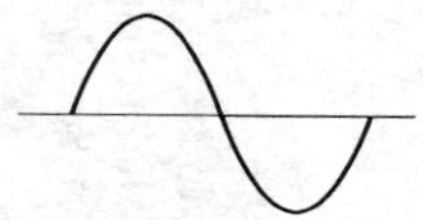	$= \frac{U_p}{\sqrt{2}}$ $\approx 0.707\ U_p \approx 0.354\ U_{pp} \approx 1.1\ \overline{U}$	$= \frac{2}{\pi}\ U_p$ $\approx 0.637\ U_p \approx 0.318\ U_{pp} \approx 0.9\ U_{eff}$
symmetrical rectangular wave (square wave)	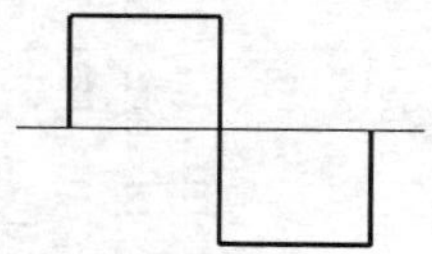	$= U_p = 0.5\ U_{pp}$	$= U_p = 0.5\ U_{pp}$
symmetrical triangular wave or balanced sawtooth	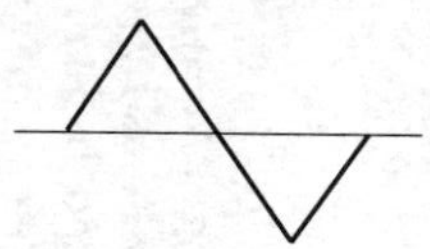	$= \frac{U_p}{\sqrt{3}}$ $\approx 0.577\ U_p \approx 0.289\ U_{pp} \approx 1.1547\ \overline{U}$	$= 0.5\ U_p = 0.25\ U_{pp} \approx 0.866\ U_{eff}$
half-wave rectification **)		$= 0.5\ U_p = 0.25\ U_{pp}$ *) $= 0.707\ \overline{U_{sine}}$	$= \frac{U_p}{\pi} \approx 0.318\ U_p \approx 0.159\ U_{pp}$ *) $= 0.5\ \overline{U_{sine}}$

*: U_{pp} refers to the rectified sine wave.
**: *Caution:* the DC component.

Table 4.8 *Effective and DC values of important alternating-current waveforms.*

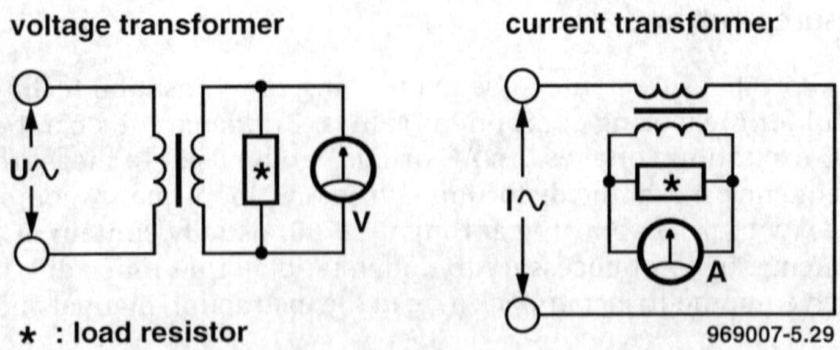

Figure 4.29 *Voltage and current conversion transformers.*

the secondary of a current transformer – such as a multimeter or an oscilloscope – is practically an open circuit!) For more details, see the relevant literature (Appendix 8).

Measuring power in AC circuits

The product of the measured values of the voltage and current in an AC circuit is only the *apparent* power. The *effective* or *real* power can only be determined if we also know the *power factor* cos φ (see Appendix 5).

The following types of instruments can be used for measuring AC power:

- Electrodynamic meters (wattmeters). These meters have two coils, one of which is fixed and the other moveable (as in a moving-coil meter). One coil measures the current and the other one the voltage. There are two sets of terminals: one for the current and the other for the voltage. *Caution:* be careful not to confuse the two sets of terminals!
- Induction meters. These are the normal electricity meters, which are found in every building. An induction meter contains a rotating disc, whose speed of rotation is proportional to the effective power. The cumulative power consumption is measured by counting the number of rotations of the disk. *Tip:* electricity meters that have been taken out of service and are no longer approved for official use can be obtained from electricity boards at reasonable prices. Such meters can be used to directly read the amount of energy consumed by whatever equipment is connected to them; the power consumption can then be calculated by dividing the energy by the time over which the measurement is made.
- Electronic power meters. These are based on either (a) Hall-effect multipliers or (b) measuring the phase angle (the phase displacement of the zero-crossings of the current and voltage waveforms). The latter type of instrument typically contains a microcontroller and a ROM-based conversion table for determining the power factor from the phase angle.

4.2.9 Measurement bridges

Bridge circuits allow measurements to be made using compensation techniques and are especially useful for measuring component values. Bridge circuits can be powered by either DC or AC excitation voltages. An AC bridge can be used to measure inductance, capacitance, frequency, harmonic distortion and so on. Good-quality component testers and RLC bridges (with prices starting at roughly £700) usually contain AC bridges with automatic balancing. It is not necessary to understand all that much about the internal operation of such equipment: just unpack it, read the instruction manual and try it out.

We devote our attention here to two types of DC bridge circuits (see Figure 4.30):

a) *Wheatstone bridge.* This is the simplest form of a bridge circuit. As can be seen, no current flows through the null-indicator meter if the voltage drops across the two voltage dividers are equal, so that the voltage at point A is the same as the voltage at point B (that is, the balance condition is $U_{AB} = 0$). In this case the following must be true:

$$\frac{U_1}{U_2} = \frac{U_3}{U_4} \text{ and thus } \frac{R_1}{R_2} = \frac{R_3}{R_4}$$

The divider ratio R_3 / R_4 must be adjustable to allow the balance condition to be achieved. If R_1 is replaced by the the resistor R_x that is to be measured, then:

$$R_x = R_2 \cdot \frac{R_3}{R_4}$$

In practice, the voltage divider R_3 / R_4 is made up of a number of individual resistors (resistance decades) that can be incorporated into the voltage divider using switches or shorting plugs. The resistance R_2 is often a precision potentiometer that can be adjusted to achieve an exact balance. High-accuracy bridges allow resistance values between 10 Ω and 1 GΩ to be measured with an error on the order of 0.001% to 0.1%.

b) *Thomson bridge.* This circuit can be used for measuring very low resistances (in the range of 0.1 µΩ to 100 Ω). Lead resistances (labeled R_{L1} through R_{L6} in Figure 4.30) cause problems when such low resistance values must be measured. Both the resistor to be measured (R_x or R_1) and the balance-adjustment resistor (R_2) are connected to the second voltage divider using the four-wire technique. The second voltage divider consists of two matched branches (R_3 and R_3', R_4 and R_4'). The bridge is first (roughly) balanced by changing the corresponding resistance values of the second divider in matching steps using mechanically-coupled switches. A mathematical analysis of this circuit (which is not presented here) shows that the effects of all lead resistances are canceled out, so that the value of the resistor to be measured can be determined in exactly the same manner as with the Wheatstone bridge.

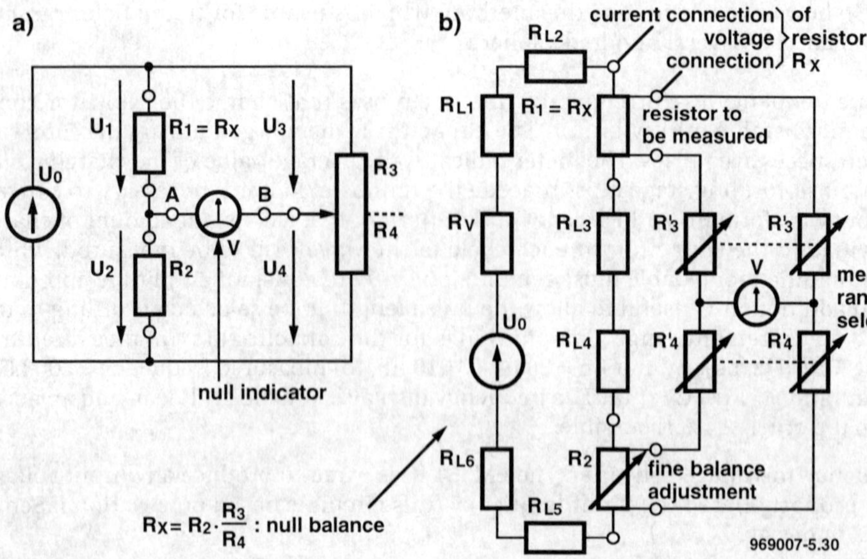

Figure 4.30 *DC bridge circuits. a) Wheatstone bridge, b) Thomson bridge*

4.2.10 Frequency measurements

Frequency is most commonly measured in modern practice by counting. Direct or 'true' frequency measurements can be made by counting the number of oscillations that occur in a given length of time. A frequency can also be calculated as the inverse of the measured duration of one cycle (period measurement). The following equipment is usually used for measuring frequency:

1 Oscilloscopes: with simple types, the period must be measured 'manually' by counting the number of scale divisions for one or more complete cycles; more sophisticated types automatically calculate and display the period and frequency of the displayed waveform. See Section 4.3.5.

2 Frequency counters or universal counters: such equipment is necessary for the exact measurement of very high frequencies (in the GHz region); see Section 4.4.5.

3 Digital multimeters with a frequency-measurement ranges. The cutoff frequency of such instruments is typically several hundred kilohertz. Their operation is otherwise the same as that of a frequency counter or universal counter.

Frequency measurement using the capacitor integration method. The waveform whose frequency is to be measured is converted into a train of fixed-length pulses. This pulse train periodically recharges a capacitor. The average value of the charging current (which can for instance be measured by a moving-coil meter) is directly proportional to the pulse repetition frequency. This method is particularly suitable for monitoring purposes or for trend displays (such as a computer activity indicator); see Volumes 2 and 3.

Figure 4.31 shows two versions of this method, which is usable for frequencies ranging from several hertz to several hundred kilohertz:

a) A simple frequency-to-current converter. The pulses that charge the capacitor come from a monostable multivibrator. The capacitor is discharged during the intervals between successive pulses. The meter indicates the average value of the charging current. A pulse-forming stage must precede the monostable if it is necessary to convert the input waveform into a TTL signal. It is important that a constant amount of charge is provided to the capacitor for each cycle of the waveform to be measured, which means that the monostable must generate pulses with constant amplitude and duration. In addition, it is useful to allow the pulse length to be selected according to the desired measurement range. Sample values for the capacitor are: (a) measured frequency = 10 kHz, pulse duration = 20 µs: C = 10 nF; (b) measured frequency = 200 kHz, pulse duration = 1 µs: C = 1 nF. The frequency display is practically linear, and an accuracy on the order of 1% is possible.

b) A frequency-to-voltage converter. The LM331 IC is wired to produce an output voltage that is proportional to the input frequency (this circuit is based on a National Semiconductor design).

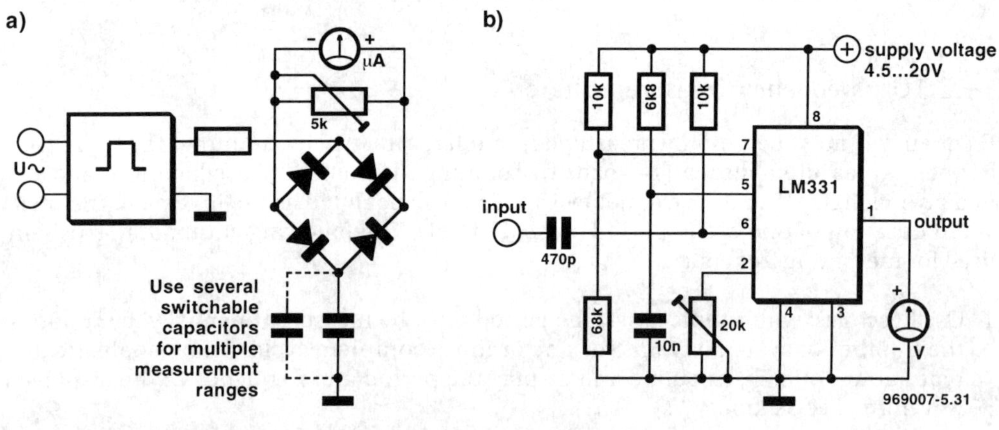

Figure 4.31 Frequency measurement using the capacitor-integration method.

4.2.11 Analogue stimuli

The measurement techniques that are described in this chapter are predominately related to digital circuitry. Systems based on processors or computers usually come with their own test stimuli (in a manner of speaking), so that you can simply measure whether everything is order without having to apply any sort of external test stimulus. With such systems, test setups that take the form of '(analogue) stimulus → test object → measurement equipment' (as shown in Figure 2.4) are typically only necessary for testing individual components or subassemblies. An additional application for this sort of testing is the verification of measurement equipment and measurement setups.

DC voltage sources

DC voltage sources are used for two purposes: (1) to power the circuit under test and (2) to provide input voltages or currents for static measurements.

Power supply assemblies and laboratory power supplies

Ordinary power supplies. When no unusual requirements are placed on the power supply, it is usually sufficient to perform a cursory test of the power supply in question (which is usually contained in the device under test). This means checking that all output voltages and any superimposed disturbances lie within their specified acceptable ranges (see Section 5.2.2).

Replacement testing. It is sometimes worthwhile to check a fault suspicion by replacing the power supply with a known good unit. Watch out for the following potential problems:

- Some mass-produced power supplies (including the usual PC types) require a certain minimum load, so that it may be necessary to provide a suitable supplementary load.
- Sometimes problems occur because the power supply is not suitable for operating the circuitry in question. In other words, the power supply is not actually faulty, but it is the wrong type or it is not properly integrated into the equipment (incorrect voltage tolerance, excessive ripple voltage, excessive radiated interference and so on). Possible remedies are: (1) use a high-quality power supply, (2) locate the power supply outside of the equipment and connect it via an extender cable (to verify the suspicion of interference problems).

Limit-value testing. If the equipment is to be tested at the limits of specified supply-voltage tolerance ranges, the power supply must be precisely adjustable. An obvious solution is to use a laboratory power supply. The following problems may be encountered:

- Logic-voltage supplies must often provide large currents (10–20 A for a PC).
- An operator error can result in a 'smoked' piece of equipment (for instance, selecting 16 V instead of 5 V).
- It is sometimes necessary to supply several different voltages at the same time.
- Several different voltages must be switched on and off simultaneously, and in addition a common 'ready' signal (Power Good) must be provided.

These problems can be solved using commercially available equipment (programmable laboratory power supplies or remotely-controlled current sources, plus a control computer), but the costs are significant. See Volume 3 for work-around solutions (which require a certain amount of handiwork and are still not all that inexpensive).

Battery substitution. It is a natural idea to use a suitable mains-powered supply in place of the battery when testing battery-operated equipment. *Caution:* with switch-mode power supplies, the superimposed voltages (AC ripple, switching noise and other disturbances) may be excessive (batteries do not have AC ripple). This means that you should never use a 'generic' switch-mode supply in such situations; always use an adequate-quality laboratory supply. (In this case such equipment is affordable, since the current required is rarely more than 1 to 2 A.) For practical considerations see Section 5.2.3.

Providing voltages and currents as measurement stimuli. Suitably adjustable mains-powered supplies are required for this purpose. The main question is whether the power supply has built-in meters (and whether they are sufficiently accurate), or whether external meters must be used.

Requirements for laboratory power supplies:

- coarse and fine voltage and current adjustments,
- adjustable current limiting,
- built-in voltage and current meters. Meters with measurement errors that are significantly more than 1% are not 'meters' but at best 'indicators'. It does not matter all that much whether the meters are digital or analogue. Digital meters should display at least 3½ digits, and analogue meters should have an accuracy grade of 1.5 or better. Manufacturers of inexpensive equipment often economize on the resistors used for current sensing, range selection and the like. *Practical tip:* test such equipment using a sufficiently accurate digital multimeter. It is possible to calibrate an analogue meter that is not sufficiently accurate. You can also consider which of the following options is more economical: (1) a high-quality power supply or (2) a relatively low-quality power supply plus a an additional digital multimeter. Figure 4.32 shows a simple outboard circuit that allows a single meter to be switched between current and voltage measurement, using only a single switch and without requiring any connections to be changed.
- *Operation:* equipment is available that is exceptionally easy to use, with features such as an operator keyboard, programmable operation and so on. Some laboratory power supplies can even be used to a certain extent as arbitrary waveform generators. If you do not plan on using such capabilities, which means that you intend to use conventional 'manual' measurement procedures, then a potentiometer is better than a keyboard for setting the output voltage. The output voltage can be easily varied by turning a potentiometer knob back and forth, which is useful if (for instance) you want see how a particular signal responds to such variations.
- *Selection:* in most cases, a dual supply that can provide up to 30 V at 2 to 3 A for each voltage is adequate (for powering logic circuits, checking out drives (which require +5 V and +12 V), experimenting with operational amplifier circuits and so on). *Practical tip:* if you frequently need 'standard' voltages (for logic circuits, operational amplifiers and the like), then fixed-voltage supplies are often more economical. You can also assemble such supplies from standard subassemblies at a reasonable cost (see Volume 3).

Batteries. It is sometimes worthwhile to supply the device under test from a battery in place of its own mains-operated supply, since a battery does not produce any interference signals. (If the circuit works properly when powered from the battery, then the suspicion of coupled interference coming from the power supply is usually confirmed. If using a battery to supply the circuit does not help, then the fault must be in the circuit itself.)

Practical tips:

1 Adapters: it is not always easy to make connections to individual cells or battery assemblies. Commercially-available battery holders and clips, which can for instance be built into an enclosure provided with test jacks, are a possible solution. See also Volume 3.

2 Battery-operated test and measurement devices have two advantages: (1) they do not introduce noise (earth loops, pulse and radio-frequency disturbances) and (2) they can readily be used for 'floating' measurements (given a suitably insulated construction).

3 Using rechargeable batteries as sources of test signals or for powering test aids: the most important selection criteria are (low) cost, obtainability and the availability of suitable chargers and adapters. (Rechargeable batteries for laptop computers and mobile telephones do not always meet these criteria.) There are two good candidates: (1) ordinary NiCd cells that are intended to be used to replace primary cells (they can be used with standard battery holders and adapters, and chargers are available in every hardware shop) and (2) rechargeable battery packs used for remote-controlled models (see Volume 3), which are powerful, relatively inexpensive and usually provided with suitable connectors; there is also an extensive selection of chargers for such batteries.

4 NiCd versus NiMH: NiCd cells have a better price/performance ratio and a lower self-discharge rate, and they can also supply high currents. The disadvantages of NiCd cells are (1) the memory effect, (2) the fact that they contain heavy metals and thus require special disposal and (3) their somewhat lower capacity compared to NiMH cells.

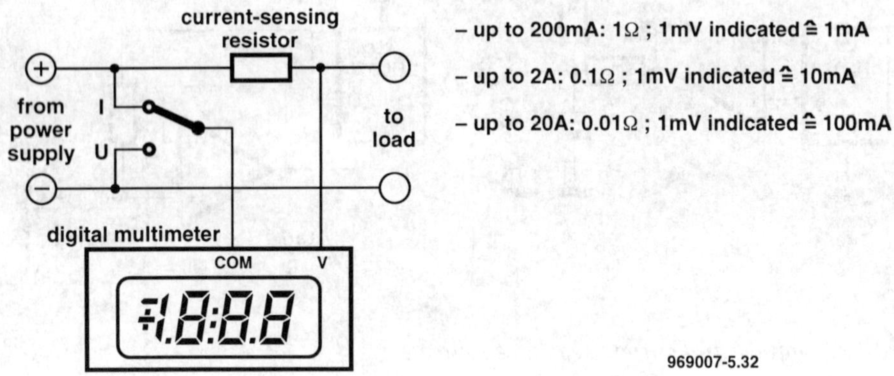

Figure 4.32 *Switchable voltage/current meter.*

Notes:

1 *An autoranging meter is the most convenient.*

2 *This circuit is an ideal accessory for instruments that do not have any current-measurement ranges (such as the Fluke Model 10 or Model 70 multimeters).*

3 *If the current is positive (as shown in the illustration), the indicated value of the measured current will be negative.*

5 Never short-circuit a battery! Using banana plugs attached to the leads of a battery pack is poor practice. It is better to build the battery into an enclosure with a suitable connector. To protect against short circuits or reversed-polarity connections to the circuit being tested, use two or three more cells than are needed for the required voltage and connect a low-drop regulator between the battery and the circuit that it supplies. See Volume 3 for more information.

Reference potential sources. Reference potential sources (as illustrated in Figure 4.33) supply very accurate voltages at low currents (in the milliampère range). Their primary

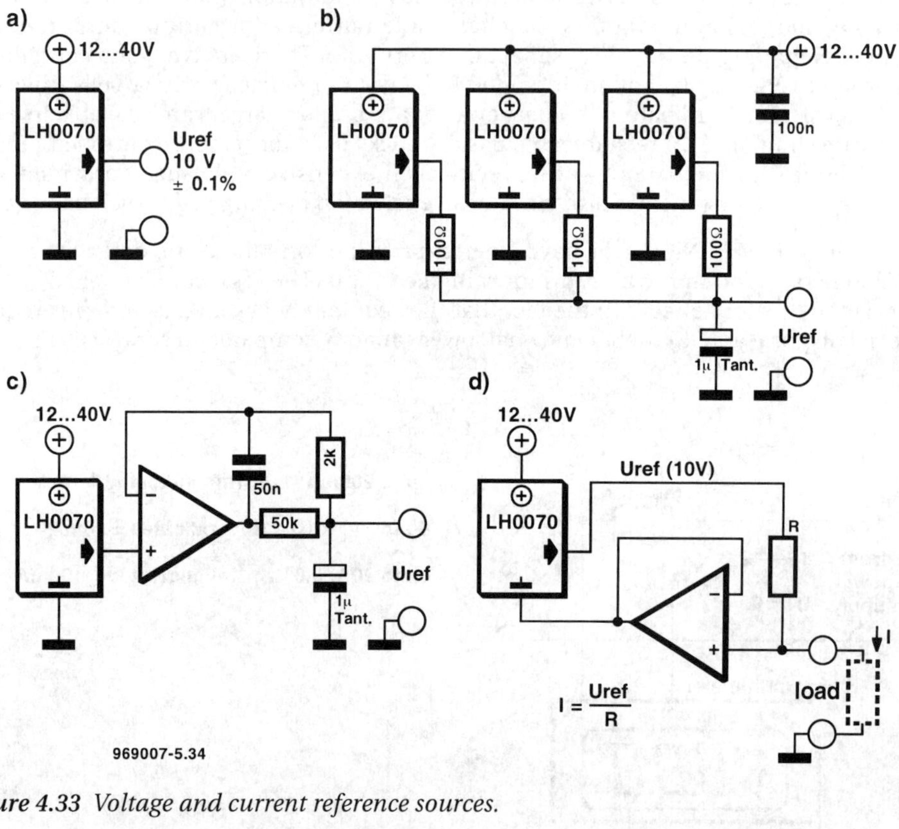

Figure 4.33 Voltage and current reference sources.

a) The simplest application circuit, which in this case provides 10.000 V with 0.1% accuracy. The LH0070 is short-circuit protected and can provide 20 mA maximum.

b) A 'statistical' reference potential source can be constructed by connecting several LH0070 ICs in parallel (more than three can be connected in this manner). All reference potential sources generate a certain amount of noise; when several are connected in parallel the individual noise voltages tend to be averaged out.

c) The noise voltage can also be filtered out. This is an example of such a filter circuit (source: Burr-Brown).

d) An example of a constant-current source.

applications are checking instruments and providing comparison voltages. They can also provide reference potentials for precise voltage measurements, as shown in Figure 4.14c.

Note: a reference potential source can be used to make a DIY test aid with high inherent accuracy (that is, which does not have to be calibrated using precision measuring equipment). The circuit of Figure 4.33a may already be sufficiently accurate; voltage dividers and buffer stages can be added to provide several different reference voltages. Such a reference source can reasonably be powered from a battery. See also Volume 3.

AC voltage sources

Mains transformers. If you need a 50-Hz sinusoidal test signal in a hurry, you can often make do with a mains transformer (for example, a suitable AC mains adapter).

Signal generators. In the ideal case, you should be able to apply any desired waveform (which you have sketched on a computer screen, for instance) to the test object. Suitable signal generators (arbitrary waveform generators) are commercially available. Low-resolution models are even quite affordable (see Figure 4.34b and Volume 3). However, in most cases a regular waveform (such as a sinusoid, square wave or triangular wave) is adequate. Frequently, a rectangular pulse train is all that is needed to allow the signal-handling behaviour of the device under test to be carefully investigated. Suitable signal generators and pulse generators are commercially available.

With regard to DIY equipment, there is an abundance of designs for basic oscillator circuits and complete application circuits. The practical difficulties lie in particular in the areas of frequency stability, amplitude stability, calibration and range switching. Given the prices of ready-made equipment, only very simple designs based on modern integrated circuits are worth considering for DIY construction (see Figure 4.34).

As a further refinement of the circuit of Figure 4.34b, the waveform table can be stored in a FIFO that is loaded by a microcontroller or PC, instead of in a ROM. See Volume 3 for more details.

Loads

Sometimes an artificial load for the device under test forms part of the test setup (a 'dummy' load, for instance for testing an audio amplifier or a power supply). There are various ways to provide such loads.

Fixed resistors. Obviously, the power produced by the test object can be dissipated in a suitably-dimensioned resistor, which means that the power is converted into heat. High-power resistors are available with standard ratings of 10, 15, 25, 50, 100, 200 and 500 watts.

Practical tip: always multiply the calculated power dissipation by a factor of two (the dissipation is $P = U \cdot I = I^2 \cdot R = U^2 / R$) and then choose a resistor with the next-higher power rating.

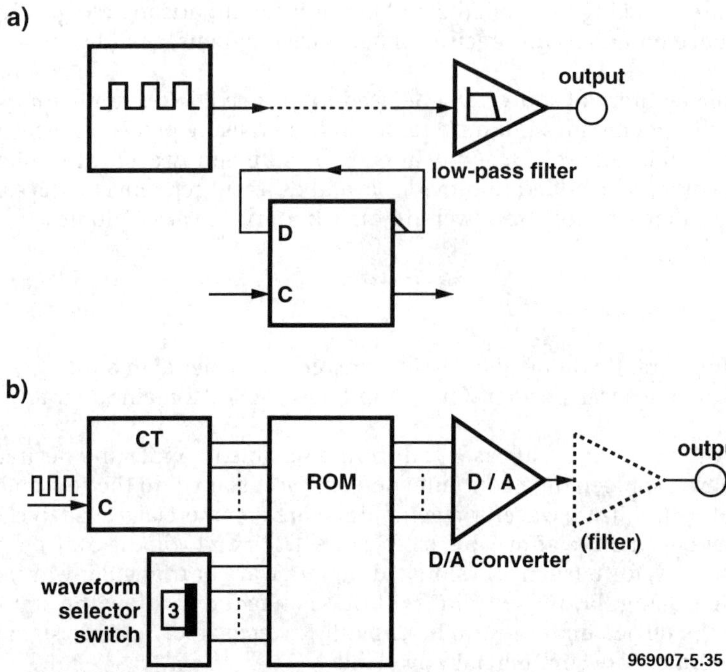

Figure 4.34 Simple signal generators: two sample circuits.

a) A sine wave generator can be made by connecting a low-pass filter to the output of a square-wave signal generator. A harmonic-free signal with quartz-oscillator accuracy can be achieved with a suitable choice of components (such as an analogue universal filter). Tips: (1) a pulse train with any arbitrary duty cycle can be converted to a symmetrical square wave by passing it through a 2:1 divider (flip-flop); (2) rectangular pulse trains reaching into the megahertz range can be generated by microcontroller program loops (see Volume 2).

b) An arbitrary waveform can be generated by using a recirculating counter to address a ROM table that stores successive samples of the desired waveform, together with a digital-to-analogue converter that produces an analogue output signal from the resulting series of digital values. The illustration also shows how a number of different waveform tables stored in a single ROM can be selected via a thumbwheel switch.

The power dissipation ratings of high-power load resistors in metal packages are only valid when the resistors are mounted on suitable heat sinks. The allowable dissipation without a heat sink is 50% of the rated dissipation for 10 to 100 W resistors and 25% of the rated dissipation for 200 or 300 W resistors.

Adjustable resistors. Clearly, adjustable resistors can be used instead of fixed-value resistors for dummy loads, to provide a continuous range of load resistance. High-power wire-wound variable resistors (rheostats) are available with standard ratings of 4 W and 20 W. *Caution:* with a constant applied voltage, the power dissipation increases as the resistance

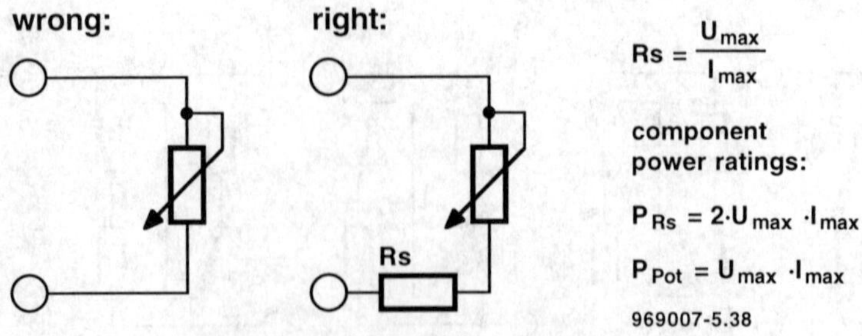

Figure 4.35 Using a variable resistor as an adjustable load.

is reduced (by turning the knob to the left), since the the power goes as 1 / *R*. Near the limit of travel, there is almost no resistance left to limit the current. This means that a variable resistor should only be used in combination with a fixed-value series resistor (see Figure 4.35) that can withstand the maximum applied voltage and the resulting current, unless the range of motion of the variable resistor can be mechanically limited.

Switchable resistors. An obvious idea is to construct a given dummy load by connecting a number of standard resistors in series and/or parallel, in order to achieve the desired resistance and power dissipation values. In this way it is possible to realize a variety of different load resistances using relatively few components (and the prices of high-power resistors are not exactly negligible). Improvised arrangements, such as using test leads and alligator clips to interconnect a bunch of loose resistors lying on the workbench, are not recommended. The resistors should be solidly mounted on some sort of surface (preferably inside an enclosure). There are several approaches to constructing switchable dummy loads, as follows:

1 The terminals of the resistors can be individually connected to 4-mm test jacks. These can be interconnected using either test leads or – given a suitable mechanical layout – shorting jumper plugs.

2 Switches can be used to make the necessary connections without using jumpers or test leads. With high load currents (10–20 A) only adequately-rated toggle switches can be used for this purpose. Figure 4.36 shows a suitable circuit design, which allows a pair of resistors to be used individually, as a series-connected pair or as a parallel-connected pair. An example of such a design in described in Section 5.2.3; see also Volume 3. *Note:* normal rotary switches are only suitable for low currents (< 1 A). High-current rotary switches are rather expensive.

Electronically-switchable resistors. Power supplies must be operated with varying loads in order to test their dynamic behaviour and voltage regulation. Quickly-changing load conditions, which essentially amount to switched loads, are especially important. (*For example:* the current consumption of a modern processor IC changes within a few tens of nanoseconds from almost nothing to several ampères when it switches from the low-power mode to the normal operational mode.) An obvious way to emulate such load conditions is to use transistors to switch suitable resistors in and out of the load circuit (see Figure 4.37, and Volume 3 for more information).

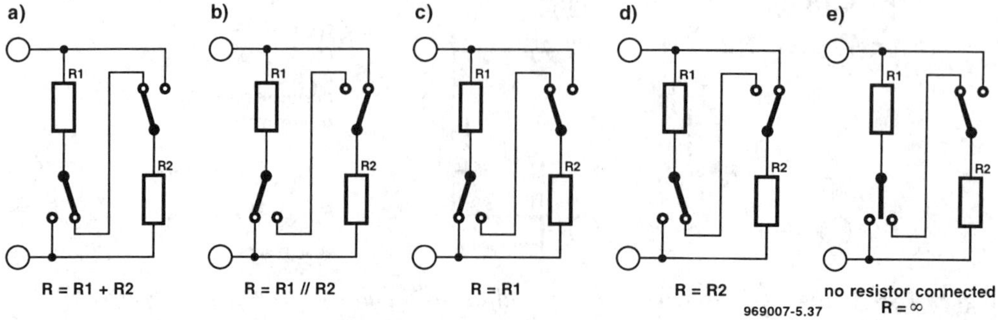

Figure 4.36 Switchable load resistors.
a–d) illustrate the individual operating configurations.
e) If one of the switches is a centre-off type, then an 'infinite' load can also be selected.

Electronic loads. An arbitrarily variable load can be realized by using a power transistor to replace the load resistor. Power FETs are particularly suitable for this purpose. The principle is illustrated in Figure 4.38. The load current flows through a FET that acts as a variable resistor; the drain-source resistance R_{DS} depends on the control voltage applied to the gate. The operational amplifier is connected in a feedback loop that maintains the potential difference between its two inputs at zero volts. If the voltage that the load current produces across the shunt resistor is greater than the reference voltage, the output of the operational amplifier goes more negative, which causes the drain-source resistance of the FET to increase and thus reduce the load current. If on the other hand the voltage across the shunt is lower than the reference voltage, the output of the operational amplifier

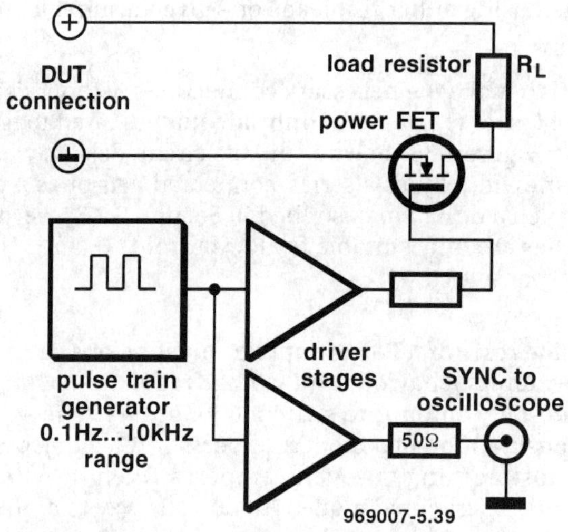

Figure 4.37 Switched load test circuit.

becomes more positive and the drain-source resistance of the FET is reduced. This circuit has several working modes, depending on how the operational amplifier is driven:

1 Constant current: if a fixed reference voltage is applied to the operational amplifier, the load current is held constant (I = constant, $R_{DS} = U / I$).

2 Constant resistance : if the reference voltage is proportional to the load voltage, the load current depends on the load voltage and the circuit behaves like a 'true' resistor (R_{DS} = constant, $I = U / R_{DS}$).

3 Constant voltage (not illustrated): the operational amplifier compares the reference voltage to the load voltage instead of the voltage across the current-sensing shunt resistor, so the load resistance depends on the load voltage (U = constant, $R_{DS} = U / I$).

4 Variable load: the reference voltage is varied; it can for instance be provided by a computer-driven digital-to-analogue converter. Any desired load characteristic can be simulated in this mode.

Switched resistors versus electronic loads. Obviously, an electronic load is much more versatile than electronically-switched load resistors. However, it is also a fair amount more costly, since the transistor is operated in continuous mode instead of in switching mode. In the circuit of Figure 4.37 the power is dissipated in the load resistor, while in the circuit of Figure 4.38 it is dissipated in the transistor – and this can be 100 or 200 W! This means that a passive heat sink, forced-air cooling or similar means of cooling the transistor is mandatory. Also, an electronic load is slower than a switched-resistor load in switching mode, when a sudden change in the load must be simulated (the rise time of the FET operated as a switch is less than 1 µs, while the rise time in the electronic-load configuration is more than 10 µs).

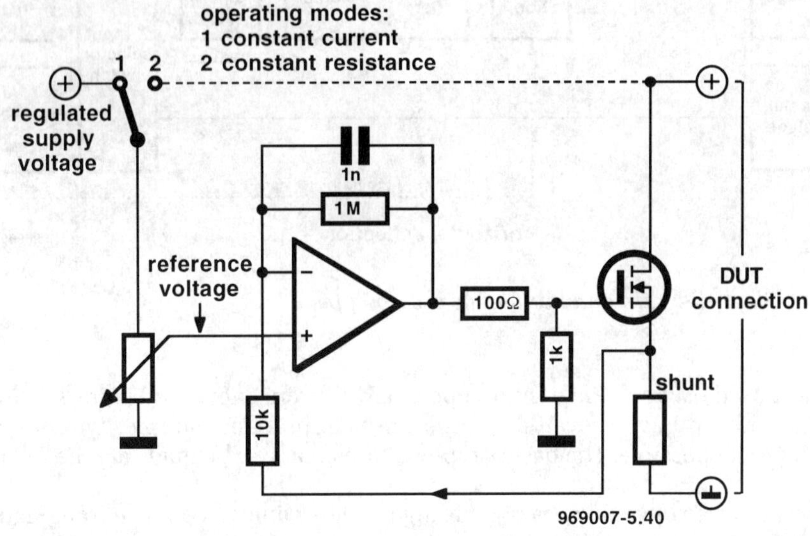

Figure 4.38 Electronic load (illustrative circuit).

4.3 Oscilloscopes

4.3.1 The conventional (analogue) oscilloscope

Construction

An oscilloscope is, in a manner of speaking, built around a cathode ray tube (CRT) (see Figure 4.39). Cathode ray tubes for oscilloscopes use electrostatic deflection. The voltage to be measured is applied to a set of deflection plates that control the up-and-down motion of the electron beam (vertical or Y-axis deflection). To display the behaviour of the measured voltage over time, the beam is swept at a constant rate from left to right (horizontal or X-axis deflection). A sawtooth (ramp) deflection voltage is generated by the oscilloscope's internal timebase circuitry to drive the horizontal deflection plates.

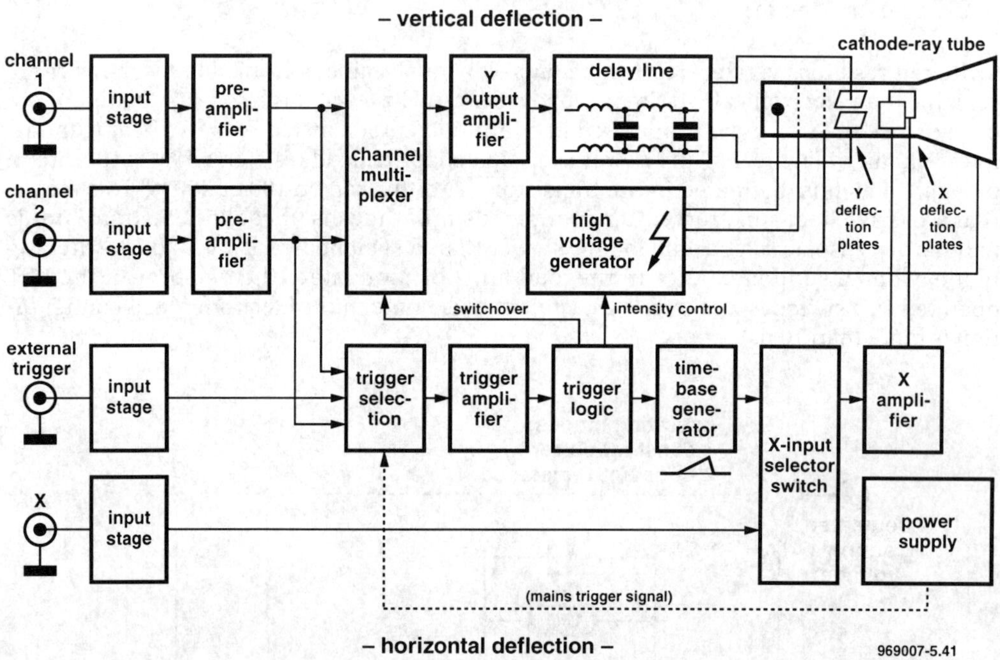

Figure 4.39 Block diagram of an analogue oscilloscope.

The simplest oscilloscope. The oscilloscope block diagram shown in Figure 4.39 includes several features that have proven their usefulness in practice. Simpler types of oscilloscope are however possible. The basic simplifications that can be made are the following:

1 No delay line: the consequences of omitting the delay line can be seen from Figure 4.41 and the accompanying explanation.

2 No external X-axis input: this makes certain measurement techniques fundamentally impossible (see Section 4.3.5). In any case, these techniques (such as measuring with Lissajous figures) are rarely used in modern practice and are certainly not essential.

3 Only one Y channel (single-channel oscilloscope): this means that the input stage and preamplifier for the second channel are omitted, as well as the channel switch. This yields the 'classical' form of the oscilloscope, which can display only a single waveform.

How many waveforms should be concurrently displayable? Sometimes it is sufficient to display only a single waveform:

- if the objective is to quickly evaluate a signal (level, pulse edges, absence of under- or overshoots, and so on), so that the actual waveform of the signal is not particularly important,
- if an individual signal is to be carefully checked or measured (waveform, frequency, amplitude and so on). This applies primarily to typical analogue circuits with linear signal flows ('front to back'), in which it is desired to trace specific signals through the various subassemblies (a certain stimulus is present at the input of a subassembly, and the output signal is observed on the oscilloscope). *Note:* individual signals in digital circuits that can be observed in this manner are (among others) reset signals, clocks and supply voltages.

In other words: in the case of an individual signal, its exact behaviour with respect to time is either not all that important (*for example:* checking whether the signals on a bus look 'reasonable') or it is already known (*for example:* a 1-kHz rectangular pulse train with an amplitude of 25 mV is applied to the input of an audio amplifier, and we observe the amplifier's output signal on the oscilloscope).

However, in modern fault-finding practice it is frequently necessary to observe the waveform of a particular signal with respect to that of some other signal (for example, a data line signal with respect to the associated control or enable signal). This means that both signals must be displayed at the same time. This is possible with either a dual-beam oscilloscope or a dual-trace oscilloscope.

A *dual-beam oscilloscope* contains a dual-beam CRT. Such a CRT has two independent beam-generation assemblies, so that two different signal waveforms can be displayed at the same time. Two vertical amplifiers are required. There is usually only a single common horizontal deflection system for the two beams, although there are dual-beam oscilloscopes that have two independent horizontal deflection systems. Dual-beam oscilloscopes are no longer manufactured.

A *dual-trace oscilloscope* contains a CRT that has only a single beam-generation assembly. The two signal waveforms are displayed by rapidly switching the input of the final vertical deflection amplifier between the outputs of the vertical input stages or preamplifiers of the two channels. The visual effect is that two independent waveforms are concurrently displayed on the screen.

Multichannel oscilloscopes can be produced by extending the principle of the dual-trace oscilloscope to more than two channels. Commercial multichannel oscilloscopes typically have three or four channels.

Channels and traces. The number of channels and the number of traces (displayed waveforms) need not be the same. Sophisticated oscilloscopes can display the signals applied to the input channels from different 'points of view', in a manner of speaking. *Examples:* (1) all channels individually, (2) the arithmetic sum or difference of a pair of channels, (3) a single channel displayed as two traces with different horizontal deflection rates (two different time scales) – the lower rate displays a larger number of pulses or cycles, while the higher rate shows individual details (such as pulse edges) more clearly. For instance, there are '4-channel, 10-trace' oscilloscopes and '3-channel, 8-trace' oscilloscopes.

Modern analogue oscilloscopes are typically two-channel instruments, so that we can limit out attention to this type of oscilloscope.

Functional elements

Input stage (see Figure 4.40). Oscilloscope input stages can typically be switched between AC coupling, DC coupling and earth (0 V or *GND*). They include switchable attenuators (voltage dividers) for selecting the desired vertical sensitivity (measurement range). These attenuators must maintain a constant division ratio (attenuation ratio) over the entire usable frequency range. Since parasitic capacitances cannot be neglected at high frequencies, the attenuators are constructed as frequency-compensated voltage dividers. Such voltage dividers have capacitors connected in parallel with the resistors. The division ratio is determined by the resistors at low frequencies and by the capacitors at high frequencies. The desired frequency compensation (constant division ratio for all frequencies) is achieved when the relationship shown in Figure 4.40 is satisfied.

Notes:

1 With AC coupling, only the AC portion of the input signal is passed through (with a low-frequency cutoff point that is usually on the order of a few hertz).

2 The input can be switched to earth at any time, even when a signal is applied to the input, since the input to the subsequent preamplifier stage is connected to earth rather than the oscilloscope input. This can be used to quickly determine the location of the zero-voltage level (baseline) for measuring the amplitude of an observed signal.

3 Frequency-compensated attenuators must be adjusted (calibrated) from time to time; see Section 4.3.5.

4 Selectable input coupling and range switching are not necessarily provided for all inputs (for instance, only a potentiometer is provided for the trigger input, and the X input has only a limited number of selectable ranges). *Caution:* some oscilloscopes with more than two channels have 'full' range selection for only two of the channels; the remaining channels have only a few ranges (2 to 4), or (in many cases) only a single range. The utility of these channels is correspondingly limited (to observing digital logic signals, for example).

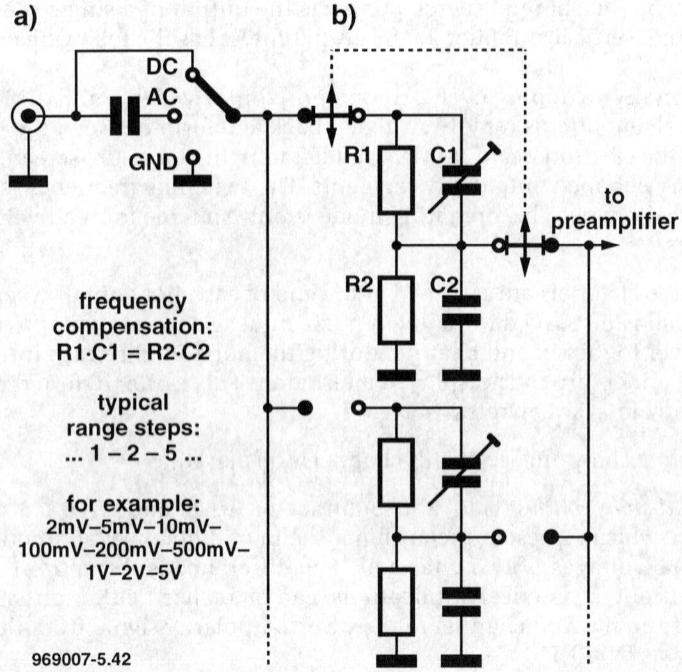

Figure 4.40 Construction of an oscilloscope input stage.
a) input coupling switch (simplified example)
b) frequency-compensated input attenuator (range switch).

5 The circuit shown in Figure 4.40 is only a basic design. Actual oscilloscopes can have the following variations: (1) complicated resistor/capacitor arrangements, (2) multi-stage range switching (an input attenuator with relatively coarse steps combined with switchable gain for the amplifier stage(s) to further refine the range selection). *Caution:* range selection by means of switchable amplifier gain is an economy measure; with a good amplifier the measurement range should be selected exclusively by the input attenuator. Since this requires high-quality components for the attenuator, the manufacturer may be tempted to save money by reducing the number of attenuator steps. The disadvantage of a switched-gain amplifier is that increased gain (to provide increased sensitivity) results in a decreased frequency range (bandwidth). This is typically not mentioned in the brochure. In case of doubt, test!

6 The input ranges of modern instruments are typically switched in a 1-2-5 sequence.

Preamplifiers. The preamplifiers increase the amplitudes of the input signals to levels that allow them to be used elsewhere in the oscilloscope. A preamplifier's input impedance must be high enough that it does not excessively load the input attenuator or the input signal source. The output impedance must be low enough to allow the amplified signal to be supplied to the following stages. The gain of the preamplifier can usually be continuously adjusted (via a potentiometer).

Channel switch. The channel switch connects the output of a single preamplifier to the input of the final vertical amplifier at any given time. It has the following operating modes:

1 *Chopped:* many switching cycles in one horizontal sweep interval. The input to the final vertical amplifier is rapidly switched back and forth between the two input channels while the electron beam moves from left to right across the screen. Each displayed trace is thus 'chopped' into many segments. The switching frequency is typically several hundred kilohertz. This operating mode is only suitable for relatively low horizontal sweep rates.

2 *Alternate:* the channels are switched at the end of each horizontal sweep interval. If the first channel is displayed for a given horizontal sweep interval, the channel switch changes over to the second channel during the horizontal retrace interval and so on. The two channels are thus displayed alternately. This is the preferred operating mode for high horizontal sweep rates.

3 *Single-channel:* only one (selected) channel is displayed.

4 *Combined display modes* ('Add' and 'Subtract' or 'A + B' and 'A – B'): a single waveform is displayed which represent the arithmetic sum or difference (addition or subtraction) of the signal voltages of two channels. The difference (subtraction) mode, which is quite important in practical applications, can be realized either directly or by adding the signal from the first channel to the inverted-polarity signal from the second channel (ADD plus INVERT).

Final amplifier. Electrostatic deflection requires relatively high deflection voltages. An amplifier that can provide the necessary voltages must be connected to each pair of deflection plates. The Y-axis or vertical amplifier is the actual measurement amplifier; the signal to be observed is applied to its input (Y input). Most oscilloscopes also allow the input of the horizontal (X-axis) amplifier to be switched between the internal timebase generator and an external X input.

Delay line. The function of the delay line shown in Figure 4.39 is illustrated in Figure 4.41:

a) We want to see this pulse edge on the oscilloscope. This requires the horizontal time base to be suitably triggered, which means that we must define a specific reference voltage for triggering the horizontal deflection circuit. If the only signal that can be compared to this reference voltage is the signal to be observed (which is typically the case), then the time base must be triggered 'internally'. This means that the horizontal sweep will be triggered when the signal to be observed crosses a particular trigger level U_{tr}. A certain amount of time t_d passes between the trigger time t_1 and the time t_2 at which the horizontal sweep actually starts.

b) An observable trace is displayed starting at the time t_2. Without a delay line, only the portion of the waveform following t_2 can be observed; almost all of the rising edge of the pulse is lost.

c) To allow the complete pulse edge to be seen, the signal must be delayed on its way to the vertical amplifier by at least the time interval between its zero-crossing time and t_2. The delay line provides a delay equal to t_{dl}.

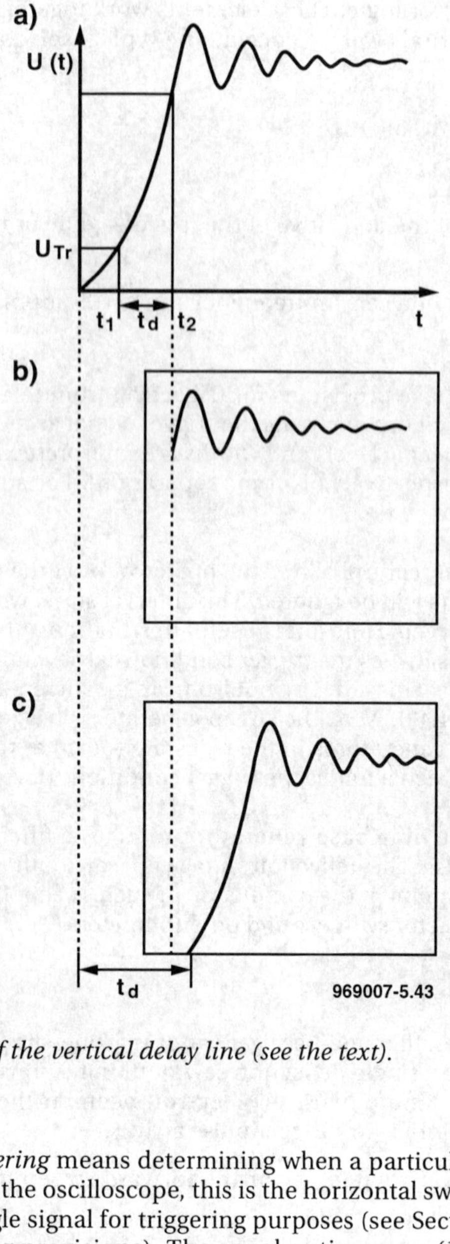

Figure 4.41 The effect of the vertical delay line (see the text).

Trigger selection. *Triggering* means determining when a particular measurement process is started (in the case of the oscilloscope, this is the horizontal sweep). Conventional oscilloscopes evaluate a single signal for triggering purposes (see Section 4.3.5 regarding more sophisticated triggering provisions). The usual options are: (1) internal triggering, in which the trigger signal is derived from the signal to be displayed (which means that it is tapped off from the signal in the vertical amplifier), (2) external triggering, in which the trigger source is a separate external signal and (with some oscilloscopes) (3) mains triggering, in which the trigger signal source is an AC signal at the mains frequency. See Section 4.3.5 for more information about trigger selection.

Trigger amplifier and trigger logic. These elements work together to evaluate the trigger signal and cause a horizontal sweep to occur. The typical selection options and adjustments are:

- trigger signal coupling: AC or DC.
- trigger level;
- trigger slope: the sign of the derivative of the trigger signal, or the effective pulse 'edge' (positive or negative);
- selectable suppression of a certain frequency range (LF suppress, HF suppress, noise suppress).

Some oscilloscopes also have provisions for triggering from television synchronization signals. In this case the oscilloscope treats the trigger signal as a composite colour television signal. Pulses with specific levels and widths are interpreted as horizontal or vertical synchronization pulses respectively (TV sync separation). For additional trigger options see Section 4.3.5.

Hold-off setting: a 'hold-off' control allows the operator to set the length of a time interval during which trigger events will be ignored. This interval starts with the trigger event that initiates the horizontal sweep. Hold-off is useful in situations in which the waveform of the selected trigger signal satisfies the trigger conditions at several different points, so that a stable display cannot be achieved. The hold-off can be adjusted to exclude 'unsuitable' trigger events (see Figure 4.42). *Note:* the sweep generator is triggered under the following conditions: (1) the sweep generator is in the rest state (beam at the left-hand edge of the screen) and (2) the trigger signal fulfills the trigger conditions (level and slope).

Time base generator. The time base generator produces a linear deflection voltage in response to the trigger pulse. The deflection voltage is a sawtooth, since the electron beam must return to its starting point at the completion of each sweep. The sweep time or speed can be set by means of selector switches and potentiometers.

Notes:

1. In modern oscilloscopes, the time-base generator is a 'one-shot' pulse generator, which means that it produces a single deflection sawtooth in response to a trigger pulse and then returns to the rest state (with the electron beam at the left-hand edge of the screen), where it stays until a new trigger pulse arrives.
2. The electron beam is only gated on for the forward sweep (from left to right); it is blanked during the retrace.
3. Two operating modes are available: 'normal' and 'automatic' triggering. In the normal mode (NORM) the oscilloscope behaves as just described (no trigger = no display). In the automatic (AUTO) mode, by contrast, the time base generator 'free runs': it is periodically retriggered even in the absence of a trigger signal. If no input signal is applied a horizontal line is displayed; as soon as a signal that provides trigger pulses is applied, the oscilloscope operates in the normal mode.

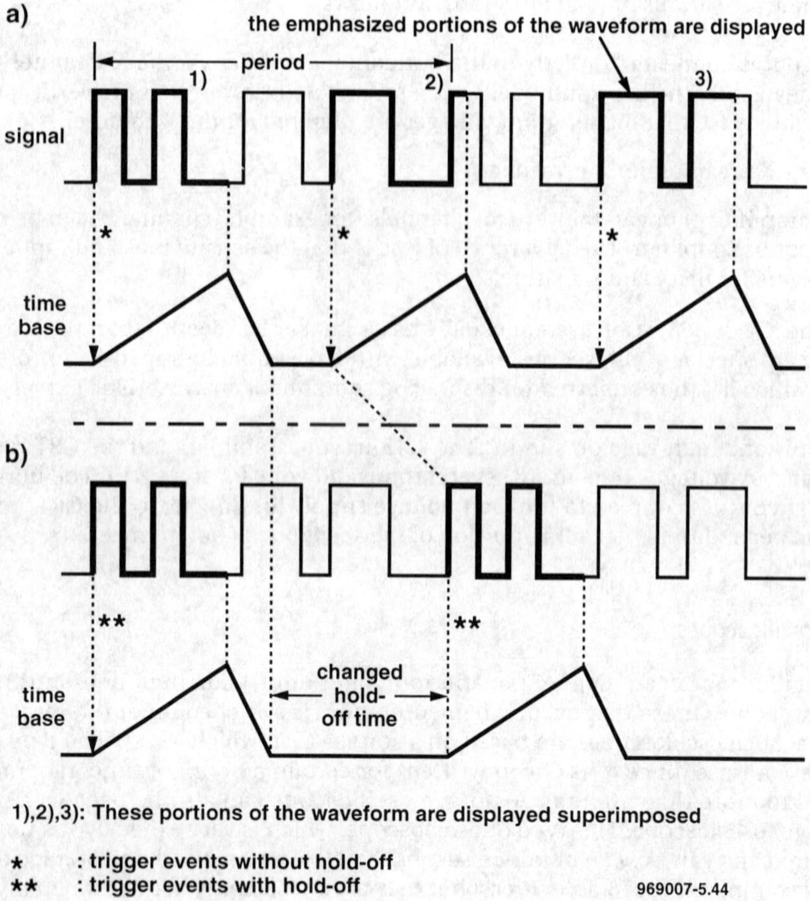

Figure 4.42 Using hold-off. (source: Hameg)

a) A second trigger event occurs following the completion of the first sweep. However, a different portion of the signal waveform is displayed if the oscilloscope is triggered at this point. This results in an unstable or unclear display that cannot be evaluated, since different portions of the observed signal are displayed on top of each other.

b) With a suitable hold-off time setting, the undesired trigger events are suppressed, so that the same portion of the signal waveform is displayed each time.

4 Very old and very simple oscilloscopes have a free-running time base generator. This cannot be triggered, but must instead be *synchronized* to the signal being observed.

X-channel switch. The X-channel switch selects either the internal time-base generator or the external X input as the signal source for the horizontal amplifier. See Section 4.3.5 regarding various measurement techniques that make use of the external X input.

X input. There are various implementations, as follows:

1 The X input is connected directly to the horizontal amplifier via the X-channel switch, which means that the horizontal deflection sensitivity is lower than the vertical deflection sensitivity (so that higher input voltages are required for the X channel).

2 A separate X preamplifier is provided.

3 The preamplifier of one of the vertical channels (for example, channel 2) can be used as an X-input preamplifier. The advantage of this is that the sensitivities and input ranges are the same for the X and Y channels.

Caution: the X-axis deflection system usually has a smaller bandwidth than the Y-axis system. There are special oscilloscopes available with have equal bandwidths for both systems (and which are thus preferred for displaying component characteristic curves).

Power supply and high-tension supply. The various subassemblies and the CRT require a variety of supply voltages (including several thousand volts for the CRT). The horizontal deflection circuitry is coupled to the high-voltage supply (or directly to the CRT) to blank the electron beam during the falling portion of the sawtooth signal (retrace).

Storage oscilloscopes

Storage oscilloscopes can display non-periodic and non-recurring ('one-shot') signal waveforms, such as those that occur when equipment is switched on or off, reset signals and the like. Such oscilloscopes are based on a storage CRT, which is a special type of CRT that can save a trace (once it has been written) for certain amount of time, ranging from 10 minutes to more than an hour. A storage oscilloscope can also be used as a normal (non-storage) oscilloscope. This type of oscilloscope, which is very expensive, is no longer manufactured and can only be obtained second-hand. Even second-hand storage oscilloscopes still command respectable prices, but they are significantly better than many models of digital storage oscilloscopes for observing non-recurring events.

4.3.2 The digital storage oscilloscope (DSO)

Construction

In a digital storage oscilloscope (see Figure 4.43), the CRT is purely a display device. The actual measurement device is a fast digital-to-analogue converter, whose output data are stored sequentially in a memory. The data thus stored can be evaluated and displayed.

The functions of a digital storage oscilloscope can be implemented in various ways. Figure 4.43 shows a relatively simple and easily-understood design. The input stages, preamplifiers, channel switch and trigger selection elements are constructed in practically the same manner as for a conventional analogue oscilloscope. The additional functional elements can be briefly described as follows:

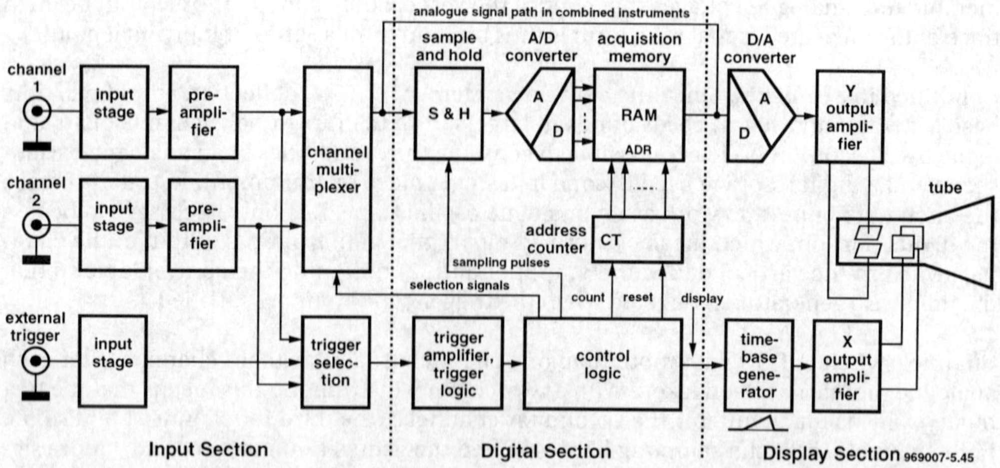

Figure 4.43 Block diagram of a digital storage oscilloscope (DSO).

- Sample and hold: the signal that is to be measured is periodically sampled. Each sampled analogue value must be held until it has been digitized.
- Analogue-to-digital converter: each analogue sample is converted to a binary number that corresponds to the amplitude of the sampled voltage (digitization).
- Acquisition memory: the digitized amplitude data are sequentially stored in the acquisition memory, which is continuously addressed by a counter.

Acquisition process:

1 The system is initially at rest; the address counter stands at zero.

2 The selected trigger condition is satisfied. As a result, the control logic starts the acquisition process: sample the signal, convert the sampled voltage to a binary number, enter the binary into the memory and increment the address counter. This sequence is repeated until the memory is full.

Display. Once the acquisition memory is full, the data can be used in almost any desired manner, including transferring them to a PC for further processing or displaying them as a signal waveform. A CRT of the type used in an analogue oscilloscope is not required for the display – some DSOs have a built-in VGA monitor and some have LCD screens. A microprocessor can be used to generate the display from the data that it reads from the acquisition memory. Figure 4.43 shows a simple solution that manages without a microprocessor: an actual oscilloscope-type CRT is used, with the associated deflection amplifiers and so on. The data to be displayed are read from the acquisition memory, starting from address zero, and a horizontal deflection signal is generated in parallel by a time base generator. A digital-to-analogue converter transforms the binary data read from the memory into analogue voltages that control the vertical deflection of the electron beam. A trace of the measured signal waveform is thus built up from a series of individual points.

Combined instruments. Since the above arrangement (a CRT, deflection amplifiers, time base generator and so on) corresponds in large part to the functional elements of an analogue oscilloscope, it is clearly possible to combine the capabilities of an analogue oscilloscope and a digital storage oscilloscope in a single piece of equipment. In order to make the circuit of Figure 4.43 work as an analogue oscilloscope, it is only necessary to bypass the digital portion, which means (1) providing a signal path from the output of the channel switch to the vertical deflection amplifier and (2) designing the control logic so that the time base generator is triggered when the trigger conditions are satisfied.

Channel switch. A DSO with more than one channel usually switches channels after each analogue sample has been taken. With a two-channel instrument, this means that the first binary value is for channel 1, the second for channel 2, the third for channel 1 and so on. Both channels share the sampling hardware and the acquisition memory and – more significantly – the sampling rate, which is critically important for the correct reproduction of the sampled waveforms. You must carefully read the specifications. A better approach is to provide separate conversion and memory hardware for each channel.

X input. A separate X input is not common. Instead, an X/Y operating mode is provided in which the X values are acquired via one channel and the Y values via the other channel.

Basic sampling theory

The objective of sampling is to be able to reproduce the actual signal waveform with as high a degree of fidelity and accuracy as possible. The following factors impede the realization of this objective:

- The signal is sampled only at discrete points in time; the parts of the waveform that lie between the sampling times cannot be observed.
- The capacity of the acquisition memory (the memory depth) is limited; it can only store a finite number of sample values.
- When the acquisition memory is full, it must be read out so that the data can be displayed. It is not possible to observe the signal that is being measured while the acquisition memory is being read.
- The analogue-to-digital conversion process is fundamentally subject to digitization errors (see Section 4.1.3).

Sampling rate and reproduction fidelity. Figure 4.44 illustrates the obvious fact that details of a waveform can be lost if it is not sampled at a high enough rate. The density of the sampling points is determined by the sampling rate or sampling frequency f_s, which is measured (as a train of sampling pulses) in hertz or in samples/s (which amounts to the same thing).

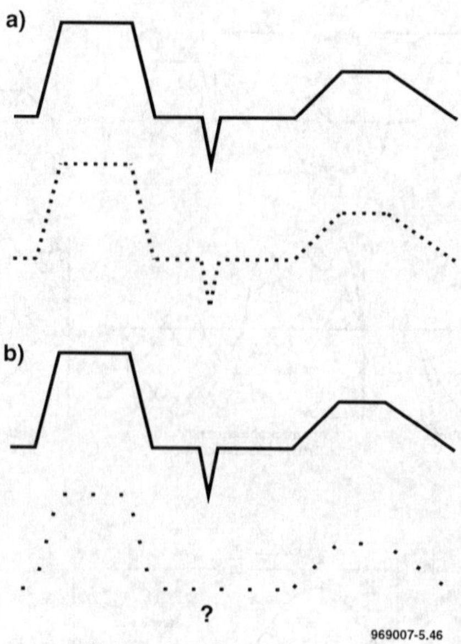

Figure 4.44 Reproduction of a sampled signal waveform (a) with a high sampling rate and (b) with a low sampling rate. In each case, the actual waveform is shown at the top and the sample values are shown at the bottom.

The sampling theorem (Nyquist criterion) describes the relationship between the highest frequency f_u of a bandwidth-limited periodic signal and the minimum sampling frequency f_s that is required to allow the sampled waveform to be exactly reconstructed:

$$f_s \geq 2 \cdot f_u$$

This means that the waveform must be sampled at a rate that is at least twice the upper limit of its bandwidth. This relationship is well known, but it is not always correctly interpreted. Figure 4.45 should help to clarify what this relationship means.

Aliasing. If a sinusoidal signal with a frequency f_u is sampled at a rate $f_s < 2 \cdot f_u$ (or put another way, if the bandwidth of the signal to be sampled is not limited to $f_u \leq 2 \cdot f_s$), then a signal will appear at the output of the arrangement of Figure 4.45d at the alias frequency $f_a = f_s - f_u$. This cannot be removed by any sort of low-pass output filter!

Oversampling. This term is commonly used to emphasize that a bandwidth-limited signal (with a cutoff frequency f_u) has been sampled at a rate that is greater than the minimum sampling frequency ($f_s > 2 \cdot f_u$). The *oversampling rate* is the ratio of the actual sampling frequency to the minimum required sampling frequency, or in other words the value of $f_s / 2 \cdot f_u$). *For example:* if a signal with a cutoff frequency f_u = 100 MHz is sampled at a rate f_s = 500 megasamples/s, the oversampling rate is 500 ÷ (2 · 100) = 2.5.

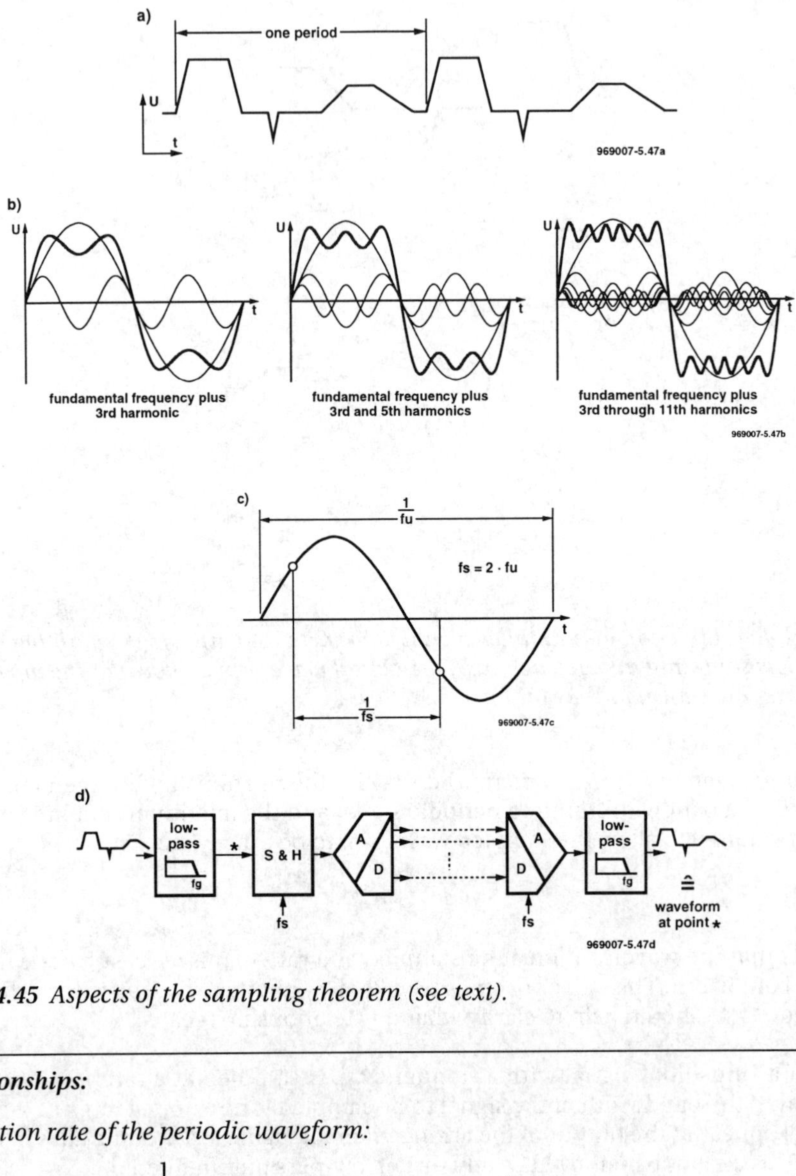

Figure 4.45 Aspects of the sampling theorem (see text).

Relationships:

Repetition rate of the periodic waveform:

$$f_0 = \frac{1}{t_p} = \text{fundamental frequency (first harmonic)}$$

The full waveform can be represented as infinite series of sine waves (Fourier series):

$$u(t) = a_1 \cdot \sin\omega t + a_2 \cdot \sin 2\omega t + a_3 \cdot \sin 3\omega t + \ldots = \sum_{i=1}^{\infty} a_i \cdot \sin i\omega t$$

where $\omega = 2\pi f_0 = \dfrac{2\pi}{t_p}$

Explanation of Figure 4.45:

a) *The waveform of Figure 4.44 is shown here as a periodic signal. The fundamental or repetition frequency can be calculated from the period* t_p *as* $f_0 = 1 / t_p$. *Every periodic waveform can in principle be constructed by the superimposition of sinusoidal waveforms whose frequencies are integer multiples of the fundamental frequency* f_0 *(that is,* f_0, $2{\cdot}f_0$, $3{\cdot}f_0$ *and so on). The fundamental frequency is also called the first harmonic,* $2{\cdot}f_0$ *is called the second harmonic or first overtone, and so on. In the exact mathematical sense, such a sequence (a Fourier sequence) has an infinite number of terms.*

b) *Since infinitely high frequencies do not exist, the sequence must be truncated after a certain number of terms. The signal waveform is thus approximately reproduced by the combination of a finite number of sinusoidal waveforms. Here, for example, we can see that a symmetrical square wave can be approximated reasonably well by a relatively small number of sinusoidal waveforms. (Note: the Fourier series of a symmetrical square wave contains only odd harmonics.) Even using only the first three harmonics gives a 'usable' pulse waveform that certainly could be applied to the input of a normal logic IC. If the series is extended to the* 11th *harmonic, the resulting waveform is quite acceptable, with steep edges and only a small amount of ripple on the flats.*

c) *A sinusoidal waveform can be exactly reconstructed from just two sample values that are separated by one half of the period of the waveform – as long as we know that the two samples represent a sine wave with a frequency* f_u. *This means that a set of sample values that have been taken at a frequency* $f_s = 2{\cdot}f_u$ *can be used (with respect to the frequency* f_u*) to reconstruct a sinusoidal signal, but not an arbitrary waveform.*

d) *This is the 'technical environment' in which the sampling theorem is valid: we satisfy ourselves with approximating the waveform with a finite number of harmonically related sine waves. The frequency of the highest harmonic that is included in the series is then the limiting frequency* f_u. *We must therefore sample at a rate* $f_s \geq 2{\cdot}f_u$. *The signal that is to be sampled must be bandwidth limited, since we are not allowed to process signals with frequencies higher than* f_u. *To this end, the signal is passed through a low-pass filter with a cutoff frequency* f_u *before it reaches the sample-and-hold stage. The sampled data can then be stored or processed in any desired manner.*

   To reconstruct the sampled signal, we feed the sample values to a digital-to-analogue converter at the same rate as the original sampling frequency f_s. *The output of the converter is again passed through a low-pass filter with a cutoff frequency* f_u *in order to to limit its bandwidth. The signal that appears at the output of this filter is the reconstructed input signal, at least in the form that it had after passing through the input low-pass filter. For example: suppose that we want to sample a symmetrical square-wave signal whose fundamental frequency* f_0 *is 1 MHz, and we wish to retain all harmonics up to and including the eleventh. This means that* $f_u = 11$ *MHz, so our low-pass filter must cut off all frequencies above 11 MHz and we must sample at 22 MHz. The reconstructed signal that appears at the output will look like that on the right-hand side of Figure 4.45b.*

Theory, practice and advertising. You must pay attention to the following factors:

1 The sampling theorem says only that sampled data that have been taken at a rate f_s can be used to reconstruct a sinusoidal waveform with a frequency of $f_s/2$ – but not just any arbitrary periodic waveform with a repetition rate of $f_s/2$.

2 The sampling theorem is only exactly valid when the signal to be sampled has been subjected to a precise bandwidth limitation, which means that it has been passed through an ideal low-pass filter with a cutoff frequency $f_u = f_s/2$ before being sampled. An ideal low-pass filter, however, distorts all non-sinusoidal signals (see Section 4.3.3 and in particular Figure 4.55a). Non-ideal low-pass filters allow signals to pass that unavoidably result in the generation of alias signals.

3 Any desired periodic waveform can be exactly reconstructed from an infinite series of sine waves (and approximately reconstructed from a finite series of sine waves). The Fourier decomposition of a waveform is however a mathematical idealization that is only valid if the individual sine waves are in phase (see Figure 4.45: the zero crossings of all of the harmonics must coincide with the zero crossing of the fundamental frequency). The sine waves of Figure 4.45 are lines on paper and can easily be made to add up to the desired approximation of a square-wave signal, but real signals need a certain amount of time to arrive at a given point (cable and circuit propagation delays). A time delay is however the same as a phase offset between the input and output signals. See Section 4.3.3 and in particular Figure 4.56 regarding the consequences of such delays.

4 The specified cutoff frequency of a digital oscilloscope, as given in brochures and data sheet specifications, usually refers to the maximum frequency of a sinusoidal signal that can be correctly processed by the equipment. Also, a sampling rate that is more than twice the specified upper frequency is often promoted as 'oversampling' for advertising purposes.

5 Regarding the cutoff frequency, our actual requirement depends on the types of measurements to be made. For cursory checks of rectangular waveforms (such as clocks and bus signals), a bandwidth that includes the third harmonic (or better yet, the fifth harmonic) of the base frequency is usually adequate. However, in this case noise pulses and so on will be lost. The important point is that the narrower the pulses and the shorter the signal transition times that we wish to observe, the higher the maximum sinusoidal frequency must be (or the larger the number of included harmonics). *For example:* the minimum cutoff frequency for observing a 10-MHz digital signal should be at least 30 MHz, and preferably 50 MHz. Now suppose that we also want to see 10-ns noise pulses. Let's assume that such pulses occur periodically at an interval of 10 ns, which corresponds to a repetitive waveform with a period of 20 ns or a frequency of 50 MHz. Here again, we want to capture at least the third harmonic, and preferably the fifth harmonic. This means that our equipment needs to have a cutoff frequency of 150 MHz (or better yet 250 MHz). With a cutoff frequency of 150 MHz (or 250 MHz), the sampling rate must be at least 300 MHz (megasamples/s) or 500 MHz respectively. Another way of putting this is to say that a DSO with specifications that appear to be quite impressive – 150 MHz upper frequency and 300 megasamples/s – is only suitable for making measurements in logic circuits running at a maximum clock rate of 10 MHz (which is relatively feeble by current standards), and even then you will not see everything that may be important for recognizing and tracking down faults.

6 Fundamentally, although the cutoff frequency is a commonly used specification (and is therefore popular with the advertising crowd), it is totally inadequate as a means of characterizing the suitability of an oscilloscope for a particular task. Instead, specifications that relate to the time domain instead of the frequency domain (in other words, the impulse response instead of the frequency response) are more suitable. The most important characteristic is this regard is the rise time (see Section 4.3.3).

7 Although the usability of an analogue oscilloscope for a given task can usually be judged from its signal-handling characteristics (such as rise time and sensitivity), the usability of a DSO is also dependent on specific features that are included in its digital and display hardware. These are discussed below and in Sections 4.3.3 and 4.3.4.

Fundamental time-domain relationships

Let's first investigate the basic questions of how a representation of a signal waveform can be displayed and what the worst-case signal forms are that we can expect to encounter (independent of what the signal-handling capabilities of the oscilloscope may be). The following questions arise naturally in this regard:

- What can actually be portrayed on a display screen?
- What types of waveforms can actually be expected, given the electrical and physical characteristics of the circuits being investigated, and in particular, what are the details of these waveforms?
- How many sample values can be stored?

The image raster. The relationship between the sampling rate and the fidelity of the reproduced signal can be investigated in a very simple manner. We can consider sampling to be equivalent to transforming a pictorial representation of the signal waveform into a raster image. We are all familiar with rasterization; it is used to reproduced photographs in newspapers and books, to generate laser printer outputs and to display images on a PC monitor. Obviously, a waveform of any desired degree of complexity can be displayed or printed, but the quality of the resulting image depends on the resolution of the image raster (see Table 4.9).

A resolution of 24 lines/cm is so low that hardly any newspaper today would dare to impose it on its readers, while photographs printed at 60 lines/cm look quite handsome. We can thus set a minimum requirement of 30 lines/cm. This means that a display whose resolution is lower than that of the VGA standard (640 × 480 pixels) is in practice hardly usable (no matter how attractive the screen pictures may be in the advertising literature).

An obvious approach is to take one waveform sample for each pixel of the image display. If we want the length of one cycle of the displayed waveform to be 10 cm, we thus need 300 samples per cycle. A periodic signal with a 50 MHz repetition rate (a typical value for the bus clock of a modern PC) would thus have to be sampled at 50·300 = 15,000 megasamples/s (ie, 15 GHz), just to achieve a display quality comparable to that of an analogue oscilloscope. However, modern DSOs have a sampling rate that is at best 1/10 to 1/3 of this rate. We thus have to be satisfied with (for instance) ten samples per period. Obviously, it

Reproduction medium	Resolution in lines/cm or pixels/cm (reference values)
newspaper and book printing; photo reproduction	24–60
300 dpi computer printers	118
600 dpi computer printers	236
laser image setters (> 2000 dpi)	≈ 800
17" computer monitor, 800 × 600 pixels	30
17" computer monitor, 1600 × 1200 pixels	60
high-performance display, 4096 × 4096 pixels 1)	> 130
10.4" monitor, 800 × 600 pixels 2)	30
7" monitor, 640 × 480 pixels 3)	35
7" monitor, 320 × 240 pixels 3)	17
8.1-cm flat-screen display, 240 × 240 pixels 4)	29 (25 pixels/div)
8.5-cm flat-screen display, 140 × 180 pixels 4)	16 (20 pixels/div; see also Table 4.13)
resolution capability of the human eye	approximately 20 (only about 10 lines/cm can be distinguished in 'line on/line off' patterns)

1: Vector display (not raster-scanned); every pixel can be directly written by deflecting the electron beam to its location. This resolution was already achieved in the 60's!
2: Flat-screen displays of modern portable PCs.
3: CRT or flat-screen display of a DSO.
4: Flat-screen displays of portable DSOs (for example).

Table 4.9 Examples of raster resolutions.

is not possible to generate a usable image if these ten samples are spread out over ten centimeters on the display. It is only possible to create a 'nice' image under the following conditions: (1) the physical size of the image is reduced (to one centimeter per period, for example) (2) the signal waveform is very regular and (3) we view the display in the form of a 'screen shot' or printout, or with data acquisition halted (the displayed waveform will always jump around if data acquisition is continuous and cyclic).

Waveform details. The shortest-duration details of a waveform that we actually should be able to see (because they may help in fault-finding) typically resemble noise (spikes, glitches and the like), which means that they are very short-lived departures from the normal waveform (see for example the negative spike in Figure 4.44). In practice, it is enough to be able to recognize and/or evaluate the following: (1) the actual presence of such disturbances, (2) the time(s) at which they occur and (3) their amplitudes. What sampling rate is needed to capture pulse-shaped disturbances (glitches) with a given duration t_g? At a sampling rate of $1/t_g$, it can happen that the samples are taken at exactly the times when no glitches occur (see Figure 4.46). Clearly, we must sample faster than this ($f_s > 1/t_g$), and to be on the safe side the sampling rate should be at least twice the glitch rate.

What can we expect? Clearly, very 'wild' waveforms can occur, with all sorts of positive and negative spikes, sharp bends, over- and undershoots and the like (see Figure 4.47). However, there is an upper limit to the wildness of signal waveforms, since switching times and rise times cannot be reduced to arbitrarily small values:

1 The absolute limit is set by the fact that signal conductors act as low-pass filters with cutoff frequencies that are typically greater than 1 GHz. A low-pass filter limits the rate

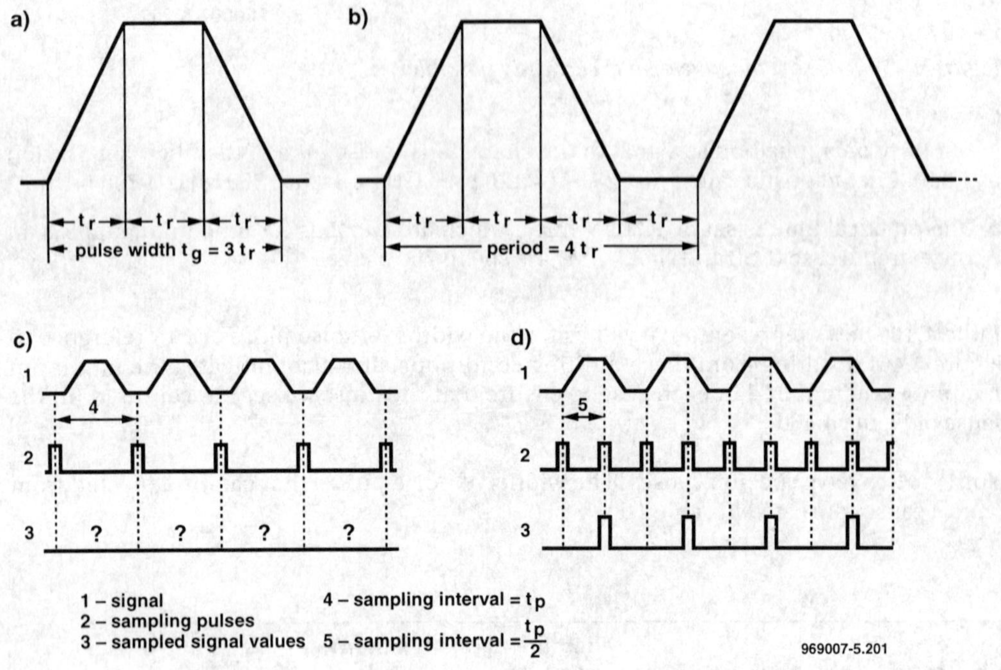

Figure 4.46 Effect of noise pulses on sampling.

a) Simplified representation of a noise pulse: finite rise and fall times mean that 'ideal' (zero-width) spikes cannot occur, since any change takes time (even if it is only a few picoseconds). If we assume that the rise time or fall time t_r of the noise pulse is as short as is possible given the physical characteristics of the circuit, and in addition that the pulse stays high for the same interval t_r, then the total pulse duration is $t_p = 3{\cdot}t_r$.

b) Clearly, the worst-case situation is a continuous train of such noise pulses, with a repetition frequency $f_g = 1 / (4{\cdot}t_r)$.

c) If we sample at an interval of $t_g = 4{\cdot}t_r$, the noise pulses may not be not detected at all.

d) If we sample at an interval of $t_g/2$, we can be sure of detecting the noise pulses. Note: this is an idealized representation. In reality the sampling pulses will not have a fixed phase relationship to the noise pulses. If we sample at an interval of $t_g/2$, we can detect that the signal has been disturbed, but we cannot obtain any exact information about the exact times and widths of the noise pulses, nor about their amplitudes (since the samples may be taken on the edges of the noise pulses).

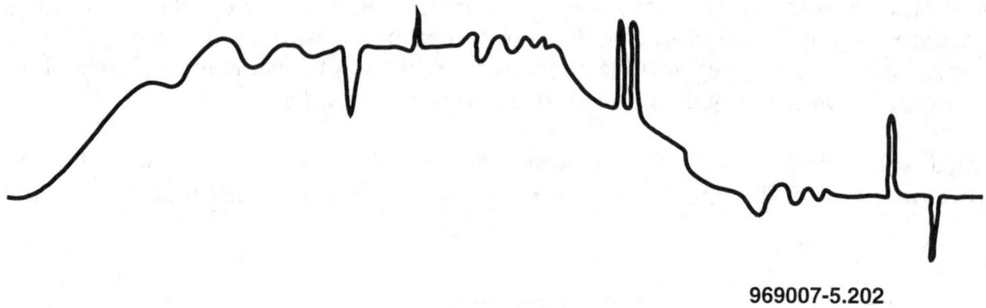

Figure 4.47 *This sort of waveform is certainly possible.*

of rise of an applied pulse signal to the characteristic rise time of the filter (see Section 4.3.3). Characteristic values are: 2 GHz ≙ 170 ps, 1 GHz ≙ 340 ps, 500 MHz ≙ 700 ps.

2 The practical limit is set by IC rise times, which are 1 to 2 ns for modern digital ICs; in the extreme case (ECL and GaAs) they are 200 to 700 ps.

Table 4.10 shows representative values for the widths of noise pulses (with reference to Figure 4.46) resulting from these limiting conditions, and also indicates the minimum sampling rate required in each case to ensure that the noise pulses are captured (in the sense of Figure 4.46d).

Noise pulses have specific causes. The widths of noise pulses that can be expected from typical causes are listed in Table 4.11.

System environment	Rise time t_r	Minimum glitch width $t_g = 3 \cdot t_r$	Maximum glitch repetition rate $f_g = 1/(4 \cdot t_r)$	Sampling rate $f_s = 2 \cdot f_g$
general wiring network: 2 GHz; GaAs logic	170 ps	510 ps	1.5 GHz	3 Gigasamples/s
general wiring network: 1 GHz	350 ps	1 ns	600 MHz	1.2 Gigasamples/s
ECL	600 ps	1.8 ns	400 MHz	800 Megasamples/s
CMOS, BiMOS, GTL	1–2 ns	3–6 ns	125–250 MHz	250–500 Megasamples/s
'garden-variety' logic (TTL, HC)	5 ns	15 ns	50 MHz	200 Megasamples/s

Table 4.10 *Widths of noise pulses*

Cause of the disturbance	Approximate pulse width (representative values)	Remarks
propagation delay effects in logic ICs	1–20 ns	the minimum pulse width is 1 gate delay time
signal-line reflections and short-range crosstalk	< 2 to > 40 ns	1 pulse ≙ 2 propagation delays (5 ns/m for 'open' wiring; 10–20 ns/m for bus conductors)
long-range crosstalk	several hundred picoseconds to 5 ns	the pulse width ≙ the rise time of the interfering waveform
radiated interference (EMI)	from several hundred picoseconds upwards	frequencies ≤ 2.5 GHz (electromagnetic waves are sine waves!)

Table 4.11 Important noise sources.

Actually, the situation cannot be all that bad: waveforms such as that shown in Figure 4.47 are only possible if the period of the signal is relatively long compared to the smallest rise time of the circuit or device. A mains-voltage waveform with superimposed 1-µs spikes is certainly possible, and quite bizarre waveforms can occur in switch-mode power supplies running at frequencies on the order of 100 kHz. However, a waveform like that of Figure 4.47 practically cannot occur if the fundamental signal frequency is (for instance) 200 MHz: the spikes would have to be a fraction of a picosecond wide, which is not possible if the shortest possible rise time is more than 100 ps. Figure 4.48 shows that a 'fat' glitch with a rise time of 500 ps (as shown in Figure 4.46) just fits into one pulse of a 200-MHz pulse train. Table 4.12 lists reference values for noise pulses that can be expected at various pulse repetition rates. *Note:* in principle, every circuit can be affected by every type of noise. However, if we regard the most likely types of noise to be propagation-delay effects, crosstalk and signal reflections, the minimum noise pulse duration can be approximately determined from the usual values of gate propagation delay times and conductor lengths.

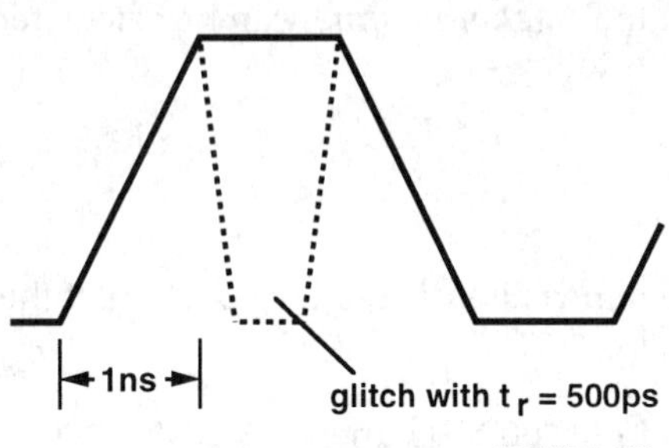

Figure 4.48 A glitch in a 200 MHz pulse.

Pulse repetition frequency	Period	Typical rise time t_r	Shortest noise pulse (glitch)	1 pixel represents [1]
200 MHz	5 ns	500 ps	1.5 ns	16 ps [2]
100 MHz	10 ns	1 ns	3 ns	33 ps [2]
50 MHz	20 ns	2 ns	6 ns	66 ps [2]
20 MHz	50 ns	2 ns	6 ns	166 ps [2]
10 MHz	100 ns	5 ns	15 ns	330 ps [3]
1 MHz	1 µs	5 ns	15 ns	3 ns [3]
100 kHz	10 µs	5 ns	15 ns	33 ns [4]
200 kHz	50 µs	5 ns	15 ns	160 ns [4]
100 Hz	100 µs	5 ns	15 ns	33 µs [4]
1: assuming that one period is displayed with 300 pixels over a length of 10 cm 2: unrealistic sampling rate 3): glitches displayable 4): glitches not displayable				

Table 4.12 Typical noise pulse widths (representative values).

It is clearly not reasonable to demand very high display resolution when measuring very-high-frequency signals, since even the shortest noise pulses will occupy many pixels. In the low megahertz region (1–10 MHz), glitches can be detected with manageable sampling rates and can also be displayed within the bounds of the available image resolution. However, with signal frequencies under 1 MHz the image resolution is not good enough to unambiguously display the noise pulses that are commonly encountered in practice (an individual pixel is 'broader' than a glitch).

Memory depth. The memory depth, or in other words the number of binary sample values that can be stored, is fundamentally limited. (Inexpensive oscilloscopes typically have a memory depth of 2048 samples per channel, while the acquisition memories of high-quality equipment can be extended to several hundred thousand or even a few million samples per channel). If the maximum sampling rate is fixed, then the memory depth is the limiting factor in the fidelity of the displayed waveform.

The maximum acquisition time t_c (*capture time*) can be calculated from the sampling rate f_s and the memory depth M as follows:

$$f_{su} = \frac{M}{t_c}$$

This means that for any given measurement, only a portion of the signal can be observed, with a duration of t_c.

For example: if M = 2048 and f_s = 100 MHz, then $t_c = 2048 \div 100 \cdot 10^6 / \mathrm{s} = 20.48\ \mu\mathrm{s}$. We can thus capture a segment of the signal being observed that is at most approximately 20 µs long. If we assume that the signal in question is periodic and that we want to display a full cycle, then the maximum period of the signal cannot exceed t_c (in this case roughly 20 µs). The repetition frequency of the signal must therefore be at least 1 ÷ 20 µs = 50 kHz. If relatively low-frequency signals are to be captured and displayed, the sampling rate must be reduced. Some equipment reduces the actual sampling rate, and some equipment samples at the same rate but stores only every n^{th} sample. The usable sampling rate f_{su} for a desired capture time t_c and a memory depth M can be calculated as follows:

$$t_c = \frac{M}{f_s}$$

For example: to store one mains cycle (t_c = 20 ms) with a memory depth of 2048 samples:

$$f_{su} = 2048 \div 20\ \text{ms} = 102.4\ \text{kHz}$$

We therefore sample at around 100 kHz, which means one sample every 10 µs. The displayed waveform can look quite good – it is after all made up of 2048 discrete points, which is a much higher resolution than most monitors can display.

Caution:

1 *In the time domain:* what we see does not necessarily faithfully represent the actual signal. There is a good chance that noise pulses shorter than 20 µs will not be captured if the sampling rate is 100 kHz (a sampling interval of 10 µs) – see Figures 4.46c and d.

2 *In the frequency domain:* if the sampling rate f_s is reduced, then according to the sampling theorem the cutoff frequency of the oscilloscope must also be reduced (to f_s / 2). However, the bandwidth of the input low-pass filter is not reduced (in the example it should be limited to 50 kHz to satisfy the conditions of the sampling theorem). This means that aliasing products are generated that systematically distort the displayed waveform.

Table 4.13 lists the sampling rates and associated upper cutoff frequencies of a typical inexpensive digital storage oscilloscope.

Remedies:

- An extremely large acquisition memory: 'lots of RAM', possibly supplemented by data compression. This is the obvious and theoretically best solution, but it is expensive (the prices of suitable equipment lie in the five-figure range).
- Sophisticated triggering provisions: these allow specific portions or features of the signal being examined to be selected. (*For example:* you can trigger on all pulses whose amplitude exceeds a certain level and whose duration is less than a certain interval, in order to investigate mains disturbances.) The practical problem with this approach is that you must first have some idea of what to look for.
- Peak-value detection: the basic idea is to monitor the signal in the intervals between sampling points and to register any unusual events that occur in these intervals. There are various ways in which this can be implemented.

Time base (time/div)	Sampling rate (samples/s)	Capture time [1] (rounded off)	Cutoff frequency for arbitrary waveforms, assuming 10 samples per period (Hz)	Cutoff frequency according to the sampling theorem (Hz)	Number of sample points per scale division
2 s	10	200 s	1	5	20
1 s	20	100 s	2	10	20
0.5 s	40	50 s	4	20	20
0.2 s	100	20 s	10	50	20
0.1 s	200	10 s	20	100	20
50 ms	400	5 s	40	200	20
20 ms	1 k	2 s	100	500	20
10 ms	2 k	1 s	200	1 k	20
5 ms	4 k	0.5 s	400	2 k	20
2 ms	10 k	0.2 s	1 k	5 k	20
1 ms	20 k	0.1 s	2 k	10 k	20
0.5 ms	40 k	50 ms	4 k	20 k	20
0.2 ms	100 k	20 ms	10 k	50 k	20
0.1 ms	200 k	10 ms	20 k	100 k	20
50 μs	400 k	5 ms	40 k	200 k	20
20 μs	1 M	2 ms	100 k	500 k	20
10 μs	2 M	1 ms	200 k	1 M	20
5 μs	4 M	0.5 ms	400 k	2 M	20
2 μs	10 M	0.2 ms	1 M	5 M	20
1 μs	20 M	0.1 ms	2 M	10 M [2]	20
0.5 μs	40 M	50 μs	4 M	10 M [2]	20
0.2 μs	50 M [3]	40 μs	5 M [2]	10 M [2]	10
0.1 μs	50 M [3]	40 μs	5 M [2]	10 M [2]	5

1: Acquisition memory depth = 2048.
2: The analogue circuitry is specified for 10 MHz.
3: The sampling rate is halved if two channels are used.

Table 4.13 Sampling rates for a digital storage oscilloscope rated at 50 MS/s. (source: Hung Chang)

Peak value detection

Digital detection. This technique is illustrated in Figure 4.49. The signal is sampled at the maximum available rate, but sample values are only stored at an interval (the sampling interval) corresponding to the selected capture time and the depth of the acquisition memory. For each sampling interval, the minimum and maximum values of the signal during the interval are stored (two sample values are stored for each sample interval). With a signal that normally changes slowly (which is why a relatively long capture time

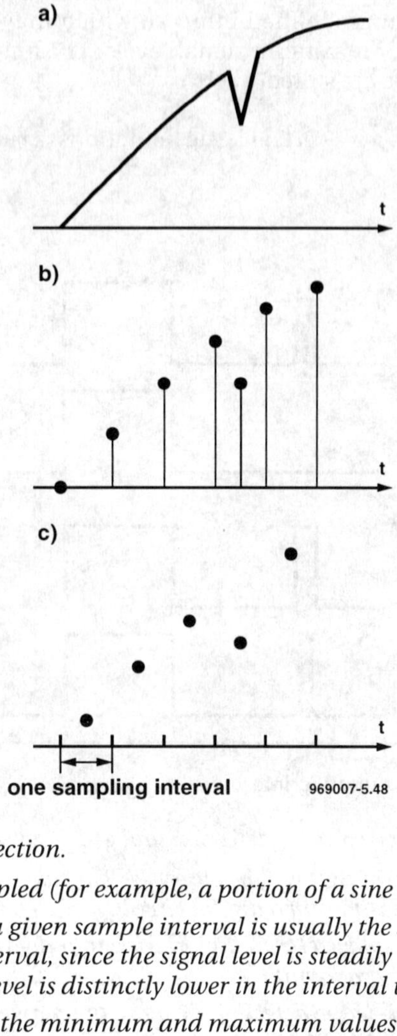

Figure 4.49 Peak value detection.

a) The waveform to be sampled (for example, a portion of a sine wave).

b) The minimum value in a given sample interval is usually the same as the maximum value of the previous interval, since the signal level is steadily rising (except for noise pulses). The minimum level is distinctly lower in the interval with the noise pulse.

c) If the difference between the minimum and maximum values is small, their average value can be calculated and displayed as a single pixel on the screen. If the difference exceeds a certain level, the average value is not calculated and the displayed pixel shows the measured value instead of the average value. Sometimes the 'exceptional' pixels are optically emphasized (displayed with higher intensity or a different colour, for example).

has been selected), the difference between the minimum and maximum levels during the sample interval will usually be small, so that the average value of the signal during the sample interval can be displayed. A relatively large difference between the minimum and maximum values for a given sample interval is an indication of some sort of disturbance.

Analogue detection. In this method, an analogue peak-value detector is connected ahead of the analogue-to-digital converter. The output of the peak detector is evaluated at the end of each sampling interval.

Glitch capture. This method is based on the detection of narrow pulses with durations that are less than a certain predefined interval, which may be selectable. It does not depend on the minimum or maximum signal levels. The amplitude and time of occurrence of each detected glitch are stored.

All peak-detection methods have certain basic limitations, as illustrated in Figure 4.50:

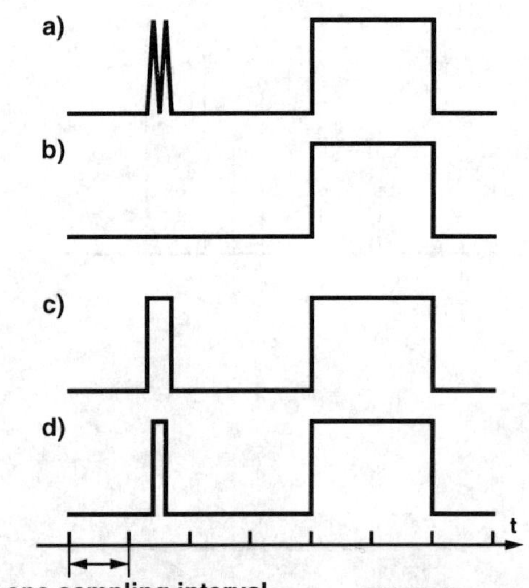

***Figure 4.50** Limitations of peak-value detection and glitch detection.*

a) This is the actual waveform that is to be observed. The two closely-spaced noise spikes could certainly trigger flip-flops in a digital circuit.

b) This waveform will be displayed by an oscilloscope that does not have peak-value detection or glitch capture capability.

c) An oscilloscope with peak-value detection displays the maximum signal level in the sampling interval in question. We can see that 'something is wrong', but we cannot see that there are two pulses, and we also do not know the exact timing of the disturbance.

d) An oscilloscope with glitch capture can register the exact time at which the first noise pulse occurs. However, we still cannot see that there are two pulses.

1 The precision with which the time of occurrence of a given noise pulse can be fixed is determined by the effective sampling interval and not by the actual sampling rate. *For example:* suppose that we are displaying one cycle of the mains AC voltage. With a memory depth of 2048 locations, the effective sampling interval is 20 µs (since two values must be stored for each sampling interval). Even if the oscilloscope has a sampling frequency of 1 GHz, the timing of noise pulses can only be determined to within 20 µs. A more sophisticated implementation of glitch detection (with a sampling clock counter) could register and display each glitch with a resolution of ±1 ns.

2 You can basically see the presence of at most one glitch in any one sampling interval, even if several closely-spaced glitches are actually present.

3 The minimum/maximum method amplifies the effect of a noise signal superimposed on the signal being observed (since the noise signal increases the difference between the minimum and maximum values).

Sampling methods

Real-time sampling (RTS). The signal to be measured is sampled at a particular rate, and the samples are converted to binary values that are sequentially stored. The advantage of this method is that it can be used to acquire non-recurring or non-periodic signals (single-shot operation).

Random interleaved sampling (RIS). Each time that the trigger conditions are satisfied (a trigger event), the signal is sampled at a particular rate as for real-time sampling. Trigger-condition detection is not synchronized with the sampling clock, so that the time delay between each trigger event and the first sampling clock is different each time. This time delay is precisely measured and used to determine an offset value for storing the sampled data in the acquisition memory. If recurring trigger events occur, the sampled data for each trigger event are stored in the acquisition memory in an interleaved fashion according to the delay between each trigger event and the first sample clock. In this manner an increasingly detailed representation of the waveform being measured can be built up over time. *For example:* suppose that 500 sample values are stored for the first signal period; this gives a raster image with 500 pixels. The next period is again sampled 500 times with a certain (random) time offset, and the resulting sample values are stored 'in between' the first 500 values, with a offset that is proportional to the sampling time offset (see Figure 4.51). The advantage of this method is that it is much easier (and less expensive) to precisely measure the length of the delay between the trigger event and the first sampling clock (for example, with a resolution of 1 ns) than to sample the signal at a rate that provides the same number of sample values for a given capture interval (for example, 1 GHz).

Effective sampling rate. An oscilloscope corresponding to the example illustrated in Figure 4.51 can display periodic waveforms with a resolution of 1 ns, even though the actual sampling rate is only 125 MHz (8 ns period). It thus has the same capability as an oscilloscope with an actual sampling rate of 1 GHz (1 ns resolution). This means that the reciprocal of the resolution with which the interleaving time is measured can be taken to be the effective sampling rate for periodic waveforms.

Equivalent-time sampling (ETS). This method can only be used with periodically repetitive waveforms. Only one sample is taken for each cycle of the signal being measured (one sample per period), which allows the sampled values to be digitized and stored at a 'comfortable' pace. The displayed waveform is constructed from samples that are taken from a large number of successive periods of the signal (see Figure 4.52).

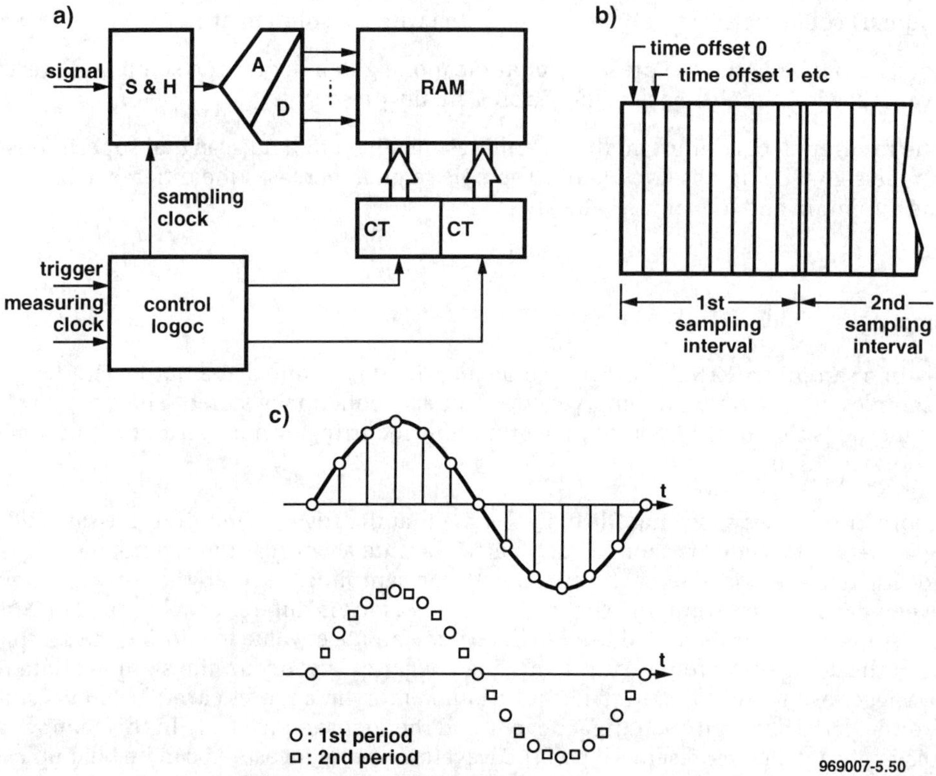

Figure 4.51 Random interleaved sampling (RIS).

a) Block diagram. A very fast counter is added to the arrangement shown in Figure 4.43. This counter provides part of the acquisition memory address. The exact implementation depends on the resolution with which the sampling delay is measured and the rate at which the signal can be sampled. In this example, the time resolution is 1 ns and the sampling interval is 8 ns.

b) Sample memory organization: in this example, eight memory locations are required for each sample interval. The data that are sampled with a delay time of 0 ns are stored in the first location, those sampled with a delay time of 1 ns in the second location and so on.

c) The displayed waveform becomes increasingly detailed. The upper waveform shows a sine wave that has been sampled once at the actual sampling rate. In the next period the signal is sampled at a (randomly-determined) offset to the first set of samples, so that the second set of sample points fills in the first set.

Note: Since the difference in the time delay between the trigger pulse and the sampling pulse has a fixed value from one cycle to the next, this is a *deterministic* sampling method. The difference between successive time delays (the incremental delay) determines the effective sampling rate. *For example:* an incremental delay of 1 ns corresponds to a effective sampling rate of 1 GHz. Sampling rates that are significantly higher than this can be achieved with analogue circuit techniques (that is, not using a counter).

The advantage of equivalent-time sampling is that very high effective sampling rates can be realized with relatively slow hardware (sampling oscilloscopes are available with cutoff frequencies up to 50 GHz); only the input amplifier stage and the sample-and-hold circuit have to be designed to handle the actual 'fast' signal waveform. (Even this is not particularly easy, and the equipment is correspondingly expensive). The only prerequisite for this method is that the waveform being measured must be truly repetitive (although the actual period does not matter). It is not essential to take a sample in every period; as much time as is desired can elapse between successive samples. It is even possible to directly output the reconstructed waveform to a plotter.

Note: sampling oscilloscopes that achieve upper cutoff frequencies in the gigahertz region at an affordable cost have also been built. They use analogue techniques to augment the effective sampling rate.

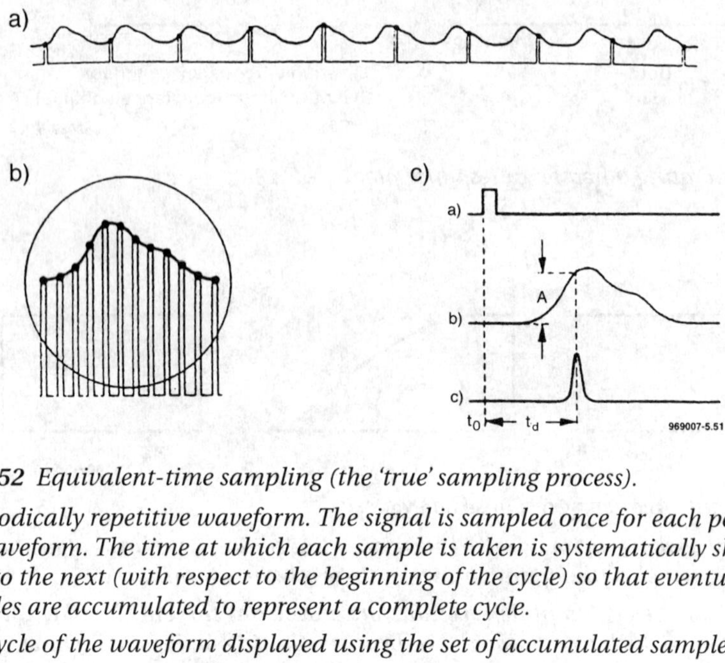

***Figure 4.52** Equivalent-time sampling (the 'true' sampling process).*

a) A periodically repetitive waveform. The signal is sampled once for each period (cycle) of the waveform. The time at which each sample is taken is systematically shifted from one cycle to the next (with respect to the beginning of the cycle) so that eventually enough samples are accumulated to represent a complete cycle.

b) One cycle of the waveform displayed using the set of accumulated sample values.

c) Determining the sampling time: (a) is the trigger pulse, which occurs at the beginning of each cycle, (b) is the (delayed) waveform which is to be displayed and (c) is the sample pulse. The sample pulse is generated with a time delay t_d with respect to the trigger pulse. This delay is increased by a certain amount from one cycle to the next.

4.3.3 Specifications

General specifications

Frequency response. A frequency response curve is a graphic representation of a device's behaviour in the frequency domain. Figure 4.53 illustrates the appearance of typical frequency response curve and shows how such curves are generated. A signal generator is connected to the input of the test object (such as the vertical amplifier of an oscilloscope). The signal generator produces a constant-amplitude sinusoidal signal whose frequency can be set to any value in the frequency range of interest. The amplitude of the signal at the test object's output is measured while the signal generator is swept through this frequency range, and the result is plotted as an graph of amplitude versus frequency. (The output amplitude is usually normalized with respect to the input amplitude, so that the vertical axis represents the ratio U_{out} / U_{in}.) Several types of frequency response are shown in Figure 4.54.

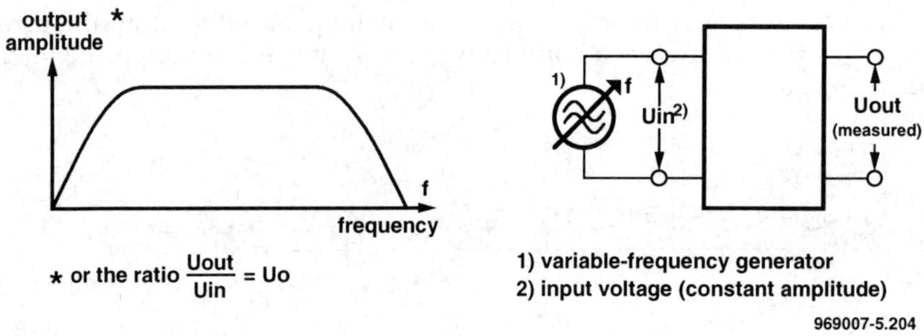

Figure 4.53 Generating a frequency response plot (see text).

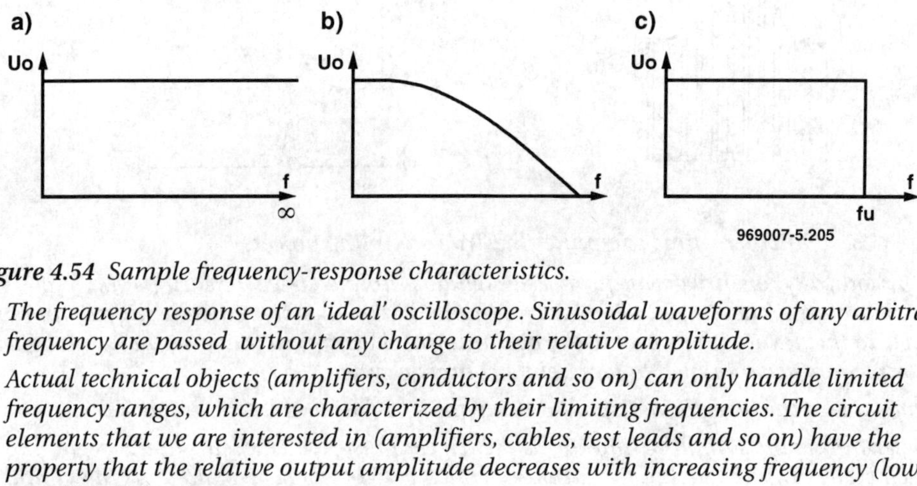

Figure 4.54 Sample frequency-response characteristics.

a) The frequency response of an 'ideal' oscilloscope. Sinusoidal waveforms of any arbitrary frequency are passed without any change to their relative amplitude.

b) Actual technical objects (amplifiers, conductors and so on) can only handle limited frequency ranges, which are characterized by their limiting frequencies. The circuit elements that we are interested in (amplifiers, cables, test leads and so on) have the property that the relative output amplitude decreases with increasing frequency (low-pass filter behaviour).

c) The frequency response of an ideal low-pass filter.

Given that an oscilloscope must exhibit a low-pass response, what is the most useful sort of frequency response? An obvious first choice is a very simple response: the oscilloscope acts like an ideal instrument (with a constant response for all frequencies) up to a certain limiting frequency f_u, but does not respond at all to signals above his frequency. However, an oscilloscope with such a frequency response cannot be expected to faithfully reproduce arbitrary non-sinusoidal waveforms. This becomes clear when the relationship between the frequency-domain response and the time-domain response is more closely investigated. Figure 4.55 shows the time-domain response to a step change in the input voltage (zero rise time) for devices with several different types of frequency response characteristics. The upper illustration in each case shows the frequency response and the lower illustration shows the associated response to the step pulse (the step response).

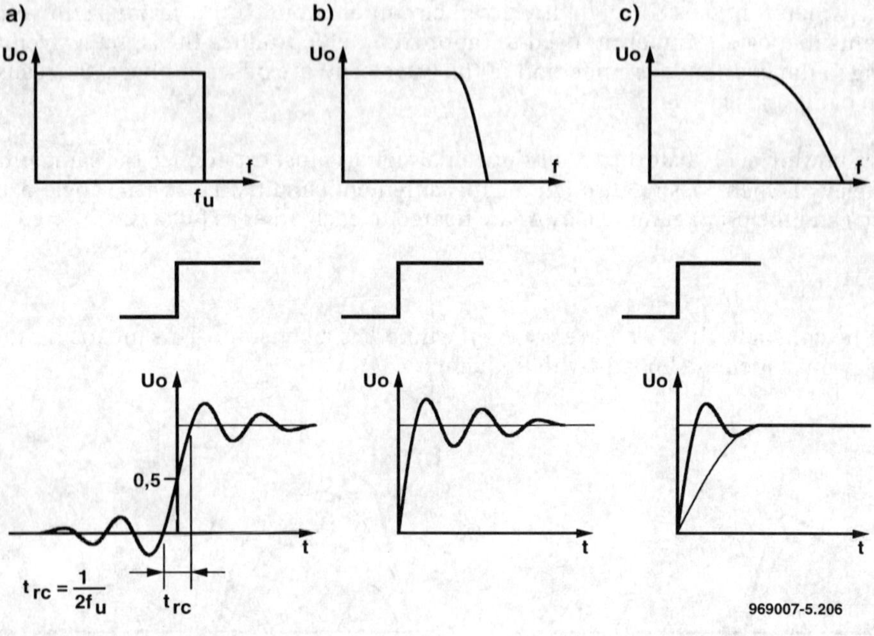

Figure 4.55 Frequency response characteristics and related time-domain responses. Top: frequency response ; middle: input step; bottom: output signal (time-domain response)

a) An ideal step pulse at the input of a device with an ideal low-pass response causes overshoots in the output signal. The mathematical analysis yields overshoots both before and after the leading edge of the output signal. This is a physical impossibility that results from the combination of two idealizations. It is in any case evident that an ideal low-pass filter responds to an ideal step pulse (zero rise time) with an output pulse that has a finite (non-zero) rise time.

b) A real low-pass frequency response characteristic that closely approximates that of an ideal low-pass filter also produces overshoots in response to an ideal step input.

c) A low-pass filter with a more gradual cutoff produces either a lower level of overshoots or no overshoot at all, depending on the exact shape of the frequency response. An excessively gradual cutoff however causes the characteristic rise time to be too long.

Notes:

1 An oscilloscope with the frequency response shown in Figure 4.55b would display waveforms that do not actually exist.

2 An ideal low-pass filter is required if the conditions of the sampling theorem are to be satisfied (see Section 4.3.2). If a filter is used that essentially meets this requirement (so that alias frequencies are eliminated), the reconstructed signal waveform will be distorted by the overshoots resulting from the low-pass filter (as shown in Figure 4.55b).

Frequency response and phase response. Figure 4.56 shows that the frequency response alone (or just the cutoff frequency specification) is not enough to characterize the ability of an oscilloscope to faithfully reproduce the measured signal. Any signal experiences a time delay when it passes through any given circuit element. If the various sinusoidal components that can be superimposed to (approximately) produce the signal waveform (according to the Fourier decomposition of the waveform) are offset in phase, this causes distortion of the signal waveform.

In order to prevent such distortions, all sinusoidal signals must experience the same propagation delay Δt when passing through the measurement circuitry. The phase angle φ, the frequency f and the propagation delay Δt are related to each other as follows:

$$\varphi = 2\pi\,\Delta t \cdot f$$

If Δt is to be constant, then we have $\varphi = k \cdot f$, where k is a constant. This means that the phase delay must increase linearly with frequency.

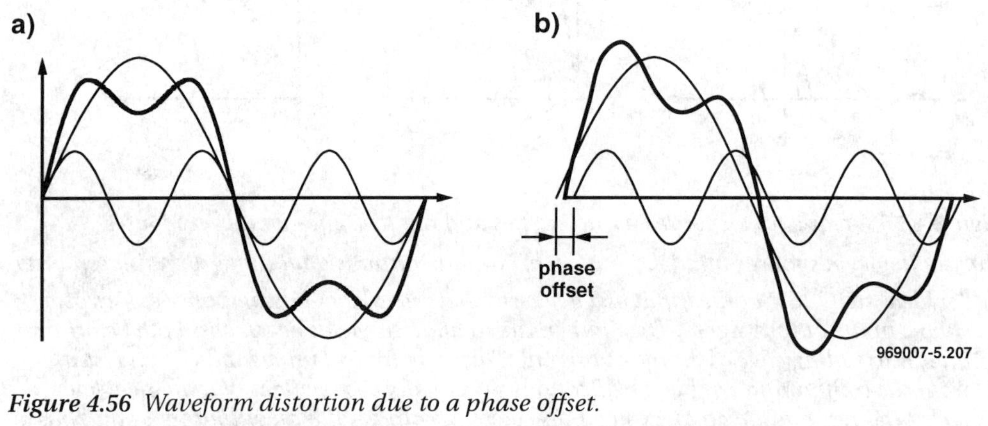

Figure 4.56 *Waveform distortion due to a phase offset.*

a) A square-wave signal can be approximately represented by the superimposition of two sine wave signals. Since there is no phase offset between the sine waves, the representation is correct (compare this to Figure 4.45b).

b) The sine waves have a relative phase offset. This causes the representation to be strongly distorted.

Frequency limits and bandwidth. A cutoff frequency specification relates to the point at which the amplitude response drops off by a factor of $1/\sqrt{2} = 0.707$, or –3 dB. *For example:* if we apply a 20-MHz sine wave with a peak-to-peak value U_{pp} of 1 V to the input of an oscilloscope with a cutoff frequency of 20 MHz, the displayed peak-to-peak amplitude will be 0.707 V (see Figure 4.57). Such behaviour cannot always be ignored.

Upper and lower frequency limits. Modern oscilloscopes have DC-coupled amplifiers, so that it is not necessary to specify their bandwidths in terms of both lower and upper cutoff frequencies. (The lower cutoff frequency is only relevant for operation with an AC-coupled input; it is typically around 10 Hz).

The cutoff frequency specifications in the equipment brochure refers to the vertical (Y-axis) signal path. Sometimes the horizontal (X-axis) path has a lower limiting frequency (which is specified in the data sheet or operator's manual).

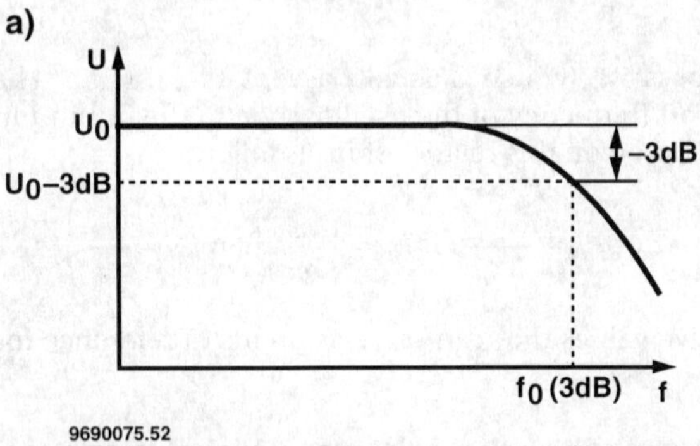

b)

relative frequency[1]	attenuation[2]	
1.0 · fo	–3dB	–29%
0.5 · fo	–1dB	–11%
0.1 · fo	–0.1dB	– 1%

1): referred to the -3 dB limit frequency f_0
2): referred to nominal amplitude ratio U_0

Figure 4.57 Upper cutoff frequency specification.

Rise time. If an ideal step pulse (with zero rise time) is applied to the input of an oscilloscope with an upper cutoff frequency f_u, the displayed waveform will be a step with a rise time t_{rc} (the oscilloscope's *characteristic rise time*). The characteristic rise time is related to the cutoff frequency by the following approximation:

$$t_{rc} = \frac{0.35}{f_u}$$

where t_{rc} is measured between the 10% and 90% amplitude points of the step waveform.

When we observe a real pulse, its own rise time t_{rs} contributes to the net rise time of the displayed waveform. The rise time t_{scr} that we measure on the screen is given by:

$$t_{scr} = \sqrt{t_{rs}^2 + t_{rc}^2}$$

If more than one low-pass device (such as a probe assembly, probe lead, amplifier and so on) is connected in series between the signal to be measured (with its rise time t_{rs}) and the display, then the net rise time of the displayed waveform is calculated in the same manner from the signal rise time and the individual rise times t_{r1}, t_{r2}, t_{r3} and so on as:

$$t_{scr} = \sqrt{t_{rs}^2 + t_{r1}^2 + t_{r2}^2 + t_{r3}^2 + \ldots}$$

Selection criteria

Characteristic rise time. If the signal that we wish to observe has a rise time t_{rs}, then obviously the characteristic rise time of the oscilloscope should be significantly less: $t_{rc} < t_{rs}$. The percentage error p due to the effect of the oscilloscope's characteristic rise time t_{rc} can be calculated from the ratio $n = t_{rs} / t_{rc}$ by means of the following formula:

$$p = \frac{t_{scr} - t_{rs}}{t_{rs}} \cdot 100\,\% = \left\{ \sqrt{\frac{n^2 + 1}{n^2}} - 1 \right\} \cdot 100\,\%$$

From Table 4.14 we can see that the error is 5 % (which is usually acceptable) if n is 3. The usual rule of thumb that relates the cutoff frequency of the oscilloscope to the minimum rise time of the signals to be observed is based on this relationship, as follows:

$$f_u \geq \frac{1}{t_{rs}} \quad \left(t_{rc} = \frac{0.35}{f_u} \Rightarrow f_u = \frac{0.35}{t_{rc}}, \text{ or } f_u \approx \frac{1}{3\,t_{rc}}. \text{ If } t_{rs} = \frac{t_{rc}}{3} \text{ then } f_u \approx \frac{1}{t_{rs}} \right)$$

Table 4.15 lists a number of representative values that can serve as points of reference for selecting an oscilloscope.

Note: Section 4.3.4 includes several examples of how an oscilloscope responds to extremely short pulses and signals with extremely short rise times.

Deflection sensitivity. The deflection sensitivity is specified in terms of the deflection factor or deflection coefficient, in volts/cm or volts per scale division (volts/div), for the various positions of the input attenuator (range switch). A scale division is the distance between two lines of the grid on the face of the CRT (see Section 4.3.4 for sample illustrations) and is equal to 1 cm for most instruments. The deflection sensitivities provided by a typical selector switch range from 5 mV / div to 5 V / div in a 1-2-5 sequence (*for example:* 5 mV – 10 mV – 20 mV – 50 mV – 100 mV and so on).

Horizontal sweep (time base range). The time base range is usually specified in terms of time coefficients, and only rarely in terms of the sweep rate. A time coefficient (or horizontal deflection factor) has the form of time/cm or time/div. A typical time base covers the range from 0.1 s / div to 0.2 µs / div (again in a 1-2-5 sequence).

The advantage of such units is that a voltage or time interval can be easily measured by counting the number of scale divisions that it covers and then simply multiplying by the selected vertical deflection coefficient or the time base coefficient, respectively.

Input impedance. The input impedance is basically frequency-dependent. It is usually specified as a resistance in parallel with a capacitance (see Figure 4.58, which shows values that are practically an 'industry standard' for oscilloscopes). *Note:* the higher the frequency, the more the input impedance is determined by the capacitance.

Rise time ratio $t_{rs}:t_{rc}$	Error due to the oscilloscope's rise time
1:1	41 %
2:1	12 %
3:1	5 %
4:1	3 %
5:1	2 %
6:1	1.3 %
7:1	1 %
8:1	0.7 %
9:1	0.6 %
10:1	0.5 %

Table 4.14 Rise time measurement error due to the oscilloscope's characteristic rise time.

Cutoff frequency	Characteristic rise time	Suitable for observing pulses with a minimum rise time t_{rs} of ...	Maximum viewable pulse repetition frequency f_p [1]
5 MHz	70 ns	200 ns	1.25 MHz
10 MHz	35 ns	100 ns	2.5 MHz
20 MHz	17 ns	50 ns	5 MHz
40 MHz	9 ns	25 ns	10 MHz
60 MHz	6 ns	17 ns	15 MHz
100 MHz	3.5 ns	10 ns	25 MHz
150 MHz	2.3 ns	7 ns	35 MHz
250 MHz	1.4 ns	4 ns	60 MHz
500 MHz	700 ps	2 ns	125 MHz
1 GHz	350 ps	1 ns	250 MHz

1: $f_p = 1/(4\,t_{rs})$. This is the frequency of a sequence of 'just-observable' pulses, based on the oscilloscope's characteristic rise time (see Figure 4.46b).

Table 4.15 Representative values for selecting an oscilloscope.

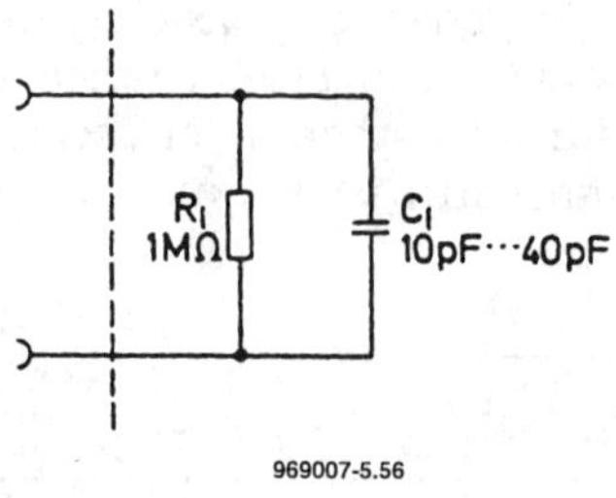

Figure 4.58 *Oscilloscope input impedance.*

Specifications of digital storage oscilloscopes

Digital oscilloscopes are in the first instance specified by the same parameters as analogue oscilloscopes. *Note:* any oscilloscope can only be as good as its analogue signal paths (which precede the digital portion; see Figure 4.43). The specific characteristics of the sampling process are addressed by a number of additional parameters.

Sampling rate. The sampling rate is specified in samples/s (S/s) or in MHz. Typical values range from 5 MS/s (megasamples/s) up to (currently) 10 GS/s (gigasamples/s). In the low-cost range – up to £700 – you can expect rates of 10 to 100 MS/s; instruments that are still affordable (say up to £1800) offer roughly 500 MS/s to 1 GS/s. Instruments that reach into the high GS/s region cost about as much as an auto in the mid-price bracket.

The sampling rate is often used to distinguish a particular instrument from its competition. However, the sampling rate alone is not everything. The digitization resolution (in bits) and the size of the acquisition memory (in samples) are also important. Additional factors to consider are the specifications of the analogue circuitry (input stages and preamplifiers) and the instrument's triggering and display capabilities. Manufacturing quality and accurate assembly are also decisive factors (which can easily result in price ratios of up to 10:1 for two different instruments whose specifications are at first blush the same).

Cutoff frequency. This specification is also extensively quoted in the adverts. You should carefully check what it actually means! Many DSOs have two cutoff frequencies: one for periodic signals and another one for single-shot signals. The cutoff frequency for periodic signals may even be higher than the sampling rate (*for example:* sampling rate = 20 MS/s, cutoff frequency = 100 MHz). If a 'true' sampling technique is used (such as RIS or ETS), the cutoff frequency for the signal to be displayed is essentially determined by the bandwidth of the preamplifier.

If the instrument in question does not have real-time sampling (RTS) capability, you must look for its single-shot cutoff frequency specification. If this is not stated, you can reasonably assume that the single-shot cutoff frequency is no more than one-tenth of the specified sampling rate – and don't forget that this applies to a sine wave signal. (*For example:* a 20 MS/s sampling rate implies a single-shot cutoff frequency of 2 MHz for a sine wave signal. If you are observing signals in digital logic circuits, this is suitable for signals whose pulse repetition rates are less than 600 to 800 kHz.)

The signal rise time is often a more relevant specification than the cutoff frequency (see Figure 4.59).

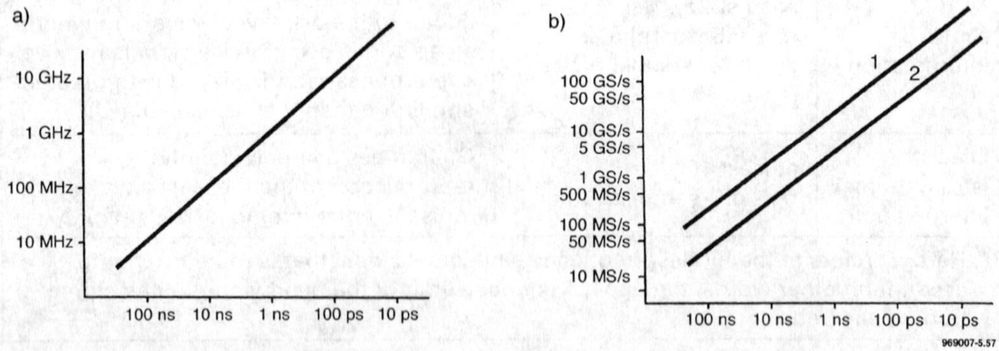

Figure 4.59 Rise time (horizontal axis) versus cutoff frequency (vertical axis). (source: Tektronix)

a) for periodic signals

b) for single-shot signals (1 – sin(x)/x interpolation, 2 – linear interpolation).

Dividing the sampling rate over several channels. You should also take this possibility into account if the sampling rate per channel is not explicitly specified. Whenever several individual signals are simultaneously acquired, the instrument's sampling rate may be 'divided' among the various channels (the sampling hardware is in a sense time-shared). In the previous example, if the single-shot bandwidth is 2 MHz and two channels are being used, then only 1 MHz of bandwidth is available for each channel. In practice, this means that you should try to use only one channel to acquire a single-shot signal (this may require external triggering).

Digitizing errors. Based on practical experience, a resolution of 8 bits per sample is fully adequate. An 8-bit sample can represent 256 levels, so ±1 bit represents 1/256 or an error of around 0.4%. *Caution:* this error calculation is valid when the full dynamic range of the A/D converter is utilized (see the discussion of 'range error' in Section 4.1.3). For lower-amplitude signals, the relative error is correspondingly higher.

Note: 8 bits/sample represents common practice, but some high-quality instruments work with 11 to 13 bits per sample. A few old-model DSOs use 6 bits, which is by no means adequate for the faithful reproduction of complex waveforms.

Interpolation. A display that consists of a relatively small number of sample points can obviously be made more attractive by either inserting intermediate points or connecting the sample points to each other with straight lines or curve segments. Two techniques are commonly used for this (see Table 4.16). With a given sampling rate, the frequency of the signal being measured can be higher with the $\sin(x) / x$ interpolation technique, but the fidelity of the displayed waveform is better with linear interpretation ($\sin(x) / x$ interpolation is only acceptable when the measured waveform is more or less sinusoidal).

Method	Minimum sampling rate *)	Remarks
sin (x)/x interpolation	at least $2f_u$; $4f_u$ is better ($f_u \leq 0.5 \cdot$ sampling rate; $f_u \leq 0.25 \cdot$ sampling rate is better)	Artificial under- and overshoots may be added to the displayed waveform where steep edges occur; noise signals may be overemphasized (displayed with greater amplitude than they actually have).
linear (straight- line) interpolation	at least $10f_u$ ($f_u \leq 0.1 \cdot$ sampling rate)	Guarantees a largely faithful reproduction of the measured waveform (noise is not overemphasized, etc.).
*: Here, f_u refers to the highest-frequency sinusoidal signal that is to be taken into account (in other words, the upper frequency limit of the bandwidth-limited signal to be measured).		

Table 4.16 Comparison of interpolation methods.

Display resolution. The display assembly is sometimes the weakest link in the chain (especially for instruments that use an LCD screen). LCD screen resolutions typically range from 240 × 240 pixels to 320 × 240 pixels; some (inexpensive) instruments have only 140 × 128 pixels (for example). With low screen resolutions it is certainly not possible to display 8-bit amplitude values! The consequence of insufficient screen resolution is that the displayed waveforms appear to be worse than they actually are (that is, they are always somewhat jagged). They also jump around more or less if the display is periodically updated (due to the extra digitization error that comes from converting the stored sample data to the display format, and with some instruments also due to the none-too-sophisticated trigger hardware). *Practical tip:* such instruments usually have an interface for connecting a printer or a PC. Most signal waveforms will appear the way they should be (the way they have actually been digitized) on a printout or on the PC monitor. However, you cannot view a repetitive process in this manner, so using a printer or PC only makes sense when you know exactly what to expect (application examples: exact evaluation of stored waveforms, documentation, training).

Note: See Section 4.3.4 regarding the signal acquisition and display rate (update rate).

4.3.4 Selecting and testing an oscilloscope

Analogue or digital?

There is no simple answer to this question. If we leave aside extremely demanding situations and concentrate on fault-finding in commonly-used hardware, the following aspects should be taken into account.

Application scenario 1: development, initial operation and fault-finding down to the lowest possible level. Every possible type of fault can be encountered in these situations. Although we often know how the signals are supposed to appear (based on the design objectives), we do not know their actual appearance, and the oscilloscope must be able to provide us with this information.

Application scenario 2: general service practice. In this situation the objective is usually to identify a particular item that should be replaced (an FRU), and to do so as quickly as possible. Measurements often are made to simply check whether or not a particular signal is correct ("does that look OK or is there something suspicious?") in order to decide whether or not to replace a particular item. It is not always essential that the oscilloscope shows every last detail of the signal being observed, as long as it allows us to judge whether the signal is correct or somehow different from what it should be.

The human factor. Naturally, a technician cannot be expected to make measurements with his or her bare hands or with truly unusable equipment. On the other hand, resourcefulness and technical knowledge cannot be replaced by some set of 'special features' of the test equipment, since (1) the technician must know what to look for and how to use the equipment and (2) 'effortlessly' identifying the cause of a fault from a screen display alone only happens in the adverts. In practice, it is frequently necessary to first devise fault hypotheses and then either confirm or refute them. There are several well-proven ways to do this, including (1) using conventional test equipment (multimeters and oscilloscopes) augmented by simple test accessories (see Section 4.4) and (2) using several different types of equipment in combination (oscilloscope, universal counter, logic analyzer and so on). We also must not forget 'tricks' such as pulling signal lines HI and LO, applying a cyclic stimulus, modifying the software (breakpoints and the like), writing small test programs and so on. All of this argues for giving our preference to equipment that has excellent fundamental characteristics but which can otherwise be fairly basic.

Cost. Digital instruments are significantly more expensive than analogue types with the same cutoff frequency and similar features and accuracy. If you want to be able to make useful measurements on (for example) a Pentium motherboard or a PCI bus but have a limited budget, then your only real choice is an analogue oscilloscope, since it will allow you to see more of the electrical details of the signals (pulse edges, noise and so on).

Cutoff frequencies of analogue instruments: truly high-performance oscilloscopes (which are definitely affordable as second-hand equipment) have cutoff frequencies that range up to 500 MHz; only a few types are specified for 1 GHz. (*Caution:* (1) sometimes a very high cutoff frequency specification applies to the direct connection of the signal to the CRT deflection plates (see Section 4.3.5), in which case the deflection sensitivity is only 2–10 V/cm (which is however adequate for making measurements in digital circuitry); (2) look carefully – the instrument in question may be a sampling oscilloscope.)

Sampling oscilloscopes. 'True' sampling oscilloscopes are outstanding instruments, but they are of only limited use in the service environment. They presuppose that the signal to be measured (or the test stimulus) is periodic and perfectly repetitive. Such signals can be relatively easily produced in test setups (for instance, using a signal generator), and sampling oscilloscopes are very well suited to tasks such as measuring signal rise times and the switching times of individual components (from diodes to processors). However, they are not suitable for routine service measurements of poorly-documented circuits that cannot be easily 'manipulated' by the user (such as those found on motherboards).

Functions and ease of use. Digital instruments have outstanding advantages with regard to functions and ease of use:

- automatic range selection (autoscale),
- stable display of long-duration processes,
- storage of single-shot processes,
- capturing a portion of the signal prior to the occurrence of the trigger event (pre-triggering; see Section 4.3.5),
- the digitized waveforms can be evaluated directly, subjected to further processing, printed out or transferred to a PC, and so on.

The state of the art and the market. High-performance oscilloscopes are presently manufactured almost exclusively as DSOs, with the development trend running towards extremely high (real-time) sampling rates (500 MS/s to more than 1 GS/s) so that both repetitive and non-repetitive processes can be directly sampled. The actual cutoff frequency is thus largely determined by the analogue 'front end' stages. The only analogue oscilloscopes that are still manufactured are specifically designed to be used as service equipment and can typically handle frequencies ranging up to 20–100 MHz (up to 250 MHz in some cases).

Guidelines (1): analogue instruments. Dual-trace oscilloscopes with a 20 MHz bandwidth cost roughly £350. Significantly less expensive (very simple) equipment is practically useless, although it may well be suitable for training and general 'hobby' use. Instruments with bandwidths in the range of 60–100 MHz, which is adequate for many purposes, can be obtained starting at roughly £700.

Guidelines (2): digital instruments. Truly useful DSOs cost at least £1700. 'Cheap' DSOs, regardless of whether they are self-contained instruments or plug-in cards for a PC, are largely useless (except for cursory plausibility checking of well-known waveforms with repetition rates of at most a few megahertz – but in many cases such instruments are not even suitable for typical hobby applications). *For example:* try displaying the servo pulse of a remote-control transmitter for a model aeroplane or car on an inexpensive DSO – which is a task that even the most primitive analogue oscilloscope can easily handle.

Special features (selected):

- More than 2 channels (3 or 4 is common). *Caution:* the extra channels often have a limited selection of measurement ranges. They are however adequate for measuring signals in digital circuitry, which is where the additional channels are the most useful. *Practical tip:* most people cannot hold more than two probes at once. If you have to use more than two channels, you should consider using some sort of test adapter (see Section 4.3.5).
- Digital readout of vertical and horizontal scale parameters: this feature comes practically 'for free' with DSOs. It is hardly worth the extra cost for analogue instruments, unless you need to make screen photos (some types are available with or without this feature).

- On-screen measurement functions (using a cursor): certain calculation functions and the like can be worthwhile. Measuring amplitude and time using adjustable cursors (see Section 4.3.5) at least saves counting scale divisions and (sometimes) a bit of mental arithmetic. Basically, if this feature involves an extra cost, you should carefully consider whether or not it is really necessary (in normal service practice it is not essential).
- Dual time base. Have a look at the possibilities provided by the oscilloscope in question (highlighting a portion of a displayed waveform, expanded-time display, displaying the same signal as two traces with different time scales and so on), and consider whether or not they are really necessary. (The pretty pictures in the brochures and operator's manuals can only be obtained if the oscilloscope is driven by signals that contain cyclically repetitive waveforms. Typical examples are television composite colour and combined sync signals.)
- Glitch capture (DSO). This is essential! A DSO that does not have some provision for detecting the behaviour of the signal being measured between the sampling points is almost useless for making critical measurements. Examine the specifications carefully. It must be possible to enable glitch capture for every available sampling rate setting. The specified minimum glitch duration should be ≤ 10 ns (≤1 ns is better). Some instruments can only detect glitches that are longer than 50 ns (for example), but a glitch that is only 10 ns long can easily clock a flip-flop or increment a counter.
- PC interface. For the types of measurements that we have in mind, the possibility of remotely controlling the oscilloscope or transferring the display data to a PC does not by itself provide any practical benefits. Storing the display data in a PC is only necessary if the waveforms are to be saved for documentation or something similar. In actual practice, the only advantage of being able to display the waveforms on a PC monitor is that the higher resolution of the PC display sometimes allows more details to be recognized. *Practical tips:* (1) a PC is no substitute for an oscilloscope with an inadequate display. Connecting the oscilloscope to the PC is awkward, and the update rate of the resulting display is so low that it hardly justifies the name. (2) 'Useful' software provides the display data in a format that can be imported into standard application programs for further processing. An even better solution is 'modular' software that you can incorporate into your own programs (for instance, using Delphi or Visual Basic).

What the oscilloscope shows

The information content of the screen image. An analogue oscilloscope displays an image that is basically three-dimensional: its three axes are time (X or horizontal axis), voltage (Y or vertical axis) and brightness (Z axis). Some instruments have a special Z input (often located on the rear panel) that allows the brightness of the beam to be modulated. (See Appendix 8 for references to literature that describes what can be done with this feature.) It is sometimes possible to glean useful information from the brightness variations of 'normal' waveform displays. For instance, pulse edges are usually characterized by lower brightness (very steep edges are often not even visible), portions of the displayed waveform that are not as bright as the rest occur less frequently (that is, not in every sweep cycle) and so on. With a bit of experience (acquired through practice, by making measurements on working hardware), it is not difficult to correctly interpret waveforms that 'swim' across the screen, jittery waveforms and so on.

A DSO, on the other hand, provides no brightness information, and the way in which the waveform is displayed can encourage incorrect interpretations, especially to the extent that a low sampling rate and a low screen resolution are employed. *Examples:* (1) an absolutely repetitive periodic signal appears on the screen of an analogue oscilloscope as a perfectly clean and stable waveform. If you expect this and instead see an unstable or jittery display, you instinctively know that something is wrong (supply voltage unstable? phase modulation? metastability problems? coupled-in interference? and so on). By contrast, the waveform displayed by a DSO is always a bit 'restless'. (The conversion of the sample data to the raster format of the LCD or monitor represents a second digitization. If the resolution of the display is less than that of the sample data (which can happen – see Table 4.9), then even copy-book sine waves and rectangular waves will look quite irregular, especially at low amplitudes.) (2) With an analogue instrument, you can easily detect small notches in the waveform, weak superimposed ripple voltages and the like, and you can judge whether they represent harmless disturbances or true faults. With a DSO, by contrast, the (two-fold) digitization of the waveform means that a minute amplitude variation results in either a distinct pixel (the variation is displayed in exaggerated form) or no pixel (nothing at all is displayed).

Signal details. An analogue oscilloscope has the same signal bandwidth for all time bases (sweep rates) and passes the smallest signal details through to the CRT – although you do not see anything useful at very low sweep rates (see Figure 4.67). A DSO does not 'see' the signal at all in the interval between each sample point, and when long-duration processes are to be captured the sampling rate must be drastically reduced (see Table 4.13). Many details of such long-duration processes (such as noise pulses that occur during a switch-on process) are thus not observable. The remedies are (1) glitch-detection capability that is as effective as possible (in order to see whether something suspicious is happening) combined with (2) very flexible trigger provisions (so that subsequent measurements can be triggered as accurately as possible on the fault event – with some luck it will recur and you can capture it with finer time resolution).

Basically, short-lived signal variations (such as 2-μs spikes superimposed on a full mains cycle) either cannot be displayed at all, or cannot be displayed in their true forms, whenever measurements are made with a 'long' horizontal time base, regardless of whether we use an analogue or a digital oscilloscope. A digital oscilloscope with peak-detection or glitch capture capability is however superior in such cases, since it can at least direct your attention to potential problems. (You should regard it as a sort of signal analyzer with a graphical display, rather than a waveform-display instrument – see also Figure 4.50).

Note: the way in which oscilloscopes typically react to glitches (even without glitch capturing) can be seen from a number of examples in Figures 4.66 through 4.71.

Measurement time (acquisition time). It is often necessary to continuously monitor a waveform in the course of fault-finding, to see whether anything suspicious occurs (departures from the normal waveform, glitches and so on). The problem is that the entire waveform cannot be acquired; this is illustrated in Figure 4.60.

What you can see is ultimately affected by several time intervals that depend on the characteristics of the oscilloscope:

- With an analogue oscilloscope, the sequence of events is: (1) the trigger event occurs, (2) the beam is swept across the screen, (3) the beam returns to its starting position (retrace) and (4) the oscilloscope waits for the next trigger event.
- With a digital oscilloscope the sequence is: (1) the trigger event occurs, (2) the signal is digitized (the acquisition memory is filled), (3) the data rae read from the acquisition memory and displayed (4) the oscilloscope waits for the next trigger event. (Here the ability to digitize and store signal data before the trigger event can be an advantage.)

The signal is only actually measured in the second step of each of these processes, which means that only the portion of the waveform that occurs during this time (the measurement time) is observable. The retrace time of an analogue oscilloscope is typically on the order of 20% of the measurement time, but the time required to read and display the data from the acquisition memory of a digital instrument can be many times longer than the measurement time. The measurement time of an analogue oscilloscope amounts to somewhere between 20% and 70% of the total time between trigger events (the elapsed

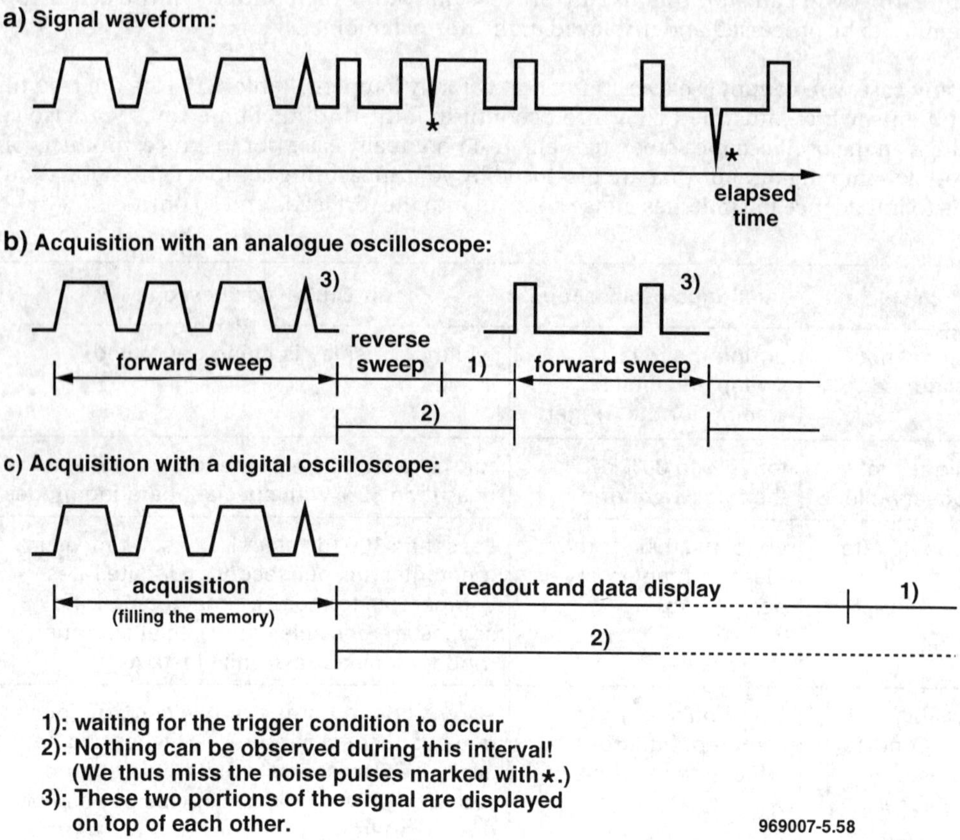

Figure 4.60 Measurement or acquisition time.

time), since the time spent waiting for the next occurrence of the trigger event must also be taken into account. With digital instruments the measurement time is often less than less than 1% of the elapsed time (the acquisition rate is at most several cycles per second, since many milliseconds are required to prepare the display following each measurement interval). *Note:* the number of measurement intervals per second is specified in the data sheets as the *update rate*. The update rate naturally depends on the selected time-base factor. If you consider the smallest possible time-base factor (for example, 100 ns/div , corresponding to a 1 µs measurement interval), analogue oscilloscopes can achieve update rates of more than 100,000/s. Normal digital oscilloscopes, on the other hand, have update rates ranging from less than 100/s to a few thousand per second.

Special data acquisition techniques (such as InstaVu from Tektronix) can increase the measurement time factor to more than 20%, which allows several hundred thousand waveform acquisitions per second. The update rate can thus reach the same level as that of an analogue instrument (and can even exceed it, with pre-triggering, since it is not always necessary to wait for the trigger event to occur before acquiring data). The basic idea of pre-triggering is that the memory depth is restricted, and the acquired data are immediately read out and displayed without extensive processing. If you see something interesting, you can stop the display process and allow the contents of the acquisition memory to be processed and displayed in the normal manner.

In any case, you cannot see everything that actually happens. Table 4.17 lists the resulting problems, which must be taken into account in fault-finding. Sometimes you also see things on the oscilloscope screen that either do not really exist (for instance, products of a too-low sampling rate) or that are produced by your measuring setup (such as overshoots due to impedance mismatches, noise coupled into the test leads and so on).

Problem	Analogue oscilloscope	Digital oscilloscope
signal not acquired ...	during the reverse sweep and while waiting for the trigger	while the display is being generated.
signal not observable ...	for 20% to 80% of the total waveform	for more than 95% of the total waveform (less than 80% with special operating modes)
update rate	up to 100,000 (or more) measurement intervals per second	Less than 100 to several thousand measurement intervals per second. (Update rates comparable to those of analogue oscilloscopes are possible with special operating modes, such as Tektronix' InstaVu.)
visibility of short noise pulses (glitches)	Noise pulses are acquired, but are difficult to recognize.	Acquisition requires suitable circuitry (peak detection, glitch capture). Glitches can be easily recognized with suitably prepared displays, but no precise conclusions can be drawn regarding durations and waveforms.

Table 4.17 *Problem areas with fault-finding using an oscilloscope.*

Testing oscilloscopes

Not surprisingly, oscilloscopes are tested using known signal waveforms. In principle, any source of signals that match the amplitude and frequency-range characteristics of the oscilloscope in question can be used for such testing. There are various reasons for testing an oscilloscope:

1 Rating and selecting an oscilloscope: the preferred test stimuli are those that address the presumed weak points of the oscilloscope(s) in question.

2 Gaining practical experience: this aspect is often overlooked, but it is an important prerequisite for successful fault-finding. The ability to interpret oscilloscope displays must be learned. This applies to all types of oscilloscopes, both high-quality and inexpensive. With the first type you must be familiar with the multitude of their features, and with the second type you must know their performance limitations. (The operator's manual and the data sheet do not always reveal everything.) The way to get relevant experience is to practise viewing characteristic signal waveforms. A good introduction consists of making measurements on relatively easily-understood circuits. Later, you can work with 'real', functioning hardware, such as PCs and PC interfaces, to develop a feeling for the correct appearance of various types of signals.

3 Functional checkout: it is certainly not a bad idea to make sure that the oscilloscope is in good working order before using it for a major project.

4 Calibration: this relates to the accuracy of the amplitude display and the time base, and frequency compensation of the probes.

All tests are based on applying known waveforms to the inputs of the oscilloscope and observing what it displays. In addition, practically all tests can be accomplished with a limited number of stimulus signals (which – for calibration purposes – must meet relatively high accuracy standards).

Note: the operator's manuals of many oscilloscopes include prescribed measurement and adjustment procedures, which describe how to check the instrument's specified performance and how to adjust or calibrate it if necessary.

Calibration signals. Calibration signals can be simple (pulses are sufficient), but they must be accurate enough for the specific parameters being tested. For more detailed information, refer to the literature listed in Appendix 8.

Amplitude calibration. Square-wave signals with defined amplitudes are sufficient. Simple reference voltage sources can also be used for preliminary DC calibration checks.

Time-base calibration. Square-wave signals with defined periods are sufficient. Needle-pulse trains (time-marker signals) with a spacing equal to one horizontal scale division are also 'tried and true' test signals.

Frequency compensation adjustment. The most common application for calibration signals in everyday practice is adjusting the frequency compensation of probe assemblies (the procedure is described in Section 4.3.5). A square-wave signal with a repetition rate of several kHz, a maximum amplitude of 1–2 V and sufficiently steep edges is adequate. Most oscilloscopes have built-in signal sources that are suitable for probe compensation adjustment. *Note:* these are usually only intended to be used for frequency compensation adjustment and are not sufficiently accurate for voltage or time-base calibration.

Special waveforms. Suitable signals can be generated by quite simple circuits. Certain oscilloscope manufacturers have published suggested circuits, and equipment that is suitable for calibrating oscilloscopes is commercially available.

Test circuits

A very simple test generator. A simple calibration generator that provides sufficient frequency and amplitude accuracy can be constructed by taking advantage of the fact that CMOS ICs switch almost exactly between the supply voltage level and the earth level (rail-to-rail switching). All that is needed is to use a reference voltage source to provide the operating voltage (see Figures 4.61 and 4.62, and Volume 3 for more detailed information).

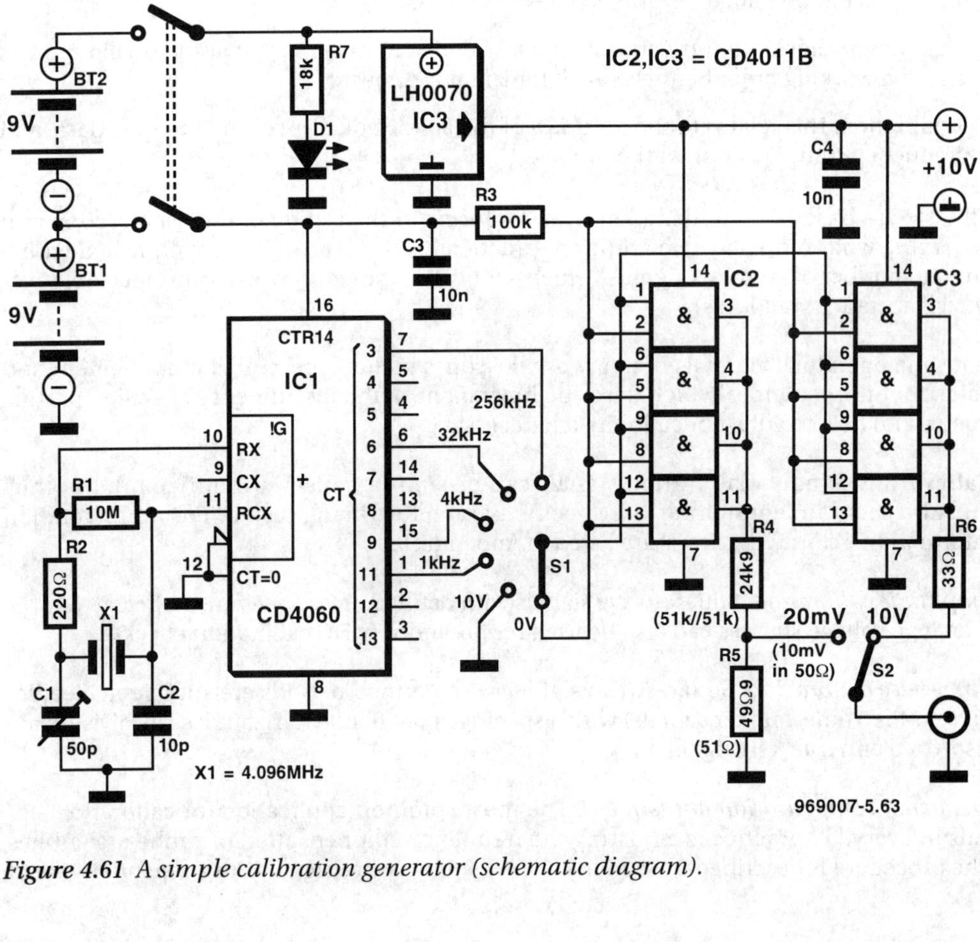

Figure 4.61 A simple calibration generator (schematic diagram).

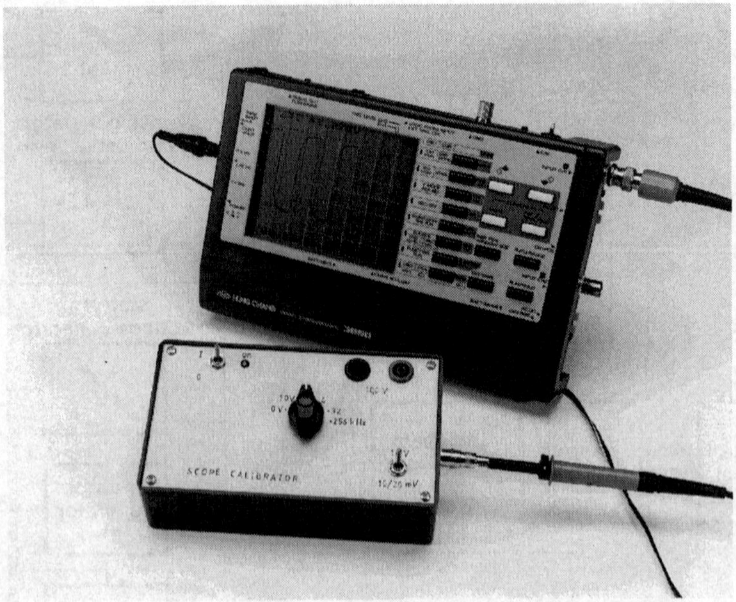

Figure 4.62 A simple calibration generator (sample construction).

A universal oscilloscope tester. Test equipment can obviously be designed to perform a variety of functions, so that it can be used for several different tasks. Test aids should however not be too complicated. Figure 4.63 shows the block diagram of a oscilloscope test generator that meets these criteria (see Volume 3 for more detailed information). This tester is based on a switchable frequency divider, laid out according to the 1-2-5 scheme, which is complemented by a one-shot circuit, a pulse generator for narrow pulses (which can be used as glitches, time markers and so on) and several other test circuits. Such an arrangement can be used as a general-purpose pulse and signal generator.

Scope exerciser. Figure 4.64 shows the schematic diagram of a tester that can be used to check several essential operating characteristics of analogue and digital oscilloscopes (see Volume 3 for more detailed information). The operation of this tester is described by means of typical measurement examples (following Figure 4.64).

Notes:

1 uncommon components: SW = DIL slide switch, Δt = delay line

2 Every signal of the circuit can be used as a measurement signal; separate test jacks are only provided for especially important signals.

3 The arrangement of the asynchronous counter and D/A converter is based on a similar Tektronix test accessory.

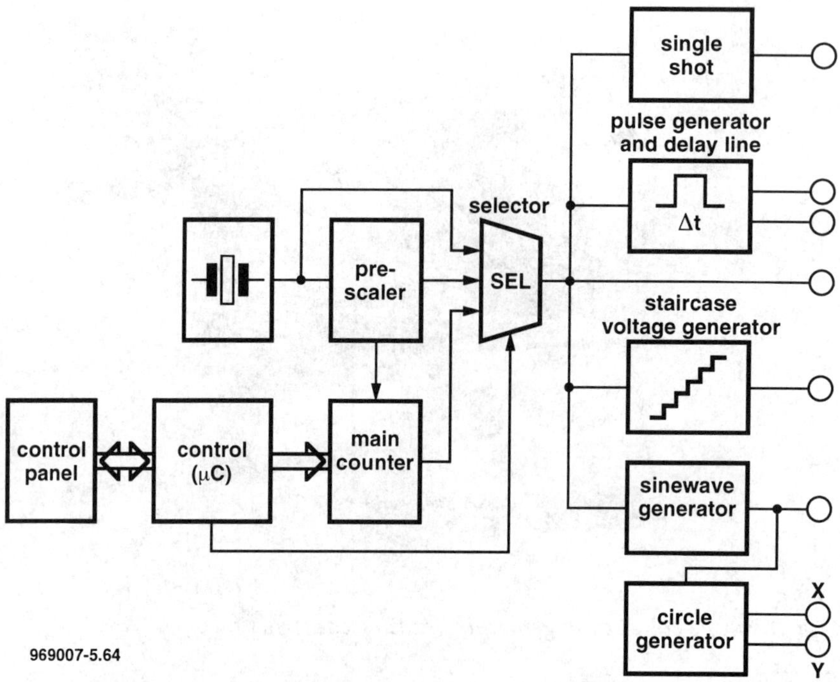

Figure 4.63 A 'universal' oscilloscope tester.

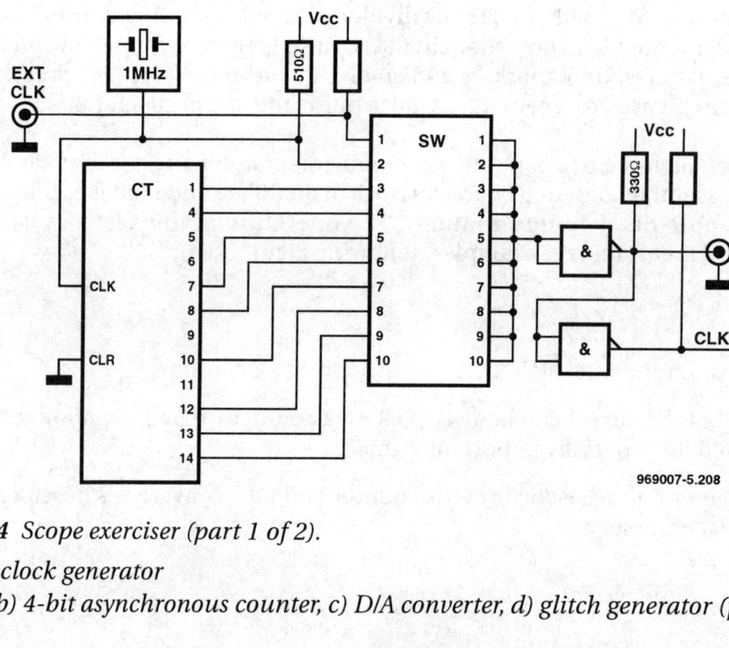

Figure 4.64 Scope exerciser (part 1 of 2).

above: a) clock generator

opposite: b) 4-bit asynchronous counter, c) D/A converter, d) glitch generator (part 1)

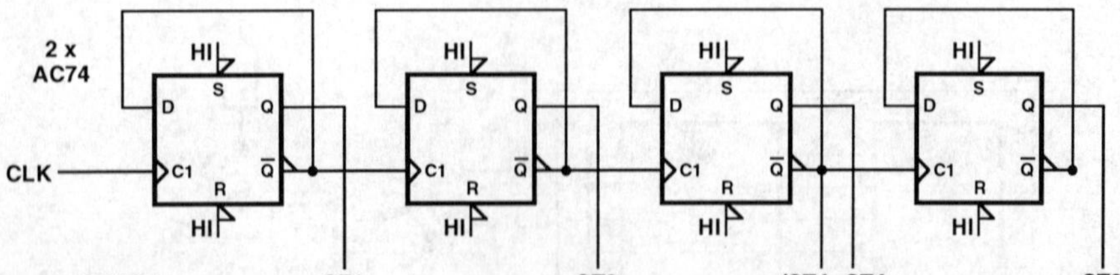

969007-5.209

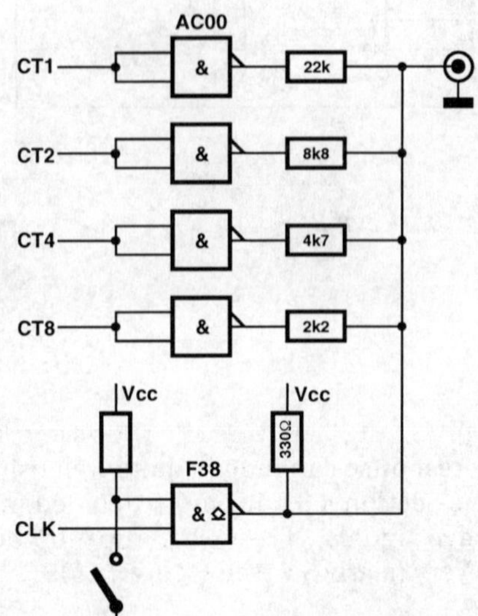

969007-5.210

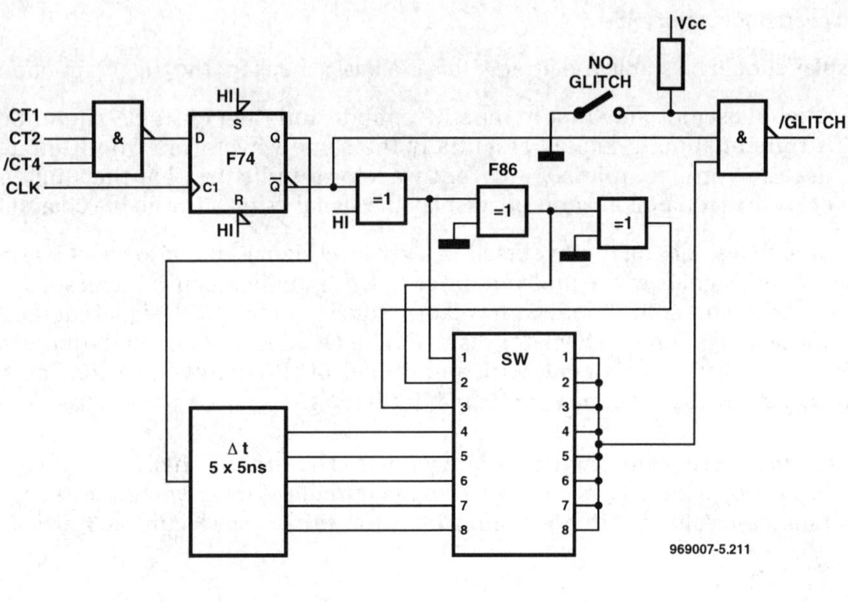

969007-5.211

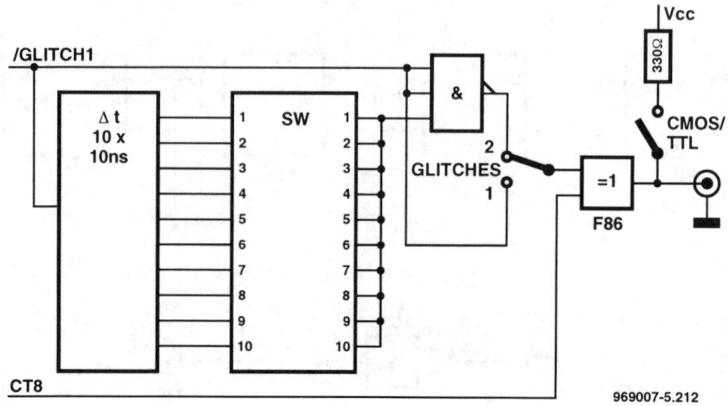

Figure 4.64 *Scope exerciser (part 2 of 2).*
e) Glitch generator (part 2).

Measurement examples

Bandwidth and frequency response can judged quite well using square-wave signals with sufficiently steep edges (see Section 4.3.5 for more detailed information on making measurements with square-wave signals). The CMOS flip-flop outputs deliver steep-edged pulses whose amplitude is very nearly 5 V (see Figure 4.65).

Notes:

1 A small overshoot is normal.

2 The display should be stable (no jitter) – this is a basic check for the trigger functions.

3 Check a digital oscilloscope first in repetitive mode and then in single-shot mode. A good instrument should display essentially the same waveform in both modes (a longer rise time or a completely different waveform indicates that the single-shot bandwidth is inadequate). In addition, display the signal using all available channels.

4 For a thorough test, do not limit yourself to logic-level signals, but also check the oscilloscope (1) with signals in the millivolt range (if the amplifier gain is increased for the high-sensitivity ranges, the bandwidth will be reduced) and (2) with signals at the limit of the input voltage range (check for distortions). Of course, it is not all that easy to generate clean, fast-rising signals with amplitudes of 10–20 mV or 10–20 V and rise times less than 5 ns; see Volume 3 for more information.

Triggering in the presence of high-frequency noise. The (intentionally primitive) D/A converter, formed by the resistors connected to the outputs of the asynchronous counter, generates a staircase voltage with noisy and glitchy transitions (see Figure 4.66). The oscil-

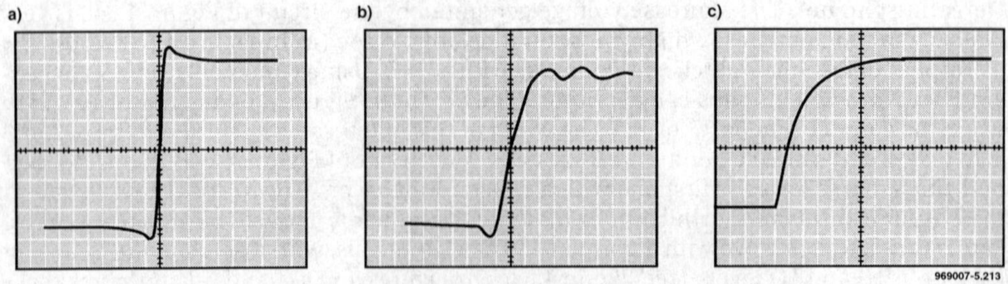

Figure 4.65 *Judging bandwidth and frequency response.*

a) 20-MHz analogue oscilloscope, 50 ns/div (with 10:1 magnification): the rise time (25 ns) is nearly the same as the characteristic rise time (18 ns). The probe assembly has a relatively small effect.

b) 50 to 150-MHz analogue oscilloscopes and DSOs, 10 ns/div (with 10:1 magnification): the display on an analogue oscilloscope is completely stable with a sharp, thin trace, while the DSO display always appears somewhat noisy. The characteristic rise time of the probe assembly significantly affects the rise time of the displayed waveform.

c) An inexpensive DSO with 5 MHz analogue bandwidth and 50 MS/s, 100 ns/div: the characteristic rise time of the oscilloscope (70 ns) gives the impression that even fast CMOS ICs can only deliver 'soft' pulses. Notes: (1) the displayed trace jumps up and down and can only be evaluated if the display is frozen and (2) a 5-MHz analogue oscilloscope would display the same waveform, but completely stable and sharp.

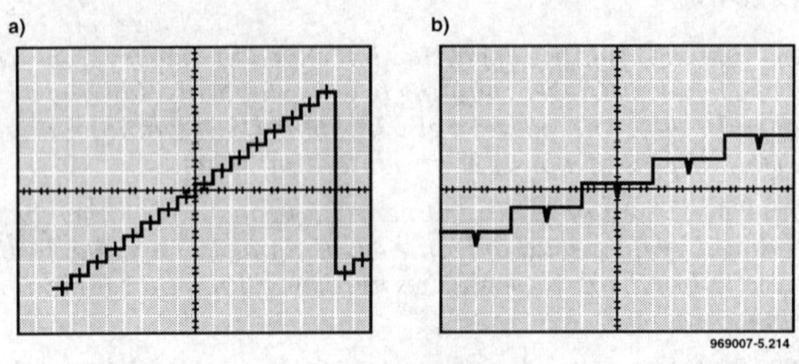

Figure 4.66 *Triggering and glitch test with a staircase voltage. (source: Tektronix)*

*a) Staircase waveform displayed on an analogue oscilloscope or DSO.
Clock rate 1 MHz, time coefficient 2 ms/div.*

*b) A portion of the staircase waveform displayed on a DSO.
Clock rate 1 MHz, time coefficient 500 ns/div.*

loscope display should be stable (select 'HF reject' trigger coupling if necessary). The glitches can be easily recognized. The waveform displayed on an analogue oscilloscope is completely stable, but a DSO displays a waveform that jitters and is somewhat fuzzy.

Detecting glitches. The staircase voltage generated by the circuit of Figure 4.64c is very likely to contain glitches, which you should be able to see (as in Figure 4.66) – possibly using peak detection or glitch capture mode with a DSO. Using circuit delay effects to generate well-defined glitches is even better. The circuit of Figure 4.64d generates a glitch when the counter state is 3_H or B_H. The width of this glitch can be selected to be 1, 2 or 3 gate delays and can be extended in 5-ns steps. The circuit of Figure 4.64e produces a second glitch that follows the first one with a selectable delay of 10 to 100 ns. The XOR gate at the output inverts the fourth counter bit when the glitch is active. The signal at the test jack is thus a square wave with a period of 16 clock intervals, with a glitch in the middle of each of its HI and LO levels (see Figure 4.67g–i). Figures 4.67a–f and j–l show how these glitches appear on various oscilloscope displays.

Explanation of Figure 4.67:

Diagrams a–f show single and paired glitches displayed on various types of oscilloscopes. Glitch width = 25 ns, time between paired glitches = 100 ns, repetition rate = 62.5 Hz (16 µs period). Note: Figure 4.68 shows why the glitches appear to be wider than they actually are.

a) *20-MHz analogue oscilloscope, 20 ns/div (magnified): 1 glitch*

b) *20-MHz analogue oscilloscope, 200 ns/div: 2 glitches*

c) *inexpensive DSO with 5 MHz analogue bandwidth, 50 MS/s, 100 ns/div. The glitches jump around on the screen and can only be evaluated if the display is frozen.*

d) *high-quality 100-MHz analogue oscilloscope, 50 ns/div*

e) *same as (d) with the bandwidth reduced to 20 MHz*

f) *150 MHz DSO, 20 ns/div*

Diagrams g–i illustrate the visibility of glitches imbedded in longer pulses. Note: you may have to select 'AC' or 'HF reject' trigger coupling to get a stable display (the oscilloscope must trigger on the 'real' pulse edge and not on a glitch). Some DSOs cannot manage even an approximately correct display of a pulse train containing glitches.

g) *time coefficient 2 µs/div. Analogue instruments and good DSOs show the waveform that is actually produced by the circuit of Figure 4.64d. Some DSOs show the basic pulse waveform correctly, but display the glitches as vertical lines that jump up and down, instead of stable spikes.*

h) *150 MHz DSO, 500 µs/div, pulse repetition rate 500 Hz. For the HI level the presence of glitches is suggested by the occasional occurrence of individual pixels; but the glitches in the LO level are totally obscured by a cloud of noise pixels resulting from 'ground bounce' in the test generator circuit. Note: you can determine whether something in the displayed waveform is due to a glitch by comparing it to the waveform that is displayed with glitch capture disabled.*

i) *analogue oscilloscope, 100 µs/div, pulse repetition frequency ≈ 1 kHz. The glitches can be seen as notches in the waveform, which are occasionally extended to form very narrow, low-intensity spikes. (The spikes can sometimes be made more visible by expanding the time base, but this will not make them any wider).*

Diagrams j–l show details of glitch waveforms displayed by a 20-MHz analogue oscilloscope (time coefficient = 50 ns/div).

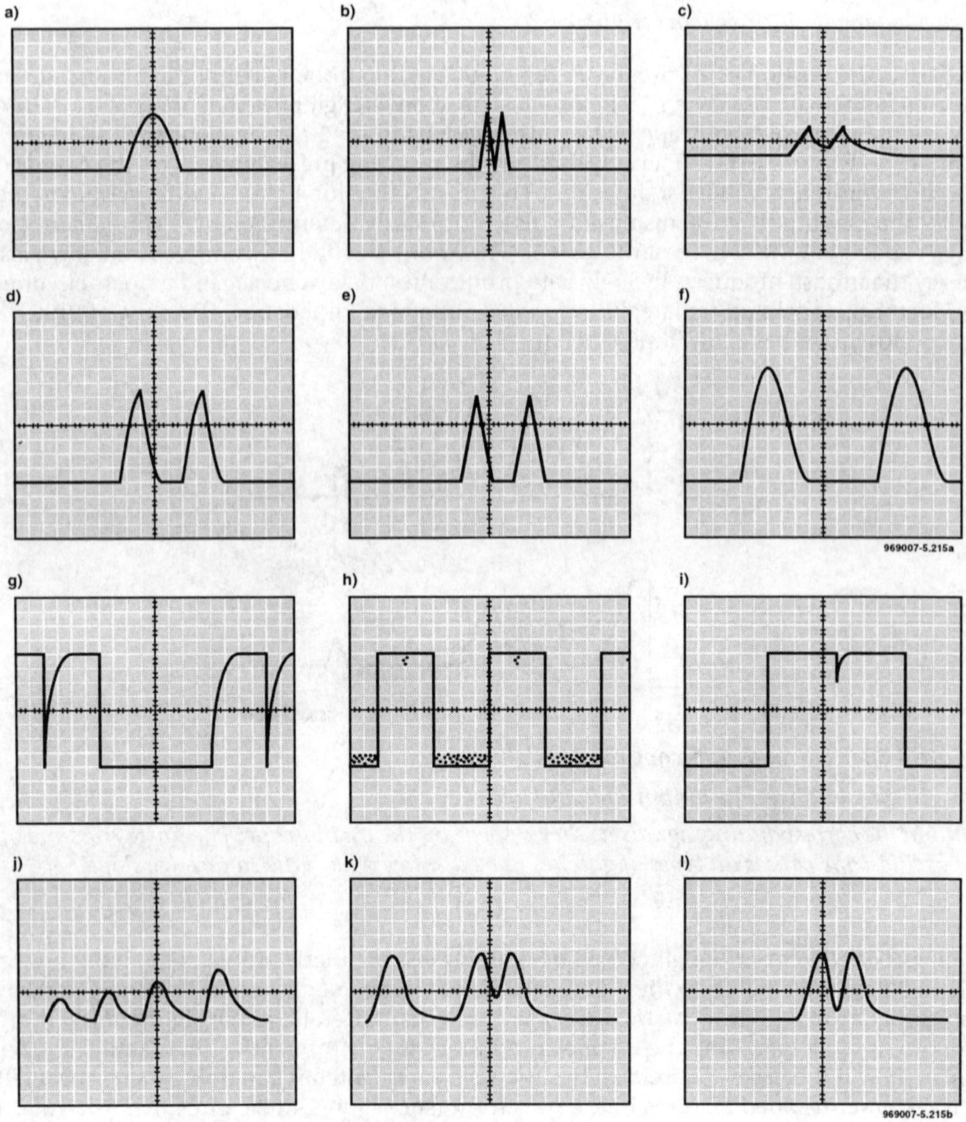

Figure 4.67 *Sample glitch waveform displays.*

---

j) Glitches of various widths, displayed together for comparison by combining several oscillograms. The glitch durations are (from left to right) 5, 10, 15 and 20 ns (according to the corresponding taps of the delay-line IC in Figure 4.64d).

k) Similar to (j). Left: a single 25 ns glitch, right: two 25-ns glitches separated by 60 ns.

l) Two glitches generated by the circuits of Figure 4.64d/e: 25 ns width, 100 ns delay.

How does an oscilloscope react to glitches?

The analogue portion of an oscilloscope (including the probe assembly, cable and so on) has a characteristic rise time t_{rc}. The oscilloscope converts glitches that are shorter than t_{rc} into roughly pyramidal or trapezoidal pulses with widths $> 2 \cdot t_{rc}$ (see Figure 4.68, and compare this to Figure 4.67j–l). The amplitude of the resulting pulse depends on the energy of the glitch (current × voltage × duration). This means that for a glitch with a certain actual width, the amplitude of the displayed pulse is primarily determined by the impedance or drive capacity of the source of the glitch. (*For example:* glitches from crosstalk have less energy than those produced by logic gate propagation delay effects, and a open-circuited conductor that picks up radiated interference has a high impedance, while the output of a gate or bus driver has a low impedance.)

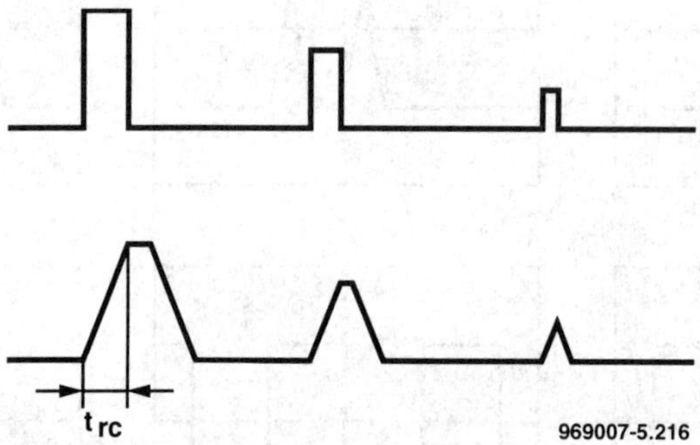

Figure 4.68 *How an oscilloscope reacts to glitches.*
top: glitches with various amplitudes and widths
bottom: the corresponding signals as they appear on the oscilloscope. The finite rise time of the oscilloscope causes all types of glitches to be displayed with the same edge slope.

Analogue oscilloscopes: the glitch will be more or less distinctly visible, depending on the selected time coefficient. In the worst case, the presence of glitches is only suggested by very faint vertical lines. (*Practical tip:* such lines can often be made significantly more prominent by defocussing the display and increasing its intensity – their actual shapes can anyhow not be seen.) At relatively low sweep rates, glitches can only be seen (if at all) as weak notch-shaped features in the waveform (see Figures 4.67a and 4.69). As a rule, a glitch with a duration of $2 \cdot t_{rc}$ can still be reliably recognized if its width on the screen is 0.5 mm (0.05 division). This corresponds to a time base setting for which one division equals $40 \cdot t_{rc}$. With a greater time coefficient, very little of the glitch will remain to be seen. *Note:* in experiments with the test generator of Figure 4.64, it is still possible to clearly see glitches using significantly larger time coefficients (see Figure 4.68g/i). However, in this case we know in advance where to expect the glitches, which is rarely true in real fault-finding.

Practical tip: don't deceive yourself. If a circuit does not behave as it should, there must be a reason. If a glitch seems to be a plausible fault hypothesis (based on experience and analysis) then you should not casually dismiss minor waveform notches as 'uninteresting'.

DSOs (without glitch capture): the analogue input circuits also have a characteristic rise time t_{rc} and thus form a pulse as shown in Figure 4.68. If this is captured by the sampling process, then will appear on the screen (more or less jittery, depending on the quality of the instrument, and as a single pixel for long time scales – see Figure 4.67h). With short time scales the pulse may appear to be fuller (broader or more energetic) than it actually is. If it is not sampled, it will not be visible at all.

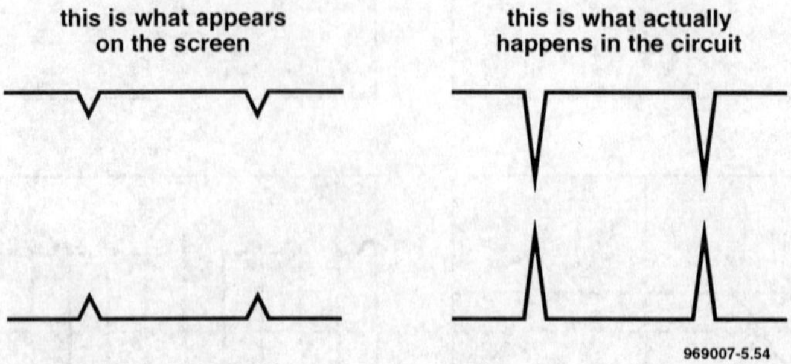

Figure 4.69 *Incorrect interpretation of narrow pulses due to rise-time limitations.*

Detecting signal variations. The oscilloscope's update rate should be fast enough to allow signal changes to be observed practically 'live'. This can be checked by (for example) periodically varying the amplitude of a rectangular waveform (see Figure 4.70). (The D/A converter circuit of Figure 4.64c can be used for this purpose; its output is pulled low at the clock rate via an open-collector gate.) In Figure 4.70, the upper portion of each pulse should appear to be 'painted in'. Depending on the trigger setup, some oscilloscopes show several distinct lines instead of a solid area. Sometimes the displayed image can be resolved into a number of superimposed waveforms with different amplitudes. A good DSO should at least show several stacked lines. Some instruments, however, only show a series of pulses with stepwise increasing amplitude, and some only show 'jumping' pulses with constantly varying amplitude.

An additional form of universal test circuit for oscilloscopes is a feedback shift register. If the shift register output does not show a typical 'eye pattern' as illustrated in Figure 4.71, the update rate is too low (see Section 5.3.3 for more information).

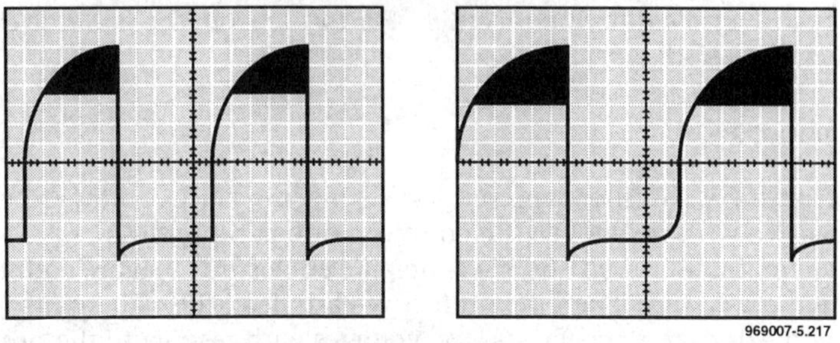

Figure 4.70 *Checking the update rate (see the text). (source: Tektronix)*

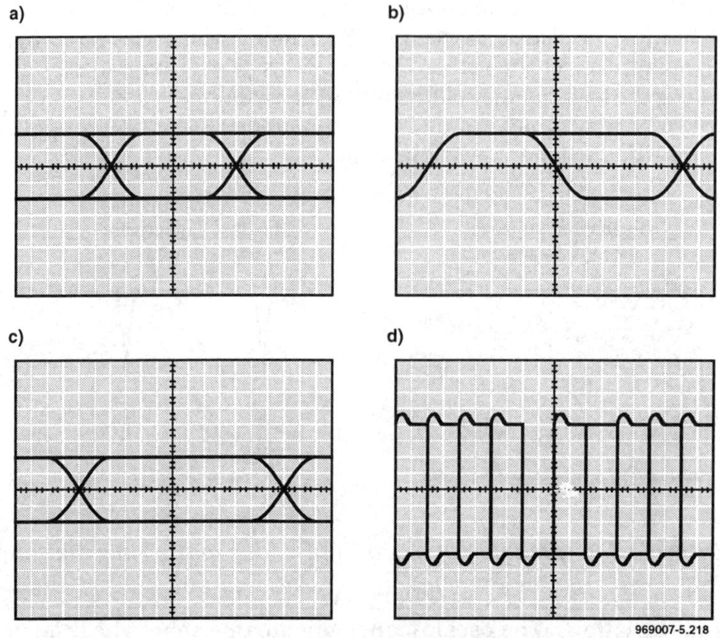

Figure 4.71 Eye -pattern diagrams.

a, b & c) Eye-pattern diagrams displayed by various analogue oscilloscopes. You must experiment with the trigger level and time scale to obtain 'textbook' eye -pattern displays (sometimes it helps to use the shift register clock as an external trigger source). Diagram (b) is an magnified display, which clearly shows the clock signal that triggers the display.

d) The display of a good DSO. Clock frequency 10 MHz , time coefficient 100 ms/div. The clock signal that triggers the display is here also not overlaid by other waveforms. If the time scale is too long, only 'jumping' pulses can be seen. Some DSOs display complete rubbish with such an input signal, regardless of their settings.

4.3.5 Measurement practice

As long as you take the basic selection criteria into account (bandwidth, sampling rate and so on), there is virtually no difference between analogue and digital oscilloscopes as far as their practical application is concerned. For this reason, the material in this section applies equally well to both types.

Basic configurations

Measurement setup. An oscilloscope fundamentally measures voltages, which may be positive or negative with respect to the common or earth potential (COM = 0 V). Most oscilloscopes are mains-powered and have metallic enclosures that are connected to both the measurement earth (COM) and the protective earth lead (PE). All inputs to the oscilloscope (X, Y, Z and external trigger) thus 'see' voltages with respect to the protective earth or mains earth potential (see Section 5.2.1). *Caution:* as illustrated in Figure 4.72, you must keep this fact in mind – avoid situations in which the instrument earth is connected to some other potential that is also referenced to the mains earth!

Figure 4.72a shows the usual situation when a measurement is made using an oscilloscope. The circuit and the instrument are connected to a common earth potential via their respective protective earth leads. If you attempt to make a measurement as shown in Figure 4.72b, you will at best measure rubbish. You cannot make measurements 'inside' a circuit using oscilloscope as easily as you can using a multimeter (see Figure 4.14d for comparison). The point where the earth lead of the probe cable is attached (marked with a *) will be shorted to earth. In this example, the transistor can clearly no longer work at all – and sometimes such an earth connection can cause real damage (for instance, if the resistor is connected to the supply line at the point *).

Possible remedies:

- Float the circuit being tested – which means that its earth must not have any connection to the oscilloscope earth or the mains earth. This is for example possible with experimental circuits that are powered from a laboratory power supply or batteries.

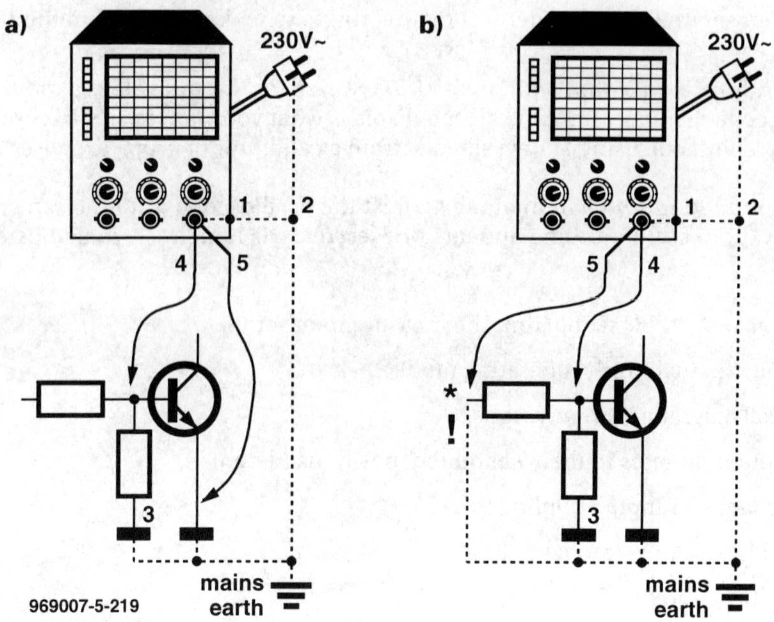

Figure 4.72 The basic measurement setup.

1 – connection between the oscilloscope's internal circuit earth and the enclosure

2 – connection between the enclosure and the mains protective earth line

3 – earth of the circuit under test

4 – test signal lead

5 – earth lead

Caution: check very carefully that there is no connection between the circuit earth and the mains earth (including via a capacitor!).

- Float the oscilloscope. *Caution:* see Appendix 3.
- Battery-powered oscilloscopes are available that are specifically designed to be used for floating measurements.
- Use a differential voltage measurement technique (refer to Figure 4.76).

Basic instrument setup. There are certain tried and true rules, although you should consult the operator's manual for detailed instructions regarding the use of the equipment.

You must always:

1 Check that the measurement setup matches the one illustrated in Figure 4.72a – that is, make sure that none of the oscilloscope inputs somehow has its earth lead connected to a point that is not at earth potential. Also check that all voltages to be measured lie within the input range of the oscilloscope (or will you need a high-voltage probe?).

2 Switch on the oscilloscope and let it reach its normal operating state before making any connections to the circuit under test or attempting to make any measurements.

With relatively modern equipment: press AUTO SCALE (or the equivalent), and then adjust the oscilloscope manually until the screen displays what you want to see. In case of doubt, check the oscilloscope using known signals (from a calibrator or scope exerciser).

With conventional equipment: manually adjust the oscilloscope until the screen displays what you want to see. If nothing appears or everything is hopelessly maladjusted, do the following:

- disconnect the oscilloscope from the measurement setup,
- select the 'channel 1 only' operating mode,
- turn off all magnification functions,
- set all potentiometers to their 'calibrated' positions (detents),
- set the channel-1 input coupling to 'DC',
- set the trigger source to 'channel 1',
- turn off trigger hold-off,
- set the trigger mode to 'automatic',
- select vertical and horizontal deflection factors (input sensitivity and time coefficient) that are in the middle of the respective available ranges,
- a trace should now be visible. If necessary, adjust the vertical (Y) position control until the trace is vertically centered.
- Adjust the intensity and focus until the trace is sharp.

Calibrated and variable settings. Be careful! If fine-adjustment controls are present, the marked deflection factors of the vertical and horizontal selector switches (such as V/div and ms/div) are valid only if the associated fine adjustment control is in the calibrated position (which can be felt as a 'click' detent when the control is moved). You must also take selectable magnification functions and the divider ratios of the probes into account.

Magnification functions. Most oscilloscopes can magnify the display in the vertical and horizontal directions by fixed factors (Y and X magnification). For instance, dividing the selected deflection factor by 10 corresponds to a 10-fold magnification for the affected axis. This allows details to be more closely examined. *Caution:* magnification can significantly degrade the linearity of the displayed image.

Operating mode 1: Y–t operation (waveform display)

The signal amplitude is displayed on Y axis and time on the X axis (via repetitive sweeping or digitizing). The amplitude and time values can be directly read from the scale divisions.

Measuring amplitude and time. It is only necessary to read the quantity in question from the scale (or count the number of scale divisions) and then multiply this by the selected deflection coefficient (see Figure 4.73).

For example: suppose that you want to measure the amplitude and duration of the undershoot in the waveform of Figure 4.73. The measured scale amplitude is 1.4 divisions; if the vertical deflection factor is 1 V / div then you can calculate (1.4 div)·(1 V / div) = 1.4 V. The measured duration on the horizontal scale is approximately 0.6 divisions (estimated); if the horizontal deflection factor is 0.1 µs / div then you can calculate (0.6 div)·(0.1 µs / div) = 0.06 µs or 60 ns.

Calculations. Amplitude values must be multiplied by the divider ratio of the probe assembly, and time values must be multiplied by any magnification factor that is used.

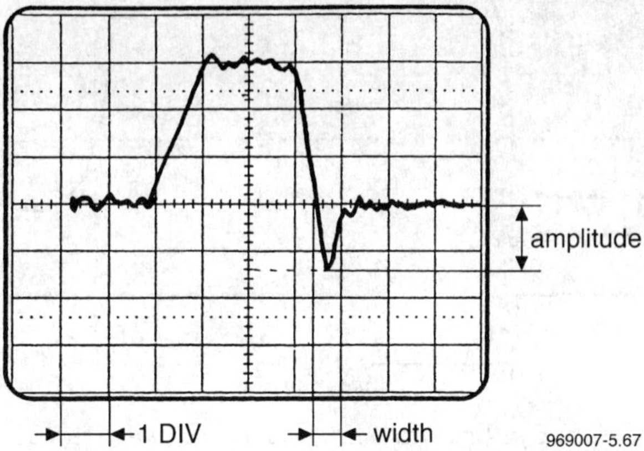

Figure 4.73 Amplitude and time measurements (see text).

Example 1: amplitude measurement with a probe assembly: displayed amplitude = 2.6 scale divisions, deflection factor = 20 mV/div and the probe assembly has a ratio of 1:10 (ie, a division factor of 10). The measured value is (2.6 div)·(20 mV / div)·(10) = 520 mV.

Example 2: time measurement with magnification: displayed time = 5.2 scale divisions, deflection factor = 0.1 µs/div and 10× horizontal magnification is enabled. The measured value is (5.2 div)·(0.1 µs / div)·(10) = 0.052 µs or 52 ns.

Caution: the accuracy of the amplitude measurement decreases with increasing frequency. It is by no means possible to accurately measure the amplitude of a 100 MHz sine wave with a 100 MHz oscilloscope! (The cutoff frequency is specified as the frequency at which the amplitude response falls off by 3 dB, or in other words at which it is 0.7 of the value at lower frequencies, which corresponds to a 30% error.) If the frequency response is known exactly, the measurement error can be taken into account.

Cursor measurements. Some oscilloscopes have moveable vertical and horizontal cursor lines that can be used to mark the limits of the quantity to be measured. These can be positioned using cursor buttons or knobs (for analogue instruments). Horizontal cursor lines are used to measure amplitude, and vertical lines are used to measure time (see Figure 4.74). The corresponding values can be directly read from the screen in numerical form (although in some cases it is still necessary to multiply the amplitude reading by the division factor of the probe assembly).

Some oscilloscopes can automatically measure amplitude, frequency and the like. Some also have probe interfaces that allow the division factor of the probe assembly to be automatically sensed.

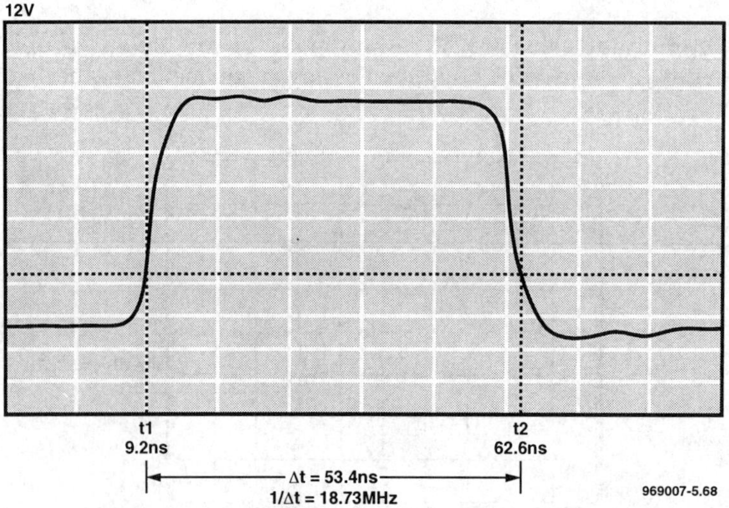

Figure 4.74 *Measuring pulse width with cursors. The value can be read directly (53.4 ns). The horizontal cursor marks the desired amplitude level. (source: Hewlett-Packard)*

Notes:

1 You basically measure voltages, and instantaneous or peak voltages at that. Effective values can be easily calculated only if there is a simple relationship between the effective and peak values (such as for sine waves, where $U_{eff} = 0.35\ U_{pp}$; see Appendix 5). Be careful when selecting the input range, since in many cases the peak-to-peak value is more than twice the effective value (U_{pp} = 650 V for the mains voltage, for instance).

2 Check the manual regarding the accuracy of cursor-based measurements. It is sometimes significantly lower than you might expect. Above all, keep in mind the screen resolution with a digital oscilloscope: the cursor can only be moved in steps of at least one screen pixel. *For example:* suppose a LCD screen has a vertical resolution of 128 pixels, and the signal covers 50 pixels peak to peak. One pixel then corresponds to $U_{pp}/50$, which means that the cursor resolution is 2% of U_{pp}. The error is proportionally larger if even smaller amplitudes are measured. *Practical tip:* if you transfer the displayed waveform to a PC, you can usually measure it more accurately (with suitable software).

3 Measure the waveform at the right places (and in case of automatic measurements, be aware of what is actually measured). See Figure 4.75 for an example.

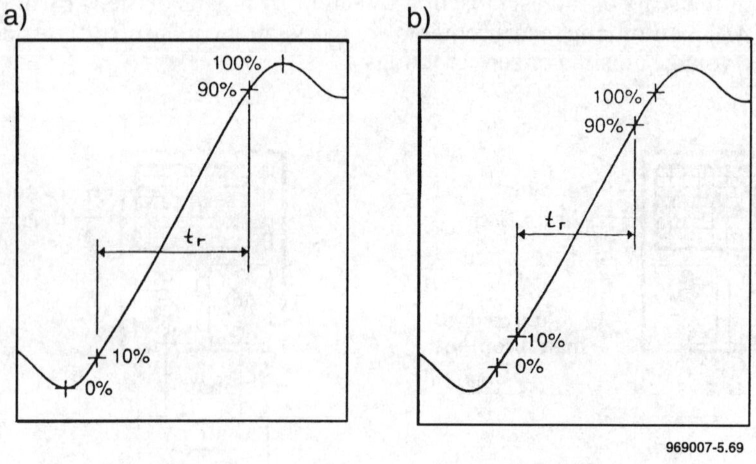

Figure 4.75 *Measuring the risetime t_r of a pulse. (source: Tektronix)*

The rise time should be measured between the 10% and 90% amplitude points.

a) Wrong! The measurement is made with reference to the visible peak values of the waveform (the instrument interprets the minimum and maximum voltages as the 0% and 100% levels respectively).

b) Right! Small undershoots and overshoots are normal. The 10% and 90% amplitude levels must be chosen with reference to stable signal levels. If you make the measurement manually you can easily recognize these levels, so that it is not difficult to read the correct amplitude from the scale or to correctly set the cursor lines. An automatic measurement process would have to calculate the stable levels from the sampled signal values.

Current measurements. Although oscilloscopes measure voltages, it is possible to measure currents indirectly. There are in principle two ways to do so:

1 Measure the voltage drop across a sensing resistor (*Caution:* normally, the voltage can only be measured with respect to earth (refer to Figure 4.72); one way to work around this is to make a differential measurement.)

2 Use a current-to-voltage converter.

Differential voltage measurements. Here the displayed waveform represents the difference between two signal voltages (see Figure 4.76). A differential amplifier is necessary (this is available as an accessory device or a differential probe assembly). If high accuracy is not required, a differential measurement can be made by operating the two input channels in differential mode (SUB or ADD + INV). *Practical tips:* (1) set both channels to the same sensitivity, (2) use identical probes and (3) check the common-mode rejection. (If both probes are connected to the same signal, then strictly speaking only a horizontal trace should be seen. Adjust the compensation of each of the probe assemblies to come as close as possible to this condition.).

Frequency measurements. The frequency is the inverse of the period, which can be read directly from the scale of the oscilloscope screen or by means of the vertical cursor lines (see Figure 4.77). Automatic measurements are based on the measured time between voltage minima, voltage maxima or zero crossings.

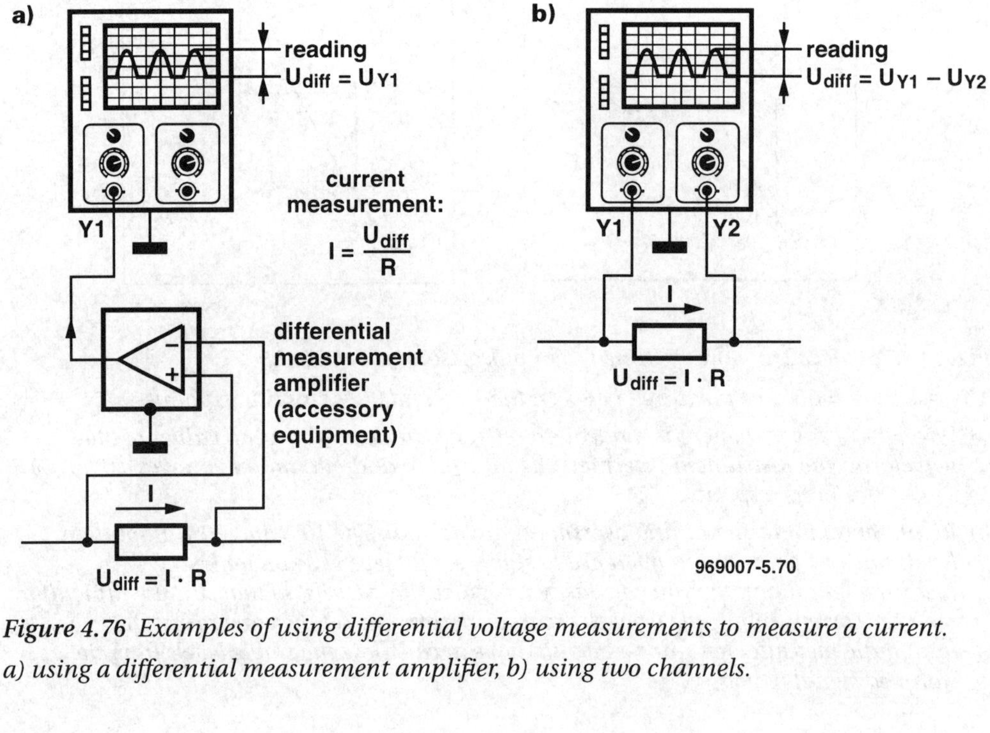

Figure 4.76 Examples of using differential voltage measurements to measure a current. a) using a differential measurement amplifier, b) using two channels.

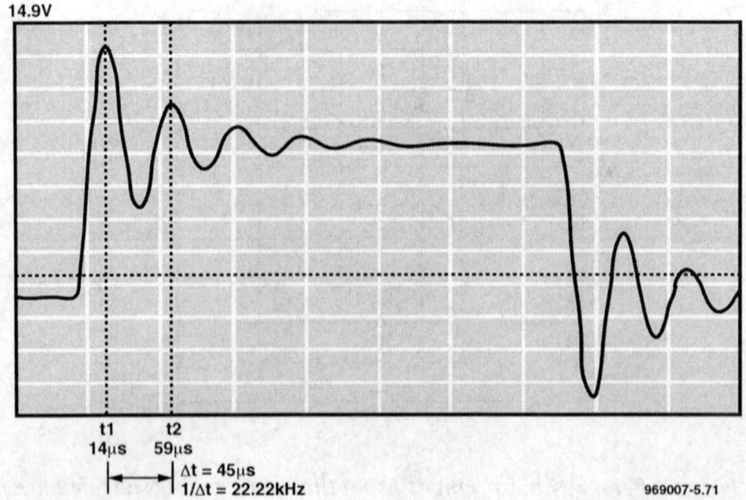

Figure 4.77 *Measuring the characteristic frequency of a an overshoot. The period and frequency can be directly read (Δt = 45 μs, f = 22.22 kHz). (source: Hewlett-Packard)*

Practical tip: always choose the most distinctive features when making frequency measurements. Sometimes peak values (maxima or minima) are the most suitable and sometimes zero crossings; with pulse waveforms the edges may be the most suitable. In any case, measure the time between two successive, identical peak values (maxima or minima) or two successive pulse edges with the same slope, and then calculate the frequency as the inverse of the measured time.

Phase-difference measurements. The phase offset between two signals with the same frequency can be directly read from the oscilloscope scale (see Figure 4.78) or measured using the vertical cursor lines.

First measure the period T (as for a frequency measurement) and then the offset t between two identical reference points for the two waveforms (peaks, zero crossings or pulse edges). The period T represents a phase interval of 360°. The phase angle φ can thus be calculated as follows:

$$\varphi = \frac{t}{T} \cdot 360°$$

Sign of the phase angle (leading or lagging phase). Be careful! Elapsed time increases from left to right on the screen, so that the further to the right a given waveform is displayed, the later in time that it occurs. *For example:* suppose that we take signal *A* as the reference signal in Figure 4.78. Point *b* lies further to the right than point *a*, which means that signal *B* occurs later than signal *A* (it is delayed with respect to signal *A*, which means lagging phase or a negative phase angle). The phase angle would be positive (leading phase) if point *b* would occur earlier than *a*, which means that it would lie to the left of point *a*.

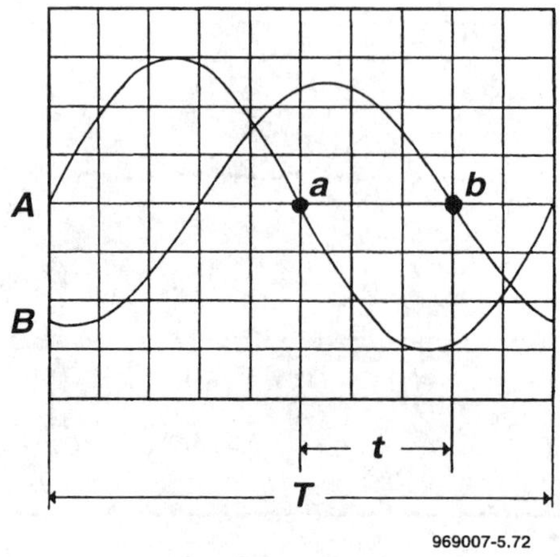

Figure 4.78 Phase difference measurement using the zero crossings as reference points. (source: Hamag)

Operating mode 2: X-Y operation

In this mode, the signals to be measured are applied to both the X and Y axes. The oscilloscope is thus operated as a sort of plotter (see Figure 4.79).

Caution: if only a DC voltage is applied to each of the X and Y inputs (or if there is nothing applied to either input), only a single spot is displayed. In such a situation there is a risk of damage to the screen of an analogue oscilloscope.

Lissajous figures. Screen images that result from the X-Y display of sinusoidal signals are called Lissajous figures. Figure 4.80 shows two examples of very simple Lissajous figures.

If the signals applied to the two channels have a non-zero phase difference, then in general an ellipse is displayed. A phase difference of 90° causes a circle to be displayed, while phase differences of 0° and 180° produce straight line segments. (The phase angle can be calculated by measuring the ellipse.) If sinusoidal signals with different frequencies are applied to the X and Y inputs, the resulting Lissajous figure rotates (for non-integer frequency relationships) and forms loops (if the frequency ratio is greater than 1:2). The number of intersections in the figure indicates the frequency ratio (no intersection = 1:1, one intersection = 2:1, two intersections = 3:1 and so on). Lissajous figures can be used for frequency comparisons: if the figure does not rotate then the two frequencies are related by an exact integer ratio.

Note: this type of measurement produces pretty screen patterns, but it has gone somewhat out of fashion for two reasons: (1) it is only usable for frequencies up to a few hundred hertz at most (at higher frequencies, unacceptable errors result from the non-equal phase delays of the two channels) and (2) with modern equipment, many measurements can be made more conveniently and more accurately using Y-t operation. For this reason, you are referred to the literature listed in Appendix 8 for more information.

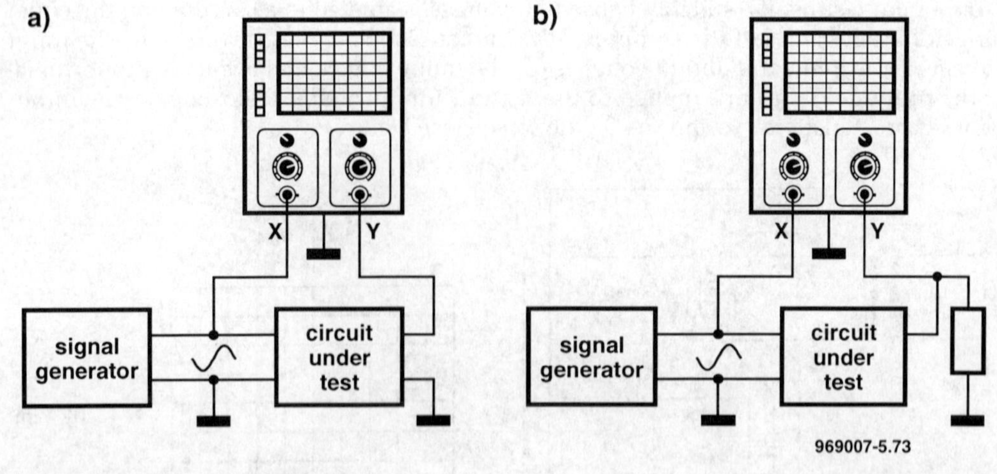

Figure 4.79 Measuring in X-Y mode.

a) The X-Y display of two voltages. The display shows the dependence of the output voltage of the device under test to the (time-varying) input voltage. Applications include measuring frequency response and measuring characteristic curves of components (component testing).

b) The X-Y display of a voltage and a current. The voltage is measured directly, while the current is measured as a voltage drop across a resistor. Applications include measuring characteristic curves of components (component testing) and measuring phase angles.

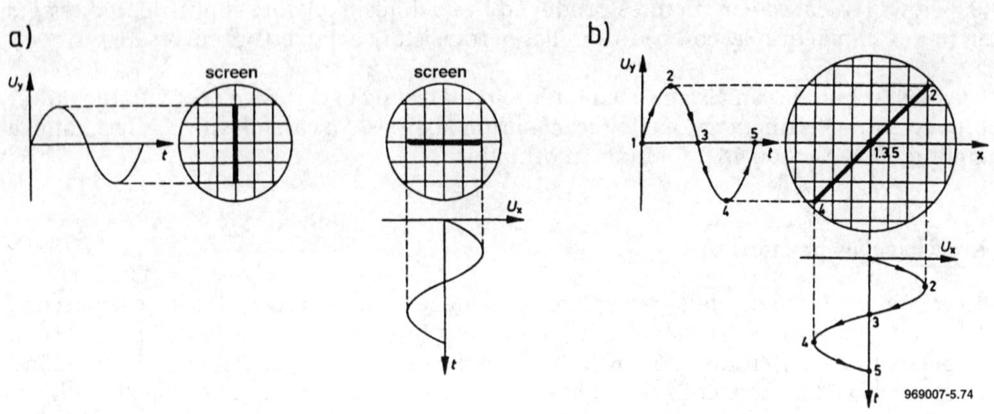

***Figure 4.80** Simple Lissajous figures.*

a) If a sinusoidal voltage is applied to only one channel and no signal is applied to the other one, only a straight line segment will be displayed, either vertical (if the signal is applied to the Y axis) or horizontal (if the signal is applied to the X axis).

b) If the same signal is applied to both channels at the same time, a diagonal line segment is displayed , running from the lower left to the upper right.

Component testing. If a suitably-chosen AC voltage is applied to a component, the entire range of input values of the component's characteristic curve is traversed by the input voltage. If the input and output voltages (or the input voltage and a voltage proportional to the output current) are applied to the X and Y inputs of an oscilloscope, the component's characteristic curve appears on the screen (see Figure 4.81).

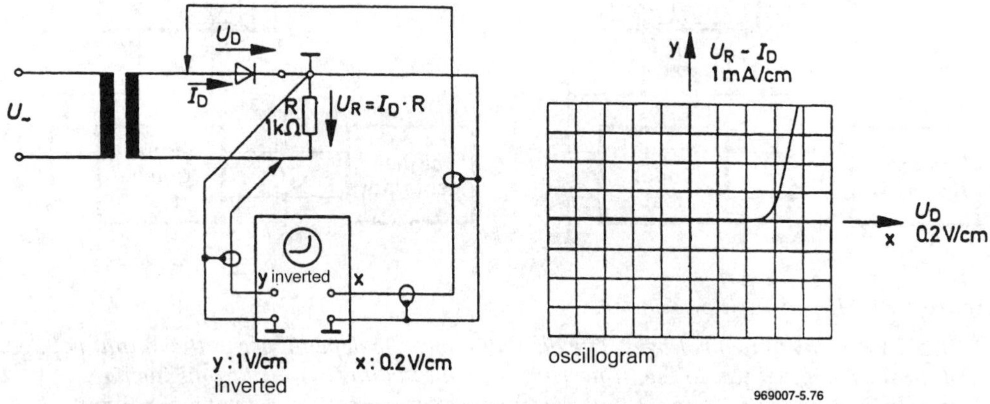

Figure 4.81 Measuring and displaying the characteristic curve of a diode.

Component testers based on this principle are available for some oscilloscopes (either built-in or as accessory devices). Simple equipment uses an AC voltage derived from the mains, which is applied to both the component being tested and the X channel. Figure 4.82 shows typical screen pictures produced by such testers. More sophisticated testers can (for example) display complete families of transistor characteristic curves.

Note: this technique can also be used for in-circuit testing of components (with the power off!) as a sort of 'elementary' analogue signature analysis. Be careful with the test voltage amplitude! See Section 5.1.2 for more information.

Miscellaneous practical tips

Direct connection to the deflection plates (analogue oscilloscopes). Modern oscilloscope CRTs are quite sensitive. They typically have deflection factors ranging from 2 to 15 V / cm. A signal with an amplitude of 5 V will thus produce a clearly measurable deflection, and you could consider connecting the signals to be measured directly to the deflection plates. The advantage of this is that it avoids the effects of the amplifier stages with regard to loading, bandwidth and frequency response. It requires almost no power to drive the deflection plates! Some high-grade oscilloscopes allow the signal to be measured to be connected directly to the deflection plates, thus allowing measurements to be made with a cutoff frequency of more than 1 GHz. Even instruments that are unquestionably in the hobbyist class can be used in this manner for frequencies of 30 MHz and more, as long as the deflection plate terminals are accessible (switched jacks are sometimes provided for this purpose on the rear panel).

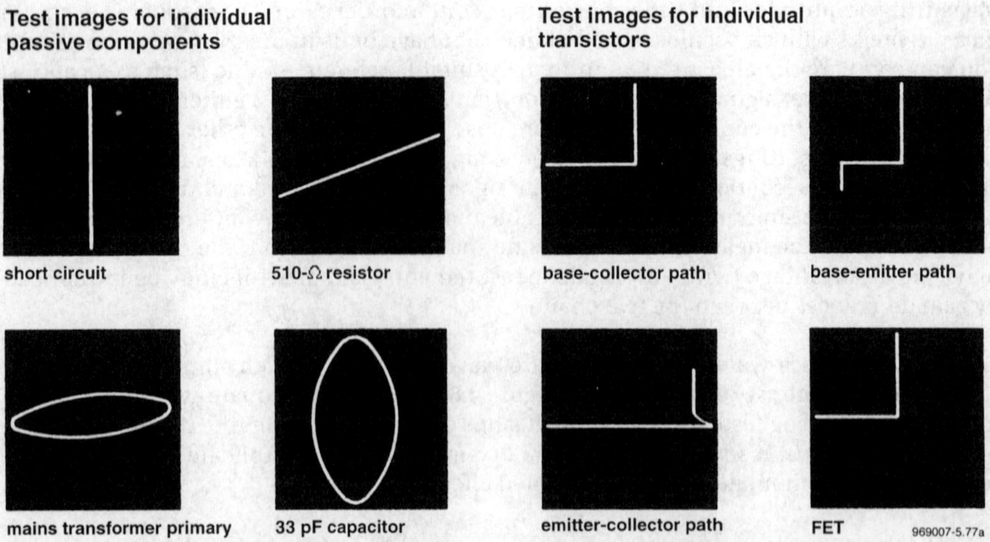

Figure 4.82 Examples of screen displays from a simple component tester. (source: Hamag)

Accuracy. With very simple analogue oscilloscopes (with all measurements made directly from the face-plate scales) you cannot expect anything better than ±5 %, all things considered. High-quality instruments with digital readouts have accuracies of 1 % or better.

Triggering versus synchronizing (antique or very cheap analogue oscilloscopes). Formerly, inexpensive oscilloscopes had free-running horizontal deflection systems that had to be synchronized with the signal being measured. (The horizontal deflection frequency was set by means of a potentiometer control and range selector switch until it was nearly an integer multiple of the signal being observed, so that the sawtooth generator could then be 'pulled' into synchronization with the external signal.)

Invisible disturbances. Sometimes the suspicion arises (as a fault hypothesis) that disturbances are present in a circuit, but none are visible on the oscilloscope (refer to Figure 4.69). Such situations always test the technician's resourcefulness, since there is no fool-proof way to deal with them. Several well-tested tricks can be suggested:

1 Try to utilize the special features of the oscilloscope (such as peak detection or glitch capture, as well as special triggering modes). (*For example:* trigger on all impulses that are shorter than 50 ns. If necessary, try triggering on various impulse widths to see if you can thereby get a handle on the problem.)

2 Attempt to verify the presence of the disturbance by using additional test aids, such as a sensitive pulse detector or a suitable supplementary triggering aid (see Volume 3).

3 Attempt to verify the disturbance hypothesis by experimentally 'shorting out' the suspected noise pulses in selected locations. You can for example connect a ceramic capacitor (20 pF to 2.2 nF) between the signal line in question and earth.

Measuring amplitudes and times by comparison. In order to make reasonably accurate measurements with an oscilloscope that does not have built-in measurement provisions, you can compare the measured quantity to a suitable reference value (such as a calibration or time-marker signal). This can be done in two different ways: either (1) the signal to be measured and the comparison signal are observed one after the other without altering the oscilloscope settings or (2) the reference signal is applied to a second channel. The advantage of the second method is that the signal being measured and the reference signal can always be superimposed in a suitable manner, so that any difference is immediately visible and the measured value does not have to be read from the oscilloscope display. The disadvantage (with amplitude measurements) is that errors may be introduced by gain differences between the two channels.

Testing with square-wave signals. A square-wave signal is a proven stimulus for judging the frequency response of a circuit (see Figure 4.83). If an 'ideal' square-wave signal is fed into the circuit being tested, certain conclusions can be drawn from the distortions of the output signal. *Note:* if several phenomena occur together, a complicated waveform will result from the combination of the associated effects.

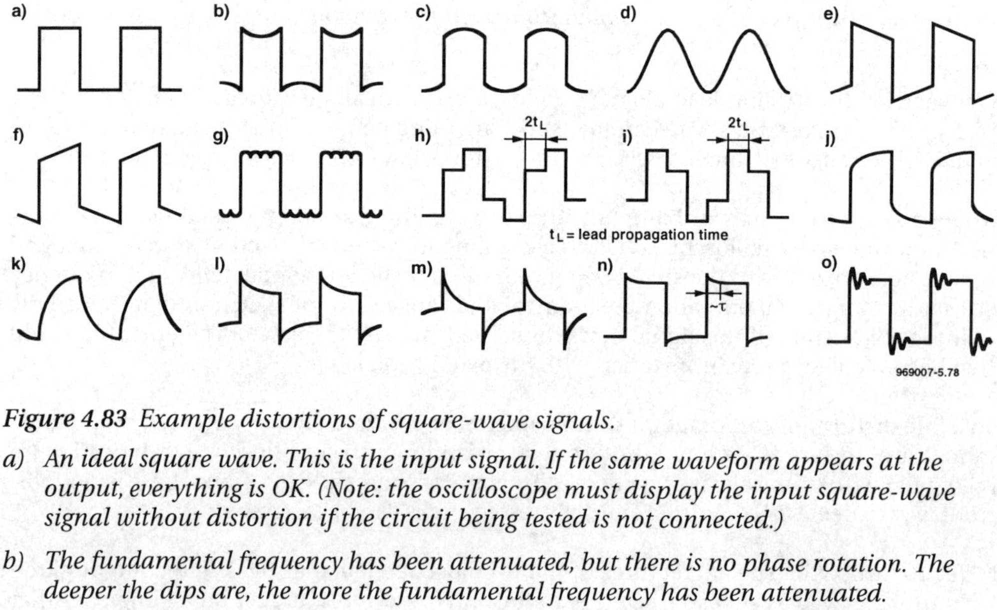

Figure 4.83 *Example distortions of square-wave signals.*

a) An ideal square wave. This is the input signal. If the same waveform appears at the output, everything is OK. (Note: the oscilloscope must display the input square-wave signal without distortion if the circuit being tested is not connected.)

b) The fundamental frequency has been attenuated, but there is no phase rotation. The deeper the dips are, the more the fundamental frequency has been attenuated.

c) The fundamental frequency has been accentuated (but there is no phase rotation and no attenuation of the higher harmonics).

d) Higher harmonics have been attenuated (without phase rotation). If only the fundamental frequency is passed through, the result is a pure sine wave.

e) Lagging fundamental frequency (leading phase for the higher harmonics).

f) Leading fundamental frequency (lagging phase for the higher harmonics).

g) *Resonance effects. (The frequency of the oscillation can be measured from the displayed waveform.)*

h) *Impedance mismatch (of a conductor): positive reflection factor (not terminated, or excessive termination impedance).*

i) *Impedance mismatch (of a conductor): negative reflection factor (shorted, or insufficient termination impedance).*

j) *Low-pass behaviour: high frequencies are suppressed; lagging phase.*

k) *Extreme low-pass behaviour (integrator). A typical cause is a capacitive load to earth due to incorrect component insertion.*

l) *High-pass behaviour: higher frequencies are accentuated; leading phase.*

m) *Extreme high-pass behaviour (differentiator). Low frequencies are suppressed.*

n) *Overshoots in the form of single peaks with exponential decay. Suspicion: capacitive effects. The RC time constant can be approximately determined from the decay time.*

o) *Overshoot in the form of a damped sine wave. Suspicion: inductive effects (including earth loops, among other things). The resonant frequency can be measured from the displayed waveform (compare to (g)).*

Rise time measurements. The scale markings typically include a set of lines that mark the 10% and 90% amplitude levels (see Figure 4.84). Adjust the amplitude and position of the displayed waveform so that the minimum and maximum levels (excluding under- and overshoots, see Figure 4.75) lie on the horizontal scale lines *a* and *b*. You can then read the rise time between the two dotted horizontal lines.

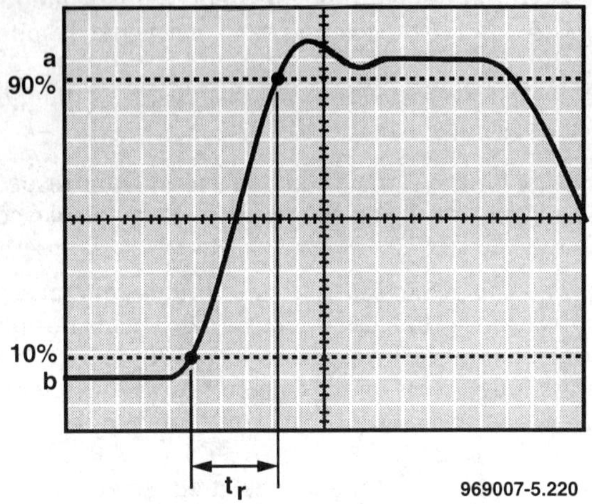

Figure 4.84 *Measuring the rise time.*

Recording waveforms. Photographing the oscilloscope display is tedious (and the necessary equipment is expensive), and it is not always possible to save or print the data. One way around this is to copy the displayed waveform by hand onto a prepared grid (see Figure 4.85). It takes a certain amount of skill to draw a complete waveform 'freehand', but it is relatively easy to examine the waveform square by square and copy each section to paper (*tip:* make a copy of Figure 4.85). The scale divisions of this sample screen are subdivided to further ease the copying process (it is relatively simple to observe the shape of the waveform in each of the grid cells and to sketch its shape: is it located (for example) in the lower left corner or the upper right, does it have a rising or falling slope, and so on).

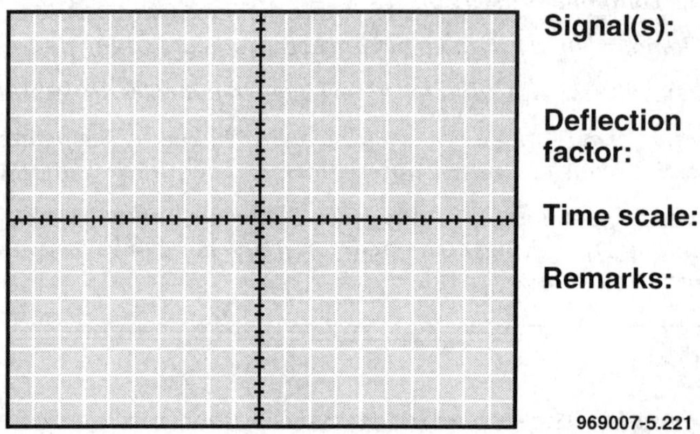

Figure 4.85 Blank screen grid for sketching waveforms.

Triggering

Analogue oscilloscopes without delay lines. Inexpensive instruments do not have delay lines. There are two ways to observe the leading edge of a signal in spite of this limitation:

1 Display two periods of the waveform (with the disadvantage that details may be lost or the measurement accuracy may be reduced).

2 Provide an external delay for the signal to be measured (with respect to the trigger signal), so that the trigger signal for the oscilloscope leads the displayed signal.

In the basic measurement setup in Section 2.4, the stimulus generator must provide a separate trigger signal that precedes the actual test signal. This is not particularly difficult, and many test generators include such provisions (see Volume 3 for sample circuits). Software test loops (stimulus loops) can be written so that they generate a trigger output before executing the actual test accesses (see Volume 2 for details).

If the trigger must come from the signal being measured, an adjustable monostable pulse generator can be used to delay the trigger point until just before the start of the following period of the signal (see Volume 3 for more details).

All in all, this means that oscilloscopes without delay lines can still be used for many measurements on interfaces, bus systems and the like.

Triggering modes

Normal triggering. In 'normal' mode, a trace is only displayed if the signal waveform matches the trigger settings (level, slope and coupling). If positive-slope triggering is selected, a trigger only occurs if the signal reaches and passes the trigger level on a rising slope (negative → positive). If negative-slope triggering is selected, a trigger only occurs if the signal reaches and passes the trigger level on a falling slope (positive → negative).

The advantage of the normal triggering mode is that you can display the details of complicated signals by properly selecting the trigger coupling and slope, and (the critical factor) carefully adjusting the trigger level and hold-off.

Automatic triggering. In 'automatic' mode a trace is always displayed, even if no signal is applied to the oscilloscope, and it is not necessary to be concerned with the triggering level. The disadvantages of this mode are: (1) it does not work with low-frequency signals (typically below 10–20 Hz) and (2) the trigger usually occurs at the zero-crossing point of the signal, which means that a complicated signal will often not be displayed as you might want to see it, and that sometimes the displayed waveform is rather unstable.

Trigger coupling. Experiment! The labels are self-explanatory, but you sometimes have to consult the manual to determine the specific characteristics of the various options:

- DC: direct-current coupling of the trigger signal,
- AC: alternating-current coupling of the trigger signal (ranging from the lower cutoff frequency (typically 10 Hz) to the upper cutoff frequency of the oscilloscope),
- HF: high-pass AC coupling; the low-frequency cutoff is typically several kilohertz,
- LF: low-pass AC coupling (typical bandwidth 20 Hz to >50 kHz).

Trigger source

Internal, each channel individually. It should be possible to select any individual channel as the trigger source, regardless of what is being displayed.

Internal, alternating (ALT or COMPOSITE). The trigger signal is taken alternately from the channels that are selected to be displayed (the channel that is currently connected to the input of the vertical deflection amplifier also provides the trigger signal). This mode is practically useless in normal fault-finding, since we basically want to investigate all signals with reference to a common time base (see also Section 5.5). *Caution:* with some oscilloscopes, alternate triggering mode is selected whenever alternate display mode is selected (in which each channel is displayed in turn for one horizontal sweep).

External. This mode allows a reference signal to be used for triggering without applying it to one of the Y channels, and enables the use of test aids (triggering accessories) that allow very selective definition of the triggering conditions (such as triggering on an address match, on pulses shorter than a defined interval and so on).

Triggering modes of digital oscilloscopes. Acquisition of signal data in a digital oscilloscope is based on an addressable storage memory. This makes it possible for signal data to be acquired *before* the trigger condition occurs: in this case data acquisition is stopped instead of started when the trigger event occurs. (This requires somewhat more complicated addressing logic than that shown in Figure 4.43 – the address must return to zero after the maximum address is reached (wrap-around addressing) and registers must be provided to store the start, end and range addresses.) Figure 4.86 illustrates the various operating modes.

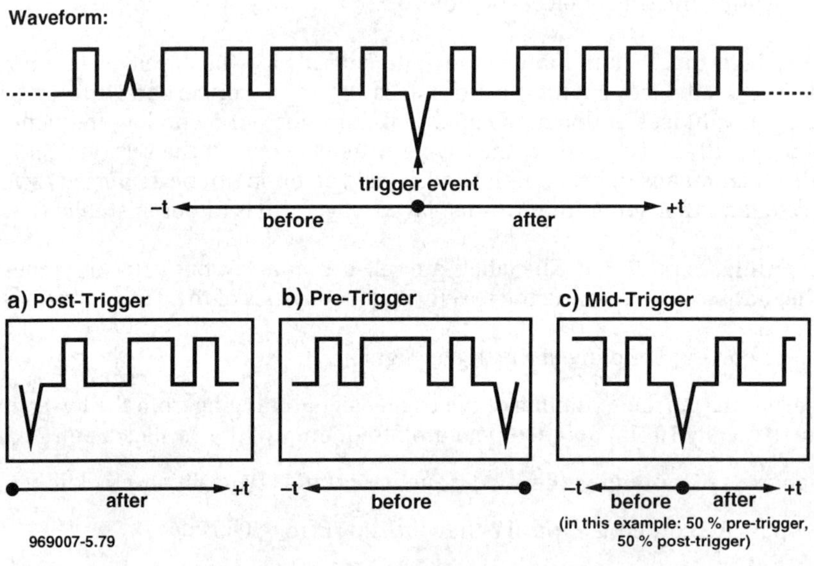

Figure 4.86 *Triggering modes.*

a) Data acquisition is initiated by the trigger event (post-trigger): this corresponds to the operation of an analogue oscilloscope. The acquisition memory is filled starting immediately after the trigger condition is satisfied. Sample data are stored from the starting address until the acquisition memory is full.

b) Data acquisition is stopped by the trigger event (pre-trigger): the acquisition memory is continuously refilled in a cyclic manner. The storing of sample data stops immediately when the trigger condition is satisfied. The entire acquisition memory then contains the 'pre-history' of the trigger event. The end address, which is the address at which data acquisition stopped, marks the location of the trigger event.

c) Data acquisition brackets the trigger event (mid-trigger): the acquisition memory must be split for this mode. Some oscilloscopes have a fixed relationship between the pre-trigger and post-trigger portions (such as 10% pre-trigger and 90% post-trigger), while others have selectable ratios (so that you can for example select 30-70 or 50-50).

Special trigger conditions. Modern digital oscilloscopes provide a variety of selectable trigger conditions (see Table 4.18).

Trigger condition	Trigger event	Selection or adjustment options
signal edge	the signal level rises above or drops below a defined threshold level (trigger level)	trigger level, slope (rising or falling); various types of coupling (DC, AC); basic filter functions (HF reject, LF reject etc.)
pulse width	a pulse is wider than a defined threshold value	minimum pulse width, polarity (positive/negative/both)
edge steepness	a signal edge is steeper or less steep than a prescribed value	edge steepness (slew rate) and condition (less steep or more steep)
glitch	a pulse is narrower than a defined threshold value	maximum pulse width, polarity (positive/negative/both)
timeout	a signal stays HI or LO for longer than a specific time	timeout interval, event selection, level monitored (HI or LO)
logic signal combination	a particular combination of binary signals occurs (and sometimes, it also stays active for a certain length of time)	combinational conditions (in the sense of a boolean equation), minimum duration (if applicable)
logic state combination	a particular combination of logic signals occurs, with reference to a clock signal	combinational conditions (in the sense of a boolean equation), clock edge (positive/negative)
a pulse in a given interval (setup and hold times)	a trigger signal occurs in the vicinity of a clock edge (shortly before or shortly after)	time range before and after the clock edge, clock edge (positive/negative)
signal pattern	a portion of the waveform matches a predefined pattern (*for example:* horizontal or vertical sync pulses in a composite colour signal)	pattern selection, level settings

***Table 4.18** Selected triggering conditions.*

Notes:

1 The more sophisticated the triggering provisions, the better your chances of detecting tricky faults (glitches, signal level changes during flip-flop setup or hold-time intervals, metastable states and so on) – and naturally, the more expensive the equipment.

2 Certain conditions can also be detected with suitable accessory devices.

3 The art of triggering, and of detecting unusual conditions in general, is one of the basic problems of practical testing. This applies to logic analyzers and signature analyzers as well. Moreover, triggering techniques can be used as the basis for simple but effective test aids. For more information, see Section 4.4.4 and Volume 3.

One-shot versus cyclic stimuli. This question relates to stimuli (in the sense of Section 2.4) that you generate yourself, using either hardware or software. Is a single execution of the test enough, or is cyclic operation (using a test loop or cyclic stimulus) better? In spite of the progress of measurement technology, the cyclic approach still has its advantages:

1 Not all oscilloscopes have equally good performance for single-shot and periodic measurements (many digital types, especially somewhat older or less expensive models, have quite good specifications for cyclic operation and lousy specifications for single-shot operation).

2 It is easier to see noise pulses (and other fault symptoms that do not always occur predictably) with cyclic operation (and even if they are not immediately visible, it may be possible to trigger on them).

3 Single-shot measurements require solid measurement setups. If (for example) you are probing an IC pin, you will recognize an intermittent contact almost right away with cyclic operation. But how should you interpret the results of a single-shot measurement with a hand-held probe? In order to distinguish between a true disturbance and a poor connection between the probe and the signal lead, you must repeat the test a few times – and in this case you could just as well have used cyclic testing from the start.

4 Even 'general' disturbances can be correctly interpreted with a cyclic display without too much effort. (With some experience, you can for example immediately recognize that certain overshoots always occur or always affect all signals, so that they do not have anything to do with the fault symptoms in question.)

5 A series of test points can be checked more quickly when a cyclic stimulus is used (*for example:* start a test loop program and then check each bus line in turn.) See Section 5.5 for more information.

Probes and test cables

If you're not careful, you can introduce disturbances when you connect the oscilloscope to the signals to be observed (see Figure 4.87). Improvised connections using (for example) bell wire, multimeter test leads or ordinary coaxial cables will load both the oscilloscope's input and the circuit under test with their parasitic capacitance and inductance. In addition, unscreened cables can pick up interference from their surroundings. For these reasons, only screened cables are used to connect the oscilloscope to the device under test. There are two reliable ways of doing this: (1) using a probe assembly and (2) using 50-Ω coaxial cables.

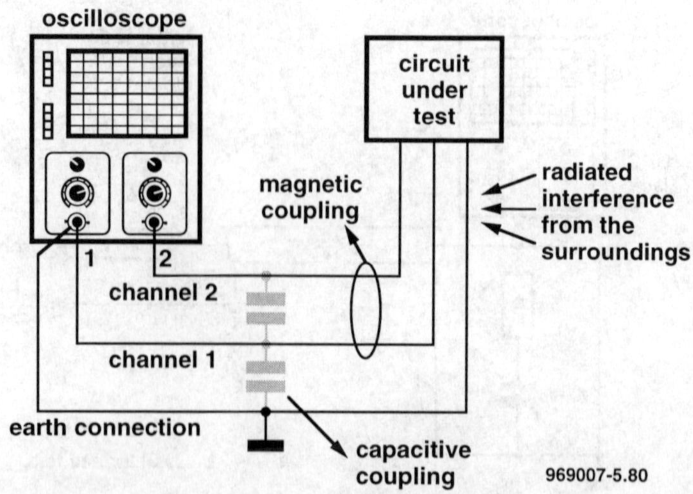

Figure 4.87 What can happen: the connections between the oscilloscope and the circuit under test are susceptible to various types of interference.

Passive probes. A passive probe is effectively a frequency-compensated voltage divider that is connected to an oscilloscope input (see Figure 4.88).

Divider ratio. This specification describes the voltage division that the probe assembly provides (the ratio of the voltage at the oscilloscope input to the voltage applied to the probe tip). Typical divider ratios are 1:1, 1:10 and 1:100. The reciprocal of the divider ratio is the attenuation factor (which for example can thus be 1, 10 or 100).

A 1:1 probe does not cause any voltage division. The signal is conducted to the oscilloscope via a screened cable and is not attenuated. The disadvantage of this is that the full capacitance of the cable directly affects the circuit under test. A voltage-divider connection increases the input impedance of the oscilloscope with regard to the device under test and allows higher voltages to be measured. The voltage divider is formed by a resistor in the probe body and the input impedance of the oscilloscope.

Frequency compensation. A voltage divider can be compensated by connecting capacitors in parallel with the resistors, so that the divider ratio is essentially constant over the entire usable frequency range. The capacitors must occasionally be adjusted (See Figure 4.89).

In order to calibrate the voltage divider, connect a 'good' square-wave signal to the oscilloscope input via the probe assembly. If the displayed waveform shows peaks (Figure 4.89a), this means that the high frequencies are over-emphasized (or the low frequencies are excessively attenuated); if it has rounded corners (Figure 4.89b) then the opposite is true. The compensation must be adjusted until a clean square wave is displayed (Figure 4.89c). See also Figure 4.83.

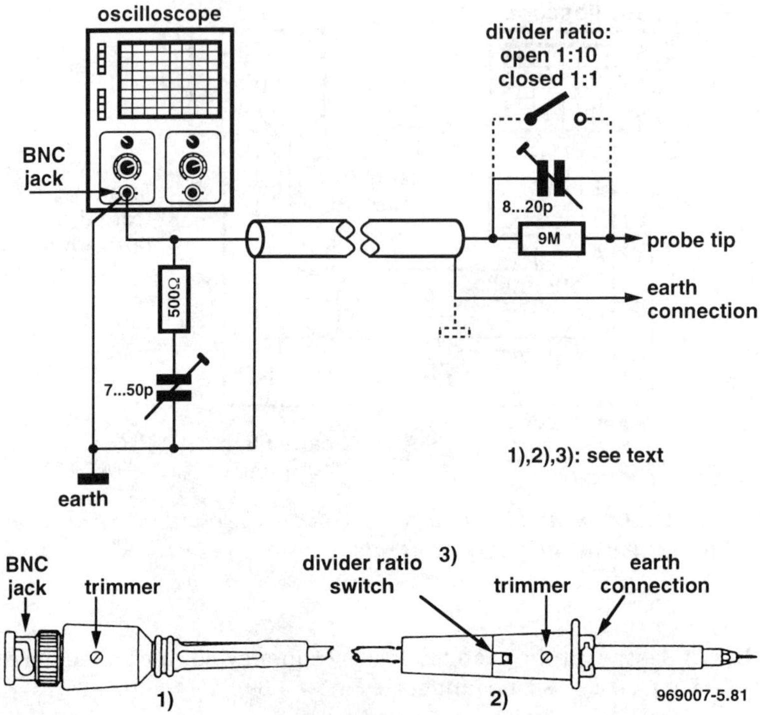

Figure 4.88 *Passive probe assembly (from a Tektronix example).*

1) & 2) Simple types of probes have only one trimmer capacitor (either in the probe head or in a small enclosure next to the BNC connector assembly). Some types have a trimmer capacitor at the BNC assembly but no resistor.

3) Normally, probe assemblies of a given type are designed to be built with two possible divider ratios (such as 1:1 and 1:10 or 1:10 and 1:100) and are available in the following versions (whose specifications are otherwise the same): (1) with only the lower ratio (eg 1:1, in which case the RC combination is omitted), (2) with only the higher ratio (eg 1:10) and (3) with two switch-selectable ratios.

Practical tips:

1 Many oscilloscopes have their own 'calibrators' in the form of built-in square-wave generators to which a probe can be connected.

2 The best results are obtained if the compensation is adjusted at two different frequencies (suggested values: 1–2 kHz and several hundred kilohertz to 1 MHz). Use a non-conductive screwdriver to make the adjustments (a suitable screwdriver is usually provided with the probe).

3 The compensation trimmer is located in the probe body or in a small enclosure at the BNC connector, depending on the construction of the probe assembly. Some probes

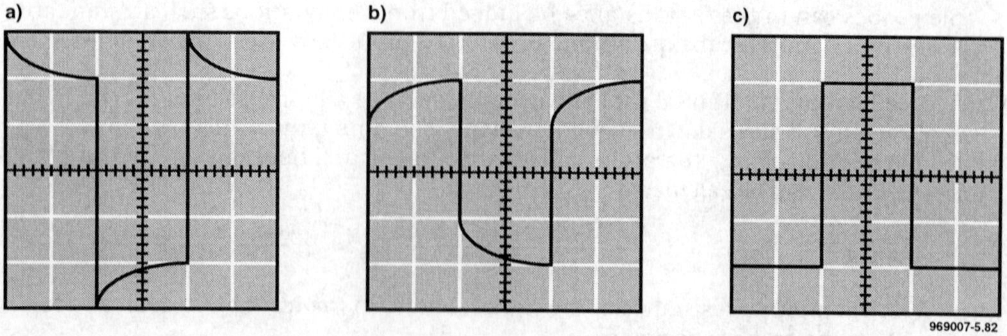

Figure 4.89 *Adjusting the compensation of a wideband voltage divider with the help of a square wave signal.*

have trimmers at both locations (see Figure 4.88). The compensation is best adjusted with a test signal in the kilohertz range for a trimmer located in the probe body, and in the megahertz range for a trimmer located at the BNC connector.

4 Switchable probes should be adjusted at the higher divider ratio.

5 When should you calibrate a probe? (1) when it is new, (2) whenever it is connected to a different oscilloscope, (3) after it has been stored for some time and (4) whenever the displayed signals appear to be a bit strange (especially in the case of undershoots or overshoots on logic signals).

6 Pay attention to the bandwidth of the probe and the effect of its input capacitance on the signal being tested. A 1:1 probe cannot provide sufficient bandwidth for making usable measurements in logic circuits. Table 4.19 lists typical probe specifications, and Table 4.20 shows the effects of capacitive loading on gate propagation delays.

7 You should avoid probes with switchable divider ratios in situations where an incorrect selection could result in damage (such as mains-voltage measurements). By the way, an incorrectly selected divider ratio is frequently the cause of incorrect measurements.

8 Don't save money in the wrong place. (This applies not only to cheap products but also to attempting to build your own probes). Quality always has its price. A certain amount of knowledge (based on experience) is required to be able to produce good probes, and there are also various technical tricks of the trade. (The quality of commercially avail-

Divider ratio	1:1	1:0	1:100
bandwidth (MHz)	10–30	150–250	≥ 300
input capacitance (pF)	40–50	10– 5	3–5
input impedance (MΩ)	oscilloscope input (typically 1)	10	100

Table 4.19 *Summary probe specifications.*

able probes can in the first instance be judged from designations such as 'standard', 'professional' and 'HF'; the quality and price of the probe increase in the order listed.)

Capacitive loading. Additional capacitive loading acts as a low-pass filter for the signal being observed and slows down signal transitions. The time t_r that is required for the signal to charge or discharge the probe capacitance (measured between the 10% and 90% amplitude levels) can be calculated as:

$$t_r = 2.2 \cdot R_s C_p$$

where R_s is the internal resistance of the signal source (in the test object, such as a logic gate) and C_p is the probe capacitance.

The combined characteristic rise time t_{rc} of the oscilloscope and probe can be calculated from their respective characteristic rise times t_{rO} and t_{rP} as:

$$t_{rc} = \sqrt{t_{rO}^2 + t_{rP}^2}$$

Logic circuits become slower with additional capacitive loading, as shown in Table 4.20.

The higher the frequency, the lower the input impedance of the probe (the capacitive portion dominates) and the greater the load placed on the circuit being tested (*for example:* 10 pF represents an impedance of around 300 Ω at 50 MHz). Table 4.21 shows the effects of different types of probes on measurements in digital circuits, in terms of the number of

Capacitive load	Standard CMOS Δt	High-speed CMOS Δt
15 pF	25 ns	2.5 ns
6 pF	13 ns	1.3 ns
2 pF	3 ns	0.3 ns

Table 4.20 Increased gate propagation time (Δt) due to probe capacitance. (source: Hewlett-Packard)

Probe type	Capacitive load	Representing ...
passive 1:1	100 pF	around 30 CMOS inputs
passive 1:10	10 pF	3 to 4 CMOS inputs
active	1.5 – 3 pF	0.5 to 1 CMOS input
50-Ω divider *)	< 1 pF	around 0.25 CMOS input
*: not all signal sources can drive a 50-Ω divider probe		

Table 4.21 Additional circuit loading due to probe capacitance. (source: Tektronix)

equivalent standard loads that each type of probe places on the circuit. For comparison, a typical CMOS input is assumed to have a capacitance of 3 pF.

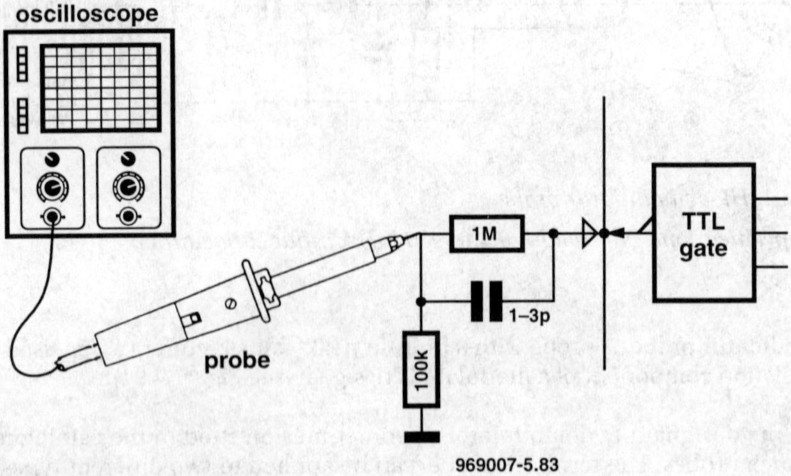

Figure 4.90 Supplementary filter circuit for making measurements at TTL outputs. (source: EDN)

Connecting a probe to the output of a TTL gate in the usual manner can cause overshoots due to the probe capacitance. The additional voltage divider reduces the capacitive load on the logic gate to roughly 1 pF.

Additional compensation for measurements in logic circuits. Even with a properly-compensated probe, overshoots may be seen on signals that are measured at TTL gate outputs, although such overshoots are not present when the probe is not connected. One remedy for this is a supplementary filter network as shown in Figure 4.90 (see Volume 3 for more information).

Cable propagation time. Typical probe cables have propagation times of roughly 4 ns/m.

Earth connections. Avoid creating earth loops! If you have connected several probes to a circuit, all the earth leads should be connected to a single point. Sometimes it is better to connect only one of the earth leads. See Appendix 3 with regard to floating measurements (in case the oscilloscope must be isolated from the earth potential).

Note: Earth loops produce under- and overshoots in the form of damped oscillations (refer to Figure 4.83o). These typically have resonant frequencies between 50 and 80 MHz.

Active probes. An active probe contains a preamplifier (which must be powered by a separate supply). Sample specifications for an active probe are: cutoff frequency 500 MHz, input impedance 1 MΩ || 3 pF, divider ratio 1:1 / 1:10 (switchable).

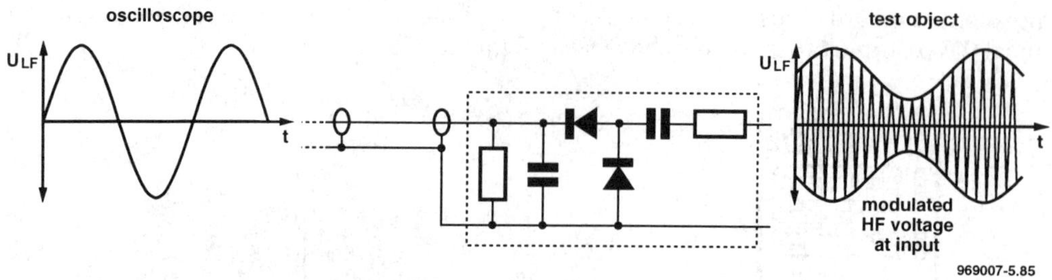

Figure 4.91 HF demodulator probe.
(typical specifications: bandwidth 0.1 to 900 MHz, input capacitance ≈ 5 pF).

HF demodulator probe. A probe with a built-in peak-value rectifier can be used to display the modulation component of a modulated HF signal (see Figure 4.91).

Caution: the designation 'demodulator' is sometimes omitted in the catalogue descriptions of such probes. The term 'HF probe' can be applied to two different types of probe: (1) a DC-coupled probe (eg 1:10 or 1:100) with an especially high cutoff frequency (250–300 MHz) or (2) a demodulator probe. (HF probes of the first type typically have extra compensation components in the small enclosure attached to the BNC connector.)

50-ohm probes. These typically have divider ratios of 1:1, 1:10, 1:20 and 1:100, with an input capacitance of <1 pF. (Some models have interchangeable tips that provide different divider ratios.)

50-ohm test cables. The outputs of many types of test equipment have an impedance of 50 Ω, under the assumption that 50-ohm coaxial cables will be used for interconnections. Such cables can be easily attached to the BNC jacks at the oscilloscope inputs, but they will not be properly terminated by the oscilloscope input impedances. This means that a 50-ohm cable must always be connected to an oscilloscope input via a feedthrough terminator (which is a sort of BNC coupler with a built-in resistor to earth).

Special features

- *Trace finder.* Some probe assemblies have a pushbutton that switches the oscilloscope input from the measured signal to earth.You can identify the trace associated with the probe by momentarily switching the input to earth while observing the screen.
- *Probe interface.* Some probe assemblies (used with specific series of oscilloscopes) can indicate the divider ratio and probe type to the oscilloscope via special contacts. Some probes also have a programmable pushbutton (softkey) that the user can assign to the 'most important' measurement function in a given situation. (A particularly important function is 'store measurement'. You can make a series of measurements, storing each one in turn by pressing the pushbutton. At the end of the test you will have the results of the whole set of measurements on a diskette or PCMCIA memory card.)

Connecting the oscilloscope to the DUT

Probe assembly. Figure 4.92 shows what is normally included with a probe assembly. *Note:* this illustrates a modular probe assembly. Items *9* and *10* are usually available separately, and the remaining items are available as an accessory set (the illustrated accessories represent a sort of 'industry standard'). Some types of especially inexpensive probes cannot be dismantled exactly as illustrated.

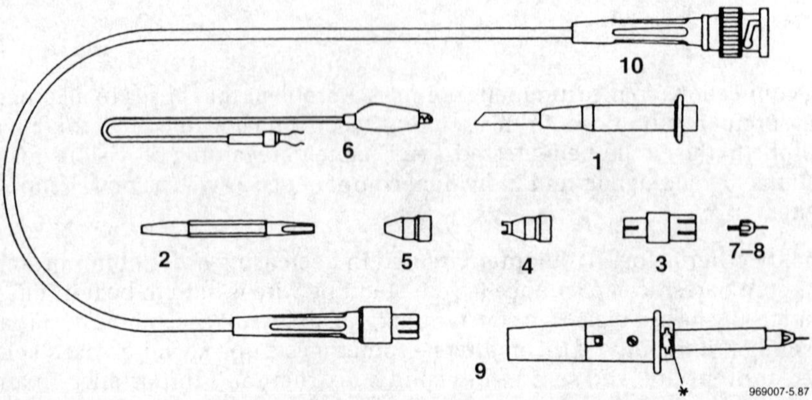

Figure 4.92 Probe assembly and accessories (excerpt from an operator's manual).

*1 – spring-hook adapter (this is slid over the probe tip (9); the spring hook is exposed when the sleeve is pulled back), 2 – compensation screwdriver, 3 – BNC adapter (for connecting the probe tip to a BNC jack), 4 – IC test adapter (this is fitted over the end of the probe tip and allows individual IC pins to be probed; the insulation at either side prevents the probe tip from sliding off the IC pin), 5 – insulating collar (covers the earth ring of the probe tip), 6 – earth lead with insulated crocodile clip (the spring clip is attached to the probe tip by inserting it into the opening marked with a *), 7 & 8 – screw-on tips made of various materials (such a beryllium-copper or vanadium steel), 9 – probe tip, 10 – probe cable with BNC connector (items 9 and 10 are normally connected together in use; the cable-end connector also mates with a standard BNC jack).*

Making connections in modern circuitry. Mechanical access to modern hardware for measurement purposes poses certain difficulties:

- It's desirable to be able to make fixed connections to the signals to be tested, to avoid always having to manually hold the probes to the pins of the ICs being tested.
- If several signals must be checked, it would be better to avoid having to rearrange the connections. In most cases, the device under test must be switched off before any connections can be changed, due to the risk of inadvertent short circuits. Even if brief shorts between signal lines do not cause any lasting damage, they may cause malfunctions that can only be remedied by switching the equipment off and on again. However,

repeatedly switching the equipment off and on degrades its reliability, and it may take a while for the device to be usable after it has been switched on (a PC, for example).

- It would be nice to be able to measure anything at all with modern components (SMD packages and very narrow pin spacings).

Suitable accessories are commercially available (in particularly, you should look at what is available from Tektronix), but the prices are rather steep. We therefore offer some suggestions for DIY solutions.

Notes:

1 Making connections to measurement points is a problem that arises in all types of test and measurement situations. In this section, we only deal with connecting an oscilloscope probe to the circuit being tested. See Chapter 5 regarding provisions for making connections to logic signals and individual components, as well as power supply measurements.

2 The 'quality criteria' for DIY adapter circuits (in decreasing order of importance) are: (1) minimum parasitic capacitance (a free-standing wire is hard to beat in this regard, since air has the lowest dielectric constant), (2) reliability, (3) acceptable cost (parts and labour) and (4) durability (if this matters – sometimes a quickly improvised solution is the most appropriate, and sometimes solid construction is important). In any case, prettiness is totally unimportant.

Measurement points for use with oscilloscope probes. An obvious solution is to add test points to the printed circuit board so that the probe's spring-hook adapter (item *1* in Figure 4.92) can be clipped onto them. Connector pins and ordinary solder terminals are suitable for this. *Notes:* (1) leave at least 5 mm between any two adjacent test points (and even more between adjacent points to which probes are to be attached at the same time) and (2) take care that the test points are not rendered inaccessible by adjacent components (such as large electrolytic capacitors).

Excessively long earth leads. Sometimes even the relatively short lead that connects the probe body to the circuit earth (see Figure 4.92) can be too long, so that it picks up interfering signals. The solution is to use the shortest possible connection to the earth ring located immediately behind the tip of the probe body (see Figure 4.93).

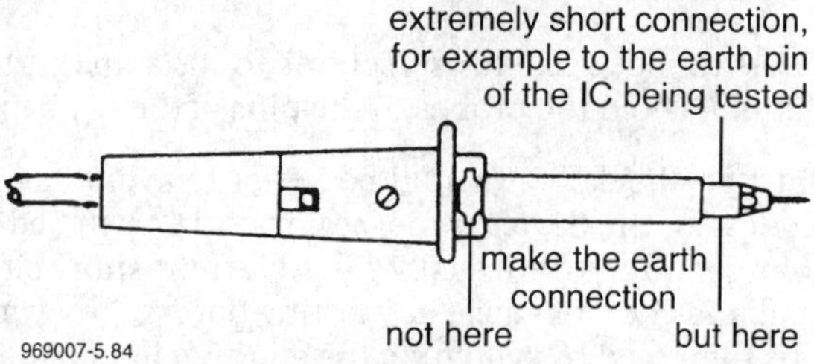

Figure 4.93 *Making the shortest possible earth connection to the probe tip.*

Possibilities:

- Short spring clips that can be slipped over the earth ring are included in some accessory kits and with some modular probe assemblies.
- Tektronix has an extensive selection of accessories, which includes not only various types of earth contacts but also complete jacks designed to accept a probe, including a signal lead and an earth connection. These are available in versions for soldering onto a circuit board (Circuit Board Test Points) and for panel mounting (Chassis Mount Test Jacks).
- Work-around 1: ordinary fuse clips will also make a secure connection to the earth ring of the probe body (see Figure 4.95).
- Work-around 2: provide BNC jacks in the measurement adapter and use the probe with its BNC adapter (see Figure 4.96).

Probe plus miniature spring clip. Miniature spring clips have proven their worth for making connections to component leads. However, inexpensive spring clips are not compatible with inexpensive probe assemblies. Some suggested solutions are:

1 For a single connection, construct a probe-tip adapter. You will need the following items from the probe accessory kit: (1) an insulating collar (item *5*) and (2) a screw-on tip (items *7* & *8*) that can be soldered (a beryllium-copper type). Connect a short insulated wire to the spring clip and then slide the insulating collar over the wire (enlarge the opening if necessary). Then solder the end of the wire to the point of the screw-on tip. This can now be screwed onto the end of the probe body and the collar slid down to cover the earth ring (see Figure 4.94). *Caution:* never attempt to directly solder to the tip of the probe body. The materials used for the tip are not suitable for soldering, and the probe body may be damaged by the heat.

Figure 4.94 *A spring clip connected to an oscilloscope probe.*

2 For multiple connections, make up an adapter using a piece of prototyping board. The individual test points can be connected to spring clips or a suitable mass connector via insulated leads. The connection to the oscilloscope probe body for each test point is provided by the combination of a 1-mm test jack (which will make a reliable connection with the tip wire of most probe bodies) and an ordinary fuse clip (for connecting to the earth ring). The whole assembly can be mounted inside a plastic enclosure, so that it can simply be 'dropped' into the equipment under test (see Figure 4.95).

Figure 4.95 *Multiple test-point adapter for an oscilloscope probe.*

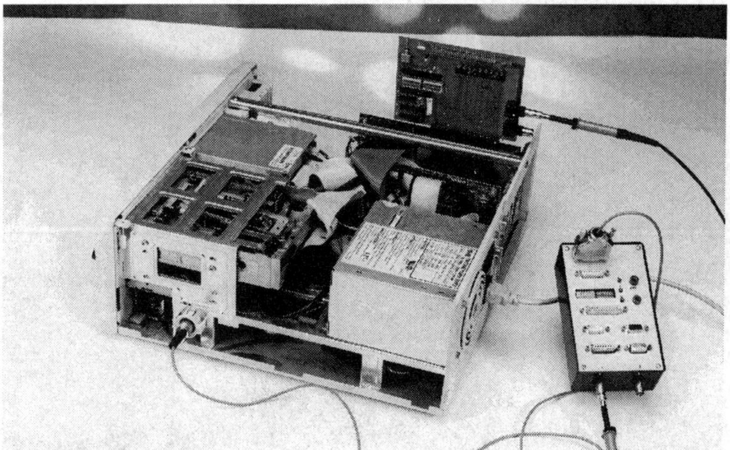

Figure 4.96 *Diagnostic adapter (left, as a PC plug-in card) and cable-type measurement adapter in use. The probes are connected via BNC adapters, and the signals to be measured are connected to the measurement adapter via a short cable.*

An alternative arrangement is to provide a fixed connection for the probe body together with selectable connections to a number of test points via a short interconnecting cable (see Figure 4.96).

Multiplexer accessories. The objective of such devices is to make the selection of individual measurement points more convenient. There are three possibilities:

1 a selector switch ahead of the probe,

2 a selector switch at the input of the oscilloscope (with multiple probe cables),

3 a fast multiplexer for the simultaneous display of multiple signals. An extremely high switching rate (at least several hundred times per second) is used to connect each of the signals to the oscilloscope in turn. The objective is to allow a large number of waveforms to be displayed at the same time on the screen. *Note:* fast multiplexers cannot be used with DSOs whose update rate is too low.

There are several ways to make such multiplexers (see Volume 3 for details).

Mechanical switching. An obvious solution is to use rotary switches to connect a probe, or the oscilloscope itself, to the individual signals to be measured. The disadvantages of this approach are (1) the signals are no good if an inexpensive switch is used, (2) the capacitive loading is relatively high and (3) the signals cannot be freely chosen (which is however exactly what is sometimes necessary in fault-finding: you want to (for example) first see signal 5 together with signal 7, then signal 3 together with signal 5, and so on). Figure 4.97 shows a probe-input signal selector based on miniature slide switches. *Note:* non-shorting (break-before-make) switches must always be used.

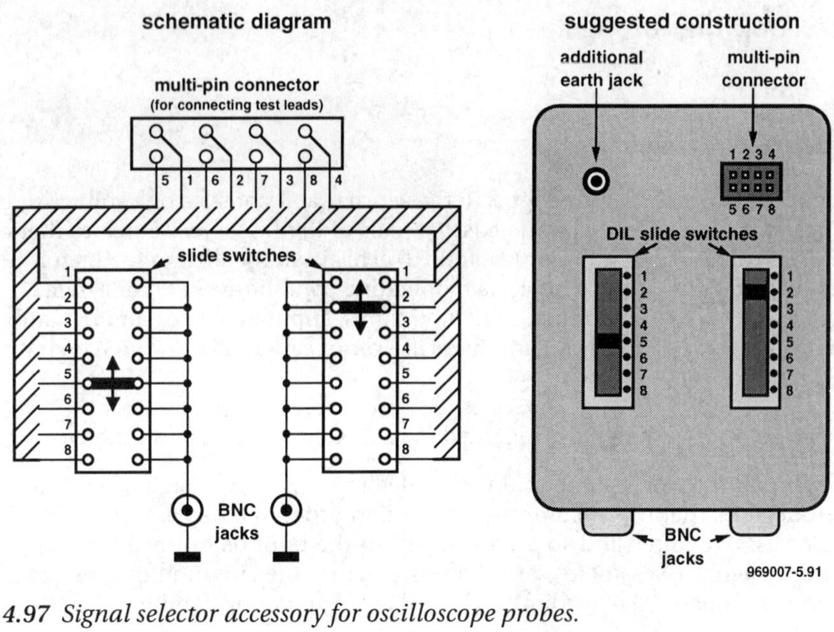

Figure 4.97 Signal selector accessory for oscilloscope probes.

Electromechanical switching. Relays can be used instead of switches. Miniature reed relays have quite small parasitic capacitances and only slightly affect the characteristic impedance of the signal path. They have the additional advantage of providing complete galvanic isolation between the control signals and the signals to be observed. Their rather high cost is however a disadvantage. *Note:* selector-switch relays (multiple-way relays) for coaxial cables, with typical impedances of 50 Ω and 75 Ω, are commercially available.

Electronic switching. Analogue multiplexers can naturally be used as signal selectors. Furthermore, with digital circuits it is not essential to treat the waveforms as 'analogue' signals. For example, with a bus system you can initially check only the 'digital' behaviour of the signals (and only later look more closely at any signals that appear to be behaving in a suspicious manner). A digital multiplexer arrangement is sufficient in such cases (this is the serialization technique; see Section 3.3).

Does it make sense? Even if the cost of a certain switching arrangement is acceptable, the question remains of how useful it is in actual practice. The deciding factor is the total time spent on making the measurements in question (connecting the test leads or probes and measuring the signals). The advantage provided by a 'convenient' multiplexer device is totally lost if a large number of connections must be made before it can be used (for example, to all 64 lines of a data bus: for quick checking it is usually faster to just touch a probe to each bus line in turn). Attempting to construct your own 'all-singing, all-dancing' equipment is not to be recommended. A better approach is to build a number of relatively simple test aids, each of which can be used for a particular test or measurement task. These test aids can employ mechanical switching, electromechanical switching, electronic switching or just simple test points, according to what makes sense in each case.

4.4 Testing digital logic

4.4.1 Basics

Analogue and digital testing

Every signal is fundamentally analogue, in the sense that it consists of a voltage or current that changes with time. However, the basic signals of digital logic are pulses that change between fixed values ('0' and '1', or LO and HI). This is an idealization, which is justified as long as the circuit complies with certain operating conditions that relate to factors such as the signal levels, edge steepness, switching times, supply voltages and the ambient or IC package temperature. In this regard, there are various manners in which digital circuits can be tested:

- *Analogue testing.* Every signal is regarded as an analogue signal and tested accordingly.
- *Assessment testing.* We assume that we are dealing with digital signals, but we want to assure ourselves that this assumption is valid. In order to do so we check whether the signal levels, rise times and so on all lie within the bounds of the allowed operating conditions (signal assessment), or we check whether the circuit in question still works properly if it is operated at the limits of the allowed operating conditions (limit testing).

- *Digital testing.* We the test the behaviour of the circuit exclusively on the basis of binary signal values ('O' and '1', or LO and HI). In this regard we rely on the signals being truly digital, which means that their levels, rise times and so on are all within the bounds of the allowed operating conditions.

Thorough analogue testing is very time-consuming. The *characteristic curves* of digital ICs, as found in the manufacturer's application handbooks, are examples of measurement results from analogue testing processes. There are only two situations in which such test processes would be used: (1) during IC development and (2) when a particular type of IC is to be characterized, for instance in order to investigate its suitability for specific types of applications. In normal service practice, analogue testing is limited to the careful examination of signals using an oscilloscope (with occasional static measurements of signal amplitudes or currents), in order to draw conclusions (from the observed waveforms or measured values) about possible causes of a particular fault.

Pure digital testing, by contrast, requires an absolute minimum of effort (even when built-in self-test provisions, test interfaces and test software are utilized (see Chapter 3), the testing is exclusively digital). The vast majority of failure mechanisms affect the digital or switching behaviour of the circuitry in question, so that in most cases digital test procedures are adequate for verification. Also, most hard faults (failures in equipment that has previously functioned) can be successfully localized by digital testing.

The disadvantages of digital testing are the following:

- It can occasionally lead to incorrect conclusions. In particular, if you attempt to localize the cause of a fault by a process of logical elimination based on incorrect circuit behaviour, you can be misled in situations where the incorrect behaviour is due to the operating conditions. (*For example:* you replace an IC that is behaving incorrectly – but in fact the supply voltage is too low, although all other ICs just barely work as they should.) For this reason, you should always check the 'infrastructure' (supply voltages, ambient temperature, clock system, reset lines and so on) before starting any sort of wide-ranging test process. Built-in fault-finding aids should have suitable provisions for this (displays, test jacks and the like) or should perform such checks automatically (service processors; see Volume 3).

- Purely digital testing can in general lead to incorrect conclusions (when you are searching for the causes of faults in newly-developed hardware or hardware that has just been assembled). *For example:* an IC behaves incorrectly and is thus regarded as suspect, when in fact the problem is due to an incorrectly mounted filter capacitor, or a signal that is connected to too many inputs (design error). For this reason, you should make 'analogue' checks of suspect signals (using an oscilloscope). Sophisticated digital test equipment (logic analyzers, automated testers and the like) is designed to be used for assessment testing. Such equipment includes circuitry to determine whether the signals being examined comply with the allowed operating conditions of the logic series in question (a signal is only regarded as being LO or HI if it falls within prescribed limits; otherwise it is considered to be 'undefined'). In addition, it allows test signals with specific levels, edge steepness and so on to be generated.

Test equipment

Figure 4.98 shows the basic configuration of a piece of digital test equipment. This simple block diagram applies to a variety of types of test equipment (as described in Table 4.22), which differ considerably with regard to their cost, construction and utilization (ranging from simple lash-ups, consisting of a few toggle switches and LEDs, to computer-based automatic testers).

In the remainder of this chapter we devote our attention to the basic design features of such test equipment. Refer to Chapter 5 for its use and to Volume 3 for concrete examples.

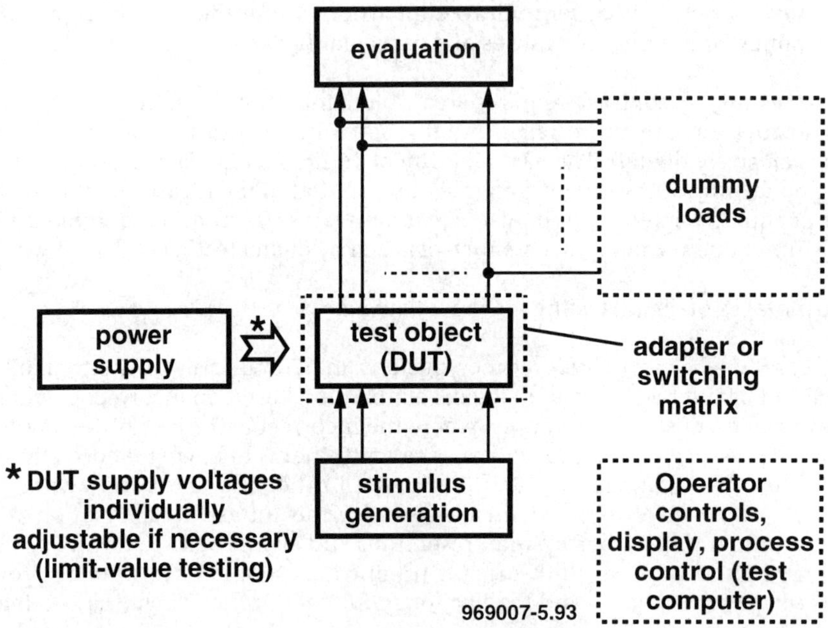

Figure 4.98 The basic structure of digital test equipment.

Connecting the test equipment to the DUT

Either the test equipment must 'accept' the DUT (IC testers and expansion card testers) or it must somehow be connected to the DUT (IC emulators, interface testers), which requires sockets, test cables, adapter plugs and the like. The basic aspects of making connections (which are largely independent of what is being tested) are treated in Section 5.1.4; here we limit our attention to specific issues related to digital test equipment.

Universal connectors. In the general case, it must be possible to connect any contact of the test equipment (that is, any stimulus-generator output, any input for data evaluation and if necessary any power supply terminal) to any contact of the DUT. The simplest solutions (for manual connections) are (1) patch panels and cross-connection arrangements together with standard test cables and (2) specially-prepared cables or wire bundles, which are attached to the test equipment via connectors and to the DUT via individual test clips or the like. Automated connection requires the use of a bed-of-nails adapter (or a suitable universal test socket) and a switching matrix, constructed using (for example) reed relays or CMOS analogue switches.

Special arrangements. The objective is to be able to connect all test signals 'in one go'. Typical solutions are (1) connector(s) in the DUT in which the arrangement of the test signals matches that of the test equipment connectors (so that the connection can be made with a 1:1 cable) and (2) suitable 'hard-wired' test adapters, plugs and the like.

Manual or automatic connections? In the service environment, automatic connection is virtually out of the question. Bed-of-nails adapters and switching matrices are very expensive; they represent a considerable portion of the price of a modern automated tester (which can lie in the 6-figure range).

Type of test equipment	Test object(s)	Test adapter	Number of inputs and outputs
IC tester	digital ICs belonging to specific logic series	IC socket(s)	typically < 40; > 100 for processors, SIMMs etc.
breadboard	experimental circuits	patch panels etc.; 'free' connections	20 to 50 is usually adequate
IC emulator (mimics an IC, such as a processor)	any digital circuit that employs the emulated IC	a plug with the same dimensions and pin arrangement as the emulated IC	typically 40 to > 100
expansion card tester	expansion cards with specific interfaces	a socket according to the applicable interface standard	typically around 100 (ISA or PCI bus)
interface tester	peripheral equipment, subassemblies and so on with specific types of interfaces	interface connectors according to the applicable interface standards	typically < 100 (8-bit SCSI: 18; Centronics: 17; serial interface: 8 etc.)
special-purpose test beds and test assemblies	complex ICs (such as processors); subassemblies, assemblies etc.	sockets and interface connectors that match the item being tested; 'free' connections	very few (clock plus reset, for example) to several hundred

Table 4.22 Digital test equipment.

On the other hand, making a large number of individual connections by hand is time-consuming and prone to errors. The number of signals in question can range from under 20 (for simple logic ICs, parallel and serial interfaces, 8-bit SCSI buses and so on) up to well over 200 (high-performance processor ICs such as the Pentium and Pentium Pro, motherboard ICs and the like). But even making connections to (for example) the 17 lines of a parallel interface costs time and offers ample opportunity for mistakes.

Remedies:

- prepare in-line adapters that can be used together with 1:1 cables to make the necessary connections,
- test using a cyclic stimulus, and briefly probe each of the output signals in turn (for example, it takes much less time to quickly touch a probe to each of 64 data lines than to attach 64 spring clips (test clips or 'grabbers') to the data lines – not to mention the inevitable clips that make poor contact, fall off or are connected to the wrong lines). The prerequisites for cyclic-stimulus testing are that (1) it is possible to generate a suitable cyclic stimulus in a sufficiently simple manner (that is, without having to attach umpteen test clips) and (2) the resulting test signals can be evaluated individually (for example, using signature analysis; see Section 4.4.3) or can be temporarily stored and subsequently evaluated as a group.

Dummy loads. Dummy loads are used to provide typical operating load conditions for the outputs of the DUT. An IC tester without dummy loads is practically worthless (see Section 5.1.7 regarding standard dummy loads). Dummy loads are not necessarily needed with experimental circuit constructions, but in general it is important to reproduce the loading conditions of the interface being tested (such as terminating resistors, pull-up resistors and the like). If signals are 'tapped off' from a circuit for testing purposes, the additional loading must be kept to an absolute minimum (you should pay particular attention to the capacitive loading).

Partial power-down. If the DUT is not directly installed in and powered by the test equipment (as with an IC tester or an expansion-card tester), there are two special operational situations that must be taken into account: (1) the test equipment is switched on and the DUT is switched off and (2) vice versa. These situations are not limited to particular types of faults or the times when you intentionally create such conditions (for example, by switching off the DUT so that test clips can be repositioned or an IC can be exchanged). Rather, they can be expected to occur quite regularly whenever when the overall test setup (test equipment plus DUT) is switched on or off.

It is important to guarantee two different things: (1) the hardware that is switched off must not interfere with the operation of the hardware that is switched on, and (2) the hardware that is switched off must not be damaged by any stimulus coming from the hardware that is switched on.

There is no problem if the hardware that is switched off presents a high impedance to the remaining hardware. Since we cannot alter the equipment being tested, this requirement must be satisfied by provisions in the test equipment alone.

To determine whether a particular pin of a non-powered IC has a high or a low impedance, you can examine the relevant internal circuitry, assuming that there is effectively a short circuit between the supply pin (V_{CC}) and the earth pin (*GND*) if no supply voltage is present. Based on this assumption, you must determine what current paths can be created if an external voltage is applied to the pin in question. The diode structures at the IC inputs and outputs play a significant role here (see Figure 4.99 and Table 4.23).

Table 4.24 lists the maximum allowed operating conditions for the commonly used types and series of logic ICs.

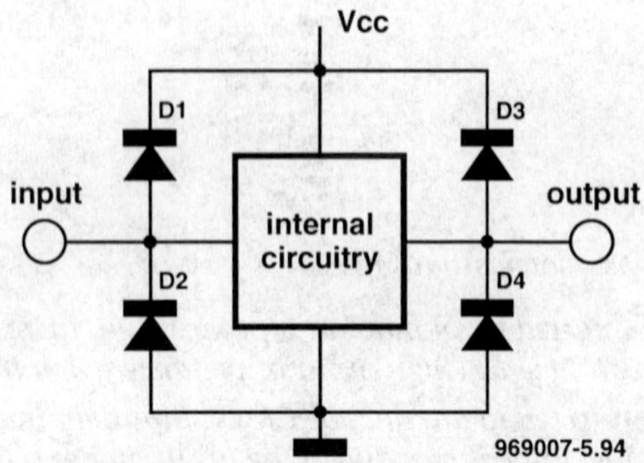

Figure 4.99 *Diode structures at the inputs and outputs of ICs. (source: Texas Instruments)*

Diode	Explanation	Available in ...	Not available in ...
D1	ESD protection (diverts positive voltage peaks)	CMOS	TTL, BiMOS
D2	parasitic structure; bipolar ICs also have an intentionally fabricated parallel diode to clip negative peaks (undershoots)	CMOS, TTL, BiMOS	
D3	parasitic structure *); some CMOS ICs have an intentionally fabricated parallel diode for ESD protection	bipolar totem-pole outputs, CMOS	bipolar open-collector and tri-state outputs
D4	parasitic structure; some ICs also have an intentionally fabricated parallel diode to clip negative peaks (undershoots) or for ESD protection	CMOS, TTL, BiMOS	
*: In CMOS ICs, the upper p-channel FET of the output stage forms a parasitic diode that is forward biased if the output potential is greater than V_{DD}.			

Table 4.23 *Descriptions of the diode structures shown in Figure 4.99.*

Figure 4.100 illustrates the specific situations that must be considered, with regard to bipolar and CMOS integrated circuits.

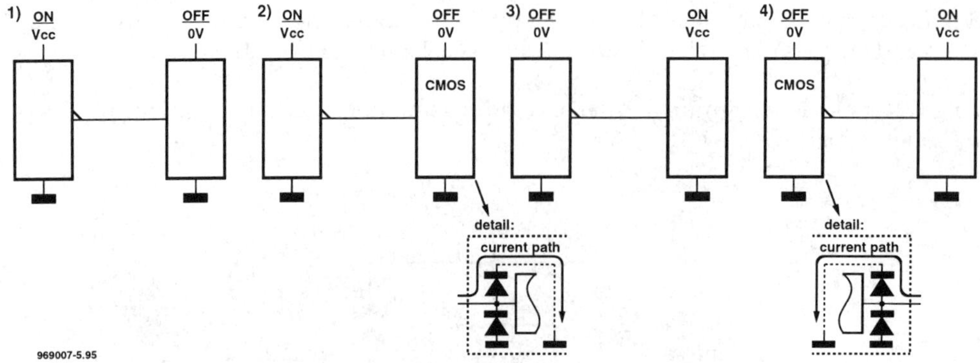

Figure 4.100 *Partial power-down situations.*

1 *An active output is connected to an inactive bipolar input: there is no problem, since bipolar inputs have a sufficiently high input impedance when the IC is not powered.*

2 *An active output is connected to an inactive CMOS input: an inactive CMOS input fundamentally has a low impedance, due to the input diodes (D1 and D2 in Figure 4.99). This is in principle the worst possible situation: the input diode connected to the* V_{CC} *line can conduct when* $V_{CC} = 0$*, and there is nothing to limit the current flow. The inactive IC can be destroyed by the resulting current.*

3 *An inactive bipolar output is connected to an active input: a tri-state bipolar output (such as with a bus driver) has a high impedance when the IC is inactive, but a normal bipolar output stage has a low impedance.*

4 *An inactive CMOS output is connected to an active input: an inactive CMOS output has a low impedance due to the output diodes (D3 and D4 in Figure 4.99). For this reason, CMOS ICs with diodes connected to the* V_{CC} *line cannot be used in interfaces that require high impedance.*

Latch-up. Latch-up is an additional failure mechanism in CMOS ICs (see also Section 2.2.2). If the voltage applied to a pin is higher than the positive supply voltage (V_{CC} or V_{DD}) or lower than the negative supply voltage or earth potential (V_{SS} or *GND*, as the case may be), it can trigger a parasitic thyristor that is formed by the PNPN structure between the supply voltage and *GND* connections. This causes a current to flow directly from V_{CC} to *GND*, which can destroy the IC. If no power is applied to the IC there is no danger of latch-up (a current cannot flow when there is no voltage), but there is definitely a risk of latch-up during transitional states (such as when power is first applied).

Clearly, if you are not careful the test equipment, the DUT or both can suffer damage (especially with DIY test equipment). Figure 4.101 illustrates the operating situations that have to be considered.

Family	Technology	Maximum input voltage	Maximum totem-pole output voltage	Maximum open-collector output voltage	Maximum tri-state output voltage	Maximum clamp current (one input)
74	bipolar (TTL)	5.5 V	7 V	7 V	V_{CC} + 0.5 V	–12 mA
74S	bipolar (Schottky TTL)	5.5 V	V_{CC} + 0.5 V	7 V	5.5 V	–18 mA
74AS	bipolar (Advanced Schottky TTL)	7 V	V_{CC} + 0.5 V	7 V	5.5 V	–18 mA
74F	bipolar (Fast Schottky TTL)	7 V	V_{CC} + 0.5 V	7 V	5.5 V	–18 mA
74LS	bipolar (Low-power Schottky TTL)	7 V	V_{CC} + 0.5 V	7 V	5.5 V	–18 mA
74ALS	bipolar (Advanced Low-power Schottky TTL)	7 V	V_{CC} + 0.5 V	7 V	5.5 V	–18 mA
74BCT [2]	BiMOS	7 V	— [3]	5.5 V	5.5 V	–18 mA
74ABT [2]	Advanced BiMOS	7 V	— [3]	— [3]	5.5 V	–18 mA
74HC/HCT	CMOS	V_{CC} + 0.5 V	V_{CC} + 0.5 V	V_{CC} + 0.5 V	V_{CC} + 0.5 V	±20 mA
74AC/ACT	CMOS	V_{CC} + 0.5 V	V_{CC} + 0.5 V	— [3]	V_{CC} + 0.5 V	±20 mA
74LV	CMOS [4]	V_{CC} + 0.5 V	V_{CC} + 0.5 V	V_{CC} + 0.5 V	V_{CC} + 0.5 V	±20 mA
74LVC	CMOS [4]	6.5 V	V_{CC} + 0.5 V	— [3]	5.5 V	–50 mA
74ALVC	CMOS [4]	4.6 V	— [3]	— [3]	V_{CC} + 0.5 V	–50 mA
74LVT/LVTZ [2]	BiMOS [4]	7 V	— [3]	— [3]	5.5 V	–50 mA
74ALB [2]	BiMOS [4]	4.6 V	— [3]	— [3]	4.6 V	–50 mA

1: The maximum input clamping current is given here. The maximum output clamping current is sometimes higher and sometimes lower, and sometimes both positive and negative currents are allowed. The maximum continuous current limitations for the V_{CC} and earth pins must also be observed.
2: The BiMOS and LVZ families have built-in voltage monitors (Power Up Tri-State Circuits) that switch the outputs to a high-impedance state when the supply voltage is lower than a defined threshold value.
3: This type of output is not available in this logic family.
4: Low-voltage technology (3.3 V).

Table 4.24 *Operating-condition limitations of integrated circuits. (source: Texas Instruments / National Semiconductor)*

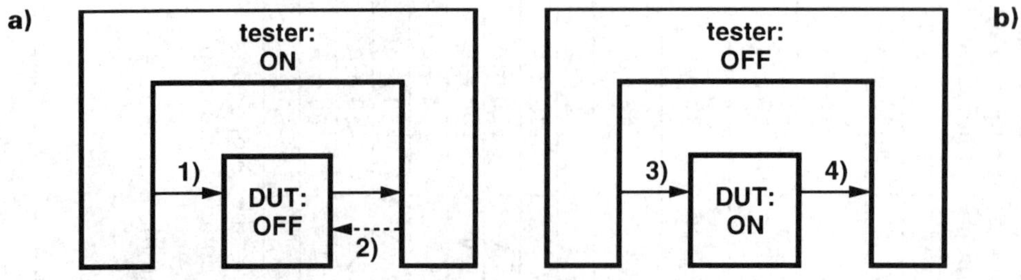

Figure 4.101 *Critical operating relationships between the tester and the DUT.*

a) Tester on, DUT off. Both connections have to be considered separately:

- *Tester output → DUT input: any sort of input stage can be present. Typical CMOS inputs (as shown in item 2 of Figure 4.100) form a particular risk.*
- *DUT output → tester input: there is a chance that the DUT has low-impedance outputs, as shown in item 4 of Figure 4.100. In such a situation, dangerous currents can flow if the tester input is held at a particular voltage (for instance, via a pull-up resistor).*

b) Tester off, DUT on. Both the inputs and the outputs of the tester must have high impedance.

Damage-proof circuitry. The choice of methods that should be used to avoid damage in partial power-down situations depends on what is actually to be tested, or in other words, what type of circuitry is needed to provide particular test stimuli or to evaluate the signals from the DUT. You can either populate the tester interfaces with ICs that have the required properties, or employ additional circuit elements. A few general principles are briefly described below. For examples of real circuits that are used in digital test equipment, refer to Volume 3.

The protection tasks are:

1 Protect the powered-down DUT from the powered-up tester (see Figure 4.101a). The DUT may have CMOS ICs at its interfaces. The simplest solution is current limiting.

2 Protect the powered-down tester from the powered-up DUT (see Figure 4.101b). The simplest solution is to use bipolar or BiMOS ICs for the tester's interfaces.

Solutions

Use a series current-limiting resistor. Choose the resistor value to limit the short-circuit current to a non-damaging level (as a guideline, several milliampères at most). The disadvantages of this approach are: (1) it can only be used with high-impedance inputs, since excessive voltage drops will occur if significant currents must be supplied in normal operation (calculate the voltage drop for normal operation and decide whether it is acceptable) and (2) the combination of the resistor and the parasitic capacitance forms a low-pass filter that spreads out pulse edges and affects the general frequency response.

Populate the interfaces with ICs that prevent or withstand critical operating conditions (bipolar and BiMOS digital ICs, bipolar comparators, fault-protected analogue multiplexers and so on). Refer to the data sheet of the item in question for guidance, but be sure to very carefully read the 'fine print'. In case of doubt, check out the item in question using a simple test setup (see Figure 4.102). Ideally, nothing should happen if a voltage is applied to a pin of the unpowered IC. A current ≤ 1 mA normally will not do any harm. A 50 kΩ variable resistor that is initially set to maximum resistance can be used to limit the current. Gradually reduce the resistance and check whether the resulting current reaches a hazardous level (guideline value: > 5 mA).

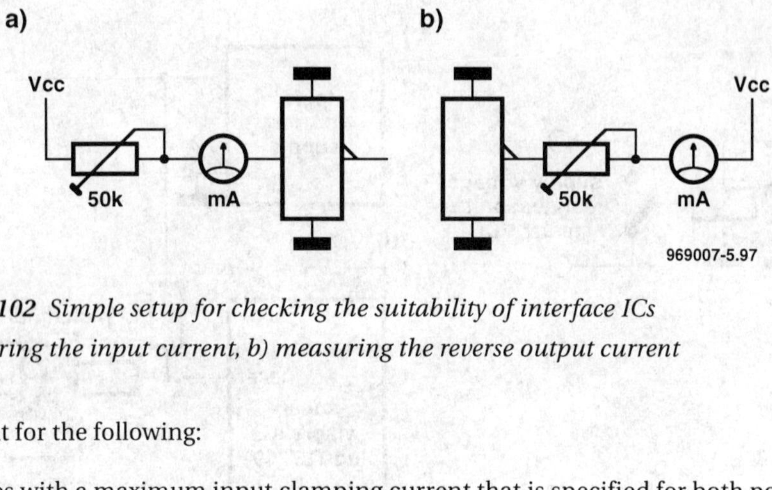

***Figure 4.102** Simple setup for checking the suitability of interface ICs*
a) measuring the input current, b) measuring the reverse output current

Watch out for the following:

- Devices with a maximum input clamping current that is specified for both polarities (±) cannot be used.
- Devices with a maximum specified input voltage that is less than the supply voltage, or at most 0.5 V greater than the supply voltage, cannot be used.
- Devices with I/O equivalent circuits (sometimes found in the data sheets or associated application notes) that include a diode connected between the pin in question and the supply voltage cannot be used.
- Protection against latch-up ('latchproof') does not necessarily mean protection against overvoltage.

Use a common power supply (tester powered from the DUT or vice versa). This is a reliable way to avoid the operating situations shown in Figure 4.101. It can be used with small test aids (logic probes, pulse detectors and the like), experimental setups, ICs, expansion card testers and IC emulators. *Note:* any protective measures that are present (such as a forced shutdown of the power supply in case of a short circuit) must affect the entire setup. A fuse (on an expansion card, for instance) that removes the supply voltage from the DUT without cutting off the power to the tester at the same time (if it blows) can create an unallowed operating condition as shown in Figure 4.101.

Hard-couple the tester power supply to the DUT power supply. With this technique the tester – or at least its interfaces to the DUT – is only powered when the DUT supply voltage has reached a suitable value (see Figure 4.103).

Note: if a portion of the tester is still powered when the DUT is switched off, then a partial power-down situation will be present inside the tester. In this case you at least have the advantage that you have control over both sides of the interface (powered and non-powered), so that it is less difficult to devise a solution. You could (for instance) use BiMOS ICs for the 'internal' digital interfaces, with the external DUT interfaces implemented in whatever way is most appropriate for their specific functions.

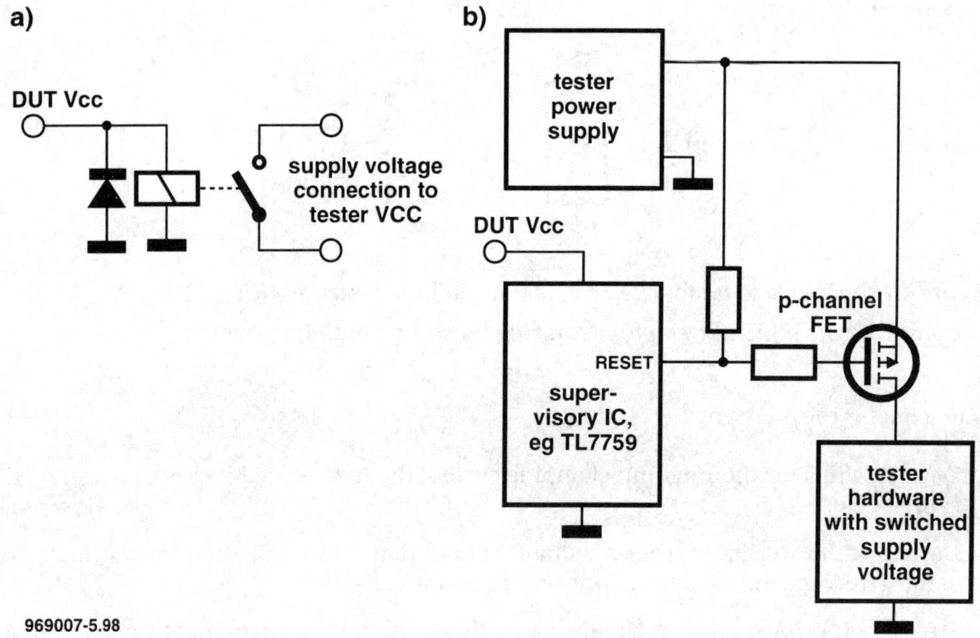

Figure 4.103 *Forced supply-voltage coupling.*

a) the simplest solution: the relay contacts switch the tester's power supply on and off

b) a more modern solution, based on a supervisory IC.

Use a blocking diode to prevent undesired current flows (Figure 4.104). Clearly, the diode is reverse-biased when the driver is switched on and the receiver is switched off, so practically no current will flow. This arrangement has certain problems and particular features:

- It can only be used with pure binary signals – it is not possible to apply selected input levels to the DUT or to evaluate the voltage levels of output signals.
- The voltage drop across the diode increases the LO signal level in normal operation (*tip:* use a Schottky diode).
- The pull-up resistor must be connected to the DUT supply voltage (receiver side).
- It is not suitable for bi-directional signals.
- It offers protection for only one condition (driver on, receiver off).

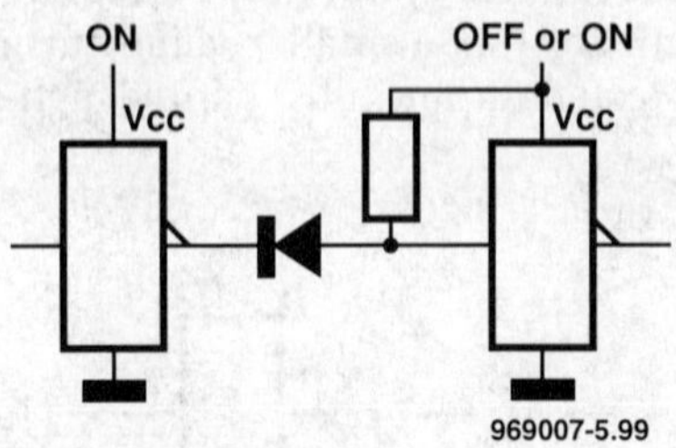

Figure 4.104 Blocking diode in a signal lead.

Connect the DUT and tester only after both devices have been switched on (Figure 4.105). This approach is possible when a matrix switch is already present (using reed relays or analogue switches, for example). The switches are only enabled when the DUT and the tester are both operational.

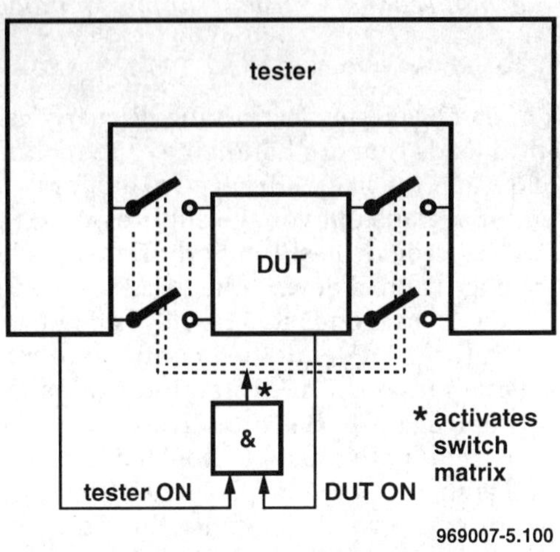

Figure 4.105 Connecting the tester and the DUT via a switching matrix.

Note: modern semiconductor switches (such as Quality Semiconductors' QuickSwitch devices and Texas Instruments' CBT-series devices) guarantee the required isolation between powered and non-powered hardware. (These ICs do not have any p-channel transistors in the signal path, so they will not be damaged if a HI-level signal is applied in the absence of a supply voltage.) Their drawbacks with respect to mechanical switches (reed relays) are (1) the on-state series resistance is roughly 5 Ω and (2) the output voltage cannot be higher than V_{CC} plus 1 to1.5 V.

Use external protection diodes. Schottky diodes connected as shown in Figure 4.106a will clip excessive positive or negative voltages to V_{CC} + 0.5 V and *GND* – 0.5 V respectively. The associated IC is thus protected, but current limiting may be necessary since any short-circuit currents will flow through the external diodes. Diode arrays as shown in Figure 4.106a are commercially available (for example, Texas Instruments Bus Termination Arrays). These devices are intended to be used for clipping transient overshoots and undershoots rather than for long-term voltage limiting, but they can tolerate relatively high currents (on the order of 50 mA). This means that a smaller-value current limiting resistor can be used (but also that the DUT or tester outputs must handle higher short-circuit loads).

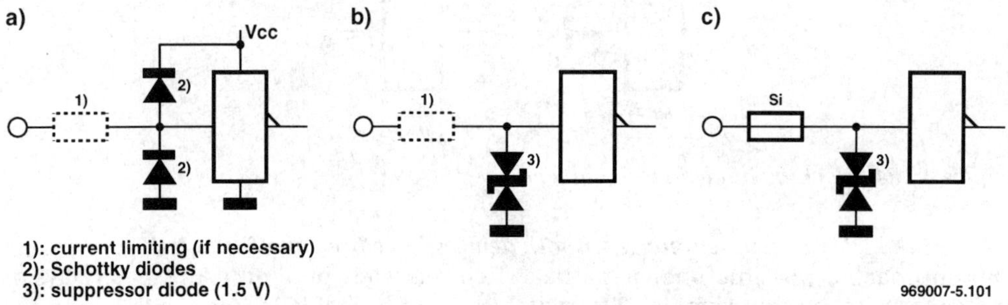

***Figure 4.106** Protection diodes.*
a) Schottky diodes, b) and c) bidirectional transient suppressor diodes.

ESD protection. The risk of damage from electrostatic discharge is especially high for circuit elements connected to leads that are handled by the technician (such as adapter cables with probes, test clips and the like). If protective diodes as shown in Figure 4.106a are not adequate, bidirectional transient voltage suppressors (TVSs) can be utilized as shown in Figure 4.106b. A TVS (trade names: TranSorb, Transil and so on.) creates a short-circuit path to earth as soon as its breakdown voltage is exceeded, and it can switch very quickly (on the order of a few picoseconds to 5 ns). TVSs can handle very high impulse currents (ranging from < 10 A to > 100 A), but their continuous current ratings are significantly lower (< 1 A to several ampères). If protection against long-term overloads is desired (for instance, protection against accidental contact with the mains voltage), some means of interrupting the current path must be provided ahead of the TVS (an ordinary fuse, a polyfuse or the like) as shown in Figure 4.106c. *Note:* limiting the current that can flow through the TVS is in any case a good idea, since the required level of protection can then be achieved with a smaller type of TVS.

Capacitive contact-loading disturbances. Each time a contact is made between the tester and a signal in the operating DUT (for instance, by touching a probe to a signal lead), the DUT must charge or discharge the input capacitance of the test probe. The resulting current can create a disturbance in the DUT (see Figure 4.107). This typically takes the form of a damped sinusoidal oscillation at a frequency determined by the parasitic inductance and capacitance of the probe (roughly 100–200 MHz). A remedy is to precharge the tester input to a voltage corresponding to the threshold voltage of the logic gates (roughly 1.5 V). Although this does not completely eliminate the oscillation, it significantly reduces its amplitude.

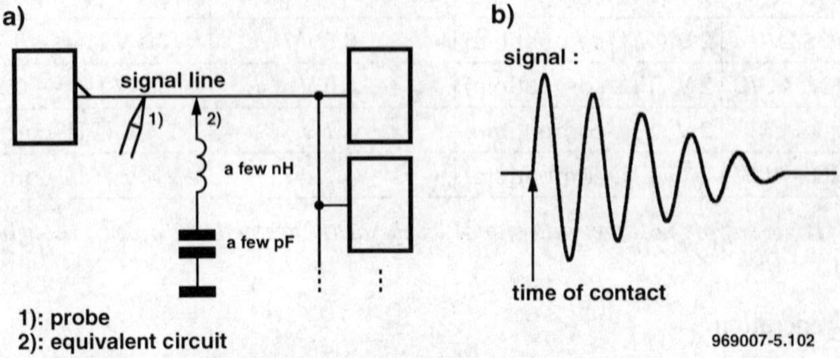

Figure 4.107 Capacitive contact loading disturbances.
a) equivalent circuit, b) a typical disturbance.

Logic supply voltages. There are two 'industry standard' voltages that have to be dealt with: 5 V and 3.3 V. If the tester is built entirely using ICs with supply voltages that match those of each of the test objects, then there basically cannot be any problems. However, the construction of such a 'universal' tester requires special provisions. The most important requirements are summarized in Table 4.25.

An obvious solution is construct the tester using 5-V logic and to provide switchable level converter stages as necessary. Such level converters must satisfy the following conditions:

- 5-V CMOS logic inputs need a HI level of ≥ 3.5 V.
- All other logic families have TTL-compatible inputs, but it is a good idea to be able to test with a minimum HI level (2 V).
- Some 3.3-V logic families cannot withstand typical CMOS output levels (≈ 5 V).

Level selection can be implemented by using open-collector gates or by using switching elements (CBT, QuickSwitch or the like). Refer also to Appendix 4 and (for actual designs) Volume 3. *Note:* tester outputs should also be short-circuit proof, so that they can survive shorts in the DUT (to ground, to the supply voltage and between signal lines).

Supply voltage and technology	Minimum HI level at DUT input (tester must provide)	V_{max} at DUT input (tester may not exceed)	DUT output delivers:
5 V TTL	2 V	5.5 V	2.4 to 3.6 V with totem-pole outputs; ≈ 5 V if a pull-up resistor is connected
5 V CMOS	3.5 V (0.7·V_{CC})	5.5 V (V_{CC} + 0.5 V)	≈ 5 V (V_{CC} – 0.1 V)
3.3 V CMOS LV	2 V (TTL-compatible)	3.8 V (V_{CC} + 0.5 V)	≈ 3.1 V (V_{CC} – 0.2 V)
3.3 V CMOS LVC	2 V (TTL-compatible)	6.5 V	≈ 3.1 V (V_{CC} – 0.2 V)
3.3 V CMOS ALVC	2 V (TTL-compatible)	4.6 V	≈ 3.1 V (V_{CC} – 0.2 V)
3.3 V BiMOS LVT	2 V (TTL-compatible)	7 V	≥ 2.4 V (TTL-compatible)
3.3 V BiMOS ALB	2 V (TTL-compatible)	7 V	≥ 2.4 V (TTL-compatible)

Table 4.25 *Logic supply voltages and signal levels for the tester-DUT output interface.*

Stimulus generation

Manually-generated stimuli. Manually setting input values (using toggle switches, for example) is by no means outdated. The simplest test equipment is often the best for trying out experimental circuits and microcontrollers, and for investigating certain fault situations. Figure 4.108 shows a number of basic circuits. (*Note:* protective measures relating to ESD, partial power-down operation and so on are not shown.)

Component values:

- Pull-up resistors at tester outputs must satisfy two conditions: (1) the minimum HI level must be maintained with the maximum allowed load and (2) sufficiently steep edges must be guaranteed when the signal level switches (the speed of the low-to-high transition is determined by the value of the pull-up resistor, since the charging current for the parasitic capacitance flows through it alone). For static signals, (1) CMOS devices require 'almost nothing' and (2) the worst case input HI unit load of TTL

Explanation of Figure 4.108:

a) A binary stimulus produced by a switch contact. The LO level is provided by the circuit earth and the HI level by a pull-up resistor. The HI level can be matched to the requirements of a particular logic family, or adjusted for limit-value testing, by adjusting the value of V_{OUT}.

b) An output stage using an open-collector gate. The gate itself needs a supply voltage V_{CC} (5 V). The value of the HI-level output voltage V_{OUT} can be set within a wide range, since it need not be the same as the gate's supply voltage.

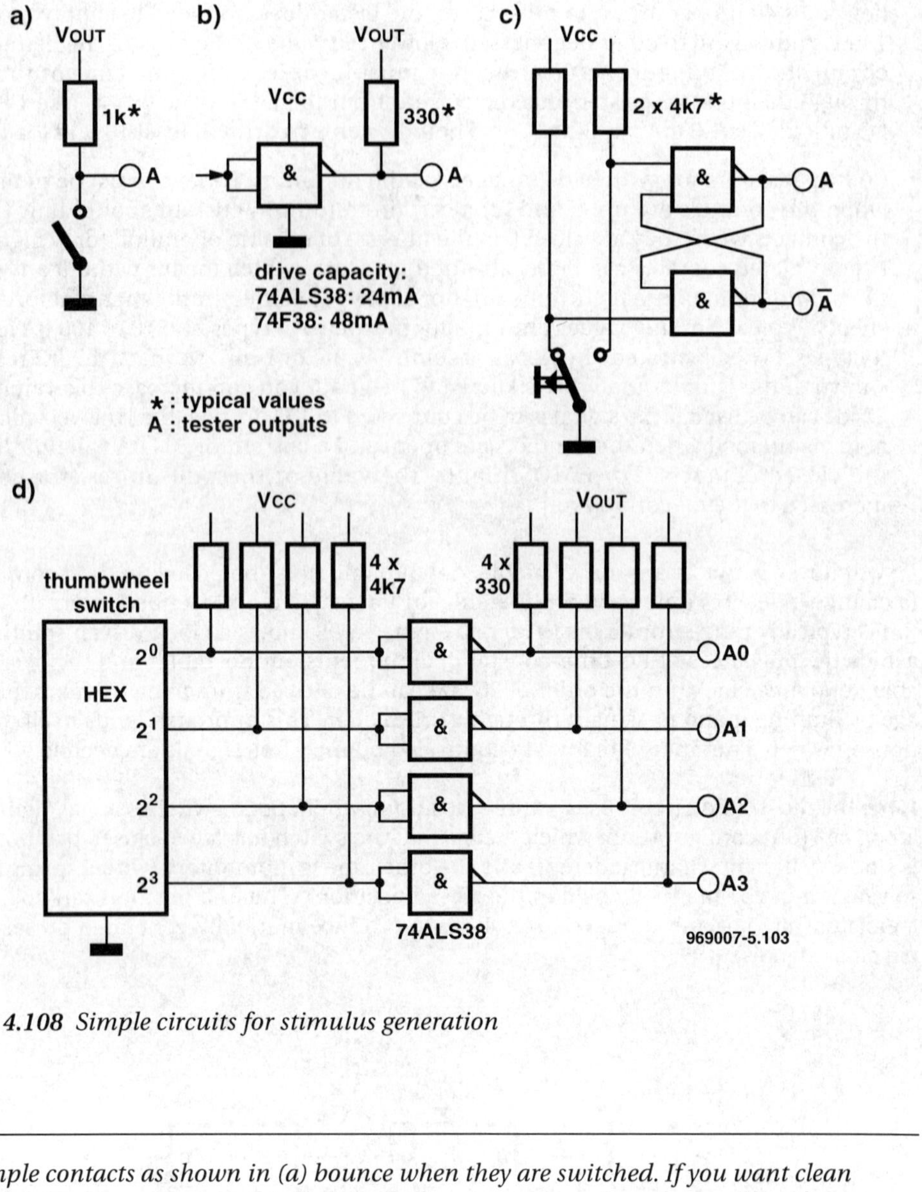

Figure 4.108 Simple circuits for stimulus generation

---

c) *Simple contacts as shown in (a) bounce when they are switched. If you want clean signals that are guaranteed to be bounce-free, you can produce them with a changeover switch and a latch. Another way of generating bounce-free signals (debouncing) is shown in Figure 4.109.*

d) *A hexadecimal switch. Various types are available (rotary and thumbwheel switches, sometimes with pushbutton actuation) with16 hexadecimal-coded positions. Such switches can be used to set four bits at a time. The switch itself represents a '1' by a closed contact and a '0' by an open contact, so inverters must be used to obtain 'positive' logic values.*

devices is 40 µA per input. In most cases, the tester does not need to supply more than 1 mA. With regard to edge steepness, the lower the value of the resistor the better. In the circuit shown in Figure 4.108b, the gate must be able to sink the current that flows through the pull-up resistor plus the current from the DUT. (The worst-case TTL input LO unit load is 1.6 mA, and the tester should be able to drive at least 10 unit loads.)

- Pull-up resistors for switched contacts: (1) the minimum HI level must be maintained when the contacts are open and (2) a certain minimum current should flow through the contacts when they are closed, in the interest of reliable operation (only high-quality gold-plated contacts can be reliably operated 'dry', which means without any current flow). Minimum contact currents are not always specified in the manufacturer's data sheets. Typical working values are 1 mA for inexpensive types and 10 to 100 µA for quality types. (The illustrated circuits all assume a value of 1 mA, so that a 4.7 kΩ resistor is shown. If the contact material is known to be good, you can increase this value. Up to 22 kΩ can be used if the signal is to be connected to TTL inputs; the HI level will be reliably maintained when the contacts are open, and a current of 200 µA will flow through the closed contacts. For CMOS inputs, the value of the pull-up resistor could be increased to the megohm level).

If switchover contacts are not available, debouncing must be based on time comparison (a change in level is only valid if it is stable for longer than a prescribed interval). Bounce times typically range from 5 ms to 30 ms. Figure 4.109 shows a clock-driven solution that avoids the nuisances of RC time-constant components, monostables and the like. A suitable debounce clock (on the order of 25 Hz) can be derived from a mains-frequency voltage by limiting it and passing it through a 2:1 divider. This approach lends itself to situations in which a number of manually-controlled bounce-free signals are needed.

Note: the illustrated circuit uses a three-position switch. Such switches are available with an on-off-(on) configuration, which means that the switch has a two closed positions: one is stable (with a mechanical detent) and the other one is momentary (with a spring return, so that it must be manually held in the closed position). The first position can be used for 'persistent' signals and the second position to produce manually-generated pulses (manual clock signals).

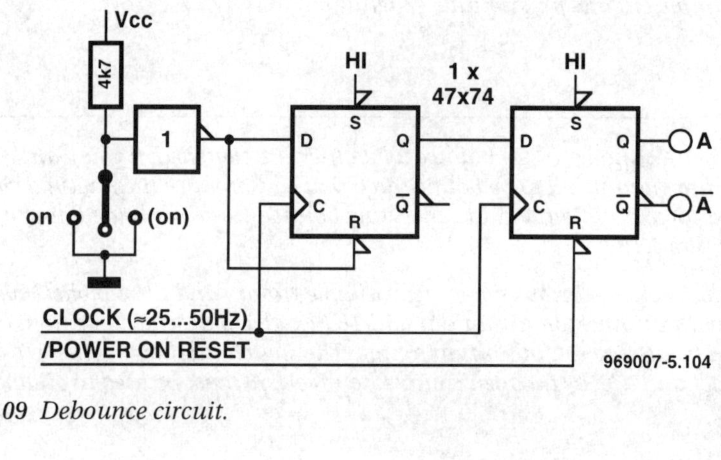

***Figure 4.109** Debounce circuit.*

The limitations of manual operation are basically determined by questions of practical utility. *For example:* consider a simple tester for ISA-bus cards. How long would it take to check all 65,536 I/O addresses using a set of 16 manually-operated toggle switches?

Basically, simple switch arrangements are worthwhile test aids in situations where a relatively small number of signals (typically 16 to 32) have to be held at essentially static levels. For all testing tasks that go beyond this, and especially in conjunction with a large number of test stimuli, you should look for an 'automated' solution.

Automatic and software-controlled stimulus generators (Figure 4.110). In terms of functional complexity, this type of test equipment ranges from simple software-driven substitutes for toggle switches and LEDs, through device-specific testers (for ICs, plug-in cards and specific interfaces) and up to freely-programmable test data generators. There are two obvious ways to build such equipment in a cost-effective manner: (1) by coupling it to a PC and (2) by utilizing microcontrollers. See Volume 3 for examples of practical circuits.

The electrical construction of the tester → DUT interface. The basic consideration is a certain degree of robustness. Do not connect signals that are used 'inside' the tester directly to external interfaces that will be connected to relatively long external cables. This also applies to the outputs of standard flip-flop, counter and shift-register ICs. Always buffer such signals by placing driver stages between them and the external interfaces. *Practical tip:* use suppressor diodes to protect interface components against ESD.

Selecting interface components:

- IC testers and plug-in card testers: in the usual implementation, in which the tester and the DUT adapter (IC test socket or card-edge connector) form a single unit and the tester and the DUT have a common power supply, the same types of interface ICs can be used in the tester as are used in the DUT (including CMOS types).
- Interface testers: it depends on the interface in question. For example, RS232, SCSI and (IEEE-1284 compatible) parallel interfaces can all basically withstand partial power-down operating conditions. This means that with these interfaces the same types of interface ICs can be used in the tester as are used in the DUT.
- For all other interfaces: determine which operating situations can occur and avoid possibly damaging conditions by selecting suitable interface components or by using additional circuit elements. Some devices that can be used as 'universal' interface components are (1) bipolar open-collector gates (see Figure 4.108b), (2) BCT or ABT series ICs and (3) switch components (QuickSwitch, CBT and the like). For practical circuit examples see Volume 3.

Signal evaluation

The simplest DUT → tester interface. If all that is needed is a purely digital test, then buffer ICs that load the DUT as little as possible can be used at the DUT/tester interface. BiMOS devices (BCT or ABT) are particularly suitable; they have input currents in the microampère range and high input impedances when they are not powered.

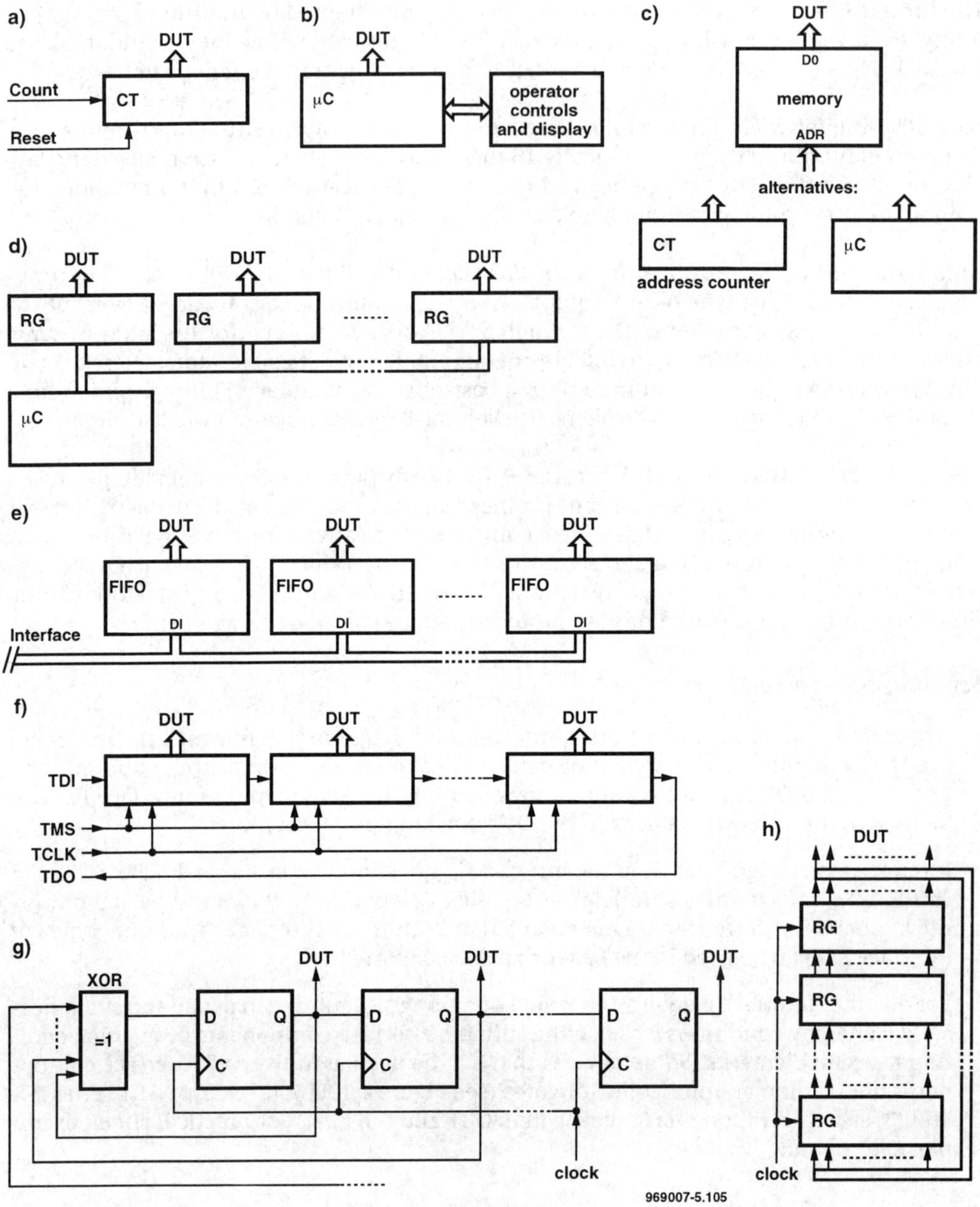

Figure 4.110 Automatic and software-controlled stimulus generation.

a) A simple counter (controlled manually or via software) is sometimes a useful stimulus generator, especially when all possible sets of input values have to be applied to the DUT (2^n for n inputs).

b) *Test vectors can be provided directly by a microcontroller. The prerequisites are that the microcontroller has (1) a sufficient number of I/O lines and (2) sufficient processing power to satisfy the demands of real-time operation. The advantages of this approach are (1) the simplicity of the circuit and (2) application flexibility. It can be used as an interface tester (for example).*

c) *The test vectors can be stored in a continuously-addressed memory (using a counter, for example). This yields the highest possible speed (up to the level of dynamic testing), since the duration of each test step can be reduced to a single memory cycle time, but it is also limited to a rigid counting process. An obvious way to make the process more flexible is to use a microcontroller. The advantage of this approach is that an extensive set of test vectors can be managed by a simple controller. An application example is a tester for digital ICs (in which the test data memory consists of an array of flash ROMs containing test vectors for all standard IC types).*

d) *The stimulus can be provided by registers that are directly addressed by a microcontroller. The advantage is a high degree of flexibility; this approach is particularly useful when the test vectors cannot be stored in advance but must be calculated. A typical application is a memory tester. (Test vectors for memories on the order of several megabytes must be generated 'on the fly', since it would be far too expensive to permanently store them. See Volume 2 regarding the basic aspects of memory testing.)*

e) *FIFOs can be used to store stimuli. Test vectors that have been loaded into the FIFOs can be repeatedly read out in a cyclic fashion. The duration of an individual test step can thus be reduced to the minimum FIFO cycle time (roughly 20–30 ns). The advantages of this arrangement are (that (1) the FIFO array can supply test stimuli made up of any desired number of bits in parallel to the DUT, but it can be loaded via a relatively narrow interface (such as a byte-parallel interface) and (2) a fast test data generator, which is also suitable for dynamic testing, can be implemented at a relatively small additional cost. This approach can be used for PC-controlled universal testers. See Volume 3 for more information.*

f) *The stimulus signals can be provided by a serially-loaded shift register. The boundary-scan interface (in line with the IEEE/JTAG 1149 standard) is especially suitable for this purpose. The advantages of this approach are that: (1) it needs a very simple interface consisting of only four lines, by means of which (2) any desired number of test signals can be generated or sensed, and (3) there is a useful range of suitable ICs available (the cost per bit with such ICs is significantly less than the price of a toggle switch). This approach can in principle be used for any desired type of static or quasi-static test equipment. See Volume 3 for more information.*

g) *The stimulus can be produced by a feedback shift register (pseudo-random test vectors). See Section 3.4.2 and Appendix 6 for more information.*

h) *A stimulus generator or test data generator can be based on a recirculating register. The test vectors are first loaded into an array of shift registers, which runs in a cyclically recirculating manner during the test. This approach can be used to realize very high pulse repetition rates (100 MHz or more) and pulse sequences with a time resolution on the order of < 5 to 10 ns. Such capabilities are typically required for testing DRAM arrays (see also Volume 3).*

Simple LED indicators. Figure 4.111 shows how LEDs can be connected to the outputs of a buffer IC and indicates typical component values.

Practical tip: Sometimes it is desirable (for instance, in experimental circuits) to be able to display a number of signals with a minimum of effort (and preferably directly on the circuit board) using LEDs. Bar-graph display components, which contain a number of LEDs in a single DIL package, are particularly useful for this purpose.

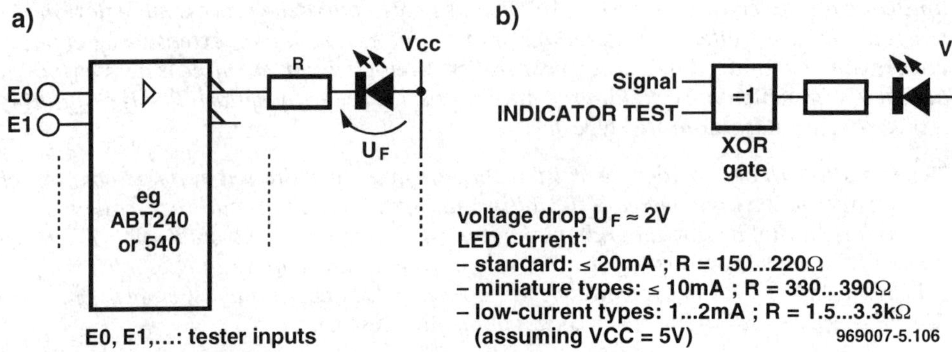

***Figure 4.111** Elementary LED indicators.*
a) Multibit display via a driver IC, b) LED driver with provision for display testing.

Display test. LEDs are much more reliable than incandescent lamps, so test provisions for them are usually not worth the effort. (A simple example of a LED test provision is shown in Figure 4.111b. The LED is driven by an XOR gate; if INDICATOR TEST is made active then a previously dark LED should glow and a previously glowing LED should go dark). In circuits such as that shown in Figure 4.111a, open inputs are interpreted as being HI, so that all LEDs connected to the buffer IC should be on when no DUT is connected. This arrangement can be tested by briefly pulling each input to earth.

Hexadecimal displays. Simple decoders that convert 4-bit hexadecimal values into control signals for 7-segment displays are not available in all logic families (an example is the MC14495). There are however LED and LCD driver ICs that can be used to build multiposition displays. The arrangement shown in Figure 4.112, for example, allows the values of up to 32 bits to be displayed in the form of eight hexadecimal characters. Four bits at a time are selected via a multiplexer and passed to the display driver IC. This type of arrangement can be used as a sort of 'logic monitor' for experimental purposes, or it can be employed in microprocessor breadboard constructions. See Volume 3 for more details.

Interfaces with signal evaluation. In most cases it is not necessary to measure the exact amplitude of a digital signal. Instead, it is usually sufficient to determine whether the signal level is in the LO range, the HI range or somewhere outside of these two ranges. Comparators are the proper components to use for this (see Figure 4.113). Figure 4.114 shows two practical examples of comparator circuits.

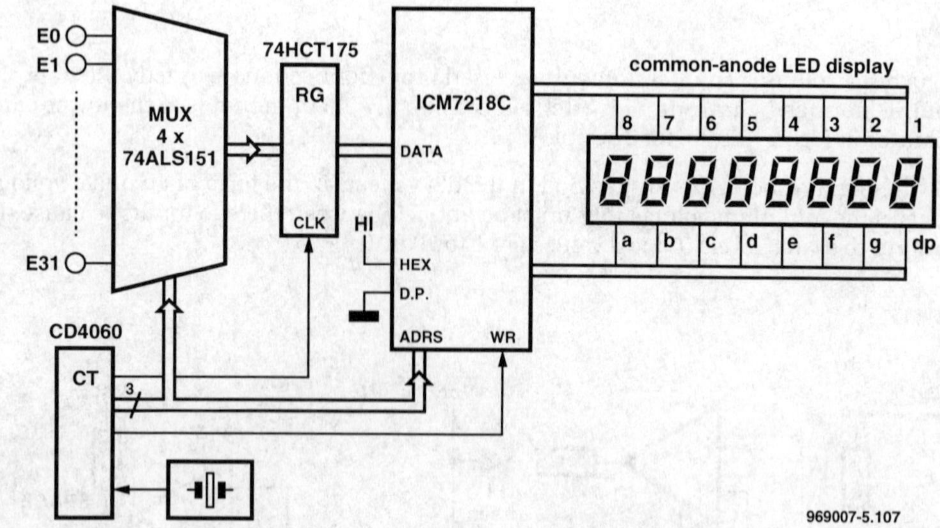

Figure 4.112 Multiposition hexadecimal display (logic monitor).

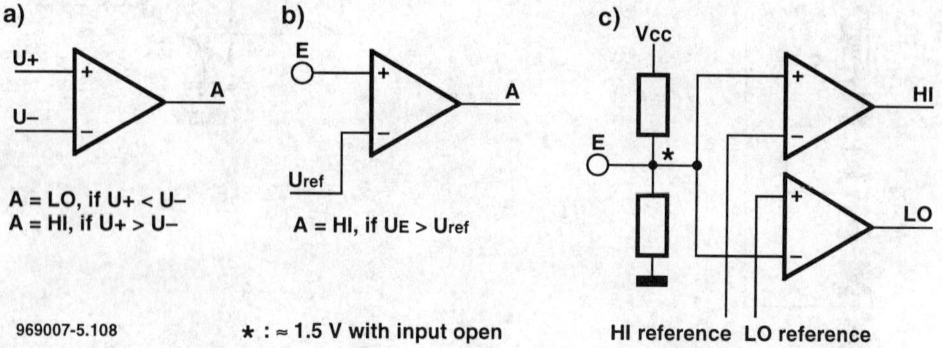

Figure 4.113 Comparator operation, and using comparators to evaluate a digital signal.

a) *A comparator is an operational amplifier that is used as a differential amplifier without any feedback. Its output level is HI when its '+' input is more positive than its '–' input.*

b) *Signal evaluation: the signal to be evaluated is compared to a reference voltage. For example: if a 2 V reference potential is applied as shown, then a HI level at the comparator output indicates that the input signal represents a valid TTL HI level.*

c) *More precise signal evaluation: the input signal is compared to a LO level reference voltage (0.7 V) and a HI level reference voltage (2.0 V or 3.5 V). If neither of the comparator outputs is HI, either the input signal lies in the 'forbidden' region between the maximum LO level and the minimum HI level or the input is open.*

Notes:

1. The behaviour of a comparator with reversed input connections is exactly the opposite of the 'normal' behaviour. This trick is occasionally encountered (see the lower comparator of Figure 4.113b, for example).
2. Switching threshold: this is specified in the data sheets in the form of an offset voltage. This is the minimum voltage that must be applied between the two inputs to cause the output to switch levels. Typical values are 1 to 10 mV.

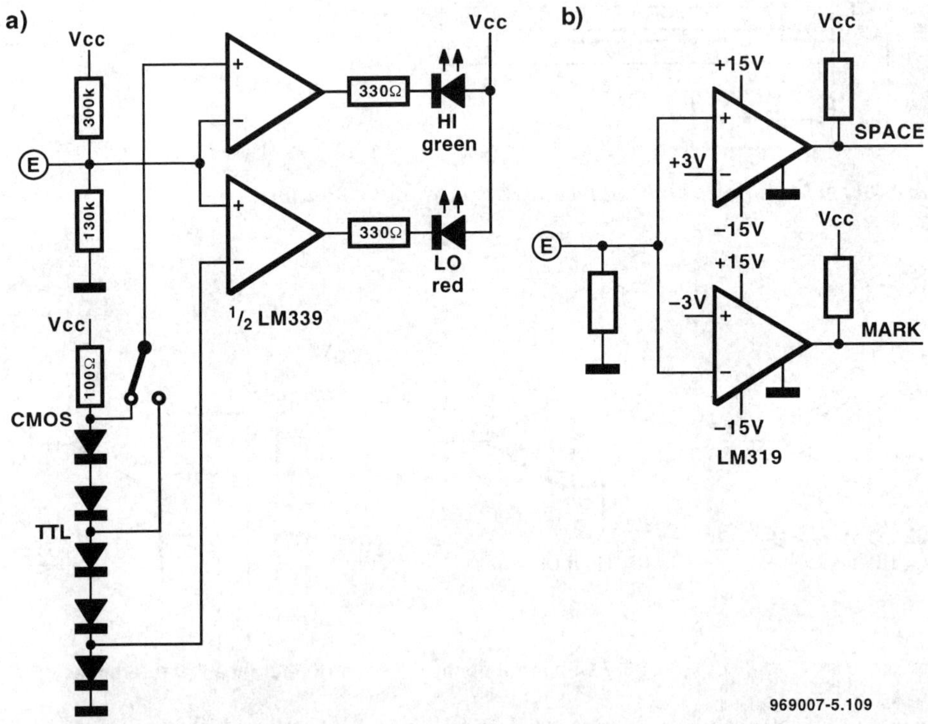

***Figure 4.114** Practical circuits for signal evaluation.*

a) The basic circuit of a hand-held logic probe or a logic monitor. The reference voltages are provided by forward-biased silicon diodes. This is usually sufficiently accurate (and the accuracy can easily be improved by using a 'true' reference voltage source; see Section 4.2). A single reference voltage source can be used for a large number of comparator arrays (for a multibit display).

b) A test circuit for an RS232 interface. The input signal is evaluated to see if it represents a MARK level (–15 to –3 V), a SPACE level (+3 to +15 V) or an unallowed level, or if it is not connected.

3 Output stage: many ICs have open-collector or open-drain outputs, which require pull-up resistors. This feature allows the outputs of several comparators to be connected together in a wired-AND configuration, in which the common output level is HI only if the '+' input of every comparators is more positive than its corresponding '–' input.

4 Speed: the speed of a comparator is specified as its *response time*. Relatively slow types are fully satisfactory for static displays (such as the LM339 with a response time of 1.3 µs). For other applications a sufficiently fast comparator must be selected (for extreme cases, ECL comparators have response times of 1 to 4 ns).

5 CMOS types: keep in mind the 'tester off / DUT on' situation. CMOS comparators can be destroyed if they are used without input protection measures. (This means, for example, that an LM339 cannot simply be replaced by a TLC339.)

6 If a comparator is operated in its linear region for an extended length of time, its output can start to oscillate. Not only does this invalidate the signal evaluation (we see pulses that do not exist in the DUT), it can sometimes also destroy the comparator. Input voltage differences on the order of the offset voltage must therefore be applied only briefly, especially if the comparator output is heavily loaded. One remedy is to use hysteresis, which means that output of the comparator circuit switches HI when the input voltage rises above one threshold voltage (E_{on}) and switches LO when the input voltage drops below a second threshold voltage (E_{off}). (Typically, $E_{on} > E_{off}$.)

Figure 4.115 shows a fast comparator circuit that can be used (for example) in logic analyzers and signature analyzers. The reference voltages are derived from the supply voltage via a voltage divider consisting of R3 through R6. The voltage divider also holds an open input at roughly 1.4 V, which has the advantage that the tester's input capacitance does not have to be fully charged or discharged when the test lead is connected to a signal line (since it is already charged halfway between the LO and HI levels). R1 limits the input current, and with diode D1 it protects the comparator inputs. R2 increases the hysteresis and C1 couples fast changes in the input signal to the comparator inputs.

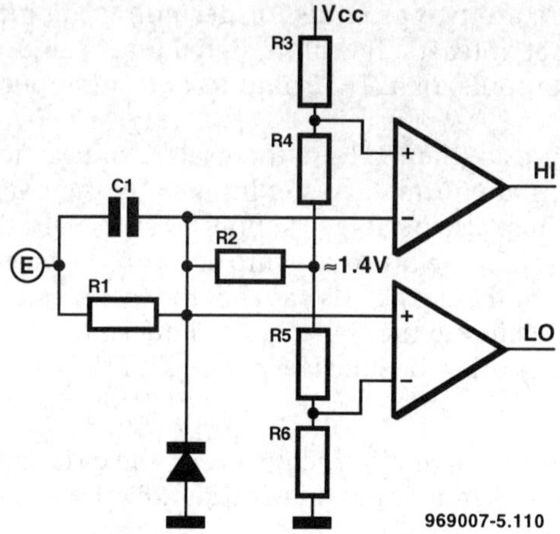

Figure 4.115 *Fast comparator circuit. (source: Hewlett-Packard)*

Multibit displays. A summary indication of whether the levels of a group of signals are correct (such as those of a computer interface) can sometimes significantly reduce the time required to locate a fault. One approach is to build as many copies of the window comparator in Figure 4.114b as are needed. Alternatively, a multibit display based on the multiplexing principle can be constructed, as shown in block diagram form in Figure 4.116. The signals to be evaluated are connected to the inputs of an analogue multiplexer array (a two-stage selector, which can for example be based on DG406 devices or Quick-Switch QS3251 multiplexers). The LEDs are connected directly to the outputs of two M5451 LED driver ICs, which have serial data inputs and do not require external series resistors. The control functions – which are not exactly trivial – can be taken care of by a microcontroller, such as an inexpensive PIC device. For more details see Volume 3.

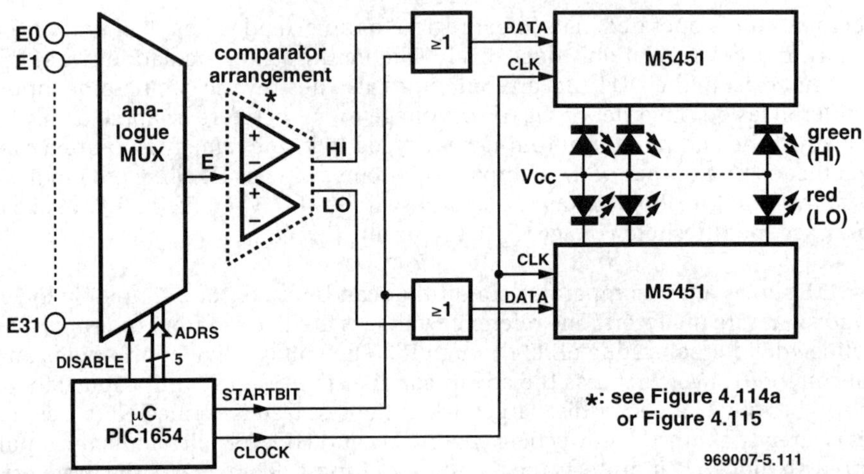

Figure 4.116 Multibit digital display with signal evaluation (logic monitor).

Bidirectional signals. There are two grounds for designing the interface between the tester and the DUT such that at least a few of its signal lines can optionally function as either DUT inputs (= tester outputs) or as DUT outputs (= tester inputs):

1 If the associated signal lines of the DUT are themselves bidirectional (for example, in bus systems). Such signals are normally tester inputs and their values are determined by the DUT. If the tester must also be able to supply these signals, then special interface arrangements are necessary. These must take into account the particular features of the DUT (such as bus request and busy signals), so that the tester can determine whether it may make an active connection to the signal lines in question. The tester then behaves practically the same as any other bus device (such as a PCI card or a SCSI peripheral device).

2 If the tester is to be 'universal', then in principle every one of its connections should be selectably configurable as either an input or an output. Figure 4.117 shows several ways in which this can be realized.

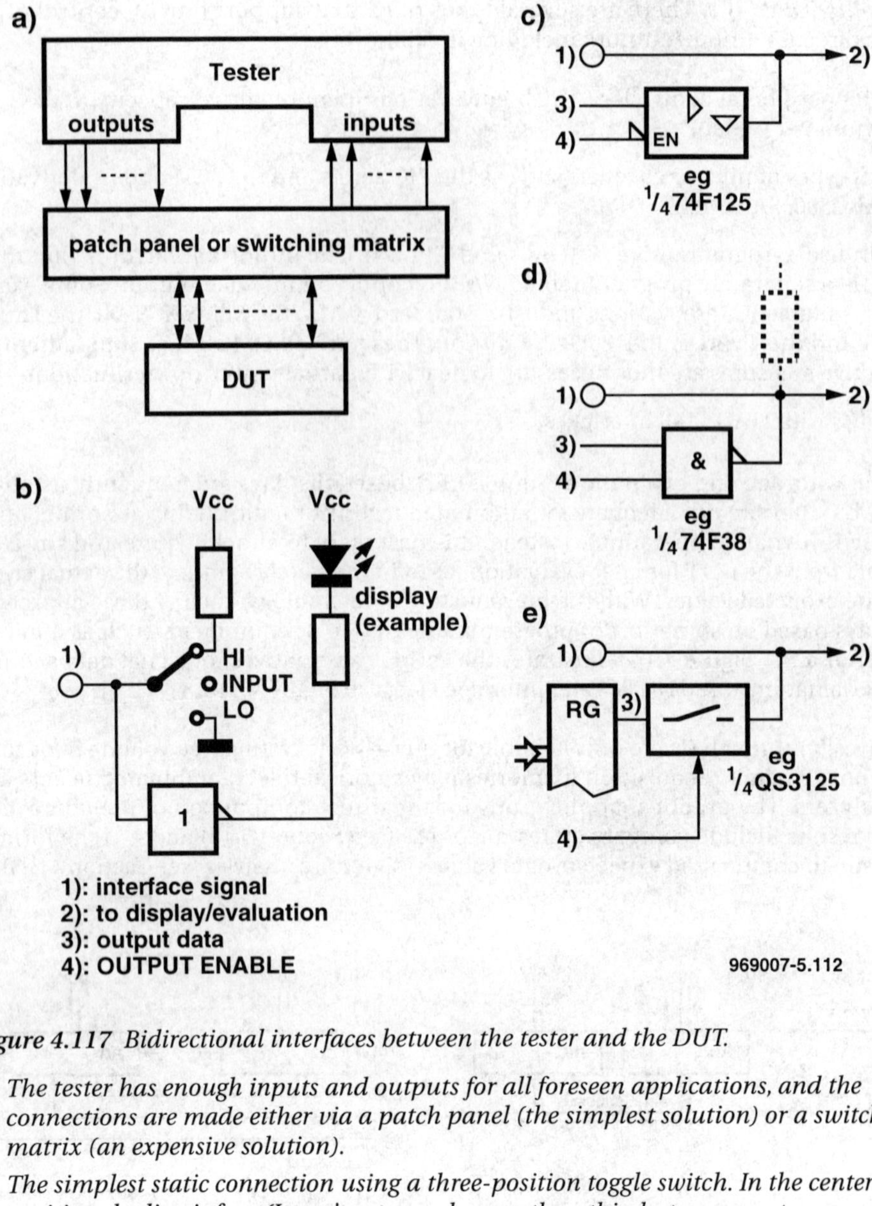

Figure 4.117 Bidirectional interfaces between the tester and the DUT.

a) The tester has enough inputs and outputs for all foreseen applications, and the connections are made either via a patch panel (the simplest solution) or a switching matrix (an expensive solution).

b) The simplest static connection using a three-position toggle switch. In the center position the line is free. (It can't get any cheaper than this, but an operator error can have serious consequences.)

c) Basic circuit using a tri-state driver (suitable for use with 74x125 ICs).

d) Basic circuit using an open-collector driver (suitable for use with 74x38 ICs).

e) Switched tester output.

More sophisticated ICs. There are several types of IC that support bitwise control of the interface direction (input/output), including the following:

- Z80 PIO peripheral controller (which however has a limited drive capacity: $I_{OL} = 2$ mA maximum per signal),
- various types of microcontroller (such as the PIC series and the 8051 and its derivatives (eg SABC500, SABC16x)),
- certain user-programmable IC (GALs and CPLDs). The important factor is that there must be a separately programmable ENABLE condition for each output. Some examples of suitable ICs are: various 'industry-standard' GALs, the Xilinx XC9500, the Lattice ispLSI and the Cypress PLD20G10. *Caution:* these are all CMOS ICs. Supplementary protective measures are thus necessary to deal with partial power-down situations.
- some ICs with JTAG 1149 interfaces.

Signal data storage and evaluation. Simple LED-based displays are frequently useful as test aids, but they are not adequate for automated testing or fault-finding at normal operating speeds (dynamic operation). Instead, it is necessary to either acquire and store the signal data from the DUT for later evaluation, or to immediately compare the actual signal data to the expected values. With the present state of technology, each of these approaches is always based on some sort of programmable device or equipment, such as a microcontroller or a PC. Figure 4.118 illustrates the various alternatives for signal data acquisition and evaluation at the DUT/tester interface (see also Figure 4.110).

Alternatives (a) through (f) are only suitable for quasi-static testing; see Volume 3 for sample circuits. 'Dynamic' acquisition memories play a critical role in automated testers and logic analyzers. The practical applications for the direct comparison of the actual and expected results include comparison testing of ICs (see Section 5.14) and – in the form of comparison to compressed expected-data values – signature analysis (see Section 4.3.3).

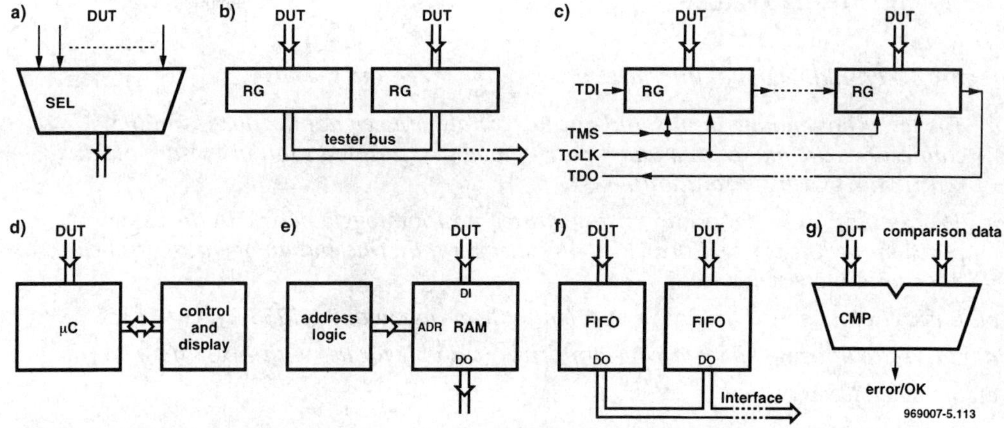

Figure 4.118 Basic techniques for signal data storage and evaluation.

4.4.2 The logic analyzer

The logic analyzer is essentially a digital data-acquisition memory that can be used to capture and display the behaviour of digital signals over time.It continuously monitors a certain number of these signals for the occurrence of pre-defined sets of values (trigger conditions). If the trigger conditions are satisfied, the signal values are subsequently sampled and stored at a certain regular interval. The sampling and storage timing can be determined either by an externally-supplied clock signal (*state analysis*) or by a preselected internal clock signal (*timing analysis*). The content of the acquisition memory is processed (usually by a microprocessor) and displayed (usually on a CRT or monitor). Figure 4.119 shows the block diagram of a typical logic analyzer.

Functional elements

Signal adapter. A logic analyzer acquires binary signals. A separate data channel is necessary for each signal to be acquired (the number of channels is typically 8, 16, 24, 32 and so on). In contrast to a DSO, each channel is normally only 1 bit wide. The test signal adapter usually contains evaluation circuitry for each signal to determine whether it represents a logical 0 or a logical 1 (see Figures 4.113, 114 & 115 for typical circuits). The general technical requirements for the signal adapter are: (1) it must not impair the operation of the test object, (2) it must not couple disturbances into the test object and (3) it must not distort the signal on its way from the test object to the acquisition memory. These requirements must be met under the condition that (in contrast to other types of equipment such as an oscilloscope) connections must be made to a relatively large number of signals.

Explanation of Figure 4.118:

a) The signals can be sensed via a selector circuit (the serialization technique; refer to Section 3.3.3 as well as Figures 4.112 and 4.116).

b) The signals can be temporarily stored in registers, or sensed via bus interface ICs that can be directly addressed by a microcontroller.

c) The signals can be read in via a shift-register interface. Compare this to the description of Figure 4.110f.

d) The test results can be directly sensed by a microcontroller. Compare this to the description of Figure 4.110b.

e) The signal values can be stored in a RAM that is incrementally addressed. This allows data to be acquired for a complete test process or portion of a test process with the DUT running at normal speed (dynamic testing). (This technique is used in logic analyzers.)

f) The acquisition memory can consist of FIFOs. If the required memory size and cycle time are relatively modest (a few kilobytes at most, and on the order of ≥ 20 ns), then FIFOs are an inexpensive alternative to RAM. In particular, this approach allows 'dedicated' logic analyzers to be built into interface testers and similar equipment.
(See Volume 3 for details.)

g) The actual signal values can be directly compared to the expected values. (The expected values can be produced in the same way as the test data, as shown in Figure 4.110.)

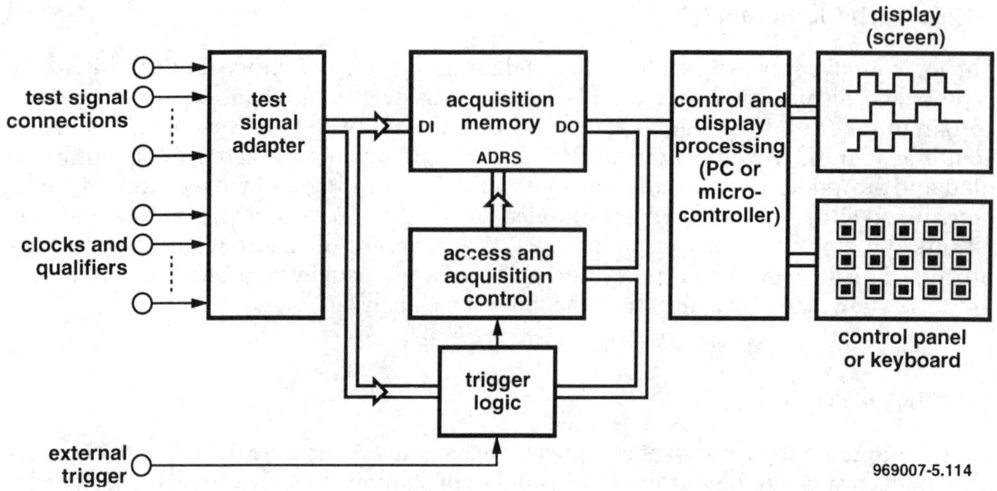

Figure 4.119 Logic analyzer block diagram.

The signal-evaluation circuitry is normally housed in a small enclosure (*probe pod*) that is separate from the main body of the logic analyzer. The actual connections to the test object (DUT or target system) are made by short flying leads attached via spring clips or plug connectors (refer to Figure 5.20). The probe pod is connected to the main body of the logic analyzer via a flat or round cable.

Notes:

1 The probe pod and the logic analyzer are often connected via a flat cable. The connection is (in the mechanical sense) not always particularly robust.

2 Some logic analyzers have several probe pods, with a separate one for each subset of the available channels (for example, a probe pod for every 4 or 8 channels).

3 Special accessory adapters (preprocessors) are available for various logic analyzers. These allow the logic analyzer to be quickly connected to particular types of target systems (for example, various types of microprocessors).

4 The practical utility of a logic analyzer largely depends on how quickly and easily it can be connected to a given test object. An important consideration is that it must be possible to build and use DIY adapters. This means that (1) the interfaces must be completely and accurately documented (in the operator's manual or service manual for the logic analyzer), (2) the probe pods must be constructed using standard, commercially-available connectors, cables and so on, and/or (3) it must be possible to purchase individual parts for the probe pods (cables, connectors, empty enclosures and so on).

Acquisition memory and access/acquisition control. In its simplest form, the acquisition memory consists of either a FIFO or a RAM with an address counter, in which the binary signal data are stored according to a particular clocking scheme. The depth of the acquisition memory determines how many zeros and ones can be stored per channel (according to the clocking scheme). Typical memory depths range from 512 bits to 8 Mbits or more. Inexpensive logic analyzers usually have a memory depth of 1 kbit to 16 kbits. Some instruments trade off the number of channels used against the memory depth per channel (the more channels that are used, the less memory there is per channel).

Trigger logic. The trigger logic works in the same way as the trigger logic of a digital storage oscilloscope. It determines when signal data storage starts (post-trigger mode) or when ongoing data acquisition stops (pre-trigger mode). It is usually possible to select part of the acquisition memory to be used for data before the trigger event and the rest for data after the trigger event (mid-trigger mode; refer to Figure 4.86c). The flexibility of the triggering system is essentially the deciding factor with regard to the usefulness of a logic analyzer in difficult fault-finding situations, just as with a DSO.

Operator controls and display. These functions are almost always provided by microprocessor-based arrangements. There are significant differences in the types of displays used (LCD or CRT), the display resolution, the features and layout of the control panel and the ease of use. Interfaces for connecting the logic analyzer to other equipment are standard features; these start with simple serial or parallel printer ports and can extend to serial interfaces for connecting the logic analyzer to a PC, or interfaces that allow a group of instruments to be connected together in a network (such as IEEE 488).

Practical tips:

1 When you are looking at the display, you often will not be able to recall which signal is connected to which channel. The logic analyzer should allow you to assign labels to the individual channels and to save these assignments (in the form of setup data) before the instrument is switched off (on a floppy diskette, for instance).

2 A PC interface is essential. However, the accompanying software is not always particularly helpful. The ideal situation is that (1) the logic analyzer can be completely controlled by the PC, (2) the data files have standard formats (that can be imported into a word-processing or graphics program, for example) and (3) the software has a (properly documented!) interface (Application Program Interface) to a widely-used programming environment (such as Visual Basic) so that you can develop your own test functions with a minimum of additional effort.

3 Especially with inexpensive logic analyzers, the ease of use and display resolution are often improved if the instrument is used via a PC instead of on its own. However, if you need a portable logic analyzer, do not rely too much on using a PC to make up for the shortcomings of a marginal instrument, since this can be quite awkward in practice.

Display formats. The stored signals can usually be displayed as pulse trains (see Figure 4.120) or in tabular form (see Figure 4.121). In most cases, you can select various representations for data displayed in tabular form (see Table 4.26).

Mnemonic displays. Mnemonic displays are important for working with microprocessor systems or standard interfaces (such as SCSI). In the mnemonic display mode, the stored bit patterns from the various channels are interpreted according to the assembly language of a particular microprocessor system or the functional concepts of a particular interface

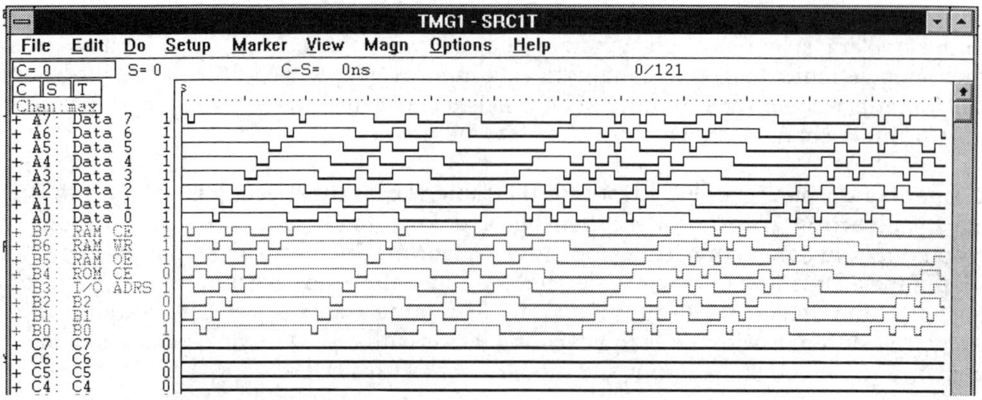

Figure 4.120 Example of a timing waveform display.

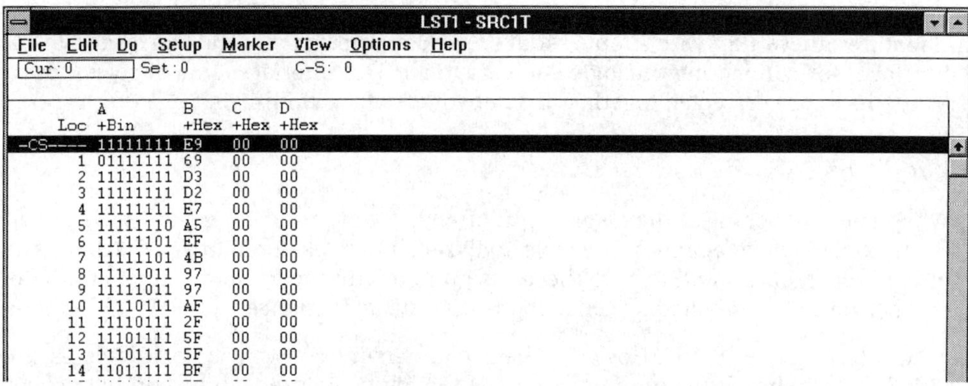

Figure 4.121 Example of a tabular display.

Type of representation	Symbols	Number of channels represented in one table column
binary	0 and 1; sometimes other special characters	1
ASCII	all characters of the ASCII character set	up to 8
octal	numerals 0–7	up to 3
hexadecimal	0–9, A–F	up to 4
decimal	numerals 0–9	Typically, 4, 8 or 16 channels are collectively displayed as a single (multidigit) number.

Table 4.26 Tabular display modes.

standard, and they are displayed in the form of their equivalent mnemonics or names. *For example:* a particular bit pattern on a microprocessor data bus will be displayed as an ADD instruction, or a set of signal values on a SCSI interface will be displayed as a RESELECTION bus phase.

Data acquisition techniques

Timing analysis (asynchronous sampling). With this technique, the timing for sampling and storing signal data is provided by an internal clock. In this mode, the logic analyzer operates in essentially the same way as a digital storage oscilloscope with a resolution of 1 bit per channel.

Edge-driven sampling (transitional sampling). This is a form of asynchronous data acquisition that makes more effective use of the acquisition memory. At relatively high sampling rates, the memory is normally filled in a short time. With transitional sampling, the signals to be acquired are monitored for changes, and data are only stored when a change is detected (the signal states and the elapsed time since the last change are stored). This technique is especially advantageous for acquiring data for burst processes, in which periods of relatively intense activity are separated by intervals of inactivity. This type of operation is characteristic for modern systems. (*For example:* a microprocessor with a built-in cache memory will take control of the memory bus only from time to time, but it usually executes a sequence of fast memory accesses – burst cycles – each time it does so.)

State analysis (synchronous sampling). With this technique, the timing for sampling and storing the signal data is determined by a clock that comes from the DUT. In most cases the logic analyzer has separate inputs for the clock signal and a number of clock-enabler signals (*clock qualifiers*). Data is sampled and stored when the selected edge of the clock (positive or negative) occurs at the same time that the associated clock qualifiers are valid (see Figure 4.129). The advantage of this method is that data are only stored for the processes that you actually want to see. The difficulties are that (1) you have to more or less understand the relative timing relationships of the various signals (including the clock and its qualifiers) and to trust that the actual signal forms roughly correspond to our assumptions (since otherwise the logic analyzer will store rubbish), and (2) you do not learn anything about the behaviour of the signals between the sample times (which is not infrequently responsible for the DUT's faulty behaviour).

Dual-mode analysis. Some instruments can sample and store data asynchronously and synchronously at the same time. This saves a lot of time spent making measurements. If you have detected a fault by examining the state-analysis data, you can then directly examine the timing-analysis data to see if there is anything that looks suspicious (such as a glitch). With a conventional instrument, by contrast, you would have to repeat the measurement using the timing-analysis mode, and the likelihood of again capturing the same fault indications under the same conditions is naturally not all that great.

Fundamentals of data acquisition for binary signals

The crucial requirement is that the signal being tested is evaluated in exactly the same way in the digital test equipment and the DUT:

1. A signal that is seen as a logical 1 by the gates and flip-flops of the DUT must not be seen as a logical 0 by the test equipment, and vice versa.
2. A signal that is regarded as valid with respect to a clock signal in the DUT and that is (for example) stored as a logical 1, must also be regarded by the test equipment as valid with respect to the same clock signal and must also be stored as a logical 1.
3. A pulse that has some sort of effect in the DUT must also be recognized by the test equipment as a valid pulse.
4. A pulse that for whatever reason has no effect in the DUT (for example, it is too narrow or it has too little amplitude) should not be seen as a valid pulse by the test equipment.

Digital test equipment, such as logic analyzers, signature analyzers, automated testers and test aids (in the sense of Section 4.4.4), is itself typically based on synchronous (clock-driven) digital circuitry (see Figure 4.122). The problem areas for data acquisition are (1) signal evaluation (that is, deciding whether a signal level at a particular time should be regarded as a logical 0 (or LO) or a logical 1 (or HI)) and (2) synchronization (that is, the clock-driven transfer of signal values into the acquisition hardware).

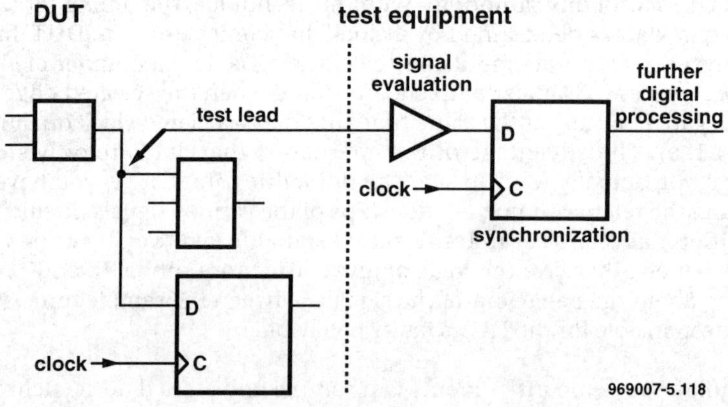

Figure 4.122 Binary signal data acquisition in synchronously-driven test equipment.

Signal evaluation. Figure 4.123 illustrates the essential problem areas:

a) The switching threshold of the comparator can be higher than that of the DUT ICs: the test equipment either totally fails to see signals that are effective in the DUT, or it sees pulses as being shorter than they 'really' are (as seen by the DUT).

b) The switching threshold of the comparator can be lower than that of the DUT ICs: the test equipment either sees signals that have no effect in the DUT, or it sees pulses as being longer than they 'really' are (as seen by the DUT).

c) The response time of the comparator can be too long: the test equipment is unable to capture short pulses that are nonetheless effective in the DUT.

d) The comparator can be too sensitive: its response time is short but its threshold is too low. Waveforms that have no effect on the DUT are seen as 'real' pulses.

The load that the test lead and comparator impose on the signal being tested also affects signal evaluation. This load primarily takes the form of additional capacitance, which causes signal amplitudes to be reduced and transitions to be broadened. This can have two effects:

1 The fault symptoms may be noticeably different when the test equipment is connected. (In many cases, nothing works properly when the test equipment is connected.)

2 The fault may disappear when the test equipment is connected. (In this case, the typical fault suspicion is that there are noise impulses present and that their amplitudes are reduced by the additional loading of the test connections.)

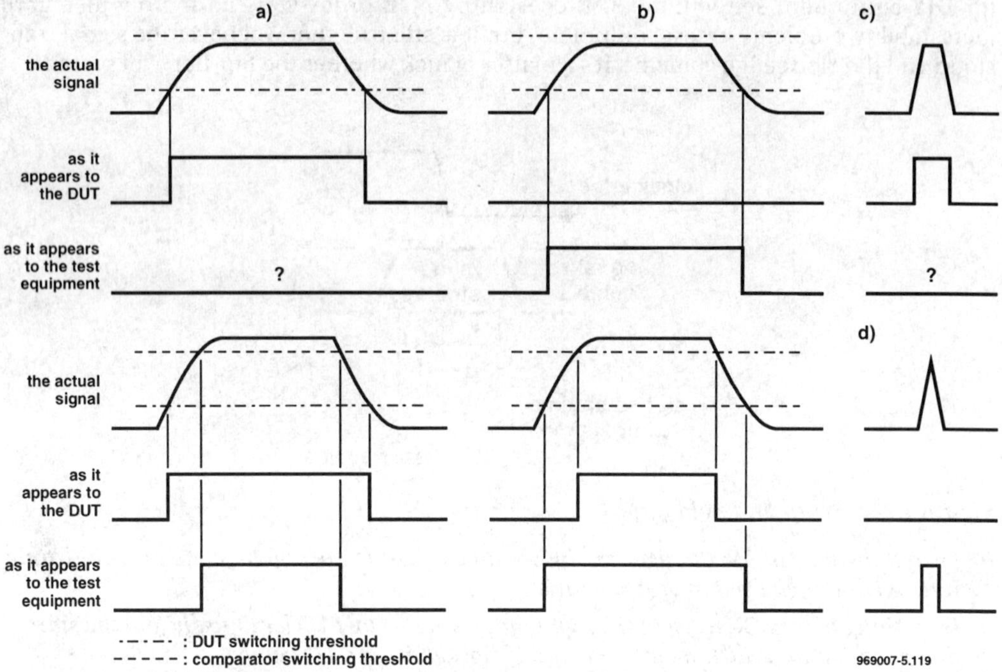

Figure 4.123 Problem areas in signal evaluation (see text).

Synchronization. *Synchronization* means transferring a signal value to a flip-flop with reference to a specific edge of a clock signal ('clocking in' a signal). This results in a clock-synchronous signal at the output of the flip-flop, which can then be further processed. There are two specifications of the flip-flop that are critical here: (1) its *setup time* and (2) its *hold time* (see Figure 4.124). The signal that is to be acquired must be stable in advance of the active clock edge for an interval that is at least as long as the setup time, and it must remain stable after the active clock edge for an interval that is at least as long as the hold time. Figure 4.125 illustrates certain typical situations.

Note: if the input to a flip-flop changes within the setup and hold-time interval, the response of the flip-flop is entirely a matter of chance. It will either change state or remain in its previous state. There is however a critical time window within the setup and hold-time interval (10 to 150 ps, depending on the logic family). If the change occurs during this window, the flip-flop will enter a *metastable* state, during which its output value is undefined; strong oscillations are present on the output for a certain length of time. It is fundamentally not possible to completely prevent metastability. Instead, one has to be satisfied with being able to prove – by means of analysis and measurement – that the probability of malfunctions due to metastability effects has been reduced to a level that is much lower than the probability of malfunctions and failures due to other causes. Metastability can only be dealt with by measures that reduce the likelihood of its occurrence or limit its consequences; it is not possible to constrain a flip-flop to always exhibit perfectly predictable behaviour. With regard to commercial test equipment, you must trust that its designers have taken adequate measures to deal with metastability problems; for DIY equipment see Volume 3. If everything is in order you can mainly disregard metastability problems in test equipment, but it is still true that whenever the signal transition and the clock edge coincide, it's a matter of luck whether the flip-flop will switch.

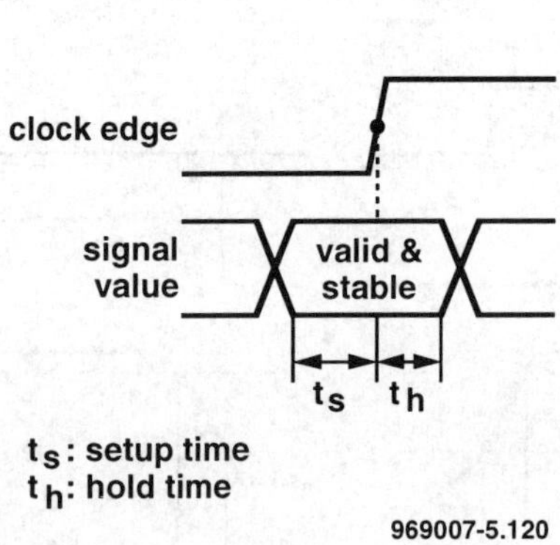

Figure 4.124 *Setup and hold times.*

a) Everything here is OK: the signal holds a stable HI value during the setup and hold-time interval and is clocked in as a HI value.

b) Everything here is OK as well: the signal has switched to a LO level just before the start of the setup and hold-time interval and is clocked in as a LO value.

c) The signal changes state within the setup and hold-time interval: the value that will be clocked in is therefore uncertain.

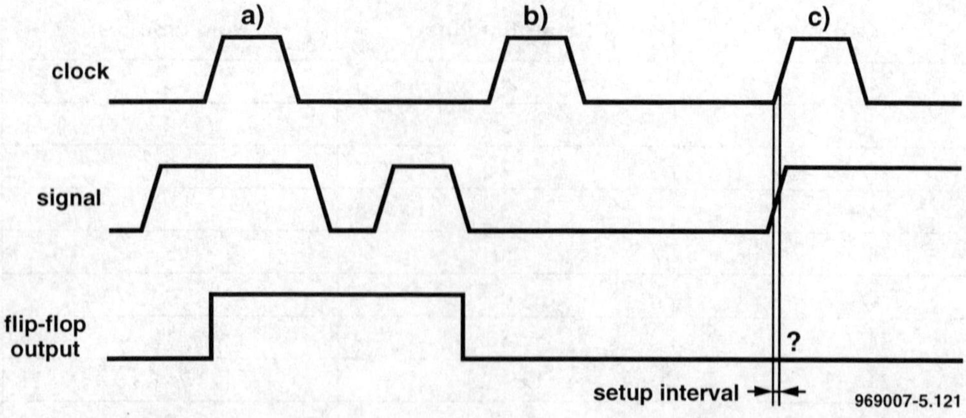

Figure 4.125 Signal synchronization situations.

In practice, this means that you have to pay close attention to the setup and hold time specifications of the test equipment. This applies in particular to any sort of reference to clock signals coming from the DUT (for state analysis and signature analysis). Only signals that are guaranteed to be stable during the setup and hold-time interval can be reliably acquired or evaluated. In all other cases you will either see signal waveforms that do not actually exist (with state analysis) or you will obtain spurious error indications (with signature analysis; see Section 4.4.3). *For example:* with state analysis, it could happen that a flip-flop in the DUT switches while a flip-flop in the test equipment that is driven by the same signal and clock does not switch, or vice versa (refer to Figure 4.123). Therefore, you must consider whether the signals that are to be checked meet the setup and hold-time specifications of the test equipment. In case of doubt, check your assumptions (with an oscilloscope, for example).

Table 4.27 lists the setup and hold-time characteristics of commonly used logic series ICs.

Note: it is evident from Figure 4.122 that the signal experiences a delay due to the response time of the comparator, so that (if the clock remains the same) the test equipment will sample and store delayed signal values. To remedy this, the test equipment should always delay the clock signal by an interval equal to the comparator response time (or in general, equal to the total propagation delay of the connection and signal evaluation circuitry).

Time offsets between various signals (skew). If several signals are acquired simultaneously, the propagation delays for the various signals will never be exactly the same. The individual synchronization flip-flops will also not behave identically in the setup and hold interval. The time offsets in the test equipment add to any offsets that are already present in the DUT (due to differing lengths of bus lines, for example). This means that (1) the skew inside the test equipment should be a small as possible and (2) the worst-case skew value must be taken into account in determining whether the setup and hold-time requirements are met.

Series	Setup time (ns)	Hold time (ns)
74	20	5
74S	3	2
74AS	2	0
74F	3	1
74LS	20	5
74ALS	15	0
74BCT	3	1
74ABT	1.8–2.5	0.7–1.3
74HC/HCT	20	5
74AC/ACT	3.5	0
74LV	15	3
74LVC	4	2
74ALVC	1.1	1.5
74ALVT/LVTZ	2.7	1

Table 4.27 *Setup and hold times for commonly-used logic series ICs. (sources: National Semiconductor, Philips and Texas Instruments)*

Disturbances in the test signal paths. The leads between the DUT and the probe pod are particularly susceptible to disturbances (ground shift, mutual crosstalk and radiated interference). To counter this, the earth lead or leads that are provided for each channel or group of channels must be connected to the DUT circuit earth as close as possible to the point(s) where the connections are made to the signal lines. *Note:* the separate earth lead of the probe pod itself should also be connected to the DUT earth in the immediate vicinity of the connections for the signal channels. If you connect the probe-pod earth lead to the DUT cabinet earth via a 1-metre length of ordinary wire (or don't connect it at all!) you should not be surprised if you measure rubbish.

Sampling errors. If the signals are sampled synchronously (that is, using a clock generated by the test equipment), then there is always a chance that the signals that you see (or the test equipment sees) are not identical to those in the DUT (see Figure 4.126).

Glitch detection. Glitch detection capability is provided to allow events that occur between sampling points to be recognized. As a rule, digital test equipment regards any variation in the signal level that occurs between any two sample points and which exhibits at least two edges to be a glitch (see Figure 4.127). Glitches can be registered and displayed, but neither the exact time of occurrence nor the exact behaviour of the signal can be determined from the stored data. *Note:* the logic analyzer will indicate a glitch for both Figure 4.127c and d, but the actual signal behaviour cannot be seen. The glitches shown could have quite different effects in the DUT!

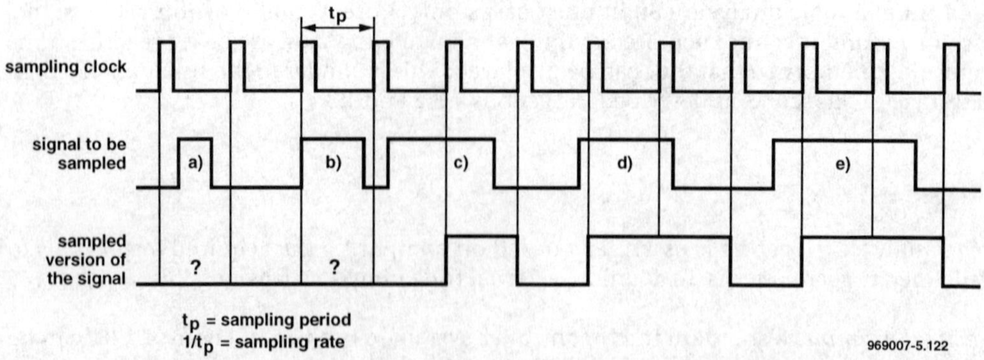

Figure 4.126 *Inaccuracies in (synchronous) sampling.*

a) A pulse that is shorter than the sampling interval can be lost. Note: this is a problem with all sampling techniques, since you cannot see what happens between samples (see also Section 4.3.2).

b) The longest pulse that can be lost is roughly one clock period (t_p) long. It will be lost if its edges exactly coincide with two successive set up and hold-time intervals and the synchronization flip-flop does not switch. Note: the minimum pulse width that can reliably be captured is $t_p + 2 \cdot$(setup + hold time); to be on the safe side it should be at least $2 \cdot t_p$. See also Section 4.4.4 regarding pulse evaluation and pulse stretching.

c) This pulse appears to be shorter than it actually is. If the leading edge is immediately after a setup and hold-time interval, the initial portion will not be sampled. The apparent start of a pulse can thus be delayed by up to one sample period (t_p).

d) This pulse appears to be longer than it actually is. If the trailing edge falls within a setup and hold-time interval, the final portion will be sampled an extra time. The apparent end of a pulse can thus be extended by up to nearly one sample period (t_p).

e) All acquired pulses are shifted more or less in time due to their being sampled by the synchronization clock. The maximum shift is approximately one sampling clock interval (t_p); the average (statistical) shift is $t_p/2$.

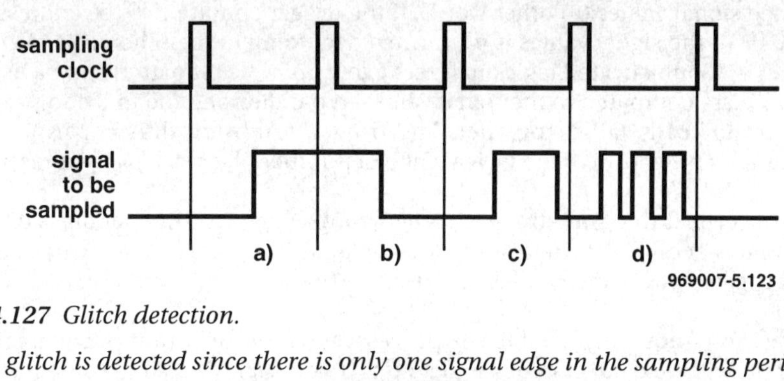

Figure 4.127 *Glitch detection.*

a, b) no glitch is detected since there is only one signal edge in the sampling period.

c, d) glitches are detected since there is more than one edge in the sampling period.

Dead time. A logic analyzer can in most cases only store a small portion of the signal waveforms, and it cannot acquire anything at all while the data are being read from the acquisition memory so that they can be displayed. This is similar to the situation in a digital storage oscilloscope, as described in Sections 4.3.2 and 4.3.4.

Triggering

Logic analyzer triggering is essentially based on comparing the acquired test signals to predefined trigger patterns. In addition, external triggering is usually possible.

Practical example: a standard microcontroller system, as shown in Figure 4.128, can be serve as a simple practical example to illustrate how the various triggering provisions can be used. We assume that all 20 signal lines of the microcontroller system have been connected to the data inputs of a logic analyzer and that (for convenience) the microprocessor's signal names can be used instead of logic analyzer channel numbers to identify specific signals.

Bit pattern comparison. The trigger event occurs when the signal values and the trigger match in every bit position. Bit patterns for defining trigger conditions are usually entered using three different values: a '1' means that the signal value must be a logical 1, a '0' means that the signal value must be a logical 0 and an 'X' means that the signal value does not matter (a 'don't-care' condition).

Example 1: you want to see what happens when a particular subroutine runs, and its starting address is $3F01_H$. Enter the trigger condition as follows: AD7–0 and A15–8, according to the binary equivalent of the starting address, PSEN = '0', ALE = '1', and WR and RD = 1 (PSEN, RE and WR are active LO!).

Example 2: you want to see whether location 2015_H is ever addressed. Again enter the trigger condition for AD7–0 and A15–8, according to the binary equivalent of this address, ALE = '1', and the memory access signals PSEN, WR and RD = 'X' (to exclude them from the comparison).

Multiple bit patterns. A single bit pattern as a trigger condition represents the logical AND of the specified signal values (in other words, if the trigger condition is the address 2 ... $_H$, the logical AND of the signal values is performed according to the scheme /A15 & /A14 & A13 & /A12 & ...). Sophisticated logic analyzers allow multiple bit patterns to be used concurrently. These are compared to the signal values in parallel, and the individual comparison results are logically ORed together. The trigger condition thus corresponds to a boolean equation in the form of the logical OR of a number of logical AND expressions.

Example 3: we suspect that something is going wrong whenever either address $023FA_H$ or address $023FF_H$ is accessed. With a suitable logic analyzer, we can specify both of these addresses as trigger conditions, as in Example 2.

Range entries. You frequently need to see processes that occur within a specific range of addresses, data values, control codes and so on. If you exclude a single signal from the trigger condition evaluation by using a don't-care entry ('X'), this means that the trigger

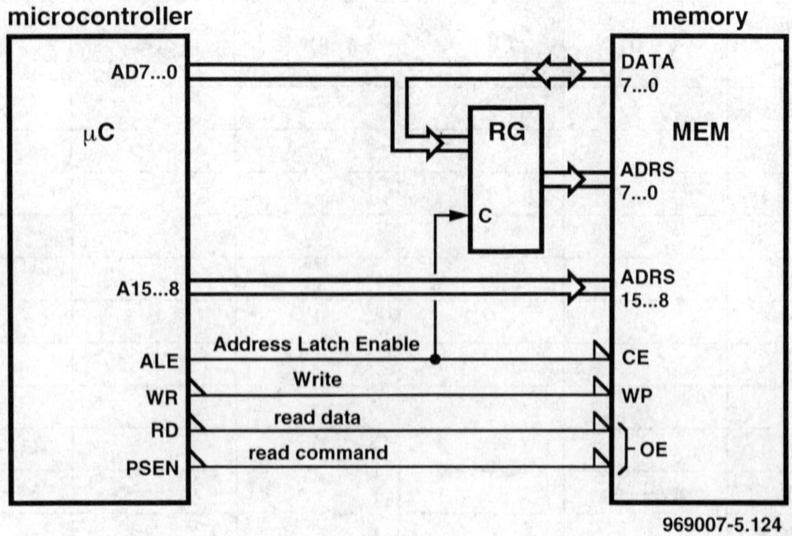

Figure 4.128 A practical example for explaining triggering provisions.

condition can be satisfied for all possible values of this signal. Triggering ranges can be defined in this manner, by marking particular data or address signals with don't-care entries. This is illustrated by Table 4.28 for a set of four signals. These signals can have a total of 16 different sets of values (column 1). The topmost cells of the remaining columns show examples of trigger conditions, and the arrow in each column indicates which combination(s) of signal values meet the associated trigger condition.

In general, a set of n signals can take on 2^n different sets of values. A trigger condition in which none of these signals is marked as 'don't care' selects exactly one set of values, and this condition is only satisfied when the exactly this set of values is present (*for example:* the trigger condition '1001' is only satisfied if the signal values are '1001'). A trigger condition with one 'X' entry can be satisfied by two different sets of signal values (*for example:* the trigger condition '100X' is satisfied by the signal values '1000' and '1001'). A trigger condition with two 'X' entries can be satisfied by four different sets of signal values (*for example:* the trigger condition '10XX' is satisfied by the signal values '1000', '1001' '1010' and '1011'). In general, a trigger condition with k 'don't-care' entries can be satisfied by a set of 2^k different sets of signal values. If you take these signals to represent an n-bit natural binary number (such as an address), then k 'don't-care' entries in the least-significant bit positions select a range of 2^k numbers (addresses) starting with the number whose k least-significant bits are all zeros (see also the third and four examples in Table 4.28). This means that 'don't care' entries can only be used to define triggering regions whose boundaries and lengths are exact multiples of 2. This is however still quite useful, although the boundaries of the regions within which you want to trigger are not always so favourably located.

Signals \ Trigger	1001	100X	10XX	1X01	1X0X
0000					
0001					
0010					
0011					
0100					
0101					
0110					
0111					
1000		←	←		←
1001	←	←	←	←	←
1010			←		
1011			←		
1100					←
1101				←	←
1110					
1111					

Table 4.28 Examples of using don't-care conditions to define triggering ranges.

Example 4: you want to see if any accesses occur in the address region 221_H–229_H. This range cannot be defined by the use of 'don't care' entries alone. *Practical tip:* you can trigger on $022X_H$ by making the four least-significant bits 'don't care'. You can then scan through the acquired data to see if any addresses occur that fall within the region of interest (the software sometimes provides a scrolling function). You might be unlucky, though: for example, if one or more addresses in the acquired region lies in a frequently-called program loop, the memory will be stuffed so full of 'unwanted' trigger events that it will be extremely difficult to see whether the addresses that you are looking for are also present.

Some logic analyzers have special trigger possibilities in which the signals are interpreted as binary numbers. These allow the trigger condition to be specified in the form of 'from *n* to *m*' or 'outside of *n* to *m*'. See also Section 4.4.4.

Time evaluation. Simple logic analyzers always generate a trigger whenever the overall set of trigger conditions is satisfied during the setup and hold-time interval of the clock in question. This can cause the logic analyzer to trigger on events that are of absolutely no interest to us, or missing events that we do want to see. For this reason, relatively sophisticated instruments can also evaluate the *duration* of the trigger condition (see Table 4.29).

Sequential evaluation (multi-level triggering). With multi-level triggering, trigger conditions can be defined in terms of a sequence of events.

Example 5: you want to trigger when data $4F_H$ is written to address $1AA0_H$. This is not possible with the previously-described triggering provisions, since the address and data are placed on a shared set of lines one after the other. You need a triggering function that can wait for ADDRESS = $1AA0_H$ & ALE = 1, and then trigger on AD7–0 = $4F_H$ & WR = 0.

Some logic analyzers have suitable multi-level triggering capability. Once the first step of the trigger condition is satisfied, the logic analyzer waits for the next step to be satisfied and so on. Time constraints as shown in Table 4.29 can also be applied. *Note:* the major consideration is how long it takes the logic analyzer to advance from each step to the following step. Only the best equipment manages to do this within one clock cycle.

Responding to trigger events. The simplest response to a trigger event is to either start or stop data acquisition. The more sophisticated types of logic analyzers can also be programmed, using a number of 'logical constructions' borrowed from high-level programming languages (such as "trigger on ... followed by ... " – which could be used for the previous example – or "trigger if ... else do ... "). These can be filled in via suitable menus.

State triggering. Triggering in the state analysis (synchronous) acquisition mode is essentially the same as for the time analysis mode. The only difference is that the trigger conditions are evaluated with respect to the clock from the DUT. Some logic analyzers do not allow certain types of triggering in state analysis mode (such as time evaluation and sequential evaluation).

Clock qualifiers for state analysis. Some simple types of logic analyzers treat the clock qualifiers as individual signals that are logically combined with the clock signal from the DUT (the qualifiers must either all be HI (AND combination) or all be LO (OR combination) for the clock to be effective). More sophisticated equipment allows 'don't-care' conditions to be assigned to individual qualifier signals. It is often possible to utilize the acquisition memory more effectively by the judicious use of qualifiers (since data then need not be acquired on *every* clock edge).

Time parameter	Additional entries	Requirements for the trigger event to occur
width	time value; selection of 'shorter than' or 'longer than'	The trigger condition is satisfied for a time that is either shorter than or longer than the specified time.
range	time interval (start and stop times); selection of 'shorter than' or 'longer than'	The trigger condition is either satisfied within the specified interval, or it is satisfied for a time that is either shorter than or longer than the specified interval.
timeout	time value	The trigger condition is not satisfied within the specified interval.

***Table 4.29** Selected time-based evaluation options for trigger conditions.*

Notes:

1 The trigger conditions must be satisfied and the qualifiers must be valid at the exact instant that the selected clock edge occurs – otherwise nothing at all will happen.

2 The clock and the qualifiers are processed by combinational logic, so there is no setup and hold-time requirement for the qualifiers with respect to the clock. However, keep in mind that (1) the clock that actually samples the data in the logic analyzer (and which thus is the reference for the setup and hold time requirements) only occurs when the external clock signal is active and the qualifier conditions are satisfied, and (2) you must avoid situations in which a qualifier can cut off the clock signal in the middle of a pulse, such that the resulting pulse width is less than the allowable minimum clock pulse width (see Figure 4.129).

Memory depth and triggering provisions. Why go to so much trouble with the triggering? An obvious alternative to sophisticated triggering provisions is to have a generous amount of acquisition memory. The idea here is to first acquire 'everything there is', and then use

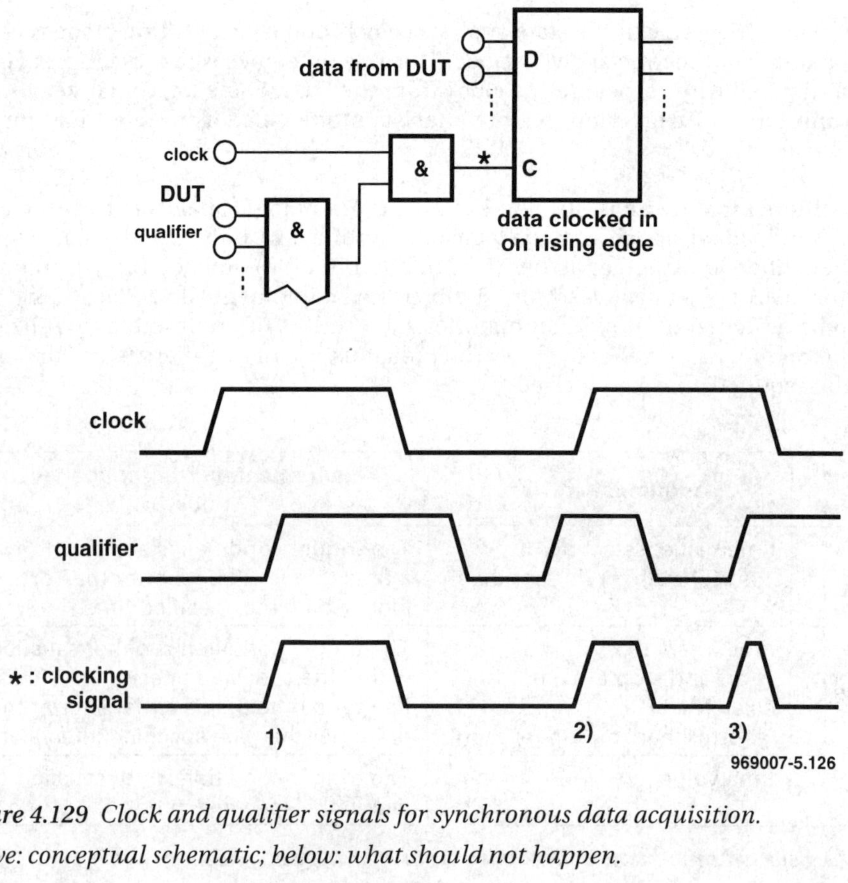

Figure 4.129 Clock and qualifier signals for synchronous data acquisition. Above: conceptual schematic; below: what should not happen.

software to comb through the entire volume of stored data to see if anything interesting turns up. Such an approach is not exactly inexpensive, but it has its advantages. You can only successfully use pattern-searching and triggering techniques if you first have some idea of what to look for. In practice, you usually have some sort of working hypothesis that you want to either confirm or refute, and you devise a search pattern or trigger conditions on this basis.If you have a large amount of data in memory you can repeatedly search through it for a variety of different patterns, but if a carefully-formulated trigger condition'misses the mark then you have no choice other than to repeat the measurement – and it's often a matter of luck whether exactly the same fault situation recurs.

Practical aspects

Types of logic analyzers

- Universal logic analyzers: these are available as self-contained ('stand-alone') units with their own control panels and displays, as plug-in cards for PCs and as elements of complex measuring systems.
- Multi-function equipment: logic analysis is one of several functions of a compact instrument (which can for example also be used as a DSO and a multimeter).
- Mixed-signal equipment: this combines the functions of a DSO and a logic analyzer, such that both functions can be used together (for example, for the concurrent measurement of two analogue channels and 16 digital channels).
- Special-purpose analyzers such as bus-system and interface analyzers for ISA, PCI, VME and SCSI. These are available as plug-in cards or self-contained equipment.

Explanation of Figure 4.129:

1 *The qualifier occurs later than the clock signal, so that the data are clocked in late. This can result in the data being sampled after it has already switched or while it is switching (changes in the data levels occur during the setup and hold-time interval). Note: many modern circuits are basically edge-driven state machines, in which almost all signals change immediately following the 'driving' clock edge. The sampling clock timing shown here could capture 'undefined' signal levels that are in the process of changing, or perhaps the stable levels before the next state transition instead of the stable levels just prior to the current transition. (The latter case will not be not so obvious, since the resulting signal waveforms will look 'clean' and thus less suspicious.)*

2 *The overlap of the clock and the qualifier is so small that only a very short sample clock is generated. There is a risk that either no sample will be taken or the sample data will be rubbish (some equipment can crash, just like a PC, if it is driven by non-compliant clock pulses).*

3 *If the qualifier switches during an active clock pulse, it can create additional sampling clocks. In principle, the logic analyzer should not take the next sample before the next rising edge of the 'true' clock signal. If it does not have suitable provisions to ensure 'one-shot' sampling, you will measure rubbish in this situation.*

Using a logic analyzer in the service environment. Logic analyzers are in principle versatile and powerful items of measuring equipment. They are however not inexpensive, and it takes some experience to be able to use them effectively. Many technicians prefer oscilloscopes for these reasons. (Some manufacturers lay out their mid-priced logic analyzers so that they have the same appearance and operator controls as an oscilloscope.) An additional hindrance is fact that connecting the logic analyzer to the DUT is time-consuming and prone to errors.

The most widely used types of logic analyzers are those for bus systems, interfaces and the like. In such cases the technician knows in advance what he or she wants to see (such as the sequence of SCSI bus phases), the controls and display are designed for the specific item to be tested and it is easy to connect the analyzer to the DUT.

For other types of test situations, you need to achieve a similar level of operational convenience. The measures that you can use are:

- purpose-built adapters (see Volume 3),
- special triggering and accessory hardware. It is frequently possible to use relatively simple aids to improvise a solution to a given test problem, so that it can be solved using inexpensive equipment instead of the quite costly equipment that would otherwise be necessary (see Section 4.4.4).

  Example 6: In the circuit shown in Figure 4.128, you want to trigger on a specific address in order to closely examine write accesses. This is not all that easy. The first possibility is to use asynchronous sampling, which allows you to see the address and the immediately following write access (with the data on the AD7–0 lines). The problem is that the sampling interval should be at most one half of the length of the smallest pulse to be observed. The accesses that are acquired then occupy a varying number of sampling intervals (depending on the relative timing of the sampling clock and the sampled signal). In addition, you have to search for the accesses that interest you (which consumes a lot of memory and makes the assessment difficult). The second possibility would be to use synchronous sampling with the WR signal – but the address bits have disappeared from the AD7–0 lines when WR is active! You could work around this by taking the address signals A7–0 from the address latch, but your logic analyzer may not have enough channels to allow this. A remedy in this case would be to use external triggering with an accessory address comparator (see Volume 3).
- 'dedicated' analyzers that are designed for specific tasks. It is hardly possible to make your own general-purpose, high-performance logic analyzer using readily available components (the critical functions of modern logic analyzers are implemented in special-purpose large-scale integrated circuits). On the other hand, quite effective acquisition memories can be 'custom-made' for specific testing tasks using modern components (such as FIFOs and CPLDs). In addition, it is often enough to be able to prove that certain conditions occur in the DUT, so that it is not even necessary to store a large number of signals (event analysis - see Section 4.4.4).

One-shot versus cyclic stimuli. This question has already been raised with respect to the oscilloscope (see Section 4.3.5). It relates to all stimuli (in the sense of Section 2.4) that you generate yourself using either hardware or software. Just as with the oscilloscope, a cyclic stimulus has its advantages with the logic analyzer. The most important of these is that if you see that the data are constantly changing (with synchronous data acquisition), this can be an indication that there is something wrong with the test arrangement or the logic analyzer setup (switching thresholds? skew? data changing during the setup and hold-time interval? problems with the qualifiers? intermittent contacts? and so on). On the other hand, in most cases it will take you significantly longer to assess the data captured by a logic analyzer than to evaluate the display of an oscilloscope, so that you will anyhow have to interrupt the data acquisition process. *Practical tip:* before beginning a 'rigorous' measurement procedure, use a simple (cyclic) test to check whether everything is properly set up and connected.

Testing logic analyzers. Logic analyzers, like oscilloscopes, are tested using known signal waveforms. The situations in which a logic analyzer should be tested are also the same as for an oscilloscope (refer to Section 4.3.4).

A universal data generator with enough channels to match the logic analyzer's capability would be the ideal tester, but such equipment is rather expensive. On the other hand, every properly-working digital circuit is in principle suitable for testing an logic analyzer. Inexpensive and/or readily-available test circuits include the following:

- feedback shift registers,
- counters,
- simple microcontroller circuits (similar to that shown in Figure 4.128), possibly with special ROM-based test routines (see Volume 3),
- a PC with a diagnostic adapter and corresponding test software.

Key aspects of logic analyzer testing:

- Glitch detection: the tester circuit must be able to insert glitches in the data stream.
- Satisfying special trigger conditions (time evaluation, multi-level triggering). Suitable test vectors must be provided to test these capabilities.
- Clock and qualifier relationships for state analysis. The tester circuit must produce various types of 'critical' signal sequences (as shown in Figure 4.129) that allow the correct operation of the logic analyzer to be verified from the acquired data. (*For example:* start with a counter and fashion a test clock that samples every second counter state. You then generate signal sequences (like those shown in Figure 4.129) that occur exactly when the counter changes to a state that should not be sampled. Invalid data sampling can be recognized by the presence of incorrect count values in the stored data.
- Comparator reference voltages: the test circuit must be able to provide signals with adjustable levels. This can be achieved by using open-collector drivers (for example).
- Checking the behaviour of the logic analyzer when acquiring tri-state signals.

4.4.3 Signature analysis

Signature analysis is a (digital) testing technique in which the comparison data are stored as compressed *signatures* instead of in their full form. Signatures are constructed using feedback shift registers, in exactly the same way as CRC sequences are generated. A signature that is (for example) 16 bits long can represent a signal waveform containing many thousand clock cycles. Reference signatures are taken from known-good hardware or generated by circuit simulators, with a separate signature for each signal. Signatures can be quite easily documented (as listings, as annotations on service schematics or in data files). When a device for which reference signatures are available is tested, the signatures are again acquired and compared to the reference values. Any mismatch indicates a fault.

There are three different applications for the signature technique: (1) test data compression, (2) built-in self-test provisions and (3) self-contained testers (signature analyzers). The last of these is described first, since it best illustrates the principles involved.

The signature analyzer

Construction and operation. Figure 4.131 illustrates the logical structure and operation of a signature analyzer, and Figure 4.130 shows a typical implementation.

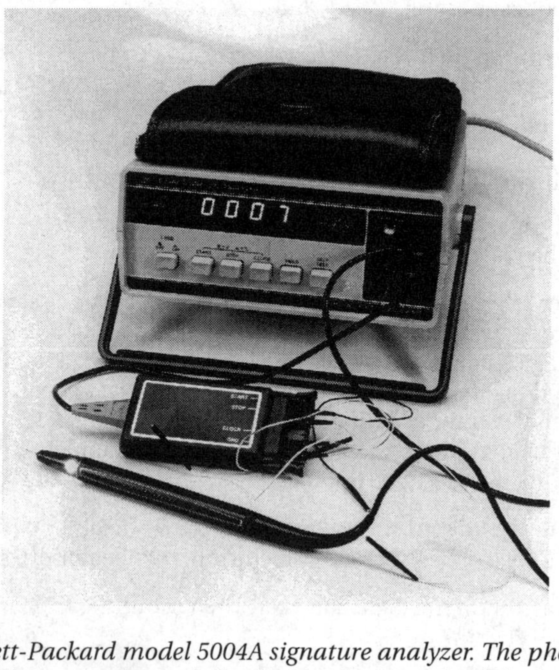

Figure 4.130 *Hewlett-Packard model 5004A signature analyzer. The photograph shows the analyzer itself, the adapter (probe pod) for the Start, Stop and Clock signals and the probe for sampling the signals to be tested.*

This sort of signature analyzer is intended to be used as a service instrument, in much the same way as a multimeter. The probe pod leads are connected to the DUT's Start, Stop and Clock signals, the proper clock edge and the signal levels for Start and Stop are selected, stimulus generation is initiated in the DUT and the data probe is touched to each signal in turn whose signature is to be checked. For each signal, the displayed signature is read and compared to the expected value. (One suggestion is to note the expected signature values alongside the signal test points on the service schematic diagram, in the way that voltage readings and oscillograms are traditionally shown on service diagrams for radios and televisions. This is why it is important to be able to easily remember a signature value for a short time.)

Error detection probability. An error is indicated whenever the actual signature is not the same as the expected signature. All single-bit errors (in which only one bit position has an incorrect value) will be detected with 100% certainty. The probability that a multibit error will cause an incorrect signature to be displayed lies around 99.94%. If we assume that in most cases the faults that have to be located correspond to rather crude fault models (such as 'stuck-at'), this is practically perfect.

Problems:

1 A signature is not a unique representation of the data: although a signature can be generated from a test data sequence, it is not possible to reconstruct the original data sequence from the signature.

2 The result of a signature comparison can only be a yes/no or go/no-go decision (either the actual signature and the reference signature are the same or they are not the same). It is practically useless to try to draw any conclusions about the nature of the test signal from its signature, with two exceptions: (1) if the signal being tested is stuck at LO or at earth potential, it yields a null signature ('0000' display) and (2) if the signal being tested is stuck at HI (or at V_{CC}), it yields a characteristic signature (which should be documented for reference purposes; see also item 4 below).

3 Signature analysis is a purely digital technique that is strictly clock-synchronous. It can only be used if the DUT provides a suitable clock signal and proper Start and Stop signals; if any of these is missing then signature analysis is not possible. The input setup and hold-time specifications of the test equipment with respect to the Start, Stop and test signals must be satisfied (see also Figure 4.124); if they are not then only erroneous signatures will be seen (which renders any further fault localization impossible).

4 A signature is only valid if the number of clock pulses between Start and Stop is constant (see Figure 4.131b). Even one clock too many or too few makes the result useless. This means that processes that, although standardized, are subject to device-specific or situation-specific timing variations cannot be characterized by a consistent set of signatures. (Think of bus systems with a variable number of wait states.) *Note:* you can verify that the Start, Stop and Clock signals are valid (correct connections, correct selection of signal levels and clock edge, correct number of clock pulses in the Start-Stop interval) by checking the HI-level signature (touch the data probe to a V_{CC} connection while stimulus generation is active). This is why the expected value of the HI-level signature should be documented for each test setup.

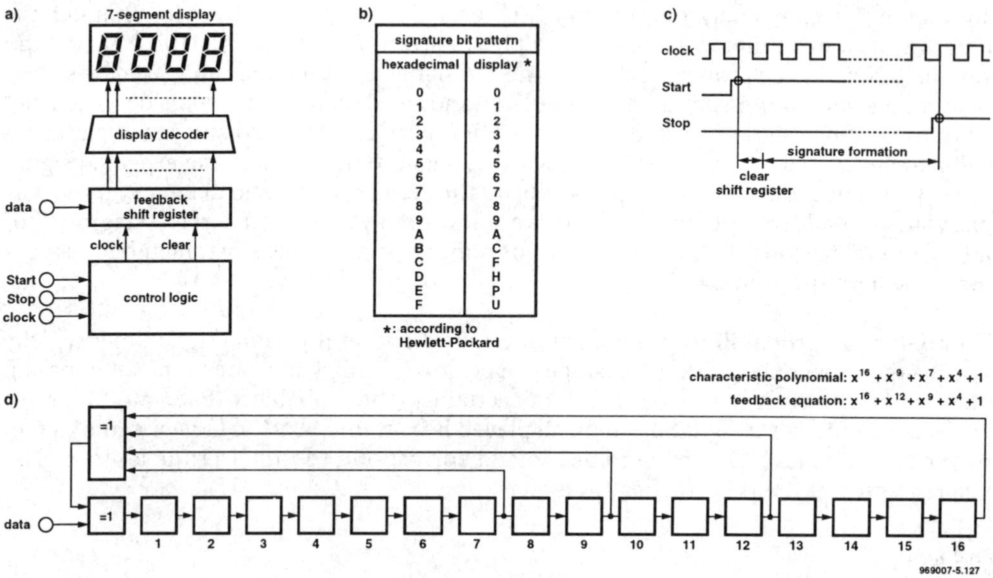

Figure 4.131 Construction and operation of a signature analyzer.

a) Simplified block diagram. Signature analysis is fundamentally based on a feedback shift register with an attached display. This can form one signature at a time from a given waveform. The signal to be checked is connected to the input of the signature analyzer. The DUT must also provide a clock signal plus Start and Stop signals, which together provide the required time reference.

b) Display. The example shows a 16-bit shift register whose content is displayed by four 7-segment indicators. In principle, many different types of display are possible (such as octal, hexadecimal or even binary, using 16 LEDs). The display scheme shown here was introduced by Hewlett-Packard in 1979/80. The reasons for choosing it were: (1) to allow 7-segment displays to be used, (2) to distinguish signatures from hexadecimal data (so that nobody would try to attribute some 'deeper' significance to the signature values), (3) to limit the signature to four characters (since a sequence of four characters is significantly easier to note and remember than a five-character sequence) and (4) to limit the characters to numerals and capital letters, so that the signatures can be easily documented and pronounced (which is important for remote servicing by telephone, for example).

c) Acquisition timing. Everything is based on a suitable test stimulus, which must be generated in the DUT in strict synchronization with a clock signal. The DUT must mark the limits of the measurement window with suitable Start and Stop signals. The first valid clock edge after the Start signal goes active clears the shift register to all zeros. The data bits are shifted in starting with the next following clock edge, and the signature is formed via the feedback circuitry. This process stops as soon as an active Stop signal is sensed by a valid clock edge. The content of the shift register at this time is the signature.

5 Signature analysis can only be used for detailed fault localization if the conditions described under points 3 and 4 are satisfied. The equipment to be tested must be designed with signature analysis capability in mind if these conditions are to be consistently satisfied. The following recommendations are important in this regard:

- The circuit should be divided into subcircuits for testing purposes.
- Stimulus generation (in particular, the generation of the clock and the Start and Stop signals) for a particular subcircuit should be performed by another part of the circuit that cannot be affected by a fault in the subcircuit being checked.
- It should be possible to open feedback loops to create linear signal flows in subcircuits to be tested. (The preferred signal tracing method is thus backtracing, in which you start with the subassemblies or ICs located at the end of a signal path and work back toward the signal sources until you find the first item that displays incorrect output signatures and has correct signatures at its inputs. See also Section 2.5.3.)

Typically, testing using signature analysis amounts to a strictly incremental process, which (for example) starts with the DUT's microprocessor operating in the 'free-running' mode (see Volume 2 for more information). The very high error-detection probability of signature analysis can also be exploited by constructing composite signatures from the results of several signature measurements. This can be extended to identify specific fault conditions within the DUT. Composite signatures can (for example) be produced by adding up a series of measured signatures (which in practice is done by the signature analyzer itself).

The measures needed to make a given device suitable for testing by signature analysis are part of the art of test- and service-oriented system design. They are also useful for other testing techniques. In particular, distinctly identifying specific test processes by means of Start and Stop signals has proved to be generally useful. (For instance, it can be used to

---

d) The shift register recommended by Hewlett-Packard. In principle, signatures can be generated by any desired feedback shift register that generates a maximum-length sequence (see Appendix 6). The shift registers that are used for generating CRC strings could thus also be used for signature analysis. The following factors relate to the selection of the shift register:

- *The longer the shift register, the better the probability of detecting an error.*
- *The longer the shift register, the longer the signature. Long signatures are however unwieldy; they are difficult to read and to document. A 16-bit signature represents an optimum length.*
- *There are 2048 different feedback equations that produce maximum-length sequences with a 16-bit shift register. The illustrated arrangement was chosen because its probability of not detecting an error is distributed as evenly as possible over all normally-occurring bit patterns. (The requirements are: (1) it must represent an irreducible polynomial and (2) it must not include any feedback terms that are separated by 4 or 8 bits (since modern microprocessor-based circuitry frequently produces bit sequences that repeat at an interval of 4 or 8 bits).*

Note: such a shift register is also a quite useful circuit for testing oscilloscopes, logic analyzers and cables (refer to Figure 5.112).

distinguish test processes from control processes in a test program, and a logic analyzer or oscilloscope can be triggered by the Start signal at the beginning of each test process. See Chapter 5 and Volume 2 for more information.) Many technicians gladly utilize test loops designed to be used for signature analysis (SA loops) as stimuli for signal tracing with an oscilloscope. The features of the two different techniques can perhaps be clarified by means of an example: suppose you want to locate a fault on the ISA bus of a PC, and you use a cyclic test loop that periodically writes to a specific I/O address as a test stimulus. Table 4.30 describes what you can expect.

Test example	**Fault-finding using a signature analyzer**	**Fault finding using an oscilloscope**
A bus clock is used as the test clock signal.	Since different types of PCs have different instruction execution times, the number of bus clocks during a test loop is different in each case (and you also have to take other types of bus accesses into account, such as memory refresh and interrupt handling). The net result is that stable signatures cannot be expected.	With a bit of practice, you can judge whether the process in question is correct, even if the timing relationships are different from one PC to the next.
The I/O Write signal *IOW#* is used as the test clock, but it behaves incorrectly or is completely inactive.	You see either invalid signatures or none at all.	You see that the behavior of *IOW#* is somehow incorrect and can draw your own conclusions.

***Table 4.30** Comparison of fault-finding methods.*

Other signature generation methods. In principle, any process that has the following properties is suitable for generating signatures:

1 An unambiguous signature (that is significantly shorter than the original) is generated for any given bit pattern or bit sequence (a particular bit pattern or bit sequence always produces the same signature).

2 There is a sufficiently high probability of a difference between the signature of the correct data and the signature that is generated when there is any sort of error in the bit pattern or bit sequence.

The following methods (among others) are possible:

- checksum generation (adding up successive bit patterns),
- parity checking (in the general sense of XOR combinations of bit patterns),

- running-total checks (adding up the number of ones in a bit pattern),
- transition counting (adding up the number of HI/LO or LO/HI transitions).

The above-listed methods are technically simpler than CRC generation, but they have lower error-detection probabilities.

Signature analysis as a built-in self-test provision

Self-contained signature analyzers are practically only available as second-hand equipment. By contrast, the use of signature analysis as a method for built-in self-testing is currently standard practice.

Serial signature analysis. If serial signal paths are already present (as with the serialization or scan technique described in Section 3.3), then it is a natural idea to build in permanent signature-testing circuits similar to that shown in Figure 4.131 (but without the display). The detected signatures are either evaluated by a service processor or directly compared to expected values (see Figure 4.133).

Parallel signature analysis. A signature can be generated for a group of signals in parallel if they are applied 'broadside' to a suitable feedback shift register (see Figure 4.132). The resulting 'combined' signature can be used to make a go/no-go decision regarding the overall set of tested signals. Any linear feedback shift register (in which the feedback terms are generated exclusively by XOR combinations; see Appendix 6) can be configured for parallel signature analysis by inserting XOR gates in the shift path in order to couple in the signals to be tested (see Figure 4.132).

Notes:

1 The feedback shift register must be set to a predefined initial state before the signature is generated.

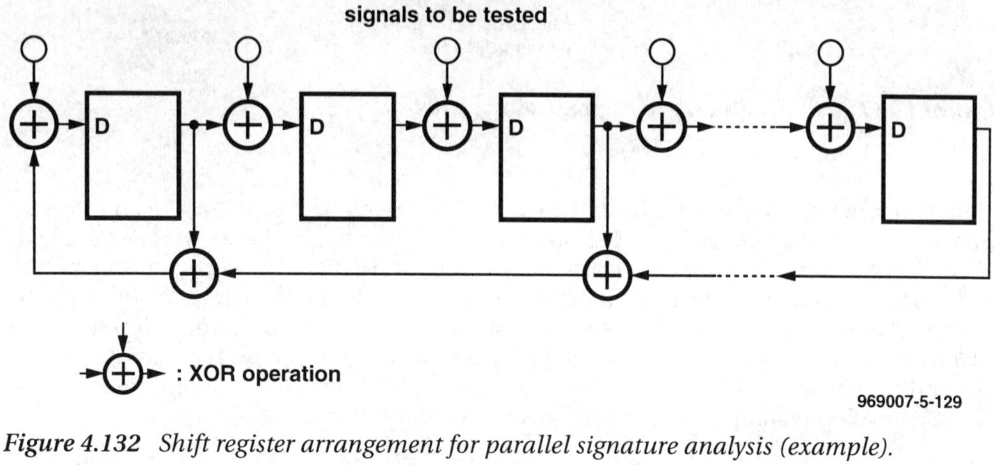

Figure 4.132 Shift register arrangement for parallel signature analysis (example).

2 The actual signature is generated by clocking the shift register a predefined number of times after it has been placed in the initial state, while the signals to be tested applied to its parallel inputs.

Signature analysis in highly integrated ICs. Built-in self-test provisions are usually based on using feedback shift registers for both test stimulus generation and signature testing (the BILBO technique; see Section 3.4). Figure 4.133 depicts the built-in self-test provisions (BIST) of the Intel 386DX processor, as an example. The hardware to be tested consists of the microprogram control memory (control ROM) and the PLA structures of the process control circuitry (which practically amounts to its combinational logic). These elements are tested using a pseudo-random bit sequence, and a single signature is generated that is compared to a predefined value by bitwise exclusive-NOR operations (so that the result is all zeros if the two signatures match). The result of the comparison is loaded into the EAX processor register, where it can be read by software.

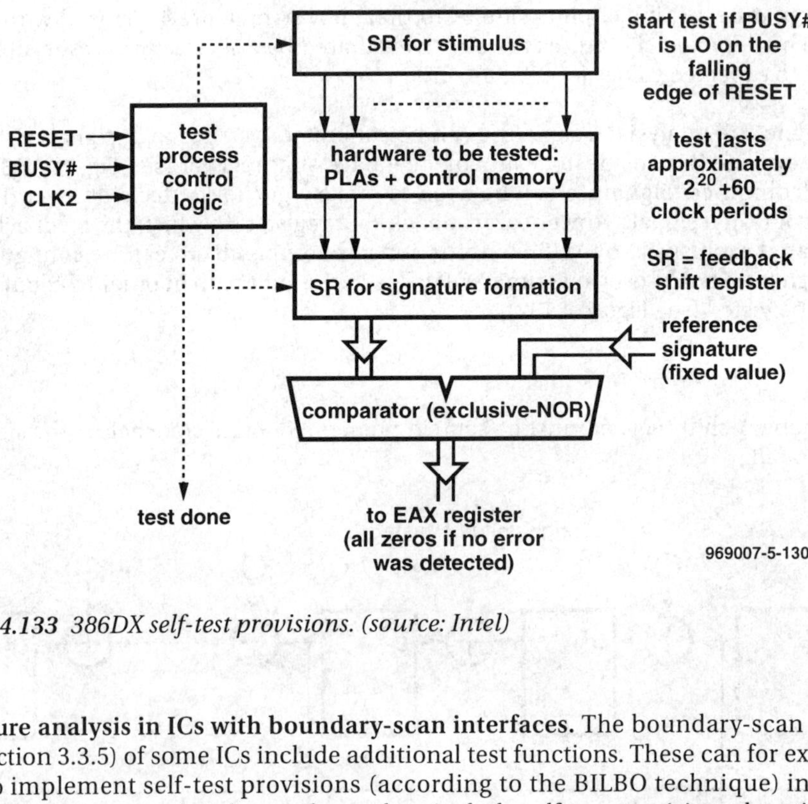

Figure 4.133 386DX self-test provisions. (source: Intel)

Signature analysis in ICs with boundary-scan interfaces. The boundary-scan interface (see Section 3.3.5) of some ICs include additional test functions. These can for example be used to implement self-test provisions (according to the BILBO technique) in a circuit based on components that do not themselves include self-test provisions, by utilizing the bus-interface ICs that are already present in the circuit. The following short description is based on the use of suitable Scan Test Device ICs (manufactured by Texas Instruments). The additional functions are not part of the official JTAG/IEEE 1194 standard, so you must always consult the manufacturer's data sheets regarding specific details.

Circuit structure. Figure 4.134 shows the block diagram of one of these ICs, which is a bidirectional bus driver with 2 × 18 data pins. It is a bus transceiver that performs same functions as a 74x245 IC, arranged as two 9-bit channels.

An additional control register (the Boundary-Control Register), that can be accessed via a special test instruction (Boundary-Control Register Scan), is essential for the test functions. A suitable operation code must be loaded into this register before a test process can be executed. A total of five test functions are available:

1 Pseudo-random pattern generation (PRPG). The 36 boundary-scan cells of the data pins are operated as a feedback shift register, which generates a pseudo-random test pattern that appears at the IC's outputs at every falling edge of the test clock (TCK). *Note:* an initial value (*seed value*) must be loaded into the boundary-scan cells before the pseudo-random number can be generated. The initial value can be any desired bit pattern except all zeros (since this would cause only zeros to be generated).

2 Signature formation (Parallel Signature Analysis, PSA). The 36 boundary-scan cells of the data pins are operated as a feedback shift register that is affected by the values applied to the inputs of the IC each time that a rising edge of the test clock occurs. The input data contribute to parallel signature formation via XOR gates, as shown in Figure 4.132. *Note:* an arbitrary initial value (seed value) must be loaded into the boundary-scan cells before the signature is formed.

3 Combined PSA and PRPG. The 36 boundary-scan cells of the data pins are configured as two 18-bit feedback shift registers. One of these produces a pseudo-random bit sequence (synchronized to the falling edge of the test clock TCK) that is provided via the data outputs, and the other forms a signature for the signals present at the data inputs, which are clocked in on the rising edge of the test clock. *Note:* an initial value (seed value) must be loaded into all 36 boundary-scan cells before the test is started. The initial value can be any desired bit pattern, but a pattern of all zeros in the test pattern generator shift register will cause only zeros to be generated.

4 Combined PSA and counting (PSA/COUNT). The boundary-scan cells of the data pins are configured as an 18-bit binary counter together with an 18-bit feedback shift register. The test pattern from the binary counter (available via the data outputs) is a binary count that increases by 1 on each falling edge of the test clock TCK. The shift register forms a signature from the signals present at the data inputs, which are clocked in on the rising edge of the test clock. *Note:* an initial value (seed value) must be fed into all 36 boundary-scan cells before the test is started. The initial value is arbitrary; the initial value that is placed in the counter cells forms the starting value of the counting sequence.

5 Toggle Outputs, Sample Inputs (TOPSIP). On every rising edge of the test clock TCK, the output data are inverted ($0 \rightarrow 1$, $1 \rightarrow 0$) and the input data are transferred to their associated boundary scan cells.

Note: the test provisions just described apply only to the data pins. The test functions can only be executed if all 18 data bits are (1) enabled and (2) configured to transfer data in the same direction (either from left to right or from right to left in Figure 4.134). The data pins are configured via additional boundary scan cells.

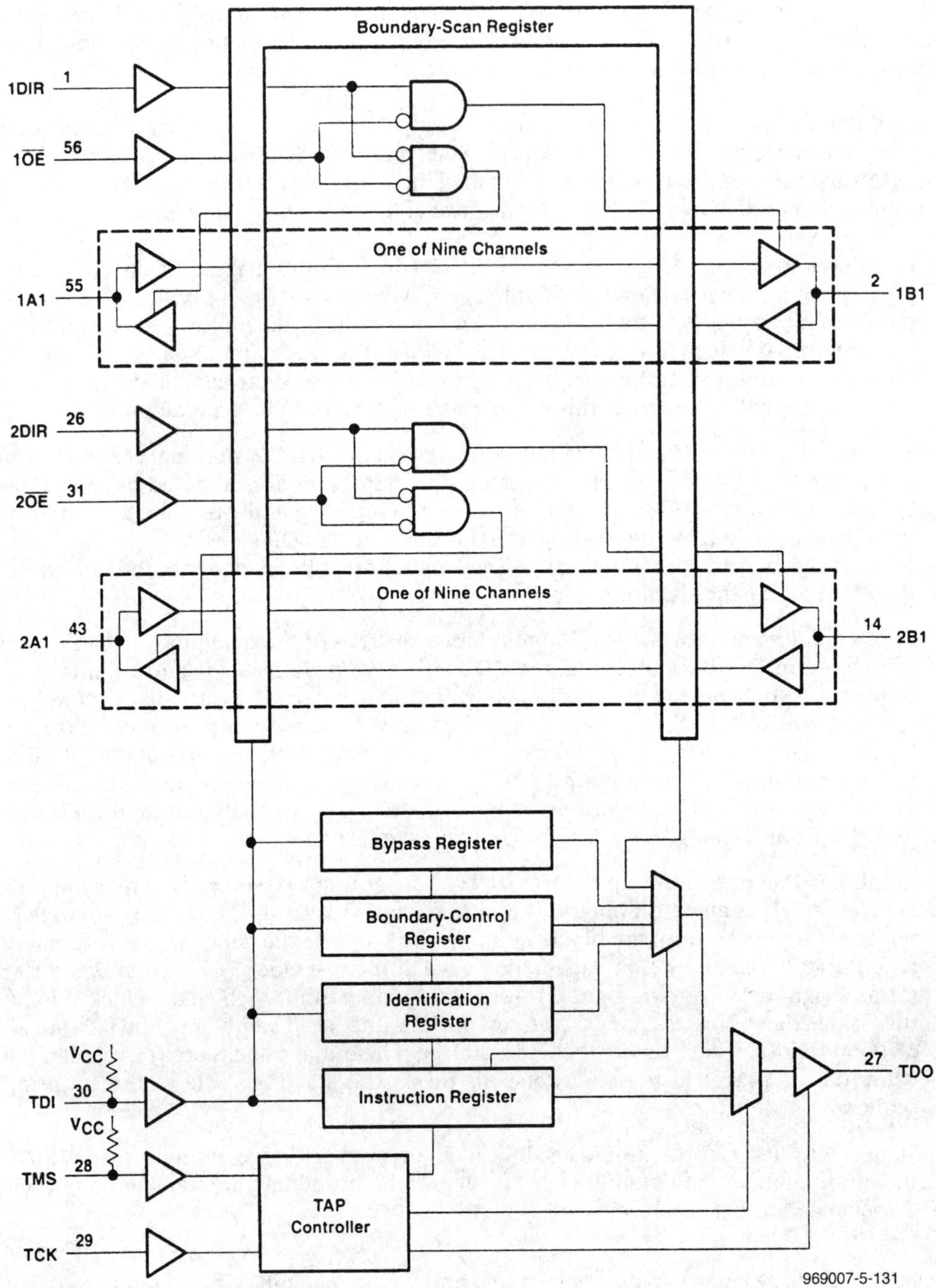

Figure 4.134 Block diagram of the type SN74ABT18245 IC. (source: Texas Instruments)

Signature analysis in programmable ICs. Figure 4.135 shows a representative application example, based on a Xilinx XC9500 CPLD IC. This field-programmable IC contains a DRAM controller with an address multiplexer, control signal generation logic and a refresh-interval counter. These circuit elements are all connected to a feedback shift register to support signature-analysis testing. In the example circuit, the test stimulus comes from a control processor (which means from a suitable test program). The processor must also fetch and evaluate the signature once it has been formed. When designing a circuit such as this, you must keep the same considerations in mind as for a circuit that will be used with an external signature analyzer (with respect to the clock window, setup and hold times and so on).

Note: the logic elements required for signature analysis testing are sometimes only present 'on demand' in integrated circuits that can be loaded or reprogrammed in the application environment. This means that certain parts of the IC that are part of the 'functional' logic in normal operation can be programmed to function as part of the test logic whenever this is required (in test mode), and then reprogrammed to perform their normal operational functions.

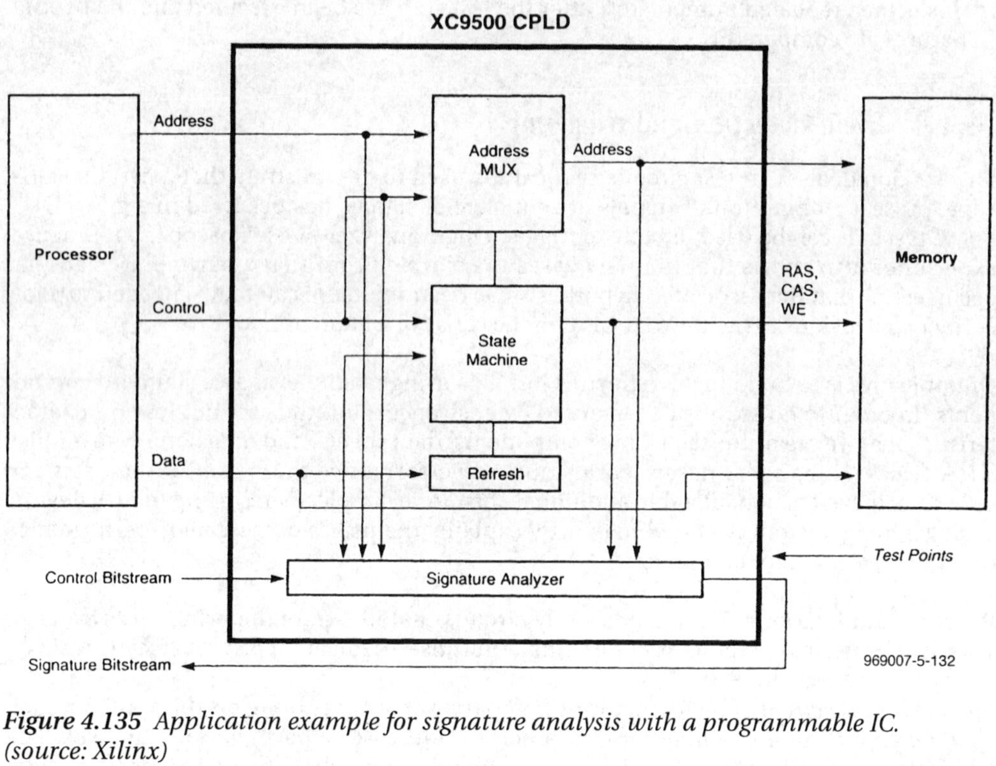

Figure 4.135 *Application example for signature analysis with a programmable IC. (source: Xilinx)*

Using signatures to compress test data

It is obvious that large files of comparison data (which are primarily found in test software and automated testers) can be stored in the form of relatively small signatures. In the case of test software, this applies primarily to the following:

- Computation results (comparison data for mathematical and logical instruction tests).
- Data streams (for peripheral device testers). *Note:* built-in peripheral-device test functions that are based on stored error-checking data (such as the VERIFY function of a disk drive) are in themselves a sort of signature analysis. Such tests do not however check the data paths.
- Register and memory contents. Some well-known examples are ROM checksums (to verify the contents of ROMs) and file checksums (to test files for corruption by a virus). In test programs, this technique is primarily useful for checking the 'integrity' of registers or memory regions. (*For example:* only certain registers or certain memory locations are supposed to be modified in a particular test step; any other changes indicate a fault. Just before the test is executed, a checksum (or other type of signature) is formed from the contents of all registers or memory locations that are not supposed to change; this is then repeated immediately after the test step has been executed and the two signatures are compared.)

4.4.4 Event detection and triggering

This section deals with test circuits that can be used to demonstrate that a certain signal edge, pulse, combination of signals or sequence of signals has occurred in a given DUT. These test aids can be used directly to trigger a logic analyzer or oscilloscope, but practical experience also shows that in many cases the simple proof that a particular event has occurred (or that the associated hypothesis has been refuted) is all that is needed to allow correct conclusions to be drawn regarding the cause of a fault.

Our objective is to build highly-effective test aids using readily-available standard components. In order to do so, it is necessary to extensively exploit the capabilities and characteristics that we are promised by the components' data sheets, and occasionally to employ a few tricks (since every nanosecond counts). Performance that is better than this can only be achieved at considerable additional cost (for example, by resorting to ECL devices or developing custom ICs) – which partly explains the prices of truly high-performance test and measurement equipment.

Practical applications. We consider only circuits that illustrate the principles involved. They can be used in practice as either 'single-purpose' (dedicated) or 'universal' test aids.

Single-purpose circuits: the detector or comparator circuit is built 'on the spot' for each specific application, in a manner that precisely addresses a particular testing task. The advantages of this approach are that (1) you are not limited by the restrictions of a ready-made test aid and (2) very short switching times and/or high upper frequency limits can be realized due to the simple construction (limited circuit depth). You can choose from the following approaches:

- 'On-thy-fly' construction on a prototyping board or solderless breadboard using basic IC types (gates, comparators and so on). This has the disadvantages of being labour-intensive and error-prone.
- 'Semi-permanent' construction using a fixed array of gates, flip-flops and the like that are interconnected as needed via connectors or switches (see Volume 3).
- Utilizing ICs (CPLDs, FPGAs) that can be programmed in the application hardware. In many cases a single IC, suitably packaged (for instance in a probe pod), can be used for event-recognition and triggering functions that are almost unimaginably diverse (the trigger logic of some logic analyzers works on this basis). The drawback is that a PC and a suitable development system (which can be fairly expensive) are required.

Universal test aids. Event detector circuits form the basis for a variety of test aids, including pulse detectors, triggering aids and event analyzers. They can be implemented as self-contained devices, computer-controlled devices or plug-in cards that can be connected to a suitable bus system (such as the ISA or PCI bus). See Volume 3 for more information.

Edge detectors. A signal edge (that is, a LO-to-HI transition or a HI-to-LO transition) can be detected by connecting the signal line in question to the clock input of an edge-triggered flip-flop (see Figure 4.136). The function of an edge detector is to verify that the signal under test (*P*) has either switched from LO to HI or from HI to LO. The number of times that this has occurred, and the particular sequence of transitions, are in this regard unimportant. The detector flip-flop must be cleared by the reset signal */CLR* before the start of the test. If at least one edge event occurs, the flip-flop switches to the set state and remains in this state until it is again reset.

Flip-flop selection. There are only two basic IC types that can be used as edge detectors: the 74x74 D-type flip-flop and the 74x112 JK-type flip-flop (a single IC package contains two flip-flops). The general characteristics of these devices are listed in Table 4.31, and their critical specifications are given in Table 4.32. Since we want our tests to be as 'rigorous' as possible, we consider only the fastest possible devices. *Notes:* (1) the F, AC/ACT and LVC series are modern types, (2) if a fixed value is applied to the data input (as in Figure 4.136a), the hold-time specification is not significant. Otherwise, a hold time of 0 reduces the risk of metastability; (3) Flip-flops in the F series are the least inclined to show metastable behaviour.

Device	Type	Advantages	Disadvantages
74x74	D-type flip-flop, positive-edge triggered	slightly better setup times than 74x112	may need additional logic (feedback) to maintain its state if a fixed level is not applied to the D input; hold time is not always zero
74x112	JK flip-flop, negative-edge triggered	state held if J = 0 & K = 0 (no feedback needed); zero hold time for all series	slightly longer setup times than 74x74

***Table 4.31** Flip-flop types that can be used in detector circuits.*

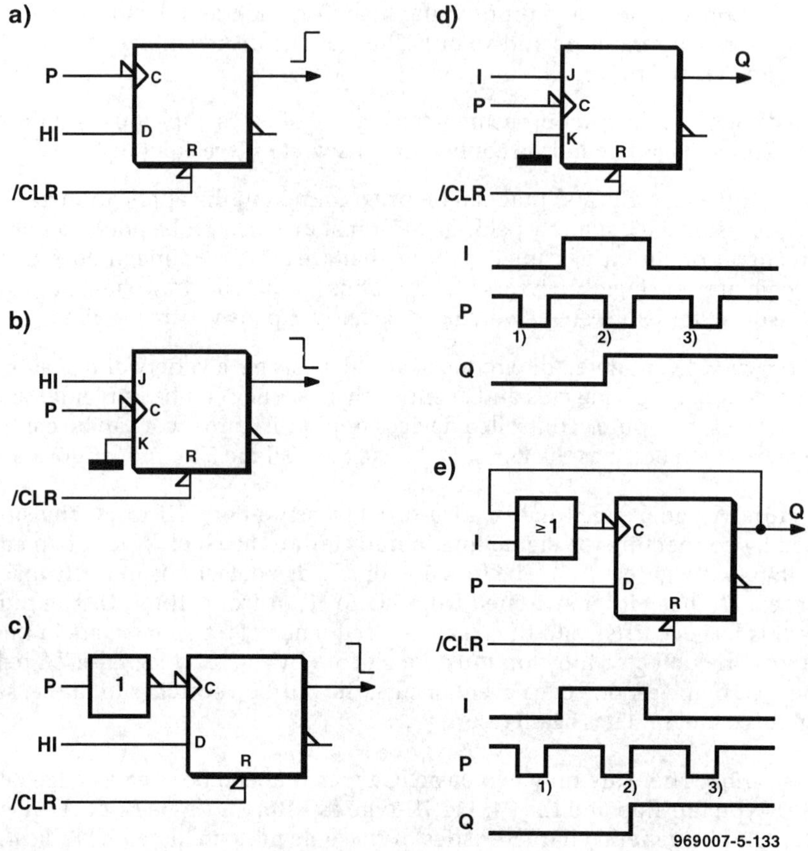

Figure 4.136 *Basic edge detector circuits.*

a) Detecting a LO-to-HI transition ('positive' edge) with a D-type flip-flop.

b) Detecting a HI-to-LO transition ('negative' edge) with a JK-type flip-flop.

c) Detecting a HI-to-LO transition with a D-type flip-flop. Since D-type flip-flops are clocked on the positive edge, the signal P must be inverted. In a similar manner, positive edges can be detected with a JK-type flip-flop by inverting the signal P.

d) Proving that P switches from HI to LO in the interval defined by signal I. Edges 1 and 3 have no effect, since the flip-flop can only be clocked when I is HI (edge 2). A JK-type flip-flop does not need any additional circuitry to hold its state when I is LO.

e) Proving that P switches from LO to HI in the interval defined by signal I. Edges 1 and 3 have no effect, since an active clock edge will cause the flip-flip to change states only if I is HI (edge 2). A D-type flip-flop needs feedback, since it would otherwise be reset by edge 3 when I is LO.

Device type	Setup time (ns)	Hold time (ns)	Minimum clock pulse width (ns)
74S74	3.0	2.0	7.3
74AS74	4.5	0	5.5
74F74	3.0	1.0	5.0
74AC74	3.5	0	4.0
74ACT74	4.5	0	5.0
74LVC74	3.0	2.0	4.0
74S112	7.0	0	7.5
74F112	5.0	0	5.0

Table 4.32 Flip-flop specifications.

Pulse detectors. A pulse has two edges and thus can be detected by a combination of two edge-detector circuits as shown in Figure 4.136. See Figure 4.137 for two examples. A problem with the circuit of Figure 4.137b is that it needs both the true and the inverted forms of the signal to be tested (*P*).

Inverting a signal without time offset. Various types of circuits require both inverted and non-inverted versions of certain signals (Figure 4.137b is only one example), with minimal time offset between the two versions. Figure 4.138 shows various ways to generate both true and inverted forms of signals.

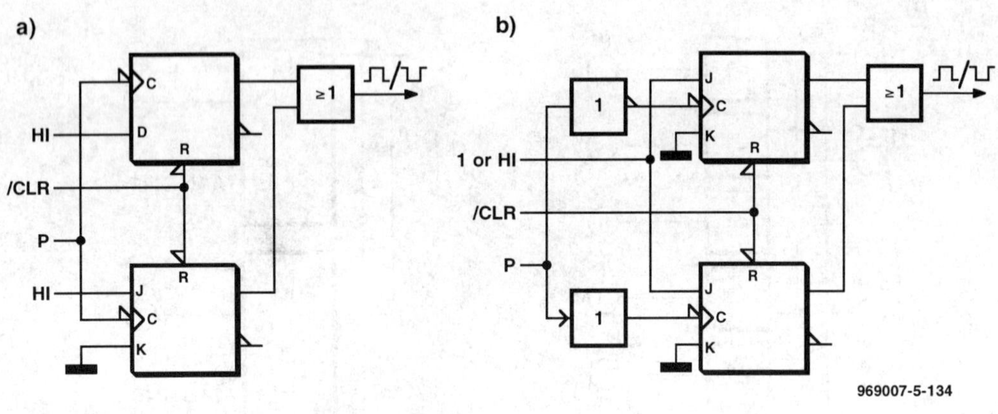

Figure 4.137 Pulse detector circuits.

a) A pulse detector using two different types of flip-flops.

b) A pulse detector using two flip-flops of the same type. The advantage of this is that both flip-flops are contained in the same IC package, so that the differences in their characteristics can practically be ignored.

Pulse-type detectors. The circuits of Figure 4.137 can only demonstrate that at least one pair of edges has occurred. The sequence analyzer circuits in Figure 4.139, by contrast, can distinguish between positive and negative pulses. A positive pulse is characterized by a rising edge followed by a falling edge, while a negative pulse is characterized by the opposite sequence of edges. The edge-detector flip-flops are connected in series according to the edge sequence to be detected.

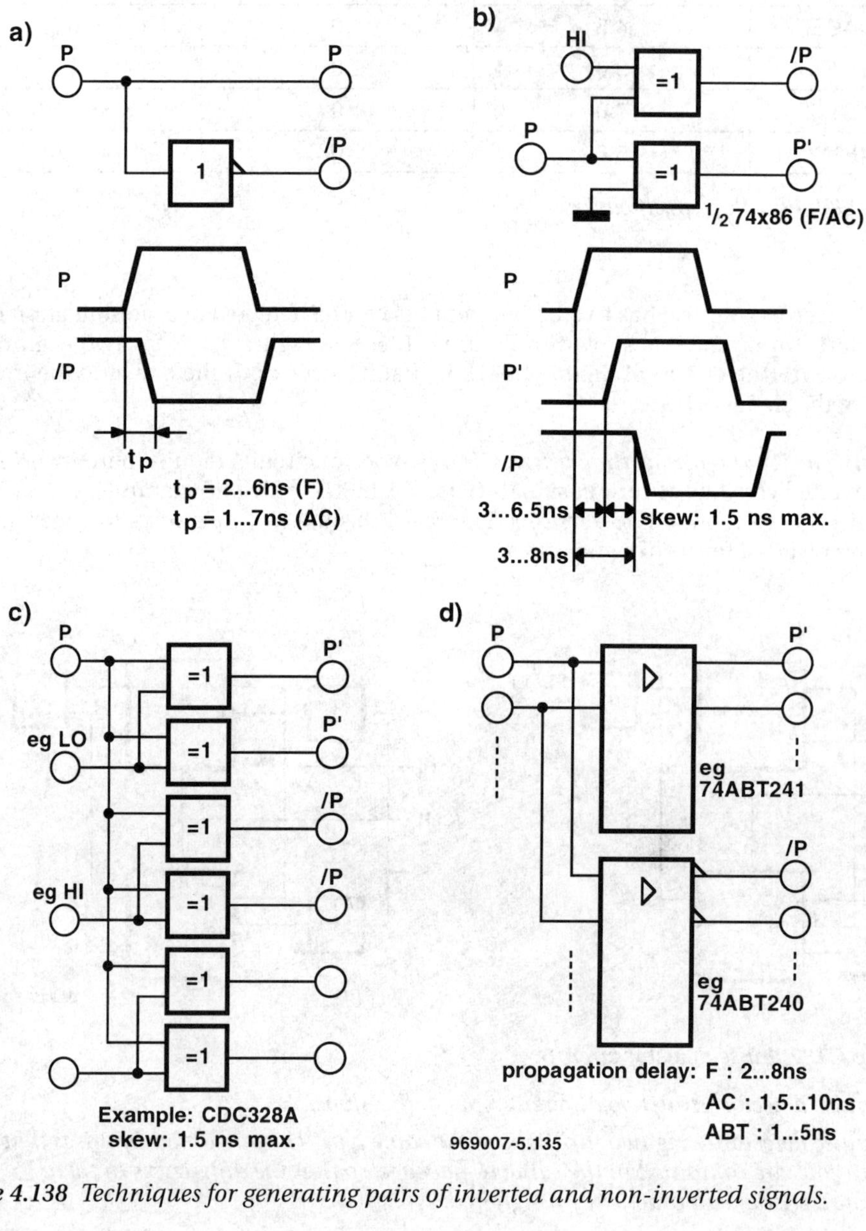

Figure 4.138 Techniques for generating pairs of inverted and non-inverted signals.

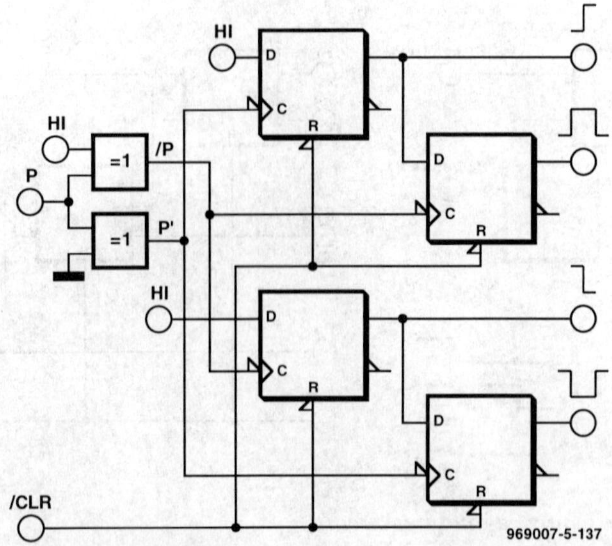

Figure 4.139 A pulse-type detector circuit.

Sequence analyzers. The objective of a sequence analyzer is to determine whether a set of signals *P1, P2, P3* ... switches in a specific sequence; the time intervals between the various transitions do not matter. Figure 4.140 shows several basic circuits.

Sequence analysis with qualifiers. Validity signals (qualifiers) can be incorporated into flip-flop chains similar to those shown in Figure 4.140b, as shown in Figure 4.141. The signals *C1, C2* and *C3* define time intervals during which transitions of the associated signals *P1,*

Explanation of Figure 4.138:

a) Simply connecting an inverter to the signal line does not meet the objective, since the inverted signal is delayed with respect to the non-inverted signal by the propagation delay t_p of the inverter (typically 2 to 7 ns).

b) A pair of XOR gates ensures approximately equal delays for both the true and the inverted signal. The time offset (skew) between the two signals is typically in the range of 1 to 3 ns (for 'F' or 'AC' devices with both gates in the same IC package). The inverted signal is delayed slightly more than the non-inverted signal.

c) There are special clock-driver ICs that are specifically designed to produce minimum skew. The Texas Instruments type CDC328A is illustrated here. The input clock signal (in this casc the signal to be tested, P) is available at six outputs that can be pairwise configured as inverting or non-inverting by means of XOR gates. The skew between any two outputs does not exceed 1.5 ns.

d) If many signals are involved, an economical solution that still yields a tolerable amount of skew (roughly 4 ns with ABT-series devices) is to use inverting and non-inverting versions of bus-driver ICs. See Volume 3 for sample circuits.

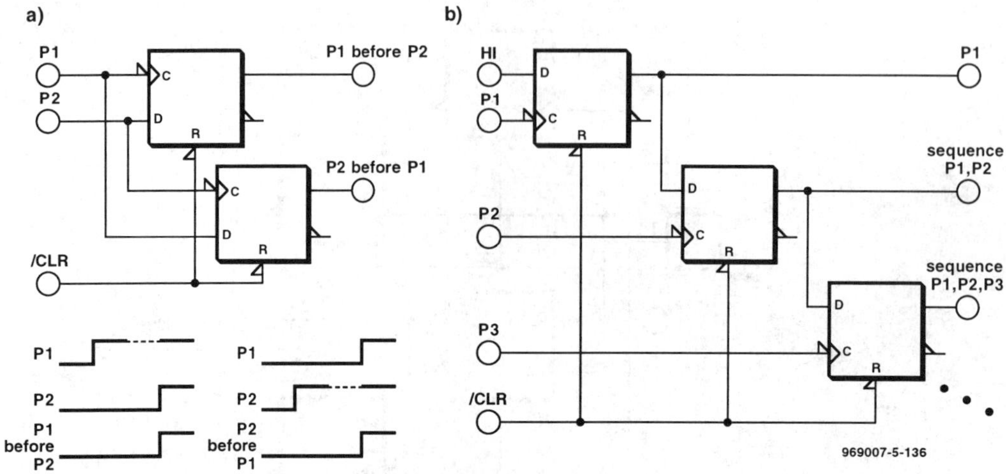

Figure 4.140 *Basic sequence-analyzer circuits.*

a) *This circuit exploits the time-dependent relationship of the data and clock inputs of a flip-flop. If the signal P1 is applied to the data input and the signal P2 is applied to the clock input, the flip-flop will only be set by the edge of P2 if P1 has been HI for at least the duration of the setup time. This arrangement can thus be used to verify the sequence 'first P1 and then P2'. Two such pairs of flip-flops, suitably connected, can check for both of the two sequences that are possible with two signals (P1 → P2 and P2 → P1).*
Notes: (1) it may be necessary to use additional feedback logic (see Figure 4.136e);
(2) JK flip-flops can also be used in a similar manner.

b) *Sequence detection with a chain of flip-flops. Each successive flip-flop can only be set if the previous one has already been set. If the third flip-flop is set, this means that a sequence of LO-to-HI transitions has occurred in the order P1 → P2 → P3.*

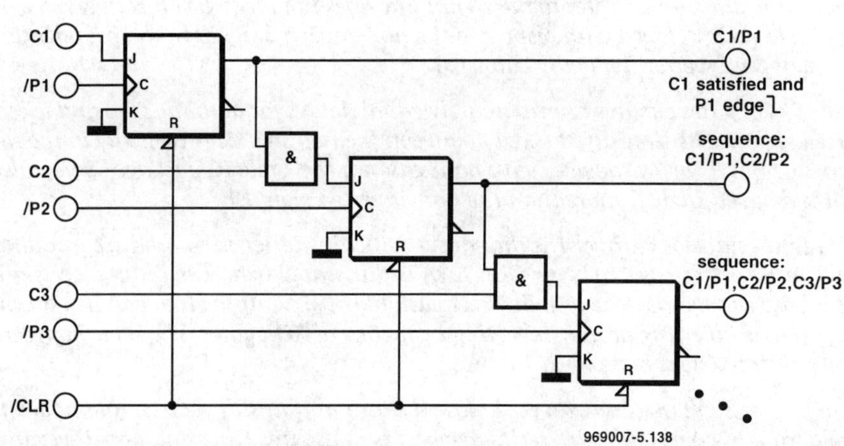

Figure 4.141 *A sequence analyzer circuit with qualifiers.*

P2 and *P3* are valid. The signals *C1*, *C2* and *C3* can also be products of combinational-logic circuits, such as comparator output signals and the like (refer to Figure 4.151).

For example (see also Figure 4.128): suppose that *C1*, *C2* and *C3* come from three address comparators and */P1*, */P2* and */P3* are tied together and connected to the ALE line. It is then possible to verify whether or not a series of memory accesses has occurred in the sequence address 1 → address 2 → address 3. If */P1*, */P2* and */P3* are separately connected to ALE, WR, RD and/or PSEN according to a particular test objective, it is possible to check for sequences such as 'instruction read from address 1, then data written to address 2' and so on (corresponding to multi-level triggering in a logic analyzer).

Additional examples of sequence analyzer circuits

1 Resource utilization on the ISA bus. The circuit shown in Figure 4.142a can indicate whether an interrupt request line has been used. In a similar manner, the circuit of Figure 4.142b indicates the usage of the DRQ/DACK line pair. A suitable test aid based on such circuits consists of a plug-in card with the appropriate circuit for each IRQ line and each pair of DRQ/DACK lines. Insert the card and allow the PC to run for a while, and then check to see which LEDs are lit. You can draw additional conclusions from the states of the individual LEDs. For instance, with reference to Figure 4.142a: (1) if only LED *A* is on, an interrupt request has occurred that was not acknowledged; (2) if both LEDs *A* and *B* are on, an interrupt request has occurred and has been acknowledged.

2 Handshaking signals on the parallel interface. Modern parallel interfaces (which comply with the IEEE 1284 standard) can be used in various modes that employ different types of handshaking (standard mode, nibble mode, byte mode, EPP mode and ECP mode). To determine the mode in which the interface is currently operating, you could connect a logic analyzer to it. However, a suitably-designed sequence analyzer circuit can also yield the desired information. A circuit such as that shown in Figure 4.142c, for instance, can verify that write accesses have been executed in the ECP mode.

Note: see Volume 3 for details of practical circuits.

Pulse length evaluation and pulse discriminators. Suppose that you have to determine whether a given pulse is shorter or longer than a certain predefined interval. In this regard, very short intervals are particularly significant (for distinguishing between glitches and 'real' signals). Such intervals cannot be defined by means of clocked circuit elements, since the narrowest pulse that can be detected by such an element must be longer than one clock period. There are various types of delay lines available that are suitable for defining the time intervals in question (see Table 4.33). *Note:* a delay line is not the same thing as a pulse generator (as described in Section 3.2.4). A delay line does not generate a pulse, but simply passes whatever binary signal is applied to its input through to its output with a delay of Δt, without otherwise altering the signal. Figure 4.143 shows a number of basic circuit designs for pulse length evaluators and pulse discriminators.

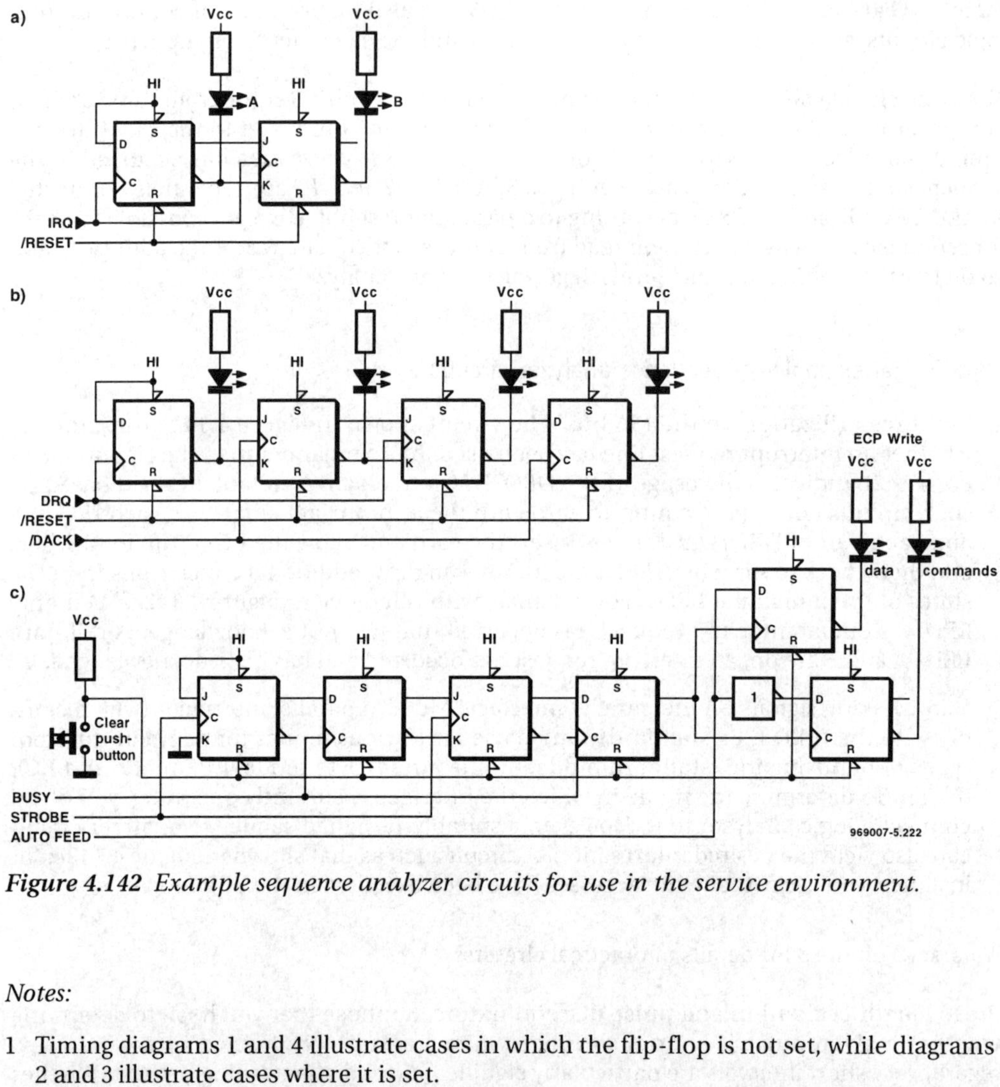

Figure 4.142 Example sequence analyzer circuits for use in the service environment.

Notes:

1 Timing diagrams 1 and 4 illustrate cases in which the flip-flop is not set, while diagrams 2 and 3 illustrate cases where it is set.

2 The behaviour of any of these circuits is unpredictable if the pulse width is nearly the same as the delay time (length of $P \approx \Delta t$) – whether the flip-flop switches then depends on component tolerances and the timing characteristics of the flip-flop in question. In such cases, you can get a more accurate idea of the pulse length by using two detector circuits together, one for pulses longer than Δt and the other for pulses shorter than Δt. If both of these give positive results at the same time, this indicates that the length of the pulse in question is equal to Δt, within a tolerance of a few nanoseconds. Of course, neither one of the flip-flops may switch in such a situation, so that you must also verify that a pulse is present in the first place (see Figure 4.144).

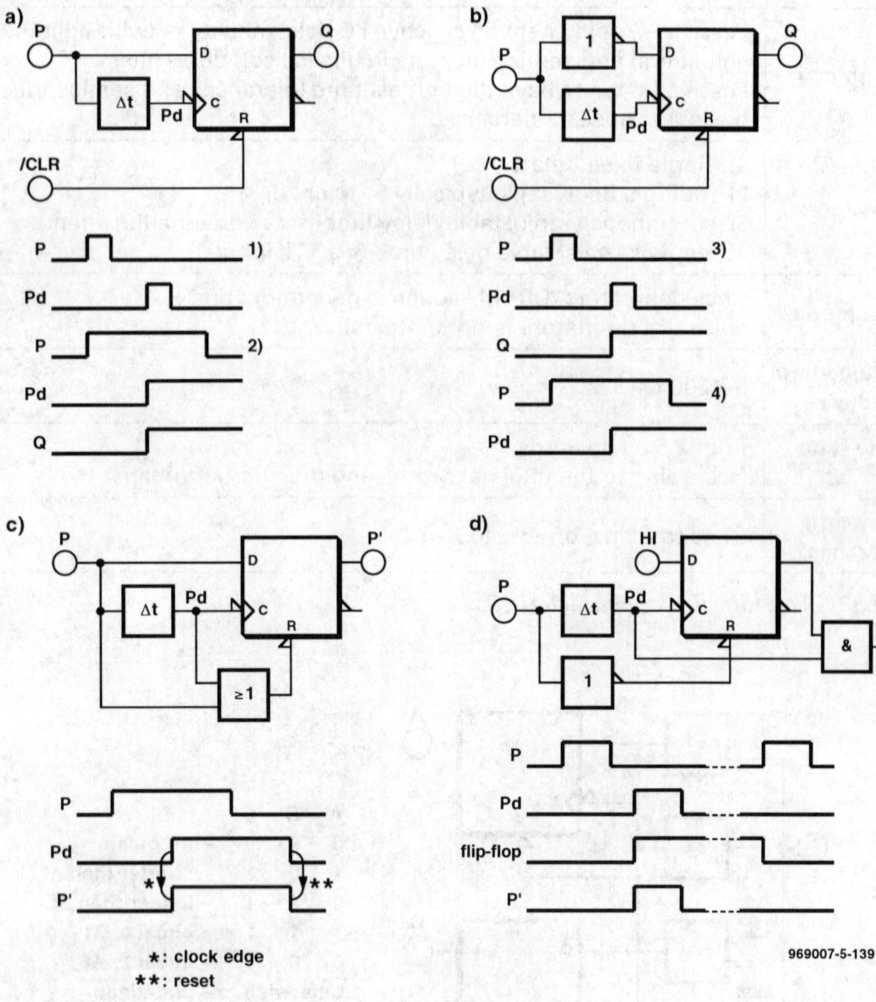

Figure 4.143 *Basic time-interval evaluation circuits.*

a) *Verifying that a pulse is longer than Δt. The flip-flop will be set if the pulse applied to the data input is still active (HI) when the delayed pulse arrives at the clock input.*

b) *Verifying that a pulse is shorter than Δt. The flip-flop will be set if the pulse applied to the data input is no longer active when the delayed pulse arrives at the clock input.*

c) *A pulse discriminator for 'longer than Δt'. Only pulses that are longer than Δt are passed through to the output, delayed by the interval Δt.*

d) *A pulse discriminator for 'shorter than Δt'. Only pulses that are shorter than Δt are passed through to the output, delayed by Δt. Note: this circuit also shows how the reset function of the flip-flop can be utilized in pulse evaluation. The reset input takes precedence over the clock input (if the reset input is active, the state of the flip-flop cannot be changed by an active clock edge). If the pulse P is longer than the time delay Δt, the flip-flop is held reset and thus cannot be set by the delayed rising edge of P.*

Technology	Passive LC delay networks, active LC delay networks (with built-in amplifiers) and semiconductor circuits (silicon delay lines). Passive LC types have the tightest time tolerances and semiconductor types the loosest tolerances.
Versions	1) single fixed delay, 2) multiple delay taps (typically 5, 10 or 20), 3) continuously-adjustable delay time (screwdriver adjustment), 4) stepwise selectable delay time (via TTL inputs).
Special models	clock generators, DRAM sequence generators and pulse discriminators (among others)
Total delay time per device	5–1200 ns
Delay time per tap	from 2.5 ns upwards (according to the total delay time and the number of taps)
Delay time tolerance	±1 ns to ±2 ns, or ±5% to ±20%

Table 4.33 Delay-line device characteristics.

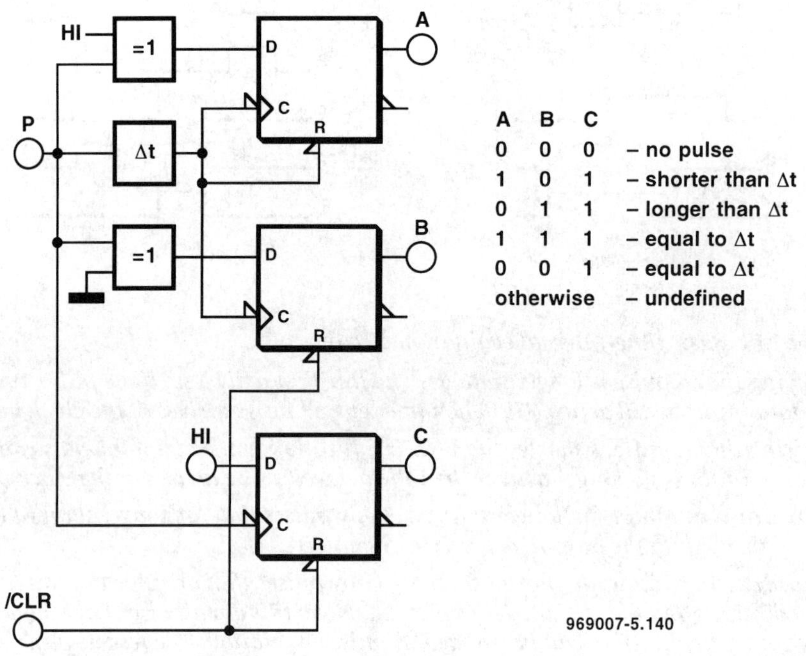

Figure 4.144 A pulse width detector.

3 It may be necessary to provide additional feedback logic for the data input of a D-type flip-flop to prevent it from being switched again by subsequent pulses.

4 Ready-made pulse discriminators (which work in the same way as the circuits shown in Figure 4.143c/d) have significantly better specifications than 'discrete' versions made from several ICs, since the entire circuit of the ready-made device is housed in a single IC package.

5 See Volume 3 for practical circuits.

Clock-driven (synchronous) pulse discriminators. Time intervals can also be measured by counting. With the present state of the technology, it is best to design a synchronous pulse discriminator as a state machine (see Figure 4.145) and to implement it using a programmable IC (since the costs can otherwise be quite considerable). Clock frequencies in excess of 100 MHz are quite manageable (with corresponding time resolutions on the order of 10 ns, since a typical digitizing error of one clock period must be taken into account). If the pulses in question are relatively long (in the millisecond range, which can be the case if you are investigating mains disturbances, for example) then a state diagram design as shown in Figure 4.145 can even be implemented in microcontroller software.

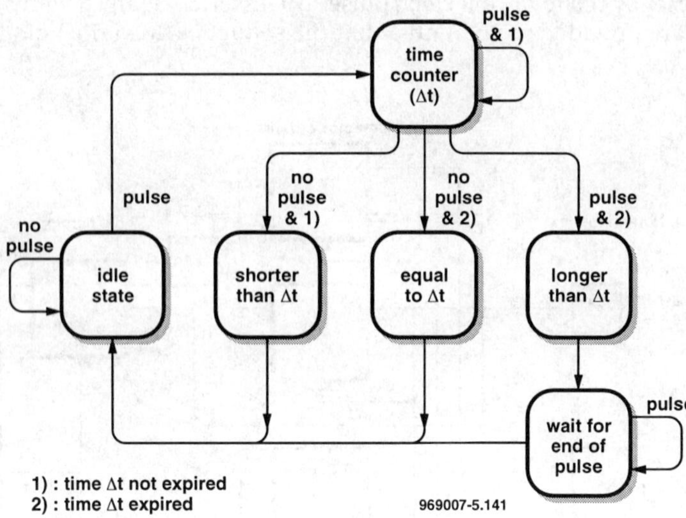

Figure 4.145 State diagram of a clock-driven pulse discriminator.

When a pulse occurs, it activates a counter that counts out the time delay Δt. The state machine has three states that can be used to classify the pulse ('shorter than Δt', 'equal to Δt' and 'longer than Δt'), each of which lasts for one clock interval. Each of these states could for example initiate a suitable logging process.

Glitch detection. There are two different definitions of what constitutes a glitch:

1 The asynchronous definition: a glitch is any pulse that is shorter than a certain predefined interval (which is itself already short). (This is how a digital oscilloscope sees glitches; see Section 4.3.) A 'shorter-than-Δt' pulse width evaluator circuit (refer to Figure 4.143) is suitable for detecting such glitches.

2 The synchronous definition: a glitch is any change in a given signal that produces at least two edges within one period of a reference clock. (This is how a logic analyzer sees glitches; see Section 4.4.2.) Figure 4.146 shows a synchronous glitch detector circuit that utilizes impulse detection circuits as shown in Figure 4.137. A difficulty with synchronous glitch detection is that the signal *P* must be monitored for an entire period of the reference clock *CLK* to see if it exhibits at least two edges, but the reference periods follow each other without any interruption. The solution is to use two detector circuits, which are alternately enabled for successive reference periods. The upper circuit could thus be active for the first period, the second circuit active for the second period and so on. Each time that a changeover occurs, the output level of the active detector is saved in a 'hold' flip-flop, following which the detector in question is reset.

Note: in the circuit of Figure 4.146, the output value of the currently active detector circuit is clocked into the hold flip-flop on the trailing edge of its enable signal, and the detector circuit is then reset by the following clock pulse. The inverters marked with an asterisk (*) in Figure 4.146 are provided to guarantee that the sequence 'load the hold flip-flop and

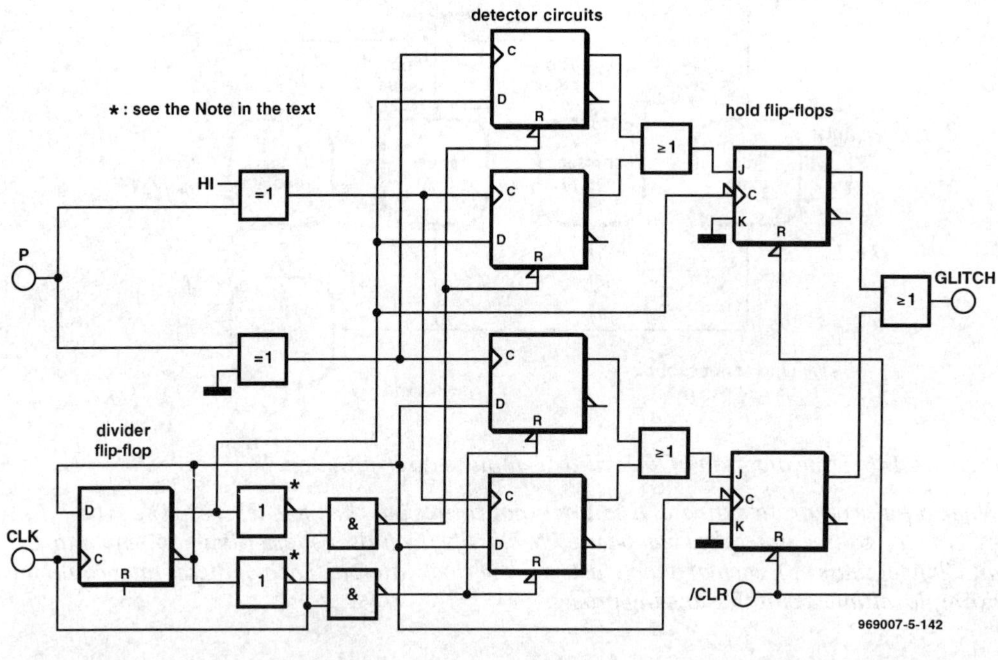

Figure 4.146 Synchronous glitch detection.

then reset the detector' is maintained under all foreseeable circumstances. (The reset signal is not taken from the complimentary output of the divider flip-flop, since one output can change before the other one.)

Event counting. The detection circuits that have been presented so far do not allow you to determine how many times the event in question has occurred. Obviously, it is possible to count the number of times that an event occurs. The counter used must be sufficiently fast, since events can occur in rapid succession, and it must not be possible for the counter to reach its maximum count and 'roll over' without this being noticed. There are three basic way to deal with counter rollover: (1) counter rollover can be registered by a supplementary circuit, (2) the circuit can be designed such that the counter cannot roll over and (3) the counter can be made so long that it will never reach its maximum count within the time intervals that occur in actual testing. Detector circuits can be readily designed to count events (that is, to act as the input stages of counters), as shown in Figure 4.147. Such stages (which function as 2:1 dividers) can be used to drive the inputs of standard counter circuits (or normal universal counters working in the event count mode). See Volume 3 for more information on counters that are very long and also fast.

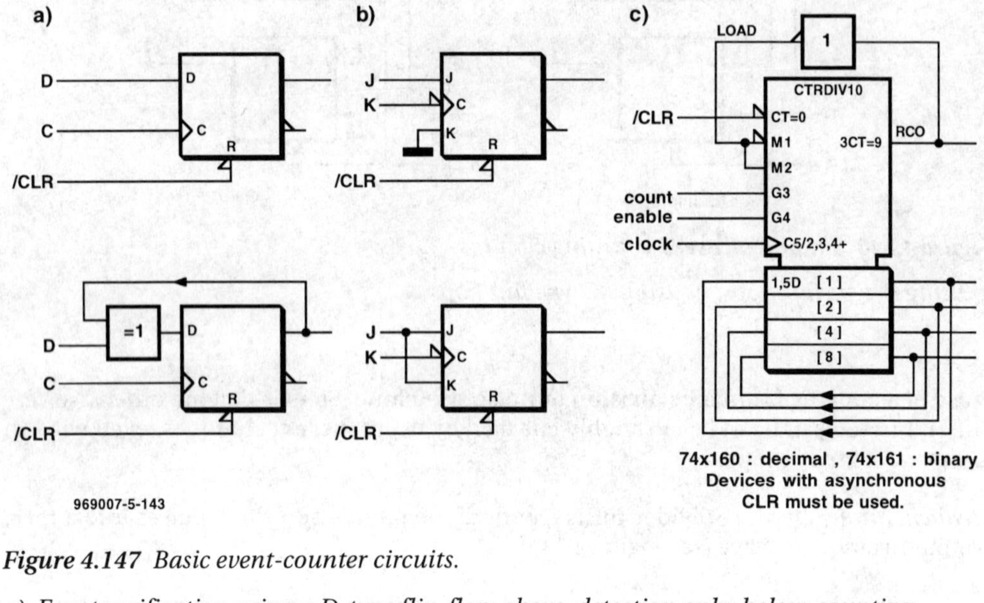

Figure 4.147 Basic event-counter circuits.

a) *Event verification using a D-type flip-flop: above, detection only; below, counting. The output of the flip-flop is inverted and fed back to its input via an XOR gate.*

b) *Event verification using a JK-type flip-flop: above, detection only; below, counting.*

c) *Counting with a counter IC. This circuit shows how rollover can be prevented. When the terminal count is reached, RCO becomes active and in turn activates the parallel-load input. Since loading takes precedence over counting, the terminal count is repeatedly reloaded 'on top of itself'. Note: several counter ICs can be connected in series (cascaded). See the relevant application notes for more information.*

Simplification: practical experience with fault-finding shows it is usually fully adequate to be able to distinguish the following three cases: (1) an event has occurred only one time, (2) an event has occurred several times and (3) an event recurs continuously. A quite simple series connection of two detector flip-flops is all that is needed to be able to make this distinction (see Figure 4.148). The signals *D*, *J* and *C* in Figure 4.148 correspond to the flip-flop input signals of the previously-described basic detector circuits (Figures 4.133 and following). If only the first flip-flop is set, then the event has occurred only one time. If the second flip-flop is also set, then the event has occurred two or more times. If the reset signal (*/CLR*) is now briefly activated and both flip-flops are again found to be set shortly after the reset signal goes inactive, this indicates that the event in question recurs continuously.

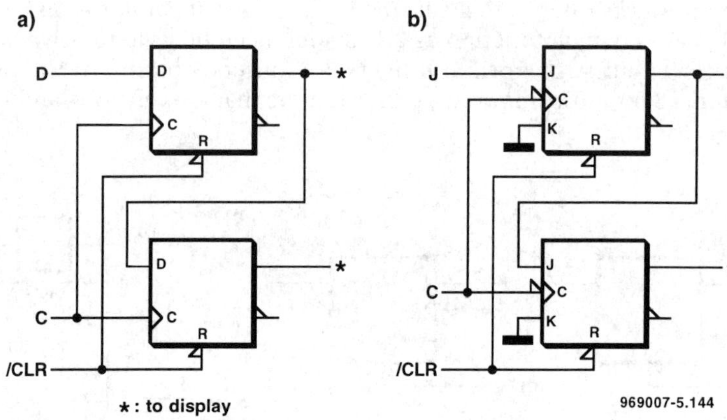

Figure 4.148 *Simplified event-counting circuits.*
a) using D-type flip-flops, b) using JK-type flip-flops

Pulse detection by frequency division or pulse stretching. Signals that are otherwise very difficult to see can be at least roughly checked by using these techniques, even with an inexpensive oscilloscope.

Frequency division: a 2:1 divider turns a periodic sequence of narrow pulses into a nice, symmetrical square wave (see Figure 4.149).

Measurement examples:

1 The refresh interval in a PC. Here you are dealing with pulses that recur roughly every 15 µs and are only several hundred nanoseconds wide. If the refresh pulses are connected to a 2:1 divider, the length of the refresh interval can be directly measured on the screen of an oscilloscope.

2 A 100-MHz clock signal. Division by 2 yields 50 MHz, division by 4 yields 25 MHz and so on. Suitable test aids are described in Volume 3.

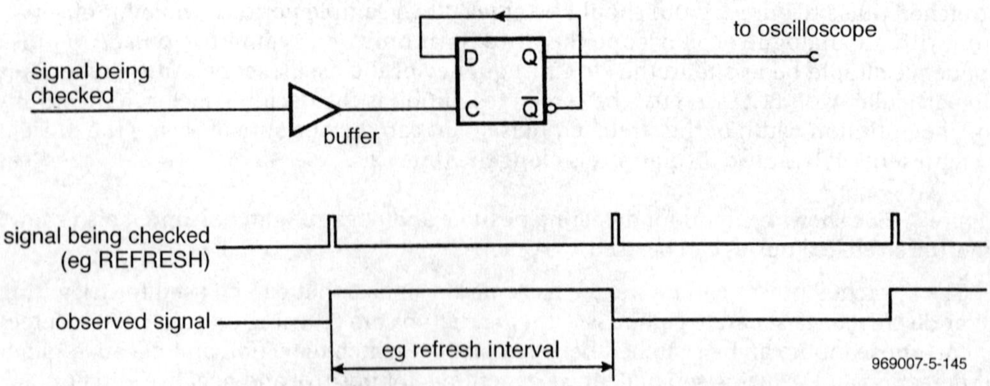

Figure 4.149 Using a flip-flop divider as a test aid.

Pulse stretching. A 2:1 division produces pulses with widths that correspond to the time between successive input pulses (the pulse repetition interval).This technique allows periodically repetitive pulse trains to be readily observed (as in the previous examples). Using a 2:1 divider for noise pulses (glitches), on the other hand, will in most cases produce a waveform that is so irregular that nothing useful can be seen on the oscilloscope screen. The remedy is to use a pulse discriminator to detect glitches and stretch the output pulses until they can be evaluated using an oscilloscope or logic analyzer (see Figure 4.150). The stretched pulses can either be viewed directly or be used as trigger signals.

The basic circuit consists of a pulse discriminator followed by a monostable pulse generator (Figure 4.150a). For sampling (with a logic analyzer or DSO), the width t_w of the

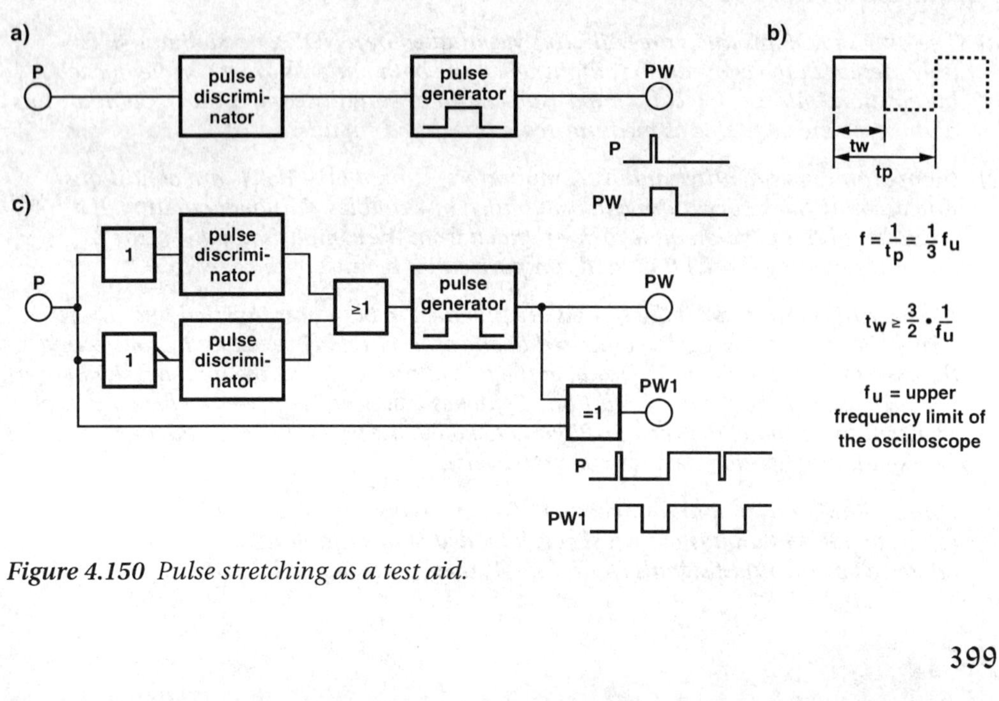

Figure 4.150 Pulse stretching as a test aid.

stretched pulse (Figure 4.150b) should be at least two sample periods, while for observation with an analogue oscilloscope the third harmonic of a symmetric pulse/no pulse sequence should be less than the cutoff frequency of the oscilloscope. *Note:* sometimes (in particular, with LCD screens) the screen resolution is the deciding factor in determining the optimum width of the stretched pulse (you can only be sure of seeing the pulse if t_w represents at least two display pixels along the time axis.

Figure 4.150c shows a circuit for detecting positive and negative glitches, and it also shows how the stretched pulses can be used:

1 The stretched pulses can be treated as separate signals that can be used for triggering or displayed via separate channels. Other variations are (1) using only one glitch detector whose input can be switched between positive-glitch detection and negative-glitch detection and (2) using separate pulse generators for positive and negative glitches.

2 The stretched pulses can be merged back into the original waveform (via an XOR gate); the resulting signal inversions can be readily seen on the screen. Other methods for optically emphasizing the stretched pulses (for oscilloscope display) are (1) intensity modulation (if a Z-axis input is available) and (2) amplitude modulation (displaying the stretched pulses with signal levels that are clearly distinguishable from the logic levels). See Volume 3 for more information.

Signal combinations (bit-pattern comparisons). A task that frequently must be performed in fault-finding is to investigate a group of signals to see whether they match a predefined bit pattern or satisfy a particular boolean equation. Figure 4.151 shows basic designs for comparator circuits.

Explanation of Figure 4.151:

a) *The bitwise XOR products are collectively evaluated by a NOR gate, so that a '1' is only present at the comparator output when the both patterns are the same in each bit position (this is a 'logical' comparison circuit or equality comparator). Comparators of this type are available as medium-scale integrated circuits.*

b) *Comparator ICs (identity/equality comparators) cannot take 'don't care' conditions into account. Such conditions can be handled by suitable additional circuitry. If a given signal (bit position a_i) is to be excluded from the comparison ('don't care'), the corresponding mask bit m_i and comparison bit b_i must both be set to '0'.*

c) *A universal comparator. If the signals are provided in both non-inverted and inverted forms, they can effectively be compared to one of three reference values by connecting the associated input of an AND gate to the non-inverted signal, the inverted signal or a fixed HI value (a 'don't-care' condition). Switches (using either mechanical contacts or semiconductor switches such as CBT or QuickSwitch devices) have the advantage of very small propagation delays (typically < 1 ns).*

d) *Using a RAM as a comparator. The signals to be compared are applied to the address inputs of a RAM that has previously been loaded with suitable data. If the location addressed by the input signals contains a '1', then the comparison condition is satisfied.*

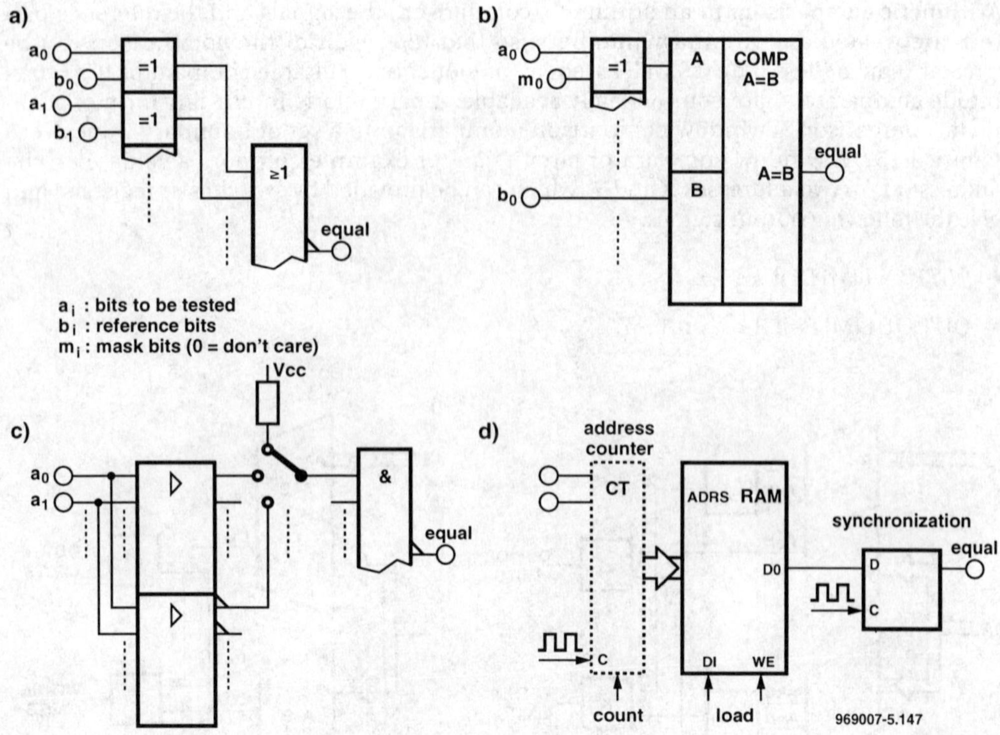

Figure 4.151 Basic comparator circuits.

The advantage of this approach is its unlimited flexibility: a given set of signals can be verified against an arbitrary boolean equation without requiring any additional circuitry to be built. The disadvantages are (1) low speed: RAM cycle times are at least an order of magnitude longer than gate propagation delays, (2) some extra circuitry is required to load the RAM and (3) the cost of the RAM itself. With the current level of technology (using fast static RAMs with 15 to 20 ns cycle times) it is possible to compare at most 20 to 24 signals.

Notes:

1 *Access to the RAM address lines is required in order to load the RAM. There are two approaches: either (1) dual-port RAMs can be used (these are relatively slow, expensive, and have relatively small capacities) or (2) an address counter can be connected in series with the signal lines (and used as a simple register in comparison mode).*

2 *Clocking: the RAM output signals must be evaluated with respect to a clock signal. Glitches can occur on the output lines every time the address changes, due to switching processes in the address decoder and the memory matrix.*

3 *See Volume 3 for practical circuits.*

Arithmetic comparison. In an arithmetic comparison, the signals and the reference pattern are treated as two binary numbers, so that the result of the comparison can be 'greater than' or 'less than' as well as 'equal' or 'not equal'. Suitable comparator ICs (magnitude comparators) are commercially available. A particularly interesting form of arithmetic comparison is window comparison (comparison to a set of boundary values); see Figure 4.152. A window comparator circuit can for example compare a value A on the address bus to two addresses X and Y (which can be provided by switches or registers) and give the following output signals:

- WITHIN LIMITS if $X \le A \le Y$,
- OUT OF LIMITS if $A < X$ or $A > Y$.

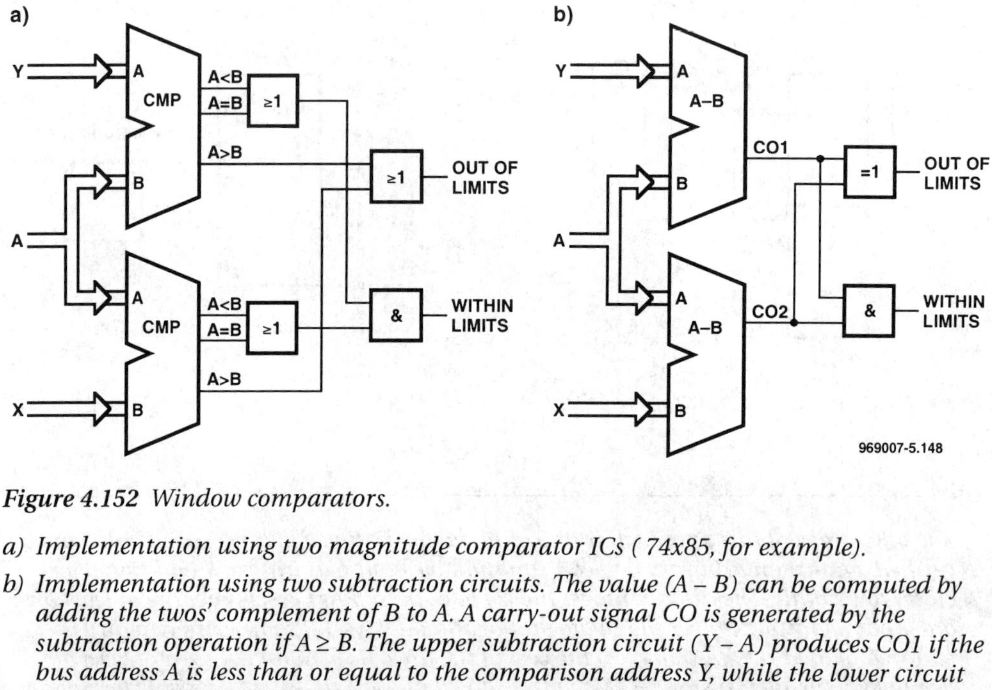

Figure 4.152 Window comparators.

a) Implementation using two magnitude comparator ICs (74x85, for example).

b) Implementation using two subtraction circuits. The value (A – B) can be computed by adding the twos' complement of B to A. A carry-out signal CO is generated by the subtraction operation if A ≥ B. The upper subtraction circuit (Y – A) produces CO1 if the bus address A is less than or equal to the comparison address Y, while the lower circuit (A – X) produces CO2 if the bus address A is greater than or equal to the comparison address X (equivalent to X ≤ A). The condition WITHIN LIMITS thus results if both carry-out signals are active (CO1 & CO2) and the condition OUT OF LIMITS results if only one of the two carry-out signals is active (CO1 ⊕ CO2).

Logging signal combinations (event analysis). Occasionally it is important to know which of a number of possible combinations of signal values have actually occurred in the circuit under test. *For example:* suppose that you want to check which I/O address ranges are used in a given PC (perhaps in order to determine which regions are free, if additional expansion cards have to be configured). An obvious approach is to observe all I/O activities over a sufficiently long period of time (for example, starting with when the PC is switched on, continuing while each available application is started up in turn, and ending after Windows has been shut down). The problem is that some addresses will be accessed literally millions of times, while others will only be addressed a few times. Trying to capture the addresses with a logic analyzer will not help much, since its acquisition memory will be repeatedly swamped by the large number of accesses to certain locations (such as the video adapter). The solution is shown in Figure 4.153: connect the signals in question to the address inputs of a RAM. The RAM is cleared before the test is started, and during the test a '1' is written to each memory location that is addressed by the applied signals. On completion of the test, read out the content of the RAM. The locations that contain '1's directly indicate which combinations of signal values (addresses) have actually occurred.

For practical circuits see Volume 3.

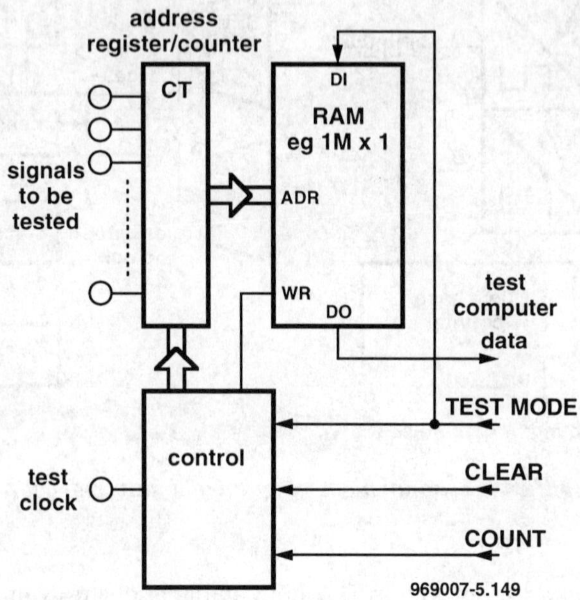

Figure 4.153 Basic event-logging circuit.

An address register that can also be operated as a counter is connected ahead of the RAM address inputs. The entire arrangement is controlled by a test computer. To prepare for a test, the test computer first makes the signal TEST MODE inactive, which causes a '0' to be applied to the RAM data input and the register to be configured as a counter. The RAM can then be cleared by a series of write accesses, starting at address 0. If the TEST MODE signal is then made active, the RAM is addressed by the signals to be checked. Each test clock triggers a write cycle, so that every applied set of signal values is marked by a '1' in the addressed location.

4.4.5 Counting

Frequency counters and universal counters. These are the most commonly used types of counting equipment. The simplest types of frequency counter can only measure the frequency of a single input signal. Universal counters have at least two input channels (A and B) and can measure frequency, period, frequency ratio and so on. *Note:* there is no standard distinction between a *frequency counter* and a *universal counter*. An instrument with a particular set of functions may be called a frequency counter by one manufacturer and a universal counter by a different manufacturer. When comparing various instruments, you must always examine the range of functions that each one offers (as specified on its data sheet). Figure 4.154 shows the functional block diagram of a typical universal counter.

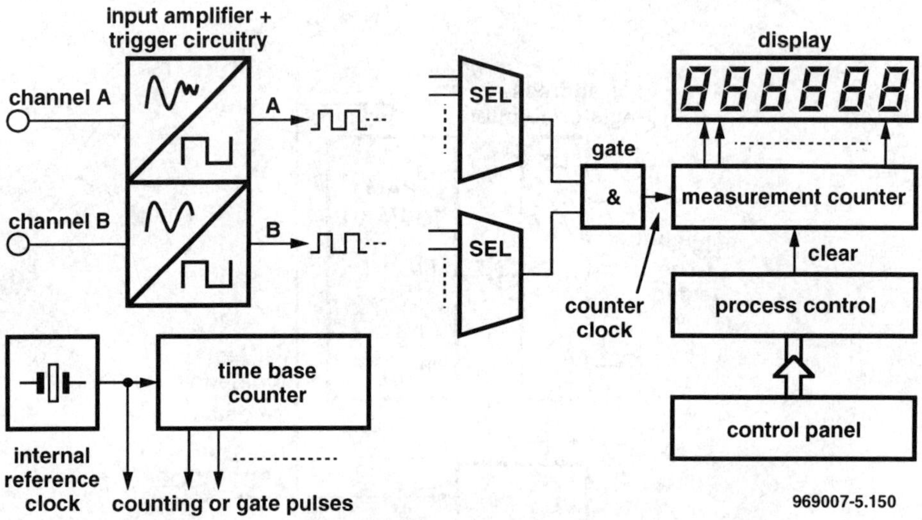

Figure 4.154 *Simplified functional block diagram of a universal counter.*

Input amplifiers and trigger circuits. The input amplifier of a truly universal instrument is practically the same as the input amplifier of an oscilloscope, with regard to its construction and use. The specifications are also similar (*for example:* input impedance 1 MΩ || 15 pF, cutoff frequency 300 MHz, characteristic rise time 1.5 ns). The trigger circuits convert the analogue input signals into rectangular pulse waveforms.

Internal reference clock and time-base counter. The internal reference clock is a crystal-controlled oscillator that provides a discrete time base for all measurements. The time-base counter provides either subdivided clock signals or gate signals, depending on the type of measurement being made. *For example:* the time-base counter can divide a 10 MHz reference clock by 10, 100, 1000 and so on, or it can generate gate pulses lasting 10 ms, 100 ms, 1 s and so on.

Measurement counter and display. The measurement counter accumulates the incoming pulses. Its content is displayed in decimal form.

Gate. The gate delivers count pulses to the measurement counter. In works on the principle of an AND gate, in which the incoming pulses from the signal to be measured are ANDed with a gate signal. The gate thus enables the actual counting process (see Figure 4.155). The various types of measurements that the instrument can make are essentially distinguished by the sources of the count pulses and gate signal in each case. This is indicated in Figure 4.154 by the selection logic ahead of the gate logic; the actual signal paths for the various types of measurements are described below.

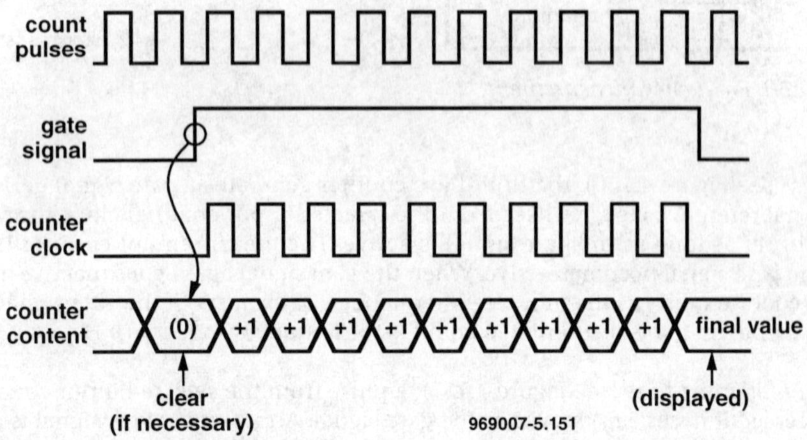

Figure 4.155 Gate circuit operation.

Process control logic and control panel. These elements select the sources of the count pulses and gate signal for each type of measurement and control the counting and display processes. Modern instruments employ a microcontroller or microprocessor for these functions. If such a processor is present, it can also be used to perform computations and thus to provide a variety of additional measurement functions. (In such instruments the display is controlled by the processor instead of being directly connected to the measurement counter.)

Measurement functions. There are only two basic types of measurement: (1) frequency and (2) pulse duration or time interval. The large variety of available measurement functions results from the fact that various types of signals can be selectively subjected to frequency or time-duration measurement. (Some instruments – referred to as multifunction counters – also include functions that in principle have nothing to do with counting in the conventional sense, but which are quite useful in regard to the signals to be measured.)

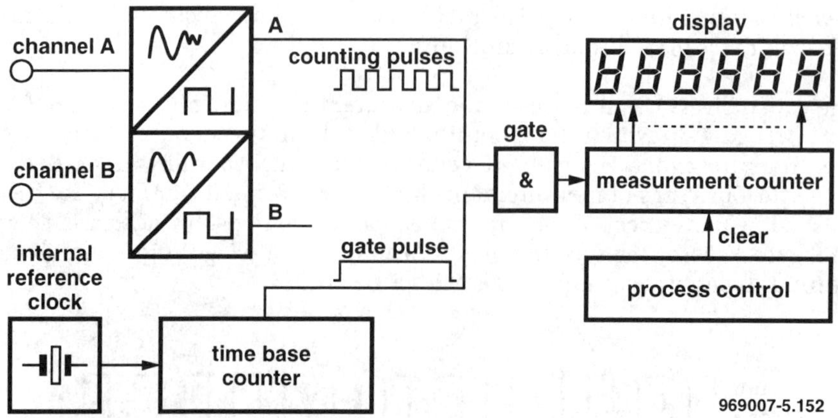

Figure 4.156 Frequency measurement.

Frequency (see Figure 4.156): the time-base counter generates a gate signal derived from the internal reference clock. Pulses from the selected input channel clock the measurement counter as long as the gate signal is active. The measurement counter is cleared before the gate signal becomes active. When the gate signal again goes inactive, the accumulated count represents the measured frequency. *For example:* if the gate signal is active for one second, then the accumulated count is the frequency in hertz (n Hz = n pulses/s).

Duration or elapsed time (see Figure 4.157): a pulse from the selected input channel (further processed if necessary) serves as the gate signal. An internal clock signal is accumulated by the measurement counter as long as the gate signal is active. This clock signal

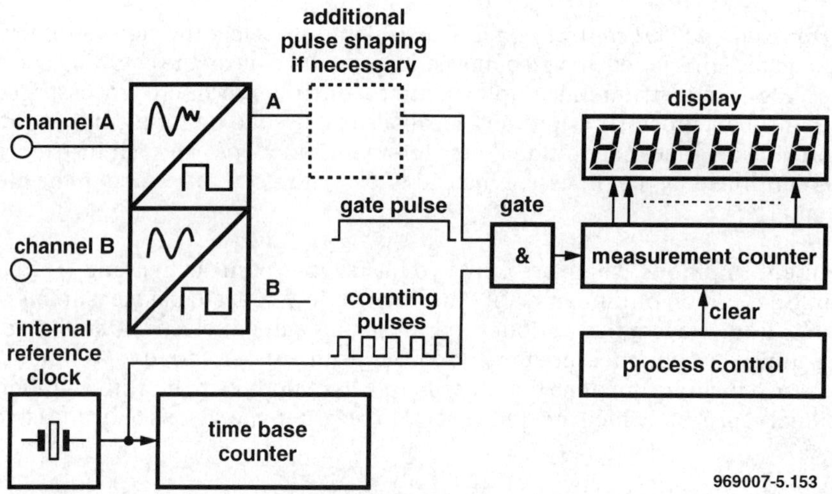

Figure 4.157 *Duration or elapsed-time measurement.*

comes either directly from the reference clock generator or from the time-base counter as a subdivided reference-clock signal. The measurement counter is cleared before the gate signal becomes active. When the gate signal again goes inactive, the accumulated count represents the elapsed time interval (the time is directly counted). *For example:* if 3244 pulses are counted at a reference clock frequency of 10 MHZ (100 ns period), then the measured duration is 324,400 ns = 324.4 µs.

Period. The signal applied to the selected input channel is converted into a pulse train in which the length of each pulse corresponds to one period of the input signal (this can for instance by done by using a 2:1 divider). The duration of these pulses is measured. With periodic signals, the measurement error can be reduced by counting the total duration of a certain number of pulses (for example, 1000).

Pulse width. Either the high or the low portion of the signal can be selected for measurement. With periodic signals the measurement error can be reduced by counting the total duration of a certain number of high or low intervals (for example, 1000).

Frequency ratio (see Figure 4.158): the ratio between two frequencies f_A and f_B (applied to the channel A and channel B inputs) can be measured by taking the counting pulses from channel A signal and the gate signal from a subdivided version of the channel B signal (the channel B signal is input to the time-base counter in place of the internal reference clock). *For example:* suppose that f_A = 1 kHz, f_B = 100 Hz and the time-base counter divider ratio is 1:10. The gate pulse is then 100 ms long, and in this interval 100 pulses from channel A will be counted. The frequency ratio is then 100 ÷ 10 = 10.

Note: the accuracy of the measurement increases with the number of pulses that are counted. The lower-frequency signal should thus be connected to channel B, since this will result in a longer gate time.

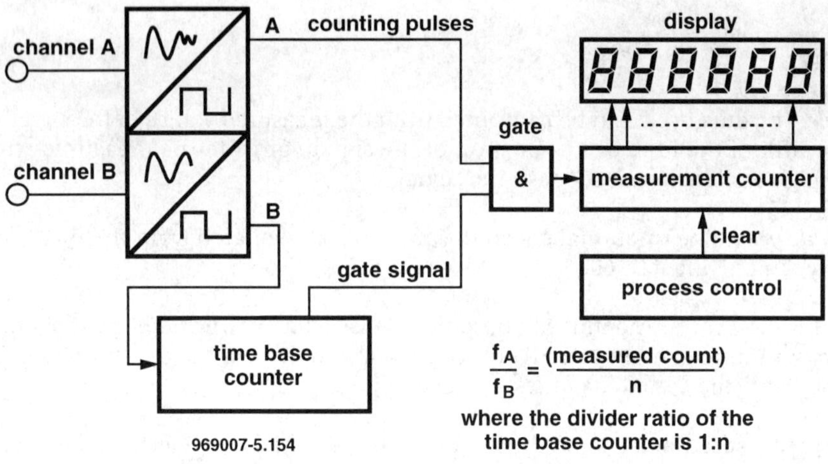

Figure 4.158 *Frequency ratio measurement.*

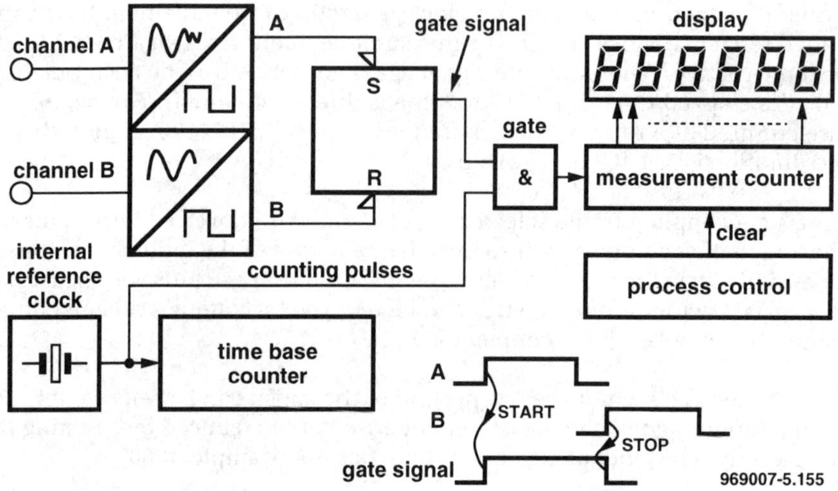

Figure 4.159 Time interval measurement.

Time interval (Figure 4.159): the start condition is applied to the Channel A input and the end condition is applied to the Channel B input (for example, the rising edge may be interpreted in each case as the start or stop condition). The time interval is measured in the same manner as a pulse period, by counting pulses from the reference clock generator or a subdivided clock from the time-base counter.

Phase angle. The phase angle of two signals that differ only by a phase offset can be calculated from the signal period t_p and the time interval Δt between two identical signal edges. One signal period represents 360°, so the phase angle φ can be calculated using the following formula:

$$\varphi = \frac{\Delta t}{t_p} \cdot 360°$$

Duty cycle. The duty cycle can be computed from the measured values of the signal period and the width of the high or low portion of the signal, or from the measured widths of both the high and the low portions of the signal.

Rotational speed. The rotational speed in rpm can be calculated from the measured frequency by multiplying it by 60.

Event counting. A running count can be kept of pulses that arrive at the input of one of the channels. Counting can be started and stopped manually (via the control panel) or by a signal applied to the second channel.

Reciprocal measurements. When the frequency is measured, the selected channel provided the pulses to be counted and the internal time reference provides the gate signal. When the period is measured, the situation is reversed. Since the measurement error becomes smaller as the number of pulses counted increases (if only because there is always a digitization error of ±1 count), relatively low frequencies are best determined by measuring their periods and relatively short periods are best determined by measuring their frequencies. In each case, the desired value can be calculated as the reciprocal of the measured value (modern instruments do this automatically).

Triggering. The job of the triggering circuitry is to derive a series of rectangular waveforms with identical periods and identical duty cycles from an arbitrary input waveform. Triggering is the weak point of any measuring instrument that is based on counting; you can measure no end of rubbish if the trigger settings are not good. The two basic types of measurement, frequency measurement and time measurement, place different demands on the triggering circuitry (see Figure 4.160).

Note: it is simply not possible to make any usable pulse width or period measurements with a universal counter if the signal waveform looks like that of Figure 4.160b/c. A remedy in this situation is to make an approximate measurement of the period using an oscilloscope (directly on the screen or using the cursors). Sophisticated counters may be able to give reasonable – and possibly more accurate – results with such signals by using averaging techniques, but it is still a good idea to check the signals with an oscilloscope to see if the measured values are plausible.

Hold-off setting. Some instruments have an adjustable hold-off setting, similar to that of an analogue oscilloscope. This can be used to prevent the occurrence of multiple triggers within a single period (see Figure 4.161).

Checking the trigger settings. The best method is to monitor the signal with an oscilloscope. You can then see if the signal waveform is problematic (for instance, if it is similar to the waveforms shown in Figures 4.160b/c and 4.161). In addition, you can approximately measure the frequency or pulse width and thus check the reliability of the values read from the counter. Some counters have connectors for attaching an oscilloscope. Some instruments can even display the signal waveform and the hysteresis band themselves (see Figure 4.162).

Measurement errors. There are essentially three error sources:

1 False triggering (see Figures 4.160 and 4.161 for the most important causes).

2 Counting or synchronization errors. All counting processes have a fundamental synchronization error of ±1 digit. As with other types of measurements, it is important to exploit the full measurement range for counter-based measurements (which means arranging for the measurement to be made with the longest possible gate time or the highest possible clock rate).

3 Time base errors. The accuracy of time interval measurements and gate times is determined by the internal crystal-controlled oscillator. There are two factors that affect this: (1) the ambient conditions (supply voltage, temperature and so on) and (2) aging. Table 4.34 summarizes the various types of reference oscillators that are available.

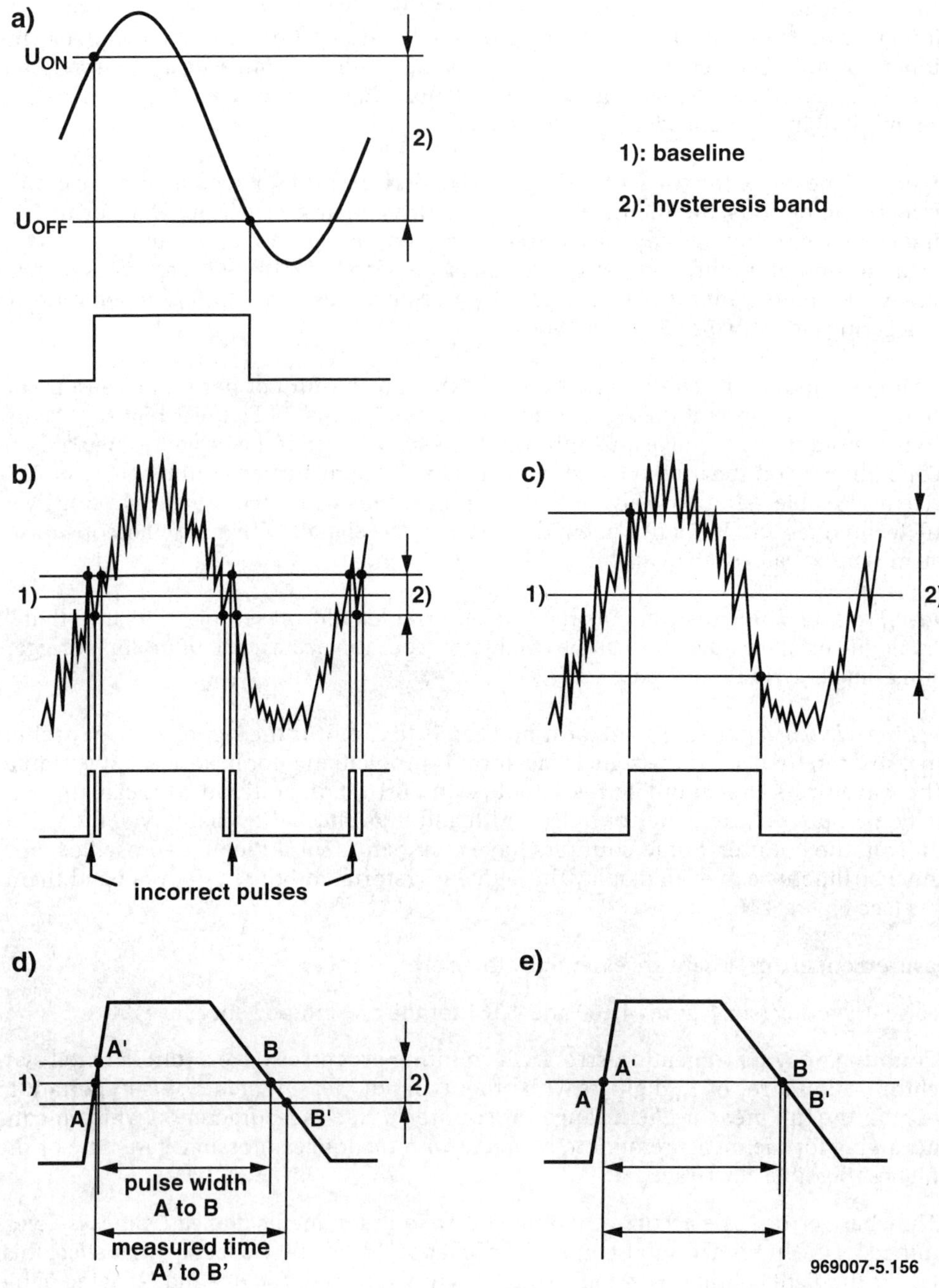

Figure 4.160 Triggering conditions for frequency and pulse-width measurements.

Explanation of Figure 4.160:

a) Hysteresis: he trigger signal switches from LO to HI when the analogue signal rises above the first threshold value $U_{ON,}$ and it switches from HI to LO when the analogue signal falls below the second threshold value (U_{OFF}. The region between U_{ON} and U_{OFF} is the hysteresis band ($U_{OFF} \leq U_{ON}$ is a fundamental requirement).

b) Frequency measurement: you need exactly one pulse for each period of the input signal. The width of this pulse and its exact location within the period do not matter, as long as there is only one pulse per period. This is not the case with the illustrated signal – it is so noisy that the trigger conditions are met more than once per cycle, so that more than one pulse is counted for each period.

c) For frequency measurements, increase the hysteresis band (in other words, reduce the sensitivity of the triggering circuit) so that in fact only one pulse is produced for each signal period. Practical tip: in many cases an AC-coupled input yields better triggering conditions for frequency measurements.

d) Pulse width measurement: the portion of the waveform that is to be measured must be determined as accurately as possible. In this example, the pulse width should be measured at the 50% amplitude level. With a relatively broad hysteresis band, the measured pulse goes HI shortly after the input signal rises above the 50% level and only returns LO well after the signal has fallen past the 50% level. The resulting trigger pulse is too long, and the measured value of the pulse width is thus too large.

e) For pulse width measurements, reducing the hysteresis band (in other words, increasing the sensitivity of the triggering circuit) will significantly reduce the measurement error. Practical tip: use DC input coupling for pulse width measurements.

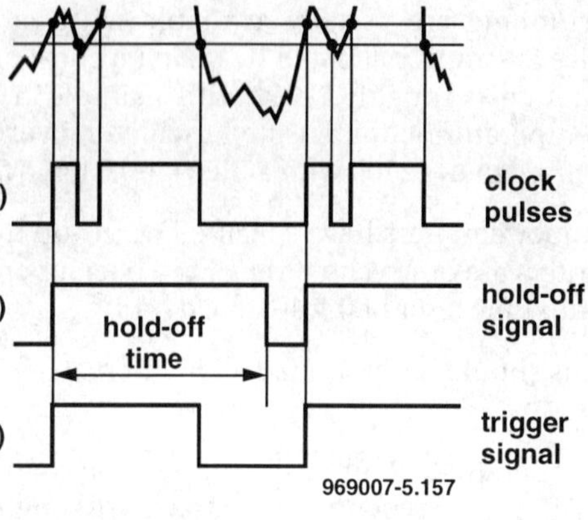

***Figure 4.161** Hold-off setting. (source: Fluke)*

a) The signal is so jagged that several trigger pulses can occur within a single period. This causes too many pulses to be counted.

b) The hold-off setting can be adjusted to ensure that the next effective trigger event does not occur until the following period.

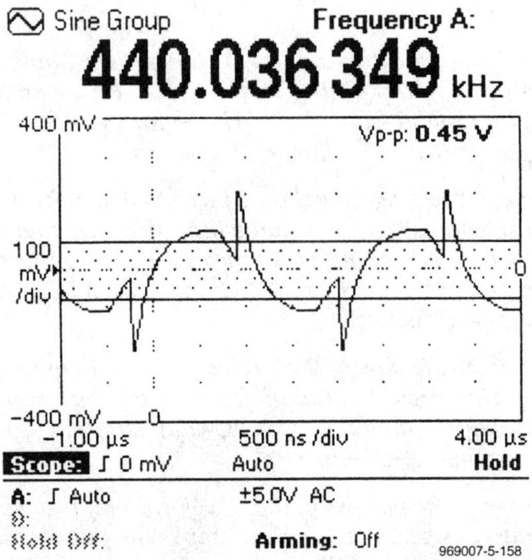

Figure 4.161 LCD display of a modern multi-function counter. (source: Fluke)

Notes:

1 With very inexpensive instruments that have 'outstanding' specifications (such as a frequency-measurement range reaching into the gigahertz region), you should beware of poor triggering performance and reference oscillators with inadequate accuracy.

2 Good-quality instruments are usually available with various reference oscillator options, and the price is strongly affected by the type chosen. (*For example:* one particular model of counter costs roughly £1200 with a simple crystal-controlled oscillator, over £1500 with a temperature-compensated oscillator, over £2000 with an oven-stabilized oscillator and more than £2700 with a high-precision oven-stabilized oscillator.)

3 Relatively high-accuracy counters have specified warm-up times. The accuracy specifications of such counters are only valid if they have been operating for at least as long as the specified warm-up time after being switched on.

4 Precision instruments should be calibrated at an interval of six months to a year, to correct for aging effects.

5 For DIY projects, temperature-compensated and oven-stabilized oscillators are commercially available. Affordable types have accuracies of roughly ±0.1 ppm to 10 ppm.

Testing counters. The basic operation of a counter can be tested with the same equipment used for testing oscilloscopes and logic analyzers. To test the characteristics of a counter's triggering behaviour, you must be able to generate signals similar to those shown in Figures 4.160b or 4.161. An arbitrary waveform generator is suitable for this, but a microcontroller-based tester that works in a similar manner can also be used (see Volume 3).

Type	Accuracy (ppm)	Measurement error at 1 MHz	Measurement error at 100 MHz
for comparison: typical microprocessor crystals and crystal oscillators in metal housings	50–200	±50–200 Hz	±5–20 kHz
crystal oscillator without special provisions	1–10	±1– 10 Hz	±100 Hz–1 kHz
temperature-compensated crystal oscillator	0.1–1	±0.1–1 Hz	±10–100 Hz
crystal oscillator in a temperature-controlled enclosure (oven)	0.001–0.1	±0.001–0.1 Hz	±0.1–10 Hz
rubidium frequency standard	$\approx 10^{-5}$–10^{-3}	±10^{-5}–10^{-3} Hz	±0.001–0.1 Hz

***Table 4.34** Reference clock sources for use in frequency meters and universal counters.*

Note: although relatively inexpensive time-marker generators (oscilloscope calibrators) are more accurate than the average oscilloscope, they are not as accurate as any reasonably useful counter. If your time-marker generator and your counter disagree by a relatively small amount, the counter is more likely to be correct. In case of doubt, check with the manufacturer's calibration service.

Using counters in the service environment. High-precision frequency measurements are hardly necessary in practical fault-finding, which is usually 'digital' in nature. However, there are certain special counting functions (which are rarely if ever found in inexpensive instruments) that occasionally would be rather nice to have:

1 Very fast, very deep counters (such as a 64-bit counter that can count at 100 MHz). Such counters can be used (1) for performance measurements and (2) as fault-finding aids for counting signal transitions (event counters).

2 Very fast, cyclically repeating counters (that can keep up with the external clock rate of a high-performance processor and work with perfect clock synchronization). Such counters can be used for generating cyclic stimuli, such as those needed for signature analysis. (*For example:* such a counter should be able to trigger a reset after every 1000th clock when driven by a 100 MHz clock signal). See Volume 3 for more information.

Certain testing problems that in principle can be mastered quite elegantly with counters require special triggering provisions (such as triggering on specific events or on a particular combination of binary signals). Accessory devices (based on the circuits shown in Figures 4.151 and 4.152, for example) are suitable for such tasks. Several testing chores can be inexpensively handled by using dedicated circuits (such as a graphics card tester; see Volume 3). Such circuits are relatively problem-free, since high-precision reference clocks are not necessary and they are driven exclusively by digital signals, so that the subtleties of analogue triggering are not a factor.

5 Introduction to practical testing and fault-finding

5.1 Component testing

5.1.1 Test methods and test equipment

Components (such as resistors, capacitors and integrated circuits) can be tested either in place (incorporated into a circuit) or separately. If you test a component in place, you check whether it behaves as it should in its particular environment, and the stimulus and load are provided by the circuit in question. If on the other hand you test a component separately, you must check whether it functions as it should or whether it meets its specifications, and the stimulus and the load must be provided externally. Clearly, testing a component in the circuit where it is located (testing in place) would appear to be the preferred method in the service environment. There are two basic techniques that can be used: (1) signal tracing and (2) in-circuit testing.

1 *Signal tracing.* You make measurements at the pins or terminals of the component and draw conclusions from the results of these measurements. The stimulus always is always provided by the 'functional environment' of the component being tested. (Sometimes all that is necessary is to simply operate the equipment or assembly in its normal manner; in other cases a special test stimulus may be more useful.) Standard test equipment is usually adequate for signal tracing (in most cases an oscilloscope is all that is needed). However, there are certain problems that must be taken into account:

   - The severity of the test is determined entirely by the component's environment (so that a particular stimulus may not applied at all, critical load conditions may not be present and so on).
   - You need accurate documentation (or at least a schematic diagram) and a wealth of technical knowledge to be able to evaluate the observed behaviour of the component being tested.
   - Certain measurements may simply not give any clear indication of the status of the component being tested, due to the way that it is connected in the circuit *(for example:* a resistor connected in parallel with the base-emitter junction of a transistor).

2 *In-circuit testing.* You test the component in the same way as it would be tested separately, although it remains in place in the circuit (*for example:* measuring the characteristics of a transistor or checking the functional behaviour of a D-type flip-flop). The test stimulus must be applied directly to the component in question, using test equipment that is specifically designed for this sort of work.

Compared to separate testing, in-circuit testing appears at first glance to require less effort and promise a better chance of success, but it requires rather expensive test equipment and involves certain practical problems (making connections to the component can be difficult, and it is not possible to identify all types of faults). Consequently, the 'pragmatic' technique of unsoldering the component and testing it separately is often the faster solution (for instance, it is clearly much simpler to test a power transistor by itself than when it is part of a circuit).

Standard test equipment (utility instruments), special component testers and test jigs can be used for component testing. Standard test equipment, which is anyhow indispensable (especially digital multimeters and the oscilloscopes), often has extra functions for testing simple types of components. Some digital multimeters, for instance, can measure capacitance as well as resistance and can test diodes and transistors, and some oscilloscopes have built-in component testers. The available measurement ranges and accuracy are of course not adequate for all types of tests. True high-quality component testers are however usually prohibitively expensive, so that in many cases the best approach is to test components in special, more-or-less improvised test jigs.

5.1.2 In-circuit testing

Obvious fault symptoms

Sometimes a fault appears to be obvious: if something appears to be scorched, or if it gives off smoke and stinks, then something must be wrong (or have been wrong). However, simply replacing a component that has been overheated does not solve the problem. Passive components (such as resistors), in particular, do not just overheat by themselves.

Practical tip: if a component is scorched or burned out, always check its immediate environment. In particular, check the following items:

- the load (including possibly jammed mechanical parts, in the case of drive circuits),
- the connections to the load (short circuits in the cabling, etc.),
- the drive source.

Faults in these areas can either create new current paths or eliminate parallel current paths (see Figure 5.1).

The shotgun procedure. If there is no obvious sign of a failure and you do not have any starting point for a systematic fault-finding process, then your only choice is to regard 'everything' as suspect and to check each component in turn. Sometimes the time required to find the fault can be considerably reduced if you test according to probability of failure (based on statistical experience) instead of following the signal flow. The most likely locations of failures (in decreasing order) are: (1) contacts and connectors, (2) fuses, (3) power components, (4) control ICs (in power stages), (5) opto-couplers, (6) discrete semiconductor devices, (7) capacitors, (8) ICs, (9) resistors and (10) inductors.

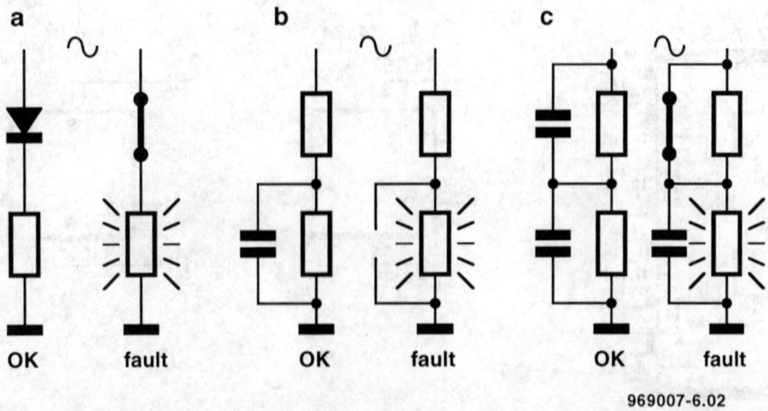

Figure 5.1 Examples of failure mechanisms that can cause component overheating. These are simple circuits driven by an AC voltage.

a) Diodes tend to fail shorted. The voltage applied to the resistor is nearly doubled if the diode fails (U_{pp} instead of U_p).

b) The capacitor has failed open, so that the parallel current path no longer exists (the current is no longer shared).

c) The upper capacitor has failed shorted, so that the voltage is no longer divided over the two resistors; the lower resistor carries the full voltage.

Signal tracing

The basic principles of signal tracing are described in Section 2.5. If a component is being tested, the question that must be answered is whether it is behaving correctly with a given set of inputs and load conditions. In this regard, it is in principle not important whether the signals in question (waveforms, voltages and so on) make sense with regard to the total circuit.

Practical tip: You should first check whether the component is possibly being adversely affected by some external factor. Check the following points (see Figure 5.2):

1 Are the operating conditions (supply voltage, ambient temperature and so on) within the allowable limits? *For example:* in Figure 5.2a, the IC supply voltage is much too low.

2 Are the inputs valid? *For example:* in Figure 5.2b, PRESET and CLEAR should not both be LO at the same time. In this case, there is no point in making any further measurements on this component.

3 Is the load within acceptable limits? *For example:* in Figure 5.2c, both outputs of a flip-flop should never be LO at the same time (assuming that PRESET and CLEAR are not both LO at the same time; if they are then the signals at the D and C inputs do not matter). Possible suspicions are: (1) the flip-flop is defective and (2) one of the outputs is being pulled LO. There are several ways to proceed in this situation: (1) examine the cir-

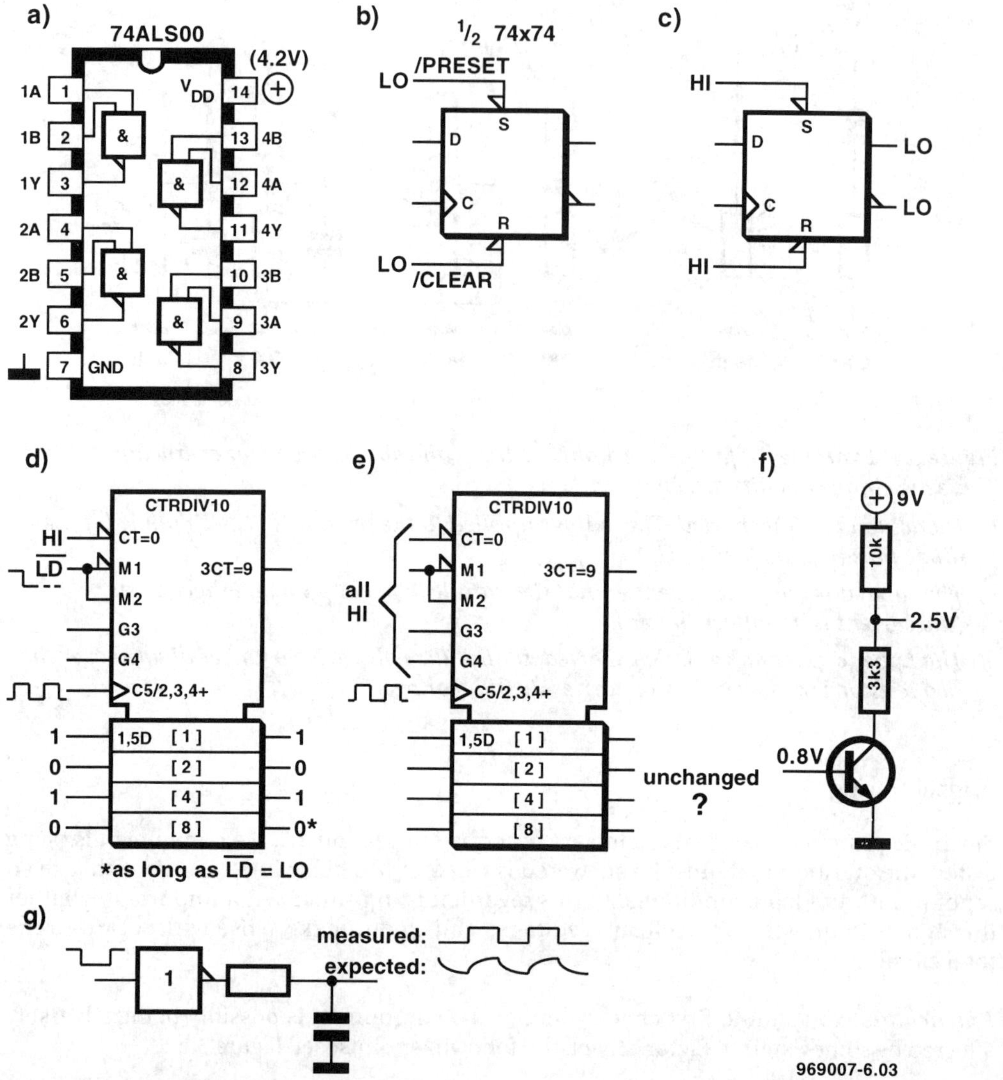

Figure 5.2 Examples of signal tracing situations.

cuit to determine which output should be HI, and cut the associated signal line, (2) use a current tracer to try to determine whether a current is flowing in the suspect signal line or (3) remove the flip-flop from the board (test it separately if it survives being unsoldered) and then check whether the signal line is still being held LO (this approach is favoured by the technician when time is short).

4 Is the component behaving correctly under the given circumstances? *Examples:*

- Figure 5.2d: the counter is receiving load pulses, and the input values appear at the outputs with each load pulse; this is OK. (It is does not matter at this point whether the load pulse is being generated correctly or whether it should even occur at all.)
- Figure 5.2e: the counter is receiving clock pulses, and it should count with the given values at the control inputs, but it does not count. This is a fault!
- Figure 5.2f: a base voltage of 0.8 V means that the transistor is turned on. The measured output voltage corresponds to the divider ratio. Everything is OK.
- Figure 5.2g: you measure a rectangular waveform, although you would expect to see exponential rising and falling edges at the capacitor. This is a fault! (The capacitor is probably open and thus not working).

The stimulus. The stimulus can either be 'artificial' (a static or cyclic stimulus) or 'natural' (a signal that occurs in normal operation). An artificial stimulus usually makes fault-finding easier, but in some cases it may not be severe enough (if it does not include the waveforms that cause the fault to occur). Sometimes it is simply not possible to generate a proper stimulus (for example, if suitable test equipment (such as a data generator) is too expensive or if the circuit in question cannot be operated in a suitable manner).

Static stimuli (DC operation). DC voltages and fixed logic levels can be checked with a multimeter or logic probe. There are various ways to generate an artificial static stimulus, including the following:

- simply connect the component to the supply voltage without applying any input signals. (*Caution:* some types of circuits cannot handle this. For instance, an assembly with CMOS IC inputs should not be operated with the inputs floating. In such cases, even a simple test aid with several toggle switches (or the like) can prove useful.),
- apply static values to the inputs,
- apply a static reset signal,
- 'halt' the clock (remove the clock oscillator and feed in a static test clock signal).

Cyclic stimuli. Many circuits can be tested under dynamic operating conditions using relatively simple test equipment (such as a service oscilloscope or signature analyzer) if they are driven by a cyclic stimulus. There are various ways to generate a cyclic stimulus, including the following:

- periodically reset the circuit,
- run suitable test software (test stimulus loops – see Volume 2),
- apply cyclic input signals generated by suitable test equipment (a signal generator or data generator).

Natural stimuli. If the equipment is tested in its normal operating mode, it is often only possible to make plausibility tests, especially in the case of complex digital circuits (*for example:* checking whether all lines of a system bus behave 'reasonably', checking whether the supply voltages and the earth appear to be 'clean' and so on). On the other hand, testing in the normal operating mode is usually an entirely adequate fault-finding technique for many analogue circuits with signal flows that are predominantly linear (from the input to the output), such as conventional radio and television sets.

Two fault-finding tricks

Opening leads. The location of a suspected fault can be narrowed down by opening leads. This trick can be used to determine whether a given waveform is being generated incorrectly at its source or being affected 'downstream' by other components, short circuits, incorrectly installed components and the like (see also Section 2.5). This technique is only usable if it is actually possible to open leads (from both a functional and a practical point of view).

- *Functional aspects:* it only makes to sense to open a signal path if doing so does not disrupt the stimulus in question. In by far the majority of cases, the following types of signal paths can be opened: (1) those that do not contain any feedback and (2) those in statically-driven digital circuits (ie, in circuits that can work with the clock halted). *Caution:* sometimes you must consider what will happen if certain inputs are left floating when a signal path is opened (CMOS ICs behave unpredictably with open inputs, for instance, and some analogue and power ICs will oscillate strongly if their inputs are open). In such cases, you must provide some sort of fixed, defined input values (via pull-up resistors to the supply voltage or by direct connection to earth, for example). If the load of a power IC must be disconnected for a test, it is sometimes necessary to provide a suitable dummy load.
- *Practical aspects:* it must be possible to open the lead without too much expenditure of time and without damaging the equipment. 'Natural' junctions in the circuit path, such as connectors and IC sockets, are favourable locations. Suitable adapters can be inserted in the signal path at such locations, so that leads can be individually opened as desired, and both sides of the connection can be made conveniently accessible via test points (see also Section 5.1.4). In some cases, all that is necessary is to simply remove a plug-in card or a socked-mounted IC. Cutting through pins or tracks, or opening leads in a cable, must be regarded as measures to be used only in extreme emergencies. *Note:* with very high-frequency signals, the circuit or equipment may no longer work properly if an adapter is inserted in the circuit path (due to the increased length of the signal leads or the additional capacitive loading).

Tying a signal line to a fixed level. This means applying a fixed value to a signal line in order to cause a particular response. *Example 1:* you can test an NPN transistor by connecting its base lead first to its emitter lead and then to a potential of a few volts (such as V_{CC}). In the first case it should be cut off, and in the second case it should be turned on. *Example 2:* with a flip-flop as shown in Figure 5.2c, you could check whether the output values change when the PRESET and CLEAR inputs are individually tied LO.

Tying a signal line HI (connecting it V_{CC}) or LO (connecting it to earth) is an old trick, but you must watch out for certain things:

- Signal lines that are driven by TTL ICs can be tied to earth without any risk (since the collector resistor in the TTL output stage will limit the resulting current). However, tying a signal line to V_{CC} can result in excessive currents. For this reason, always use a series resistor (100–220 Ω) if you must tie a signal line to V_{CC}, and do not keep the signal line HI any longer than is absolutely necessary.
- CMOS output stages have no current-limiting resistors at all.
- Some inputs may be internally connected to fixed levels. Before experimentally tying such inputs to earth or V_{CC}, check to see if the internal connection is made directly to earth or V_{CC}. A fixed LO level is usually made by a direct connection to earth; if such an input is tied to V_{CC} then the resulting short circuit will 'smoke' something. Using pull-up resistors to provide fixed HI levels (so that tying the associated input LO does not carry any risk) is good practice, but some designers rely on the power supply to never output damaging voltage spikes on the supply line, and thus use direct connections to the V_{CC} line to provide fixed HI levels.

Basically, you must always check (in advance!) what current paths will be created if a given line is tied HI or LO, and whether the component in question (and in some cases the interconnections) can withstand the resulting currents.

In-circuit testing

With in-circuit testing, the component under test remains in place in the circuit, but it is tested using an external stimulus that is applied directly to the component.

There are two different ways in which in-circuit testing can be performed:

1) testing in the non-powered state (with the operating power switched off): *analogue signature analysis*,
2) testing under operating power: *in-circuit functional testing.*

Analogue signature analysis (ASA). This technique is based on tracing the characteristic curve of the component being tested, in a circuit that otherwise has no power applied. It works the same way as component testing using an oscilloscope, as described in Section 4.3.5 (see Figures 4.81 and 4.82). One test channel is needed for each two-terminal device, where a two-terminal device is any component or circuit element that completes a path between two points. (One of these points could for example be an IC input and the other point the earth pin, or the PN junction of a transistor could be tested as a two-terminal device.) A voltage-current characteristic line (*V-I* trace) is traced by applying a defined AC voltage to the two terminals and measuring the resulting current, as illustrated in Figure 5.3. Such a V-I trace practically amounts to a signature for the component or circuit element in question (see Figure 5.4). The *V-I* traces are usually stored in digitized form; some testers also have screens for displaying signatures as illustrated in Figure 5.4.

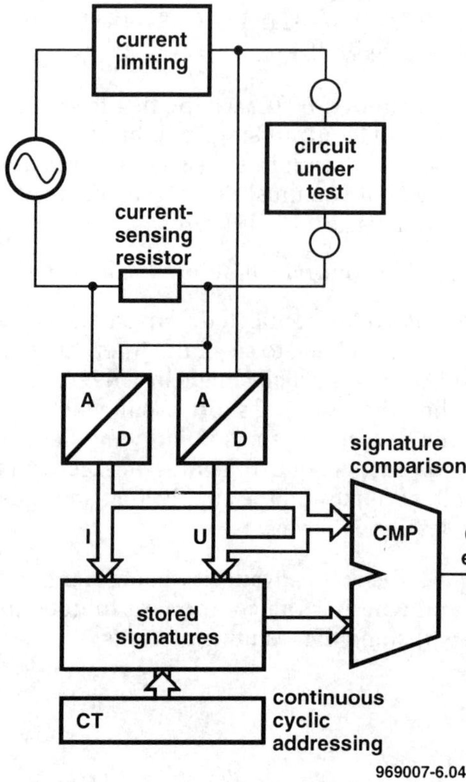

Figure 5.3 Block diagram of an analogue signature analyzer.

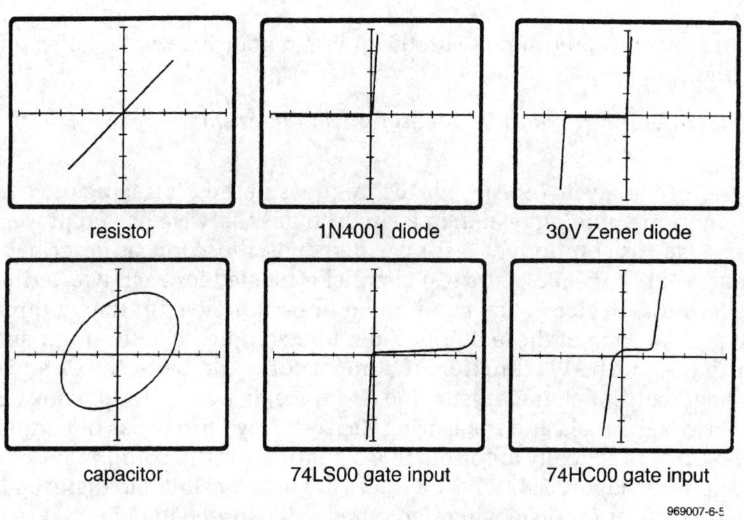

Figure 5.4 Examples of analogue signatures. (source: Polar)

Testers based on analogue signature analysis (Fault Locators) have a large number of test channels (ranging from 32 to 128) that are connected to the components to be tested via a bed-of-nails adapter or test clips (see Section 5.1.4 for more information). Such testers can 'learn' signatures from known-good test objects and can compare actual signatures to stored reference signatures. Allowed component tolerances are automatically taken into account in the comparison process. Some testers can also make direct ('live') comparisons between the measured signatures of two different circuits that are connected to the tester at the same time. Table 5.1 summarizes the test voltages and currents used by such testers.

Note: the frequency of the test voltage is normally selectable (eg 90 Hz, 500 Hz and 2 kHz).

Object under test	Open-circuit voltage (peak-to-peak)	Peak short-circuit current
PN junction	1 V	500 μA
logic circuit	10 V	5 mA
discrete component, low-voltage range	10 V	150 mA
discrete component, medium-voltage range	20 V	1 mA
discrete component, high-voltage range	50 V	1 mA

Table 5.1 *Typical test voltages for analogue signature analysis. (source: Polar)*

In-circuit functional testing (ICFT or 'backdriving'). This technique can be used for in-place testing of individual interconnections and integrated circuits in a digital circuit, regardless of the functional relationships of the items being tested. Suitable test vectors (stimuli) are briefly imposed on the circuit in question (see Figure 5.5), which has normal operating power applied.

The difficulty that in-circuit functional testing has to deal with is that quite often the inputs of the IC to be tested are directly connected to the outputs of other ICs. If a LO-level signal is required for the test stimulus how can this be provided if the input is being held HI by another IC? The answer to this question is based on the well-known fact that the output of an IC can be pulled HI or LO without damaging the IC, as long as this only lasts for a short time. This restriction does not pose a problem, since each test step can be readily completed in a few hundred nanoseconds.

ICFT is employed in high-performance automated testers as well as in service-oriented test equipment. Since bed-of-nails adapters are prohibitively expensive for ordinary servicing, they must be replaced by manual connections to the IC being tested, using an IC test clip (for example). The simplest form of in-circuit functional testing consists of switching the level of an individual signal, which can be done using a suitable hand-held tester (logic pulser or current pulser). Figure 5.6 shows the circuit diagram of the output stage of a typical logic pulser. Automated in-circuit functional testers have a similar circuit for each test channel; 40 to 196 channels are typical for such testers.

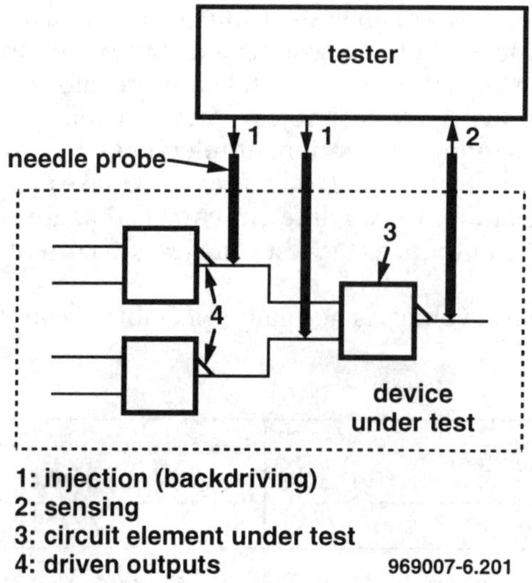

Figure 5.5 In-circuit functional testing (ICFT or backdriving).

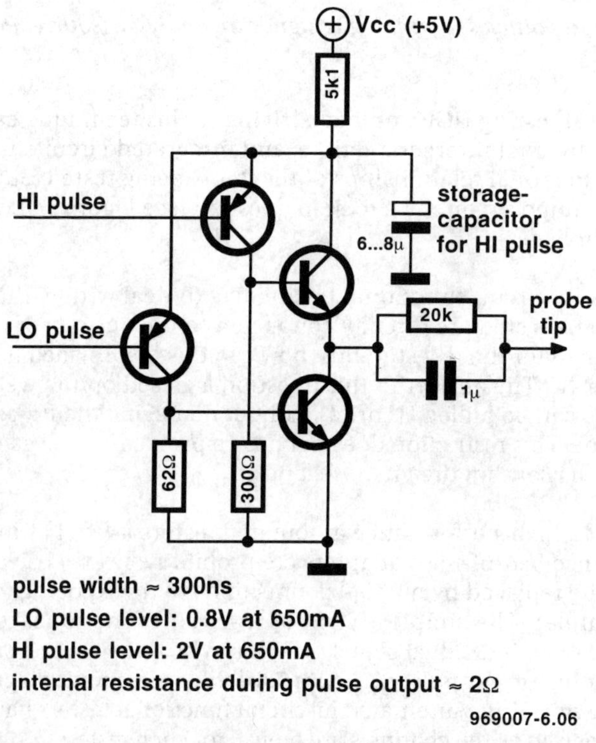

Figure 5.6 The output stage of a hand-held logic pulser. (source: Hewlett-Packard)

A logic pulser can be used to test for a short between a circuit lead and earth or the supply voltage (since the pulser cannot force the signal level to change if such a short is present). It is sometimes even possible to functionally test basic logic elements (such as gates, counters, bus drivers and the like) using a logic pulser.

Standard rules of thumb (these apply to conventional ICs, which means those (a) in DIL packages, with (b) at most 8 to 10 outputs per package and (c) having a maximum drive capacity of 40 mA per output): (1) the current limit is 0.6 to 0.75 A and (2) the duration of the excitation must be less than 5 µs.

Recommendations for modern ICs (according to Texas Instruments): (1) pulse duration ≤ 100 ms, (2) duty cycle < 1:10, (3) injected energy per IC not to exceed 25 mWs (voltage × current × pulse duration), (4) current with a forced LO level ≤ 100 mA per output, (5) current with a forced HI level ≤ 300 mA per output at a maximum voltage of 3 V, (6) wired-OR signals may not be backdriven and (7) the maximum allowable IC die temperature is 125 °C. The critical values are the total energy per IC and the maximum allowable currents. *For example:* 3 V × 300 mA = 900 mW, which means that the pulse duration cannot exceed 27 ms if the energy is limited to 25 mWs. If eight outputs of a single IC are backdriven at the same time, the maximum allowable pulse duration is 3.5 ms.

Bus contention with modern bus drivers. This applies to bus driver ICs in small packages (SSOP) with many outputs (up to 36) and high drive capacity (64 mA per output, or up to 180 mA for IWS types). Attempting to pull the level of the output signal to its opposite value represents a conflict situation (bus contention). The following facts (according to Texas Instruments) must be taken into account in such a situation: the die temperature increases at rate of 200 °C/µs and the region with an elevated temperature expands at approximately 1 mm/s. Since the die temperature must remain under 150 °C, bus contentions can just be tolerated if (1) their duration is significantly less than 1 µs and (2) they occur only infrequently.

Notes:

1 Table 5.2 describes the differences between backdriving and bus contention.

2 High-current, steep-edged pulses often cause strong overshoots, which can disturb the operation of an IC so much that it is no longer possible to test its functions.

3 IC tests that require a large number of inputs to be driven at the same time are only possible if these inputs are connected to the outputs of several different ICs (since only a few outputs of any given IC can be backdriven at any one time).

4 Bus systems can be tested if they can be put into a high-impedance state (tri-state test mode; see also Section 3.4).

5 Backdriving is without question possible if only one signal is to driven at a time. A typical application is testing printed circuit boards (see Section 5.4), but the test equipment should meet the above recommendations. Backdriving is expressly prohibited for certain application areas (such as military equipment).

Bus contention	In-circuit testing (backdriving)
All outputs of the affected IC are involved (sometimes 32 or more).	The test setup can be arranged such that only a few outputs are tested at the same time (at minimum, only one).
The signal levels that occur during the conflict are irrelevant.	Exact logic levels must be achieved.

Table 5.2 Bus conflicts versus back-driving.

In-circuit test methods in modern fault-finding practice. The utility of the two techniques (ASA and ICFT) essentially depends on how easy (or difficult) it is to make connections to the components to be tested. Although this is only a question of cost with traditional components on traditional printed circuit boards (bed-of-nails adapters versus test clips), modern component packages and construction techniques impose basic limits on the 'connectability' of components. In traditional construction, the components are soldered into holes drilled in the circuit board, and the circuit board has a component side and a solder side. All component leads (or pins) are accessible from the solder side, and they can be contacted using a bed-of-nails adapter. The pin pitch (inter-pin spacing) of traditional components is 0.1 inch (100 mil or 2.54 mm), which allows reliable bed-of-nails needle probes and affordable test clips to be built. In addition, the large majority of traditional components has a low to medium level of integration, which means that these components can be individually tested to a satisfactory level using a relatively small number of test steps. With modern technologies, particularly when they are used in combination, the situation is much less favourable for in-circuit testing. This relates to (1) surface mounting technology (SMT), (2) reduced pin pitches (1.27 mm (50 mil), 0.645 mm (25 mil) or even 0.4 mm), (3) dual-sided component mounting (see Figure 5.7) and (4) very high levels of integration, which make simple 'external' tests practically meaningless (it's just about impossible to test a Pentium processor using ICFT). The net result is that it is increasingly necessary to rely on scan techniques (possibly augmented by more extensive self-testing provisions and redundancy), since it is hardly possible to access individual components for in-circuit testing.

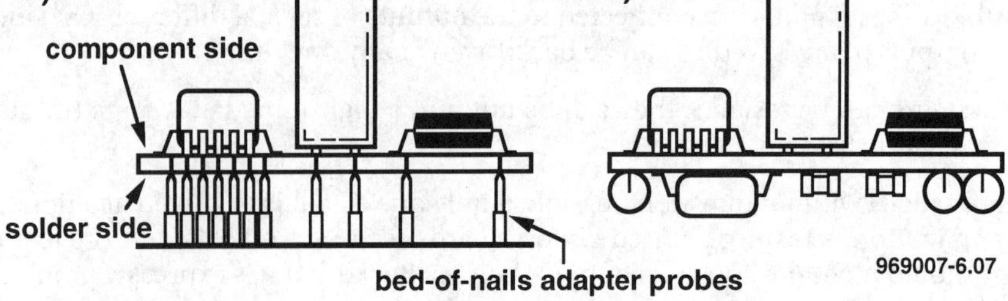

Figure 5.7 Circuit board construction: a) traditional, b) modern.

Service-level test equipment with reasonable performance (which usually combines ASA and ICFT capability) costs £750 and up. Although the adverts stress the ease of use of such equipment (above all, that it can learn correct values from the test object itself, and even generate net lists and schematic diagrams in the absence of detailed hardware documentation), it is definitely not possible to achieve useful results without a certain amount of technical knowledge and the expenditure of a certain amount of time. The typical application situation for such equipment is fault-finding in hardware that, although it may be relatively complicated, is built using traditional construction techniques. Since ICFT has become less and less usable, ASA techniques are primarily employed (even with digital hardware) to locate commonly-occurring, 'trivial' faults such as incorrect components, failed passive components, shorts between circuit board tracks and the like.

Manual in-circuit testing

Plausibility tests for components. Sometimes all that is needed to confirm or refute a particular fault hypothesis is a few control measurements in the operating circuit (using an oscilloscope or multimeter) – see Figure 5.8.

- Figure 5.8a: the measured voltage (8 V) is very close to the expected value (8.2 V) for the voltage divider formed by the two resistors — the circuit is thus OK.
- Figure 5.8b: the bus driver works in the direction $A \rightarrow B$. It is enabled by pulses applied to the *G* input. When *G* is active (its input is LO), every output should have the same value as its associated input. *Sample faults:* (1) incorrect signal behaviour: the outputs switch, but you can see that they do not always achieve the proper levels (something is 'holding back' the signal) — some other driver is trying to drive the bus at the same time; (2) there is practically no change in one or more outputs when *G* is active — the output(s) are defective or (if all outputs are affected) the IC is not being enabled; (3) a single output remains fixed at a LO level — you should suspect a short to earth.

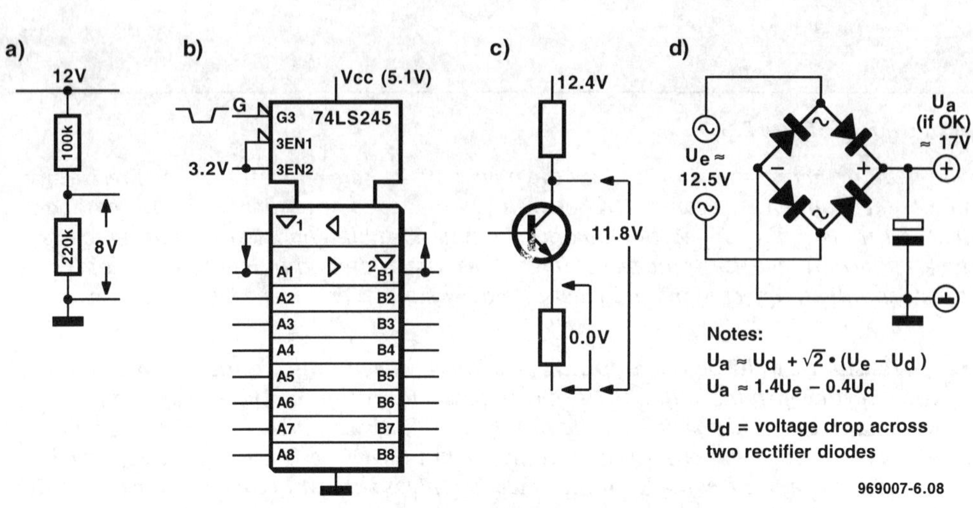

Figure 5.8 Examples of measurement circuits for plausibility testing (see text).

- Figure 5.8c: a transistor in an amplifier stage. The voltage at the emitter is too low and the collector voltage is too high. *Fault suspicion:* the transistor is open.
- Figure 5.8d: an unloaded bridge rectifier. If the capacitor is OK, then it should be charged to very nearly the peak value U_p of the applied voltage (see the note). *Sample faults:* (1) if the capacitor is ineffective (open), you will measure the average value of the input voltage (or a bit less, with a meter that is not all that accurate — see also Section 4.2.8). With an oscilloscope, you can directly see the difference between a smoothed DC voltage and a full-wave rectified AC waveform. (2) If the measured voltage is much lower than it should be (or zero), possibly in connection with overheated components, the capacitor may be shorted. Other possible faults (transformer or diode defects) can be checked by making additional measurements.

Component testing with the power off. In many cases, it is a natural idea to check components with the power switched off, using suitable test equipment connected directly to the components in question (ohmmeter, component tester and the like). Sometimes this can be done without any problems, but the measurements are often affected by other components, and in some cases it is simply not possible to check the component in question. You must always consider the effects of related circuit elements in order to avoid drawing incorrect conclusions from such measurements. (*For example:* you may have measured a resistance value that was too low and thus decided to replace the component – but in fact the measurement was affected by a parallel resistance.)

Caution: measurements made in switched-off equipment are basically 'tester on / DUT off' situations (as described in Section 4.4.1). You must be careful that the device under test will not be damaged by the level or polarity of the voltage applied by the test equipment. In addition, you should check how the switched-off power supply and other components behave (is the overall impedance high or low? — see Figure 5.9), and determine what sort of signal(s) the test equipment actually uses (DC or AC voltage, test voltage level, test current, polarity — see Figure 5.10).

Explanation of Figure 5.9:

a) *You want to measure resistor R_x with the power off. R_2 here represents the net resistance of all other components connected between V_{CC} and earth (in other words, the equivalent resistance of the V_{CC}-earth network). For now, we will ignore issues relating to the type, value and polarity of the test voltage (these are addressed in Figure 5.10), and consider only the effect of the switched-off power supply. This has a DC resistance and an AC impedance.*

   - *DC resistance: if the power supply connections of the assembly under test are galvanically open (as with a plug-in card or a battery that has been removed from the equipment, or if a switch in the supply lead has been opened), the DC resistance of the power supply is effectively infinite (∞), but the DC resistance of the V_{CC}-earth network has a finite value R_2. If the power supply is still connected, it presents a finite DC internal resistance R_i.*

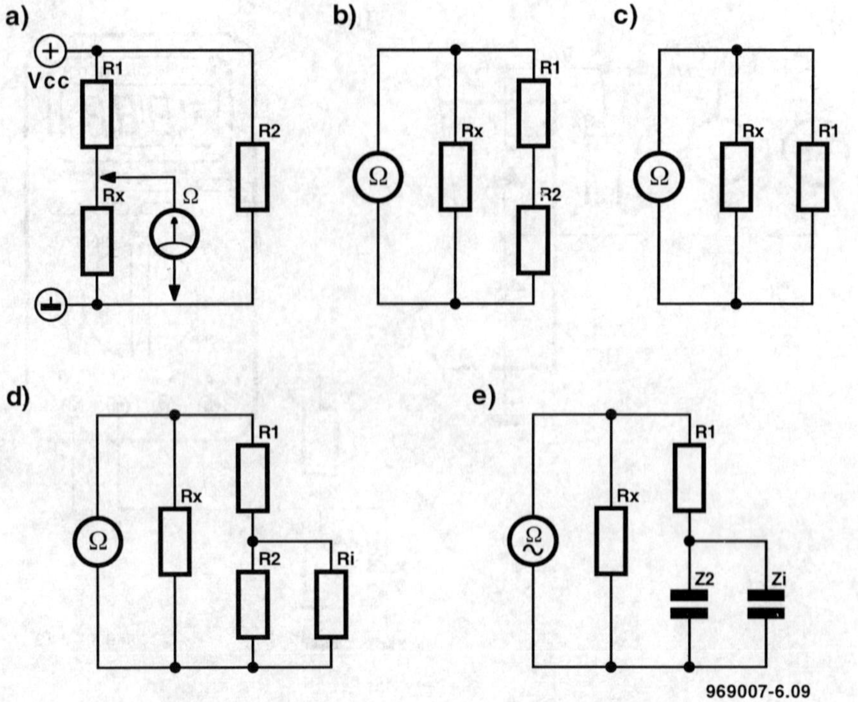

Figure 5.9 *Measurement examples showing the effect of a switched-off power supply.*

- *AC impedance: if the power supply is galvanically isolated, the AC impedance of the power supply is effectively infinite. The AC impedance of the V_{CC}-earth network is determined by the decoupling and/or smoothing capacitors. In typical digital circuits, which have a large number of decoupling capacitors, this impedance is nearly zero. If the power supply is connected, it presents an AC impedance that is also nearly zero.*

For illustrations b–d, the measurement is assumed to be made using a DC voltage.

b) Power supply DC resistance = ∞: the resistance seen by the ohmmeter consists of R_x in parallel with the series combination of R_1 and R_2.

c) Power supply DC resistance = 0: R_2 is short-circuited by the power supply's internal resistance. The ohmmeter thus sees the parallel combination of R_x and R_1.

d) Power supply DC resistance R_i is finite but not zero: R_i is effectively in parallel with R_2.

e) This is a simplified equivalent circuit for the situation in which measurements are made using an AC voltage (with an RLC bridge or a component tester). Z_2 and Z_i are the AC impedances of the V_{CC}-earth network and the power supply, respectively. These values are quite low (very nearly 0 Ω) due to the effects of the decoupling and filter capacitors.

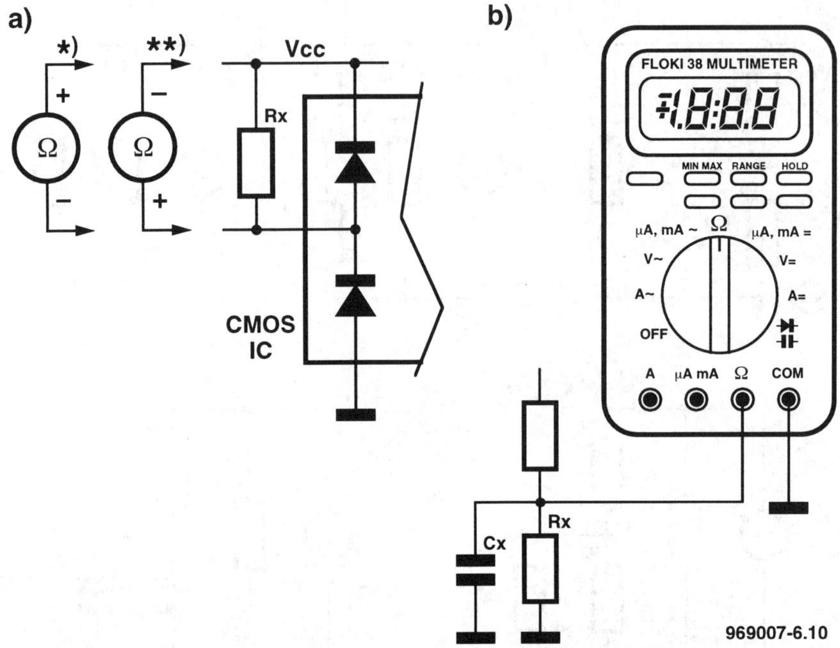

***Figure 5.10** Measurement examples with various test voltages.*

a) *Measuring the value of a pull-up resistor at the input of a CMOS gate: the measured value depends on the polarity with which the ohmmeter is connected.*

Case 1: the power supply is galvanically isolated. Polarization (*) is OK (ignoring the effects of other parallel resistances as described in Figure 5.9), but with polarization (**) the test voltage will be clipped to the forward voltage of the protective diode if it is greater than 0.45 to 0.5 V, and the current flow can damage the IC.

Case 2: the power supply is connected and has a relatively low internal resistance. If the test voltage is greater than 0.45 to 0.5 V, then a current will flow through the lower protective diode with polarization (*) and through the upper protective diode with polarization (**). If the test current is not limited to approximately 10 to 20 mA, there is a risk of damage to the IC. This means that you should not make measurements on circuit boards using test voltages higher than 0.5 V!

b) *Measuring resistance and capacitance: it all depends on what equipment you use. Component testers or measurement bridges, which apply an AC test voltage, measure the impedance of the parallel circuit ($R_x \parallel C_x$). C_x thus cannot be measured on its own, and a reliable value for R_x can only be obtained using an ohmmeter (or multimeter) that uses a DC test voltage. (The value of C_x can be calculated from the measured DC value of R_x and the measured AC impedance of $R_x \parallel C_x$ if the frequency of the AC test voltage is known.)*

SMD components. SMD components can be tested in the same way as conventional components (see Section 5.1.4 regarding making connections). One problem is that many types of SMD components look practically alike, in contrast to conventional components that can readily be seen to be resistors, capacitors, diodes and so on (see Figure 5.11). This means that you should first check to see what you should actually be measuring (using a schematic diagram, a component layout diagram, component identification labels on the circuit board or identification markings on the components themselves) before making any measurements.

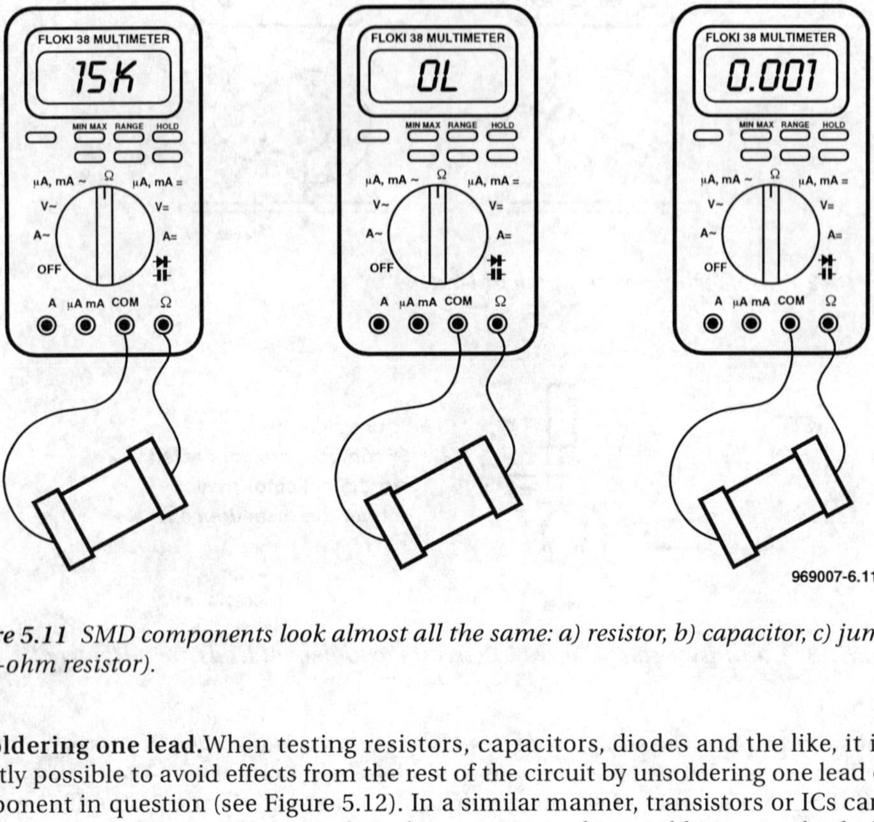

Figure 5.11 *SMD components look almost all the same: a) resistor, b) capacitor, c) jumper (zero-ohm resistor).*

Unsoldering one lead. When testing resistors, capacitors, diodes and the like, it is frequently possible to avoid effects from the rest of the circuit by unsoldering one lead of the component in question (see Figure 5.12). In a similar manner, transistors or ICs can also be made individually testable in a relatively easy manner by unsoldering one lead of each of the associated passive components (see Figure 5.13).

Practical tip: it is preferable to fully unsolder SMD components. If only one end of the component is unsoldered, and the joint insulated by carefully prising up the component and slipping a thin piece of plastic film between the component and the solder pad, the component itself and the other solder pad are subjected to mechanical stress. In addition, such a procedure takes almost as much time as unsoldering and resoldering the component in a proper, workmanlike manner.

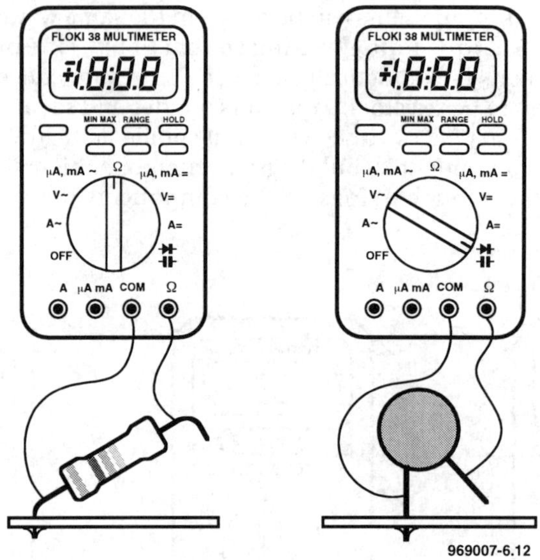

Figure 5.12 Unsoldering one component lead.

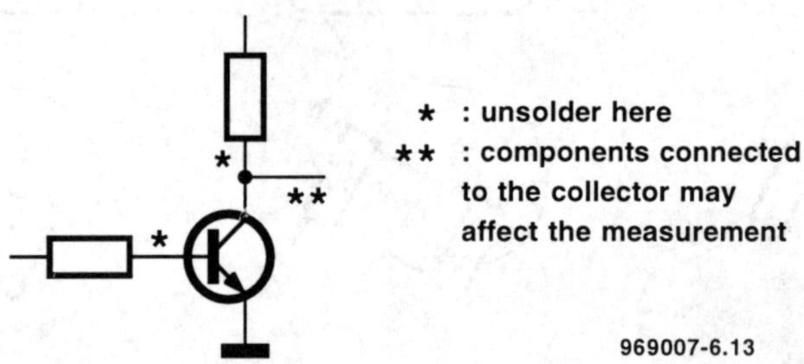

Figure 5.13 A transistor can be tested in-circuit if component leads are unsoldered.

'Simple and stupid' - the substitution test. Sometimes it is not even necessary to make a measurement; all that you have to do is to solder a component with the same value 'on top of' the component in question (a quick solder tack is fully adequate). If everything then works properly, you have located the fault.

Caution:

1 This trick only works with a 'failed-open' fault. This is the most likely failure mode for resistors, and it is also frequently encountered with capacitors. Diodes, on the other hand, most often fail shorted.

2 ***Do not*** use this technique with large-value capacitors (especially power supply filter capacitors)! Instead, first discharge the capacitor (see Section 5.1.5) and then unsolder at least one of its leads; after this you can solder in a new component for testing.

3 Diodes, quartz crystals and active components cannot be bridged by a new component, so they must first be removed. Whether the new component must be properly mounted, or whether a temporary connection can be used (possibly via short lengths of soldered-on wire), depends on specific circumstances (high working frequencies and multi-ampère current pulses, for example, will not tolerate improvised connections).

4 Pay attention to heat dissipation when mounting power-handling components. *For example:* if the component in question is a power transistor that easily manages to heat up a massive heat sink in normal operation, it is hardly a good idea to simply hold the substitute transistor in your hand.

Practical tip: first check that the component in question is not fully shorted (by making a continuity test) – if it is, then usually you have also located the fault.

Comparison testing

Comparison testing is a form of semi-automated signal tracing. It consists of applying the input signals of the suspect IC to the inputs of a known-good IC of the same type and comparing each of the equivalent outputs of the two ICs (see Figure 5.14). In practice, this amounts to a variation on the master/checker technique (see Section 3.2.1). The comparison technique has the following drawbacks under modern conditions:

1 Problems can arise due to additional circuit loading (current, capacitance, lead length), especially with high-frequency signals.

2 A suitable reference IC is always needed.

3 It is difficult to make connections to the pins of modern SMD components.

Practical tip: for a while, testers along the lines of Figure 5.14 were available from several manufacturers. If you plan on servicing somewhat older equipment and have an opportunity to acquire such a tester at a good price, it's worth giving it a try.

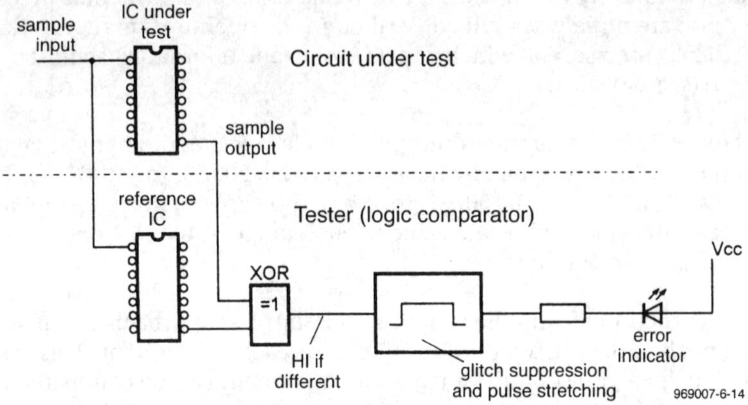

Figure 5.14 Comparison testing of integrated-circuit logic devices.

5.1.3 Reference components

Reference components are components that are expressly intended to be incorporated into a circuit for testing purposes, to be used for experimenting and to be used for checking test equipment.

Known-good or 'golden' devices. The practice of holding a certain stock of components (resistors, capacitors, transistors, SIMM modules, processors and so on), exclusively for testing purposes, has long since proved its worth.

Practical tips:

1 Store reference components separately, and if possible.mark them with coloured paint or self-adhesive labels.

2 Always check that the operating conditions are correct before installing a reference component in a circuit. (If anything untoward happens to your reference Pentium processor or SIMM when you install it without first checking the supply voltage, for example, then it's your own fault.)

3 Basically, you should always clearly mark defective components and store them apart from other components, or render them permanently unusable. It is not always possible to see that a given component is defective, and it is certainly not unheard of for a defective component to be mistakenly used in another circuit at some later date.

Checking test equipment. It's a good idea to keep a number of components on hand that have known characteristics, in order to check test equipment. You can decide for yourself whether to simply store these components somewhere or to build them into convenient test aids.

Precision components. The easiest way to determine whether a given piece of test equipment is in reliable working order, or if it perhaps needs to be calibrated or is quite simply defective, is to make a few test measurements on components with suitably tight tolerances. This is also a good way to check that you are using the equipment correctly, and sometimes also to identify the limitations or weaknesses of a particular piece of equipment (since these are not always fully described in the operator's manual). Components that are intrinsically precise, and which can be easily (and affordably) acquired, are particularly attractive (see Table 5.3).

Component values. The measurement ranges of modern test equipment (digital multimeters and component testers) typically have steps of 20/200/... (or 40/400/...); precision equipment uses 50/500/... or 10/100/.... You should be able to make test measurements at approximately 80% of the full-scale value of each range. Tables 5.4 and 5.5 list suitable resistance and capacitance values.

Mechanical construction of suitable test aids. The simplest approach is to just mount the components on a board, with a pair of test jacks for each component. This arrangement provides a high degree of flexibility (since each component can be connected in series or parallel with any other component or set of components), but it is rather awkward in use. An obvious remedy is to use selector switches, but good-quality rotary switches are not

Component type	Tolerance	Remarks
metallic-film and deposited-metal resistors	1%	inexpensive
precision wire-wound resistors	0.003% to 1%	very expensive for tolerances better than 1%
thick-film resistor networks	2%	only certain configurations and resistance values are available; inexpensive
precision measurement resistors and resistance networks	0.05% to 0.1%	only certain configurations and resistance values are available; quite expensive
precision film capacitors	1%	available values: roughly 100 pF to 10 nF
ceramic capacitors	5%	available values: roughly 1 pF to 20 nF
tantalum electrolytic capacitors	10%; 5% for precision types	available values: ≥ 1 µF
aluminium electrolytic capacitors	20%	available values: ≥ 1 µF
high-frequency inductors	5% or 10%	available values: roughly 0.1 µH to > 10 mH

Table 5.3 *Standard components that can be used for plausibility testing of test equipment.*

particularly inexpensive. A compromise solution is to connect all of the components to a single bus lead and provide jacks for making additional interconnections (see Figure 5.15). In order to keep the size of this arrangement within reasonable limits, 2-mm jacks can be used for the components, with a pair of 4-mm jacks for connecting to other equipment using standard test leads.

Note: don't overlook the fuse and the pushbutton switch(es) for discharging large-value capacitors. See Volume 3 regarding construction details.

Experimental arrangements. Small boxes containing switchable resistors, capacitors and inductors are commercially available as *substitution boxes* or *decade boxes* – at rather respectable prices. Clearly, you can also build the components that you ordinarily use into such assemblies. Figure 5.17 shows schematic diagrams of standard arrangements.

Suppose that (for example) you want to be able to select any arbitrary resistance value between 0 Ω (short circuit) and 1 MΩ, in steps of 1 Ω. A straightforward approach is to provide a set of resistors for each decade, using a rotary switch or an arrangement of slide or toggle switches to select individual resistors. Figure 5.17 shows various ways in which such a decade selection can be realized. (Individual decades are connected in series for

Range	Test Value	Range	Test Value	Range	Test Value	Range	Test Value
10	8.2	20	18	40	36	50	47
100	82	200	182	400	383	500	475
1 k	825	2 k	1k82	4 k	3k83	5 k	4k75
10 k	8k25	20 k	18k2	40 k	38k3	50 k	47k5
100 k	82k5	200 k	182 k	400 k	383 k	500 k	470 k
1 M	820 k	2 M	1M8	4 M	3M9	5 M	4M7
10 M	8M2	20 M	18 M	40 M	10·3M9	50 M	10·4M7
100 M	10·8M2	—	—	—	—	—	—

Table 5.4 Test resistor values for commonly-used measurement ranges.

Range	Test Value	Range	Test Value	Range	Test Value
200 p	180 p	400 p	330 p	500 p	470 p
2 n	1n8	4 n	3n6	5 n	4n7
20 n	15 n ‖ 3n3	40 n	33 n ‖ 3n3	50 n	47 n
200 n	150 n ‖ 33n	400 n	330 n ‖ 33 n	500 n	470 n
2 μ	1μ5 ‖ 330n	4 μ	3μ3 ‖ 330 n	5 μ	4μ7
20 μ	15 μ ‖ 3μ3	40 μ	33 μ ‖ 3μ3	50 μ	47 μ
200 μ	150 μ ‖ 33 μ	—	—	—	—

Table 5.5 Test capacitor values for commonly-used measurement ranges.

resistors and in parallel for capacitors.) Capacitors with the same rated value but different types of construction behave differently in certain circuits. To investigate such effects, you can use a multiple-section rotary switch (as shown in Figure 5.18b) to connect different types of capacitors with the same capacitance to a circuit (*for example:* aluminium versus tantalum electrolytics or plastic-film versus ceramic capacitors).

DIY construction. The working principle of decade boxes is relatively simple. However, problems can arise with the selector switches (good results cannot be achieved with inexpensive types) and the general mechanical construction. The circuits shown in Figure 5.17c and 5.17d present the fewest difficulties in this regard (since standard slide or toggle switches are adequate, and a complete decade arrangement can be built up on a single (prototyping) circuit board). To be sure, the ease of use leaves something to be desired, since a bit of mental arithmetic is always needed.

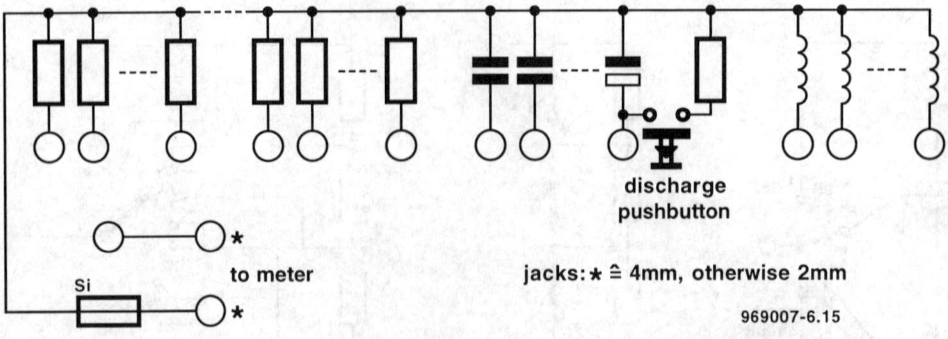

Figure 5.15 Test jig for plausibility checking of test equipment (conceptual diagram).

Figure 5.16 Resistance decade box (left) and capacitance decade box (right), with their enclosures removed. Note the intricacy of the mechanical construction.

Simplifications:

- In line with Figure 5.17b, you can use a rotary switch to select one of a number of standard resistance or capacitance values (see Figure 5.18).
- In place of switched fixed-value resistors, you can use variable resistors (potentiometers) with values that are set using an ohmmeter (digital multimeter or component tester) before they are used.

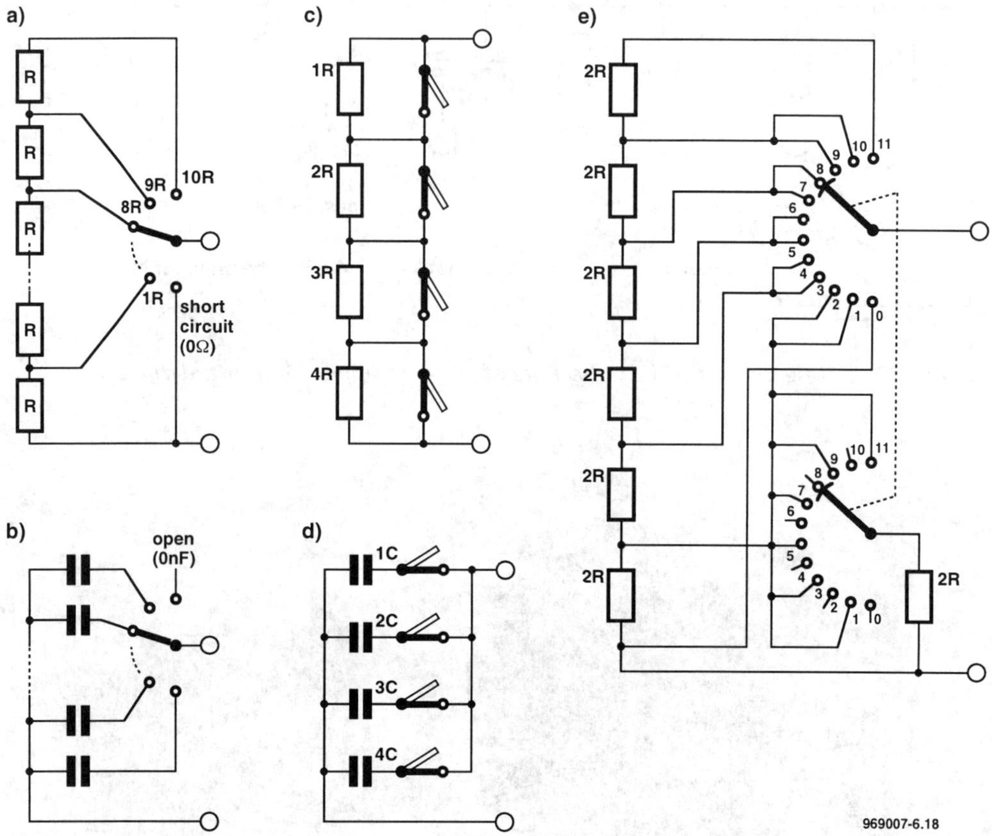

Figure 5.17 *Basic resistance-decade circuits.*

a) Series-connected components: resistance values of 0 Ω, 1 Ω, 2 Ω and so on can be selected using an 11-position switch and 10 resistors (each resistor has a value of 1 Ω).

b) Parallel-connected components: only one component is selected at a time. To be able to select any value in a given capacitance decade, for example, you would need to have ten capacitors with values of (for instance) 1 nF, 2 nF and so on (which if necessary can be made up by connecting two or more standard-value components in parallel).

c) Selective combination of series-connected components: the net resistance value is formed by individual resistors with values of R, 2 R, 3 R and 4 R. If a given switch is closed, the associated resistor is bridged over (shorted); if the switch is open then the associated resistor contributes to the total value. For example: if R = 10 Ω, the available values are 0 Ω (all switches closed), 10 Ω, 20 Ω, 30 Ω, 40 Ω, 50 Ω (= 10 + 30 or 40 + 10), 60 Ω (= 20 + 40 or 10 + 20 + 30), 70 Ω (= 30 + 40), 80 Ω (= 10 + 30 + 40), 90 Ω (= 20 + 30 + 40) and 100 Ω (= 10 + 20 + 30 + 40).

d) Selective combination of parallel-connected components: the net capacitance value is formed by individual capacitors with values of C, 2 C, 3 C and 4 C. If a given switch is

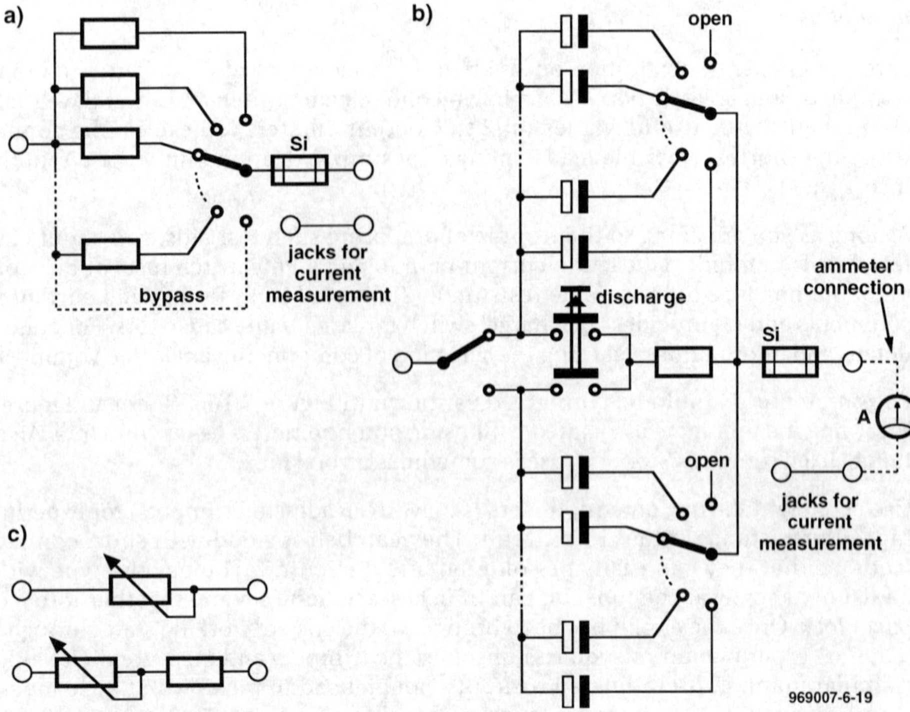

Figure 5.18 Simple test aids.
a) with resistors, b) with capacitors (see text), c) with potentiometers

open, the associated capacitor is not connected; if the switch is closed then the associated capacitor contributes to the total value. For example: if C = 10 nF, the available values are 0 nF (all switches open), 10 nF, 20 nF, 30 nF, 40 nF, 50 nF (= 10 + 30 or 40 + 10), 60 nF (= 20 + 40 or 10 + 20 + 30), 70 nF (= 30 + 40), 80 nF (= 10 + 30 + 40), 90 nF (= 20 + 30 + 40) and 100 nF (= 10 + 20 + 30 + 40).

e) *One stage of a resistance decade box (see Figure 5.16, left-hand side). A resistance of R, 2 R, 3 R … 11 R can be selected using 7 resistors with values of 2 R, and no 'gaps' or intermediate values occur when switching values. The working principle of this circuit is that an extra 2 R resistor is connected in parallel with the bottom-most resistor for odd-valued switch positions (1, 3, 5 and so on), thus reducing the combined value to R. In the first switch position the situation is clear: 2 R || 2 R = R. In the second position, only the lowest resistor of the chain is effective, so the resistance is 2 R. In the third position, the second series resistor is included and the extra resistor is connected in parallel with the first resistor, so the net resistance is 2 R + (2 R || 2 R) = 3 R. In the fourth position the parallel connection is omitted, so the net resistance is 2 R + 2 R = 4 R.*

Practical tips:

1 Large component-substitution boxes are not particularly useful. It's better to prepare a number of small plastic boxes, each having one selector switch and several well-chosen components with 'useful' values (or 2 to 4 potentiometers). These can be connected using the shortest possible leads and can be simply 'dropped into' the circuit being investigated.

2 As long as you are going to the trouble of preparing such test aids, you should keep a few details in mind: (1) fuse the current path (using a PolySwitch fuse, or an ordinary meltable-link fuse for low-value resistors), (2) provide jacks for making current measurements and (3) provide a 'discharge' switch for large-value capacitors. For additional details and information regarding the selection of component values, see Volume 3.

3 Potentiometers should be connected as shown in Figure 5.18c. (Refer to Figure 4.38 regarding the problems associated with using potentiometers as test resistors. An additional, fixed series resistor or a fuse is sometimes a good idea.)

4 Use cermet multi-turn potentiometers (screwdriver-adjusted trimpots) for experimenting with resistance values in a circuit. They can be provided with short connecting leads, so that they can easily be soldered into a circuit. If you use the type with the adjustment screw at the top, you can (if necessary) tape several together into a compact block. Once the circuit has been brought to the correct working state through (systematic!) experimentation, you can unsolder the trimpots and measure their values. In a similar manner, pre-adjusted trimpots can be used to make worst-case measurements on circuits. (A test adapter such as that shown in Figure 5.33 is a suitable aid for setting up and measuring the resistance values.)

5.1.4 Connection techniques

In this section, we give our attention to the various ways in which the hardware being tested can be connected to the test equipment. An extensive selection of test leads, cables, probes, test clips, adapters and so on is commercially available for this purpose.

Practical tip: do not over-economize on test accessories! If you try to make measurements on a Pentium motherboard using banana plugs attached to home-made 'test leads' constructed from bell wire, you will get the results that you deserve.

To be sure, highly-specialized hardware comes at a rather steep price, so in many cases the only choice is DIY construction (or 'solid' improvisation). The key idea is to utilize more-or-less standardized, commercially-available items as much as possible, in order to minimize the total effort.

Plug-and-jack connectors and test leads

Plug and jack systems. A certain few of the large variety of plug-and-jack systems have a special place in the field of measurement and testing (see Figure 5.19 and Table 5.6).

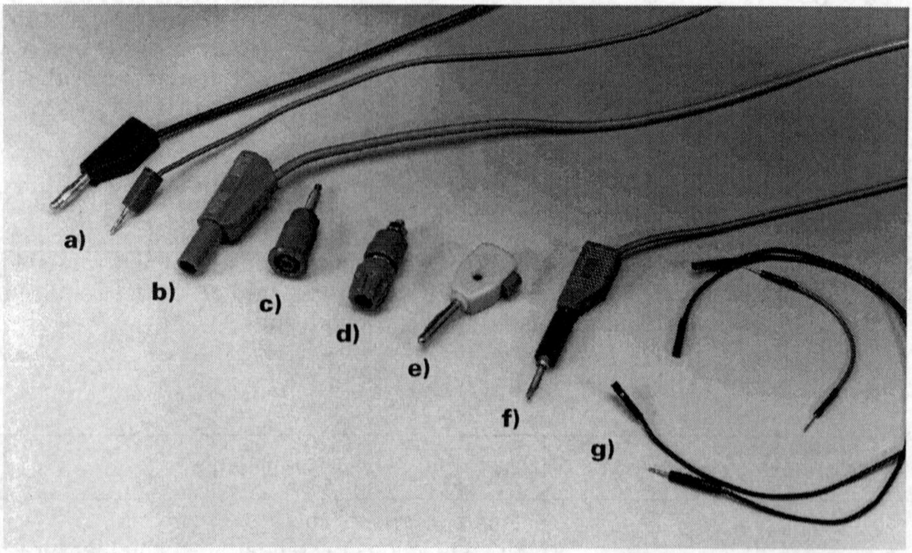

Figure 5.19 Examples of selected types of connector systems.

a) The preferred type of laboratory plug. Features: (1) spring-blade contacts and (2) stackable connectors (several plugs can be joined together, one on top of the other).

b) Safety-type laboratory plug. An insulating sleeve protects the user against accidental contact with the metallic part of the plug. There are models with fixed sleeves and ones with retractable sleeves. The first type can mate only with a safety-type jack, while the latter type is especially suitable for making measurements on the mains (since it can be inserted into the socket of an ordinary outlet while still protecting the user against accidental contact with the mains voltage).

c) Safety-type jack. Its advantages are: (1) protection against accidental contact and (2) low-profile construction (it protrudes only a small distance from the front panel).

d) Standard jack (binding post). It has the advantage that any desired wire or lead can be quickly fastened using the screw clamp.

e) Laboratory plug with a screw-clamp connection, usable as an alternative to a binding post. It can accept wires with cross-sectional areas of 0.12 to 2.5 mm^2.

f) Plug adapter (2 mm to 4 mm)

g) Various types of subminiature test leads (see also Practical Tip 2 in the text).

Practical tips:

1 Regarding miniaturized laboratory material: although the 2-mm system is inherently more robust, 1-mm system plugs mate with the sockets of standard sub-D connectors. The 1-mm system is thus a good choice for cable adapters, breakout boxes and the like.

2 Examples of simple subminiature plug-and-jack connector combinations: (1) a 0.56-mm pin in an IC socket, (2) a 0.64-mm jack on a wire-wrap post, (3) the 0.79-mm plug system (in probe pods). Test leads are available in a variety of different configurations.

The commonly-available combinations are: (a) two 0.56-mm pins, (b) a 0.56-mm pin with a wire-wrap socket, (c) two wire-wrap sockets, (d) a wire-wrap socket plus a wire-wrap pin, (e) a wire-wrap socket with a 0.79-mm plug, (f) a 0.56-mm jack with a 0.79-mm plug (for connecting an IC test clip to a probe pod), (g) a wire-wrap socket with a 0.79-mm plug and (h) a wire-wrap socket with a 1-mm plug.

3 The 0.64-mm plug-and-jack system is significantly more robust than the combination of a 0.56-mm pin and an IC socket (or mating solder jack).

4 The standard configuration for 0.64-mm test connections consists of pins (posts) located on the circuit board or test adapter (clip or probe) and at the test equipment, together with test leads that have mating sockets at both ends.

Connector system	Applications
4-mm ∅ (banana) plugs and jacks	general-purpose laboratory material; for connections to standard test and measurement equipment, power supplies and so forth. The plugs fit into mains outlet sockets!
2-mm ∅ (banana) plugs and jacks	special-purpose laboratory material; for connections to test jigs that should be small but still robust
1-mm ∅ (banana) plugs and jacks	versatile connection system for test accessories that must be as small as possible (such as breakout boxes); adapters for probes and test clips
round pins and sockets, 0.56 mm ∅ (0.022") (typical)	The pins fit in standard IC sockets and in breadboards ('Circuit Test Strips' and solderless breadboards). The sockets can be used for connecting to IC test clips, among other things.
square-section sockets and wire-wrap pins, 0.64 mm (0.025") wide (effective ∅ 0.63 mm)	inexpensive system for single and multiple connections (using flat cables), for test points on circuit boards and test adapters, for connecting test leads to DIY probe pods, and for adapters for probes and test clips.
0.79-mm ∅ (0.031") pin and socket combinations	probe pods for logic analyzers, data generators and the like. The usual arrangement is: a socket in the pod, a test lead with a pin and socket at either end, and a pin in the test clip.
BNC connectors	oscilloscope inputs, pulse generator outputs and so on. (normally for use with 50 Ω coaxial cable)
subminiature coaxial connectors (SMB, SMC)	coaxial connections in high-speed test circuits (50-Ω coaxial cable is preferred)

Table 5.6 *Selected connector systems for test and measurement equipment.*

5 The standard configuration for 0.79-mm test connections consists of jacks located in the test pod and pins located on the circuit board or test adapter (test clip or probe), together with test leads that have a plug at one end and a jack at the other end.

6 Multi-position pin-and-socket connectors are available 'by the yard' in single and double-row configurations. Mass-termination connectors for flat cables (IDC) are available with the following numbers of contacts: 10, 14, 16, 20, 26, 34, 40, 50, 60 and 64.

7 The individual leads of a stranded-wire flat cable that is connected to an IDC (Insulation Displacement Connector) at one end can be separated and connected to test clips, plugs or jacks (see the left-hand side of Figure 5.20, for example).

8 *Terminology:* socket = jack = female connector; plug = pin = male connector.

The left-hand side of Figure 5.20 shows a simple way to connect test leads using an IDC plug and matching socket. The test leads at the lower right of Figure 4.20 have 0.79-mm jacks that can be connected to test clips or adapter leads that have 0.79-mm pins (the latter are used for making connections to IC test clips). The pod itself has a 'quick-release' connector that is attached to the test leads via special 2-mm plugs (not visible here). These parts are in most cases vendor-specific, so you must purchase a certain number of pre-assembled test leads if you want to construct adapters for specific purposes. (In this case, you could also improvise with standard 2-mm plugs if you are willing to do without the convenience of the quick-release connector.)

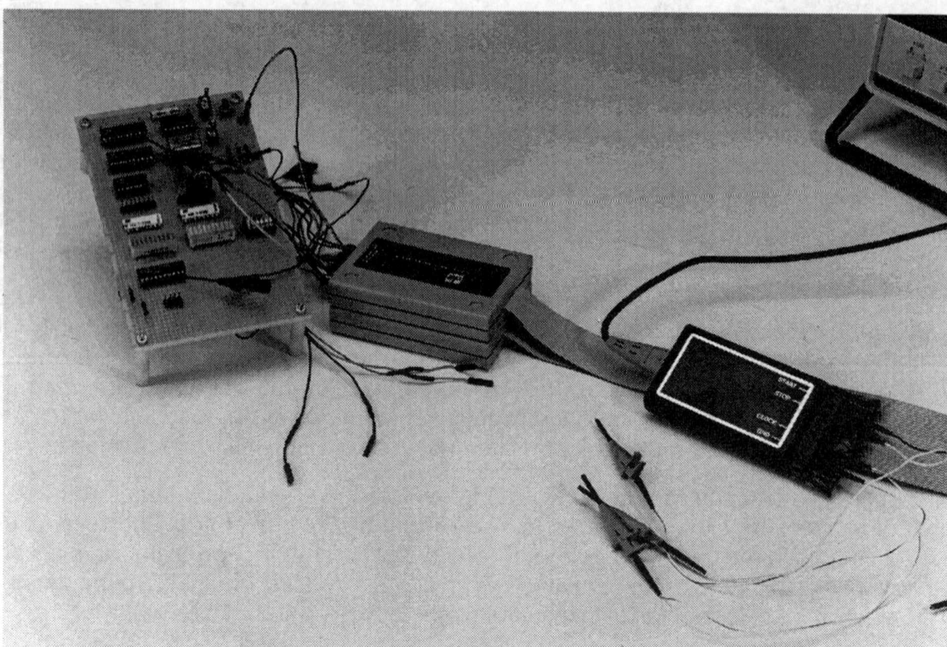

Figure 5.20 Probe pods with test leads (see text).

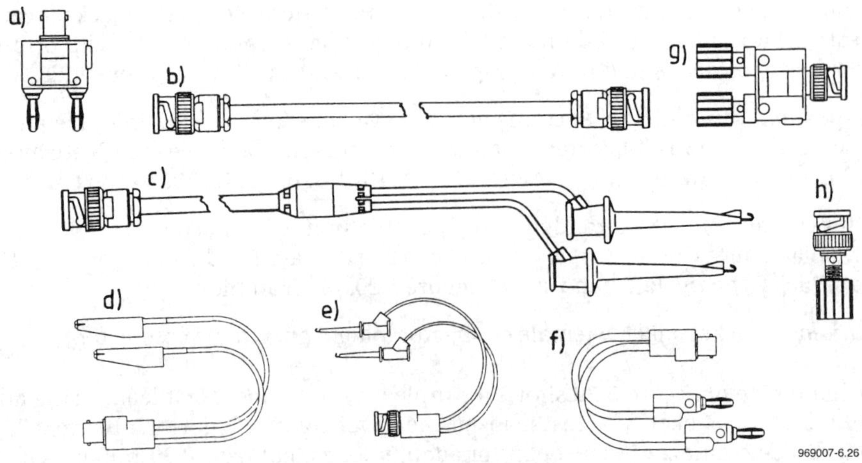

Figure 5.21 Selected BNC test cables. (source: ITT-Pomona)

a) 4-mm plug adapter with BNC jack, b) universal BNC cable, c–f) various types of BNC test leads, g) 4-mm socket adapter with BNC plug, h) BNC plug with 4-mm socket.

BNC cables and accessories. FIgure 5.21 illustrates a number of different types of test leads and adapters that use BNC connectors.

Notes:

1 *BNC test leads:* there is a large variety of test leads available with a BNC connector at one end. Probes, banana plugs, alligator clips, test clips and the like can be attached to the other end. Be careful when using such cables for making measurements with an oscilloscope: they are not the same as probes! (and not every cable that has a BNC connector is actually a coaxial cable with a well-defined characteristic impedance). In practice, using a probe with an adapter that suits the specific measurement task is usually better than using such test leads (which are anyhow not all that inexpensive).

2 *BNC connectors:* measurement equipment usually has BNC jacks, and 'universal' cables (as shown in Figure 5.21b) have BNC plugs at both ends. Adapters, test leads and the like are available with BNC plugs (for direct connection to the measurement equipment) or BNC jacks (for use with a universal cable serving as an extension cable). (See Figure 5.21e & f, for example.) Bear this in mind when selecting or ordering parts!

3 Short test leads (such as that shown in Figure 5.21d) are usually intended to be used with an extension cable. The metallic shell of the BNC connector can however cause short circuits in the device under test, if you are not careful. This problem can be avoided by using BNC plugs with insulated bayonet rings. At least one of the connectors of the extension cable should be an insulated type.

Subminiature coaxial connectors. SMB-type connectors have a snap-on fastener, while SMC-type connectors have a screw-together fastener. The plugs and jacks of both of these connector systems are small enough that a relatively large number of measurement points can easily be placed on a circuit board (for instance, directly next to an IC socket for the device under test). The SMB type is preferred in the laboratory (since it is faster to connect and disconnect), while the SMC type is suitable for portable measurement equipment that is exposed to vibration and rough handling. See Volume 3 for practical examples.

Kelvin-connection test cable. A Kelvin-connection test cable has two clips for connecting to the component in question, with two leads attached to each clip (see Figure 5.22). One pair of leads carries the test current, while the second pair is used for the voltage measurement. In high-quality versions the individual leads are gathered into a screened cable, with the screen reaching from just after the test clips to just before the banana plugs.

Practical tip: you can make your own Kelvin-connection test cable, either by soldering together test clips and flex lead or by using standard test leads with stackable plugs. The order in which the leads are soldered or connected (first the current lead and then the voltage lead, or vice versa) may affect the results of the measurement (see Section 4.24). To make accurate resistance measurements, connect the voltage lead directly to the component and connect the current-carrying lead to the voltage lead (with laboratory test leads, plug the current-carrying lead into the connector of the voltage-measurement lead).

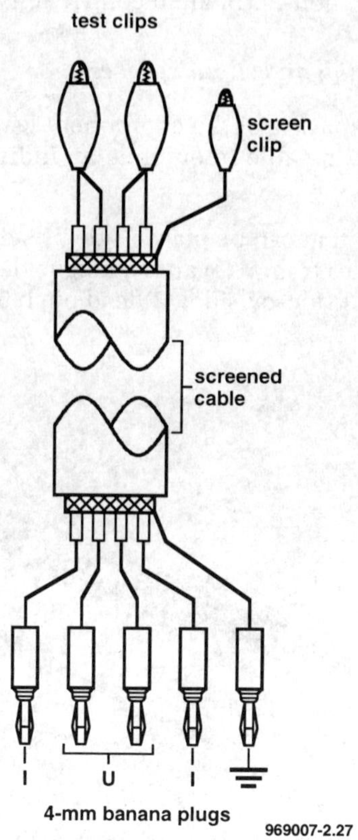

Figure 5.22 Kelvin-connection test cable.

Making your own test leads. Soldering or otherwise attaching leads to connectors is one of the basic skills of the electronics technician. Pay attention to the following:

- Use only well-insulated, sufficiently-flexible stranded wire.
- Every type of lead material has a maximum allowable operating voltage. For measurements on the mains network, the material should be specified for an operating voltage of 1000 V (RMS). Ordinary hookup wire (and especially the relatively inexpensive sort intended for automotive use) should only be used for low-voltage measurements.
- If the current is significant (battery-charging cables and current-measurement test leads, for example), make sure that the cross-sectional area is adequate.

Making connections to components in the circuit

Probes. Simple probes, as shown in Figure 5.23, are intended to be briefly touched to a measurement point. The probe tip must be truly sharp (so that it can penetrate an oxide or flux layer).

Test clips. Test clips (alligator clips, clip-on probes or 'grabbers'), as shown in Figure 5.24, are used to make fixed but non-permanent connections to component leads or pins.

Selection criteria for test clips and clip-on probes:

1 The diameter and/or spacing of the component leads to which connections are to be made: the probe must be able to encircle an individual lead without shorting it to neighboring leads.

2 The number of probes that can be placed side by side for making connections to a row of leads. Some examples from relevant catalogue descriptions: (1) lead pitch 2.54 mm to 0.8 mm, maximum 3 side-by-side; (2) lead pitch 0.5 mm to 0.3 mm, unlimited side-by-side.

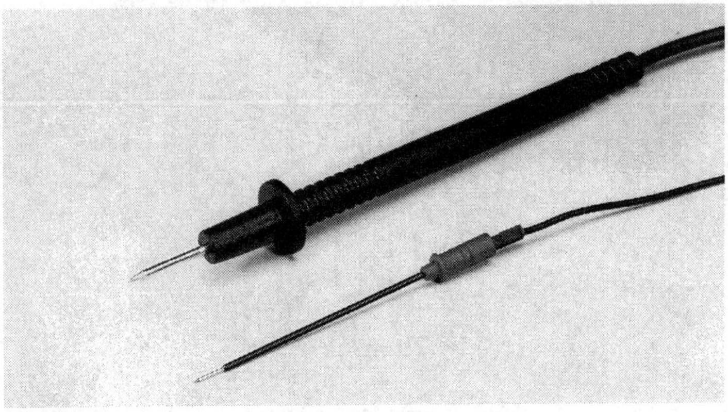

Figure 5.23 *Example test probes. Left: multimeter probe with fixed test lead; right: miniature probe (spring-loaded, with the test lead connected via a 0.64-mm plug).*

3 The type of connection to the test lead: (1) solder connection, (2) 4-mm jack, (3) 1-mm jack, (4) 0.64-mm pin, (5) 0.79-mm pin, (6) 0.56-mm pin.

4 The type of clip (see Figure 5.25): (a) hook ('grabber'),(b) spring clip ('pincher').

The 'pincher' type of clip is usually better, since it will stay firmly attached, while a hook-type clip can pop loose. However, the hook type is sometimes easier to attach and remove, so you should actually have both types on hand.

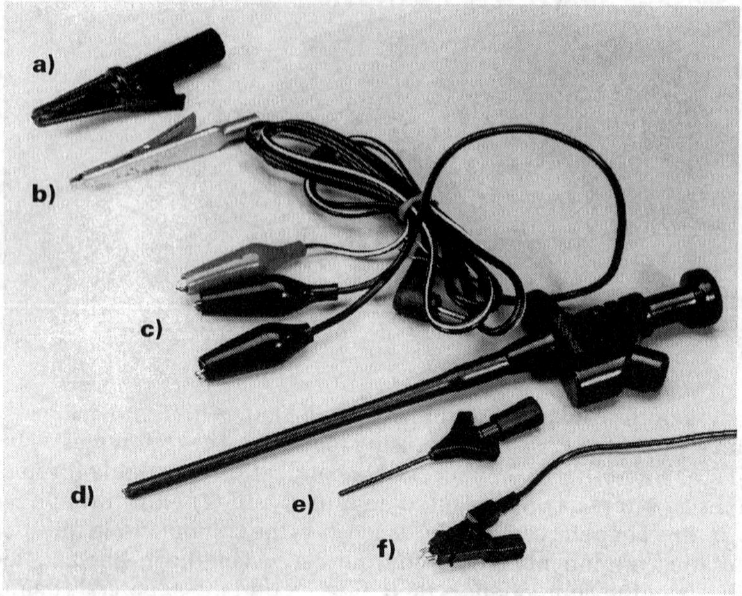

Figure 5.24 Selected types of test clips.

a) insulated alligator clip (preferred), b) alligator clip for 4-mm laboratory connectors (the simplest type), c) pre-assembled test cable with miniature insulated alligator clips (minigators), d) heavy-duty clip-on probe for 'rough' work, e) miniature clip-on probe, f) SMD alligator clip.

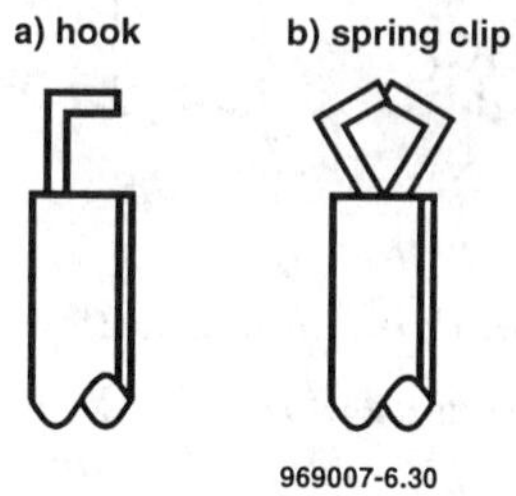

Figure 5.25 Types of test clips.

IC test clips. IC test clips (Figure 5.26) can make connections to all pins of an IC. Test clips for DIL ICs are inexpensive, but clips for SMD and PLCC packages are rather dear. Simple test clips only provide access to the IC leads, but there are also IC test aids, based on test clips, that can display the signal level of each lead using LED indicators.

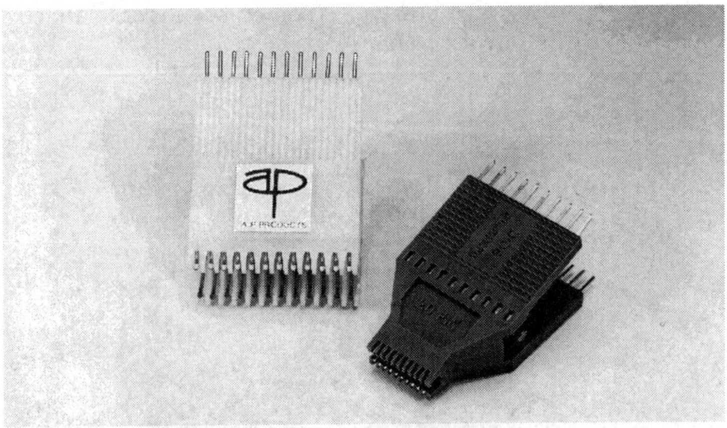

Figure 5.26 Test clips for various types of IC package (left: DIL, right: SOP).

In-line adapters. In-line adapters, as illustrated in Figures 5.27 and 5.28, are only necessary for relatively detailed fault-finding (using measurement techniques). They serve two purposes: (1) making otherwise inaccessible connections available for measurements (measurement adapters or *pin-breakout* adapters) and (2) allowing signal paths to be interrupted. In-line adapters can only be used when the component in question is mounted in a socket (the component is removed from the socket, the in-line adapter is plugged into the socket, and the component is then plugged into the adapter). Their utility in the service environment thus depends on the specific hardware that is to be serviced.

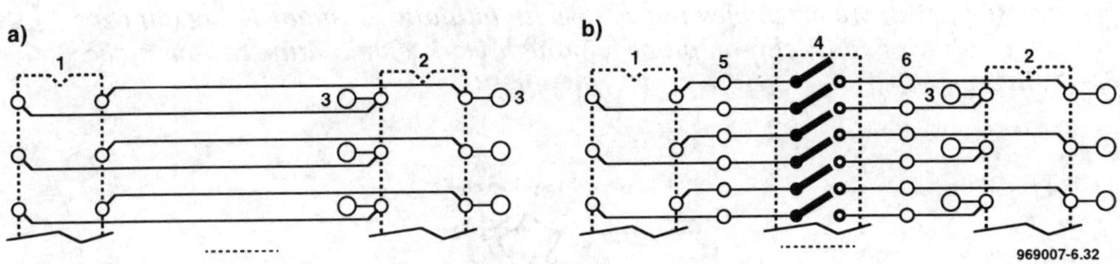

Figure 5.27 Basic types of in-line adapter.

a) measurement adapter (pin breakout), b) adapter with switches for opening signal leads.

1 – IC plug (installed in the socked in place of the IC), 2 – IC socket (receives the IC),
3 – measurement points (eg, 0.64-mm pins), 4 – DIL switch or bridging jumpers,
5 – source-side measurement points for opened signal lines,
6 – destination-side measurement points for opened signal lines.

Practical tips:

1 Test clips and adapters can only be used if there is adequate room around the IC socket in question. (Refer to the dimensioned drawings in the relevant catalogues when developing your own hardware.)

2 Before attaching a test clip to an IC, you may need to first clean the leads of the IC (to remove any residual flux, oxide layer or varnish). There are test clips that are specifically designed to penetrate such surface layers, but you should nonetheless watch out for possible problems.

3 Most test clips have 0.64-mm pins for making connections to test leads. This allows flat cables to be used for the test leads. Adapter cables with 0.64-mm sockets at one end and 0.79-mm pins at the other end (for plugging into probe pods) are commercially available.

4 It is a good idea to provide sockets for practically all ICs (and also for certain 'discrete' components) in the prototyping stage. Special prototyping adapters are available for SMD ICs. These have the same 'footprint' as an SMD IC and can be soldered to the circuit board in place of the IC, but they provide a socket for a functionally-equivalent IC in a more convenient package. *Some examples:* original SOP, substitute DIL; original PLCC, substitute DIL; original QFP, substitute PGA. The last type of adapter can be used when you plan to use a processor IC (for example) with an inexpensive QFP package,

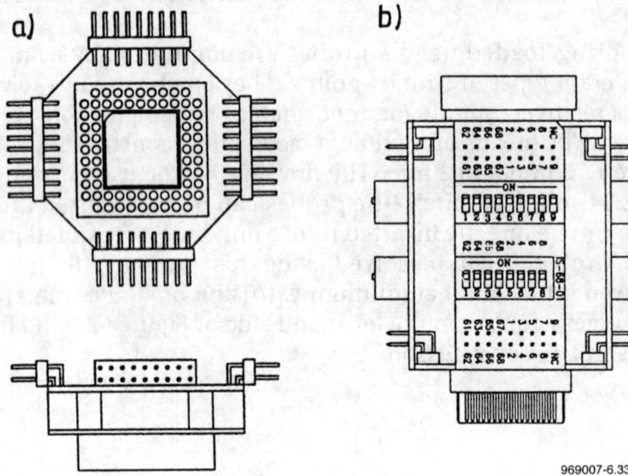

***Figure 5.28** Examples of IC socket adapters. (source: Synatron)*

a) Measurement adapter (corresponding to Figure 5.27a) for ICs with PLCC packages (plan view and side view). The measurement points are located at the sides of the adapter (using 0.64-mm pins).

b) Interrupting adapter (corresponding to Figure 5.27b) for ICs with PLCC packages (side view). This adapter is constructed in the form of a tower, with source-side measurement points at the bottom, interruption switches (DIL switches) in the middle and destination-side measurement points at the top. All four sides are the same.

which is almost impossible to remove once it has been soldered to the board. The prototyping adapter allows you to initially check out the circuit using a processor in a PGA package, which can be removed and replaced as desired.

5 DIY adapters: it is certainly possible to make your own adapters, although it can be somewhat tricky and time-consuming. You will need a mating plug for the IC socket in question, an IC socket, circuit board material (or a piece of prototyping board), DIL switches (for an interrupting adapter), multi-pin connectors, wire and epoxy glue.

6 Where can adapters be used in modern hardware? In the service environment, adapters are more useful for making measurements in the 'neighbourhood' of an IC (including opening bus lines, for example) than for testing the IC itself. If the IC itself must be tested, it is usually better to test it separately. Adapters can be used relatively easily with bus driver ICs and memory modules (and with memory ICs if they are socket-mounted). Processor adapters are usually not worthwhile, since they are too expensive and too difficult to use (see Volume 2 regarding the testing of processor ICs).

SMD test tweezers. It is quite easy to make connections to an SMD component using SMD test tweezers. Two types are available: (a) single-pole (the two arms are electrically connected) and (b) dual-pole (the two arms are electrically isolated). The dual-pole type can usually be recognized by the fact that the insulating coatings on the two arms have different colours; the single-pole type always has a single colour for both arms. Only the dual-pole type can be used to make connections to both ends of an SMD resistor (for example) at the same time. *Note:* these tweezers often have a screened or coaxial cable. The standard configuration is: red arm = inner conductor, black arm = screen.

Needle probes. Spring-loaded needle probes are commercially available with various diameters and different types of contact points. They can be used to solve special connection problems in a relatively simple manner, such as making reliable connections to individual SMD contacts (by means of a suitable mechanical construction). The standard outside diameters are 1.75 mm and 2 mm. The first size can be used to construct needle-bed adapters with a 2.54-mm lead pitch (the probes can be press-fitted into 1.7-mm holes). The 2-mm probes can be directly inserted into 2-mm board-mounted jacks. A large number of connection problems can be solved using this approach. Simple arrangements of jacks can be screwed into a bit of aluminium extrusion or similar material (see for example the battery-charger adapter on the left-hand side of Figure 5.95). Refer to Volume 3 for more information on DIY construction.

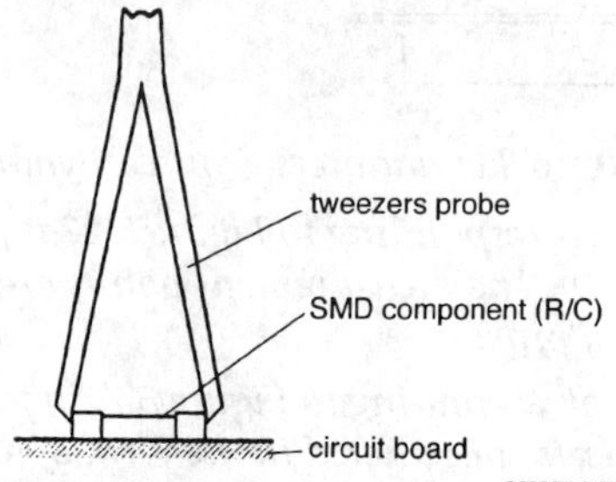

Figure 5.29 Connecting a dual-pole tweezers probe to an SMD component.

Soldered-on adapter cables. This is in many cases the simplest and most effective manner of making connections.

Practical tips:

1 To prepare the adapter cable, first remove 2 to 3 mm of insulation from the end of each lead and then carefully tin the exposed wires (be careful that no strands protrude from the tinned wire ends).

2 Attaching the cable: a simple tack joint is fully adequate in most cases. It is sometimes a good idea to slip a length of heat-shrink tubing over each lead before soldering it. Once the lead is firmly attached, the tubing can be slid over the solder joint.

3 Suggested locations for attaching the leads: (1) pins soldered into through-holes in the circuit board; (2) readily-accessible circuit-board tracks. Soldering directly to SMD contacts is a matter of practice and luck.

FlexLead adapter (Figure 5.30): this type of adapter consists of a piece of flexible circuit that has a set of multi-pin connectors at one end and a row of contacts at the other end. These contacts can be soldered to the full set of pins of one side of the IC to be tested. The adapter is unsoldered after use (and discarded – it is intended to be used only once). The advantages of FlexLead adapters are: (1) they can be used with very small lead spacings (down to 25 mil, or 0.625 mm), (2) they are relatively inexpensive and (3) they have relatively small inductive and capacitive loading. Some suppliers can provide video cassettes that describe how to solder the adapters to ICs.

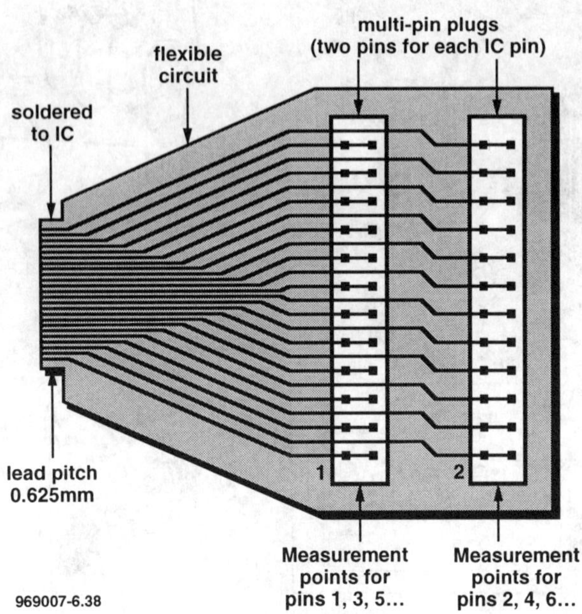

Figure 5.30 A flex-lead adapter.

Making connections to loose components

It is sometimes possible to test individual components 'loose' on the workbench. Even so, you should still use reliable probes, test clips, test tweezers and the like.

Test clips. Insulated clips are available in a variety or sizes, so there is no reason why you should use uninsulated clips for making connections to a power transistor (for example). Ready-made test leads with small, insulated alligator clips are inexpensive and suitable for all imaginable types of improvised test setups (see Figure 5.31 – these circuits are actually usable – and Section 5.1.6).

Practical tip: do not make 'benchtop' tests on a conductive surface (including an anti-static mat). Ceramic working surfaces are available (if need be); a sheet of Pertinax or an (undecorated) ceramic dinner plate can also be used as an expedient.

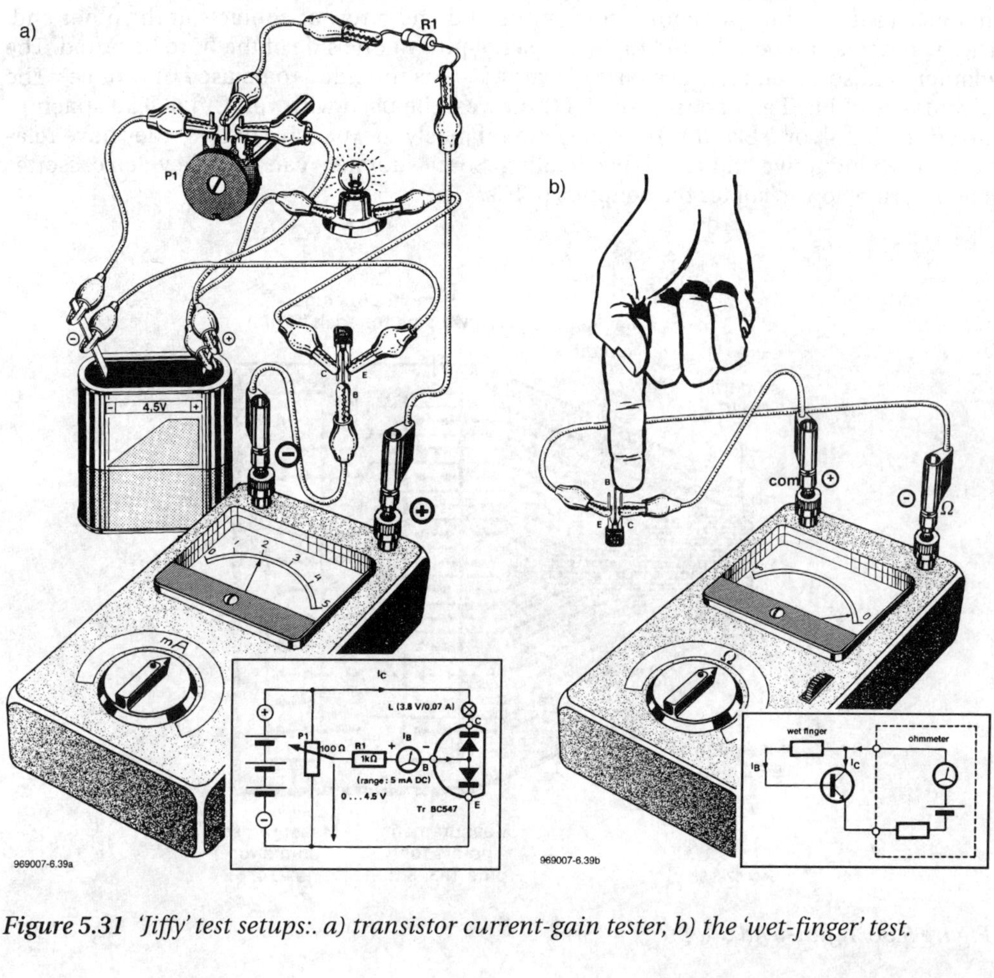

***Figure 5.31** 'Jiffy' test setups:. a) transistor current-gain tester, b) the 'wet-finger' test.*

Test boards. If you frequently have to test components, then robust test boards are a worthwhile. These need not be particularly 'pretty', but they must be reliable and durable. (By the way, using such test aids is by no means 'unprofessional'. In industry, there is a long tradition of employing competent and inventive technicians for the sole purpose of producing test jigs, special adapters and unique equipment.)

The simplest model: sockets, adapters and the like can be arranged on a piece of prototyping board (mounted on a frame) and interconnected using test leads and jacks (see Figure 5.32). Any desired type of test circuit can be 'knocked together' on this basis.

Figure 5.32 *Sample construction of a test board for component testing.*

Suitable sockets and adapters. The following items (among others) can be used: (1) zero-insertion-force sockets (with actuating levers) as shown in Figure 5.32, with their individual contacts wired to jacks; (2) zero-insertion-force sockets used as adapters for 2- and 3-lead components (see Figure 5.33); (3) test clips for two-lead components; (4) adapters for various types of two-terminal SMD components (MELF, SOD and so on); (5) transistor sockets and (6) universal breadboards (Circuit Test Strips and solderless breadboards).

Practical tip: 2-mm jacks are recommended for such test boards, since they make for compact and manageable assemblies. A few 2–mm to 4–mm connections should be provided to allow external devices to be directly connected using standard laboratory test leads. Test points for attaching oscilloscope probes (at least two) are also a good idea. See Volume 3 for more details.

Ready-made test jigs. Some types of components are best tested in typical application circuits. Such test circuits can be built in a permanent form using relatively simple materials, such as standard prototyping circuit boards (see Figure 5.34). The required components for the test circuit are permanently soldered to the board, and suitable connectors are provided for connecting the power supplies and test equipment. The component to be tested can be connected using a socket or soldered-on flying leads with test clips.

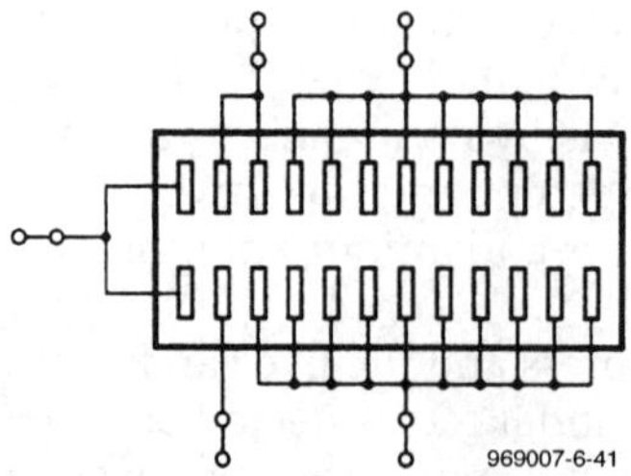

Figure 5.33 Test adapter for passive components and transistors.

A lever-arm type of zero insertion force socket can be used as an adapter. There are models available that can accept both narrow and wide DIL packages (300 mil and 600 mil); these have slot-type contacts in place of the usual pin sockets. (1 mil = 0.001"; 100 mil = 2.54 mm.) A relatively large resistor or capacitor can easily be inserted diagonally into such a socket, without requiring the leads to be bent excessively. Contacts can also be made to transistors with various types of lead arrangements (for instance, transistors in TO-220 and SOT-93 packages can be inserted without having to rearrange their leads).

Practical tips:

1 Circuit board size: the 160 mm × 100 mm Eurocard format is the most obvious choice, due to the broad availability of boards and related components at reasonable prices. If you use a 64-way DIN connector (DIN 41612), there are enough contacts for even relatively complex circuits.

2 The mechanical construction of the card receptacle must provide adequate clearance for cards using wire-wrap interconnections.

3 Certain connector pins should have pre-assigned functions, such as circuit earth, supply voltages and test equipment connections.

4 Interconnections on the circuit card can be made using insulated wire (always point-to-point) or wire-wrap techniques. Overly neat lead dress ('cable trees' formed using rectangular wiring arrangements) can result in incorrect operation (due to cross-talk and excessive lead inductance)!

5 *Test circuits (1):* one possibility is to simply copy the basic circuits that are provided in textbooks, data books, application notes and the like (the arrangement shown in Figure 5.34 should allow such circuits to be built without too much difficulty, particularly with regard to the mechanical construction). Typical test circuits can be found in data sheets and data books under the heading "Parameter Measurement Information" (for example). The values of load resistors, voltages, currents and so on can usually be found in separate columns in the data tables (under "Test Conditions") or in related footnotes. Components such as crystal-controlled oscillators, clock generators, integrated-circuit voltage regulators and voltage converters, power components, operational amplifiers and so on can be tested using such circuits. (Pay attention to the power dissipation and the current levels when constructing test circuits for power-handling components – use conductors with adequate cross-sectional area and provide heat sinks if necessary.)

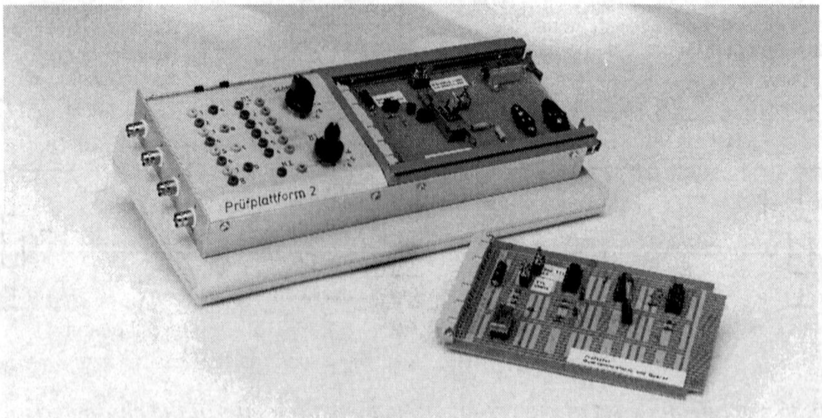

Figure 5.34 *An adapter for test circuits. This consists of a prepared test panel with jacks and related items, which accepts individual test circuits built on plug-in cards. This reduces the mechanical labour needed to construct new test arrangements.*

6 *Test circuits (2):* almost all manufacturers can provide small construction kits for certain ICs (such as operational amplifiers and voltage regulators). These 'evaluation kits', which are sometimes free of charge, can be used for experimental purposes. If you also provide a socket or some other sort of non-permanent connection for the component in question, you can build a usable test board at a relatively low cost. (The circuit board from the kit can be mounted 'piggyback' on a suitable Eurocard.)

7 Measurement equipment. Static testing (with a DC stimulus) is sufficiently rigourous for many types of components. (In contrast to the situation in a development laboratory, it is only necessary to check whether the component is in good working order, rather than thoroughly characterizing its behaviour.) How many instruments will you need to be able to work in a reasonably efficient manner? If you have only one multimeter that must be repeatedly disconnected and reconnected, testing will take far too long. As a point of reference, consider what you need to record the current/voltage characteristics of four-terminal devices (that is, devices that have an input with two terminals and an output with two terminals). Static testing in the service situation is in fact nothing more than recording device characteristics, although the measurements are made at only a few selected points rather than over the entire range of the device's behaviour (*for example:* checking whether a FET is turned on with a particular gate voltage). You must make current and voltage measurements on both the input and the output, which means that you need two voltmeters and two ammeters (see Figure 5.35).

Four instruments may be appear to be extravagant, but this is not so. First off, a portable digital multimeter is a basic necessity. Secondly, if you want to be able to make precise measurements, you should also consider an instrument with higher accuracy and more displayed digits. This is already two instruments, which could (for example) be supplemented by two panel-meter instruments, such as a digital voltmeter module and a 50-µA moving-coil meter. The last two items together would cost significantly less than £35 (or, for about the same amount, you could purchase two simple multimeters, which would

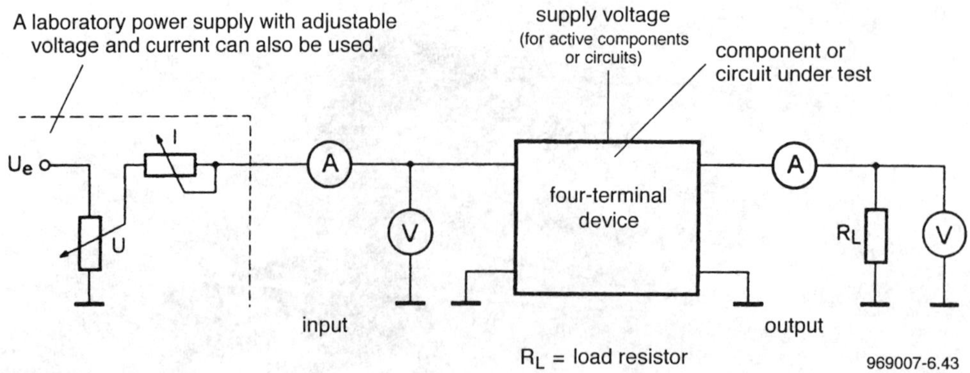

Figure 5.35 An adequate set of meters for static measurements of the characteristics of a four-terminal device.

also be adequate). The things to remember are: (1) only critical parameters need to be measured precisely and (2) the other meters perform more of a checking or monitoring function. The most convenient solution consists of a suitable plug-in measurement card for the PC or a multimeter with a PC interface, together with a selector switch for the individual measurement points.

Stimulus source and power supply. Laboratory power supplies can be used both to provide test stimuli and to power the device under test. Many tests can be performed using a high-quality dual-output laboratory power supply with built-in meters, without using any other instruments.

5.1.5 Passive components

Testing with a multimeter or an RLC component tester. An obvious way to measure passive components (or the associated electrical quantities of resistance, capacitance and inductance) is to simply connect them to suitable measuring equipment.

Practical tips:

1 Always start with the highest applicable measurement range. (Some RLC component testers, for instance, occasionally display invalid readings if the selected range is too low.) The precision and/or the number of digits of the measured value must increase each time the range is switched to a lower level. If a totally different result is obtained, something is wrong.

2 In case of doubt, disable automatic range selection (autoranging). (In particular, components with non-linear characteristics, such as forward-biased diodes, occasionally cause an unstable, constantly-changing display if autoranging is active.)

3 Make proper contact to the component being tested (to avoid measuring the effects of intermittent contacts or contact resistance).

4 Do not touch both contact points at the same time. Otherwise, the reading may be affected by the resistance or parasitic capacitance of your body.

Switch contacts. Depending on its state (open or closed), a set of contacts will exhibit either a very high resistance or a very low resistance. Normally, a continuity test is fully adequate. For more accurate measurements:

1 *Contact resistance:* check with a milliohmmeter or Thomson bridge, or measure the voltage/current characteristic (see Figure 5.36).

2 *Isolation resistance:* check with an insulation-resistance tester or a megohm bridge, or measure the voltage/current characteristic (see Figure 5.37).

3 *Insulation resistance:* in general, you must always consider the possibility of a fault if the insulation resistance drops to too low a value. This applies to printed-circuit boards and to all types of components, but naturally in the first place to high-impedance circuits. You should look for the following: (1) sweat or condensation (moisture) on component surfaces (this is one reason why equipment should be allowed to acclimatize or warm up before it is used) and (2) flux on the printed-circuit board. Flux residues, especially those located underneath surface-mounted components, can cause high leakage currents. (Whether this is actually a problem depends on the flux itself and the operating conditions.)

4 *Intermittent contacts:* try using an oscilloscope to see if you can observe anything when the equipment is operating normally. Alternatively, make a continuity test with the power off using a pulse detector (which you can construct yourself; see Figure 5.38).

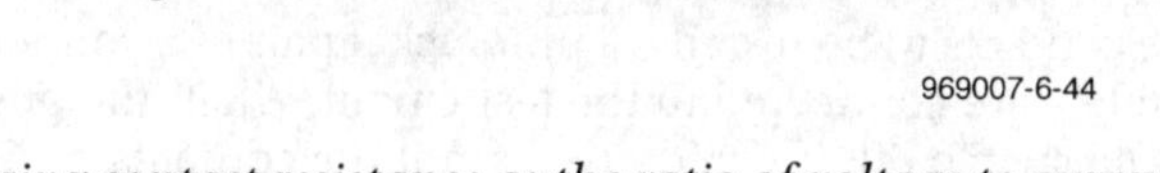

Figure 5.36 *Measuring contact resistance as the ratio of voltage to current.*

Practical tips for Figure 5.36:

1 Limit the current to the rated current of the contacts under test.

2 When testing relay contacts, apply the rated current or voltage to the relay's winding.

3 Make several measurements while repeatedly opening and closing the contacts. The contact resistance is then the quotient of the highest measured voltage and the lowest measured current.

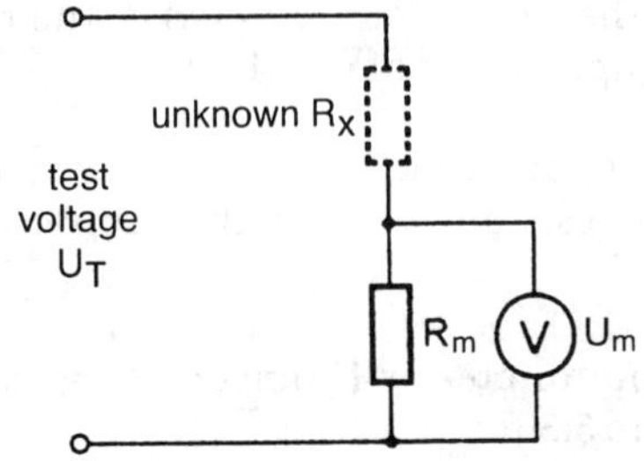

$$I = \frac{U_P}{R_X + R_m} = \frac{U_m}{R_m}$$

$$R_X = R_m \cdot \frac{U_P}{U_m} - 1$$

969007-6-45

Figure 5.37 Measuring insulation resistance via the voltage/current ratio.

4 Especially when making measurements on components that have been stored for an extended period of time, actuate the contacts several times (opening/closing) before making any measurements.

Practical tips for Figure 5.37:

1 The measured current for common values of insulation resistance lies in the picoampère range. For measurements to be at all possible, the resistance R_m must be in the megohm range. (Try using 1 MΩ and 10 MΩ resistors, starting with the lower value.) The smaller value of R_m for which usable measurements can be made, the more reliable the results.

2 The input resistance of the test equipment should be at least ten times the value of R_m.

3 Use good common sense when evaluating the results of your measurements! (The best idea is to first make test measurements on a component whose insulation resistance is reasonably well known, or that can at least be judged to be 'good', such as a freshly-unpacked relay or connector.) High-impedance measuring devices of this sort are easily disturbed, and the measurement is also influenced by atmospheric humidity.

4 Test voltage: the higher the test voltage, the more likely you are to obtain useable measurements. However, you must not exceed the component's specified test voltage.

In Figure 5.38a, the contacts to be tested – a pin/socket pair of a connector or a set of relay contacts, for example – are connected to the test circuit. When the pushbutton is pressed once, the LED should switch off. If you try to disrupt the contacts by jiggling, shaking and the like, even a very brief open contact will produce a LO-to-HI transition that will set the D-type flip-flop, causing the LED to light. Figure 5.38b shows an arrangement that can detect small variations in the contact resistance (this is one of the simplest test circuits for such purposes). The comparator produces a LO-HI transition as soon as the contact voltage exceeds the forward voltage of the diode. (The comparator output should be connected to the clock input of the flip-flop.)

Resistors. Ordinary resistance values can be adequately checked using a digital multimeter. If special test equipment for very high or very low resistance values is not available (such as a milliohmmeter or Thomson bridge), the best alternative is to measure the voltage/current relationship, as previously described.

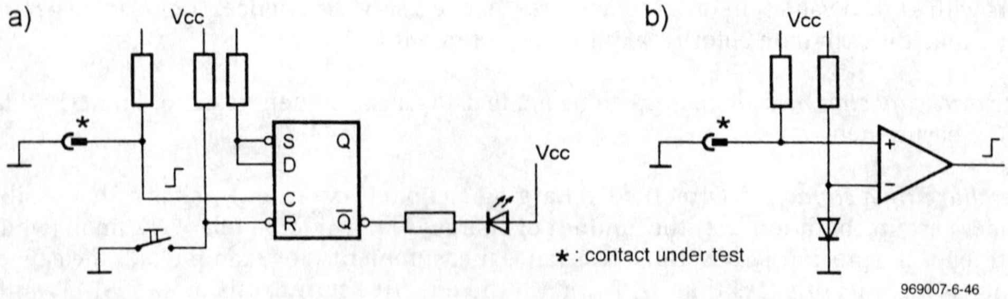

Figure 5.38 Testing for an intermittent connection.

Practical tips:

1 For non-linear resistors, voltage/current measurements yield results that are fundamentally more reliable than those obtained using an ohmmeter. (This applies in particular to 'plausibility' checks: for instance, the results of quick measurements of the resistance across the power supply terminals of a plug-in card can be different, depending on the type of ohmmeter used. A specification such as 'x mA at 5 V, y mA at 0.5 V' is significantly more exact and can be checked with any desired type of instrument.

2 If you use an RLC component tester, check to see how it measures resistance. Some instruments use an AC test voltage in the resistance ranges as well, and display the ohmic resistance as the real part of the measured impedance. This occasionally makes sense, but if you are interested in the DC resistance then the indicated value is not always usable. This applies primarily to: (1) continuity testing of inductors (you expect a DC resistance that is nearly zero, but measure a non-zero AC impedance), (2) testing capacitors for shorts (you expect a DC resistance that is nearly infinite, but measure a finite AC impedance) and (3) measuring resistors that are connected in parallel with capacitors (refer to Figure 5.10b).

Capacitors. A suspected short can be quickly checked by a (DC) continuity test. A capacitor that is not shorted but is in some way suspect is best replaced as a test. (Sometimes you can simply solder another capacitor of the same type across the suspect capacitor.)

Checking a capacitor using a multimeter. Many instruments have capacitance measurement ranges. The usual measurement process involves charging the capacitor for a defined interval. The resulting voltage is then measured and the capacitance calculated according to the formula $C = I \cdot t / U$ (in which the current is either held constant or is measured and integrated over time). Such a measurement process typically takes about one second; the result is only valid if the capacitor is fully discharged at the beginning of the measurement cycle. Capacitors of a few microfarads and more can also be roughly checked with an ohmmeter. The capacitor starts charging when the ohmmeter is connected; the resistance is initially low, but it gradually increases as the capacitor becomes charged. This effect is often more readily observed with an analogue meter than with a digital meter (the pointer quickly swings to the low-resistance end of the scale and then slowly returns to the high-resistance end). If you want to be able to better estimate the capacitance, you should practice making comparative measurements on various capacitors with known values, in order to acquire the necessary experience. (*Caution:* discharge the capacitor each time before making a measurement.)

Accurate capacitor measurements can be made using a component tester or an RLC bridge (AC measurements).

Discharging a capacitor. Directly discharging a capacitor without risk is only possible under certain conditions: (1) the amount of charge is known to be relatively small (which is true for the previously-described resistance measurements, for example), (2) the capacitance is significantly less than 1 µF and (3) the effective series resistance (ESR) is sufficiently high (this applies to energy-storage capacitors, for example – they will not be fully discharged even if they are short-circuited for an extended period of time). Otherwise, you must always discharge the capacitor using a resistor. *Note:* electrolytic filter capacitors in power supplies should have permanently-connected bleeder resistors if the capacitance is relatively large or the voltage relatively high. Select the component values as follows:

- The RC time constant should be 1 to 5 s (R = bleeder resistor, C = capacitor to be discharged); in other words $R \geq (1\text{ s})/C$ (thus 1 MΩ for 1 µF, 10 kΩ for 100 µF and so on).
- If either one of the capacitor terminals is not accessible, make sure that all circuit elements that carry the discharge current (including the circuit-board tracks) can handle the current $I = U_C/R$ (where U_C is the voltage across the capacitor).
- The power-dissipation rating of the bleeder resistor must at least $P = U_C^2/R$.

Practical tips: (1) use a resistor substitution box (pay attention to the power dissipation) and connect it using standard test leads, or (2) for a 'universal' solution, attach a pair of probes or insulated test clips to a 220-Ω, 2-W resistor via insulated leads and bridge the capacitor to be discharged for 1 to 2 minutes with this arrangement.

Inductors. A suspected open winding or short between two windings (in a transformer) can be quickly checked by continuity measurements. A partially-shorted coil will show a reduced DC series resistance. The nominal value of the coil resistance can be roughly calculated from the wire diameter, if the length of the winding is known. (This can be estimated from the number of turns and the dimensions of the coil). A simpler approach is to compare the series resistance of the coil to that of a known-good, identical inductor.

Practical tips:

1 The DC resistance will only have a significant value if the coil consists of many turns of relatively thin wire. A more accurate test is to measure the inductance directly using a component tester or RLC bridge.

2 *Caution:* do not exceed the inductor's core saturation current. Doing so can permanently change its inductance (especially with ferrite ring cores). As a general rule, use currents under 1 mA for continuity tests and resistance measurements (you can rely on with the resistance ranges of most modern digital multimeters; check in case of doubt).

Quartz crystals and crystal oscillators. Packaged crystal oscillators only need a supply voltage and a suitable dummy load (see Figure 5.39). In order to test a quartz crystal, you must construct a suitable oscillator circuit. Figure 5.40 shows a sample test circuit that operates reliably with fundamental-mode crystals up to around 20 MHz. These two circuits can be built onto a single prototyping circuit board. (Figure 5.39 shows how to connect an oscilloscope. With high-frequency oscillators, a 2:1 divider can be connected ahead of the oscilloscope.) See Volume 3 regarding construction details.

Diode components ('normal' diodes, Zener diodes and LEDs). A suitable test voltage is needed to perform continuity tests on diodes. It must be greater than the forward voltage

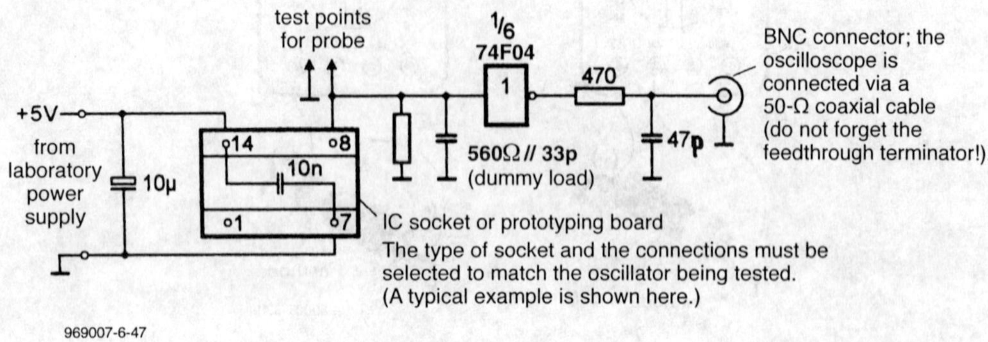

Figure 5.39 Test arrangement for packaged crystal oscillators.

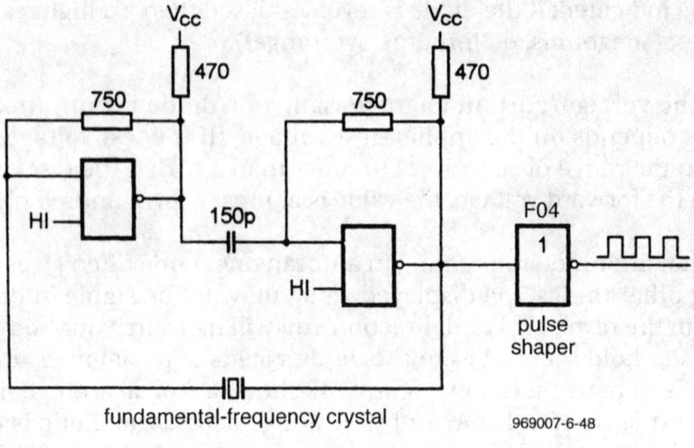

Figure 5.40 Test circuit for fundamental-mode quartz crystals.

drop of the diode for testing in the forward-biased direction, and less than the reverse breakdown voltage for testing in the reverse-biased direction.

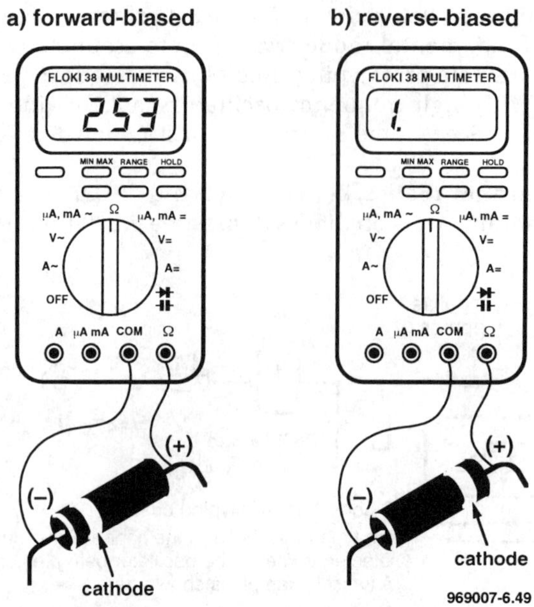

Figure 5.41 *Diode testing using an ohmmeter.*

Diodes can be roughly checked checked by making resistance measurements with an ohmmeter (see Figure 5.41). If the diode is forward-biased, a low (or at least finite) resistance should be indicated; if the diode is reverse-biased then a a high resistance should be indicated (several megohms, or 'infinite' (overrange)).

Practical tip: the voltage/current characteristic of a diode is non-linear. The indicated resistance thus depends on the applied test voltage. (If the test voltage is less than 0.5 V, the value lies in the range of several kΩ to more than 1 MΩ; if the test voltages is equal to or greater than the forward voltage, the value is at most a few hundred ohms.)

Caution: a digital multimeter operating in autoranging mode often gives a confusing indication (among other things, the displayed value may not be stable in the forward-biased direction, and in the reverse-biased direction you will measure your body resistance if you use both hands to hold the probes to the diode's leads). Special measurement ranges for diode testing offer a better solution. Figure 5.42 shows a typical arrangement: the test voltage is 3 V, the test current is 1 mA, and the voltage across the diode is displayed. In the reverse-biased direction the full test voltage should be displayed (approximately 3 V; some instruments also indicate 'overrange'); in the forward-biased direction the forward voltage of the diode should be displayed (0.4 to 0.8 V, for example, or significantly more for LEDs).

Identifying diode leads. Usually, the cathode is specially marked (with a coloured ring or a shorter lead, for example). An unknown arrangement can be sorted out by making measurements as shown in Figure 5.41 or 5.42: the lead connected to the negative terminal (the COM jack in Figure 5.41) when the diode is forward biased (low resistance) is the cathode lead.

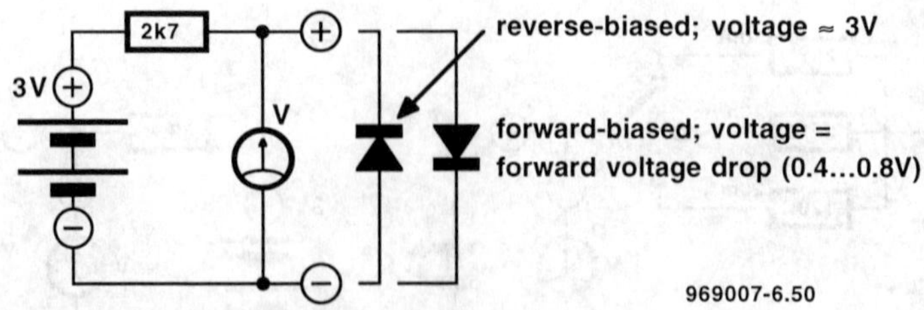

Figure 5.42 Simple diode test circuit.

Testing LEDs. LEDs have forward voltages between 1.3 V and 5 V (see Table 5.7), low reverse blocking voltages (typically 5 to 8 V, but many types can withstand up to 20 V) and can handle only low currents (guidelines: 2 mA for low-current LEDs; 10 mA for miniature types; 20 mA for 'standard' types and 25 to 50 mA for large and ultra-bright types:). A current of 1 to 2 mA is usually enough to produce visible illumination.

Colour	Forward voltage drop
infrared (IR)	1.3–1.7 V (1.5 V typical)
standard red	1.6–2.2 V (1.8 V typical)
orange & high-brightness red	1.5–3.0 V (2.2 V typical)
yellow	1.5–3.0 V (2.2 V typical)
green	1.5–3.0 V (2.3 V typical)
blue	2.7–5.0 V (3.0 V typical)

Table 5.7 Forward voltage drops of light-emitting diodes.

The large majority of LEDs can be tested using the 'diode test' range of a digital multimeter. Otherwise, it is sufficient to connect the LED to a supply voltage of 4.5 to 5 V via a current-limiting resistor (see Figure 5.43). Both test circuits shown in Figure 5.43 are designed such that the operating voltage will not exceed the allowed reverse voltage (special protective circuitry is not needed since power is provided by batteries). The LED will thus not be damaged if it is connected the wrong way round (reverse biased). In addition, it is a good idea to connect a voltmeter (digital multimeter) in parallel with the LED, so that the test corresponds to the standard diode test (as shown in Figure 5.42). In this way, it is also possible to check an IR diode without checking the actual IR emission (if you measure a voltage drop of 1.5 to 2 V in the forward direction and the supply voltage in the reverse direction, the IR diode is most likely OK). See also Volume 3.

Testing Zener diodes. Zener diodes should behave exactly the same as 'normal' diodes when forward biased; in the reverse-biased direction they must be tested with a voltage that is greater than the Zener voltage. In practice, you must check the diode characteristics at certain crucial points (see Figure 5.44).

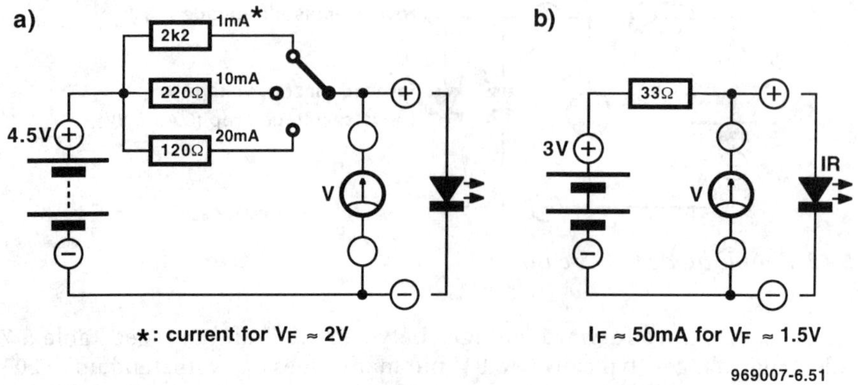

Figure 5.43 *'Simple and stupid' LED testers.*

a) LED tester with selectable forward current level. A 4.5 V supply is sufficient for testing almost all types of LEDs (the forward voltage is less than this, and the reverse blocking voltage is typically ≥ 5 V).

b) Tester for IR LEDs. These are typically operated at relatively high current levels. (The light emitted by the diode being tested should be intense enough to be detected by a very simple light-receiver test circuit; the test current is thus set to a nominal value of 50 mA.) The rated reverse blocking voltage of many types of IR LEDs is at most 3 V.

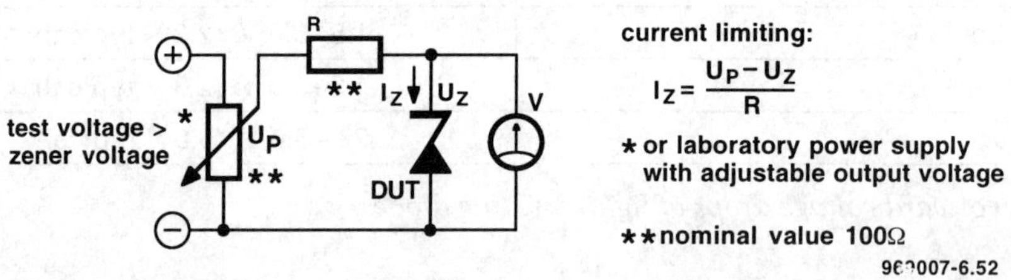

Figure 5.44 *Zener diode tester.*

The supply voltage must be greater than the Zener voltage. If the potentiometer shaft is turned from the lower limit toward the upper limit, the Zener voltage should be measured at some point. This voltage should not change when the potentiometer is turned further toward the upper limit. (A mains power supply with an adjustable output voltage can be used instead of the potentiometer.)

Testing transient-voltage suppressors (Transils, Transorbs and the like). These devices behave the same as ordinary diodes below their breakdown voltage (with bidirectional types, practically no current should flow in either direction). They can handle enormous current levels for short intervals (<10 to >100 A), but their long-term dissipation capability is no greater than that of a normal diode with the same package type. The transient-suppression function is normally tested using standardized pulse-shaped voltage waveforms, as shown in Figure 5.45.

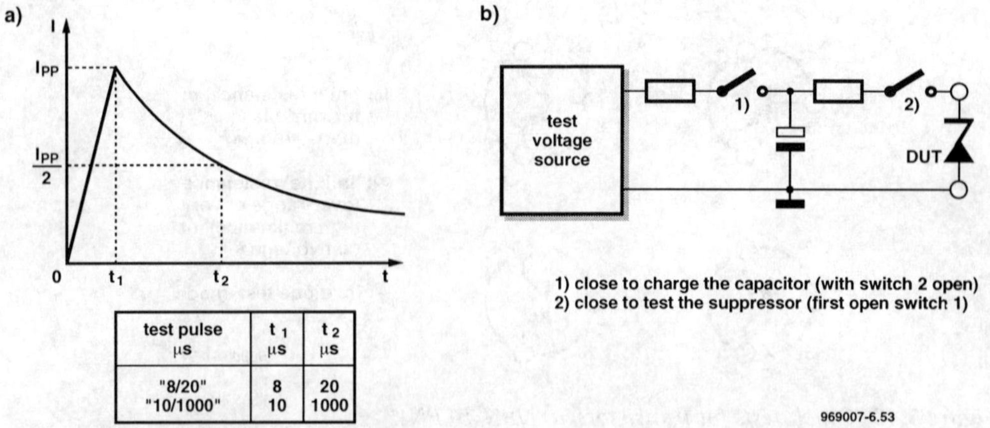

test pulse µs	t_1 µs	t_2 µs
"8/20"	8	20
"10/1000"	10	1000

Figure 5.45 Test pulse for suppressor diodes: a) waveform, b) test circuit block diagram.

Improvised DC testing. The allowed continuous current in the breakdown region is usually not specified. As an expedient, you can use the specified reverse current (I_{ZT}) of a Zener diode with the same type of package and a similar breakdown voltage as a guideline, in order to test suppressor diodes in an arrangement like that shown in Figure 5.44. (For devices in small packages and 'stacked' types, the suggested maximum current is 5 mA.)

Practical tip: the important thing is that the suppressor diode actually works if and when it should. The typical manner in which such diodes fail when they are overloaded is: (1) a short develops between the anode and the cathode, and (2) if a high current flows through the diode due to this short, it can melt internally and thus become 'open' (which by itself will not cause any other damage, but the failed device will no longer provide any protection). Since a shorted device can easily be recognized by a simple resistance test, manufacturers of suppressor diodes take pains to see that they are most likely to fail shorted when subjected to high short-term or continuous loads. A device that fails open cannot be recognized by a continuity test. The sole purpose of DC testing of suppressor diodes is to be able to detect devices that have failed open. (Such failures often produce visible indications, such as a scorched package or the like. In such cases it is recommended to replace the component as a precautionary measure.)

5.1.6 Transistors

In the service environment, it is sufficient to test transistors for basic functionality and to replace them if they are suspect (either with the same type or an equivalent type, based on a substitution chart). For a quick test, you can check the base-emitter and collector-base junctions for forward conductivity and reverse blocking using the diode-test range of a digital multimeter (see Figure 5.46). With a FET, you can similarly check whether the gate is isolated from the source and drain.

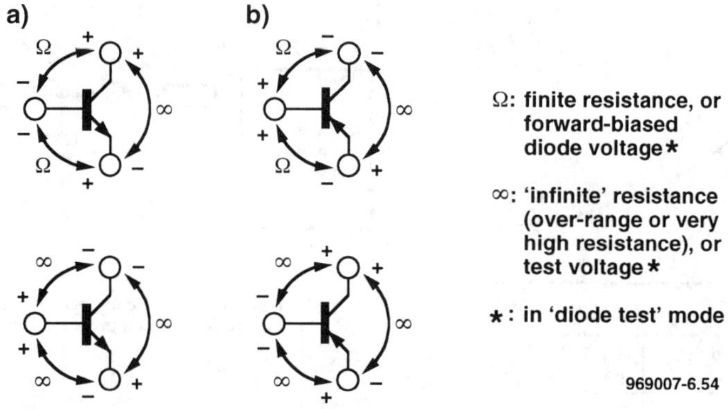

Figure 5.46 *'Quick' tests for transistors: a) NPN, b) PNP.*

A 'quick' transistor test requires a total of six measurements, since the resistance between each lead and the other two leads must be measured using both polarities. *Practical tip:* to determine the lead arrangement, you can use that fact that if the resistance between a given set of leads is 'infinite' for both polarities, then (assuming that the transistor is OK) these are the emitter and collector leads. Once you have identified the base lead, you can use a simple test circuit (as shown of Figure 5.31) to determine which of the other two leads is the emitter and which is the collector (the current gain of the transistor is the greatest when it is connected the 'right way round').

Figure 5.31 shows additional possibilities for simple plausibility testing:

a) Current-gain test: to allow the test to be made with only one meter, the collector current is 'measured' using a small lamp. If the potentiometer is adjusted from its lower limit towards its upper limit until the lamp shines with normal brightness, you can assume that it is passing its rated current (70 mA). The meter shows the associated base current.

b) Wet-finger test: this is a quick test to see whether the transistor functions at all. The meter is used as an ohmmeter. With an NPN transistor, connect the emitter to the positive terminal, and with a PNP transistor connect the collector to the positive terminal. (It is possible to determine the meter polarity with this arrangement by systematically experimenting with a known type of low-power transistor, or to determine the polarization of an unknown type of transistor.)

Transistor testers. Some multimeters have special ranges for transistor testing, and various types of transistor tester are available. Most of these can only be used with relatively low-power transistors. Such testers are actually nothing more than simple test circuits that have been more or less conveniently packaged. You can build such a tester yourself in the form of a test jig (see Figures 5.47 and 5.48). The working principle of such testers is that the various currents and voltages are measured with the transistor operating in common-emitter mode (for example). If you build your own tester, you can arrange it for testing the types of transistors that commonly fail in real life (such as the power transistors in switch-mode power supplies).

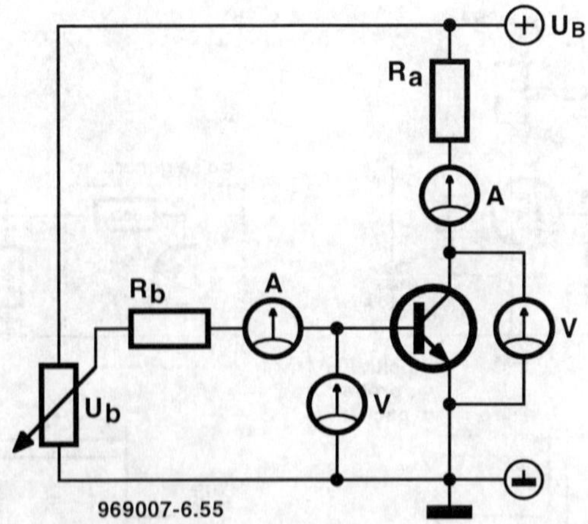

Figure 5.47 Transistor tester circuit for static measurements.

The current and voltage are measured for the transistor drive and its load. This circuit can also be used in a similar manner for N-channel FETs. PNP transistors and P-channel FETs need a negative supply voltage. The values of the load resistor R_a and the current-limiting resistor R_b must be selected according to the operating conditions of the transistor being tested (I_{CEmax} and I_{BEmax}). With a FET, practically no gate current should flow.

Figure 5.48 shows a typical circuit for testing low-power transistors. Most simple transistor testers use the working principle shown here: the base current is provided by a constant-current source (based on a junction FET in this case), and the collector current is measured. Since the base current is known, the tester can directly indicate the DC current

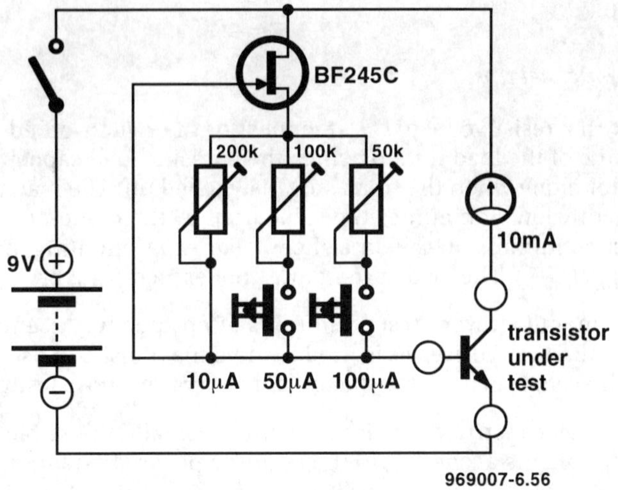

Figure 5.48 A simple transistor tester circuit.

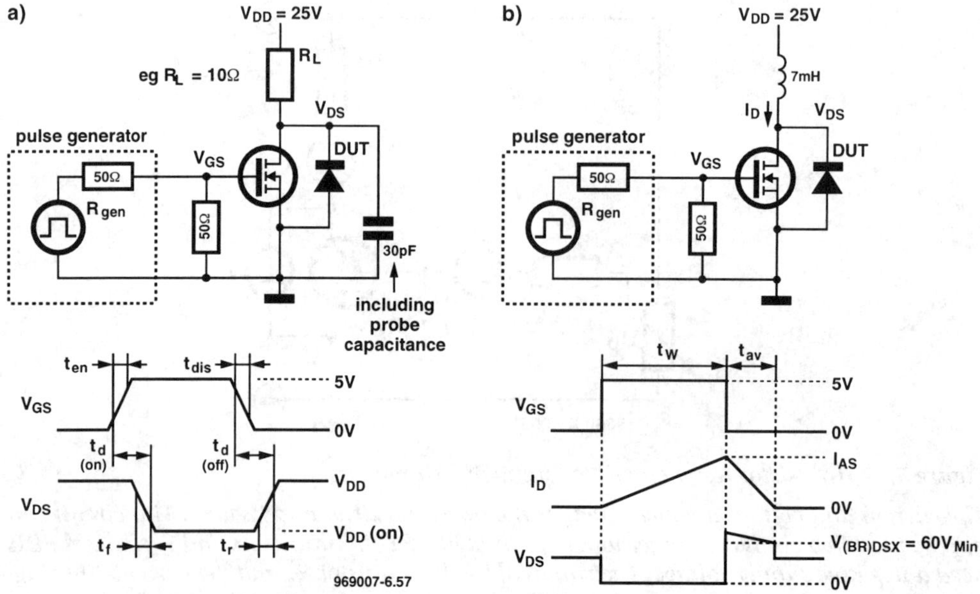

Figure 5.49 Dynamic testing of DMOS transistors: a) with a resistive load, b) with an inductive load. (source: Texas instruments)

gain. Power transistors can also be testing using such a circuit (in which case the tester should be constructed to provide base currents ranging from 1 mA to 1 A and to measure collector currents of several ampères). See Volume 3 for more information.

Dynamic testing. The dynamic characteristics of bipolar and field-effect transistors can be effectively checked by measuring their switching times using an oscilloscope (see Figures 5.49 and 5.50).

Details and practical tips:

- Testing with a resistive load: the steepness of the switch-off edge essentially depends on the value of the load resistor, since the parasitic load capacitance is discharged via this resistor alone when the transistor is switched off. The load resistance should thus be sufficiently low. Sometimes the value used by the manufacturer for testing is listed in the test conditions on the data sheet. . For experimenting, try $R_L = 2 \cdot (U_B / I_{Cmax})$ or $R_L = 2 \cdot (V_{DD} / I_{Dmax})$. The load current must not exceed I_{Cmax} or I_{Dmax}.
- Drive: the manufacturer's test conditions often specify a rise time less than 1 ns. If a gate of an ABT or AC driver IC is used to drive the transistor, rise times around 2 to 5 ns can be achieved, which are satisfactory for general-purpose testing.
- As a basic rule, construct test circuits with the smallest possible parasitic capacitance. Select components according to the required power dissipation, and use heat sinks if necessary.

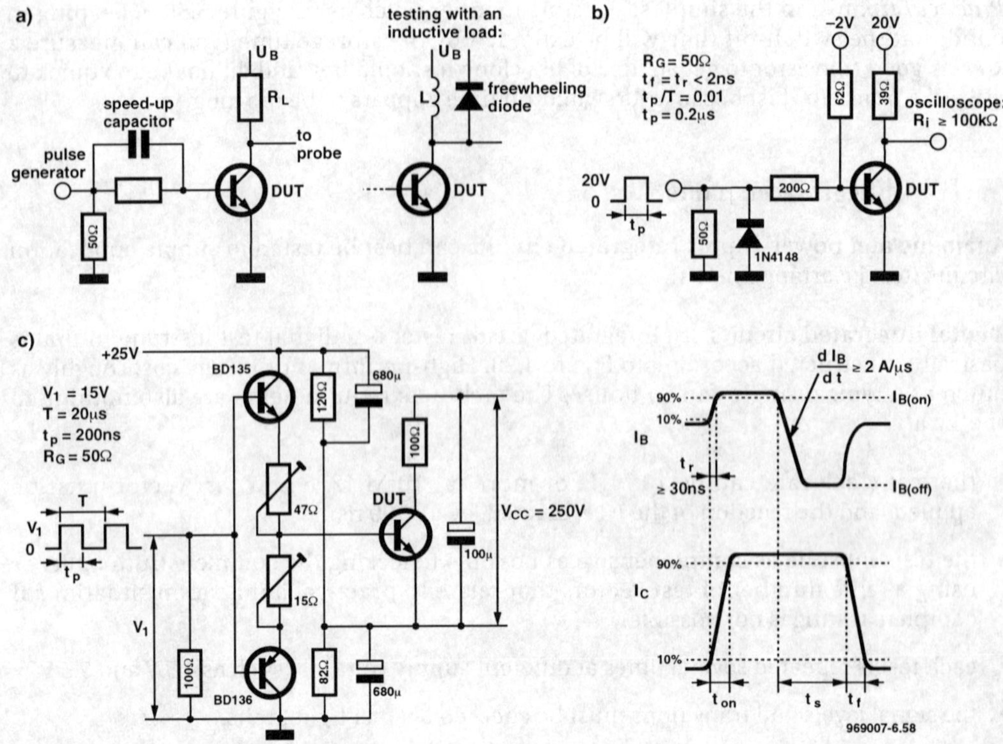

Figure 5.50 Dynamic testing of bipolar transistors (see text). (sources: AEG-Telefunken, Texas Instruments)

- Switching on a bipolar transistor: the transistor must be sufficiently overdriven if it is to switch on quickly. As a guideline for experimenting, use a base current that is five times the maximum base current shown on the U_{CE}/I_C characteristic curve chart.
- Switching off a bipolar transistor: the charge carriers must be removed from the base as quickly as possible. If this is not the case, the measured switch-off time may be considerably extended by the *storage time*. Possible remedies are (see Figure 5.50):

  a) Reduce the storage time by using only a small amount of base overdrive (the base resistor value must be selected accordingly). The speed-up capacitor in FIgure 5.50a will provide a large current pulse for switching on the transistor. Circuits that apply a negative voltage to the base when the transistor is switched off, so that the charge carriers are 'sucked out' of the base region, are even more effective (see (b).

  b) Use a supplementary negative voltage to switch off the transistor.

  c) 'Float' the transistor under test (see Figure 5.50c). When the drive pulse is LO, the lower drive transistor is turned on, so that a current can flow out of the base region.

Practical tip: test in the simplest possible manner (such as in Figure 5.50a), keeping in mind that the switch-off time will be extended by the storage time (you can measure a known-good transistor to get an idea of how long it should be), and do not be too quick to discard a transistor just because its switch-off time appears to be too long.

5.1.7 Integrated circuits

Analogue and power-supply integrated circuits can best be tested in simple application circuits (test jig arrangements).

Digital integrated circuits. An integrated circuit tester is a digital test instrument that is basically constructed according to Figure 4.98. High-performance testers cost roughly as much as a typical single-family house. The technical requirements are also nothing to sneeze at:

- the test clock rate can be 10 MHz or more (at 10 MHz, a new test vector must be applied, and the reaction of the IC evaluated, every 100 ns).
- the tests must be as comprehensive as possible (checking the complete truth table, or using a large number of test vectors that relate to practical application situations if complete testing is not possible),
- each test is repeated several times at different supply voltages (such as 4.5 V and 5.5 V),
- the signal levels and transitions must be checked against tight limits,
- the connections to the tester must be configurable as inputs or outputs according to the type of IC to be tested, and specified test loads must be connected to the outputs as appropriate,
- finally, everything must operate reliably in a mass-production environment.

At the other end of the scale, a variety of inexpensive IC testers are available (including PC cards), and some kinds of faults can even be identified using a simple 'manual' test jig (an arrangement of toggle switches and LED indicators connected to an IC socket).

Practical tips:

1. Inexpensive commercial testers are certainly suitable for locating hard faults in NAND gates, counters and the like. However, you should be skeptical if the claims made for the tester sound too good to be true.
2. There is less and less demand for testing small- and medium-scale integrated circuits, but there is a growing need to deal with programmable ICs (such as GALs), 3.3-V logic and 'wide' bus interface devices.
3. A tester that cannot apply dummy loads to IC outputs and that cannot distinguish various output levels is hardly worth its price. (Of course, such a tester will be able to recognize that a 74LS00 with an open V_{CC} bonding wire is faulty ....)

4 If you need to test ICs only occasionally, a simple test jig arrangement with an IC socket, some dummy loads and a few jacks is often the most reasonable solution. The test stimuli can be provided, and the outputs evaluated, using toggle switches and LEDs or by means of a PC-based test assembly. See Volume 3 for more information.

5 *Caution:* limit-value testing (with maximum-LO, minimum-HI input levels and maximum output loading) can lead to erroneous measurements, due to disturbances originating in the test setup. ICs are often exposed to more severe disturbances in simple or improvised test circuits than in their normal operating environments (on the circuit board of the equipment in question). Thus, an IC that seems to be faulty when tested under such conditions may work entirely correctly when installed in the actual equipment.

Dummy loads. The need to test under specific load conditions is often overlooked (with the result that you may wonder why an IC checks out OK but doesn't work properly in the circuit). The test loads or equivalent loads used for measuring device characteristics are specified in the data sheets of the IC manufacturers (see Figure 5.51).

Notes:

1 The test point for measuring the quantity in question is always the appropriate pin of the IC.

2 Where component values are not stated in Figure 5.51, refer to stated test conditions in the relevant device data sheets.

3 The specified value of the load capacitance C_L includes the parasitic capacitances of the test socket and the test adapter or oscilloscope probe.

4 All diodes shown are silicon diodes (1N4148 or equivalent) whose only function is to provide voltage drops equal to their forward voltage (0.7 V per diode).

5 Switching times for tri-state outputs: a tri-state output can be switched between the active state and a high-impedance state (Hi-Z). The switch-on time refers to the delay between the activation of the enable signal (usually identified as Output Enable) and the transition of the output to the active state. The switch-off time refers to the delay between the deactivation of the enable signal and the transition of the output to the inactive (high-impedance) state. The output can provide either of the possible active signal levels (HI or LO) when is enabled, and it can go to the high-impedance state from either active signal level. There are thus four different switching times that can be measured:

a) the output disable time from the LO level t_{PLZ} (LO switch-off time),

b) the output enable time to the LO level t_{PZL} (LO switch-on time),

c) the output disable time from the HI level t_{PHZ} (HI switch-off time),

d) the output enable time to the HI level t_{PZH} (HI switch-on time)

In addition, the gate propagation delay t_{pd} should be measured with the output active.

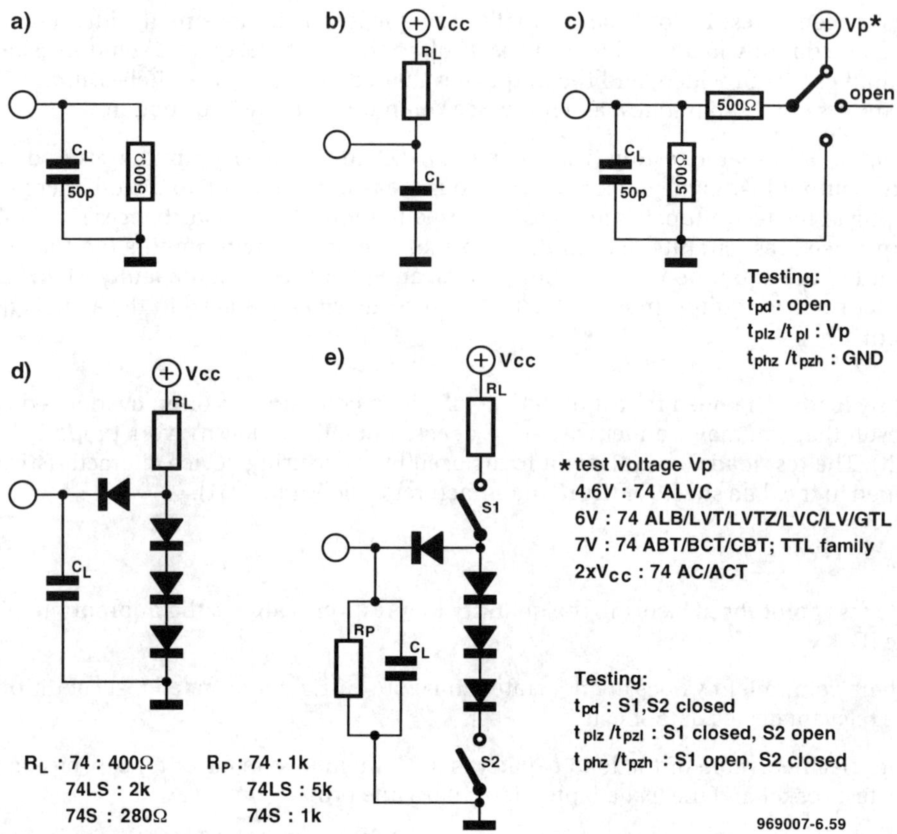

Figure 5.51 *Selected dummy loads for digital integrated circuits.*

a) the standard equivalent load for 74F/ALS, 74AC/ACT, 74BCT/ABT and low-voltage logic series,
b) equivalent load for open-collector outputs,
c) equivalent load for tri-state outputs (74F/ALS, 74AC/ACT, 74BCT/ABT and low-voltage logic series),
d) equivalent load for conventional TTL device outputs (74, 74LS/S),
e) equivalent load for TTL tri-state outputs, (74, 74LS/S).

Sources for test conditions. Most data sheets contain condensed specifications of the test conditions ('Parameter Measurement Information', 'Test Conditions'). These relate to the equivalent loads used, the signals applied to the inputs (test stimuli) and the expected switching behaviour of the outputs. Test condition specifications are sometimes provided collectively in a general-information section of the data book.

Functional testing. Functional testing is not described in detail in the data sheets. Instead, only the general conditions or – for complex ICs with built-in self-test provisions – the overall operation is described. The important information relates to how to start the self-test process and how to recognize whether the self-test process has detected a fault.

As an example, consider the following general conditions for testing basic logic ICs of the 74AC/ACT series (based on information published by Texas Instruments):

- Test patterns that ensure adequate functional testing are applied to the inputs (for simple ICs, this means exercising the complete truth table).
- 74AC-series ICs are tested with a LO level of 0 V and a HI level corresponding to V_{CC}; 74ACT ICs are tested with 0 V and 3 V.
- The test cycle frequency (application of the test vector and sensing the response) is normally around 10 MHz.
- Each test is executed three times with supply voltages (V_{CC}) of 4.5 V and 5.5 V, with the reference levels for output HI and LO levels set to V_{OH} – 200 mV and V_{OL} + 200 mV. The outputs are not fully loaded. 74AC ICs are also tested with V_{CC} = 3 V.

Test vectors and JEDEC files. The test patterns used for IC testing are usually referred to as *test vectors*. A single test vector relates to all IC pins for a single test step. The set of test vectors used for a complete test are collected together in a *test file*. The most-commonly used notation for test vectors comes from the US standards organization JEDEC (Joint Electronic Device Engineering Council). In this notation, each pin is unambiguously described by a single character. Figure 5.52 shows a sample set of JEDEC test vectors. In this example, the first D-type flip-flop of a 74x74 IC is being tested. The second flip-flop is held reset, and the test checks to see whether the reset state is maintained.

Notation

Inputs	Outputs
N - not connected	X - ignore (don't care)
C - clock	Z - high-impedance state expected
0 - LO level applied	LO - LO expected
1 - HI level applied	HI - HI expected

Example (74x74)

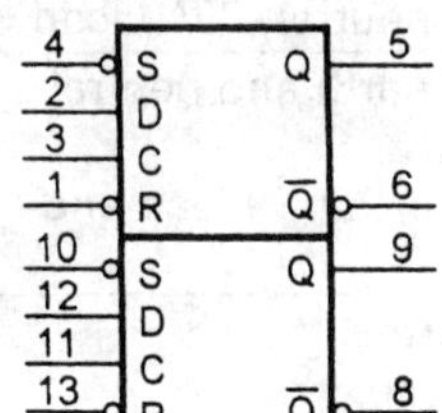

		1	2	3	4	5	6 | 8	9	10	11	12	13	
connections:		1	2	3	4	5	6 | 8	9	10	11	12	13	
test vectors:	1.	0	0	C	1	0	1 1	0	1	N	N	0	reset
	2.	1	0	C	0	1	0 1	0	1	N	N	0	set
	3.	1	0	C	1	0	1 1	0	1	N	N	0	data acquisition LO
	4.	1	1	C	1	1	0 1	0	1	N	N	0	data acquisition HI
	5.	0	1	C	1	0	1 1	0	1	N	N	0	reset

969007-6.60

Figure 5.52 Test vectors in JEDEC notation.

5.2 Testing power sources

5.2.1 Making measurements on the mains network

The mains installation

The mains installation is normally implemented as a TN-S network (see Figure 5.53, and Table 5.8 for an explanation of the terminology).

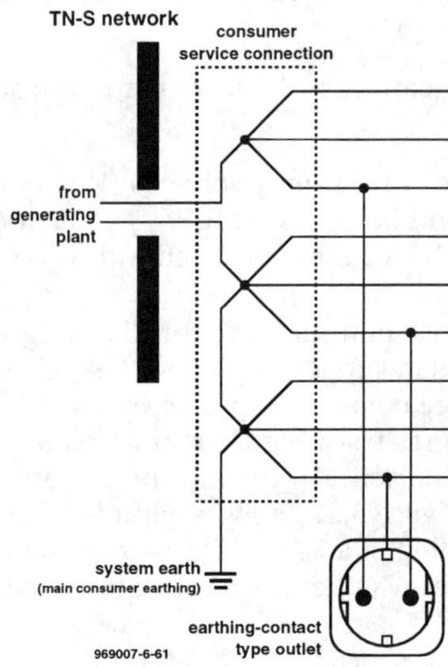

Figure 5.53 A TN-S network.

Position	Refers to	Options	Meaning
1	power source	T, I	T = star point is directly earthed; I = star point is isolated from earth
2	load equipment (enclosure or accessible metallic parts)	T, N	T = directly earthed; N = connected to the earthed star point via a protective earth lead or combined protective earth and neutral (PEN) lead
3	(TN networks only)	C, S	C = combined protective earth and neutral lead (PEN); S = separate protective earth (PE) lead and neutral (N) lead
Note: the code letters are the initial letters of the French technical terms: C = Combiné (combined), I = Isolé (isolated), N = Neutre (neutral), S = Separé (separate) and T = Terre (earth).			

Table 5.8 Mains-network type classification codes (according to VDE 0100 part 300).

End-user equipment is connected via three leads: the *live* or 'phase' lead L1, the *neutral* lead N and the *protective earth* lead PE. All externally-accessible metallic parts of the equipment must be electrically bonded to the protective earth lead (this applies to Class-I equipment). *Caution:* the earth lead and the neutral lead must not be connected to each other anywhere in the entire network, except at the main earthing terminal ('star point')!

Note: in the UK, standard mains plugs and sockets are mechanically polarized, so that the live and neutral leads have fixed assignments. Nonetheless, it is possible for wiring errors to occur, so you must pay attention to the following rules:

1 The neutral line may not be used for earthing purposes.

2 It is possible for a potential of 230 V to be present on any lead (except for the protective earth). Measurement equipment that is constructed to avoid dangerous situations when a 'live' voltage is present on the 'neutral' lead is thus to be preferred. Otherwise, it is necessary to check the actual lead assignments before making any measurements.

Additional types of networks:

- TT network: the supply source has a direct earth connection, and the installation metalwork is also directly connected to ground. An example is an overhead supply line with earth electrodes, where the mass of the earth provides the return path.
- IT network: the equipment is connected via leads that are not earthed at all; the entire network is isolated from the earth. Consequently, only a small fault current will flow if an earth connection is made (such as by touching a live lead). The equipment must be separately earthed. An IT network must be supplied by its own generator; it is restricted to emergency supply systems.
- TN-C-S network: this is the same as a TN-S network, except that the sheath of the supply cable is also used as a neutral; that is, it forms a combined earth/neutral conductor called the protective earth/neutral (PEN). This system is also referred to as *protective multiple earthing* (PME) in the UK.

A unique feature of domestic mains installations in the UK is the use of *ring circuits* instead of branch circuits for most loads (except lighting circuits), as shown in Figure 5.54. Each ring circuit can supply a larger current than a branch circuit in a standard continental European domestic installation (30 A fusing instead of 16 A, with the operating current limited to 69% of the fuse rating). In addition, the mains plug for any equipment must be a three-contact, fused type. Double-insulated (Class II) equipment (such as hand tools) also uses such plugs, although the earth pin is not connected with such equipment.

Voltages and currents in the mains circuit

The only difference between measurements on the mains network and measurements in other situations is that the mains voltage can be dangerous. It is thus important to be familiar with the basic situation and know what to expect when making measurements on the mains network. Figure 5.55 illustrates a mains circuit, both in the ideal case and under real operating conditions.

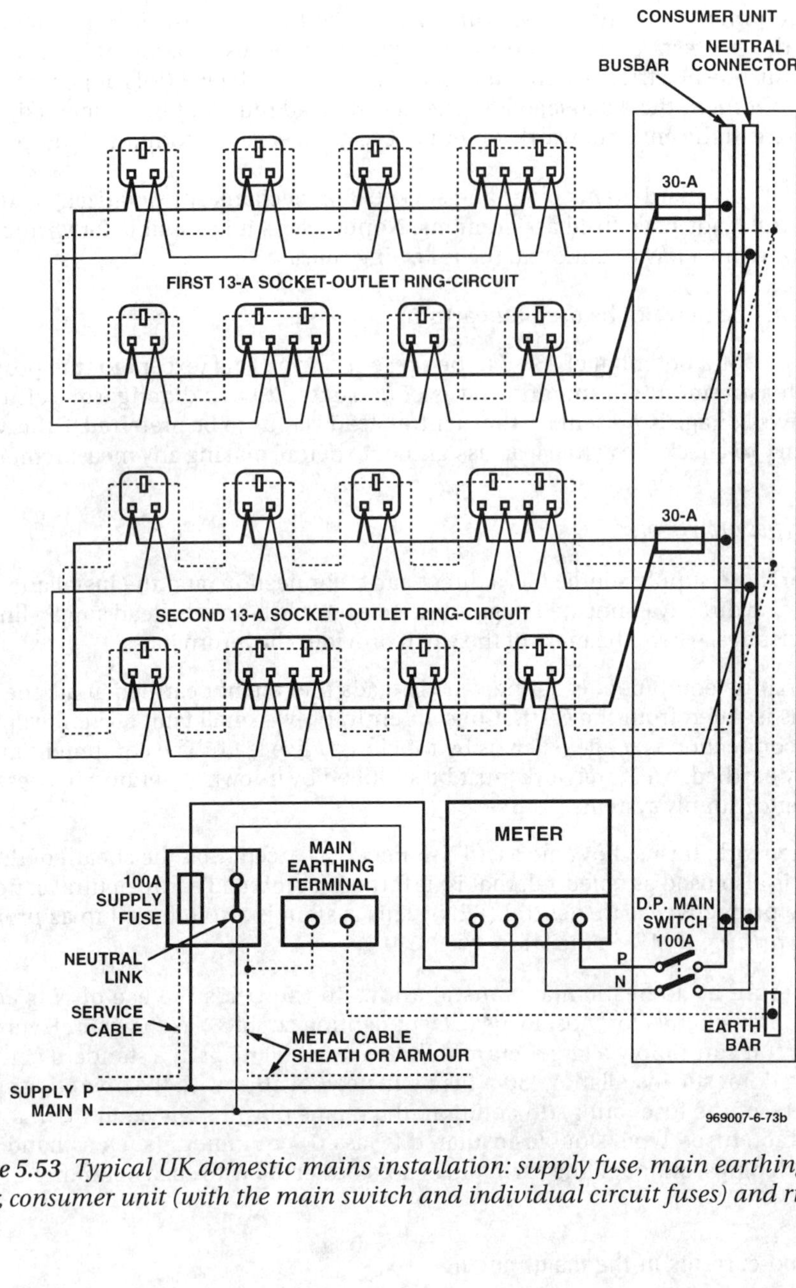

Figure 5.53 Typical UK domestic mains installation: supply fuse, main earthing terminal. meter, consumer unit (with the main switch and individual circuit fuses) and ring circuits.

Notes:

1 For a UK ring circuit, the maximum operating current for any circuit is limited to $0.69{\cdot}I_f$, where I_f is the rated current of the fuse or miniature circuit breaker (MCB) in the consumer unit. (A continental branch circuit is limited to a maximum of $0.8{\cdot}I_f$.)

2 Ring circuits are fused at 30 A in the consumer unit (lighting circuits are fused at 15 A). The available maximum load current for a ring circuit is thus 20.7 A. (A continental branch circuit is typically fused for 10 A or (at most) 16 A, yielding maximum currents of 8 A and 13 A respectively.) The maximum fuse rating for a mains plug is 13 A.

3 The maximum apparent power that can be drawn (in VA) is 230·0.69·I_f, which is 4760 VA for a ring circuit (for Continental circuits: 230·0.8·I_f, which means 1840 VA at 10 A and 2944 VA at 16 A). The true power (in watts) is equal to the apparent power times the power factor (cos φ), which typically has a value of 0.6 to 0.8.

4 In principle, voltages should be measured with respect to the earth potential. This would however require a direct, independent connection to the main earthing terminal, which is usually not available. In practice, it is usually only possible to make measurements with respect to the protective earth lead.

5 The exact value of the measured quantity is not as important as how it compares to the approximate values that are to be expected in ordinary practice (see Table 5.9). The measurement equipment should be designed to handle such voltages, with regard to measurement ranges, insulation and so forth. Specifications based on worst-case situations are therefore fully satisfactory.

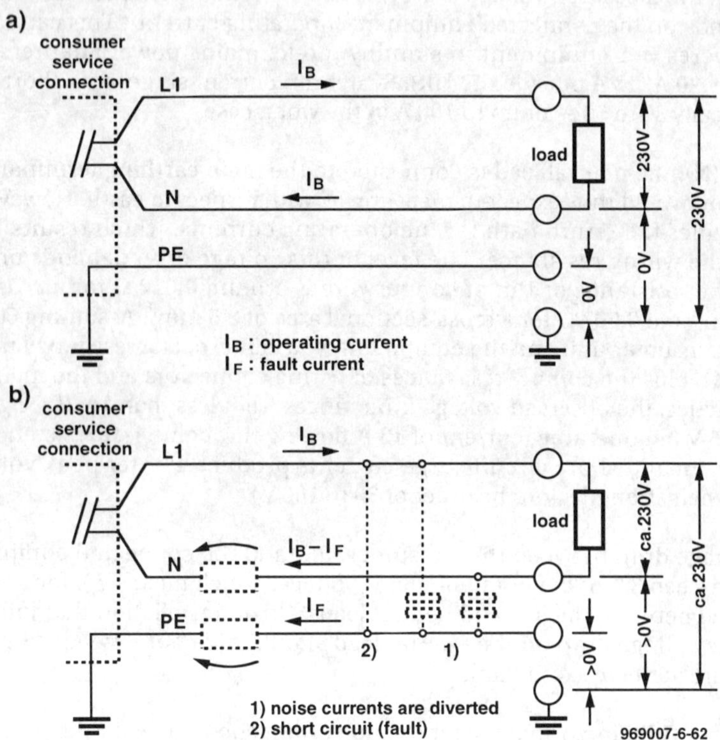

Figure 5.55 Voltages and currents in the mains circuit.
a) the ideal situation, b) the real situation.

Lead	Potential to earth	Current
live or phase (L1)	230 V ~	a few milliamperes to around 20 A (13 A for Continental circuits); short-term (switch-on) levels are higher
neutral (N)	theoretically 0 V; in actual use up to around 10–15 V; transients to 30–100 V (due to switch-on currents)	as for the live lead (carries return currents)
protective earth	theoretically 0 V; transients up to 320 V (peak) in case of a fault	theoretically none, in practice a few milliamperes; short-circuit currents around 50–80 A in case of a fault

Table 5.9 Voltages and current values in mains circuits.

Voltages and currents in the individual leads

Live lead (L1): the measured voltage is around 230 V~ (effective value), and the measured current depends on the connected equipment (operating current). This can range from a few milliampères (for equipment in standby mode, mains power adapters etc.) up to approximately 20 A (13 A outside the UK). Switch-on current surges and short-circuit currents are typically 30 to 50 A, or up to 100 A in the worst case.

Neutral lead (N): the neutral lead is connected to the main earthing terminal in the consumer unit. You would thus expect to measure 0 V with respect to earth. However, the neutral lead provides the return path for the operating currents, which results in a voltage drop due to the wiring resistance. The level of this voltage drop depends on the actual resistance. The resistance of 1 m of copper wire is around 0.012 Ω for a cross-sectional area of 1.5 mm^2 and 0.007 Ω for a cross-sectional area of 2.5 mm^2. Assuming 50 m of cable between the consumer unit and the equipment, you can expect a resistance ranging from 0.35 Ω to 0.7 Ω, plus the contact resistances of wiring connectors and the mains plug and socket. In practice, the observed voltage drop ranges from less than 1 V (for light loads) to roughly 13–15 V (an operating current of 13 A flowing through a 1-Ω resistance produces 13 V). Switch-on and short-circuit surge currents produce greater peak voltage drops, according to their strengths (on the order of 30 to 100 V).

Note: the voltage drop between the consumer unit and the connected equipment is not supposed to exceed 3% of the nominal mains voltage (which means 7 V for a 230-V mains supply, divided between the live and neutral leads). In an installation that fully meets the standards, the voltage drop on the neutral lead should thus not exceed 3.5 V, but not all installations are perfectly compliant.

Protective earth (PE): in normal operation, no current should flow in the protective earth lead. A current will however flow in the PE lead when it fulfills its intended role (protection in case of an accidental contact between the live lead and the equipment housing) or if it is used for other purposes (such as diverting noise signals).

The utilization of the protective earth lead:

1 Protection in case of an accidental contact between the live lead and the equipment housing: the original function of the PE lead is to provide a return path in case the live mains voltage comes in contact with an accessible metallic part of the equipment (due to some sort of internal fault). Since all such parts of the equipment are connected to the PE lead, in such a situation a short-circuit current will flow that will cause the fuse or MCB of the live lead to open (see Figure 5.56). A voltage of up to 320 V (the peak value of the mains-voltage half-wave) will be present on the PE lead until the fuse or MCB has opened.

2 Providing a common earth-potential reference: the PE lead is typically used as a common (0 V) reference potential in all systems in which various pieces of equipment are interconnected via signal lines, in addition to being connected to the mains network.

3 Diverting disturbances: the PE lead is used to divert induced high-frequency noise signals and overvoltage peaks to earth (see Figure 5.57).

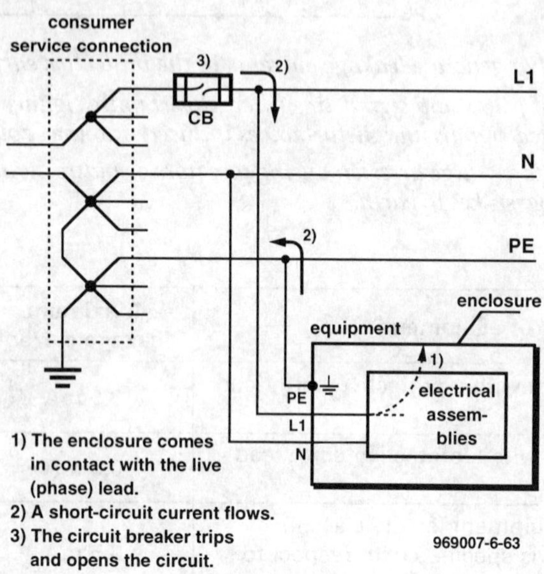

Figure 5.56 The actual function of the protective earth lead.

If noise signals and voltage spikes are diverted to earth via the PE lead, then a current (referred to as the *leakage current*) must flow in this lead. In the case of overvoltage protectors, such currents flow only intermittently, but filter capacitors produce a continuous current in the PE lead. The maximum allowed level of this current is set by the VDE 0701 standard, which also specifies the maximum allowed capacitance between the live and neutral leads and earth (see Table 5.10).

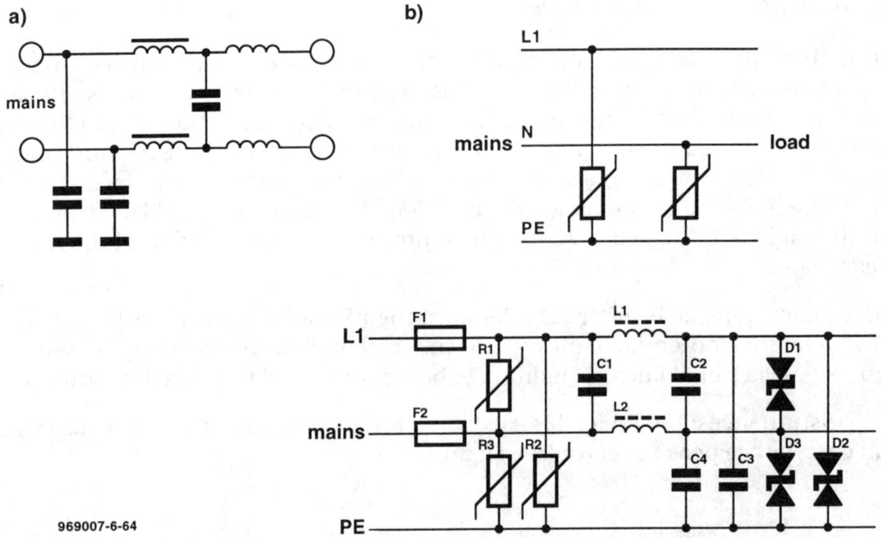

Figure 5.57 *Devices that generate leakage currents in the protective earth lead.*

a) Power-supply input filters and signal-line filters divert noise signals to earth (internally-generated signals and disturbances induced into the cable leads).

b) Overvoltage protectors (surge arresters) using varistors or bidirectional suppressor diodes divert voltage spikes to earth.

Type of equipment	Maximum leakage current	Maximum capacitance
stationary equipment with a protective earth lead (safety class I)	3.5 mA	35 nF
portable equipment with a protective earth lead (safety class I)	0.5 mA	2.5 nF
double-insulated equipment (safety class II) The leakage current is specified with respect to the equipment's internal earth reference.	3.5 mA	35 nF

Table 5.10 *Allowable leakage currents and capacitances to earth (according to VDE).*

Current consumption. Many types of equipment do not draw a steady current. In this regard, we must distinguish between switch-on behaviour and normal operation.

Switch-on current (inrush current). Incandescent lamps, all equipment with heating coils (including coffee makers), electrical motors and switch-mode power supplies (which are used in innumerable types of electronic equipment, in particular PCs) draw much more than their usual operating current for a brief interval immediately after being switched on.

Example 1: the measured cold resistance of a 100-W incandescent lamp is 37 Ω. This results in a current of roughly 6 A at 230 V. In operation, the lamp draws a current of only 0.43 A (100 W/230 V).

Example 2: the inrush current of a switch-mode power supply can range between 20 and 90 A (even if the operating current is less than 1 A). A value of 30 to 50 A for < 10 ms is typical. Figure 5.58 shows a sample inrush current characteristic.

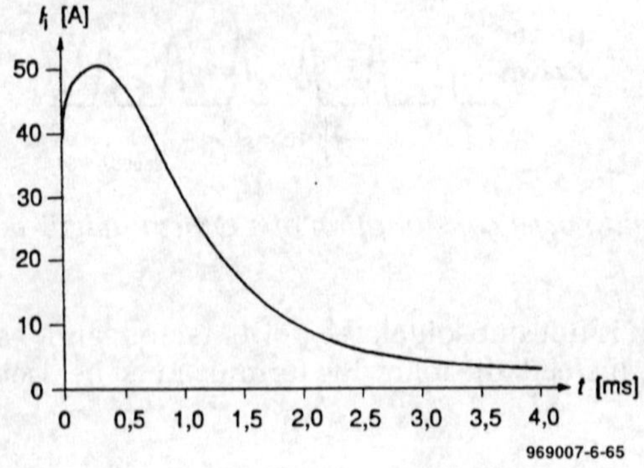

***Figure 5.58** Sample inrush current waveform. (source: Melcher)*

Notes:

1 Most switch-mode power supplies have provisions to limit the inrush current. These work on the basis of a time delay. If the equipment is switched on immediately after being switched off, these provisions may not be effective. A reasonable interval between switching off and switching on again ranges from 30 s to 2 min.

2 There are also special external devices that can be connected ahead of the equipment to limit the inrush current. They contain a resistor in series with the mains lead (typically a few ohms) that is shorted out (bridged) by a relay contact after an interval of several mains cycles following switch-on (refer to Figure 5.69).

Pulsed current waveforms in normal operation. Typical switch-mode power supplies draw current from the mains discontinuously, in the form of pulses (see Figure 5.59).

Notes:

1 The peak value of the current pulse depends on the degree of loading of the power supply — with a light load the pulses are narrow with high peak values, while with a heavy load the pulses are broad and the peak value is lower. The peak value is typically 4 to 5 times the effective value (see Figure 5.59).

2 The voltage waveform remains practically sinusoidal, due to the low impedance of the mains network. The power factor *PF* is thus determined by the current waveform alone.

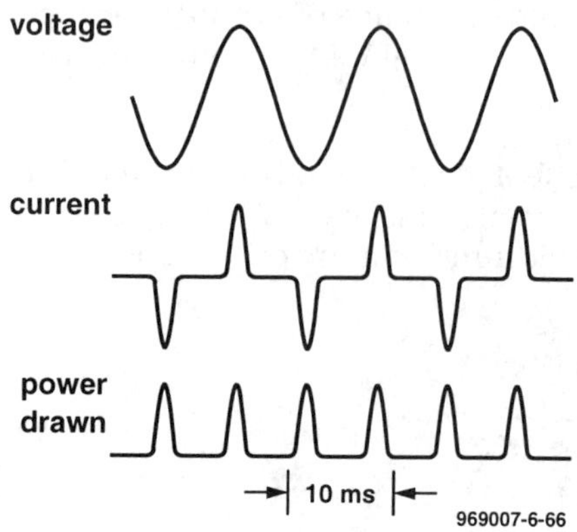

Figure 5.59 *Current and power consumption of a typical switch-mode power supply.*

Since this waveform is not sinusoidal, the power factor cannot simply determined from the phase angle (φ). Instead, the following formula must be used:

$$PF = \frac{I_1}{I_{rms}} \cdot \cos \varphi$$

where I_1 is the effective value of the fundamental term (or first harmonic) of the current waveform (50 Hz in this case), φ is the phase angle of the fundamental term of the current waveform and I_{rms} is the net effective value (real current).

In the ideal case $PF = 1$, which means that $\cos \varphi = 1$ (that is, there is no phase difference between the current and the voltage waveforms) and $I_1 = I_{rms}$ (sinusoidal current waveform). This can be approximately realized by placing a *power-factor correction* circuit (consisting of a step-up regulator or blocking converter) is connected ahead of the actual

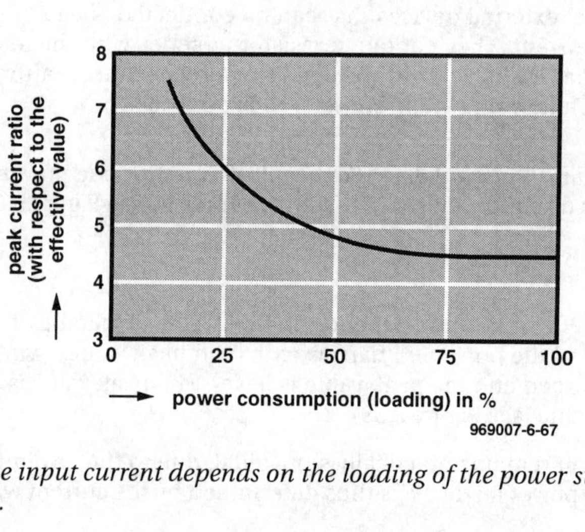

Figure 5.60 *The input current depends on the loading of the power supply (simplified representation).*

switch-mode supply (see Figure 5.61). Fundamentally, the mains voltage is passed through the input filter and immediately rectified. In a conventional power supply, the resulting pulsating DC voltage is applied to a storage capacitor. Current is only drawn from the mains when this capacitor is being charged, which means whenever the instantaneous value of the rectified voltage is greater than the capacitor voltage. This results in a charging current that only flows at the midpoint of each half-wave (see Figure 5.59). In addition, if the storage capacitor is completely discharged, as it will be on switch-on, it effectively acts as a short circuit. This explains the high level of the inrush current.

The power-factor correction circuit is a step-up converter (boost converter) that is connected between the rectifier and the DC-to-DC converter. It 'screens' the storage capacitor from the mains network, which sees a current-limiting inductor (a choke) instead of a capacitor that only draws current to the extent that it is not fully charged. The mains current waveform can be adjusted to be synchronous with the mains voltage waveform. The pulsating current waveform produced by the correction circuit approximately matches the shape of the sinusoidal mains voltage. This type of power supply can have an power factor of 0.98 (that is, almost 1). *Note:* a highly effective input filter that suppresses the higher harmonics of the current waveform shown in Figure 5.61c is essential for the proper operation of this arrangement.

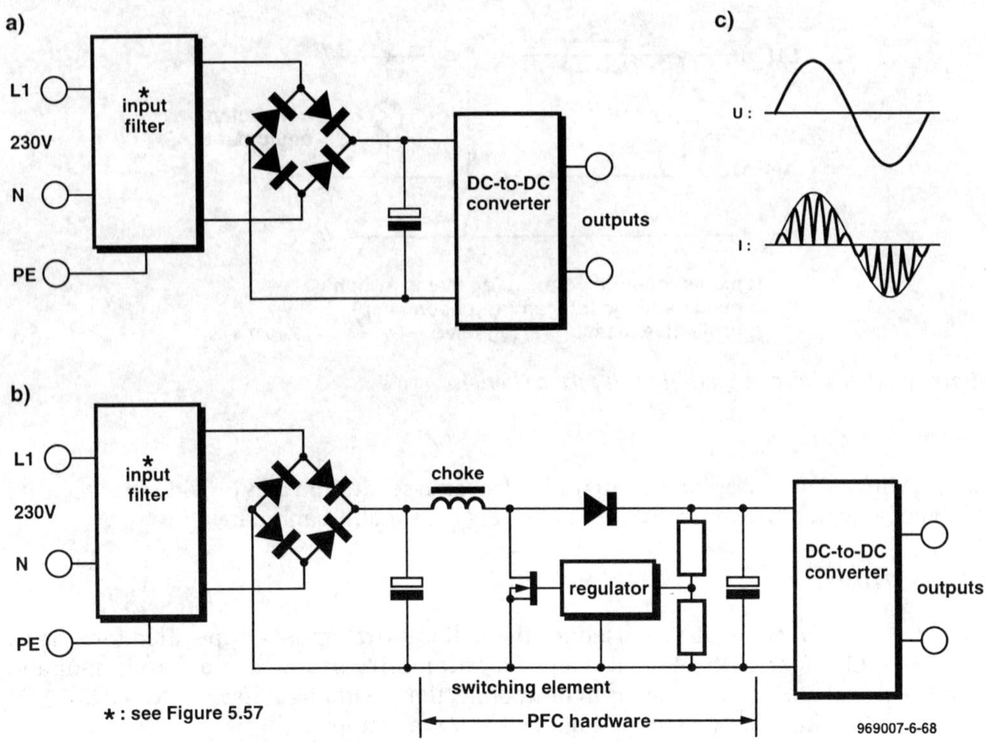

Figure 5.61 Power-supply input (a) without power factor correction (PFC) and (b) with PFC. Diagram (c) shows the input voltage and current waveforms for a power supply with PFC.

Mains connections

You will have to refer to the relevant technical literature for information regarding the fine points of mains installations. In this section, we want to examine only two things: (1) how the mains connection of the service bench should be set up and (2) the preferred form of the mains connection in the hardware being tested (this applies in particular to the usual types of PCs, which frequently have significant weaknesses in this area).

Functional requirements for the mains connection:

1 Mains switch: it should be possible to isolate the connected equipment from the mains without having to unplug any connectors.

2 Mains voltage indication: its should be readily evident whether or not the mains voltage is present.

3 Protective measures: the connected equipment should be disconnected from the mains if a dangerous situation arises, to avoid personal injury and (as much as possible) to prevent damage to equipment.

Figure 5.62 shows the necessary elements in a simple example circuit.

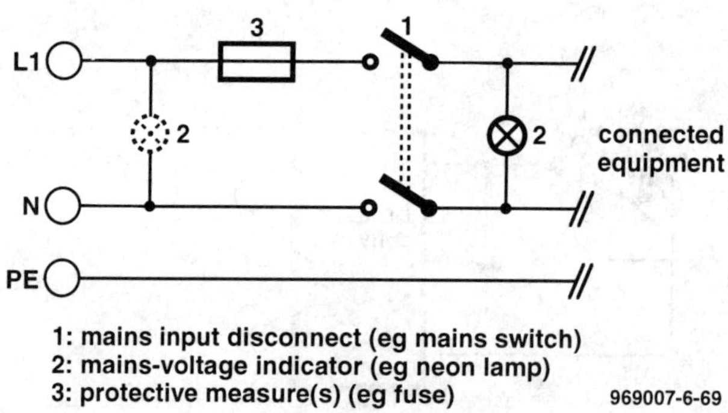

Figure 5.62 *Mains connection for a service bench.*

Mains switch. Only a two-pole switch with galvanic separation can provide true security. Mechanical switches, relays and circuit breakers are suitable candidates.

Selection criteria:

- *Switching voltage:* the device used for the mains switch must be specified for at least 230 V~. *Caution:* switches and relays for automotive use (which are abundant and inexpensive) are not suitable for mains connections. Although they are tested at 250 V, they are usually only rated for operation at 48 V maximum.
- *Switched current:* the specified switched-current rating should be greater than the rated load current to be switched. Pay attention to peak switch-on currents and pulsed

input currents! The switched-current rating may be directly specified, or it can be calculated by dividing the power rating by 230 V. Certain devices are expressly specified for high peak switch-on currents (eg 60 to 80 A for 10 ms maximum).

- *Construction:* the operating state of the switch should be clearly shown by its physical state. This means that you should choose a rotary, rocker, key-operated or toggle switch. Push-button switches may make sense for bedside lamps (since you only have to press the button to switch on the light), but they are abominable in PCs. For some sample alternatives, look at Figure 4.44. *Practical tip:* a high-quality key switch (which costs £10 or so) can provide a simple solution to a number of PC security problems (including protection against unauthorized use).

Mains voltage indicators. The basic possibilities are incandescent lamps, neon lamps and LED indicators. Neon lamps and LEDs are preferable, due to their low power consumption and reliability. They cannot however be directly connected to 230 V – see Figures 5.63 and 5.64. As an alternative or supplement to such lamps, a voltmeter (moving-iron meter) can be provided for the service bench mains connection.

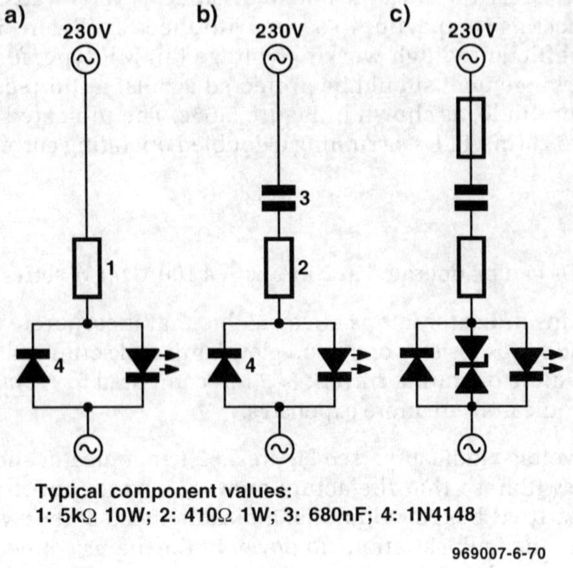

Figure 5.63 Using a LED as a voltage indicator lamp.

a) with a reverse-connected diode and a series resistor, b) connected to 230 V via a series capacitor, c) with additional protective components.

Notes for Figure 5.63:

1 *Brightness:* a LED powered by an AC current is only active on alternate half-cycles. The current must therefore be increased (as a rule, doubled) to achieve normal brightness.

$$R_S = \frac{U - U_W}{I_r}$$

U: mains voltage
U_W: working voltage
I_r : rated current

typical values

I_r	R_S
0.25mA	470...560k ; 1/10W
0.5mA	330k ; 1/8W
1.0mA	180k ; 1/4W
1.8mA	100k ; 1/3W
2.5mA	68k ; 1/2W
4.5mA	39k ; 1W

969007-6-71

Figure 5.64 Using a neon lamp as a mains voltage indicator.

2 *Reverse-blocking voltage:* LEDs have low reverse-blocking voltages. A reverse-coupled silicon diode must be connected in parallel with the LED, since the peak reverse voltage is much higher than what the LED can handle.

3 *Current and power dissipation:* the value of the series resistor can be calculated from the supply voltage (230 V) and the desired operating current. For currents in the range of 10 to 40 mA, the resistor must be able to dissipate several watts. This can be gotten around by connecting a capacitor in series with the LED (Figure 5.63b). This must of course have a sufficiently high working voltage (an MP type, for example). Strictly speaking, this arrangement should be protected against a shorted capacitor by a fuse and a suppressor diode, as shown in Figure 5.63c. The indicated component values apply to standard 20-mA LEDs, assuming a doubled operating current (see Note 1).

Practical tips:

1 Low-current LEDs can be operated at 2 mA with a 100 kΩ/1 W series resistor.

2 Ready-made mains indicator lamps are available that incorporate either a LED with a reverse diode and series resistor or a neon glow lamp and series resistor. LED indicators typically require more operating current (> 2 mA compared to < 1 mA for a neon lamp), but neon-lamp indicators are more expensive.

3 Locating mains voltage indicators (see Figure 5.62): an indicator should always be provided after the switch, so that the actual operating state (on/off) of the connected equipment is clearly visible. An additional indicator ahead of the switch has the advantage that it allows the fault situation 'no power at the mains connection' (at the mains outlet, for instance) to be distinguished from the situation 'fuse open or switch defective'. On the other hand, the cost/benefit ratio of the second indicator is fairly high, it draws a constant (if small) amount of power and it is not protected by the fuse.

Protective measures. These include provision for overcurrent protection (fuses or circuit breakers) and leakage-current circuit breakers (ELCBs).

General requirements:

1 Opening time: ≤ 0.2 s

2 Overcurrent protection should work selectively, which means that in case of a fault in a circuit only the fuse or circuit breaker immediately ahead of the circuit should open.

3 The protective measures must galvanically isolate the circuit in case of a fault and must not automatically reset themselves (which, for example, means that PolySwitch fuses are not acceptable).

Fuses and circuit breakers. Conventional fuses and circuit breakers are both available. For a fuse or circuit breaker to open, a certain level of current must first flow through it:

- With melting-link fuses, the opening current is 7 to 8 times the rated current.
- Circuit breakers are available with several different types of protection characteristics (see Table 5.11).

Trip characteristic	Range of rated current I_f	Current required for tripping in less than 1 second
B	6–63 A	3–5 I_f
C	6–63 A	5–10 I_f
G	6–32 A	6–10 I_f
K	0.2–63 A	8–14 I_f

Table 5.11 Tripping characteristics of selected circuit breaker types.

Note: the fusing or tripping current must actually be able to flow. If the load resistance is too high or the fault voltage is too low, the current will not be interrupted.

The essential specifications of overcurrent protection devices are (1) the rated current I_f and (2) opening time. Some suggestions regarding the selection of specific values:

1 The fuse or circuit breaker should not open at the maximum level of the normal operating current: $I_f \geq I_L$.

2 A given fuse or circuit breaker must open before a fuse or circuit breaker that is connected ahead of it: its rated current should be less than the rated current of the preceding overcurrent protection device.

3 Inrush currents and other peak currents must be taken into account.

4 The fuse or circuit breaker must not open as a result of commonly-occurring, short-duration peak currents (such as switch-on inrush currents). Adequately 'slow' types should thus be selected (such as circuit breakers with type C or type K characteristics).

5 Since a given fuse or circuit breaker should open before a preceding protective device, it should be 'faster' than the preceding device.

6 See Appendix 8 for references to additional literature.

Selective protection: if two fuses *F1* and *F2* (with a rated currents I_{f1} and I_{f2}) are connected in series, then *F2* will open before *F1* if $I_{f1} / I_{f2} \geq 1.6$ (this is valid for all standard mains fuses with $I_f \leq 16$ A). Selective operation is guaranteed if the rated values of the two fuses differ by two or more standard values. *For example:* (1) *F1* = 10 A, *F2* = 10 A: no selective operation; (2) *F1* = 16 A, *F2* = 10 A: selective operation is in principle assured, but problems can arise due to component tolerances; (3) *F1* = 16 A, *F2* = 6 A: selective operation is guaranteed. Selective operation of two circuit breakers, or combinations or fuses and circuit breakers, is more difficult to evaluate. For information regarding the specific case of protection for the service bench and/or the hardware being tested, see Volume 3.

Earth-leakage current breakers (ELCBs). An ELCB measures the difference between the currents in the live and neutral leads. Normally, the current that flows in the live lead should return via the neutral lead, so that the difference should be zero. If this is not the case, then some part of the current must be flowing somewhere else, such as via the protective earth lead (*fault current*). Typical rated values for leakage currents are 10 mA, 30 mA and 300 mA. An ELCB should open within 200 ms if the rated leakage current is exceed, and within 40 ms if the leakage current is more than five times the rated value. ELCBs are available as panel-mounted devices and as 'combination' devices, consisting of a mains plug, ELCB and and outlet, that can be plugged into a mains outlet.

Figure 5.65 illustrates the situations in which the previously-described protective measures are effective.

Notes:

1 A more or less dangerous voltage can be present on the enclosure of a piece of equipment without causing the overcurrent fuse or circuit breaker to open, if the resulting fault current is relatively low (a short-circuit current of 30 to 80 A is normally necessary to open a fuse). In such situations an ELCB provides better protection, since standard ELCBs trip if the leakage current exceeds 10 mA or 30 mA.

2 An ELCB cannot provide any protection against a 'hard' short between the live and neutral lines (which is not all that uncommon in experimental and fault-finding situations).

Equipment mains connections. In general, the mains connection for equipment that is plugged into an outlet is quite simple (compared to Figure 5.62). However, there are certain features that occasionally must be taken into account in fault-finding:

1 No mains voltage indication: in most cases the only power indicator is a lamp that is connected to the secondary side (output) of the power supply.

2 No true mains switch: this applies in particular to plug-in mains adapters and standby power supplies (in the latter case, the mains power is switched by a relay activated by a remote control or similar arrangement).

3 Double-insulated devices (Class-II equipment) often have only single-pole mains switches.

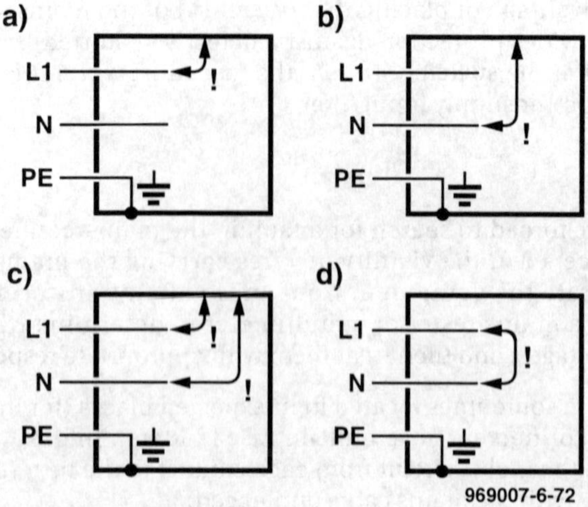

Figure 5.65 Fault situations in the mains connection (see Table 5.12).

Illustration (Figure 5.64)	Fault situation	Overcurrent protection	ELCB
a)	The L1 lead is shorted to the enclosure. The full short-circuit current flows via the protective earth lead (PE).	actuated	actuated
b)	The N lead is shorted to the enclosure. Only a small fault current flows via the protective earth lead (PE), since the potential to earth is low.	not actuated	actuated
c)	The L1 and N leads are both shorted to the enclosure. The full short-circuit current flows through the L1 lead, and it is split between the N and PE leads.	actuated	actuated
d)	The L1 and N leads are shorted to each other but not to the enclosure. The full short-circuit current flows through the L1 and N leads but not through the PE lead.	actuated	not actuated

Table 5.12 Mains connection faults illustrated in Figure 5.64.

Equipment protection. A fuse is usually provided in the mains input circuit (and in the mains plug in the UK). As a rule, this fuse should be rated at 2 to 3 times the normal operating current and should be a 'slow' type. Circuit breakers for building into equipment are also available as an alternative, including types that are combined with a mains switch.

Fuse location. The advantage of placing the fuse ahead of the mains switch (as in Figure 5.62) is that it also provides protection against a defective (shorted) switch. However, it is sometimes located after the switch. *Caution:* the fuse must without fail be located ahead of the mains filter or power supply input filter.

Practical tips:

1 Sometimes you are forced to search for a fault in the mains-connection circuitry of a switched-on device, or in the vicinity of wires carrying the mains voltage. The best practice is to operate the equipment from an isolation transformer. As a makeshift solution, first use a mains tester or a multimeter to determine which leads actually carry live mains voltages, and then treat them with appropriate respect.

2 Equipment fuses are sometimes located in unexpected places (for instance, some types of IEC-320 mains connectors have built-in fuse holders). In problem situations, you must literally trace the mains circuit from each contact to the next (using an ohmmeter to check continuity with the mains cable unplugged).

3 If auxiliary devices are connected between the mains outlet and the equipment (surge arresters, telephone-operated remote-control switch-on devices, multiple-outlet 'power bars' with filters, master/slave switching circuits and the like), then there are additional fuses or circuit breakers in the mains circuit. It is not likely that these will operate selectively, which means that if any one of these fuses or circuit breakers opens, there is a good chance that some other fuse has also opened (or a circuit breaker has tripped). Check and measure!

4 Some faults that cause a fuse to blow are difficult to locate. To keep the consumption of fuses within reasonable limits, you should employ a circuit breaker while doing fault-finding. Refer to the following description of the service-bench mains installation for more information. *Note:* an electronic overcurrent cutout can be used in a similar manner in DC circuits as a test aid. See Volume 3 for details.

Service bench mains installation. The following special requirements apply to the mains connection for the service bench:

1 Several different tools and pieces of equipment will be connected to the mains (oscilloscope, signal generator, laboratory power supply, soldering station and so on). It should be possible to switch these off as a group.

2 The test equipment should be protected against faults occurring in the item being tested. Clearly, it is not possible to do any kind of useful work if the main circuit breaker trips every time you try to switch on a suspect power supply (especially if other equipment or locations are also effected) – not to mention the nuisance of having to make a trip to the consumer unit every time.

3 Protection against bodily injury must be assured, in particular under the condition that measurements must occasionally be made on or in the vicinity of assemblies carrying mains voltages.

The following technical measures are needed to meet these requirements:

1 a master mains switch,

2 independent, selectively-operating fuses or circuit breakers in the supply circuit(s) for the item(s) under test,

3 ELCBs, possibly supplemented by an emergency switch, an undervoltage cutout and an isolation transformer.

The simplest solution: use a multiple-outlet 'power bar' with a built-in ELCB (commercially available). This satisfies the first and third requirements, but not the second one.

The perfect solution: install a permanent work station that meets the DIN VDE 0104 standard (this should be carried out by a qualified specialist firm). With such an installation, you can repair power supplies and test the protective provisions of repaired equipment against the DIN VDE 0701 and 0413 standards. Other types of fault-finding and repairs can be performed on separate, conventionally-equipped service benches.

Compromise solutions: there are various devices available that can be used to put together a service-bench mains installation that is satisfies the specific requirements listed above. Although there are commercial systems for constructing laboratory and service work stations, they have steep prices. With a DIY solution, you can make your own choices between rather modest and luxurious solutions. It is possible to put together a quite inexpensive installation if you do not put too much emphasis on external appearances and can make do with standard distribution boxes, power bars and the like (see Figure 5.66).

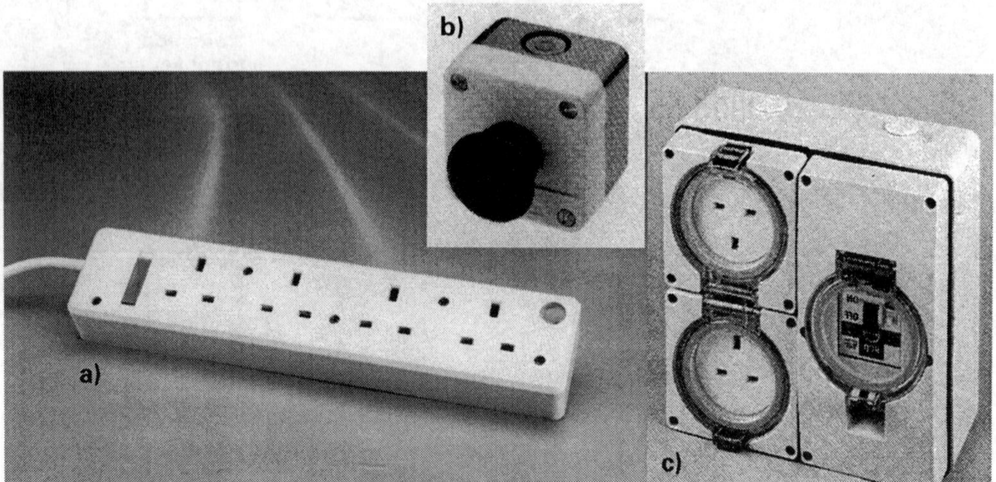

Figure 5.66 *A simple solution for the service bench mains connection.*

a) multiple-outlet power bar, b) mains switch for electrical tools, c) outlet box with its own earth-leakage circuit breaker (ELCB).

FIgure 5.66 shows a multiple-outlet power bar, an electrical-equipment switch that acts as a combined master switch, emergency shutoff pushbutton and undervoltage cutout, and an outlet box with an ELCB. Figure 5.67 shows a block diagram of the individual devices and subassemblies. See Volume 3 for practical construction details.

Notes:

1 Good-quality switches are not inexpensive, and you should not try to cut costs here.

2 Mount the power bar on the wall, or screw it to the rear panel of the work bench. Bridge over the built-in switch (if one is present), since only the power indicator is needed. Mount the box containing the master switch at the front of the work bench, underneath the bench surface (the emergency shutoff pushbutton should be within easy reach, in case it is ever needed), and wire the power bar to the master switch.

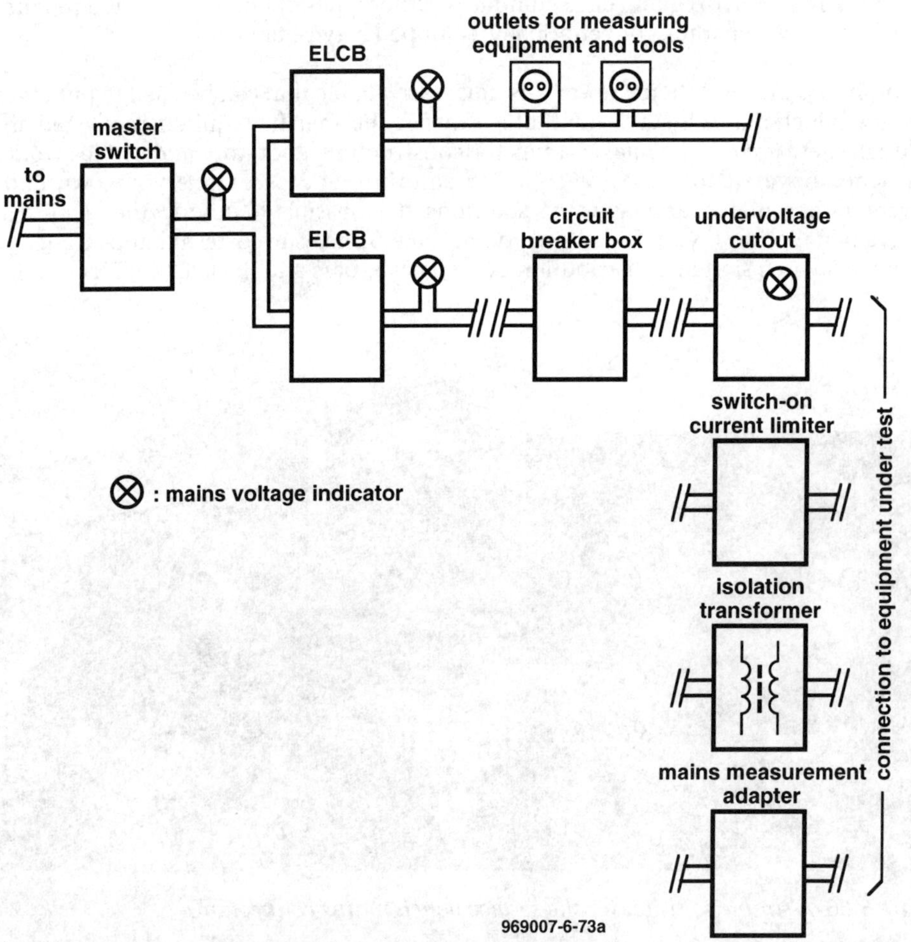

***Figure 5.67** Block diagram of the functional elements of a service bench mains connection.*

Descriptions of the items shown in Figure 5.67:

- *Mains circuit:* the mains circuit to which the service bench installation is connected should be fused at 30 A (16 A for continental branch circuits). Preferably, no other loads should be connected to this circuit.
- *Fixed equipment.* The service bench facilities should be wired as a fixed installation, connected directly to the mains circuit via wiring connectors (that is, not via a mains plug and outlet). A master switch is required with this type of connection. The advantage of a fixed installation is that the live and neutral lines cannot be inadvertently exchanged, which simplifies the construction of some of the other assemblies (such as the circuit breaker box and the mains measurement adapter).
- *ELCBs:* devices rated at 30 mA are in general use. A 10-mA type is preferred if you use an antistatic armband as an anti-ESD measure. *Note:* having two independent ELCBs, as shown in Figure 5.67, is by no means a luxury if you often service power supplies and the like. (If you accidentally short something to earth when making a measurement, or inadvertently touch a live lead, then the circuit that powers the test equipment and tools will not be cut out at the same time.) On the other hand, if fault-finding and repair of power supplies and mains connections is relatively infrequent, a single ELCB is sufficient. *Note:* if you are not careful, earth loops can be created in the test setup (if the reference potential of the test arrangement is not equal to the earth potential and it is connected to an oscilloscope (or other test equipment) that is not isolated from the mains, with the enclosure tied to the safety earth lead). The resulting earth loop current can trip the ELCB. (In other words, if an ELCB trips when there is no apparent sign of a related fault, this suggests that there is an earth loop in the test setup.)
- *Outlets for measuring equipment and tools:* master switches are not needed for 'local' circuits, since they can be isolated by manually tripping their ELCBs (for example).
- *Circuit breaker box:* this serves to prevent the fuse in the device under test from blowing every time a short circuit occurs during fault-finding (and also to prevent the main fuse or MCB in the consumer unit from opening in case of a short circuit in the device under test). Selective operation can only be assured if there is not any additional protection between the main fuse or MCB and the mains connection of the device under test, since the tolerances of fuse and circuit breaker characteristics are so large that selective operation would otherwise be a matter of luck. The circuit breaker box contains several precision cutouts or circuit breakers with various current ratings, such as 0.5 - 1 - 2 - 6 - 10 - 16 A (although selective operation is not guaranteed with the last of these). *Note:* with a fixed installation it is sufficient to interrupt the live lead (L1); otherwise two-pole circuit breakers should be used.

  Operation:

  1. Bypass the fuse in the mains connection of the equipment being tested (standard cylindrical fuses can for example be wrapped in aluminium foil – but be sure to remove the foil when you have finished with fault-finding!).
  2. Select a trip current that corresponds to the rated value of the fuse in the equipment under test. If the circuit breaker is tripped by a high switch-on current, then try the next-higher value. In case of difficulties, use an inrush current limiter.

3 After repairs have been successfully completed, test the equipment again with its own fuse.

- *Undervoltage cutout:* such a device (whose working principle is shown in Figure 5.68) is required if power should not be automatically reapplied to a protected circuit after a fuse has blown or the mains power has been disconnected. In such cases, it is necessary to ensure that the mains power is only applied as the result of some intentional operator action (such as pressing an 'ON' button). This is important for electrical tools, for example (a table saw should not just start running again after a temporary loss of mains power), but it also applies to test setups that require special caution.

  Notes:

  1 Suitable assemblies are readily available as accessories for electrical machine tools (see Figure 5.66b).

  2 In the arrangement shown in Figure 5.68, several emergency shutoff switches can be connected in series. However, such a solution is not adequate in situations that have high requirements, such as when widely-distributed emergency shutdown switches are needed in a workshop. In such cases, suitable safety cutout switch assemblies should be used.

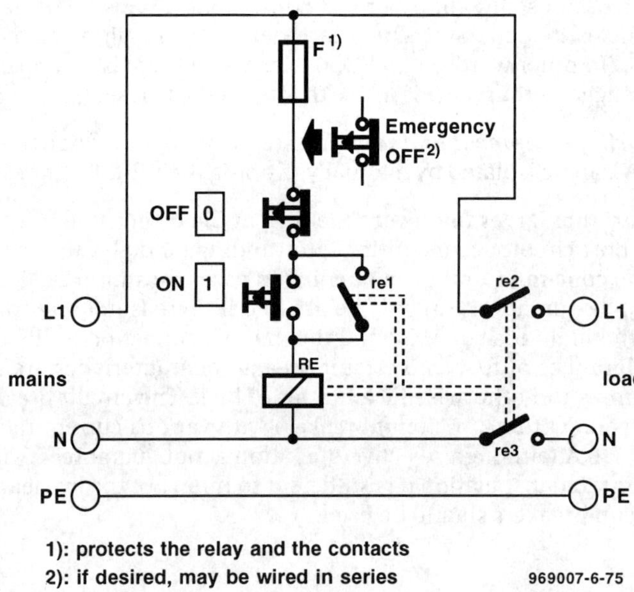

Figure 5.68 *Operating principle of an undervoltage cutout.*

The (protective) relay RE is actuated when the 'ON' button is pressed, and closes all relay contacts. These connect the mains circuit to the load and also bridge the 'ON' switch so that the relay is kept actuated. The relay is deactivated if the coil current is interrupted by pressing the 'OFF' pushbutton (or an 'Emergency OFF" switch connected in series with the 'OFF' switch), or if the mains voltage drops out.

- *Switch-on current limiter.* The schematic in Figure 5.69 illustrates the working principle of a switch-on current limiter (inrush current limiter). Such a device (which can be built as an in-line adapter that is plugged into the mains outlet) can be used when the equipment to be serviced has an unusually high switch-on current. (Based on experience, some PC power supplies can trip a 10-A circuit breaker. It is open to question whether such supplies should not just simply be replaced, but at least an inrush current limiter will allow you to investigate the fault.) *Practical tip:* the most likely causes of excessive switch-on currents are a fault in the power supply's own inrush-current limiting circuitry, problems with the storage capacitor (refer to Figure 5.61a) and a 'jittery' mains switch (which bounces so much that the power supply is effectively switched on and off several times in rapid succession).

  Notes:

  1 A current limiter that is intended to be used for testing should have several selectable resistors (see Volume 3).

  2 *Caution:* if the current-limiting resistor is located in the L1 lead, then it also limits the short-circuit current in a fault situation. The resulting current may not be high enough to cause the fuse to blow (or the circuit breaker to trip) quickly enough

  3 *Caution:* if the current-limiting resistor is located in the N lead, then practically the entire mains potential will initially appear at N-lead terminal of the load. This can be dangerous if an oscilloscope has been 'floated' on the N lead to allow the switch-on current to be measured.

  4 For the above reasons, an inrush current limiter must always be used with a fuse or circuit breaker that will reliably open the circuit in less than 0.2 s at the short-circuit current level (230 V divided by the value of the current-limiting resistor) when current limiting is active (that is, with a failed relay).

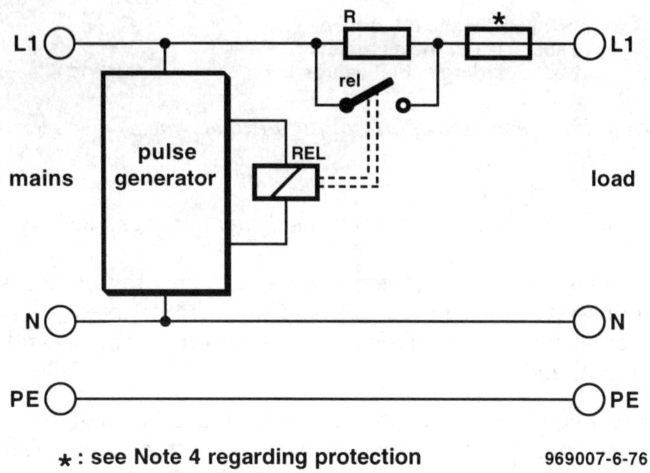

Figure 5.69 Operating principle of an inrush current limiter.

Protective isolation: this means that the load is connected via an isolation transformer, so that it is galvanically isolated from the mains network. The result is that neither of the leads that carries mains power on the load side of the transformer is referred to earth potential (see Figure 5.70).

Since a TN network is earthed, it represents a voltage source with an internal resistance with respect to earth that is nearly zero ohms. The live lead always carries 230 V with respect to earth – and if your body has a conductive path to earth and you touch a lead carrying mains potential, you will be made all too aware of this.

The fact that the leads on the secondary side of the isolation transformer have no earth connection simply means that the internal resistance of the equivalent voltage source, with respect to earth, is nearly infinite. With an infinite internal resistance, the voltage applied to the load is reduced to zero for even the smallest load (such as the internal resistance of your body). It is thus not dangerous to touch either of the secondary leads alone (but naturally not both at the same time!), and it is also possible to tie either one of the two leads to earth as desired (for instance, to make measurements with an earthed instrument).

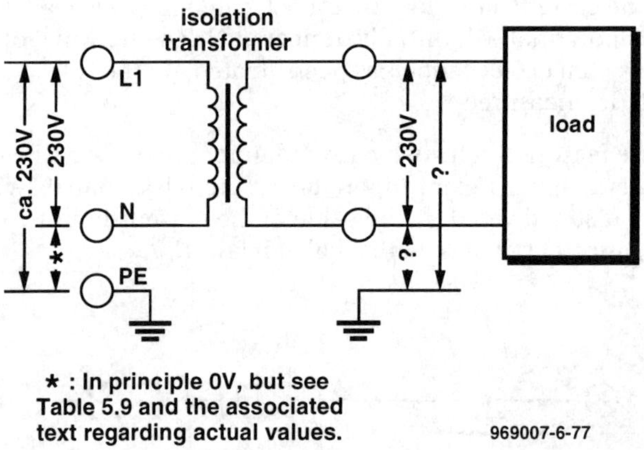

Figure 5.70 The working principle of an isolation transformer.

There are two applications for an isolation transformer in the service environment:

1 To supply the device under test: the objective is to ensure that accessible parts that carry 'live' mains-level voltages cannot supply currents to earth (this is especially important for conventional television sets with 'universal' chassis and other similarly-constructed equipment).

2 To supply the test equipment: the objective is to operate the equipment isolated from earth (so that the equipment enclosure can be connected to any desired potential, or 'floated'). See Appendix 3 regarding floating operation.

Notes:

1 The secondary winding of an isolation transformer must be short-circuit protected, either by virtue of its internal resistance or by using a suitable fuse or circuit breaker.

2 Although an earthed outlet is provided for the secondary of an isolation transformer, the protective earth contact must never be connected to anything.

3 Normally, only one piece of equipment may be connected to the secondary of an isolation transformer. However, several Class-I devices can be connected to the secondary of a single isolation transformer if their protective earth leads are all interconnected in order to equalize their earth potentials. A simple way to do this is to use a power bar.

4 Isolation transformers are available in 1:1 and adjustable versions. DIY construction based on a 'bare' transformer (plus an enclosure, fuses, plugs and sockets and so on) is possible, but the mechanical construction is quite involved (if only because of the weight of the transformer). See also Volume 3.

5 Output voltage adjustment is not necessary for simply making floating measurements and operating equipment isolated from earth; a 1:1 isolation transformer is adequate.

6 Things to keep in mind: (1) the power rating of the transformer must be adequate for the power consumption of the load(s) to be connected (do not overlook the difference between apparent power (VA) and true power (watts)!), and (2) if the equipment does not behave the same when connected to the transformer as when connected directly to the mains, even though the transformer's power rating should be adequate, the current drawn by the equipment may be very discontinuous (refer to Figure 5.59) or unbalanced with respect to positive and negative half-cycles (if the power supply has a half-wave rectifier connected directly to the mains input, for example). *Practical tip:* in case of doubt, check the secondary voltage waveform with an oscilloscope.

7 See Appendix 3 regarding the limits of floating operation (which cannot be avoided, even with an isolation transformer).

Voltage and current measurements

The following voltages and currents can be measured on the mains network:

1 the mains voltage waveform and/or differential-mode disturbances: between the live lead and the neutral lead (L1 to N),

2 common-mode disturbances (1): between the live lead and the protective earth lead (L1 to PE),

3 common-mode disturbances (2): between the neutral lead and the protective earth lead (N to PE),

4 earth-loop potentials (between various protective-earth connections),

5 supply currents (flowing in the L1 or N lead),

6 fault currents (flowing in the PE lead).

Simple (static) voltage and current measurements can be made with almost any modern multimeter without personal risk. However, it is sometimes necessary to also observe voltage and/or current waveforms (dynamic measurements, using an oscilloscope for example). Figure 5.71 illustrates typical measurement situations, and Table 5.13 describes the associated dangers.

Additional measurements. These primarily include measuring the power factor, power, harmonic content and frequency. Such measurements are either based on a set of related voltage and current measurements (power factor and power) or correspond to voltage measurements with respect to the connection of the measuring equipment (such as measuring the mains frequency with a universal counter).

Note: there are only two reasons to make measurements on the mains network: (1) to locate faults in devices (or in their power supplies or mains connections) and (2) to check

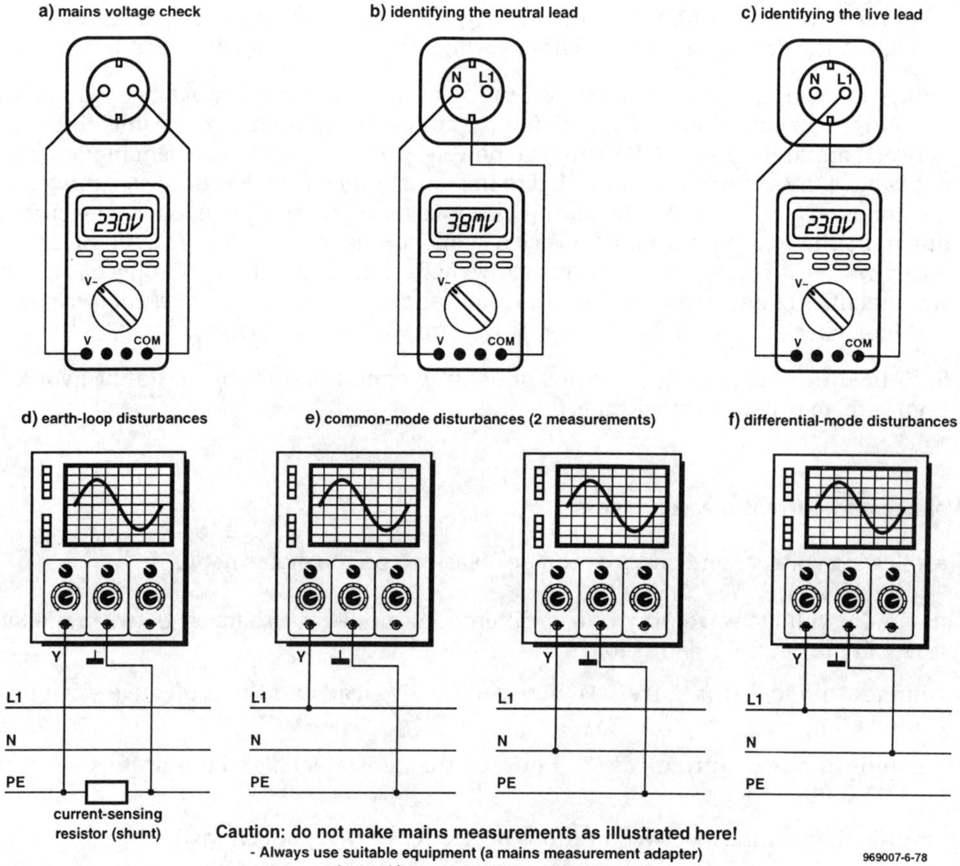

Figure 5.71 Examples of mains-circuit measurements. top: simple voltage measurements, bottom: measuring mains disturbances with an oscilloscope.

Risk	Remedy
The connections are inadequate.	Use safety probes and test leads, or use a mains measurement adapter.
The insulation of the equipment is inadequate.	Choose equipment with suitable specifications (eg, compliant with IEC 1010 or IEC 384).
The measurement setup renders the protective measures ineffective.	1) Verify that the equipment and setup are in order before starting to make measurements. 2) Arrange the measurement setup so that the protective measures can still function properly.
The measuring equipment is constructed according to Class I safety standards (the equipment earth is connected to the protective earth lead).	*In general:* be careful with the arrangement of the measurement setup. Concrete alternatives are: 1) Make measurements only with reference to earth (or the PE lead), as seen by the measuring equipment. 2) 'Float' the equipment only on the N lead (if at all). 3) Operate the equipment via an isolation transformer.

Table 5.13 Problem areas and dangers for mains-circuit measurements.

whether the mains network itself is possibly responsible for a given fault (ranging from a simple check to see whether there is any 'juice' to investigating suspicions of earth loops, inadequate earthing, excessive harmonic content (coupled into the network from other equipment) and so on). The detailed investigation of suspected problems in these areas is the proper work of the electrical technician, and you are referred to the relevant technical literature in this regard.

Phase tester. This simple test aid looks like a screwdriver and contains a neon lamp with a series resistor. It can be used to check whether a voltage with respect to earth is present on a given contact or wire (the breakdown voltage of the neon lamp typically lies in the range of 60 to 100 V (peak)). The lamp lights if the tester is touched to a live (L1) terminal and does not light if it is touched to a neutral terminal. If it lights on both terminals, then the circuit is not part of a TN network.

Caution:

1 An interruption in the neutral line cannot be detected with the phase tester (the fact that the lamp lights when the tester is touched to the other terminal is by no means a guarantee that everything is in order).

2 The lamp may not light if the resistance of the earth path is too high. The fact that the lamp does not light is thus not always proof that no dangerous voltage is present.

Practical tip: always double-check with a multimeter or a two-pole voltage tester (see the upper row of Figure 5.71). *Caution:* some phase testers draw 30 to 50 mA. This can be enough to trip a sensitive ELCB if you use a phase tester to check the voltage between the live lead and the protective earth lead.

Current measurements. You can use the following items to measure mains currents:

- moving-iron ammeters,
- multimeters with suitable current ranges,
- clip-on current meters,
- transformer or Hall-effect current converters,
- current-sensing shunt resistors.

Notes:

1 It is awkward to open a lead for current measurement. Even using a clip-on meter can sometimes be awkward, since a single conductor must be encircled by the clip; no measurement can be made on the usual three-core mains cable. A measurement adapter is the best solution to this problem.

2 Currents of 1 A or more can easily be measured using a clip-on meter, current-sensing converter or high-power, low-resistance shunt (which can also adequately handle short-term overloads). However, you will need a shunt with a resistance of 1 Ω to accurately measure currents in the milliampère range. The fact that the current can be significantly larger than its normal value when the equipment is switched on, or in a fault situation (live-to-neutral or live-to-earth short circuit), means that there are certain disadvantages to using a shunt resistor:

- the shunt resistor and/or the instrument can be overloaded or damaged,
- the voltage drop across the shunt resistor may be excessive,
- the shunt resistor will limit the current, so that in case of a short circuit it will take longer for the fuse to blow or the circuit breaker to trip.

Countermeasures:

a) To protect against the effects of excessive switch-on currents, bridge over the shunt or the instrument when switching on the equipment.

b) A short-circuit currents can only flow if there is an actual fault (a breakdown, incorrect component installation, an accidental short that occurs when you are probing the circuit during testing and so on). As a protective measure, check that everything is in order before making the measurement (for instance, before you measure the operating current, check to make sure that no live-to-neutral or live-to-earth short is present).

Using a multimeter. There is practically no danger associated with measuring mains voltages with modern multimeters and suitable safety probes (which comply with the IEC 1010 standard). Suitable meters should meet the following requirements:

1 *AC voltage range (V AC):* at least 400 V (typical upper limits are 600, 750 and 1000 V).

2 *Isolation voltage:* at least 600 V_{eff} between any terminal and earth (IEC 1010 Category III), or 1000 V peak.

3 *AC current range (A AC):* at least 10 A, fused and overload protected. (Some meters also have mechanical safety features, such as blocking the voltage-measurement jack when a current range is selected and vice versa). *Practical tip:* the switch-on current of the device under test may be too high for the instrument's fuse. As a remedy, (1) temporarily short the meter inputs when switching on the equipment or (2) measure the voltage drop across a shunt resistor instead of measuring the current directly.

4 *True-RMS capability:* if this is needed (it is not necessary for simple tests), then check the crest factor specification of the instrument (see Appendix 5). (The value of the crest factor depends on how 'jagged' the current or voltage waveform is; see Figure 5.72.) Meters with inadequate crest factor ratings give incorrect readings for heavily chopped waveforms. The recommended crest factor rating is at least 3 to 4.

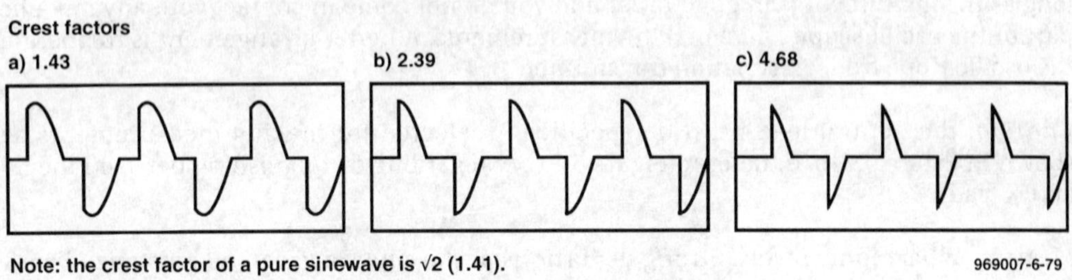

Figure 5.72 Crest factors of various waveforms. (source: Fluke)

Multi-purpose instruments. There are instruments available that have been 'optimized' for complex measurement functions on the mains network (such as 'graphic' multimeters that can display the waveform of the voltage or current being measured, like an oscilloscope (within certain limitations), harmonic-level analyzers and so on). Many difficulties associated with performing certain types of measurements on the mains network can be reduced to a simple question of making the proper connections if such instruments are used – but they are not exactly inexpensive.

Measuring with an oscilloscope

An oscilloscope can be used to measure mains voltage and current waveforms, including noise and disturbances. *Caution:* carefully review your measurement setup before making any connections to the mains – otherwise, your oscilloscope may join the ranks of those that have landed (prematurely) on the scrap heap.

The following material relates to oscilloscopes that are not specifically designed for floating operation (equipment for floating operation is isolated with respect to earth for potentials of at least 600 V_{eff}). Ordinary oscilloscopes have metallic enclosures that are connected to the measurement earth (the shells of the BNC connectors) and the protective earth lead. Thus, in normal situations (with a TN mains installation) the reference potential (common) of the oscilloscope is earthed (see Figure 5.73). *Note:* the same considerations apply to other types of instruments with similar construction (such as most universal counters).

The perfect solution (Figure 5.73): ideally, you should connect the oscilloscope in the regulation manner, which means that it is always earthed. In this case the voltage to be measured is referenced to the earth potential at the oscilloscope input. A differential measurement amplifier, a differential probe assembly or the differential mode of the oscilloscope itself must be used to measure the difference between any two voltages that are referenced to earth.

The improvised solution (Figure 5.74): you can measure voltages that are not referenced to earth by floating the oscilloscope. The oscilloscope common must always be connected to the less-dangerous voltage (which is the neutral lead in this case). To float the oscilloscope, it must be used with the protective earth lead not connected. This is acceptable as long as the instrument is in good order and you do not come in contact with any metallic part of the oscilloscope while making measurements. A better arrangement is to operate the oscilloscope from an isolation transformer.

Caution: the neutral lead must be specifically selected for making measurements as shown in Figure 5.74b/d. Refer to Figure 5.71b/c regarding distinguishing between the L1 and N leads.

Measuring harmonic levels. Strong distortions of the sine wave can be recognized with the naked eye on an oscilloscope display. The following techniques can be used to more precisely judge the amount of distortion:

1 Use an oscilloscope with a PC interface and transfer the waveform to be measured into the PC, where it can be processed by a suitable analysis program.

2 Measure the harmonics separately, by blocking the fundamental-frequency component with a 50-Hz notch filter (see Volume 3).

3 Use a harmonic analyzer.

Explanation of Figure 5.73:

a) Directly measuring the voltage between the L1 or N lead and earth (for example: measuring common-mode voltages). The voltage divider reduces the measured voltage to a level that will not damage the oscilloscope.

b) Measuring the voltage between L1 and N, using a differential measurement amplifier.

c) Measuring the leakage current with a shunt resistor in the protective earth lead.

*d) Measuring the operating current with a shunt resistor in the L1 or N lead.
The voltage drop across the shunt is sensed by a differential measurement amplifier.
Note: it is better to use a Hall-effect sensor or a clip-on current probe for currents in the 1 to 10 A range.*

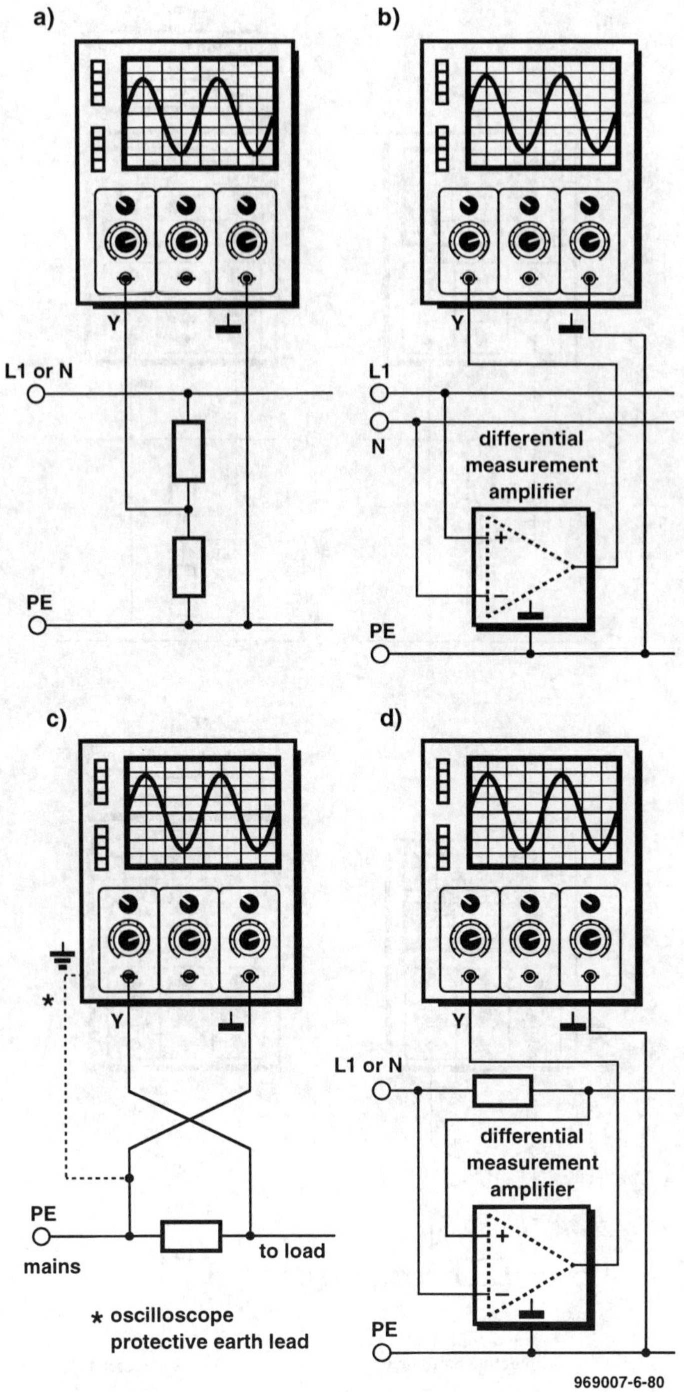

Figure 5.73 The proper way to make measurements with an oscilloscope.

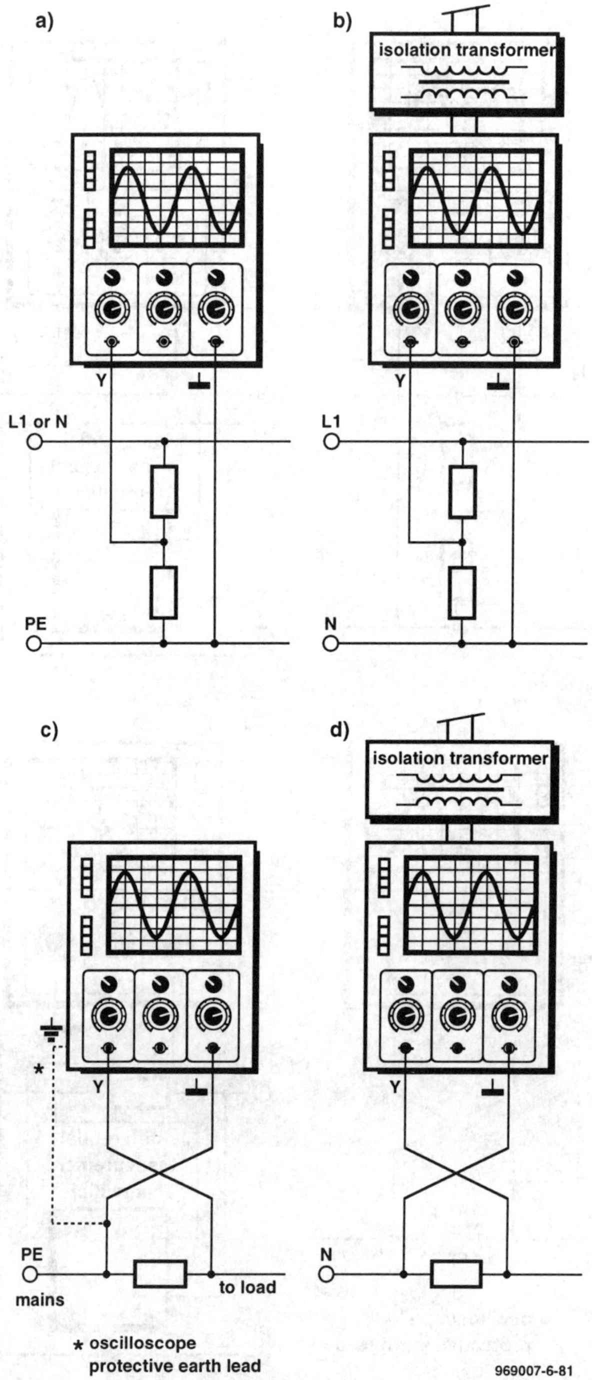

Figure 5.74 Improvised methods for making measurements with an oscilloscope.

Mains measurement adapters

Test leads. For very simple voltage measurements (as in upper row of Figure 5.71), a pair of test leads with retractable-shell safety-type banana plugs is sufficient (the plugs can be inserted in the sockets of standard mains outlets). Other useful accessories are:

- a test cable with an earthed mains plug at one end and three banana plugs at the other end, for quick voltage measurements on standard mains outlets (for simple checks as shown in Figure 5.71a, a two-terminal plug with two banana plugs is sufficient),
- a test cable with an IEC plug at one end and three banana plugs at the other end, for quick voltage measurements on equipment IEC connectors.

Note: the banana plugs are intended to be plugged into a multimeter and must be safety types with fixed shells.

Mains adapter for voltage and current measurements. Figure 5.75 shows a simple arrangement that is fully adequate for many measurement tasks.

Notes:

1 The neon lamps also act as a built-in phase tester. See the table for interpreting the lamp indications. If modern, adequately-insulated equipment is always used for making measurements (so that floating measurements do not pose any risks), or if the adapter is permanently installed, then these lamps are not essential.

2 For current measurements, the lead in question must be interrupted. Safety-type banana plugs and suitably insulated jacks provide an inexpensive and reliable means of doing this (and are better than inexpensive toggle switches or the like). *Practical tip:* arrange the jacks so that only allowable interconnections can be made using standard shorting plugs (a short-circuit connection between L1 and N, for example, should not be possible).

3 When using a clip-on current meter or current probe, replace the shorting plug with a short test lead if the meter or probe will not fit over the shorting plug.

Explanation of Figure 5.74:

a) Directly measuring the voltage between the L1 or N lead and earth (for example: measuring common-mode voltages). This is the same as Figure 5.73a, so it is not necessary to float the oscilloscope!

b) Measuring the voltage between the L1 and N leads. The earth lead of the oscilloscope is connected to the mains neutral lead.

c) Measuring the leakage current with a shunt resistor in the protective earth lead. This is the same as Figure 5.73c, so it is not necessary to float the oscilloscope!

d) Measuring the operating current with a shunt resistor in the N lead. The 'earth' lead of the oscilloscope is connected to the mains-circuit (supply) end of the shunt resistor.

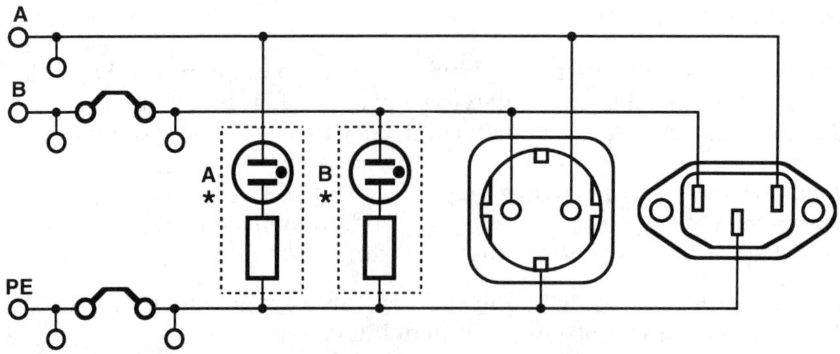

neon lamps A	neon lamps B	indication
0	0	no voltage present
0	1	Lead B is L1
1	0	Lead A is L1
1	1	no protective earth lead, or the protective earth lead is not earthed

*: not required for fixed installations

0: lamp off
1: lamp on

969007-6-82

Figure 5.75 A simple mains measurement adapter.

4 A standard mains outlet connector and a female IEC connector should preferably be provided for connecting equipment to the adapter.

5 To make measurements with an instrument that cannot easily be floated, make sure that lead *B* of the adapter is connected to the mains neutral line. Do not connect the instrument if both indicator lamps are illuminated!

Oscilloscope measurement adapters. These adapters allow measurements to be made as shown in Figures 5.73 and 5.74. Such adapters can be made more or less 'foolproof', depending on cost (see Figures 5.76 through 5.78).

The adapter shown in Figure 5.76 can be used to measure voltages with reference to earth. The neon lamps shows which lead is connected to L1 and which is connected to N. The voltage divider ratio is normally 1000:1.

In the adapter shown in Figure 5.77, the relay arrangement ensures that (1) the voltage divider is always connected between the L1 and N leads and (2) the outer shell (common) of the BNC connector is only connected if the adapter is connected to a correctly earthed mains outlet. If lead *A* is connected to L1 then relay *X* is actuated, and if lead *B* is connected to L1 then relay *Y* is actuated. If the protective-earth lead is not connected to earth, then neither of the relays will be actuated. The shell of the BNC connector is only connected if exactly one of the relays is actuated.

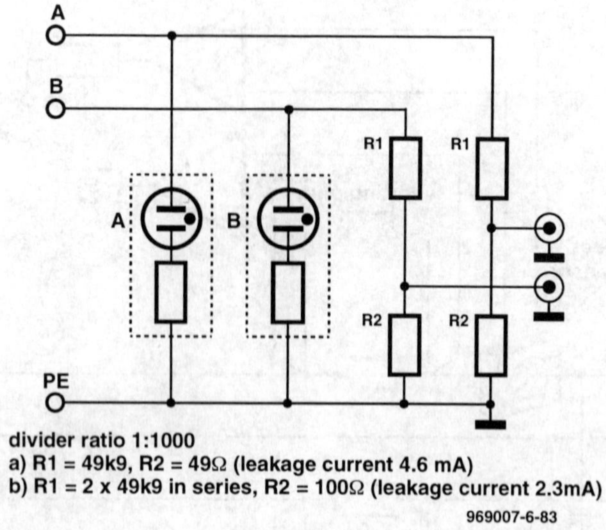

Figure 5.76 *A simple voltage measurement adapter.*

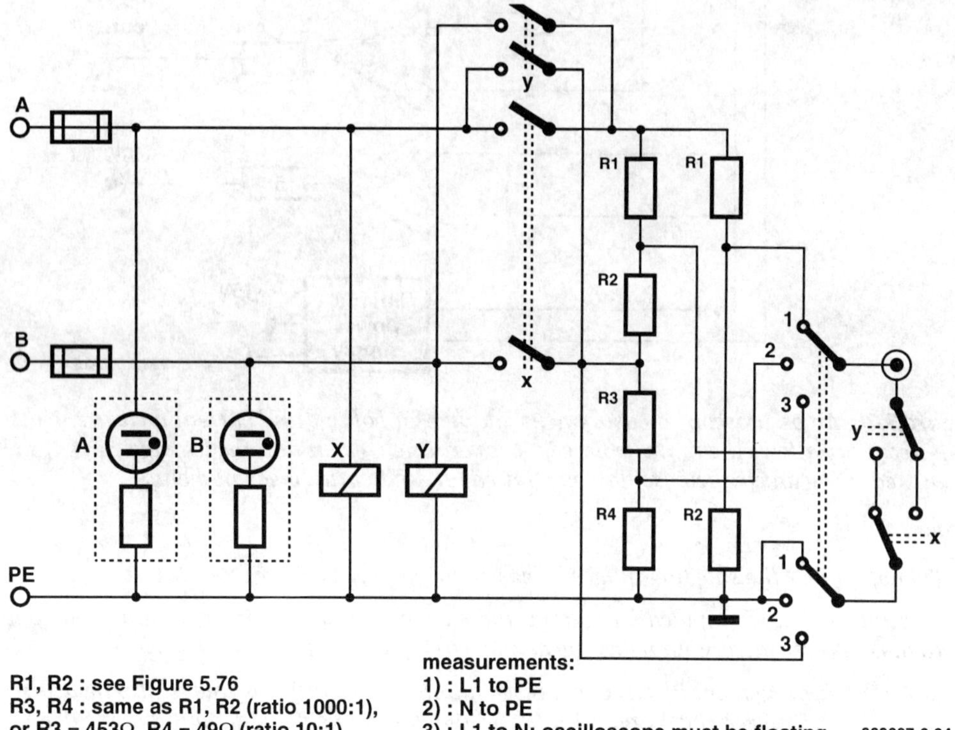

Figure 5.77 *A voltage measurement adapter with protective circuitry.*

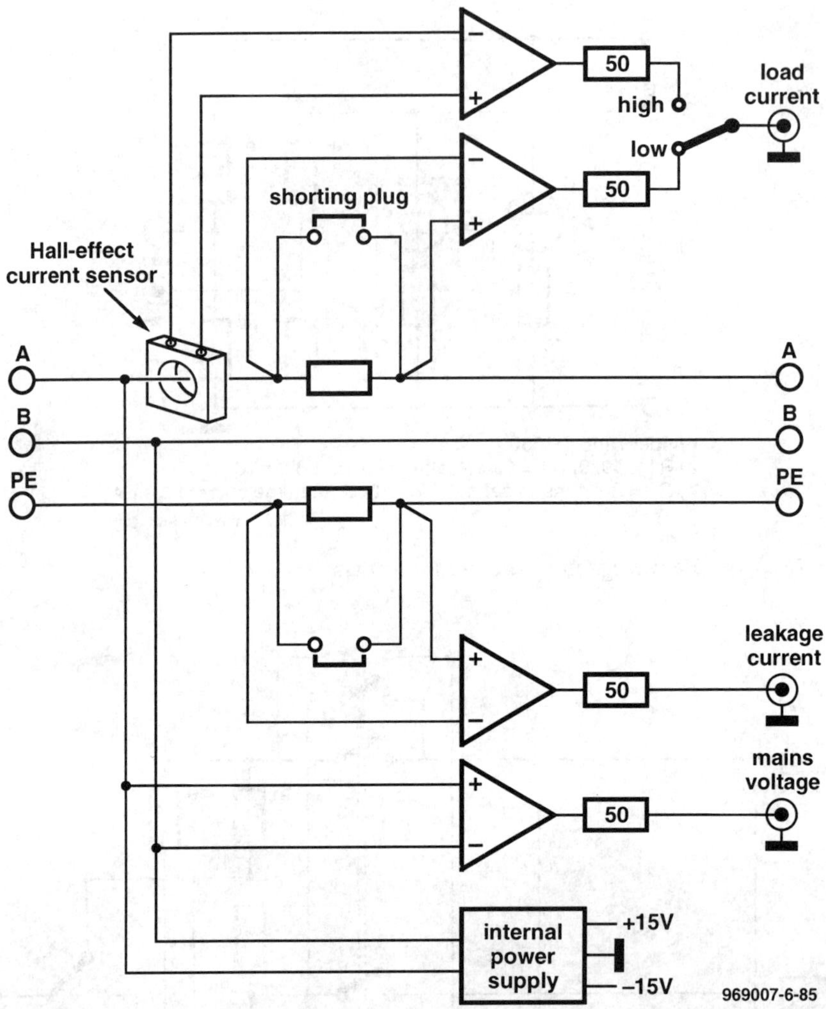

Figure 5.78 An oscilloscope measurement adapter for voltage and current measurements. This accessory allows the mains voltage, the operating current and the leakage current to be measured using an earthed oscilloscope (based on differential measurements).

Notes :

1 *Do not remove the shorting plugs if the current might possibly exceed 1 A.*

2 *Separate jacks are provided for voltage and current measurements, so that it is possible to make power-factor and power measurements (for example).*

3 *Since this arrangement has its own power supply, it is clearly possible to also measure the voltages between the A and B leads and the PE lead using differential amplifiers.*

4 *See Volume 3 for more detailed information.*

Notes for Figure 5.77:

1 Typical relays draw a current of 4 to 5 mA, to which must be added the leakage currents of the L1-PE voltage divider and the connected equipment. The resulting current may be enough to trip a sensitive ELCB (10 mA rating). *Practical tip:* clearly mark the value of the leakage current for each piece of equipment on a label attached the equipment.

2 A low-ratio voltage divider (as low as 1:10) can be used between the N and PE leads; this is important for making accurate measurements of common-mode noise.

3 If the adapter is permanently connected to the mains network, the relays are not required; only the voltage dividers and selector switch are then needed.

4 See Volume 3 for more detailed information.

Usage (Figures 5.76 and 5.77):

1 First connect the adapter to the mains, and then connect the oscilloscope to the adapter.

2 Do not connect the oscilloscope if both lamps are illuminated!

Oscilloscope connection: the oscilloscope is connected directly to the adapters shown in Figures 5.76 and 5.77 via a 50-Ω cable. If noise peaks with steep edges and the like are to be measured and the cable is longer than approximately 50 cm, it must be terminated at the oscilloscope with a 50-Ω feedthrough resistor. This will reduce the amplitude of the displayed signal by a factor of 2 if the 1000:1 voltage divider draws a current of 4.6 mA, or by a factor of 1.5 if the voltage divider draws 2.3 mA.

Detecting mains disturbances

Evaluating the mains circuit. It is frequently relatively easy to determine whether a suspect mains circuit is affected by long-term disturbances (see Table 5.14) by using an oscilloscope as shown in Figure 5.71d–f.

Observing intermittent disturbances. With infrequently-occurring disturbances, an oscilloscope – even a storage oscilloscope with one-shot mode – is not always particularly helpful, since it is necessary to trigger on the proper event in order to obtain usable measurements.

The following types of instruments are useful in this situation:

- Multimeters with minimum/maximum detection can be used to check for undervoltage and overvoltage events as well as mains dropouts. However, they cannot register very brief disturbances, since a particular level of the measured value must typically last for 1 ms (or more) for it to be detected.
- Mains disturbance analyzers: a large variety of devices is available, ranging from simple ‘indicators’ to complex measurement and analysis systems. The working principle is

that the mains voltage is rectified and converted to a suitable measurement voltage, which is compared to suitable reference values. The evaluation circuitry is normally galvanically isolated from the measurement circuitry, either by using optocouplers to pass the rectified voltage or by coupling the mains voltage to the instrument via a transformer.

Figure 5.79 shows a straightforward, simple solution, while Figure 5.80 illustrates an arrangement that allows almost all deviations from the ideal sinusoidal waveform to be detected. (Both of these figures only illustrate the working principles; see Volume 3 for additional details.)

In Figure 5.79, the mains voltage is reduced by a transformer and then rectified. The resulting DC voltage is compared to preset minimum and maximum reference levels. If the voltage exceeds the maximum level or drops below the minimum level, the associated

Type of disturbance	Causes	Measurements
earth loop noise	currents flowing in the protective earth lead (leakage currents, fault currents and coupled-in interference)	Directly measure the current flowing in the protective earth lead, or measure the common-mode noise (indirect measurement).
common-mode noise	noise injected into the protective earth lead (such as voltage spikes that are diverted to the protective earth lead by surge arresters); installation errors; overloading	Measure the voltages between the protective earth lead and both current-carrying leads (L1 & N). *Note:* an abnormally high noise level on the N lead – measured with respect to the PE lead – may indicate an earth-loop disturbance, among other things (that is, the PE terminal is carrying a voltage with respect to earth while the N lead remains approximately at the earth potential – remember that this is a differential measurement).
differential-mode noise	load switching; excessive voltages on the mains network. (Voltage spikes from external sources create differential-mode noise, since the neutral lead (N) is earthed at the entry point (consumer unit).	Measure the voltage between the live lead (L1) and the neutral lead (N).
poor power factor	pulsating current inputs; effects of connected loads	Measure the current waveform; measure the harmonic levels.

Table 5.14 Types of mains disturbance.

comparator generates a signal that can be evaluated as desired. (See Section 4.4.1 for descriptions of circuits that can be used for detecting very short pulses and classifying pulses according to their duration.) With current technology, it would be reasonable to use a microcontroller to record error events and to control the instrument's display.

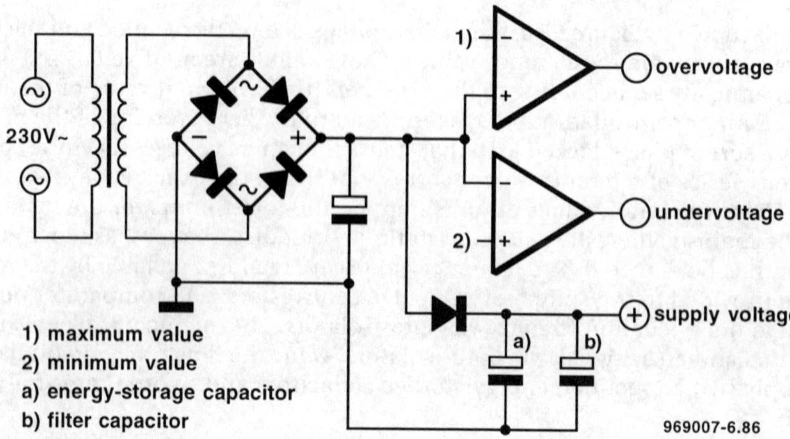

Figure 5.79 *A simple mains-voltage monitoring circuit.*

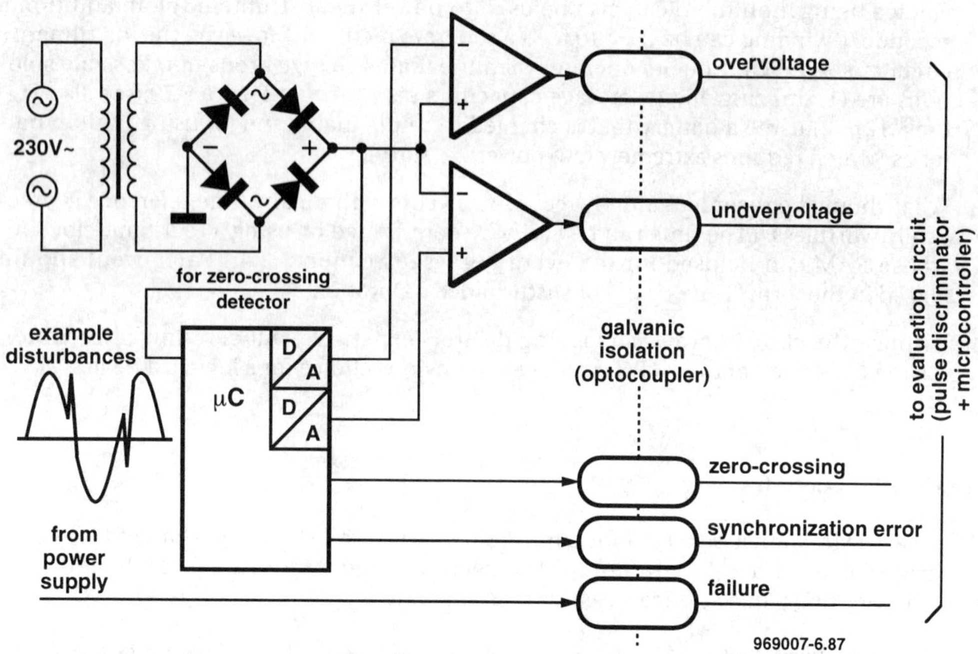

Figure 5.80 *Capturing disturbances in the time domain.*

The main shortcoming of simple types of monitoring circuits is that they cannot detect certain types of disturbances, such as dropouts and spikes that are significantly shorter than half of the mains period. Some devices analyze the harmonic content of the mains voltage in order to detect such disturbances (frequency-domain analysis, using a bank of filters followed by classification circuits). The arrangement shown in Figure 5.80, by contrast, operates only in the time domain.

In the circuit shown in Figure 5.80, the mains voltage is directly rectified and the pulsating DC voltage is compared to reference values. These values are not fixed as in Figure 5.79, but are dynamically set according to the expected (time-varying) value of the measured waveform. A microcontroller that is synchronized to the zero crossing of the mains voltage acts as a sort of phase-locked loop that tracks the mains voltage waveform, producing a continuous series of reference values corresponding to the expected level of the mains sinusoid. These reference values establish upper and lower limits for the instantaneous value of the mains voltage at each point in time. Brief disturbances, as illustrated in the figure, can thus be detected. The reference values are read from a table by the microcontroller and provided to the comparators via D/A converters. The comparator outputs are connected to the evaluation circuitry via opto-isolators. The microcontroller also provides certain status and error signals via opto-isolators. *Note:* the illustrated circuit has its own power supply (with a rectifier, energy-storage capacitors and so on) that is fully isolated from earth.

Basic problems:

a) Supplying power to the detection and recording circuitry. As shown in Figure 5.79, the voltage being monitored can also be used to power the instrument (or an additional secondary winding can be used to feed a separate rectifier). However, the instrument's circuitry should continue to operate when the mains voltage drops out. Possible solutions are (1) utilizing energy-storage capacitors (as shown in Figure 5.79; see also Figure 4.12), (2) using a battery that is charged from the mains and (3) using lithium batteries (which requires extremely low-power circuitry).

b) What should be recorded? Simply recording events without any time reference is practically worthless. (The time can be easily be determined by using a real-time-clock IC, whose RAM can be used for the event log – see Section 3.2.4.) Each event should logged in the form 'Time – Type of Disturbance – Duration'.

c) Display: the choices are: (1) a logging printer (expensive, but certainly affordable), (2) an LCD screen and (3) a PC interface (an obvious choice for a low-cost device).

Testing the safety features

Electrical devices must be tested for conformance to the relevant VDE standards (in particular VDE 0701 and 0413) after they have been repaired (see Figure 5.81). In addition, periodic checking (such as once a year) is recommended.

Safety classes. Electrical equipment is divided into four safety classes (see Table 5.15).

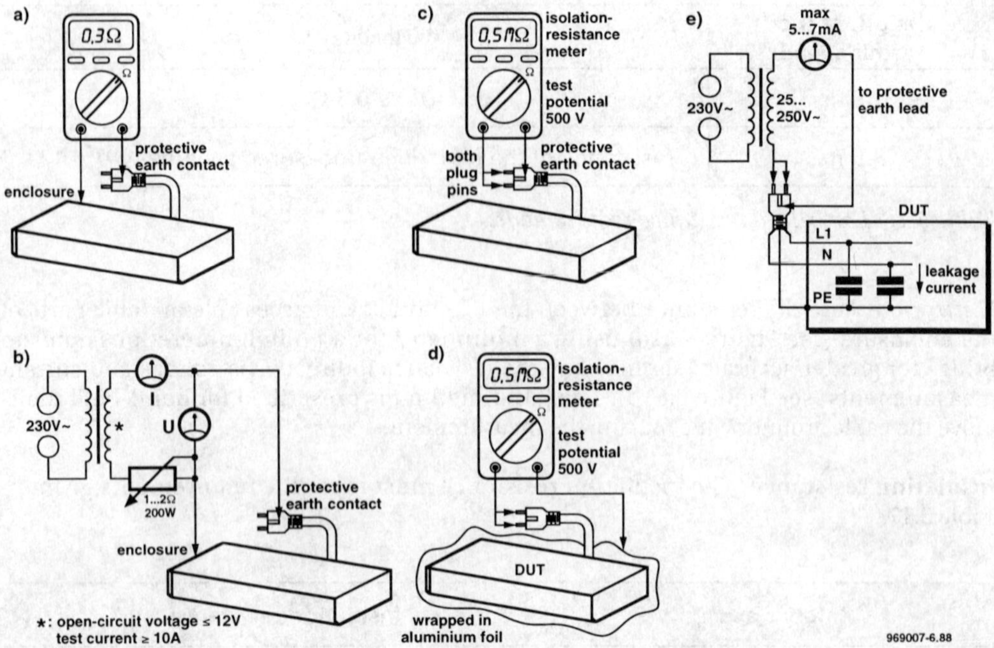

Figure 5.81 Testing safety provisions (see the text).

Class	Symbol	Construction	Remarks
0		primary insulation only	not allowed in the UK
I		primary insulation plus protective earth connection	
II		double-insulated (primary plus secondary insulation)	no protective earth connection
III	III	low-voltage connection (50 V~, 120 V⎓)	no protective earth connection

Table 5.15 Safety classes for electrical equipment according to DIN VDE 0106 part 1.

Earth lead resistance. The protective earth lead of Class I devices must be effective. The resistance of the PE lead, measured between the equipment enclosure and the PE contact of the mains plug, must meet the requirements given in Table 5.16. With a plastic enclosure, the reference point at the equipment end is the attachment of the PE lead to the interior metallic parts of the equipment (frame or chassis).

Length of the earthing lead	Resistance
≤ 5 m	≤ 0.3 Ω
> 5 m	resistance of the earth lead (measured separately) + 0.1 Ω

Table 5.16 *Earth-lead resistance requirements.*

Testing: measure the resistance between the PE contact and accessible metallic parts of the enclosure (see Figure 5.81a) using a multimeter (or a milliohmmeter or Thomson bridge for greater accuracy), or measure the resistance indirectly via voltage and current measurements (see Figure 5.81b). The latter method is prescribed for hand-held tools. Move the cable around while making the measurement.

Insulation resistance. The insulation resistance must meet the requirements given in Table 5.17.

Safety class	Insulation resistance
I	≥ 0.5 MΩ
II	≥ 2.0 MΩ
III and battery-operated devices	≥ 1 kΩ/V (1 kΩ per volt of test potential)

Table 5.17 *Insulation resistance requirements.*

Testing: a makeshift test can be made by using a digital multimeter to measure the insulation resistance (the resistance should be practically 'infinite'). More accurate results (for Class I and II) can be obtained by measuring the current with an applied voltage of 500 V DC (suitable testers are commercially available). For Class III devices, a low-resistance ohmmeter should be used (test voltage ≤ 42 V). For Class I devices, the insulation resistance is measured between the PE contact and the joined L1 and N contacts at the mains plug (see Figure 5.81c). Class II devices have a plastic enclosure; the insulation resistance is measured by wrapping the device in aluminium foil and measuring the resistance between this foil and the joined L1 and N contacts of the mains plug (see Figure 5.81d).

Leakage current. If capacitors are connected between the mains leads and accessible metallic parts of the enclosure (in a mains input filter or surge arrester; see Figure 5.57), the leakage current must be measured instead of the insulation resistance. It is measured with a test voltage of 25 to 250 V applied between the PE contact and the joined L1 and N contacts of the mains plug (see Figure 5.81e). The maximum allowed values of the leakage current are listed in Table 5.18. An improvised test can be made by measuring the insulation resistance with a digital multimeter and testing the capacitors separately, using a digital multimeter with suitable measurement ranges or a component tester.

Test with ...	Leakage current
alternating voltage U_L, 25–250 V no load, 50 Hz	≤ 5 mA for U_L > 50 V; ≤ 7 mA between accessible metallic parts and parts carrying operating voltages

Table 5.18 Leakage current requirements.

Practical tips:

1 The best way to make connections to the mains plug is to plug it into the socket of a mains adapter (refer to Figure 5.74) *Caution:* if the mains adapter has its own mains plug, dangerous voltages can be present on its contacts with certain tests! Sometimes a very simple adapter, consisting only of an outlet and test jacks, is a good idea.

2 Measurement setups as shown in Figure 5.81 should actually only be used when improvisation is necessary. If these types of measurements must be performed relatively often, you should acquire a suitable tester (VDE-701 tester), with which all the described measurements can be performed. Such a tester has a socket that accepts the mains plug of the equipment being tested, so that special adapters are not necessary.

5.2.2 Power supply testing

Basic test setup. The power supply to be tested is connected to the mains on the input side and to a suitable load on the output side (see Figure 5.82). It may be necessary to make measurements on both the mains side and the load side during testing.

The following elements are shown in Figure 5.82:

- *Mains connection:* for basic functional testing, the mains connection amounts to nothing more than plugging the mains power cord into an outlet. However, some testing tasks require more complicated arrangements (see Table 5.19).
- *Mains measurement adapter:* the mains voltage and current must be measured, both statically (with a multimeter) and dynamically (with an oscilloscope).
- *Measurement adapter (load side):* the output connectors of many power supplies cause problems for testing (this applies to standard PC power supplies – with Molex jacks for connecting to the motherboard – and also to other types of power supplies, including simple mains adapters). The measurement adapter must do the following (see Figure 5.83): (1) mate with the output connector(s) or terminals of the power supply, (2) provide connectors for the instruments used to measure voltage and current, (3) provide connectors for dummy loads and (4) provide connectors for the actual loads (so that voltages and currents can be measured under actual operating conditions).
- *Load:* testing under open-circuit conditions (unloaded) is not at all rigorous. In addition, many power supplies will not work at all without a certain minimum load. The requirements for the test load (dummy load) are: (1) to load the power supply from its

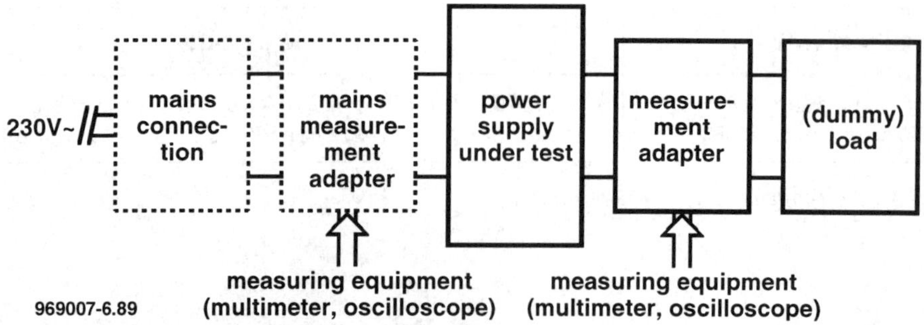

Figure 5.82 *Test setup for power supply testing.*

Test objective	Special requirements for the mains connection	Solution
checking the switch-on behavior	triggering the oscilloscope at the switch-on time	(see Note 3)
checking the input voltage regulation	varying the mains input voltage	variable transformer (see Note 1)
checking the load regulation	holding the mains input voltage constant	variable transformer (see Notes 1 & 2)
fault-finding in the power supply	selective fusing in case of a fault	test bench with ELCBs and circuit breakers

Table 5.19 *Special requirements for the mains connection.*

Notes:

1 *To accurately vary the mains voltage or hold it at a fixed value, you need a mains simulator (not to be confused with a mains emulator, which is required for EMC testing). These are however very expensive. An obvious improvisation is to use an adjustable transformer. If galvanic isolation is not required, an autotransformer type (Variac®) is sufficient (and it is lighter and less expensive than an adjustable isolation transformer with the same power rating). See also Volume 3.*

2 *The mains voltage cannot be held to a precise level with a transformer alone. Using a ferromagnetic voltage stabilizer, which might seem like a logical solution, is not always suitable, since it will increase the source impedance of the mains supply and/or increase the level of the harmonic components. You must therefore content yourself with using an adjustable transformer to set the voltage as closely as possible to the desired values (nominal value, lower limit and upper limit) .*

3 *Trigger signal: an obvious solution is to use an extra contact of the mains switch to provide a trigger signal, or to use a relay to switch the mains voltage (the relay control signal can be used as a trigger that is guaranteed to precede the application of power to the device under test). Since relay actuation times vary significantly, it is usually better to take the trigger from the applied mains voltage (for example, via an optocoupler). This approach is quite satisfactory if you use a DSO with pre-trigger capability. See Volume 3.*

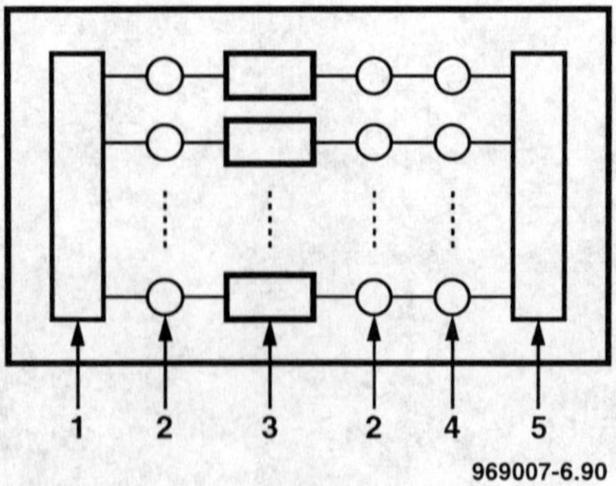

Figure 5.83 Basic construction of a load-side measurement adapter.

1 – connector for mating with the power supply output connector, 2 – measurement jacks, 3 – current measurement provisions (in this case, shunt resistors), 4 – jacks for connecting a dummy load, 5 – connector for the working load.

minimum to its maximum rated load, (2) to provide a variable load and (3) to provide some indication of changes in the load (to be used for triggering an oscilloscope, for instance).

Practical tips:

1 Make sure that the leads of the mains cable have adequate cross-sectional area (at least 0.75 mm^2 – inexpensive 0.5-mm^2 material is not suitable). As a rule, the voltage drop in the cable should not exceed 1% of the nominal mains voltage (which means not more than 0.5 % for each of the L1 and N leads).

2 Connect the power supply to the mains so that its built-in fuse (or a fuse or circuit breaker that is connected ahead of the power supply for testing purposes, as shown in Figure 5.67) is in the L1 current path.

3 If the power supply is connected to the mains via an autotransformer, connect the common terminal of the autotransformer to the N lead, not the L1 lead.

Test accessories

The simplest measurement adapter for PC power supplies. Standard PC power supplies need minimum loads only for the +5 V and +12 V outputs. The fact that both of these voltages are present at the disk drive power connectors can be utilized to construct a simple test accessory that is fully adequate for basic functional testing (see Figures 5.84 and 5.85).

PC power supply tester. Figures 5.86 and 5.87 show a simple power-supply tester that can apply three different loads to each of the four output voltages of a standard PC power supply. The loads are selected via toggle switches. The tester also includes current-sensing shunt resistors and a voltmeter.

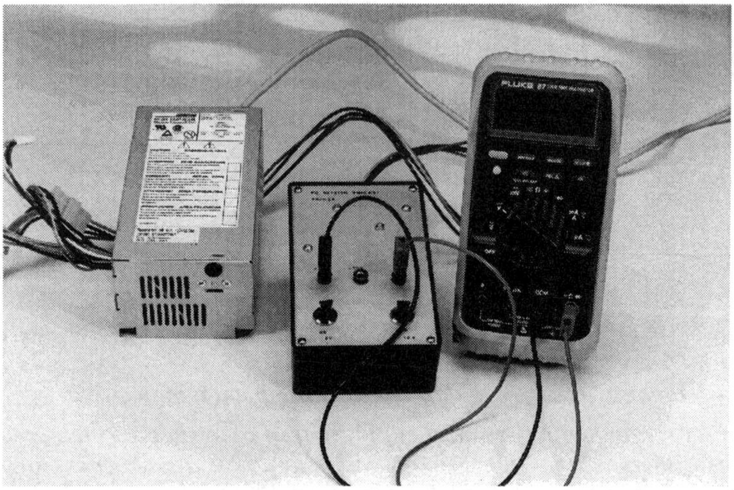

Figure 5.84 *Test aid for PC power supplies (oblique view).*

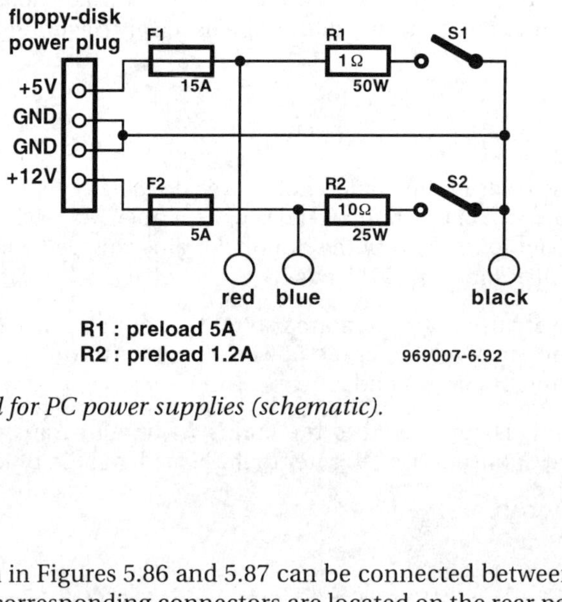

Figure 5.85 *Test aid for PC power supplies (schematic).*

Notes:

1 The tester shown in Figures 5.86 and 5.87 can be connected between the power supply and the PC. The corresponding connectors are located on the rear panel of the tester.

2 The adapter shown in Figures 5.84 and 5.85 can also be used for powering experimental circuits and the like from a PC power supply. It provides a minimum load in case nothing else is connected, allows connections to be made with standard test leads, and has fuses for the individual supply leads. (*Practical tip:* use automatic cutouts in place of ordinary fuses.)

3 See Volume 3 for additional types of power-supply test hardware.

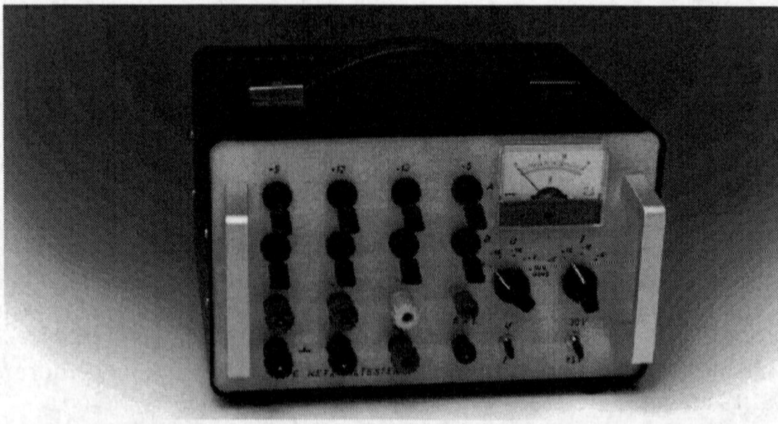

Figure 5.86 PC power-supply tester (front view).

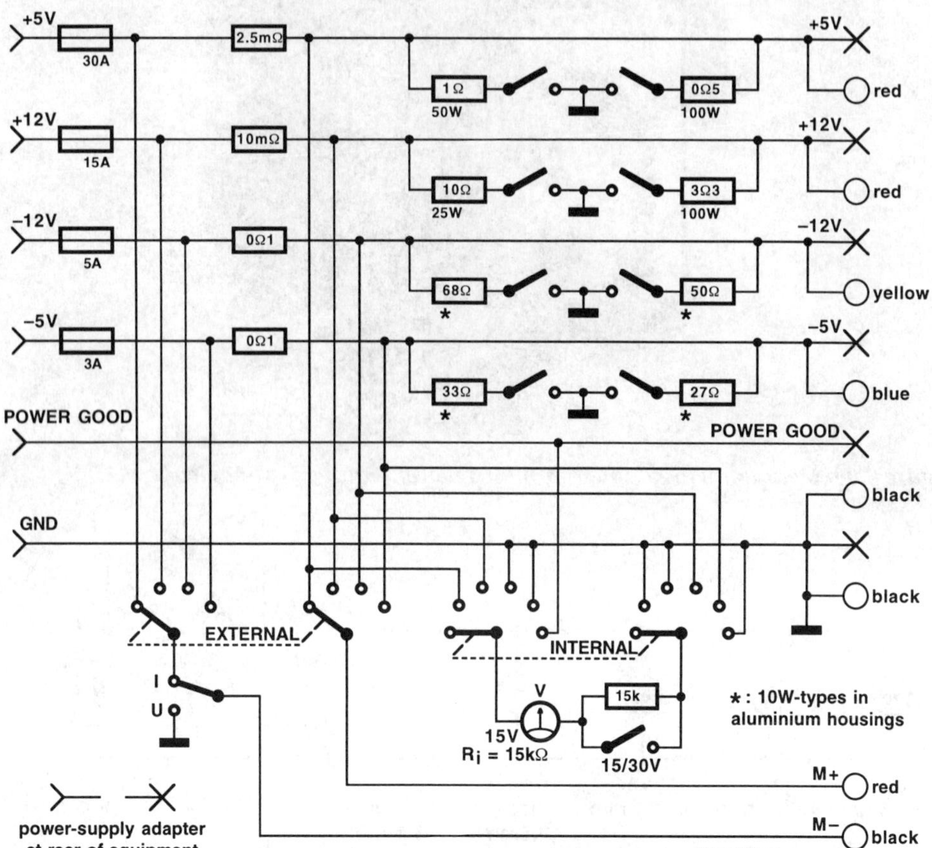

Figure 5.87 PC power-supply tester (schematic).

Measurement adapter for plug-in mains power adapters. It is difficult to connect alligator clips or the like to the round, coaxial power connectors on the output cables of mains power adapters. A simple adapter with an assortment of jacks can solve this problem (see Figures 5.88 and 5.89).

The various jacks in the adapter shown in Figures 5.88 and 5.89 mate with the various types of output connectors used with mains power adapters. They are all wired in parallel. The loudspeaker jack allows the load device in question to be connected (you must make up an adapter cable with a loudspeaker plug on one end and the appropriate coaxial or phone plug on the other end. Loudspeaker connectors are inexpensive and polarized.) The current path can be opened by a toggle switch for making current measurements. *Note:* the jack colours (red and black) correspond to the plus and minus leads in the usual manner if the plus lead is connected to the outer shell of the plug (in conformance with the DIN 45310 standard).

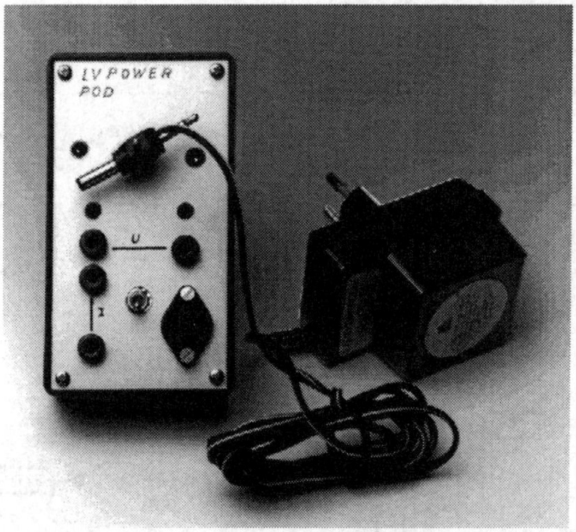

Figure 5.88 Measurement adapter for mains power adapters (front view).

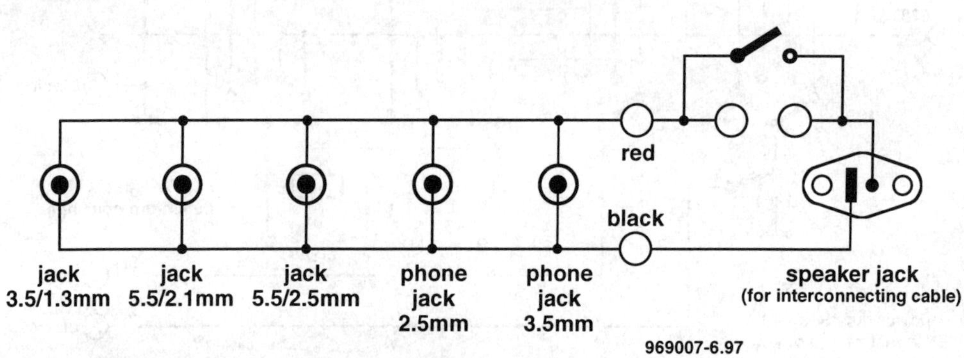

Figure 5.89 Measurement adapter for mains power adapters (schematic).

Specific tests

Input voltage regulation. The input voltage regulation of a power supply (how well it compensates for variations in the input voltage) can only be superficially tested with simple equipment. Proceed as follows:

1 Use a variable transformer to supply the mains voltage to the power supply.

2 Measure the output voltage(s) under load with the input voltage first set to the lower limit and then set to the upper limit.

3 With a DSO, you can record the output voltage level while quickly rotating the knob of the variable transformer back and forth. *Practical tips:* (1) select a suitably low sampling rate and (2) manually trigger the oscilloscope in one-shot mode just before starting to vary the input voltage. This is of course only a superficial test, since it is only possible to see whether the output voltage changes noticeably as a function of the input voltage.

Load regulation. Figure 5.90 shows the general behaviour of the output voltage when the load changes suddenly. If the load suddenly draws more current or less current, control processes in the power supply act to counter changes in the output voltage. Overshoots and undershoots can occur. For any change in the load, the recovery time of the output voltage should be as short as possible and overshoots or undershoots should not exceed the specified tolerance limits of the output voltage.

Figure 5.91 shows the measurement setup and an example of the current and voltage waveforms for a dynamic change in the load (as recorded by an oscilloscope). An electronically switchable or variable load can be used to generate periodic, rapid changes in the load current. *Practical tip:* the test load should provide a separate trigger signal for the

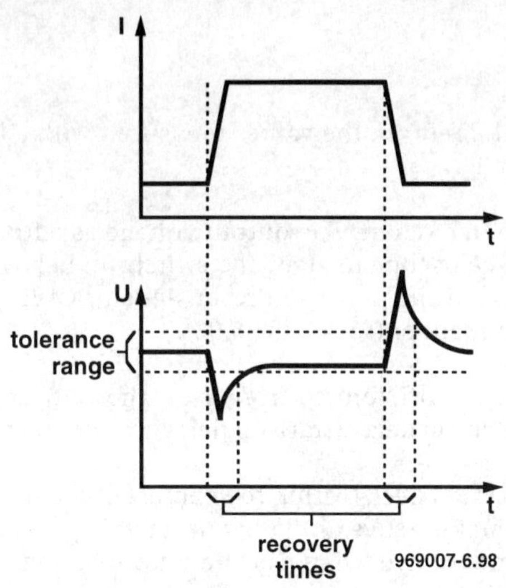

Figure 5.90 Basic output waveforms for load changes.

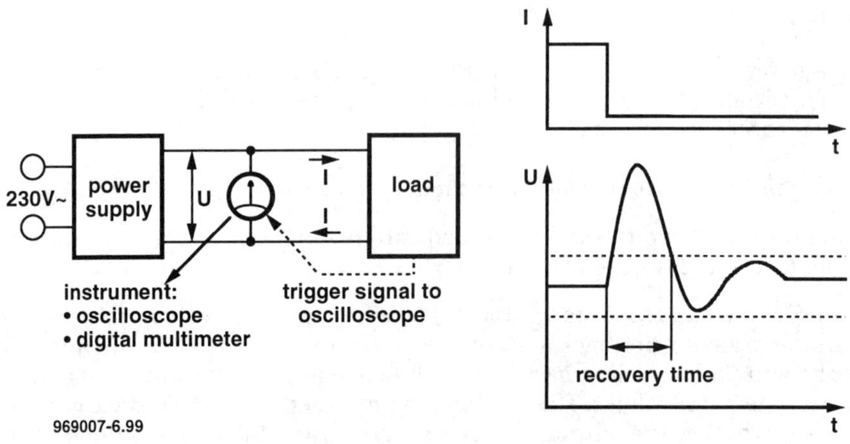

Figure 5.91 The output voltage waveform with a rapid change in the load.

oscilloscope, so that the variations in the output voltage can be observed (with a good power supply the variation in the output voltage is quite small, so that it is hardly possible to trigger on changes in the output voltage).

It is possible to make static measurements in a similar manner (using a digital multimeter), to check whether the output voltage remains within the specified tolerance band when the load current varies over its full allowable range.

Determining the effective internal resistance. The effective internal resistance R_i of a voltage source can be calculated from the voltages and currents measured with a switched load, as follows:

$$R_e = \frac{\Delta U}{\Delta I} = \frac{U_1 - U_2}{I_2 - I_1}$$

where the subscripts *1* and *2* refer to the values measured with a light load and a heavy load, respectively.

The output-side switch-on behaviour. The output voltage as a function of time must be recorded using a storage oscilloscope to allow the switch-on behaviour to be judged (see Figure 5.92). The generation of the power-on reset signal (POWER GOOD, for example) can be observed in a similar manner (see Figure 5.93).

Triggering: in principle, it is sufficient to trigger on the output voltage that is being observed. However, the switch-on delay (start-up delay) cannot be measured in this case.

An important test objective is to check (within the context of the total system) whether the hardware reset signal (1) remains active until the output voltage reaches its nominal value and (2) immediately becomes active whenever the output voltage drops below its minimum value.

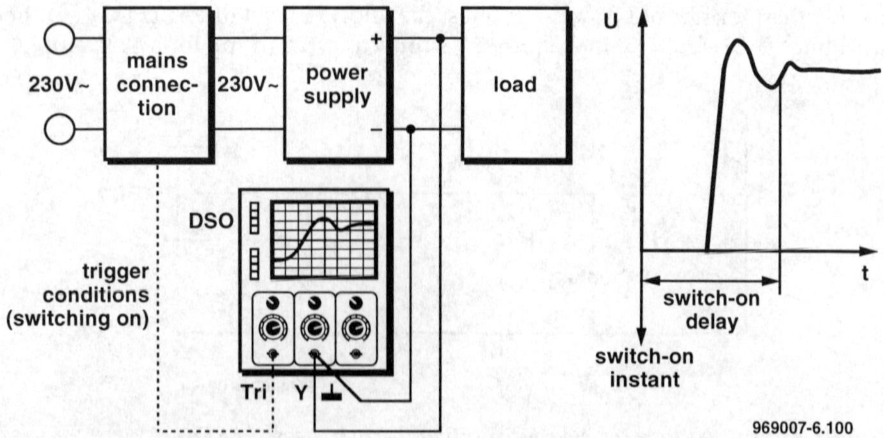

Figure 5.92 The switch-on process.
a) measurement setup. b) typical output voltage waveform.

Output disturbances (PARD). 'PARD' stands for 'Periodic and Random Deviations', which are also known as 'ripple and noise'. It is a collective term for all undesired AC voltages and high-frequency components of the output voltage (see Figure 5.94). PARD disturbances range from 20 Hz to 20 MHz ('ripple' ranges up to 50 kHz, 'noise' lies above 50 kHz, and slow variations are called 'drift'). Both the peak and the effective values should be measured. A well-screened measurement setup that is free of earth loops is required, since relatively low-amplitude AC voltages must be measured. The superimposed AC voltage is normally coupled to the test equipment by a 50-Ω coaxial cable terminated at both ends (see Figure 5.95 – the capacitor blocks the DC voltage). To avoid earth loops, float the test equipment (with an isolation transformer if necessary) or use a differential probe.

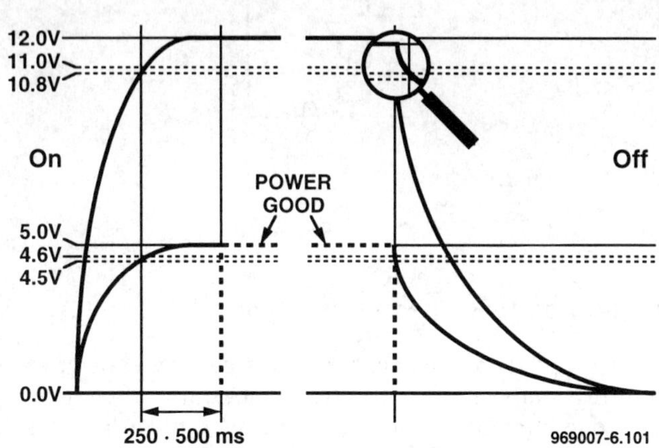

***Figure 5.93** Switch-on and switch-off transitions of a normal PC power supply.*

Practical values: for normal power supplies, the allowed level of PARD is 1% of the rated output voltage ($PARD_{max}$ = 50 mV_{pp} for a 5 V supply and 120 mVpp for a 12 V supply).

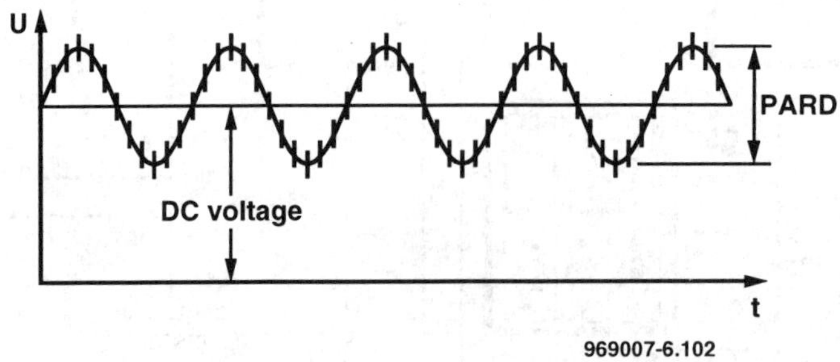

Figure 5.94 Output voltage with superimposed disturbances (PARD).

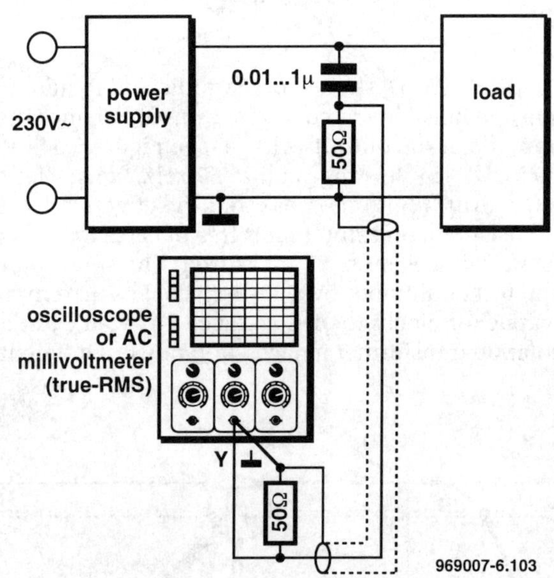

Figure 5.95 PARD measurement setup.

5.2.3 Batteries and battery-operated equipment

Batteries (both primary and secondary types) are tested by means of voltage and current measurements. Dummy loads should be selected such that the current drawn roughly matches the normal load current. However, it is more appropriate to select the dummy load for a secondary battery such that the battery's rated capacity would be fully discharged in one hour (load resistance = rated voltage ÷ CA; refer to the literature listed in Appendix 8 for basic concepts).

Connections. It is difficult to make connections for test purposes to most types of batteries. Possible (improvised) solutions are:

1 Use standard battery cases and clips (select a suitable type for the battery being tested, and solder jacks to its terminals).

2 Construct a suitable adapter (see Figures 5.96, 5.97 & 5.98).

The battery measurement adapter shown in Figures 5.97 and 5.98 can be connected between the two adapters shown in Figure 5.96. It can be used for: (1) voltage measurements, (2) current measurements (the current path can be opened) and (3) providing simple dummy loads (with a selectable resistance between 12 Ω and 51 Ω; see Section 4.2.10).

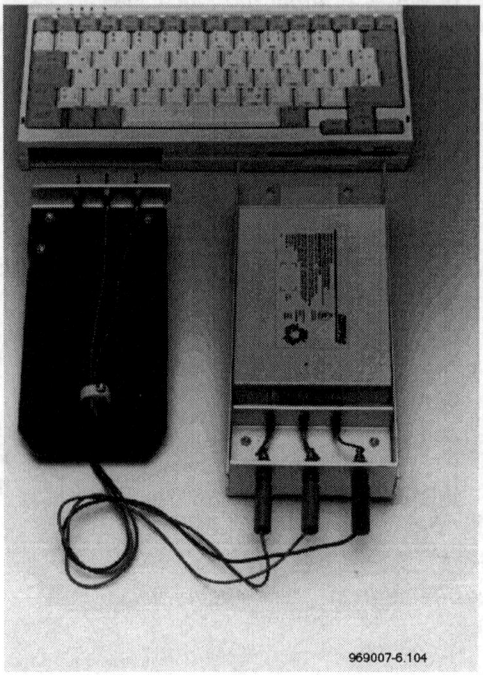

Figure 5.96 *Battery adapters for a portable PC.*

The following adapters are needed: (1) one that replaces the battery in the PC (and which allows the PC to be operated from a laboratory power supply), and (2) one that makes contact with the battery terminals. The actual construction depends on the PC in question, but you can see that useful adapters can be built from a piece of Pertinax and a short piece of aluminium extrusion (available in DIY home-improvement shops). The contacts are made using standard spring-loaded test needles mounted in insulated 2-mm jacks.

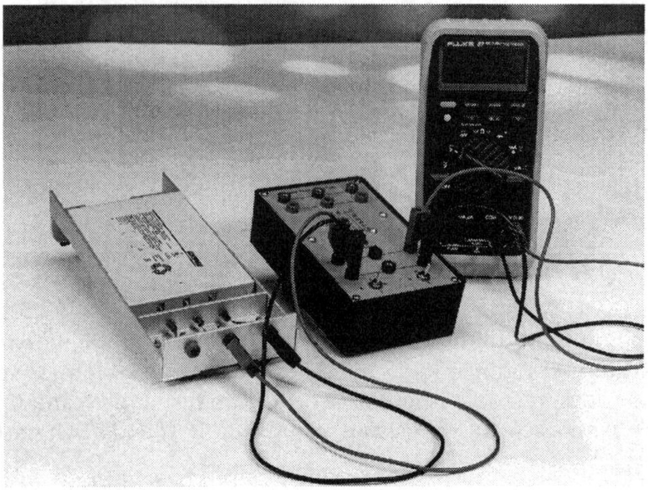

Figure 5.97 Battery measurement adapter (oblique view).

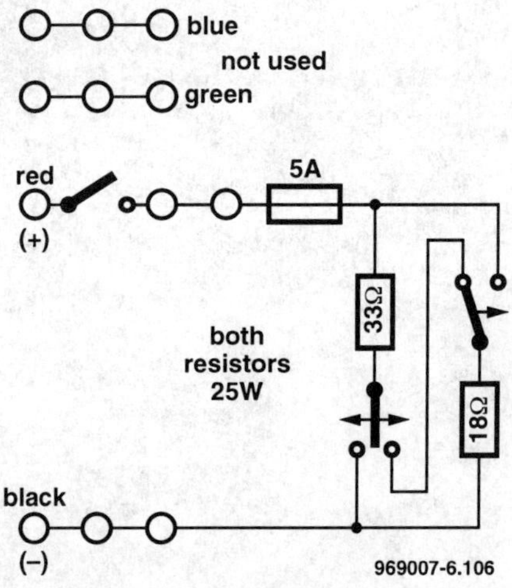

Figure 5.98 Battery measurement adapter (schematic).

Note: modern interfaces for rechargeable batteries often have additional contacts besides those for the positive and negative terminals of the battery, for instance for temperature sensors or a serial data link to the charge-state controller. These contacts must be passed through by the adapters (the measurement adapter has a free pair of jacks that can be used for making such connections).

Testing battery-operated equipment. An obvious solution is to power the equipment from a laboratory power supply. Figure 5.99 shows what is involved in such a test setup. The voltage and current are measured on both sides of the power supply. The important points are: (1) use leads with sufficient cross-sectional area for the main current paths (at least 1.5 mm^2) and keep these leads fairly short (20–25 cm maximum), and (2) connect the voltmeters directly to the input and output terminals of the device under test using separate leads (Kelvin sensing).

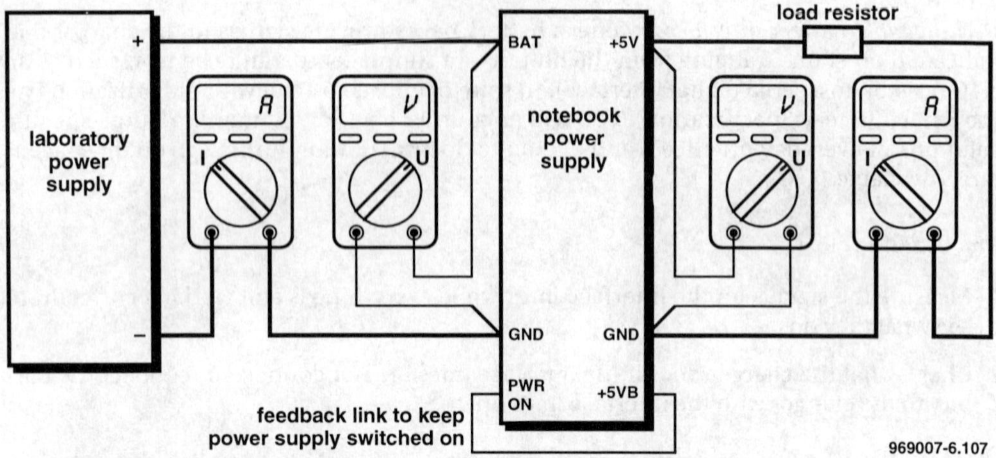

Figure 5.99 Testing the power supply of a notebook PC. (source: Electronic Design)

Testing batteries. Only spot checks are possible with primary cells and batteries; secondary batteries can be tested more thoroughly (although this takes time).

Global test: simply connecting a voltmeter to the battery is no good, since no-load voltage measurements are practically worthless. *Practical tip:* connect a dummy load that draws a 'typical' operational current (as a guideline, the resistance in ohms should be equal to the nominal voltage in volts divided by the rated capacity in ampère-hours). The cell or battery being tested should maintain its rated voltage for the duration of the test (guideline: 1 to 5 minutes). Standard commercial battery testers work on this basis; they contain of a number of different load resistors and a simple voltmeter. Since such testers are mass-produced, it is not economical to build your own.

Testing the capacity of rechargeable batteries: suitable capacity testers are available. The working principle is as follows: the battery (or cell) is discharged at a constant current to its fully-discharged (end-point) potential, and the time required for this is measured. The fully-charged capacity is then equal to the product of the discharge current and the discharge time. The battery must be fully charged before its capacity can be measured, so a complete test cycle consists of three steps: (1) discharging the battery, (2) fully charging the battery and (3) measuring the capacity.

Testing battery chargers

1 Charging test: a known-good cell or battery is first discharged, and then charged by the device to be tested. The capacity of the fully-charged cell or battery is then measured.

2 Testing as a voltage source: the charger is tested with a dummy load, just like any other power supply (as a guideline for the value of the load resistor, divide the fully-charged cell or battery voltage by the device's rated charging current).

Rechargeable-battery interfaces. Modern battery packs provide signals to the charger that indicate their status (ranging from the output of a simple temperature sensor, such as an NTC resistor, to a serial digital interface). In spite of efforts to achieve standardization (via the Smart Battery Specification), you will encounter a variety of manufacturer-specific solutions in everyday practice – and adequate documentation for any given interface is rarely available.

Practical tips:

1 Measure the signals on the interface lines while everything is still working correctly, to see what they do.

2 Check what the charger does if the interface lines are not connected (connect the battery to the charger with the interface lines open).

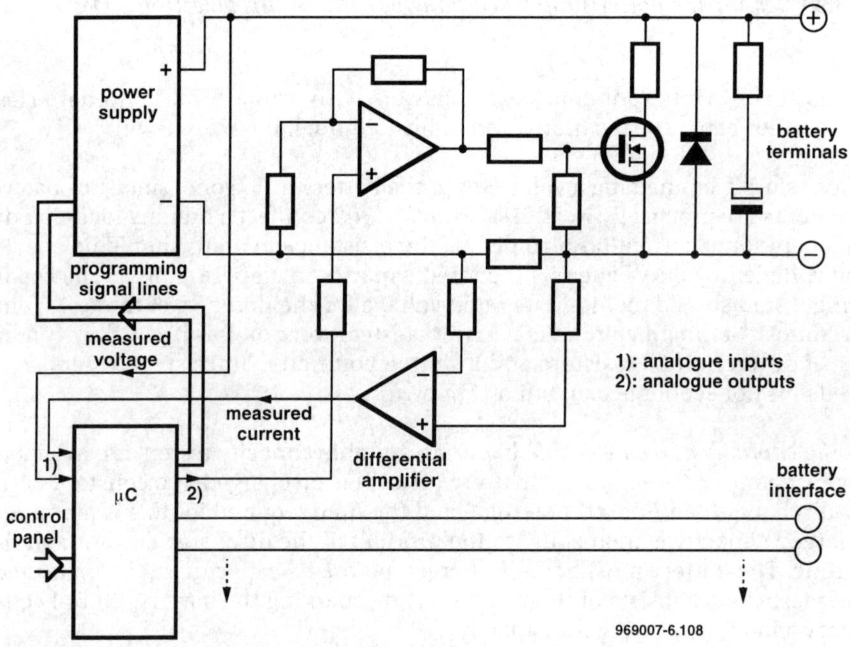

Figure 5.100 *A battery simulator (block diagram).*

3 A PTC or NTC resistor that is used as a temperature sensor can be simulated using a potentiometer, if need be.

4 It is very time-consuming to try to analyze the behaviour of a battery pack with a digital charger interface. A spare battery pack that is kept for testing purposes is usually a lot less costly. (*Practical tip:* open the battery pack and test the cells individually, or use a power supply or battery simulator instead of the battery pack.)

Battery simulator. The working principle of a battery simulator is quite simple, but the implementation is rather complicated. A device with the same electrical behaviour as that of a rechargeable battery, with regard to charging and discharging, can be built using an electronic load (refer to Figure 4.38), a programmable power supply and a microcontroller (see Figure 5.100). See Volume 3 for more information.

During charging, the DMOS transistor functions as a load resistance that absorbs the charging current. The microcontroller regulates the pass resistance of this transistor so that the resulting voltage and current characteristics match those of a rechargeable battery, including typical charge states (full discharge, reaching the terminal charging voltage and so on). When no charging current flows, the circuitry connected to the power supply mimics the behaviour of a battery connected to a load (with suitably programmed discharge characteristics).

5.3 Cable testing

5.3.1 Testing for continuity and shorts

Manual testing

Quickly testing a spool of cable. It is a good idea to perform a quick test for continuity and shorts, using an ohmmeter, before the cable is installed. For each lead, (1) there must be a conductive path (continuity) from one end to the other and (2) no short or finite (measurable) resistance to any other lead should be seen. Resistance values for the leads and insulation can be found in the technical specifications of the cable. A good approximation of the resistance of an individual lead can also be calculated from its cross-sectional area and length.

Note: the lead resistance is very low and the insulation resistance is very high. This relationship is similar to that for contact testing (see Section 5.1.5).

With certain types of cable, measuring the interlead capacitance is also a good way to make a quick quality check. Measure the capacitance between two leads using an RLC tester (for example) connected at one end of the cable; the measured value should be approximately the same as the product of the cable length and the specified capacitance per meter (as given in the data sheet; typical values are 50 to 180 pF/m).

Testing finished cables for continuity and shorts. The special considerations are: (1) the connectors also have to be taken into account (as possible sources of faults) and (2) both ends of the cable may not be accessible at the same location (with installed cables). You can connect pairs of leads together at the far end of the cable, so that the continuity of each pair can be checked at the near end; the measured resistance should then be twice that of a single lead if both leads are the same.

Semi-automatic testing for continuity, shorts and intermittent contacts. This can be achieved by monitoring all the leads while testing the continuity of each individual lead. A latch circuit is required for each lead to detect intermittent contacts or broken leads (see Figures 5.101 and 5.102).

Cable tester operation. If cable testing is only performed occasionally in the service environment, partial automation is usually adequate. Another consideration is that some cables do not have connectors on both ends (keyboards, mice and many monitors have permanently-attached cables). The cable tester shown in Figure 5.101, which contains 24 latch circuits as shown in Figure 5.102, can test up to 24 connections. It is attached to one end of the cable to be tested via an adapter and an intermediate cable, as shown in the illustration. The adapter has a mating connector for each type of connector that is used for the cables to be tested.

In operation, a hand-held probe is touched to each contact or lead in turn at the far end of the cable. The probe provides a connection to earth (logic LO) via the cable. The 'Measure' pushbutton is briefly pressed while the probe is held against each contact. If the connection is good, the 'Connected' latch will be set, and if it is open the 'Open' latch will be set. The two latches work together to detect intermittent connections. Normally, only the 'Connected' lamp of the lead being tested and the 'Open' lamps of all other leads should be lit for a given test. Shorts to other leads can thus be immediately recognized. If you wiggle the cable and pull on the connectors while the test is in progress, an intermittent connection will cause the 'Open' indicator to also light. In the same way, an intermittent short will cause the 'Connected' indicator(s) of some other lead(s) to light.

Caution: If the cable being tested is soldered to a circuit, this type of testing corresponds to making measurements in a switched-off circuit (see Section 4.4.1). A low test voltage (< 0.5 V) should be used in such cases. Figure 5.103 shows the schematic of a suitable circuit based on a simple continuity tester. (A tester like that shown in Figure 5.101 requires one circuit of this type for each input. See Volume 3 for more information.)

The continuity tester shown in Figure 5.103 uses a test voltage of 0.4 to 0.5 V provided by a voltage divider. This is compared to a reference voltage of around 0.2 V. The comparator output is HI if the test voltage is greater than the reference voltage. If the lead being tested is continuous, the test voltage falls below the reference level and the comparator output is LO. The latch arrangement corresponds to that shown for the cable tester. A continuity indicator lamp and a piezo-electric beeper are also connected directly to the comparator output. The operating-mode selector switch allows the latch or the beeper to be disabled. *Notes:* (1) the threshold level for detecting continuity (which determines the resistance level above which 'Open' is indicated) is set by the value of resistor R^*, and (2) the test current is limited to 50 µA with the given component values, which means that the continuity threshold level is 500 Ω. See also Volume 3.

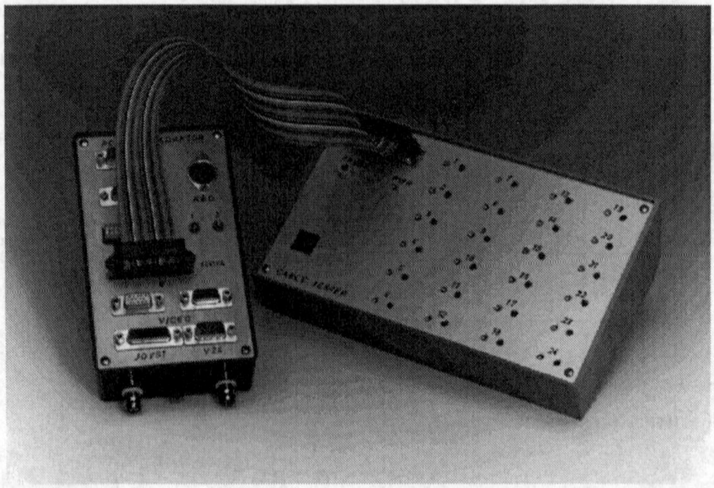

Figure 5.101 *Semi-automatic cable tester (front view), connected to a cable adapter.*

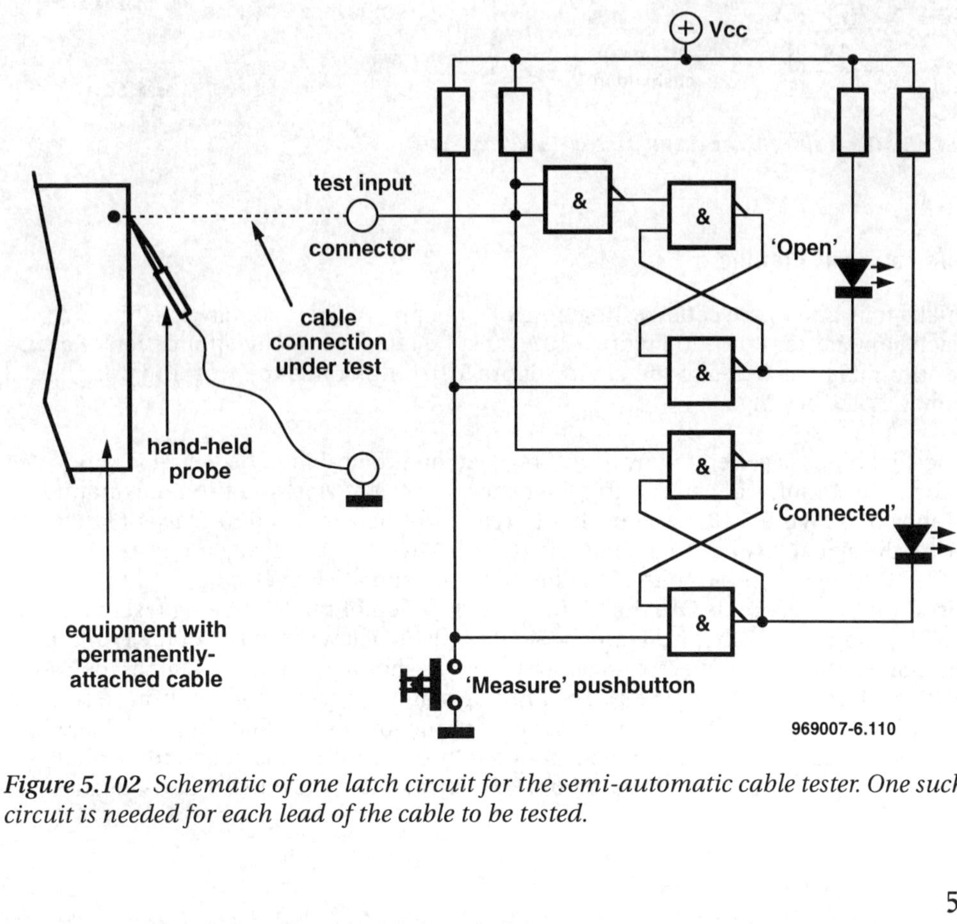

Figure 5.102 *Schematic of one latch circuit for the semi-automatic cable tester. One such circuit is needed for each lead of the cable to be tested.*

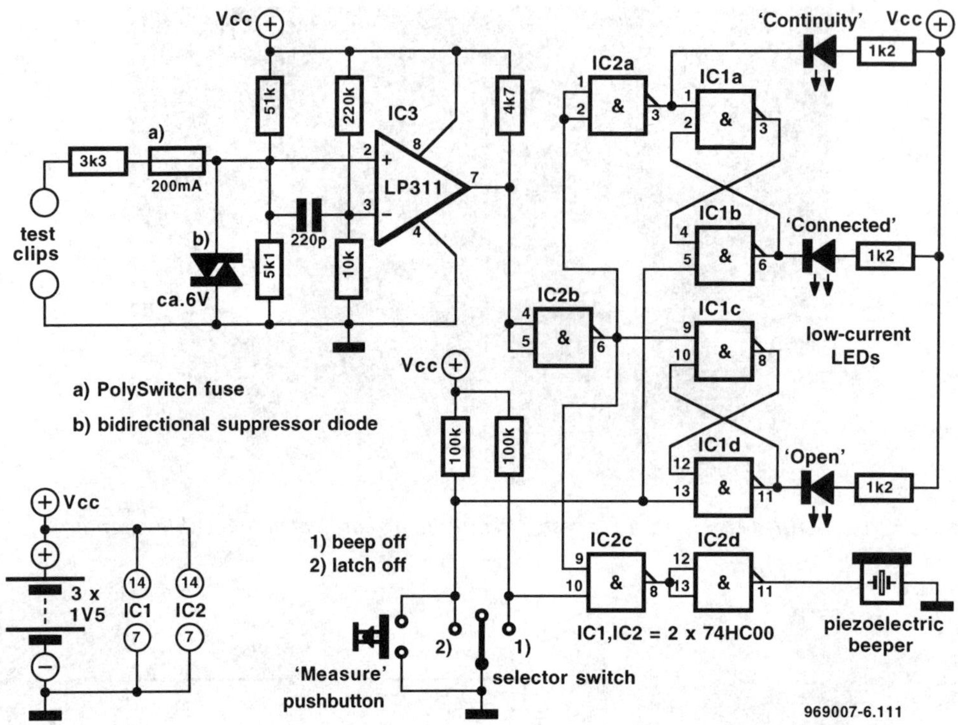

Figure 5.103 *A simple continuity tester (see the text).*

Automatic cable testing

To make fully automatic cable testing possible, the process of manually touching a probe to each contact, as shown in Figure 5.102, must be replaced by the application of a suitable stimulus to all leads of the cable. Figure 5.104 illustrates the working principle of automatic cable testing.

The test stimulus is generated by a shift register, and the cable to be tested is connected between the stimulus generator and the tester. The tester works in the same manner as that shown in Figure 5.101, except that D-type flip-flops are used in place of latches to capture the signal level of each lead (via their Set inputs). The leads are tested one at a time; a LO level is applied to the lead being tested and a HI level is applied to all other leads. If the connection is OK, the 'Connected' flip-flop of the lead being tested and the 'Open' flip-flops of all other leads will be set. In order to allow the states of the flip-flops to be evaluated, they are connected together to form a shift register (activation of the SHIFT MODE line forces all Set inputs HI). The flip-flop states can be shifted out using a suitable clock, which fills the shift registers with zeros in the process so that they are cleared. In order that the each lead of the cable is tested in turn, the stimulus generator is built as a shift register through which a single zero (LO level) is shifted (its initial state is all ones).

With current technology, a computer would be used to control the test process and evaluate the data from the latches. Using a PC for this is a natural solution, since it can also readily be used for displaying faults, storing test tables for various types of cables, 'learning' connector layouts and so on. (Memory is not a problem, and there are convenient

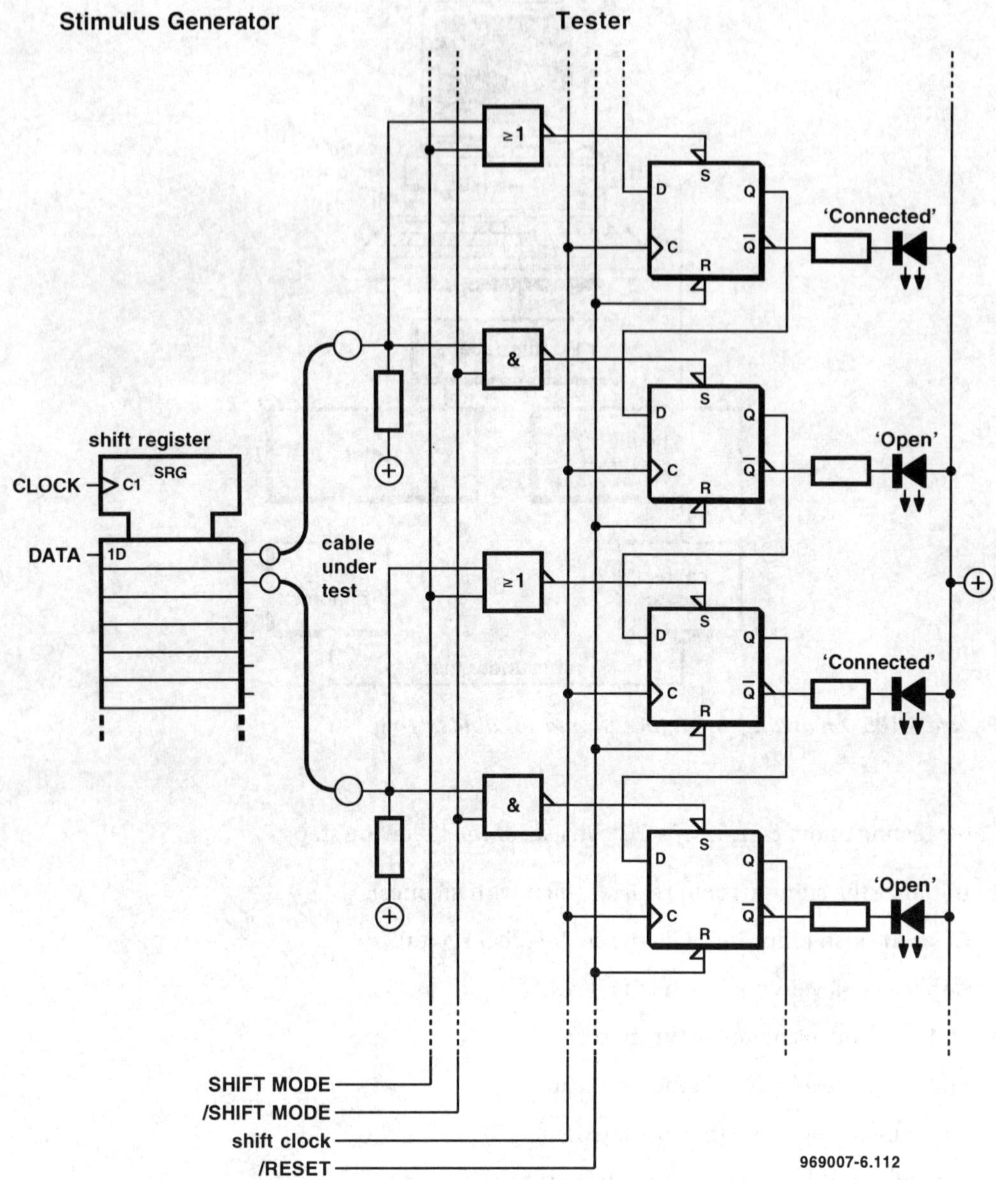

Figure 5.104 Conceptual schematic of an automatic cable tester (see the text).

development environments for creating the necessary programs. In addition, speed is not a consideration here, so the entire software could for example be written in Visual Basic to allow it to be quickly adapted to special requirements.) Figure 5.105 shows the block diagram of a suitable test setup.

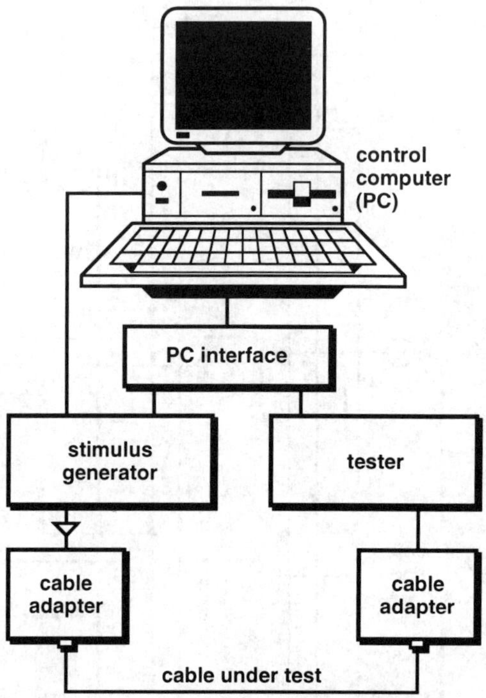

Figure 5.105 An arrangement for automatic cable testing.

Cable testing under control of a PC consists of the following steps:

1 Initialize the stimulus shift register (fill it with all ones).

2 Clear the tester flip-flops (via the shift-register path).

3 Apply a LO-level signal to the first lead.

4 Shift out and evaluate the flip-flop data.

5 Apply a LO-level signal to the next lead.

6 Shift out and evaluate the flip-flop data.

7 Repeat steps 5 and 6 until all leads have been tested.

Each test sequence should be cycled through several times, to detect intermittent faults.

Notes:

1 The cable adapter must provide connections between the contacts of the connectors used with the cable being tested and the stimulus generator and tester contacts. If a large variety of cable types have to be tested, the adapter can easily be more expensive than the actual test hardware. Two different approaches can be used to construct such adapters quickly and inexpensively: (1) use standard 19-inch rack-mount hardware and (2) use PC connector plates (with punchouts for the connectors in question) and suitable prototyping boards (which can in turn be mounted on a 'passive' ISA board). Refer to Volume 3 for further details.

2 Drivers and signal levels: since the cables are tested alone (not connected to any circuitry), relatively high signal levels can be used. Line drivers that are designed to be used with long cables are particularly suitable (an obvious choice is RS232 interface ICs). If line drivers are used in the stimulus generator, then matching receivers must be used in the tester. Standard open-collector gates or IEEE 1284 parallel interface driver ICs (which can be operated in open-collector mode) are suitable if the cables to be tested are not longer than 5 m. *Caution:* the drivers must be short-circuit proof (so that they can survive cable shorts).

Cable testing with AC signals

In principle, an AC voltage (from a simple signal generator, for example) can be used for continuity testing in place of a DC voltage. Any crosstalk between the lead(s) being tested and other leads can then be immediately seen on an oscilloscope (or an audio amplifier and headphones can be used, with a bit of practice, if an audio-frequency test signal is used). Audio-frequency signals in cables can also be detected using an inductive pickup; suitable amplifiers are commercially available as 'telephone eavesdropping' amplifiers (see Figure 5.106). *Note:* this entire arrangement should be built into a screened (metallic) enclosure. See also Volume 3.

Practical tip: the arrangement shown in Figure 5.106 can also be used for 'non-contacting' signal tracing: an AC voltage is fed into one end of the cable and the pickup is passed over or along the cable. This technique could for example be used to identify a particular cable in a cableway or conduit (see Section 5.3.2). The test tone should be distinguishable from the usual background noise. A 1-kHz tone is usable, but a pulsed tone sequence (à la Morse code, for example) is even better (as long as it is not be the same as the tone sequences commonly used on the telephone network). See also Volume 3.

Cross-connected wire pairs (split pairs)

This is a typical fault in cables that are constructed using twisted-pair leads. One lead of a twisted pair normally carries the signal, while the second lead provides the return path. The fault arises when one of the two leads belongs to a different twisted pair, so that in effect the leads of the two pairs are cross-connected (see Figure 5.107). This fault cannot be detected with DC continuity testing! If an AC signal is used for testing, on the other hand, the fault shows up as an unusually high level of cross-talk to the second split pair.

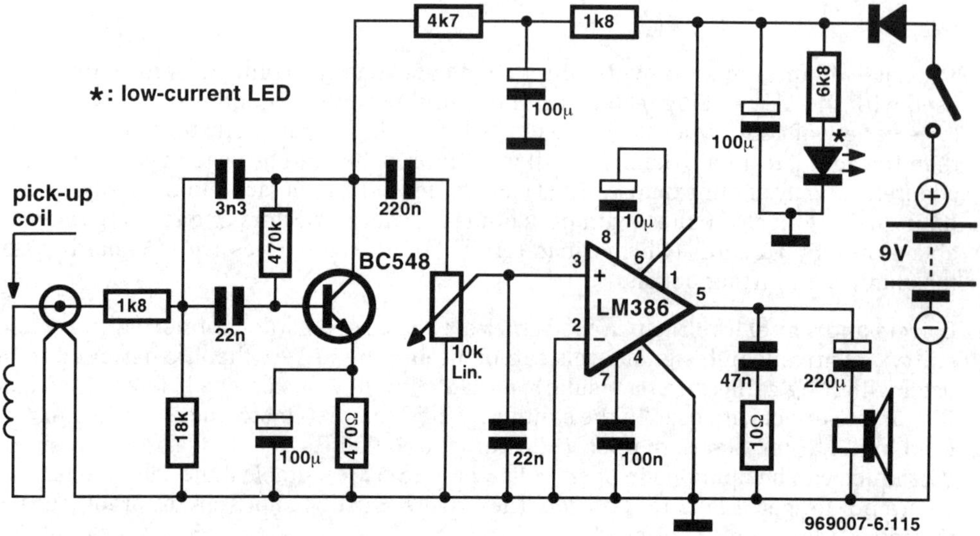

Figure 5.106 Schematic of an amplifier for use in cable testing based on a telephone pickup coil. (source: Conrad Electronics)

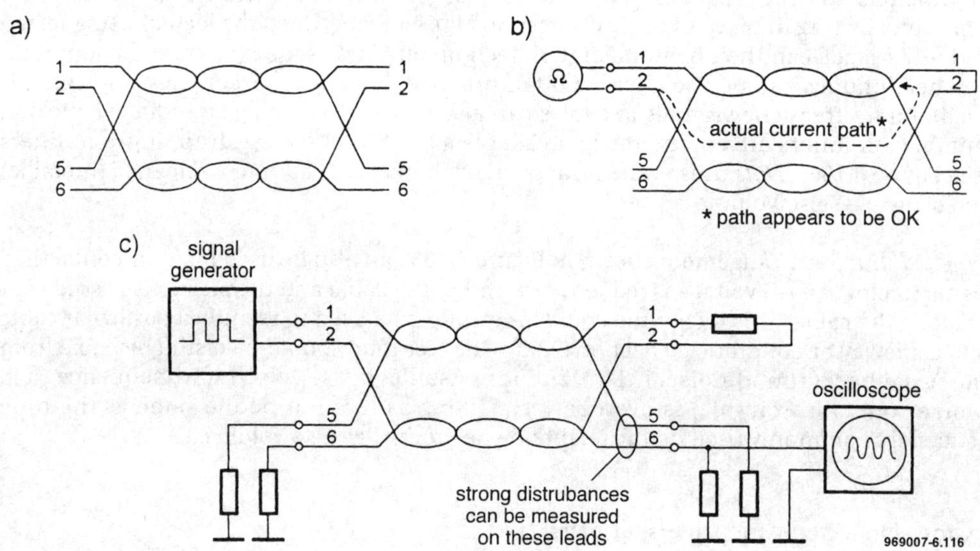

Figure 5.107 A 'split pair' fault.

a) example of an incorrectly connected cable,

b) the fault will not be detected with DC continuity testing,

c) the fault can be detected by the unusually high crosstalk level with an AC test voltage.

5.3.2 Cable identification

Figure 5.108 illustrates a typical situation in which cable identification is needed: a lot of cables have been fed into a central control room or distribution panel, and the problem is to properly connect these cables (to a network concentrator, for example). If the individual cables are not adequately labeled, then some sort of measurement procedure must be used to determine which cable goes where (or what it is attached to at the other end).

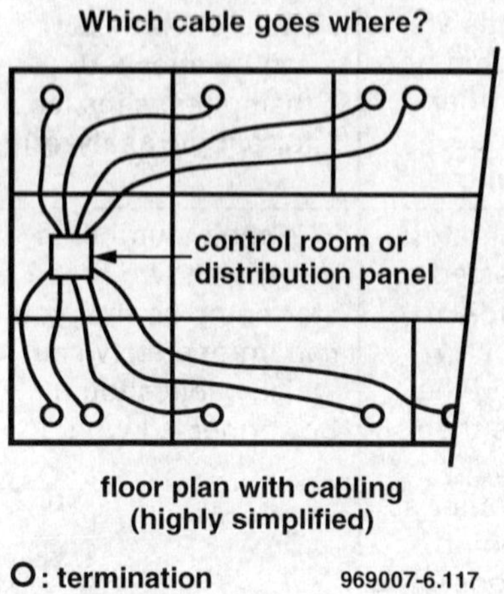

Figure 5.108 A typical cable-identification situation.

A very simple procedure is to first go through all rooms in which the far ends of the cables are located and disconnect all of the cables. Two contacts of the first cable are then shorted together, and all cable ends at the central location are tested for a shorted pair (using an ohmmeter) until the proper cable has been identified. The short is then removed from the far end of the first cable, and a short is applied to a pair of contacts at the far end of the second cable. The process is repeated until all cables have been identified.

Another situation is the following: you have a multi-lead cable with a connector at one end and you want to attach a connector to the other end, but you have no documentation that describes how the original connector is wired (ie, which insulation colour corresponds to which pin). In such a situation, it would be nice to have something that would directly indicate which lead is attached to which pin.

Two devices are needed to make possible the direct identification of individual cables (or leads in a single cable): a *code generator* and an *code display* unit. With a group of cables, the configuration of one end of the cabling installation must either be known or assumed to be fixed (either the star point or the end points). If the configuration of the star point is taken as given, then a code generator is placed at the star point to apply a unique signal to each cable, and the end points are sensed using the display unit (*centralized identification*). Alternatively, individual code generators can be connected to each of the far ends of the cables and the cable ends can be sensed at the star point using the display unit (*decentralized identification*). Table 5.20 compares the two procedures.

Procedure	Principle	Advantages	Disadvantages
centralized identification	All leads or cables are connected to a central code generator at the star point, and the far ends are then checked one after the other using a code display unit.	1 A large number of leads or cables can be identified. 2 Each far end need be visited only once. 3 Any desired type of cable or connector can be attached (using test clips, as for a logic analyzer).	Bookkeeping is necessary.
decentralized identification	Code generators are connected to the far ends of each of the cables, and the cables are then checked one after the other at the star point, using a code display unit.	The cable number is directly shown at the star point, so that you can immediately see which cable should be connected where.	1 Typically, at most 8 code generators ('Cable Identifiers') with different codes are available. 2 You must visit each far end twice (once to place the generator and once to retrieve it). 3 Code generators are often designed for a particular cabling type (such as 10-Base T or EIA-TIA 568).

Table 5.20 Comparison of cable identification procedures.

Note: an earth connection is required between the code generator and the display unit. This can be (1) a separate connection provided especially for this purpose, (2) a cable lead that is defined as an earth lead, (3) a cable shield (if present) or (4) where large physical separations are involved (such as within a building), a 'common' earth potential reference (such as the protective earth lead of the mains network or an existing, distributed conductive system (the heating system, for example)).

Centralized identification. Figure 5.109 illustrates a specific case: one end of a multi-lead cable is connected to the identification-code generator, and the individual leads are probed one after the other at the other end using the code display unit, which indicates the number of each lead.

Identification techniques. Various techniques can be used for identification. For example, signals with different frequencies or different pulse sequences can be applied to the individual cables or leads.

Practical tip: testing with a simple AC signal is often good enough. Inject a low-frequency AC signal (1 kHz, for example) into one pair of leads of each cable in turn and check the cables using an inductive pickup (see Figure 5.106) until all cables have been identified.

Quite simple tester circuits can be constructed if a certain number of pulses is used to mark each lead (in a given period of time, one pulse occurs on lead 1, two pulses occur on lead 2 and so on). Figure 5.110 shows the block diagram of an arrangement that illustrates this principle.

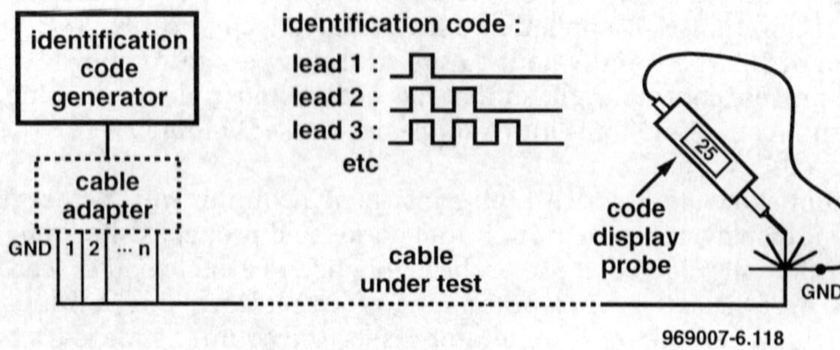

Figure 5.109 Centralized cable identification.

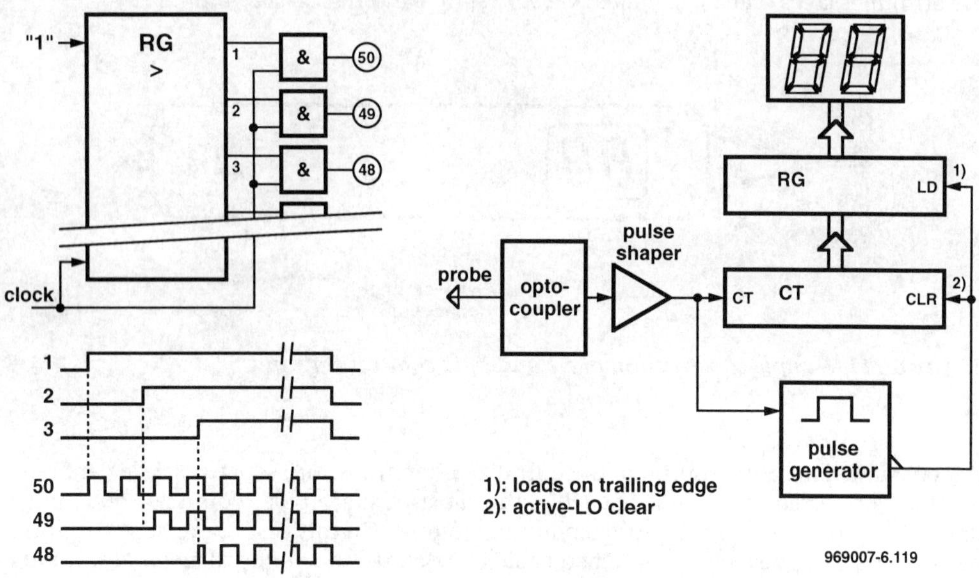

Figure 5.110 Block diagram of a centralized cable identification scheme (see the text). a) code generator, b) code display unit.

The code generator shown in Figure 5.110a is based on a shift register, where each bit position controls the pulses that are passed to a specific lead or cable. The arrangement in this example can be used for up to 50 leads, or 50 independent cables that run from a single common 'star' point. Lead (or cable) number 50 is connected to output 1 of the shift register via an AND gate, number 49 to output 2 via an AND gate and so on. Clock pulses are passed to each lead (or cable) as long as the associated shift register output is HI. The shift register is initially cleared and is then filled with shifted-in HI levels during the course of the test. After the fiftieth clock pulse it is cleared again, and a pause occurs. Following the pause, the process starts again. As a result, output 1 of the shift register is active (HI) for 50 clock intervals, output 2 for 49 clock intervals and so on, and the corresponding number of clock pulses is applied to the associated lead or cable. *Note:* the shift register arrangement is largely equivalent to the stimulus generator shown in Figure 5.104 (down to the process control level), so that it is actually possible to combine cable testing and cable identification functions into a single device (see Volume 3).

The display unit shown in Figure 5.110b consists of a counter with a storage register and seven-segment display. It can be built into a hand-held probe, as shown in Figure 5.111. The counter is initially cleared. If the probe is touched to a cable end or lead that supplies pulses, the first of these triggers a pulse generator that times out in the pause between pulse sequences. When the pulse generator resets, the counter state is transferred to the storage register and the counter is then cleared. The content of the storage register is displayed, so that the number of pulses received during one cycle can be read from the LCD screen. *Note:* this type of test probe can also be used for other purposes, such as displaying error messages that are signalled via very simple interfaces (see Volume 2).

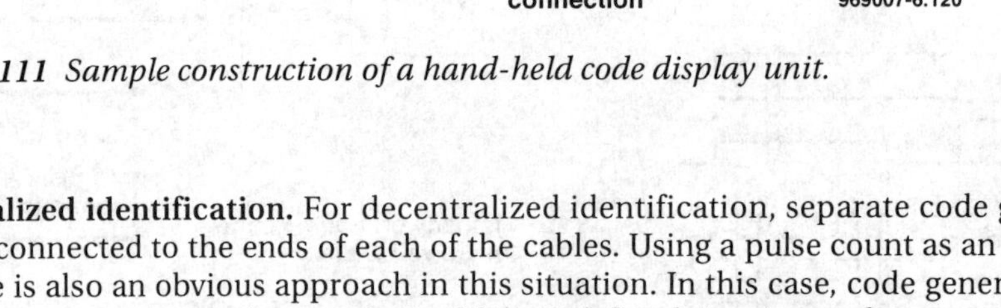

Figure 5.111 Sample construction of a hand-held code display unit.

Decentralized identification. For decentralized identification, separate code generators must be connected to the ends of each of the cables. Using a pulse count as an identification code is also an obvious approach in this situation. In this case, code generator number 1 places one pulse on the attached cable in a defined time interval, generator number 2 places two pulses on the attached cable in the same interval and so on. This can be implemented quite simply using a microcontroller. The probe shown in Figures 5.110b and 5.111 is suitable as a code display unit. See Volume 3 for more details.

Identifying cables in network and facility installations

Centralized identification:

1 Connect the code generator to all cables at the star point. The order of connection is not important.

2 Go to each of the far-end cable connectors or outlets in turn, and probe the contacts using the display unit. Note each of the indicated identification numbers (*for example:* "Room 501, code 22; Room 502, code 39" and so on).

3 Return to the star point and determine which far-end connector corresponds to which cable end (attached to the code generator), based on your notes. Disconnect the cables from the code generator and connect them to their proper locations.

Decentralized identification:

1 Go to the cable jacks or outlets in turn and connect a code generator to each one. Some record keeping may be necessary (such as noting that generator number 1 is located in Room 501, generator number 2 in Room 502 and so on).

2 Go to the star point and probe the individual cables using the display unit. The far-end location of each cable can be immediately identified from the displayed number, so that each cable can in turn be connected to its correct position.

3 Return to the far ends of each of the cables and disconnect the generator units.

5.3.3 Eye pattern diagrams

The rate at which a signal propagates along a line is frequency-dependent. If a pulse is applied to one end of the line, its various harmonics will arrive at the far end with differing delays, so that the received pulse will be more or less distorted (see also Figure 4.56). With a series of closely-spaced pulses, the harmonics from adjacent pulses may overlap each other at the receiver end. These distortions limit the data transfer rate that can be achieved. The distortions in the signal waveform can be displayed on an oscilloscope in the form of an eye pattern. Figures 5.112 and 5.113 illustrate the principles involved.

The clock signal is set to the desired bit rate (the clock period equals one bit cell or unit interval). A pseudo-random bit sequence in NRZ representation, generated by a feedback shift register, is applied to the signal path to be tested. A large variety of 1→0 and 0→1 bit transitions follow each other in the course of time (Figure 5.112 shows these transitions and how they are distorted by propagation delay). The oscilloscope is triggered at the generator clock rate, so that successive transitions are superimposed. The resulting figure resembles an eye, hence its name. Figure 5.114 shows how to interpret an eye pattern.

Note: the oscilloscope must have an adequate update rate if a true eye pattern is to be seen (see Section 4.3.4). Digital storage oscilloscopes with low update rates are practically useless for displaying eye patterns, even if their specifications are otherwise good. (This also applies to certain plug-in cards for PCs that provide 'oscilloscope' functions, and to waveform acquisition and display via a PC that is connected to an oscilloscope.)

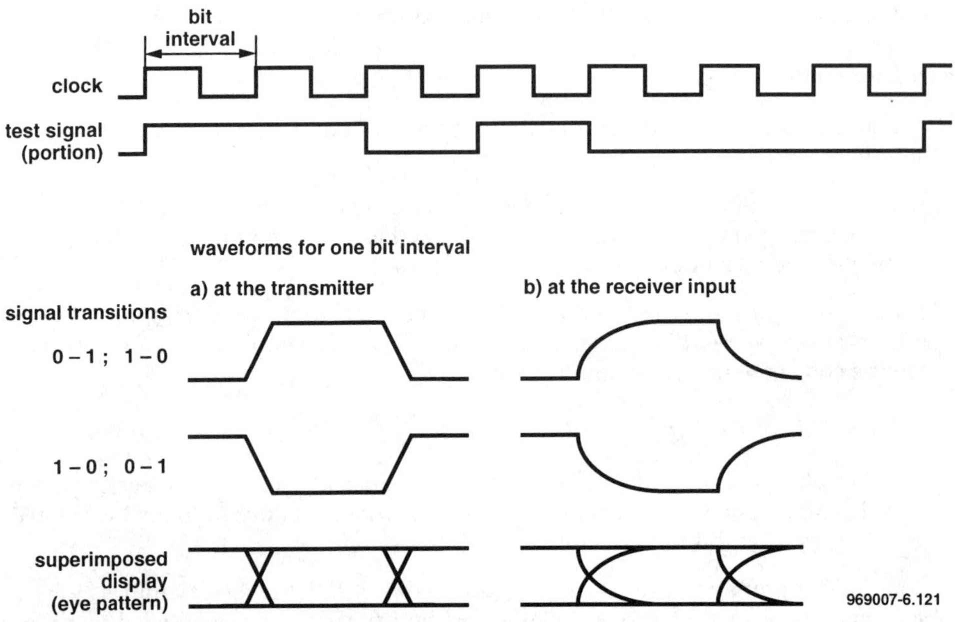

Figure 5.112 Generating an eye-pattern diagram.

Figure 5.113 shows a circuit based on a feedback shift register that generates a maximum-length cycle (see Section 4.4.3 and Appendix 6). The shift register consists of two 8-bit shift register ICs and a parity generator IC, and can be used at clock rates up to 50 MHz. The test signal consists of a sequence of pulses with widths ranging from one to sixteen clock periods. The order in which pulses of different widths follow each other, and the widths of the intervening pauses, can practically be regarded as random. This means that a large variety of signal transitions (0 to 1 and 1 to 0) occur in the unit interval.

Notes:

1 Drivers: this type of test circuit should be provided with various types of output drivers, so that it can be used in a variety of situations. In practice, the following types of drivers are important (among others): (1) CMOS, (2) TTL, (3) open-collector, (4) IEEE 1284 parallel interface, (5) RS232 and EIA-TIA562 serial interfaces, (6) differential signal interfaces (RS485, which is used for 10-Base T cable (UTP), (7) 50-Ω and 75-Ω coaxial cable (for Ethernet and video or antenna cabling). The manner in which the drivers are incorporated in the circuit is suggested in Figure 5.113 (each driver has its own cable connector (that matches its associated interface) and a test point for connecting an oscilloscope probe; each driver can be selectively enabled or disabled).These provisions make it possible (1) to drive several cables at the same time for comparison purposes and (2) to disable unnecessary drivers to minimize noise. See Volume 3 for details.

2 An adjustable clock generator (pulse train generator) must be added to the illustrated circuit to convert it into a complete tester. See Volume 3 for more information.

3 The shift register is loaded with all ones when it is reset (these are shifted in by the clock signal), so that the test cycle begins with the seed value $FFFF_H$.

4 Trigger signals: the oscilloscope is normally triggered from the clock signal. Additional trigger signals are however provided for special purposes. These include: (1) a single pulse, with a width of one clock period, that occurs every time the seed value passes through the shift register (once every 65,535 clock periods) and (2) a symmetrical square wave signal with a period equal to two complete cycles (131,070 clock periods).

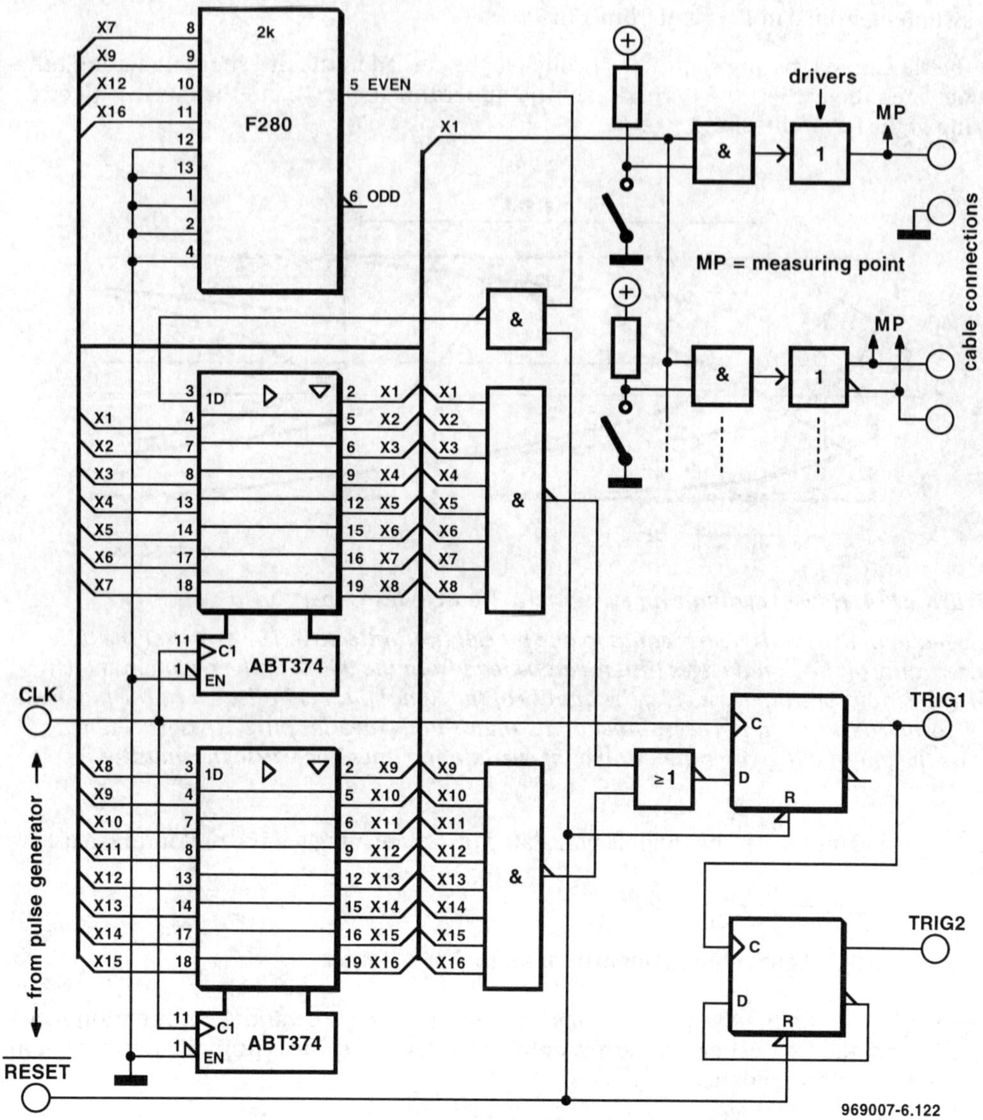

Figure 5.113 A test stimulus source for generating eye patterns.

5 An arrangement consisting of a feedback shift register and a clock generator can also be used for evaluating oscilloscopes, logic analyzers, universal counters and so on and for familiarizing yourself with the operation of such equipment.

Estimating the jitter level. The magnitude of the jitter can be estimated by measuring two features of the eye pattern at the far end of the cable (at the receiver input). These are marked in Figure 5.114, as follows:

1) the threshold range of the receiver, which is the separation between the signal levels for switching from 0 to 1 (*) and from 1 to 0 (**),
2) the threshold crossing skew (t_{TCS}), which is the spread in the threshold transition times between the earliest 0→1 crossing of the upper threshold (*) and the latest 1→0 crossing of the lower threshold (**).

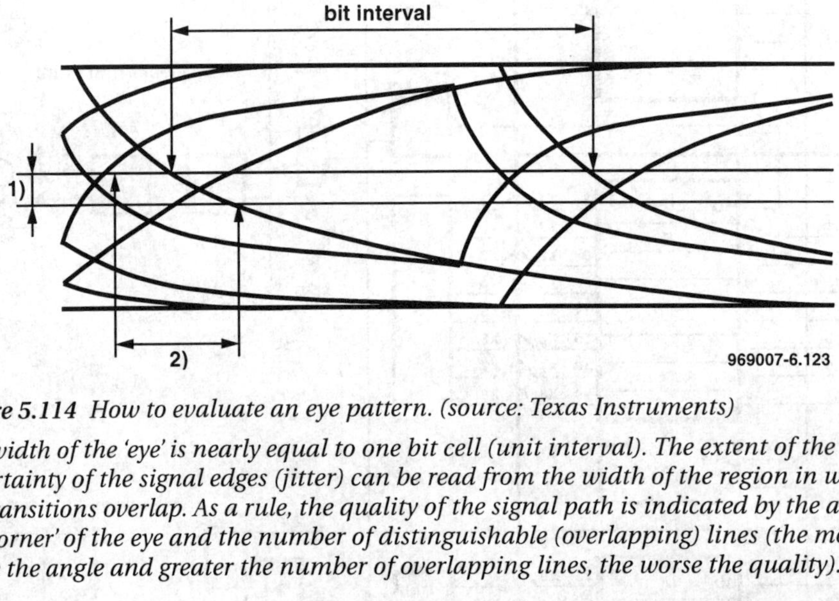

***Figure 5.114** How to evaluate an eye pattern. (source: Texas Instruments)*
The width of the 'eye' is nearly equal to one bit cell (unit interval). The extent of the time uncertainty of the signal edges (jitter) can be read from the width of the region in which the transitions overlap. As a rule, the quality of the signal path is indicated by the angle in the 'corner' of the eye and the number of distinguishable (overlapping) lines (the more acute the angle and greater the number of overlapping lines, the worse the quality).

The jitter magnitude can be roughly calculated (as a percentage of the unit interval) as:

$$\text{jitter \%} = \frac{t_{TCS}}{\text{bit interval}} \cdot 100$$

As as rule, a signal path should not exhibit more than 5% jitter.

Figure 5.115 shows the eye patterns observed for a cable with various termination resistors. Clearly, the clearest eye pattern is obtained when the cable is properly terminated in its characteristic impedance.

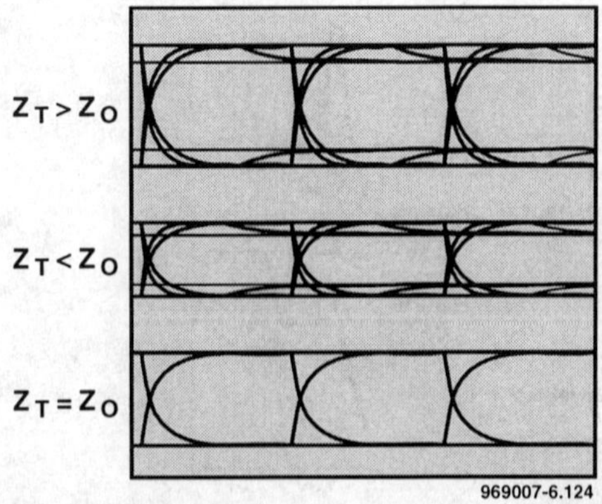

Figure 5.115 The effect of termination resistance on the eye pattern (Z_T = termination resistance, Z_0 = characteristic impedance of the cable.). (source: Texas Instruments)

The sample eye patterns in Figure 5.116 show the effect of increasing the data rate (for RS-485 signals). The eye diagram is essentially perfect at 500 kbits/s. At 1000 kbits/s it is already a bit blurry, but the jitter (at 5%) is entirely acceptable. The fully washed-out eye pattern at 3500 kbits/s shows that the signal path is not suitable for this data rate.

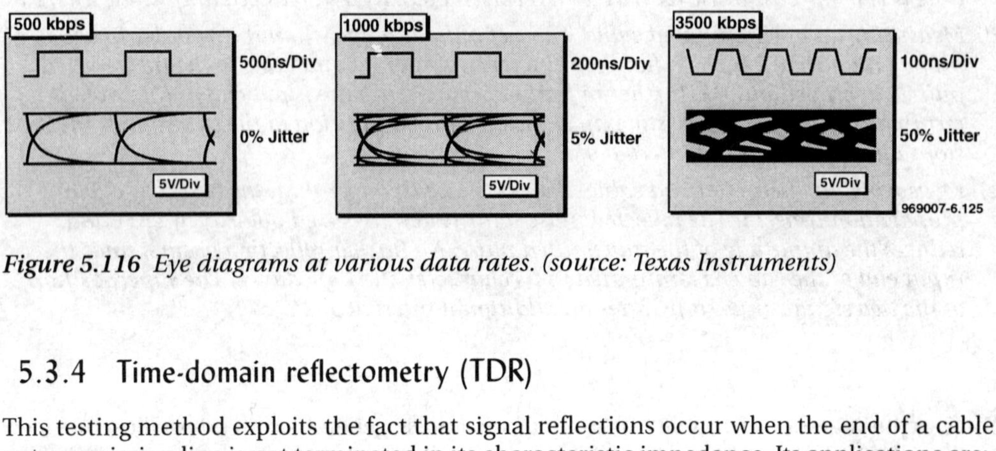

Figure 5.116 Eye diagrams at various data rates. (source: Texas Instruments)

5.3.4 Time-domain reflectometry (TDR)

This testing method exploits the fact that signal reflections occur when the end of a cable or transmission line is not terminated in its characteristic impedance. Its applications are:

1. measuring cable length, which requires an 'artificial' mistermination (usually obtained by short-circuiting the far end of the cable);
2. locating faults in the cable, by measuring the distance between the fault(s) and the near end of the cable.

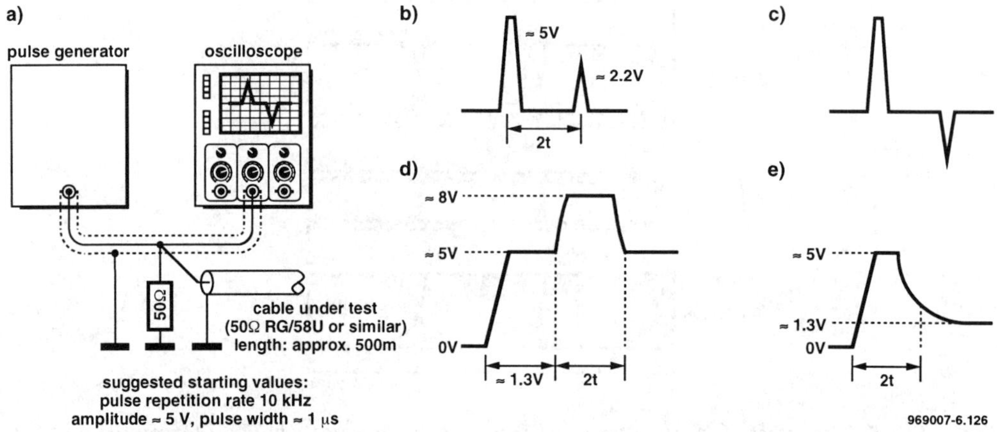

Figure 5.117 A simple, experimental TDR setup.

a) equipment configuration. The pulse generator has a 50-Ω output. The pulse generator, oscilloscope and cable under test are connected together using coaxial cables that are as short as possible. The cable under test is terminated in its characteristic impedance at the end where the pulse is injected.

b) Measuring an open-circuited cable. The test pulse is shorter than the cable propagation time t. The reflected pulse returns with reduced amplitude after 2 t.

c) Measuring a short-circuited cable. The test pulse is shorter than the cable propagation time t. The reflected pulse returns with reversed polarity and reduced amplitude after2 t.

d) Measuring an open-circuited cable. The test pulse is longer than the round-trip cable propagation time 2 t. The reflected pulse returns after 2 t and adds to the transmitted pulse (superposition). No further reflection occurs, since the input end of the cable is terminated in its characteristic impedance. The superposition of the two signals thus stops after an additional interval of 2 t.

e) Measuring an open-circuited cable. The test pulse is longer than the two-way cable propagation time 2 t. The reflected pulse returns with reversed polarity after2 t and reduces the amplitude of the transmitted pulse. No further reflection occurs, since the input end of the cable is terminated in its characteristic impedance. The superposition of the two signals thus stops after an additional interval of 2 t.

You can put together a suitable test setup using a pulse generator and an oscilloscope (see Figure 5.117). However, such an arrangement is at best only suitable for educational purposes if inexpensive service equipment is used; it would be quite difficult to use it for fault-finding in (for example) a LAN installation.

Figure 5.117d/e shows how the reflected pulse affects the leading edge of the test pulse. These effects are reversed at the trailing edge of the test pulse, as shown in Figure 5.118.

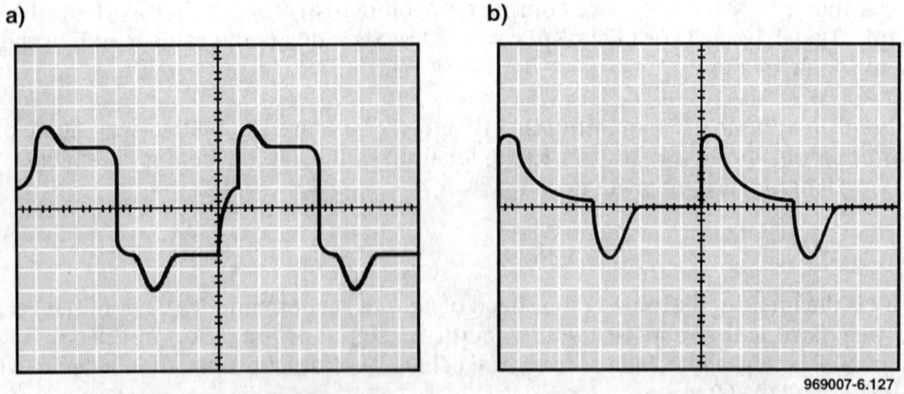

Figure 5.118 Sample TDR measurements.

a) far end of the cable open. At the start of the test pulse, the reflected wave front returns after 2 t and adds to the level of the test pulse. This is visible as a peak lasting for 2 t. At the end of the test pulse, the reflected negative wave front returns after 2 t and creates a negative pulse that also lasts for2 t.

b) far end of the cable shorted. At the start of the test pulse, the reflected wavefront returns inverted after 2 t and nearly cancels out the test pulse. (Since the cable has a finite DC resistance, the resulting signal decays exponentially after an interval equal to 2 t. The levels does not drop all the way to zero volts, since the short-circuit current flowing through the cable creates a small voltage drop as long as the test pulse remains active.) At the end of the test pulse, the short-circuit current stops. This produces an inverted wavefront that returns re-inverted after 2 t and causes a negative pulse that lasts for 2 t.

The measurements in Figure 5.118 were made under the following conditions:

- cable: RG-58/U, 100 m
- pulse generator: repetition rate 100 kHz, pulse width 5 μs, pulse source is a TTL gate
- oscilloscope: time base 1 μs/div, 50-Ω feed-through terminator connected to the input
- cable propagation time ≈ 0.5 μs

Notes:

1 Adjust the time base to obtain the most accurate measurements of pulse widths.

2 These illustrations are intended to show what can be achieved with simple test equipment. The measurements were made with a 20-MHz analogue oscilloscope and an inexpensive DSO. Both produce similar displays, but with the DSO the trace is jittery and it is scarcely possible to measure the waveforms shown in Figure 5.118.

3 It is better to measure the cable length with the far end short-circuited, since it is easier to measure the pulses that result from this arrangement.

LAN scanners. LAN scanners are compact, portable instruments that work on the TDR principle. They have LCD or LED displays that directly indicate the cable length or the distance to a fault location.

Measurement accuracy. The quantity that is actually measured is the two-way signal propagation time (from the transmitter to the location in the cable where it is reflected, and back). The length L is then given by:

$$L = 0.5 \cdot t \cdot c \cdot \mathrm{NVP}$$

where t is the measured time, c is the speed of light in vacuum and NVP (Nominal Velocity of Propagation) is the relative velocity factor for signal propagation in the cable. (The value of NVP is typically 0.7 for LAN coaxial cables. The actual propagation velocity would thus be $0.7\ c = 210{,}000$ km/s $= 2.1 \times 10^8$ m/s.)

Example: in Figure 5.117 t is very nearly 1 µs, so the length L is $0.5\ \mu s \cdot 0.7\ c \approx 105$ m.

A propagation velocity of $0.7\ c$ corresponds to roughly 20 cm/ns. If the propagation time is measured with a 100-MHz counter, then one count interval (10 ns) corresponds to a round-trip distance of 2 m, or 1 m of cable length. Since an error of ±1 digit must always be assumed with counter-based processes, the measurement error will be at least ±1 m. Inaccuracies in the counter's time base and the determination of the value of NVP will add to this error.

Practical tip: if you use a LAN scanner to measure a roll of cable whose length you know, you can refine the nominal value of NVP for this particular cable. Some instruments even have a special function for this purpose. (Even if your scanner doesn't have this function, you may not need to do any calculations — if the scanner allows the value of NVP to be specified, then enter values for NVP until the indicated length matches the actual length. Otherwise, the correction factor for NVP is equal to the ratio of the actual length to the indicated length.)

Be careful when evaluating accuracy specifications. A few typical values, collected from various data sheets, are:

- resolution: "±2 ft" to "±2 m" (very high resolution is only useful if the value of NVP is accurately known),
- accuracy: "±1%" (without any additional specification) to "±4% ±2 m ± (NVP error)" (the latter specification is more believable),
- cable length: 5 m to1000 m (with certain variations depending on the cable type). (The minimum usable cable length is important, but it is not always specified).

The effective accuracy of the measurement decreases with a long cable. Even if the instrument actually achieves ±1% accuracy, 1% of 500 m (for a 10-Base 5 Ethernet cable) is still 5 m – and with this sort of information you will need a very exact cabling plan to be able to at least identify the room in which the suspected fault is located. This means that (1) LAN scanners are indeed useful, but you cannot expect miracles from them and (2) very long individual cable runs make for problems in case of a fault.

Practical tips:

1 After acquiring a LAN scanner, practice using it in order to familiarize yourself with its operation and to get an idea of the accuracy and reliability of its indications.

2 Pay attention to the units used to display the cable length (metres or feet). This is selectable with most instruments. Mistakenly reading an indicated value in feet as a value in metres, in the throes of fault-finding, will not help your self-confidence.

3 The scanner can be connected 'in the middle' of the network (at any desired point where equipment is attached to the network cable) to obtain more accurate measurements. Remember that the 'good' end of the network cable must be terminated where it is connected to the scanner.

4 It is good idea to split up a network into several smaller sections connected via repeaters; this is beneficial for fault-finding as well as normal operation.

5.4 Fault-finding on printed circuit boards

5.4.1 Common circuit-board faults

A printed circuit board is in principle just a support for interconnections, so it can be affected by typical interconnection faults (open and shorted connections) – see Table 5.21.

Causes of open tracks:

- mechanical stress (for example, excessively flexing the circuit board),
- thermal stress (thermal expansion, hot/cold cycles),
- assembly errors,
- poor workmanship for repairs (such as cold solder joints),
- short circuits (tracks damaged due to excessive currents).

Causes of shorts:

- soldering faults (solder bridges),
- tinning faults (whiskers),
- metallic items that land on the circuit board (for example, a screw that is overlooked during reassembly or a paper clip that falls through a ventilation opening),
- mechanical stress due to improperly-performed repairs (this applies in particular to shorts between different layers of a multi-layer board),
- components that fail shorted and thus create a short between two tracks (this is especially likely to happen with capacitors and diodes).

Fault	Remarks
broken track (hairline crack)	possible causes: mechanical stress, short circuits
track broken at solder pad	possible causes: mechanical stress, short circuits
track separated from the base material	common causes: (1) overheating of a section of the track due to a short circuit, (2) ineptly performed repairs (excessive heating when soldering)
broken or high-resistance connection between a track and a plated-through hole	possible causes: fabrication faults, thermal or mechanical stress, short circuits
through-hole sleeve pulled out	cause: careless repair work (pulling out a component lead without first adequately heating the solder joint)
short between two signal tracks	For a given signal level, the driver with the greater drive capacity will dominate. Sometimes the signal level is 'undefined'. This can cause IC overloading and failure.
short between a signal track and earth	The signal is persistently LO (primary symptom: stuck-at-LO fault). IC failure is unlikely in a pure TTL environment, but CMOS ICs can be overloaded and fail.
short between a signal track and the supply voltage	The signal is persistently HI (primary symptom: stuck-at-HI fault). This can cause IC overloading and failure.
short between the supply voltage and earth	Can also be due to a defective decoupling capacitor. A short-circuit current will always flow! The typical consequence is that the weakest link in the current path is vapourized, which ultimately results in an open circuit.
shorts between tracks in power-handling circuits	Considerable currents can flow. As a result, components can fail or tracks in the current path can be vapourized, which ultimately results in an open circuit.

Table 5.21 *Typical printed circuit board faults.*

Practical tips for repairing circuit-board faults:

1 Do not use a solder bridge as a permanent repair for a broken track (this expedient should only be used to verify the correction of the fault). The proper repair is to connected a length of insulated wire between two solder holes of the affected track, close to either side of the break. Such a repair will not reduce the reliability of the board.

2 Lifted tracks should either be reattached (using an epoxy glue) or removed and replaced with insulated wire.

3 Varnished copper wire, as is used for winding coils, is suitable for making repairs. Place the wire carefully (to avoid scratching off the varnish) and secure it to the circuit board with a tacking glue.

5.4.2 Fault-finding with the power off

When performing fault-finding on a *populated* circuit board, you are dealing with a partial power-down situation (test object off, tester on) as described in Section 4.4.1. If you do not know exactly what the components can handle (if the circuit contains a mixture of component types, as opposed to a pure power stage or pure TTL ICs, for example) then to be on the safe side you should work under the conditions that apply to CMOS logic: test voltage ≤ 0.45 V or test current ≤ 20 mA.

Checking interconnections. First check suspect interconnections with the power off (continuity testing, measuring the resistance to earth and to the power supply lead(s)). Locating shorts between signal lines is always rather time-consuming.

Practical tips:

1 Not every device that is called a 'continuity tester' is suitable for checking continuity on a printed circuit board. This also applies to the 'continuity' and 'diode test' ranges of some multimeters. Check the test voltage before using the tester. (The tester shown in Figure 5.103 works with a sufficiently low test voltage.)

2 Shorted and broken tracks on the surface of the circuit board can sometimes be found by systematic visual inspection with a loupe.

Approximate track resistance: the resistance R_T of a given track can be roughly calculated if its length l, thickness t, width w and specific resistance ρ are known (l, t and w in mm):

$$R_T = \frac{\rho \cdot l}{t \cdot w}$$

Points of reference:

- standard track thicknesses: 0.035 mm (usually), 0.0175 mm, 0.07 mm and 0.105 mm
- the specific resistance at room temperature is 0.0174 Ω-mm^2/m (17.4×10^{-6} Ω-mm).

The track width can be measured at least approximately using a vernier caliper or a loupe with a graticule; the length can be either measured or estimated based on the straight-line distance between its end points (use 2 to 3 times this length as the estimated value). As a rough rule, the resistance of the narrow tracks that are normally used for logic signals is 40 to 50 mΩ per 10 cm ($\approx$ 10 mΩ per inch).

Searching for shorts. Shorts between tracks can be located in switched-off equipment by means of resistance measurements. These are typically performed by measuring the voltage drop resulting from an injected current. Figure 119a shows the working principle: if a current flows through a lead, a voltage drop can be measured over any arbitrary portion of its length. To get an idea of the magnitude of this voltage, recall that the resistance of a typical circuit-board track for logic signals is roughly 10 mΩ per 2.5 cm (1 inch), so that a test current of 50 to 200 mA will produce a voltage drop of 0.5 mV to 2 mV over 2.5 cm. A millivoltmeter with adequate sensitivity is thus needed.

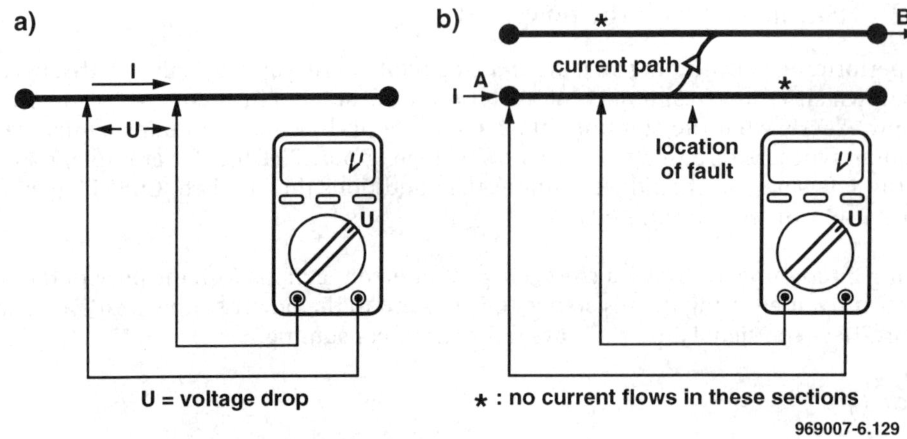

Figure 5.119 Using voltage measurements to search for shorts (see the text).

Figure 5.119b shows how to search for a short: a current source is connected to the two tracks so that the current flows through the fault location. One terminal of the millivoltmeter is connected to point *A* (where the current is fed in), and the second terminal is connected to a probe that is moved along the track. If the probe is passed along the lower track away from point *A*, the voltage will steadily increase until the probe reaches the fault location, after which it will remain constant. If the probe is now touched to the upper track at point *B* and moved away from this point, the voltage will steadily decrease from its maximum value until the probe reaches the fault location, after which it will remain constant (since a voltage drop only occurs in the current-carrying section of the track).

Test equipment. Special testers containing both a current source and a millivoltmeter are available for testing printed circuit boards. Usually, such a device also has an acoustic output consisting of a tone at a frequency that is proportional to the level of the measured voltage, so that you can concentrate on the circuit board without having to constantly watch the meter as well. A millivoltmeter that uses the four-wire (Kelvin) measurement method can also be used (it feeds a current into the resistance to be measured and senses the resulting voltage drop).

Improvised solutions. You can readily use a multimeter with sufficient sensitivity together with a current source (consisting of a voltage source and a series resister) for improvised testing. Sensitive, low-cost millivoltmeters (which should actually be called 'voltage indicators', since the objective is sensitivity and not absolute accuracy) can be constructed using modern operational amplifiers (see Volume 3). Figure 5.119 shows a number of improvised solutions.

Figure 5.119a illustrates searching for a short to earth. The combination of a 1.5-V cell and a 10-Ω current-limiting resistor supplies a current of around 150 mA. The test current is applied via a probe that is moved along the track, and the voltage drop is measured between both ends of track *1*. If the probe is touched to the part of the track to the left of the fault (point *2*), the voltmeter displays the positive voltage drop between the point

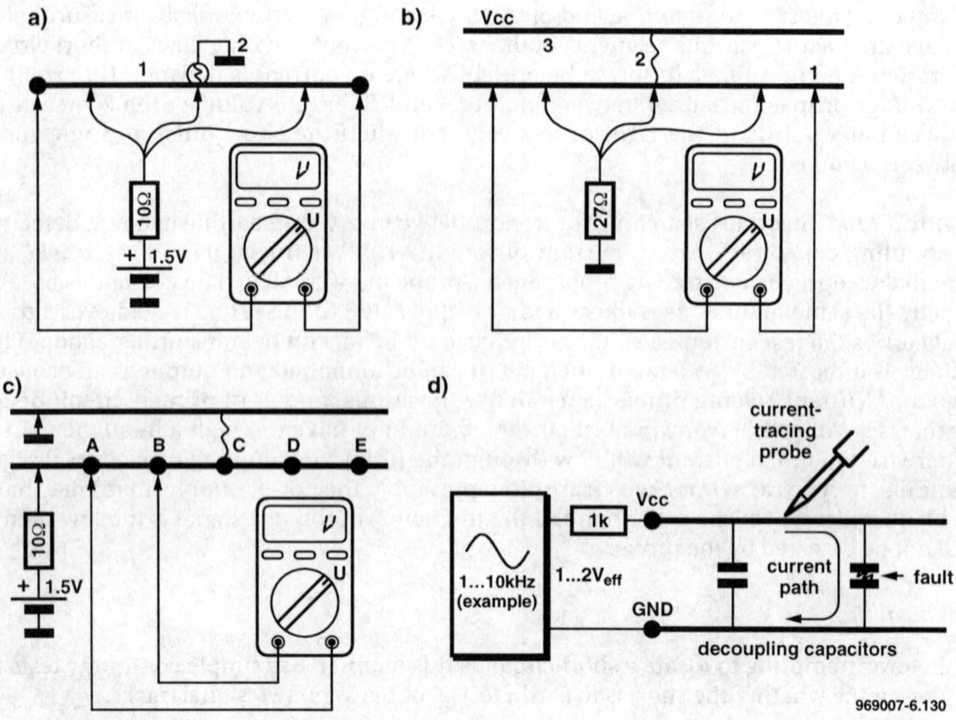

Figure 5.119 Examples of improvised methods for fault-finding on printed circuit boards.

where the current is injected and the earth potential; terminal *B* of the meter is effectively connected to earth by the fault. If the probe is touched to the portion of the track to the right of the fault, the polarity of the voltage is reversed, since terminal *A* is then effectively connected to earth. The fault is located at the point where the voltage passes through zero.

Figure 5.119b illustrates searching for a short to V_{CC}. Track *1* is shorted to V_{CC} at point *2*. The procedure is the same as just described for a short to earth, except that the test current is provided by applying the normal supply voltage to the circuit and connecting the track to earth via the probe and a series resistor (with V_{CC} = 5 V, a 27-Ω resistor will draw around 185 mA).

Figure 5.119c illustrates searching for a short between two signal tracks. One track is connected to earth and a test current is fed into one end of the other track. No operating voltages are applied to the circuit. The test current is 150 mA in this example. Since the suspect tracks are connected to earth, the test current cannot cause voltage drops that could damage CMOS devices (> 0.45 V), so it is certainly possible to work with relatively high test currents (> 200 mA). One of the meter terminals is connected to one end of the track, and a probe connected to the other terminal is then moved along the track. The voltage drop increases until the location of the short C is reached, after which it remains steady. *Note:* internal shorts in ICs that are connected to the track being checked can also be located in the same manner. You can narrow down the region in which a suspected short is located by injecting a test current at one end of the track and then systematically measuring the voltage drop across various segments of the track. A current can only flow if a short circuit is present (and no voltage drop will be present where no current is flowing). *For example:* if a voltage drop is measured between points A and B, but no voltage drop is measured between any points to the right of this segment, then the short must lie somewhere between A and B.

Figure 5.119d illustrates searching for a short between V_{CC} and earth (such as a defective decoupling capacitor). An AC current of roughly 1 mA is injected, and the tracks are scanned using a current-tracing probe. Such a probe has a small pickup coil and is used in exactly the same manner as is illustrated in Figure 5.106 (if the probe is held over a track that carries the test current, a voltage is induced in the pickup by transformer action. This voltage is indicated by an lamp or meter, or it can be amplified and output as an acoustic signal). A current-tracing probe is included with various models of printed-circuit board testers; see Volume 3. *Note:* the test current should not have too high a frequency, since otherwise too much current will flow through the (other) decoupling capacitors. Large-value electrolytic capacitors (which are often present at the power-supply input) may have to be unsoldered. On the other hand, if the frequency of the test signal is too low then it will not be detected by the probe.

Practical tips:

1 Before attempting to locate a short circuit is this manner, use simple continuity tests to determine whether the short is to earth, to V_{CC} or between two signal tracks.

2 You will have to experiment with the level of the test current. The upper limit is 180 to 200 mA; it's better if you can manage with roughly 50 mA (50 mA produces around 0.5 mV over 2.5 cm of a signal track, so you will need sufficiently sensitive equipment).

3 If you want to do this sort of fault-finding, you must first practice on discarded circuit boards with intentionally created faults.

5.4.3 Fault-finding in operating equipment

Fault-finding by signal tracing

In the course of general fault-finding, you will often be lead to suspect a faulty circuit board (for instance, if an IC pin that is connected to a particular signal line shows no sign of the expected signal, a broken track is an obvious suspicion, and a short circuit can be suspected if signals look 'strange' or two separate signal lines show identical behaviour). If you observe the suspect signal lines with an oscilloscope, you can (with some practice) readily see whether two conflicting signals are present. In addition, an unusually warm IC package is an indication of a possible short circuit.

Data lines shorted to earth or the supply voltage. If the signal in question switches in a normal manner, there cannot be a short to earth or the supply line. If the signal does not switch, then there is reason to suspect such a short if the signal level differs from the earth or supply-voltage level by only a few millivolts.

Testing with voltage measurements: a 'true' TTL LO-level signal is at least 0.1 V, and a 'true' TTL HI-level signal cannot exceed roughly 3.6 V unless it is connected to a pull-up resistor (in which case around 4.5 V can be expected). CMOS ICs *without static loads* (which means in a pure CMOS environment) have LO and HI signal levels that are practically identical to the earth and supply-voltage levels when measured with a high-impedance meter (CMOS devices switch practically 'rail-to-rail'). In such cases, a dummy load is helpful (connected to earth for a HI-level signal and to the supply voltage for a LO-level signal), or a voltage divider (refer to Figure 5.130). If the signal level changes when the load is applied, then a short cannot be present. This test can also be used when the line in question already has a pull-up or pull-down resistor.

Practical tips:

1 Watch out for consequential faults. If a section of a track turns out to be missing, then it very likely that this is due to a short (and simply bridging the missing portion of the track with a piece of wire, which will usually have a greater cross-sectional area than the track, will either not correct the fault or will cause some other part of the circuit to go up in smoke when power is reapplied). In actual service (with boards that originally worked correctly), shorts are quite often found to be due to defective components, and only rarely due to direct metallic connections.

2 Make additional tests. In modern practice, a suspected short or open connection is most often investigated by making measurements on the board with the power off. Certain simple techniques for in-circuit testing that are described in the following section have two disadvantages compared to this approach: (1) it is almost impossible to avoid a risk of damage when using in-circuit test methods with modern circuitry and (2) nowadays, suitable equipment is hard to obtain.

Fault-finding by in-circuit testing

The testing methods described here have played a certain role in the history of fault-finding practice. They are without question still useful with traditional circuit boards, preferably populated with low-scale and middle-scale-integration TTL components.

Testing with a logic pulser. A logic pulser (a probe that generates logic-level pulses) is the simplest type of test equipment for in-circuit testing (see Figure 5.121). It produces a HI-level pulse followed by a LO-level pulse each time a button is pressed. This causes a HI-level pulse to appear on a signal line that is carrying a LO level, and a LO-level pulse to appear on a signal line that is carrying a HI level. (Some logic pulsers can also generate repetitive pulse trains.) Since a signal line that is connected to earth or V_{CC} cannot be forced to change levels, a logic pulser cannot produce a pulse on such a line.

Pulse detection: the pulse from a logic pulser can be detected using (1) a logic tester with a pulse indicator, (2) a pulse detector circuit or (3) an oscilloscope. Figure 5.121 illustrates certain simple fault-finding situations.

Figure 5.121a shows a logic pulser and a pulse detector (here, a logic probe), both touched to a suspect track. If pulses can be detected, the track is in principle OK; if on the other

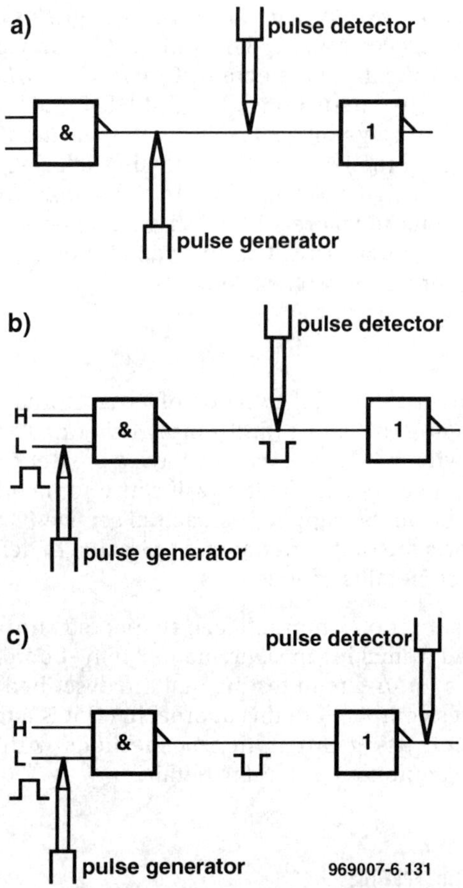

Figure 5.121 Testing with a logic pulser (see the text). (source: Hewlett-Packard)

hand the logic probe indicates a constant level, then the track has a short to earth or V_{CC}. A break in the track can also be located in this manner: the logic pulser is held on the track and the logic probe is moved along the track to detect whether pulses are present.

Figure 5.121b shows function testing of a gate: a pulse applied to the input must cause a pulse to appear at the output. Such testing is only possible if the input to which the logic pulser is connected switches 'responsibly' (if the upper input of the illustrated gate is held LO, the gate cannot be tested in this way). All types of components that have only one responsibly-switching input can be tested quite well in this manner (this applies to 'trivial' cases such as inverters, but also to bus drivers and other devices with tri-state outputs and an output-enable input, which are more significant in practice). Tri-state outputs must switch from the high impedance state to either a LO or HI level when a pulse is applied to the output-enable input (the high-impedance state can for example be detected by connecting a sufficiently low-resistance voltage divider to the output; see Figure 5.129). *Note:* see Section 2.5 for an explanation of 'responsible' switching.

Figure 5.121c shows signal tracing with a logic probe. The injected pulse is traced via the intermediate circuit elements. (This is again only possible if all signal-carrying inputs of the components in the signal path switch responsibly.)

Signal tracing with a current tracer. A current tracer (or current probe) has an inductive pickup and can thus detect the magnetic field that surrounds a current-carrying conductor. Current tracers are available as self-contained devices and as accessories for circuit-board testers (as a probe that must be connected to the tester). An important consideration is that a current tracer has an inductive pickup, not a Hall-effect sensor. It thus responds only to current changes. Such a probe cannot be used with a DC test current.

Figures 5.122 through 5.124 illustrate certain application situations for current tracers.

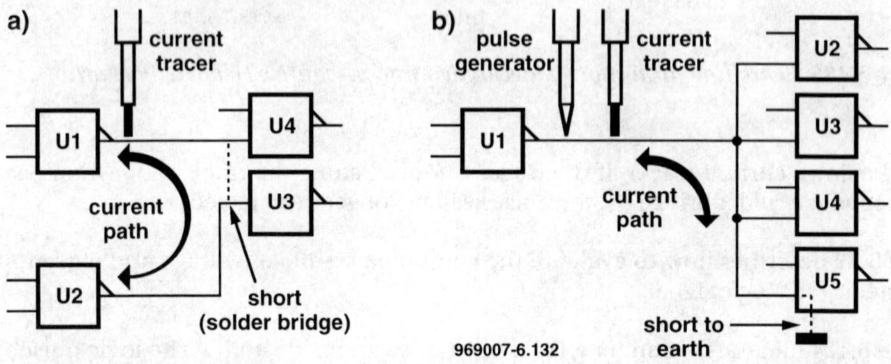

Figure 5.122 Fault-finding using a current tracer probe. (source: Hewlett-Packard)

Figure 5.122a shows how a current tracer can be used to detect current paths in a circuit. Current can only flow where a path is present. The solder bridge here forms a current path between the outputs of ICs *U1* and *U2*. If these two outputs have opposing signal levels, an unusually large current can be detected in both conductor segments. In practice, this means that if a current tracer is passed along portions of the tracks to the left of the solder bridge, it will indicate the presence of a current. If however it moved to the right, past the location of the solder bridge, it will indicate a lower current or no current at all.

Figure 5.122b shows how a current tracer can be used to search for an internal short to earth in an IC. Current pulses are injected into the track by a logic pulser and the track is scanned with a current probe. No current flows into the inputs of ICs *U2*, *U3* and *U4*, which are connected to the track, but a current can be detected on the path that leads to the input of IC *U5*. This IC is thus suspect.

Figure 5.123 shows in detail how the indicator lamp of a current tracer responds as the probe is moved along a suspect track (the lamp in the tip of the probe goes out if the probe is no longer above a current-carrying track). The solder bridge diverts the current to a second track. Pulses are injected into the beginning of the track to be checked and fol-

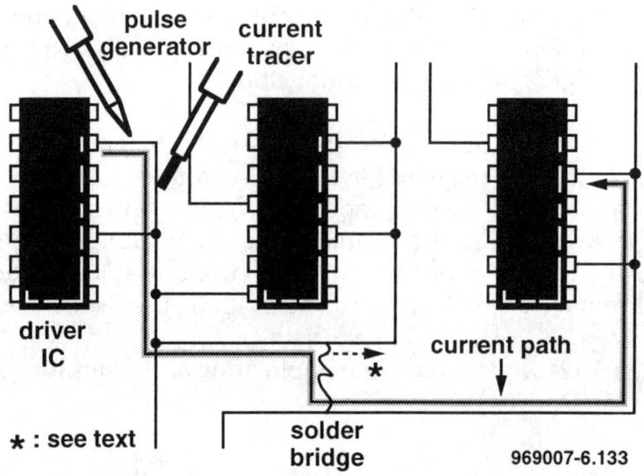

Figure 5.123 Searching for a short between two tracks. (source: Hewlett-Packard)

lowed using a current tracer. If the tracer is moved along the track in question past the location of the solder bridge (*), the pulses will no longer be detected.

Table 5.22 describes how to evaluate the combined results of voltage and current measurements for a logic signal.

Figure 5.124 illustrates searching for a short to earth inside an IC. The logic pulser produces a periodic pulse train (a rate between 100 Hz and several kHz is suitable). The brightness of the current tracer's indicator lamp indicates the strength of the current. In order to properly recognize the current path, first place the current tracer adjacent to the logic pulser and adjust the sensitivity until the lamp is just illuminated. *Note:* almost all current tracers allow the detection sensitivity to be adjusted (see Table 5.23). If you want to use them successfully, you must carefully read the documentation. Practice is essential!

Voltage testing 1)	Current testing 2)	
	Activity (switches)	No activity (does not switch)
Activity (switches)	OK	open output (no load)
No activity (does not switch)	short circuit	defective output

1: Pulse detection (logic probe, pulse detector circuit, oscilloscope).
2: Checking with a current tracer (an inductive sensor that detects current variations)

Table 5.22 Evaluating symptoms when using current tracing for fault-finding. (source: Hewlett-Packard)

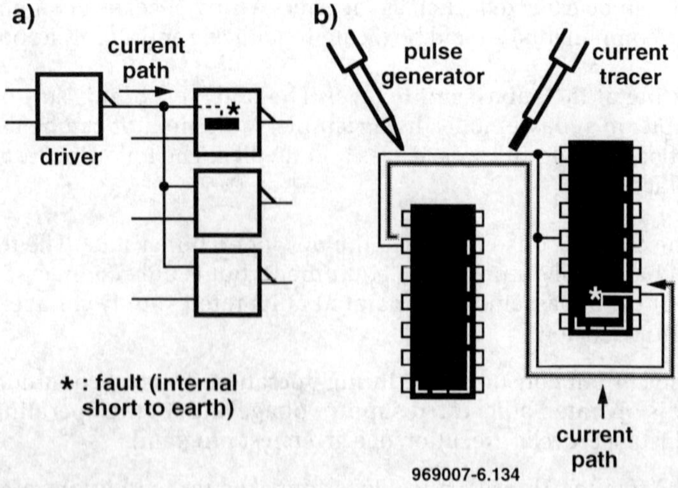

Figure 5.124 Details of current tracing: a) a portion of a schematic diagram, b) printed circuit board layout. (source: Hewlett-Packard)

Current range	Application
1–10 mA	conventional TTL and CMOS signals
1–100 mA	signal conductors driven by high-current driver outputs
100 mA–1 A	'hard' shorts (such as solder bridges) between conductors; signal conductors driven by high-current open collector outputs; conductors in power-handling circuits and so on

Table 5.23 Sample sensitivity adjustments for a current-tracer probe. (source: Hewlett-Packard)

5.5 Fault-finding in digital circuits

5.5.1 Plausibility tests

Procedures based on measurements, in which the processes taking place in the circuit are traced in detail, are frequently regarded as the only truly exact and professional methods of fault-finding. Nonetheless, such procedures are by no means always possible, even with the best of intentions and competence, since they require time, test equipment and detailed documentation of the circuit being tested (as well as the associated software if applicable). For this reason, quite 'informal' test methods also have a valid place in the service environment. This applies equally well to initial operational testing, where the trivial faults that can be expected (such as incorrect wiring, incorrect component installation and missing components) should be located as quickly and simply as possible.

Plausibility testing at the subassembly level. The objective of this sort of testing is to detect 'gross' faults in subassemblies using simple equipment. It can be used for circuit boards, motherboards, plug-in cards, drives and the like. The following tests have proved themselves in practice:

1. Measuring the current consumption in the non-operational state. The test conditions are: (1) the subassembly is not in the equipment, but is only connected to the power supply and (2) the subassembly is operated at its rated supply voltage. *Caution:* see also Practical Tip 2, below.
2. Measuring the current consumption during operation. The test conditions are: (1) the subassembly is operated at its rated supply voltage and (2) the operating state is well defined (for instance, cyclic execution of a given test program).
3. Measuring the temperature at critical locations. The test conditions are: (1) the subassembly is fitted in the equipment and connected and (2) it has been switched on for at least 15 minutes. 'Critical' locations include metallic enclosures (such as that of a disk drive), heat sinks, large IC packages and mechanical actuators.
4. Measuring the resistance between the power and earth contacts. The test conditions are: (1) the subassembly is not in the equipment and not connected, (2) the test voltage is < 0.5 V and (3) the test voltage polarity is the same as that of the operating voltage.

Practical tips:

1. These techniques are only worthwhile if you first measure the quantities in question on a known-good (new) subassembly and record them for later comparison.
2. When measuring the current consumption in the non-operational state, be careful if the device under test is a plug-in card or some other subassembly that is intended to be used in a specific environment. Simply laying such a device on the benchtop and applying power can lead to invalid measurements, due to the fact that certain inputs are 'open' (that is, they do not have defined signal levels). This applies in particular to CMOS ICs. Plug-in cards and similar items should thus only be tested when the y are fitted in suitable equipment or in a tester.
3. Measuring the resistance between the power and earth connections is simpler than measuring the current consumption, since you only have to connect an ohmmeter to the power supply connections. The reliability of the result depends however on (1) connecting the ohmmeter with the correct polarity and (2) in case of a fault, measuring with the same instrument that was used to make the original measurement (or at least the same model). *Caution:* note also that it is sometimes not possible to obtain a usable reading if autoranging is active. Combined current and voltage measurements provide more accurate results (test voltage < 0.5 V).
4. For temperature measurements, it is sometimes enough if the equipment is simply switched on, but in some cases it is necessary to keep the device under test in a certain operating state for a certain period of time, which means letting the device whose tem-

perature is to be measured actually work. Two examples are processor chips with an energy-saving mode and multi-wire dot matrix print heads. Conclusions can be drawn regarding possible faults from logical analysis of the device's temperature and its operating state (for example: a processor IC that is quite warm, even though it should be in low-power mode, suggests a fault, as well as a processor IC that remains cold when it should be operating normally).

Check for trivial faults. Experience shows that when a piece of equipment is brought to you for service, it is a good idea to first systematically check it over for trivial faults. Of course, if you don't find anything you have lost some time, but it is not all that uncommon for a technician to spend two or three hours making measurements and racking his brain over subtle problems, when the actual fault is just an intermittent contact.

Key points:

- Check contacts. This applies to all types of connectors and to all socket-mounted components. Connectors that are held together by friction alone are especially critical. This applies above all to the following:

  a) ICs (especially those in DIL packages): these can simply 'creep' out of their sockets in the course of time.

  b) Plug-in cards: these are only secured at one point in most PCs (by a retaining screw).

  c) Pin-and-socket connectors: flat cables are often used for the internal cabling of equipment, connected to the circuit boards via IDC connectors (on the cables) and multipin headers on the boards (IDC = Insulation Displacement Connection). Particularly in PCs, such connectors are frequently arranged so closely together on the motherboard or a multi-function card that it is difficult to properly install the connectors. Also, the cables are usually compressed somewhat when the cover of the enclosure is closed, which can occasionally cause connectors to loosen or even completely separate. (*Symptom:* everything worked fine when the cover was open, but nothing works after the cover has been closed.)

  d) Interface connectors: connectors that are expressly provided for attaching external devices (such as the sub-D connectors of the serial and parallel interfaces) typically have provisions for securing the plug to the jack (screws, retaining clips and so on), but these are not always used.

  e) Interface cables: a properly assembled cable that is not subject to mechanical stress will almost never fail. You should thus check whether this conditions (no mechanical stress) is actually satisfied. Cables are frequently placed where people can trip over them or step on them; although such cables may appear to be in perfect condition, one or more leads may be broken internally.

- Check the supply voltages. This is not limited to the output voltages of power supplies; it's even more important to check locally-generated supply voltages (provided by DC/DC converters or charge pumps, for example). Make both static measurements (with a multimeter) and dynamic measurements (with an oscilloscope). Some ICs have built-in voltage generators (using charge pumps); this applies to drivers and receivers for serial interfaces and to various types of power ICs, among others. To the extent that it is possible, check whether these voltage generators are still working (if the charge pump employs a capacitor external to the IC, see if anything can be measured at the capacitor lead using an oscilloscope; otherwise, try to measure signal levels).

- Measurement-based plausibility checks in the circuit: even without accurate documentation you can often locate clocks, reset signals and the like, and measure them to see whether there are any major faults (such a crystal oscillator that does not oscillate, a reset signal that is permanently active and so on). In addition, you can observe various signals with the oscilloscope and judge their general characteristics (edges, levels, distortions and so on).

Practical tips:

1 Always check whether socket-mounted ICs, plug-in cards in slots, cables attached to interface connectors and so on are properly seated or attached.

2 Disconnect and then reconnect suspect connections (remove ICs and SIMMs from their sockets, disconnect connectors, and then reinsert or reattach them). This process has a certain self-cleaning effect on the contact surfaces (any contamination is literally scraped off). Before reinstalling ICs or reattaching connectors, inspect the contact surfaces to see if they are corroded.

3 Use a fibreglass brush to clean contact surfaces that are obviously lightly corroded.

4 If an IC is actually OK but recurring errors occur due to unreliable contacts, you can simply solder the IC directly into its socket. (This may be unsightly in comparison to removing the socket and soldering the IC to the circuit board, but it is much less work. If the IC actually fails at a later date, it is still possible to remove the socket.)

5 Provide suitable securing devices, card retainers, cable conduits and the like.

6 For quickly checking whether a given cable connection is actually suspect, it is helpful to have a matching test cable that is kept on hand specifically for this purpose. The characteristics of such a cable are that: (1) it is no longer than absolutely necessary, (2) it is fitted with truly high-grade connectors and (3) the cross-sectional area of its leads is 'over-sized' for the actual application. Such cables can be purchased ready-made for certain interfaces (such as SCSI), but you will learn more if you make your own (see Volume 3).

Practical tip for the service technician: first make a few measurements when the circuit is in good working order, in order to get an idea of how it should behave (you will have to provide for well-defined operating conditions, by (for instance) running cyclic test loop programs). An oscilloscope with a PC interface is useful here, so that these samples of correct circuit behaviour can be saved.

Selected failure mechanisms in digital circuits. Sometimes all that you need is a few suggestions in order to come up with the right ideas (or more precisely, to propose fault hypotheses and then systematically verify them). To this end, Table 5.24 presents a selection of typical fault symptoms that you can expect to see in digital circuits.

Failure mechanism	Recommendations for testing and correction
'floating' signal level	Check the pull-up or pull-down resistor; check the bus termination network.
open inputs	Check the connections and any pull-up resistors. Add supplementary pull-up or pull-down resistors, as a (possibly temporary) preventative measure. (TTL: pull-up 4k7–22 kΩ, pull-down 330–512 Ω; CMOS: 100 kΩ–1 MΩ in each direction). *Note:* it's still possible to find IC inputs that have been left open, even in modern hardware.
extended-duration bus contention (during switch-on, for instance)	Widebus ICs can 'blow their tops' if they are overheated. Check the reset signals, reset hardware and controller ICs.
excessively steep signal edges at the inputs of LS TTL ICs	Signal edges that are too steep (such as from xxAC ICs) can cause malfunctions or even failure. This problem can occur with overly 'low-cost' expansion cards. Replace suspect LS ICs with equivalent ALS types.
metastable behavior	Check the clocking and drive specifications (set-up and hold times).
defective decoupling capacitors	Look for symptoms of shorted components; however, 'open' (ineffective) components are more likely. Try replacing the decoupling capacitor(s). (For a quick test, just solder another capacitor in parallel.)

Table 5.24 Selected failure mechanisms for digital circuits.

Additional possibilities:

- functional errors in ICs,
- IC faults that show up as overloaded outputs or supply voltages,
- bus-system faults (the bus lines are driven by more than one device at the same time, which is not allowed, or they are not driven at all; addressed devices do not respond and so on),
- excessive currents in protection diodes, due to very high noise levels (as a consequence of defective decoupling capacitors or the failure of external protection or termination networks). Excessive currents in the protection diodes can also be a sign of:
  - missing or incorrect termination resistors,
  - overloading by steep-edged signals or high-current signals.

See Section 5.3.3 regarding typical causes of faults in bus systems.

Component testing. See Section 5.1 regarding the basic principles of component testing. Table 5.25 describes the extent to which it is reasonable and possible to individually test modern components.

Type of component	Individual testing		Suitable test equipment
	Chances [1, 2]	Costs [3]	
discrete components (resistors, capacitors, diodes, transistors etc.)	good	relatively low	multimeters, component testers, simple test jigs
conventional ICs (low- and medium-scale integration digital ICs, operational amplifiers and the like)	good	quite acceptable	IC testers, simple test jigs
special-purpose ICs (medium-scale integration): power supply controller ICs, DC-DC converters, power ICs and the like	definitely possible	quite acceptable to relatively high	test jigs
processors	good	quite acceptable to high	test jigs (see Volume 2), master machines
memory	good	quite acceptable to high	memory testers (see Volume 2), master machines
application-specific ICs (large-scale integration)	hardly possible	very high	automated testers
expansion cards	good to definitely possible	quite acceptable	master machines, simple card testers for basic testing

1: This column relates to the situation in the service environment. The chances of being able to individually test a given component are good if its functions are either commonly known or are accurately described in sufficient detail in readily available literature.
2: Components that in principle can readily be tested individually can in practice scarcely be tested if they are soldered into a circuit and have more than three leads. (The problems are (1) unsoldering the component and (2) making connections to it.)
3: The costs of testing are judged here to be 'acceptable' if the price of the necessary equipment is at most £3500.

Table 5.25 Practical aspects of component testing.

5.5.2 Signal tracing

Signal tracing in normal operation. Fault-finding using measurement methods with the equipment operating normally is often only possible with simple circuits or with equipment or assemblies whose functions are easily understood. Furthermore, in some fault situations it is simply not possible to operate the device under test in a 'normal' mode that provides signals that are suitable for signal tracing. 'Fundamental' fault-finding requires either (1) recording processes as extensively as possible or (2) verifying that certain conditions occur, certain machine states are traversed and so on. Such procedures require suitable test equipment, in particular a sufficiently powerful logic analyzer and/or an in-circuit emulator for the processor or microcontroller in question (not to mention accurate documentation, technical knowledge and time).

Service practice: in most cases, precise fault-finding down to the last detail is out of the question. However, the normal operation of the device under test will unquestionably provide signals that are suitable as 'test stimuli' for plausibility testing. The necessary equipment consists of an oscilloscope, a logic probe, triggering aids and event detectors (see Chapter 4 and Volume 3), interface testers and (relatively simple) logic analyzers.

Examples: (1) simply switch on the PC, wait a couple of seconds, and then use an oscilloscope to check whether memory accesses occur, whether video and synchronization signals are output and so on; (2) start a print job in a Windows application and check the interface in question to see whether active signals are actually present.

Signal tracing in test mode or cyclic operation. The advantage of test mode operation is that a known stimulus is generated. Based on this, you can accurately localize faults using quite simple test equipment. Cyclic (periodically repeating) test processes are particularly advantageous, since they allow individual signals to be observed with an oscilloscope.

A proven method, which is very promising even in the absence of extensive documentation, is the following:

- Start a cyclic test program that only stimulates the hardware to be tested but does not itself test the hardware (a test loop or stimulus loop).
- Check the appropriate signals using an oscilloscope, starting in particular with standard, well-known (or otherwise adequately documented) interfaces (such as DRAM interfaces or the ISA bus). In modern hardware, there are rarely more than two or three devices or assemblies between two known interfaces (for example, there is usually only a controller IC and some easily-checked drivers between the processor and the RAM). With a little practice, you can thus identify the suspect IC with considerable certainty.

For example: suppose that the RAM complex is suspect. The first test routine only executes cyclic write accesses to a few RAM addresses (with the addresses and data selected so that at least one '1' and one '0' occurs on each of the address and data lines). The behaviour of the control and data signals for DRAM ICs is well documented in IC data books, so it is only necessary to see if the actual signals are reasonably close to what they should be (do the control signals RAS, CAS, OE and WE all switch, do all address and data signals reach the RAM ICs?). The second test routine incorporates write accesses followed by read accesses, so that you can trace the data that is read from the RAM and check whether it is correct. (Whether it is possible to trace all the way back to the processor depends on the actual circuit structure. Intermediate controllers and bus-interface ICs, external caches and the high bus clock rates used by modern processors (50, 60, 66 MHz) all considerably increase the difficulty of signal tracing.) *Note:* see Volume 2 for additional details and sample measurements (software test loops and basic diagnostic adapters), and Volume 3 regarding additional test aids.

Examples of deductions that can be made from such measurements:

- The processor is apparently largely functional (since the test loop would otherwise not be executed at all).
- Case 1: RAM accesses fail entirely — the controller circuits are suspect.
- Case 2: RAM accesses appear to function, and the signals look 'normal' (with regard to edges, levels, under- and overshoots) — the memory ICs or SIMMs are suspect.
- Case 3: errors in individual bit positions are seen: suspect memory ICs, SIMMs or bus interface ICs can be directly identified from the measurement data.

Signal tracing in static operation. Signal tracing with static signal levels is still the easiest possible technique (for which a simple logic probe is sufficient, although it should indicate whether the signal level lies in the 'forbidden' range between LO and HI, in addition to the usual LO and HI levels). Of course, it must be possible to 'halt' the circuit under test. There are two ways in which static operating conditions can be realized (see Figure 5.125):

1 stopping the clock,

2 suspending handshaking on a bus or interface, by introducing wait states or by blocking the response to a handshaking signal.

In Figure 5.125, either the halted state the normal running state can be selected using the control panel, which also allows the hardware under test to be operated a single step at a time. As indicated, the halted state can be entered as a result of certain conditions (such as "stop when address *x* is accessed". A detector or comparator circuit (designated as 'test aid' in the illustration) connected to a single-shot pulse generator can be used for this.

Notes:

1 Refer to Section 4.4 and Volume 3 regarding circuits for such test aids and certain subtleties of their use.

2 Simply stopping the clock is only possible if the circuit or IC is specified for fully static operation (for an IC, this means that the specified minimum clock frequency in its data sheet must be 0 Hz). Some ICs – in particular, many types of processors – have relatively high minimum clock frequencies (*for example:* one version of the Pentium processor has a nominal clock frequency of 60 MHz and a minimum frequency of 33 MHz). The allowable range of clock frequencies is even smaller for processors with internal clock multipliers (see Note 5).

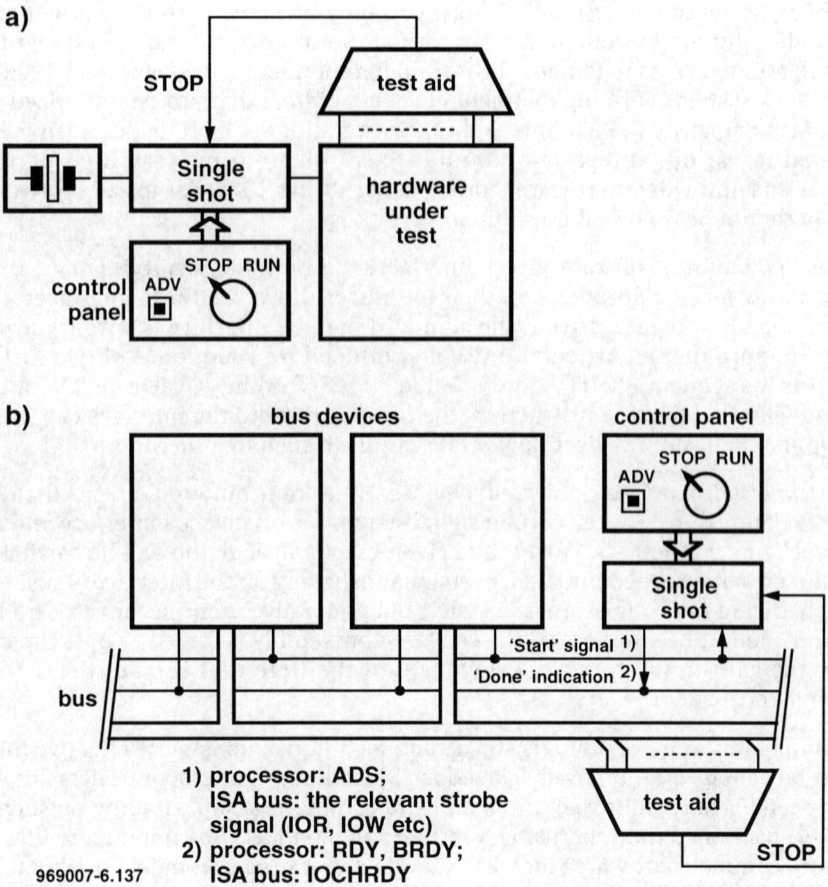

Figure 5.125 Sample arrangement for inducing static operating states.

a) Stopping the clock: a single-shot pulse generator is connected between the clock source and the circuit that is actually being tested.

b) Suspending handshaking on a bus: in 'Intel-typical' bus systems, each cycle starts with a strobe signal ('Start') that is terminated after a certain number of bus clocks unless wait states are introduced by disabling (pulling LO) the completion signal ('Ready').

3 Processors can be effectively halted via their bus signals, either by introducing wait states (with 86xx types) or by suppressing acknowledge signals. Other interfaces can be halted in a similar manner (ISA bus, PCI, SCSI and so on).

4 *Problem (1):* you cannot expect complicated hardware (including PCs) to simply carry on normal operation after it has been halted. DRAM refresh cycles may have been suspended, for example (resulting in the loss of data), or time-out errors may have occurred in certain subassemblies.

5 *Problem (2):* with current technology, you may be faced with a clock-generation or clock-distribution system makes it nearly impossible to fully stop system operation. This relates to (1) arrangements of several independent clock generators, (2) subordinate clock systems with independent clock generators that are synchronized to a central master clock by PLL circuits and (3) clock multiplier circuits (also using PLL techniques) in various integrated circuits (this applies to modern high-performance processors and video hardware, among other things). Test setups as shown in Figure 5.124a cannot be used in these situations.

6 *Problem (3):* purely software-driven interfaces. Certain subassemblies or devices expect pulses with specific durations on their interfaces. However, these interfaces are sometimes directly software-driven (the standard parallel interface is a well-known example). Suppose that a particular pulse is produced by a sequence of two instructions (such as the sequence 'OUT Address, Value = 1; OUT Address, Value = 0'). If operation is halted following the first instruction, the pulse applied to the interface can literally last for hours – and some devices do not take kindly to such treatment.

7 *Problem (4):* driving 'non-buffered' devices. Here, 'non-buffered' means that there is a direct relationship between certain signal sequences on device's interface and associated mechanical functions, such that a given mechanical action will be carried to completion once it has been initiated, even if handshaking on the interface is suspended. (A punched-card reader is a rather nostalgic but illustrative example: once a card has been fed in it cannot be stopped, so the sensor subassembly is forced to pass the data read from the card to the following stage within the time that it takes the card to pass through the reader.)

Fault-finding in static operation is still a valid technique and is quite effective for circuits that can be halted without any special effort. Most modern microcontrollers, for example, are designed for fully static operation, and you can perform fault-finding on such devices in the 'old-fashioned manner' using a test arrangement as shown in Figure 5.125a. Even with more complex hardware (such as PC systems), it is occasionally possible to use this approach, with the following restrictions:

1 Stopping the clock is hardly possible, so that you are limited to suspending handshaking on bus systems.

2 Stepwise operation after the system has been halted, or returning to normal operation, is almost out of the question. Instead, you are limited to the following procedure:

   - First devise a specific fault hypothesis (such as "expansion card X is not activating bits 8 through 15").
   - Try to verify this hypothesis (for example, by halting bus operations when an address belonging to the suspect card is accessed). This allows a static examination of the signals in question.
   - Regardless of whether the hypothesis is confirmed or refuted, it is useless to try to continue with normal operation. The only option is to reset the hardware.

Halting on a specific condition is nearly the same as triggering a logic analyzer on the same condition. The advantages of halting are that: (1) relatively simple test equipment is sufficient and (2) any desired signal can be checked, assuming that it is accessible (if the suspect card has first been mounted on an extender card, for example). By contrast, when a logic analyzer is used, it often happens that although trigger conditions correctly address the fault situation, the specific signals that are actually responsible for the faulty behaviour are not connected to the logic analyzer. The essential disadvantage of halting operation is that it can only be used for identifying 'hard' faults, so that it is not useful for problems caused by transient conditions such as spikes, state transitions and the like.

Practical problems in signal tracing

Connecting to the signals to be investigated. It is frequently necessary to check several signals together or one after another. Simply touching a probe to the signal lines is not always adequate and sometimes not even possible. 'Self-retaining' connections can be made using miniature test clips, temporarily soldered-on wires or a sort of 'artificial hand' (see Section 5.1.4 and Volume 3). With such methods, there is always a possibility of brief short circuits (between adjacent IC pins, for example) and intermittent contacts, so the device under test should always be switched off before connecting test clips or soldering on wires. With PCs in particular, it takes a while for the system to start up and load the test software after is switched on again. One remedy is to use a measurement adapter (probe pod) that allows connections to be made to a number of different signals 'in advance' (refer to Figure 4.97 and Volume 3).

Socket-mounted ICs. Whenever possible, make measurements at the IC pins and not just at the socket. Check the connections between the IC pins and the socket (sometimes probing a pin improves the contact, so that the fault disappears as long as you are looking for it).

Trick 1: verifying periodic signals. Inexpensive oscilloscopes have problems with the signals found in modern hardware. This relates to high-speed clocks (33–66 MHz and more) and narrow pulses occurring at relatively long intervals (such as DRAM drive signals, refresh signals, address decoder outputs and so on). The remedy is to use a frequency divider. A 2:1 divider converts narrow, periodic impulses into symmetric square wave signals whose durations can be easily measured (*for example:* checking the length of the refresh interval). A 16:1 divider converts a 66-MHz clock to a square wave signal with a frequency just over 4 MHz, which can be readily displayed by a 20-MHz oscilloscope. Frequency variations and the effects of noise signals can be immediately recognized. For more details refer to Figure 4.149 and Volume 3.

Trick 2: checking periodic signals. Slow down the clock ('diagnostic slowdown') – although this does not always work. Operate the hardware under test with a lower than normal clock frequency (the slow clock can come from a pulse generator or be produced by connecting a divider between the clock source and the rest of the circuit). Inexpensive oscilloscopes and logic analyzers can be readily used with clock frequencies ranging between 1 and 10 MHz, but the processor will still be running fast enough for dynamic operation.

The limitations of this technique are:

1 You have to watch out for certain situations. For instance, you should not reduce the clock rate of a video adapter if the monitor remains connected, since the monitor would otherwise be driven by invalid synchronization signals. Obviously, the monitor would not display an image under these conditions, but not all monitors will handle such situations 'correctly' (which means either disabling the drive signals to the display circuitry, or switching to internally-generated synchronization signals, if valid synchronization signals are not available).

2 In some cases, it is not possible to reduce the clock rate (for instance, if the hardware contains several independent clock generators, PLL-synchronized secondary clock generators or ICs with internal clock multipliers).

Trick 3: checking periodic signals. Observing only one signal is adequate in only a relatively small number of cases. You often need to observe the relative timing of a few (or very many) signals in order to check functional relationships. You can capture 32, 64 or more signals with a logic analyzer, but a logic analyzer that can handle this many signals (and such numbers are not uncommon in PCs – consider the processor bus of a 486 or Pentium processor, for example) has a price in the five-figure range, and it is difficult to make this many connections in a reasonably reliable manner. It is frequently sufficient to display the individual waveforms one after the other on an oscilloscope, but how can you see the timing relationships of the various signals in this case? The answer is to use a reference *trigger signal* that is specifically generated for this purpose. A D-type flip-flop (wired as a 2:1 divider) connected to an output port can provide the trigger signal. (This could be built into a diagnostic adapter for the port in question; see Volume 2). Typical software test loops (stimulus loops or scope loops) are constructed as follows:

- Start: I/O access (to the diagnostic adapter) to trigger the flip-flop,
- first test access (for example, a RAM write access with a data value of AA_H),
- second test access (for example, a RAM write access with a data value of 55_H),
- additional test accesses (as desired),
- return to Start.

You can work as follows with such an arrangement:

1 Connect the trigger signal to channel 1 of a dual-trace oscilloscope and trigger on this signal. Probe the individual signals to be checked one after the other, using channel 2. The timing relationships can be immediately seen on the display. *For example:* with the test loop just described, first look at the write pulses and note their locations with respect to the screen scale. Then probe the data and address signals one after another and check whether they have the correct values when the write pulses occur.

2 Connect the trigger signal to an external trigger input, to make channel 1 free. In our example it would make sense to display the write pulses on channel 1, so that the time relationships between the other signals and the write pulses could be directly observed.

Verifying pulse sequences. This applies to memory access signals, DRAM drive signals and bus cycle signals (among others), in addition to reset signals. A logic analyzer is the preferred instrument for this, but even simple models are rather expensive. Possible improvised remedies are: (1) pulse detectors, (2) special detector circuits, which can for example be used to verify the various pulse sequences of DRAM control logic, and (3) plug-in cards (diagnostic adapters) for analyzing busses. Figure 5.126 shows part of the circuit of a test aid (a sequence detector) that can recognize various types of DRAM control cycles. See Section 4.4.4 regarding the basic principles of detecting signal sequences, and refer to Volume 3 for additional details.

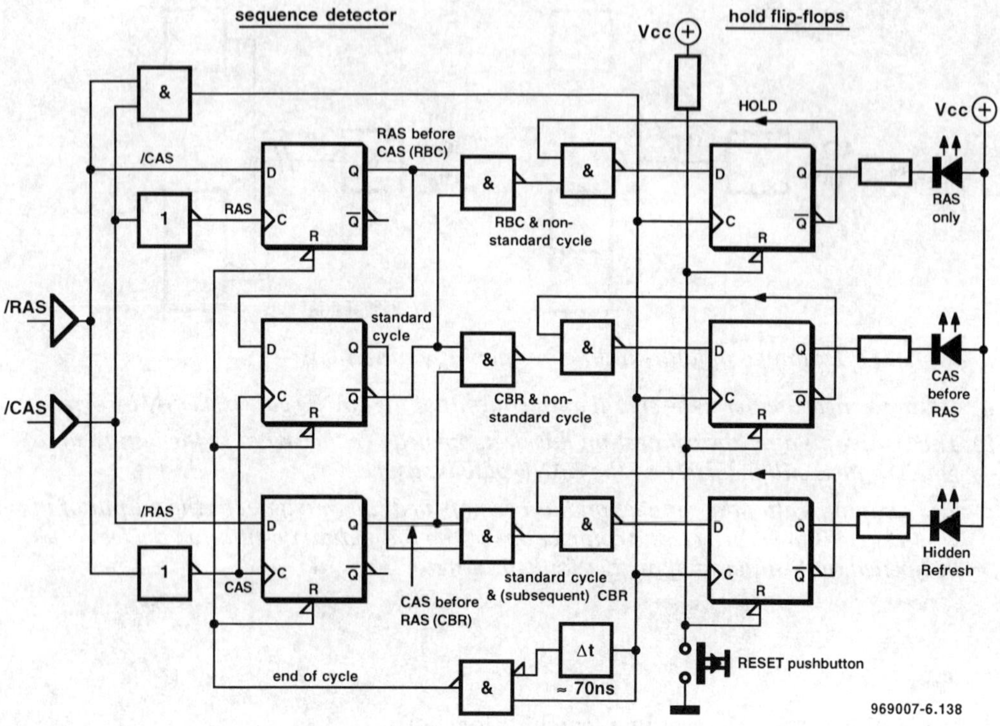

Figure 5.126 Partial schematic of a DRAM sequence detector.

This circuit recognizes various types of refresh cycles. In use, it is connected to the RAS and CAS signals of the RAM complex to be checked and the Reset button is pushed. The LED indicators then show what type of refresh cycle has been executed.

Opening connections. This is the simplest way to check for reverse coupling effects (from the destination to the source) in a faulty signal path (see Figure 5.127). Figure 5.128 shows how suspect inputs can be further investigated once they have been disconnected.

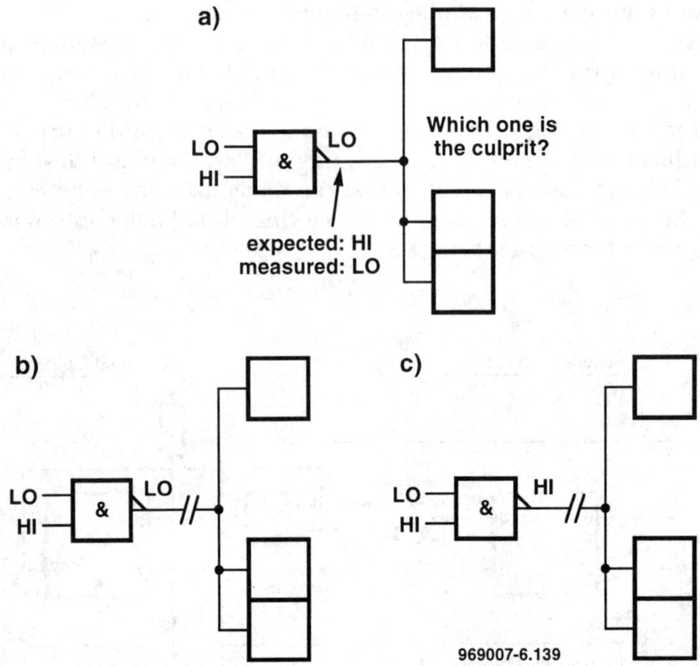

Figure 5.127 Examples of fault-finding by opening connections.

a) Example measurement: the signal at the output of the NAND gate is clearly incorrect.

b) The symptoms after the connection has been opened (1): the signal at the output of the NAND gate is still incorrect, so the NAND gate is suspect.

c) The symptoms after the connection has been opened (2): the signal at the output of the NAND gate is now correct. The connections to the individual destination devices could be opened in the interest of further fault-finding.

Notes:

1 See Section 5.4 as well regarding searching for shorts.

2 CMOS inputs draw almost no current in static operation (in the nA range), so it should be possible to pull them to either logic level with a 1-MΩ resistor (for example) if the IC is not defective.

3 The method shown can be extended in a similar manner to other interfaces with other signal levels. If it is technically possible, you can also check the values of the input currents (what you basically have to decide is whether the input current is too high).

Practical aspects of opening connections. If the source and the destination are located on different plug-in cards, opening the connection is no problem. Adapters that can be used for selectively opening connections, using DIL switches or solder bridges, are commercially available for a wide variety of interfaces (or you can make your own; see Volume 3).

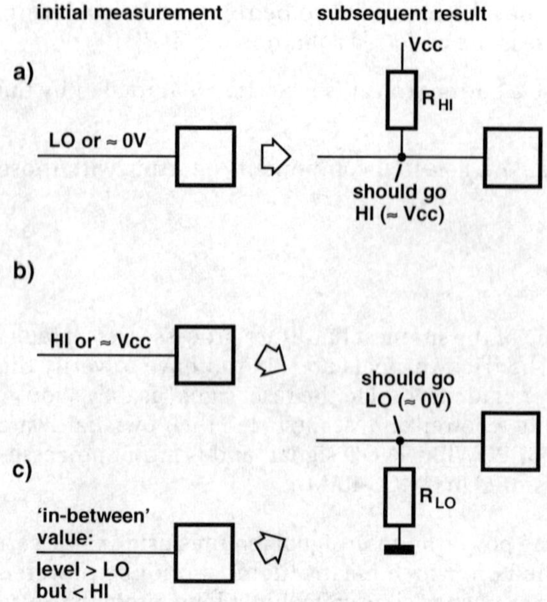

	TTL	CMOS / BiMOS
R_{HI}	1kΩ	1MΩ
R_{LO}	LS, AS: 1kΩ others: 330Ω	1MΩ
$I_{I\,HI}$*	10...50μA	1μA
$I_{I\,LO}$*	standard 1.6 mA S 2.0 mA AS, F 0.6 mA LS 0.4 mA ALS 0.2 mA	1 μA

* maximum HI and LO input currents

969007-6.140

Figure 5.128 Fault-finding with (disconnected) suspect inputs.

a) A LO level is measured. See whether the input can be pulled HI (to roughly the level of V_{CC}) by a pull-up resistor. If the measured input value is nearly 0 V and it scarcely changes when a pull-up resistor is connected, then there is a short to earth.

b) A HI level is measured. See whether the input can be pulled LO (to a level ≤ 0.4 V) by a pull-down resistor. If the measured input value is nearly V_{CC} and it scarcely changes when the pull-down resistor is connected, then there is a short to V_{CC}.

c) The measured value is in between a LO and a HI level. Perform the two tests described in (a) and (b). If correct levels are measured in both cases, the input is less suspect. Otherwise you should suspect a short to a different signal line (look for a different line with the same signal level).

Similar types of adapters are also available for IC sockets (see Section 5.1.4 and Figure 5.28 regarding test and conversion adapters). An improvised solution is to remove the IC from its socket and carefully bend the pin in question to the side before reinserting the IC (this is possible only with DIL packages) – but this is truly an emergency solution, since it is usually possible to bend each pin only once; it will break off on the second attempt. Manufacturers tend to avoid the use of sockets, in the interests of operational reliability (and cost reduction), so you will not often find them. You then have the following options:

1 Mechanically break the connection by (1) cutting through the associated track on the circuit board (only possible if the track is accessible, such as with two-sided construction) or (2) opening the connection between the IC pin and the circuit board. (*Caution:* do not use cutting pliers; instead, use a high-speed miniature drill with a milling bit or cutoff disk.) With SMD components, it is sometimes possible to remove the solder and

bend the pin aside. One problem is that the break has to be repaired by soldering, which can affect reliability. A repair wire is the preferred solution.

2 Instead of opening the connection, use a current tracer (see Section 5.4) to identify the device that is sinking the current

3 Instead of opening the connection, replace selected components (starting with those that are most suspect).

Reset faults

Although resetting would appear to be one of the simplest hardware processes, reset faults are sometimes very difficult to locate. The tricky aspects are: (1) You have to verify the power-on reset signal, (2) certain ICs may fail to respond to the reset signal as they should, (3) some plug-in cards and device generate a power-on reset signal in their own hardware (and thus do not use the general RESET or POWER GOOD signal) and (4) reset processes sometimes depend on setup parameters stored in CMOS-RAM.

Checking the power-on reset. Verifying the power-on reset signal requires using some sort of test equipment that is already switched on, which means that it cannot be powered from the PC power supply. You can use a storage oscilloscope, but a logic probe or pulse detector powered from a laboratory supply is a simpler solution.

Injecting a reset signal. Once you have verified the actual power-on reset signal, you must trace it through the rest of the circuitry to the ICs that should be reset. It is sometimes also necessary to test the *operation* of the reset process, and for this it useful to be able to intentionally inject a reset signal in order to avoid the need to repeatedly switch the equipment off and on. Figure 5.129 shows the schematic of a test aid that can drive the reset signal line. The toggle switch allows the reset signal to be held active for an extended length of time (for static testing). An externally-generated reset signal can also be injected.

Cyclic resetting. The hardware can be periodically reset by connecting a pulse-train generator to the circuit shown in Figure 5.129. This allows the startup behaviour of the processor (addressing and reading in the first few machine instructions) to be examined in detail by signal tracing. The advantage of this technique is that it can be used to check whether the processor 'hangs' on startup, which is otherwise virtually impossible with affordable test equipment. (See Volume 3 for more information.)

Clock generators

Usually, all that you can do is verify proper operation and replace suspect devices. Occasionally, however, it is necessary to precisely set the clock frequency.

In-circuit functional checks. It is frequently sufficient to look at the clock signal with an oscilloscope. You should check the signal levels, edges, general waveform and noise. A suitably fast pulse detector or a logic probe that can detect pulse trains is all you need for cursory go/no-go tests ('is it oscillating or is it not?').

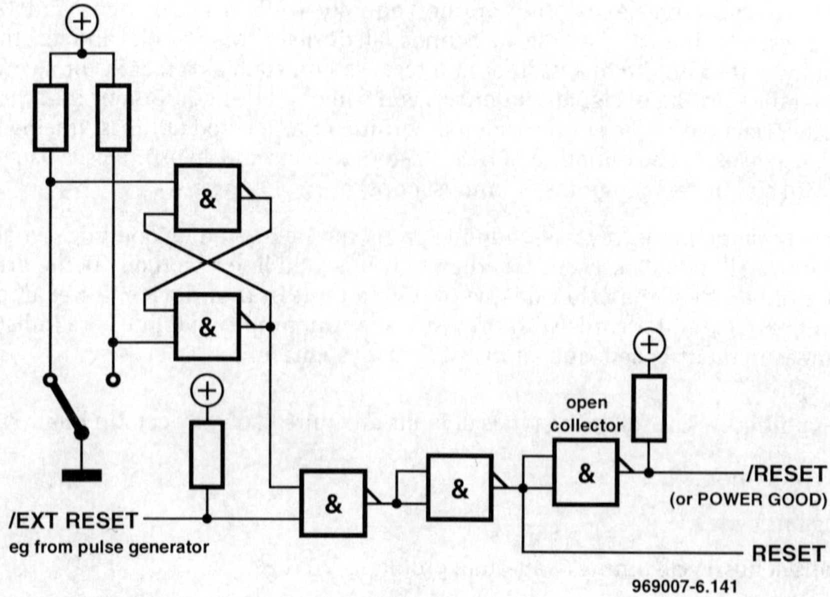

Figure 5.129 A test aid for generating reset signals.

Practical tips:

1 First measure the clock signal in the circuit. High-frequency signals will appear to be more or less distorted, especially with inexpensive instruments (slow transitions, overshoots etc), so you need to develop a feel for what can be regarded as normal.

2 Connect a divider ahead of your instrument for measuring very high frequencies. In this way you can check a 66-MHz clock using a 20-MHz oscilloscope.

3 Don't try to get by with inadequate equipment and accessories. You won't be able to measure anything useful with a length of ordinary flex lead connected between the clock generator and the oscilloscope. You should have a probe with a suitable bandwidth (guideline: DC-100 MHz).

A counter-test (for oscillators constructed from discrete components): the circuit must not oscillate if the feedback path is opened or the frequency-determining component is missing! In case of doubt (with 'wild' oscillations), check this.

5.5.3 Fault-finding in bus systems

The basic procedure. Bus systems present special problems for fault-finding, since several devices work together via a single set of interconnections. The only possible approach is to reduce the complexity of the system and to open signal paths if necessary.

1 Reducing the complexity: this applies equally well to (1) devices and (2) signal sequences. If it is at all possible, disconnect all devices from the bus and test them individually (either on the bus itself or in a test system, such as a test PC or a card tester). For detailed testing of signal sequences, you will need test equipment that can provide specific, selected cyclic stimuli (such as writing to a selected address, tracing read and write accesses, cyclic initiation of bus-master requests and so on). This is usually possible with simple test programs (stimulus loops).

2 Opening signal paths: reverse-coupling paths can be broken by opening signal leads. It is then possible to clearly see whether a given signal line is actually being driven or is being pulled by some other device (due to a fault in that device, for example). Test adapters or extender cards with provisions for opening signal lines are indispensable for measurement-based fault-finding on bus systems and interfaces.

Fault hypotheses. The following types of faults are quite likely to occur in bus systems:

- bus contention,
- hung bus cycles,
- incorrect bus-cycle timing (wait-state problems),
- signals not being driven,
- signals not being received,
- a signal line (such as an ISA-bus IRQ line) assigned to more than one device,
- system resources allocated to more than one device (for example, a set of I/O addresses or memory locations may be assigned to more than one expansion card),
- unassigned signal lines,
- multiple address decoding (slave devices),
- no address decoding (slave devices),
- multiple bus-master requests (master devices),
- no bus-master requests (master devices),
- missing, incorrect or defective bus terminations,
- incorrect configuration settings or defective Plug & Play hardware.

Bus contention. Bus contention can be caused by the following:

- faults in ICs that are connected to the bus,
- faults in the drive logic,
- incorrect resetting during switch-on or switch-off,
- using unsuitable IC types,
- human errors during measurement and testing (such as accidentally shorting signal lines when connecting ir removing test clips).

Most conventional bus drivers can tolerate bus contention,even if it lasts for a relatively long time, but they will become very warm (their internal resistance increases with temperature, which acts to limit the current). Nonetheless, you should avoid creating bus contention! (In practice, this means carefully considering what you are planning to do before starting a test or measurement procedure, and not using unduly simple accessories). *Caution:* be especially careful with modern bus drivers (such as Widebus types) – they really do not like bus conflicts! See also Section 5.1.2.

Bus contention caused by using unsuitable types of ICs: this does not refer to installing the wrong type of device, but rather to using the correct type of IC with an incorrect timing characteristic (such as installing a 90-ns RAM when a 70-ns device is required). This need not necessarily cause problems, but it is certainly possible for transient bus contention to occur due to the differences in switching times between the incorrect device and other devices, and this contention may disturb the operation of the bus. It is also conceivable that a device that is 'too fast' could cause problems in a similar manner (if it accesses the bus before a relatively slower device has become inactive). In practice, this means quite simply that if you insist on improvising in such matters, you must test the system for an extended period of time, preferably at the limits of the allowable operating conditions.

Hung bus cycles. This type of fault is caused by control or acknowledge signals that are (1) not asserted, (2) permanently asserted or (3) not received or evaluated. Signal assertion can be checked with a pulse detector (for example).

Incorrect bus-cycle timing. If a fault causes wait-state or READY signals not to be switched at the proper time, the associated bus cycle will complete with no wait states. Since the slave device will not have time to either supply or accept the data, this will sooner or later result in a malfunction whose cause is not immediately recognizable. You should check (1) whether the individual devices actually request wait states and (2) whether the READY signals switch properly on the bus.

Verification: make measurements, working backward from the drivers to the internal logic. If necessary, mount a suspect expansion card on an extender card that allows the bus lines to be opened.

Fault-finding on driven bus lines. You can check whether a given device drives the bus by opening a suitable line and connecting a test load to it (see Figure 5.130, overleaf). If the associated accesses are cyclically triggered, you can use an oscilloscope to check whether the signal in question is active.

Checking temperatures. A digital thermometer with a suitable sensor (surface-temperature probe) can be used to check the operating temperatures of ICs. (First apply a drop of heat-conductive paste to the middle of the IC package, and then press the temperature probe into the paste.) ICs that are too warm and those that are too cool are equally suspect. If bus contention is present (two drivers actively driving the same signal line), tri-state bus driver ICs will become warmer than usual and open-collector bus driver ICs will become cooler. *Practical tip:* for comparison purposes, check the operating temperatures of critical ICs in properly-working hardware after 15 minutes' warm-up time, under well-defined operating conditions (such as running a test program).

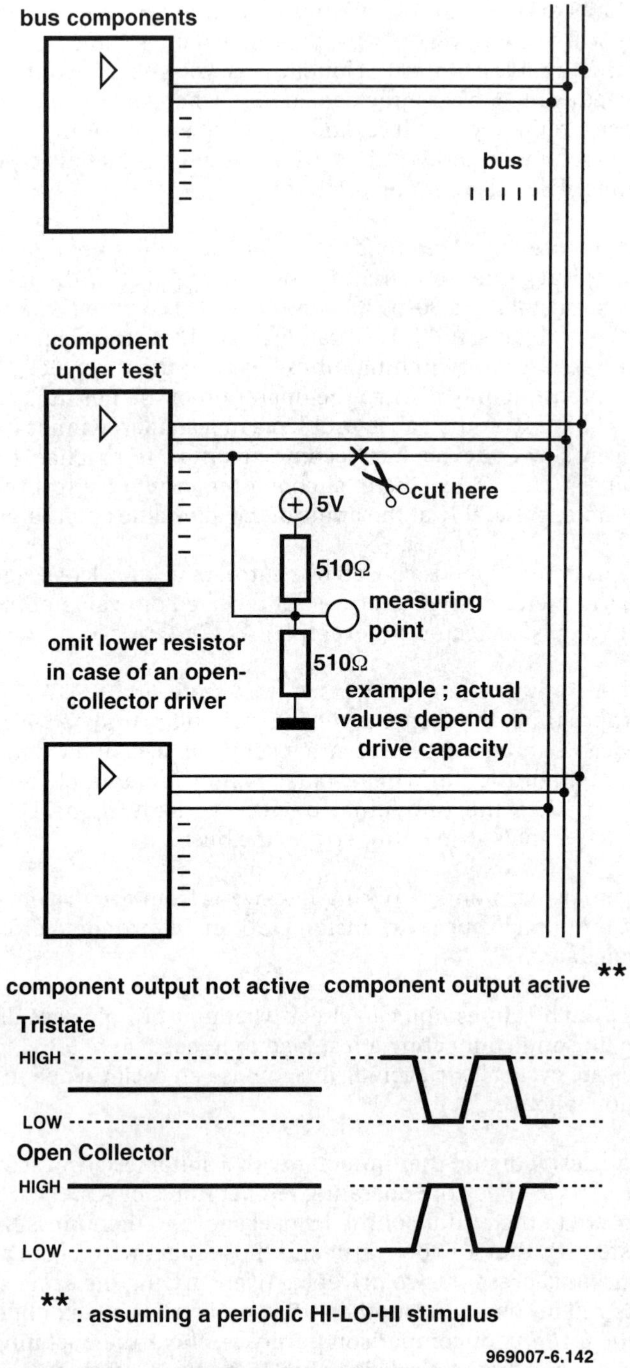

Figure 5.130 Fault-finding on driven bus lines.

Appendix 1
Anti-static measures

All measures to avoid damage due to electrostatic discharge (ESD) are based on allowing static charges to flow to earth. Figure A.1 illustrates a fully outfitted setup for an 'ESD-safe' test and service area. Simpler arrangement are often used in practical servicing, but you must observe the following basic points at all times:

- You must always be connected to a solid earth (ground) point whenever you are working with components, plug-in cards and similar items.
- The commonly used measures are (a) a conductive anti-static wrist strap, (b) an anti-static mat (connected to earth) on the work surface where you place expansion cards and components, (c) special anti-static (conductive) bags for storing components, expansion cards and similar items (do not use just any old plastic bag).

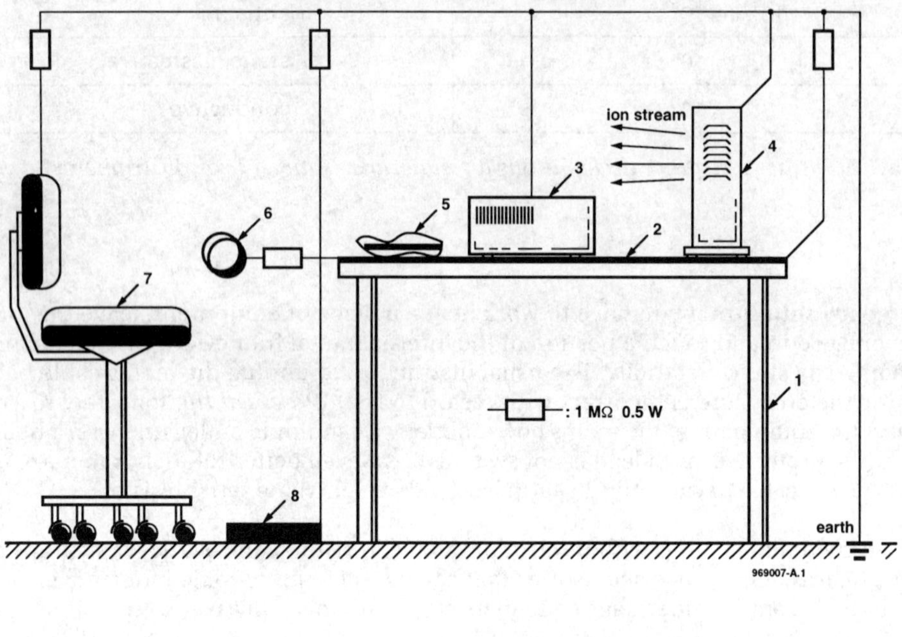

Figure A.1 ESD protection measures for the service bench. (source: Texas Instruments)

1 – work bench, 2 – conductive mat, 3 – equipment (instruments, soldering station, etc.), 4 – air ionizer (blows a stream of ionized air over the work surface to discharge static electricity), 5 – conductive packaging (for ICs, plug-in cards and so on), 6 – ESD armband, 7 – chair (with optional earth connection), 8 – earthed foot mat (optional).

- Using the protective earth connection of the mains power cable is not ideal, but it is in any case better than nothing. A professional (permanent) working area should be earthed by a direct connection to the earth rail.

Notes:

1 With regard to ESD protection, 'conductive' does not mean 'metallic' but rather 'having finite resistance'. A resistance on the order of 1 MΩ is adequate to allow static charges to bleed off. The electrical resistance of conductive plastic bags, tool grips and similar anti-static items is on this order of magnitude. Table A.1 lists approximate surface resistance values for various types of surfaces. *Note:* the important issue is that the total area in question (an anti-static mat or an anti-static coating, for instance) must have a suitable net resistance value. The actual area of the surface is not in itself important (which is why the surface resistance is stated in ohms alone).

2 Remember that moving air can also generate a static charge (this applies to compressed air sprays, cold sprays and vacuum cleaners).

Surface resistance	Designation
10^{14} to 10^{9} Ω	antistatic
10^{9} to 10^{5} Ω	static dissipative
$< 10^{5}$ Ω	conductive

***Table A.1** Surface resistance designations for materials. (source: Texas Instruments)*

Emergency solution: if you have to work inside a piece of equipment, leave the mains cable plugged in and touch a portion of the internal metal framework with one hand to discharge any static electricity. The usual instruction to unplug the mains cable before opening the enclosure is incorrect with regard to ESD! *Practical tip:* for safety reasons, connect the equipment to the mains power outlet via a multiple outlet strip with a built-in switch (the protective earth lead is not switched). An even better solution is to make up a special power cable having only an earth lead with a high-value series resistor.

Note: earthed mains plugs with built-in protective resistors, anti-static wrist straps and anti-static mats (as service accessories that can be rolled up or folded for storage), anti-static bags and anti-static storage containers are all commercially available.

Appendix 2
Working safely

Observe the following precautions:

1 Take care that you never come in contact with the earth potential *except via a series resistance* of several MΩ! The makeshift technique of using the protective earth for ESD protection (as described in Appendix 1) is only intended to be used with normal PCs and similar devices, in which the power supply assembly is almost always completely safe to touch (and it should also only to be used for the quick replacement of a sub-assembly or an expansion card). For more involved tasks, this sort of ESD protection is inadequate. Moreover, if you are working inside other types of equipment (such as a laser printer or monitor) you can easily encounter dangerous or fatal voltages.

2 There is a risk of electrical shock any time that it is possible to come in contact with parts at potentials greater than 48 V (either with reference to earth or as a difference potential). ***Always use adequately insulated test leads and probes in such cases. Use only one hand when making measurements, and keep the other one in your trousers pocket!***

3 *Accessories:* miniature probes, clips and the like can typically be used with potentials up to 50 V. Common 4-mm laboratory test leads are normally rated for up to 250 V, but there are also test leads and probes that are certified for higher operating voltages (such as 600 V and 1000 V). When you are measuring the mains voltage, you can expect a peak value of 660 V. For measuring 'true' high voltages (for example, inside a monitor) you will need special accessories (such as high-voltage probe assemblies). Don't try to cut corners here!

4 *Improvisation and DIY:* in principle there is nothing wrong with this, as long as your workmanship is tidy. The choice of material is important. Pay attention to the specifications for the allowed operating voltage and test voltage. The insulation should be rated for at least four times the actual *peak* operating voltage (not the *effective* voltage) to which it will be exposed. Inexpensive material that is commonly used in cars is only certified for use with 12–48 V and is tested at 250 V. Do not use such material to make up test leads for working with mains potentials!

Appendix 3
Floating operation

This subject of this appendix is the reference potential shared by the measurement equipment and the device under test.

Earthed operation. Normally, the reference potential is the earth potential. For mains-powered equipment, the reference potential is always that of the enclosure (the case or chassis earth), which is connected to the mains network earth via the protective earth lead.

Floating operation. In floating operation, the reference potential is not the earth potential. Two different cases can be distinguished: (1) true floating operation, in which there is a shared reference potential between the instrument's earth and the earth potential of the device under test, and (2) in-circuit measurements, in which the instrument is connected to arbitrary points within the circuit being tested (such as across a resistor or across the base-emitter junction of a transistor).

Note: refer also to Figures 4.14 and 4.72.

Practical considerations

1 A well-known improvisation is to operate the equipment with the protective earth lead not connected (experienced technicians have a stock of suitably modified mains extension cables for this purpose). This is naturally frowned on by equipment manufacturers and safety experts, but it is largely without risk as long as (1) the safety provisions of the measuring equipment (an oscilloscope, for example) are in good order (which means that they comply with the applicable standards, such as VDE 701 and 413) and (2) the reference potential is relatively low (say 24 V at most). Nonetheless, you should (1) clearly mark special mains adapter cables and (2) leave the equipment connected in this manner only as long as it takes to make the measurement(s) in question. (Using an isolation transformer, as shown in Figure 5.70, is a better solution.) .

2 There are certain unavoidable limits to floating operation, even if an isolation transformer is used. These are due to the following factors: (1) the parasitic capacitance of the instrument's enclosure will act as a capacitive load on the device under test, (2) the internal insulation of the instrument (such as the interwinding insulation of the power supply transformer) is often not designed for floating operation with a reference potential of several hundred volts and (3) many modern-day instruments have switching-type power supplies with mains input filters (refer to Figure 5.57). The last of these factors means that if the protective earth lead is not connected, the instrument enclosure will itself 'float' at one half of the mains potential (with a peak value of 330 V) and the entire setup will act as an antenna that radiates noise signals.

For example, see Figure 4.72b (Section 4.3.5): if the oscilloscope is operated 'floating', then point (*) of the circuit is connected to the oscilloscope's enclosure and is loaded by the capacitance of the enclosure with respect to earth. If the oscilloscope has a mains input filter, then the noise currents that are diverted by the filter will be injected into the circuit under test at point (*).

Safe and technically correct floating operation

- Certain types of instruments (such as oscilloscopes) are available as battery-operated models that are expressly designed for floating operation. The individual channels are isolated from each other and from the enclosure. Typically, voltages up to 600–700 V (effective value) with respect to earth are allowed, so that measurements can be made directly on the mains network (for example). *Caution:* special insulated probes must be used for such measurements.
- Differential measurements can be used in place of floating operation (refer to Figure 4.76). This can sometimes be achieved by using two channels in differential mode; otherwise a differential measurement amplifier (or differential probe) should be used with the appropriate instrument input. Such accessories can handle voltages up to around 500–2000 V with respect to earth.
- You can use galvanic isolation (of the instrument from the device under test) in place of floating operation. Isolation amplifiers, isolation probes and so on are available for voltages with respect to earth ranging from 1000 V up to true 'high tension' levels.

Appendix 4
Signal levels and transitions

TTL levels and TTL compatibility. In general, the terms *TTL levels* and *TTL-compatible* mean that the interface signals in question conform to the following parameters:

- input LO level: 0 to 0.8 V
  (the input sees any voltage below 0.8 V as a LO level),
- input HI level: 2 V to V_{CC}
  (the input sees any voltage above 2 V as a HI level),
- output LO level: 0 to 0.4 V; 0.2 V typical
  (the output LO level is usually around 0.2 V and never exceeds 0.8 V),
- output HI level: 2.4 V to V_{CC}; 3.4 V typical
  (the output HI level is usually at least 3.4 V, and not less than 2.4 V when loaded).

Traditional and modern logic families. The traditional 'industry standard' supply voltage for logic devices is 5 V (4.5 to 5.5 V range); more recent devices use 3.3 V (2.7 to 3.6 V range). Modern logic series are typically TTL-compatible, but differ in the following aspects: (1) the maximum allowed input voltage, (2) the actual output levels and (3) the input switching thresholds (which are important with regard to noise immunity). Figure A.2 compares logic levels for various logic families, and Table A.2.summarizes the interconnection possibilities between 5 V and 3.3 V families. See also Section 4.4.1.

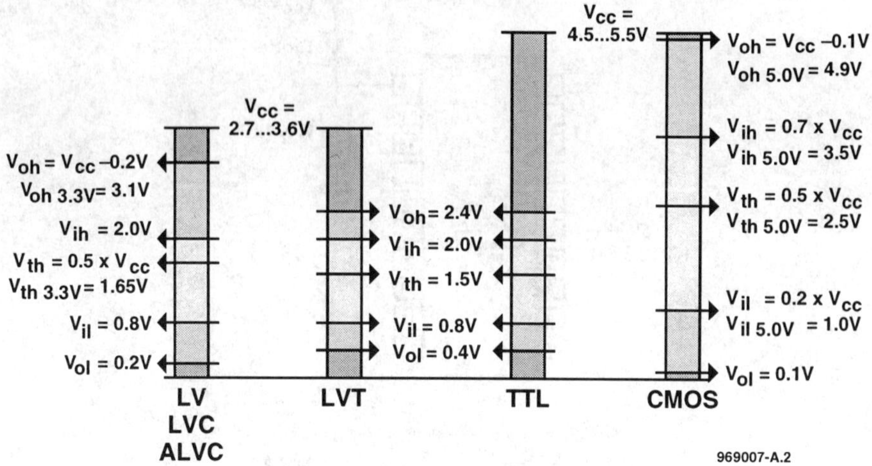

Figure A.2 *Logic levels for various IC series. (source: Texas Instruments)*
V_{il} & V_{ih} = input voltages, V_{ol} & V_{oh} = output voltages and V_{th} = threshold voltage.

Source (output)	Destination (input)	Coupling
5 V TTL	3.3 V	OK (*caution:* maximum V_{CC} + 0.5 V for LV)
5 V CMOS	3.3 V	only to LVC and LVT; otherwise see the text
3.3 V	5 V TTL	OK
3.3 V	5 V CMOS	inadequate HI level; requires special buffer ICs (Level-Shifting Transceivers) with two supply voltages (eg SN74ALVC164245)

Table A.2 *Interconnecting logic families that use different supply voltages.*

Connecting 5 V CMOS devices to 3.3 V devices: not all 3.3-V series ICs are '5 V I/O-tolerant' (this refers to ICs that can withstand 5 volts on their inputs). Figure A.3 shows one way in which the output voltage swing can be reduced to an acceptable level. The maximum level of the output signal from the switch IC (CBT, QuickSwitch and the like) is roughly 1 volt less than the supply voltage. The forward-biased silicon diode reduces the supply voltage to around 4.3 V (5 V – 0.7 V), so a 5 V signal applied to the input is dropped to around 3.3 V at the output. The special advantage of this arrangement is that the propagation delay is less than 1 ns. *Note:* switching devices with built-in diodes are available as 'level shifters'.

Signal transitions. The 'steepness' of a signal edge might seem to be self-evident, but it is specified in various ways and the definitions in the literature are sometimes ambiguous. It

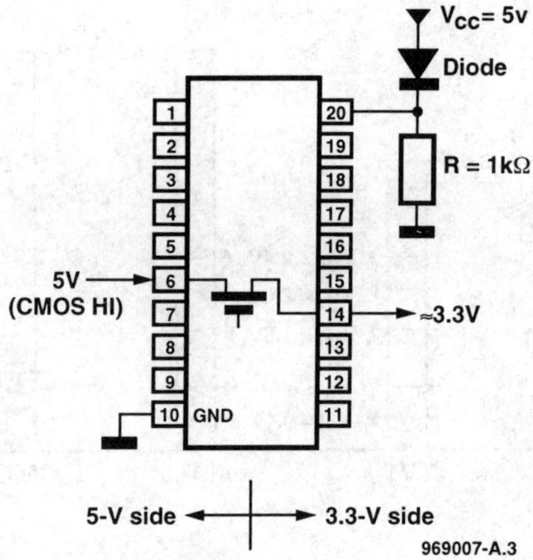

Figure A.3 *Using a switch IC to convert 5 V CMOS signal levels to 3.3 V levels. (source: Texas Instruments)*

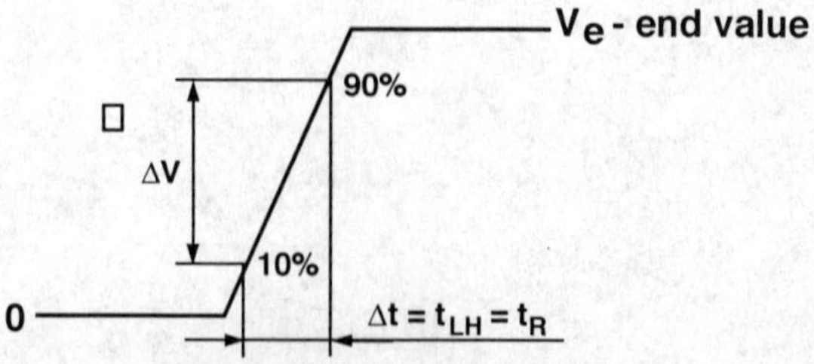

slew rate or edge steepness $= \dfrac{\Delta V}{\Delta t}$: units $\dfrac{V}{\mu s}$

approximate formula: $\dfrac{\Delta V}{\Delta t} \approx \dfrac{80\%\ V_e}{t_R} = \dfrac{0.8 V_e}{t_R}$; $t_R \approx 0.8\ V_e \cdot \dfrac{\Delta t}{\Delta V}$

rise rate : $\dfrac{\Delta t}{\Delta V} = 1$: slew rate ; units: $\dfrac{\mu s}{V}$ or $\dfrac{ns}{V}$

969007-A.4

Figure A.4 Signal transition specifications.

is thus worthwhile to define the terms *rise time, slew rate* ('edge steepness' in the true sense) and *rise rate.* The formulas and conversions are shown in Figure A.4. *Note:* the shorter the rise time, the higher the slew rate; the lower the rise rate, the steeper the edge.

Rise time and fall time. Strictly speaking, *rise time* refers to a LO-to-HI transition (t_R or t_{LH}) and *fall time* refers to a HI-to-LO transition (t_F or t_{HL}). However, *rise time* is often used to refer to both types of transitions (as are *slew rate* and *rise rate*).

Rise time. The rise time is the time that is required for a signal to make a transition between two defined voltage levels. In practice, the rise time is measured with reference to the 10% and 90% values of the stable final value. (Bear in mind that the 'final value' for logic signals is the relevant minimum HI level.)

Slew rate. The terms *slew rate* and *edge steepness* are very often used interchangeably. The value is expressed in volts/time (eg, V/ms). It is defined the same way as any other physical rate: slew rate = (voltage difference) ÷ (time difference). A slew rate or edge steepness of 1 V/µs means that the level of the signal changes by 1 volt in 1 microsecond.

Rise rate. The rise rate (or fall rate) is the inverse of the slew rate, or in other words the ratio of the time difference to the voltage difference. A rise rate of 1 µs/V means that it takes 1 microsecond for the signal level to change by 1 volt.

Typical values for signal transitions

- relatively slow logic (LS, ALS, HC): rise time 5 to 10 ns, slew rate around 0.2 V/ns
- fast logic (AC/ACT, BiMOS (BCT/ABT)): rise time 2 to 3 ns, slew rate around 1 to 2 V/ns.

Appendix 5
Alternating-current signal specifications

An alternating current periodically changes its polarity and strength (see Figure A.5). No current at all flows at the instant that the polarity changes. The current increases from this point in one direction until it reaches a maximum value (peak or crest) and then decreases until it again reaches zero. It then increases in the opposite direction until it reaches a maximum value, decreases again until it reaches zero, changes direction and continues in the same manner. The current thus swings back and forth between two maximum values, changing direction on each swing. An AC supply does not have a positive or negative terminal, since the polarity changes on every half cycle. In an AC circuit, the sign and magnitude of the voltage across a load resistor also change with the polarity and strength of the current, respectively.

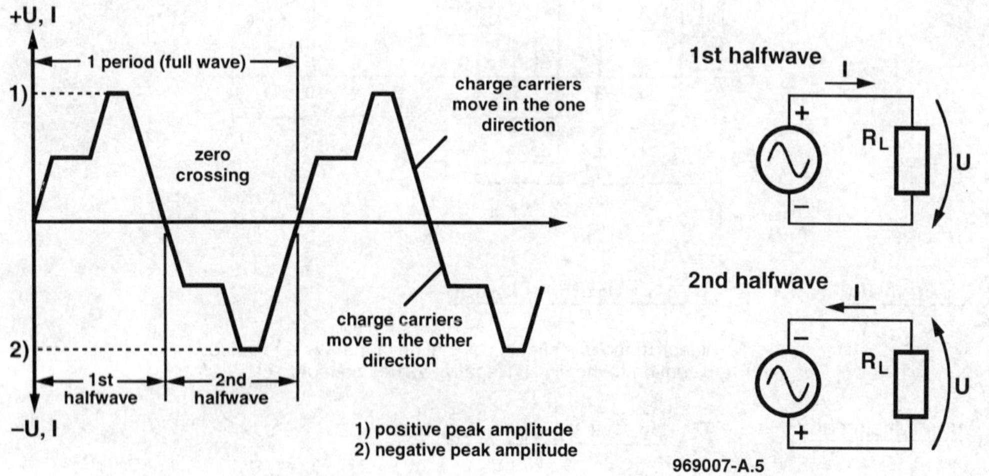

Figure A.5 Fundamentals of an alternating-current waveform.

DC component. The waveform of a 'true' AC signal (one that has no DC component) has equal areas for the portion of the signal above the baseline and the portion below the baseline (as displayed on an oscilloscope, for example), which means that the same number of charge carriers changes direction every time the current changes direction. Otherwise the AC signal has a DC component. The DC component can be so large that the waveform never passes through zero; in this case we can speak of a DC signal with a superimposed AC component (see Figure A.6).

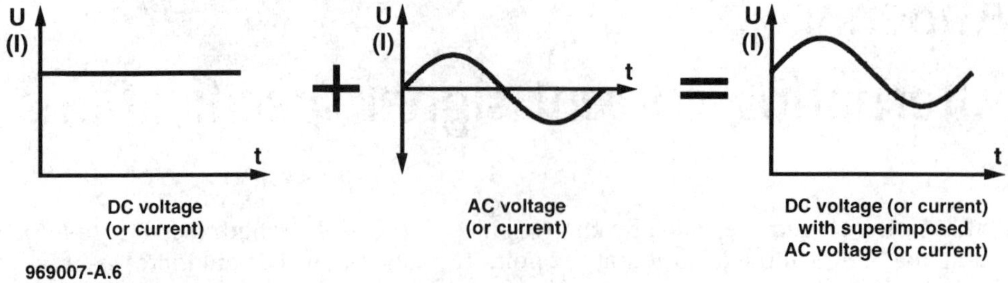

Figure A.6 *A DC signal with a superimposed AC component.*

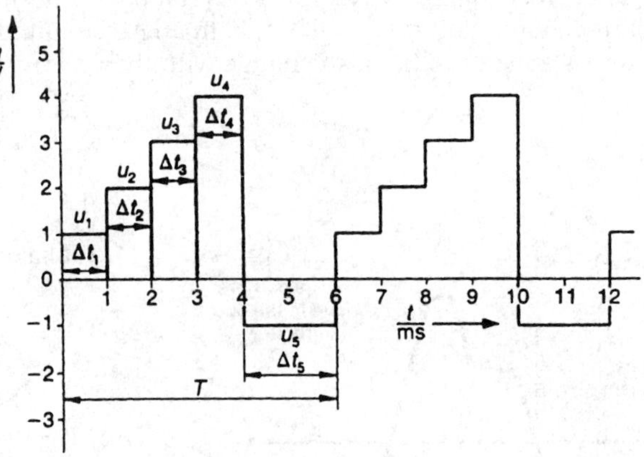

<u>linear mean</u> $\bar{u} = \dfrac{U_1 \Delta t_1 + U_2 \Delta t_2 + ... U_n \Delta t_n}{T}$

The linear mean is a simple calculation of the average value.
Pay attention to the sign! (Negative values are subtracted!)

<u>quadratic mean</u> $\overline{U^2} = \dfrac{U_1^2 \Delta t_1 + U_2^2 \Delta t_2 + ... U_n^2 \Delta t_n}{T}$

<u>effective power (true power)</u> $P = I^2 \cdot R = \dfrac{U^2}{R}$; accordingly for AC voltage or current $P = \overline{i^2} \cdot R = \dfrac{\overline{U^2}}{R}$

<u>effective value (root mean square, RMS)</u> $I = \sqrt{\overline{i^2}}$; $U = \sqrt{\overline{U^2}}$

<u>effective value of a sinusoidal AC current</u> $I = \dfrac{\hat{i}}{\sqrt{2}} \approx 0.707\,\hat{i}$; $U = \dfrac{\hat{U}}{\sqrt{2}} \approx 0.707\,\hat{U}$

$\hat{i} = \dfrac{1}{2} I_{pp}$; $\hat{U} = \dfrac{1}{2} U_{pp}$

969007-A.7

Figure A.7 *alternating-current specifications (top: measuring the mean value).*

Specifications of AC signals (see Figures A.7 and A.8):

- *Peak values:* AC currents and voltages have two peak or crest values in each cycle. The 'swing' between the two peak values is an important specification. The associated symbol has the subscript *pp* (for 'peak-to-peak').
- *Instantaneous value:* the instantaneous value of an AC current or voltage at a time *t* can generally only be determined by examining its waveform. With simple waveforms (such as sine waves), the instantaneous value can be calculated from the peak value.
- *Mean linear value:* the mean linear value is the arithmetic mean of the AC current or voltage, taking the sign into account. It is equal to the DC component of the waveform. The mean linear value of a pure AC waveform is thus always zero.
- *Mean quadratic value:* the mean quadratic value is the arithmetic mean of the squares of the instantaneous values. (A quadratic value is always positive).
- *Effective value:* the effective value of an AC voltage or current is equal to the value of the DC voltage or current that would dissipate the same amount of power as the voltage or current in question in a given resistance *R*. The effective value is calculated as the square root of the mean quadratic value (the *root mean square* or RMS value).
- *Effective power:* even if only an AC current flows through a resistor, the resistor will still become warm. The power dissipated in the resistor (the *effective power*) can be calculated from the resistance and the mean quadratic value of the current or voltage.
- *DC value:* the DC value represents the magnitude of the arithmetic mean value over one complete cycle (the sum of the instantaneous values is taken over one cycle, ignoring the sign, and then averaged). The DC value is equal to the average value of the rectified AC waveform.
- *Crest factor:* the crest factor is the ratio of the peak value to the effective value.
- *Form factor:* the form factor is the ratio of the effective value to the DC value.

Power in AC circuits. Power in AC circuits is calculated from the effective values of current and voltage. There are three related types of power specification:

1. *Effective (true) power:* the effective power is the power that is converted into heat or mechanical work. Since the current and voltage waveforms may have a phase offset, the phase angle φ must be taken into account. The formula for the effective power is:

   $$P = U \cdot I \cdot \cos \varphi$$

   The effective power is specified in *watts* (W). The cosine of the phase angle (cos φ) is called the *power factor*.

2. *Apparent power:* the apparent power P_A is the product of current and voltage without taking the phase angle into account:

   $$P_A = U \cdot I$$

   The apparent power is specified in *volt-ampères* (VA).

Average DC value

a) full-wave rectification $|\bar{U}| = \frac{2}{\pi}\hat{U} \approx 0.637\,\hat{U}$; $|\bar{I}| = \frac{2}{\pi}\hat{I} \approx 0.637\,\hat{I}$

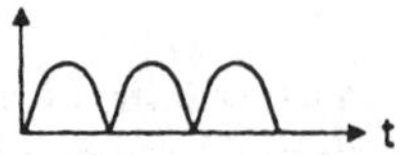

b) half-wave rectification $|\bar{U}| = \frac{1}{\pi}\hat{U} \approx 0.318\,\hat{U}$; $|\bar{I}| = \frac{1}{\pi}\hat{I} \approx 0.318\,\hat{I}$

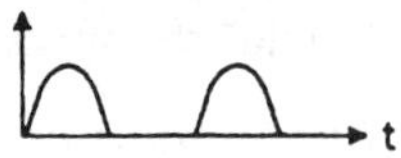

Crest factor

$\xi = \frac{\hat{U}}{U}$

crest factors

waveform	crest factor
sinusoid	1.414 – $\sqrt{2}$
squarewave	1
triangular wave	1.732 – $\sqrt{3}$
sawtooth	1.732 – $\sqrt{3}$

Form factor

$F = \frac{U}{|\bar{U}|}$

Form factors

waveform	Form factors
sinusoid	1.11
squarewave	1
triangular wave	1.1547
sawtooth	1.1547

969007-A.8

Figure A.8 *Relationships between the DC value, the crest factor and the form factor.*

3 *Reactive power:* the reactive power P_R results from the phase offset produced by reactive elements in the circuit:

$$P_R = U \cdot I \cdot \sin\varphi$$

The reactive power is specified in VA or *VA reactive* (VAr). It does not produce any heat.

The various types of power are related as follows:

$$P^2 + P_R^2 = P_A^2 \quad \text{and} \quad \frac{P_R}{P} = \tan\varphi, \quad \frac{P}{P_A} = \cos\varphi$$

Notes:

1 If an AC current passes through a resistive load, the effective power is converted to heat. The load resistor will not itself cause any phase offset, but the current and voltage applied to the resistor can already have a phase offset. The power factor of the 230 V mains network, for example, is typically 0.7 to 0.8.

2 When specifying the power dissipation capacity of a load resistor, you must take into account the possibility that the current and voltage may sometimes be in phase ($\varphi = 0°$, $\cos\varphi = 1$). This means that you must use the value of the *apparent* power.

3 A purely reactive impedance does not become warm, since it does not dissipate any power and only causes a phase shift. This results in a reactive power component that is not actually consumed, but instead is fed back into the mains circuit.

4 You pay only for effective power.

5 Watch out! The power consumption specified on the equipment's name plate can refer to either the effective power (assuming a typical value for cos φ) or the apparent power (in VA).

6 If the effective values of the current and voltage are measured separately and then multiplied together, the result is the apparent power.

Appendix 6
Feedback shift registers

The linear feedback shift register (LFSR) is a very important circuit structure for fault-finding and testing applications. It is used for error detection and correction using cyclic codes (CRC/ECC), for generating pseudo-random test data and for digital signature analysis. A shift register has *feedback* if one or more flip-flop outputs are coupled back to one or more flip-flop inputs via combinational logic. This feedback is *linear* if the combinational logic contains only exclusive-OR gates. Figure A.9 shows the basic structures of various types of shift registers.

Theory

Certain basic theoretical concepts are required to describe the operation of linear feedback shift registers. These relate to the representation of the bit pattern that is shifted, the data transformations that occur when the pattern is shifted and so on. These concepts come from mathematical coding theory. They are presented here simply as serviceable analogies (in other words, as useful descriptive tools), without any justification or proof.

Polynomials and bit sequences. A polynomial $P(x)$ is a mathematical expression that takes the general form:

$$P(x) = a_n x^n + a_{n-1} x^{n-1} + \dots + a_2 x^2 + a_1 x + a_0 = \sum_{i=1}^{n} a_i x^i$$

where x is the *variable* and the values a_i ($i = 0 \dots n$) are the *coefficients*. An expression with the form $a_i x^i$ is a *term* of the polynomial. Terms whose coefficients are zero are normally omitted. The highest power n characterizes the *degree* of the polynomial (so that we can speak of first-degree, second-degree, ..., n^{th}-degree polynomials).

In the area of coding theory that interests us, all coefficients values are either 0 or 1. The crucial idea is that any desired bit sequence can be represented as a polynomial.

For example:

$$\begin{array}{cccccccccl} 1 & 0 & 1 & 1 & 0 & 0 & 1 & 1 & \Rightarrow & x^7 + x^5 + x^4 + x + 1 \\ \downarrow & & \downarrow & \downarrow & & & \downarrow & \downarrow & & \\ x^7 & & x^5 & x^4 & & & x & 1 & & \end{array}$$

The principle is obvious: the least-significant bit position (2^0) corresponds to a_0 (= 1), and any given bit position i with a weight of 2^i and an actual value of $a_i = 1$ corresponds to the ith term of the polynomial. A sequence of n bits can thus be described by a polynomial of degree n, and vice versa.

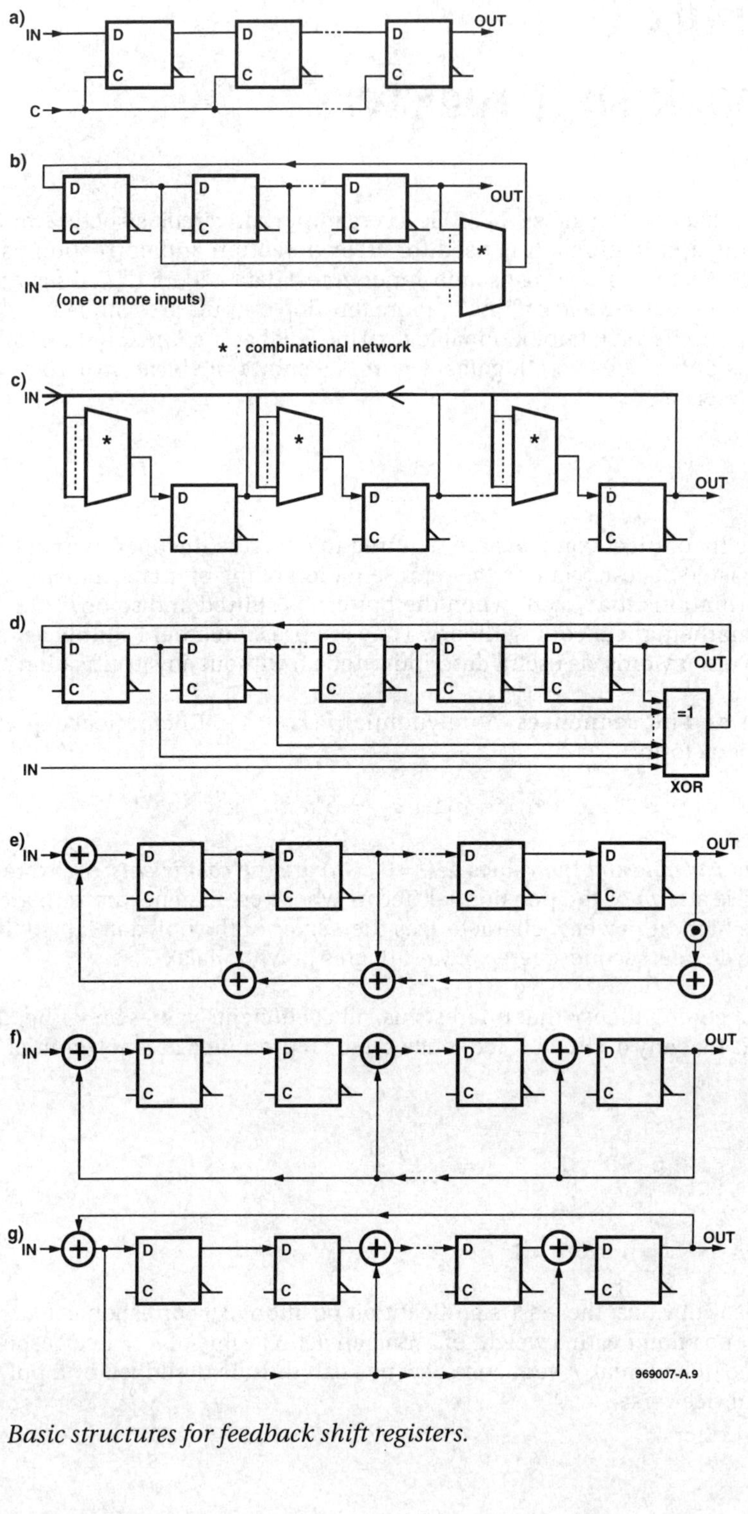

Figure A.9 Basic structures for feedback shift registers.

Two additional examples are:

$$1\ 1\ 1\ 0\ 1 \Rightarrow x^4 + x^3 + x^2 + 1$$

$$1\ 0\ 0\ 1\ 0\ 1\ 1\ 0 \Rightarrow x^7 + x^4 + x^2 + x$$

Modulo-2 arithmetic. Polynomials can be added to each other, subtracted from each other, multiplied by each other and divided by each other. Modulo-2 arithmetic forms the basis for these mathematical operations in the theory of cyclic codes and linear feedback shift registers. This may sound complicated, but in practice it is quite simple: the bit sequence representation replaces the polynomial representation and the basic mathematical operation is the bitwise XOR function. *Modulo-2* means that there is no carry between the individual bit positions. The equivalent of modulo-2 addition (or subtraction) in digital logic circuitry is the XOR gate. Figure A.10 shows some sample calculations.

The shift operation in polynomial representation. Bit shifting is a clock-driven process. If a certain input value x is applied to a shift register without feedback, then the first flip-flop takes on the value x on the first clock, the second flip-flop takes on the value x on the second clock and so on, so that the value x appears at each successive stage of the shift register with a delay of one clock period. It is illuminating to use the delay operator D to describe the clock delay. The content of the first flip-flop then corresponds to $D(x)$, the content of the second to $D(D(x))$ and so on. In other words, shifting by one clock is equivalent to multiplication by D. Shifting by two clocks can be represented by $D \cdot D = D^2$, shifting by three clocks by D^3 and so on. The next step is to simply use the polynomial representation (x, x^2, x^3 ...) in place of D, D^2, D^3 .... With respect to a specific clock interval (time) t, the final term of the polynomial (the '1') refers the input value of the first flip-flop at the time t, the x^2 term refers to the input value of the first flip-flop at the time $t - 1$ (that is, the previous clock interval), the x^3 term refers to the input value of the first flip-flop at the time $t - 2$ (that is, two clock intervals before t) and so on.

The shifting process in a linear feedback shift register. Consider the shift registers shown in Figure A.9d and A.9e. These are actually equivalent structures: in Figure A.9d the feedback logic is contained in a single circuit element (which could be a parity generator, as in Figure 5.113), while in Figure A.9e it is arranged as a chain of two-input XOR gates. Using

Explanation of Figure A.9:

a) The basic structure of a shift register. A number of edge-triggered flip-flops are connected in series (IN = input, OUT = output and C = clock). The shift direction is from the left to the right. In the remaining circuits, a common clock is assumed and is not shown.

b) The general form of a feedback shift register. At least one of the flip-flop outputs is connected via an arbitrary combinational-logic network to the input of the first flip-flop.

c) A very general form of feedback shift register. The value at each flip-flop input depends on the output of the prior flip-flop, the outputs of any number of other flip-flops and external inputs. The actual values are determined by combinational-logic networks.

d–g) Various types of linear feedback shift registers. The distinguishing feature is that the feedback paths contain only XOR gates or direct connections.

a) Addition $(x^5 + x^4 + x + 1) + (x^6 + x^4 + x^3 + 1) = x^6 + x^5 + x^3 + x^1$

```
   110011
+ 1011001
  1101010
```

'Modulo 2' subtraction is performed in exactly the same way as addition

b) Multiplication $(x^5 + x^4 + x + 1) \cdot (x^2 + x + 1) = x^7 + x^4 + x^3 + 1$

```
110011 · 111

110011
 110011
  110011
10011001
```

The same result can be achieved by fully expanding the polynomial and keeping only the terms that are present an odd number of times:

$x^7 + \cancel{x^6} + \cancel{x^5} + \cancel{x^6} + \cancel{x^5} + x^4 + x^3 + \cancel{x^2} + \cancel{x} + \cancel{x^2} + \cancel{x} + 1 \triangleq x^7 + x^4 + x^3 + 1$

(The crossed-out terms are present an even number of times.)

The basic operation of modulo-2 arithmetic is XOR addition:

0 + 0 = 0
0 + 1 = 1
1 + 0 = 1
1 + 1 = 0

c) Division $(x^7 + x^4 + x^3 + 1) : (x^2 + x + 1) = x^5 + x^4 + x + 1$ (as expected; see b)

```
    10011001 : 111 = 110011
                     1.2.3.4.5.6.
1.  111
    0111
2.   111
3.   0001
4.     010
5.      100
6.      111
         111
```

Note: As a consequence of modulo-2 arithmetic, the quotient bit is 1 whenever the current portion of the dividend has a 1 in the same position as the most-significant 1 of the divisor.

```
obviously:  ···111···  Dividend  ⇒  quotient bit = 1
               111     Divisor
and also:   ···100···            ⇒  quotient bit = 1
               111
but note:   ···011···            ⇒  quotient bit = 0
               111
```

969007-A.10

Figure A.10 Sample calculations in modulo-2 arithmetic.

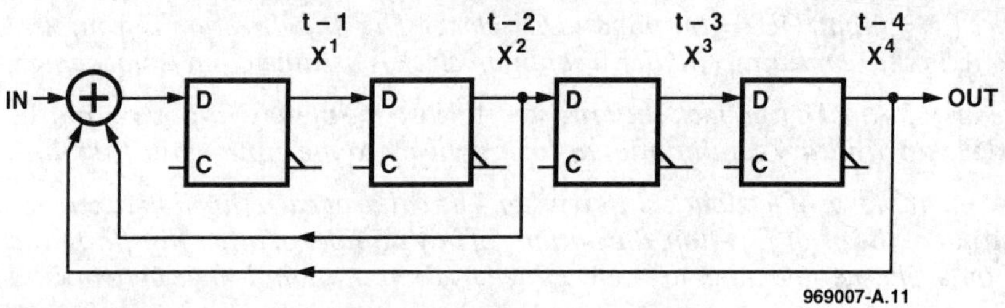

Figure A.11 A shift register with feedback.

the sample circuit of Figure A.11, it can be shown that the feedback structure and the polynomial representation are directly equivalent to each other. Since the XOR operation corresponds to modulo-2 addition, the value applied to the input of the first flip-flop at a given time (clock interval) t can be calculated as follows:

- the value at the circuit input IN (*represented as* **1**)

  $\oplus$ (XOR)

- the value at the output of the second flip-flop (the value that was present at the input two clock intervals earlier ($t-2$) (*represented as* x^2)

  $\oplus$ (XOR)

- the value at the output of the fourth flip-flop (the value that was present at the input four clock intervals earlier ($t-4$) (*represented as* x^4)

The feedback polynomial is thus $\mathbf{x^4 + x^2 + 1}$.

Applications

Maximum-length bit sequences. A shift register as shown in Figure A.11 can also be used with no input signal (see Figure A.12). Clearly, nothing will change if the initial values are all zeros, since $0 \oplus 0 = 0$. However, every non-zero initial flip-flop value will cause the content of the shift register to change with every shift clock, due to the shifting process and the presence of feedback. A pulse train or bit sequence can thus be taken from any of the flip-flop outputs.

If the shifting process continues long enough, the initial set of bit values will recur at some time. The number of clocks intervals between repeated occurrences of the initial pattern is called the *cycle length*. Shift registers that generate a *maximum cycle length* are particularly interesting. The maximum cycle length of a shift register with n flip-flops is $2^n - 1$ (all possible bit patterns except all zeros, which must not occur). Bit sequences produced by maximum-cycle-length shift registers can be used as *pseudo-random* bit patterns.

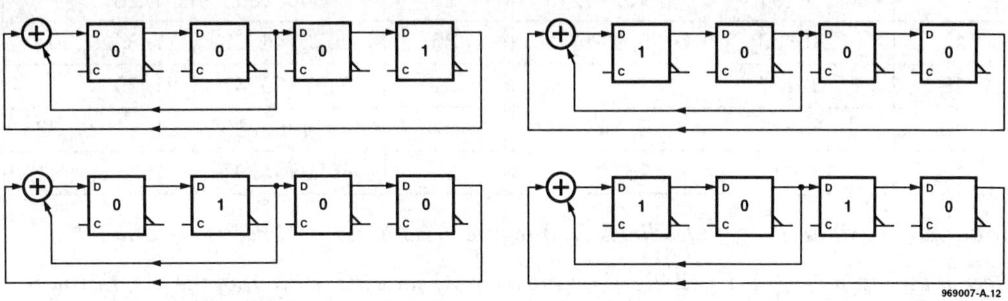

Figure A.12 A feedback shift register with no input signal. The initial state (0001) and the three subsequent states are shown.

Pseudo-random bit patterns have the following characteristics (among others):

- The number of 1's in the full cycle is one more than the number of 0's (so that both values occur with nearly equal frequency).
- A sequence of n 1's occurs exactly once in a full cycle, a sequence of $n/2$ 1's exactly twice and so on. Sequences of $(n-1)$ 0's, $n/2$ 0's and so on occur in the same manner.

Note: the shift register must be initialized with a specific bit pattern (a non-zero *seed value*) if it is to generate reproducible bit patterns.

Shift register structures that are suitable for producing maximum-length cycles are described in the technical literature. Table A.3 lists a modest selection.

Number of flip-flops	Cycle length	Feedback flip-flops [1]	Number of flip-flops	Cycle length	Feedback flip-flops [1]
2	3	1, 2	18	262143	1, 12
3	7	1, 3	19	524287	1, 15, 18, 19
4	15	1, 4	20	1048575	1, 18
5	31	1, 4	21	2097151	1, 20
6	63	1, 6	22	4194303	1, 22
7	127	1, 7	23	8388607	1, 19
8	255	1, 5, 6, 7	24	16777215	1, 18, 23, 24
9	511	1, 6	25	33554431	1, 23
10	1023	1, 8	26	67108863	1, 21, 25, 26
11	2047	1, 10	27	134217727	1, 23, 26, 27
12	4095	1, 7, 9, 12	28	268435455	1, 26
13	8191	1, 10, 11, 13	29	536870911	1, 28
14	16383	1, 5, 9, 14	30	1073741823	1, 8, 29, 30
15	32767	1, 15	31	2147483647	1, 29
16	65535	1, 5, 14, 16	32	4294967295	1, 11, 31, 32
17	131071	1, 15	33	8589934591	1, 21

***Table A.3** Maximum-length feedback shift registers. (source: IEEE Spectrum, June 1970)*

Note 1: the outputs of the listed flip-flops are coupled back to the input of the first flip-flop via an XOR gate (the flip-flops are numbered in sequence starting with '1'). The terms of the feedback polynomial can be easily derived from the indicated numbers (for instance, if flip-flops 1, 8, 29 and 30 are listed, the feedback polynomial is $x^{30} + x^{29} + x^8 + x + 1$*).*

Multiplication with a shift register. Multiplication of a polynomial X by a factor x^i is equivalent to shifting the corresponding bit sequence by i bits and then filling the i least-significant bit positions with zeros. A shift register with i flip-flops that is initially filled with all zeros can be used to perform this operation.

For example:

$$x^4 + x^3 + x^2 + 1 \Rightarrow 1\ 1\ 1\ 0\ 1$$

$$(x^4 + x^3 + x^2 + 1) \cdot x^2 = x^6 + x^5 + x^4 + x^4 \Rightarrow 1\ 1\ 1\ 0\ 1\ 0\ 0$$

Division with a shift register. The processes used to generate and check CRC codes (see Section 3.4.2) are examples of polynomial division. In these cases only the remainder of the division is important; the quotient is not used. (The remainder is either attached to the original data as the CRC term or checked to see whether it is all zeros.) Polynomial division with pencil and paper (as shown in Figure A.10c) can be directly implemented in a shift-register structure (see Figures A.13 and A.14).

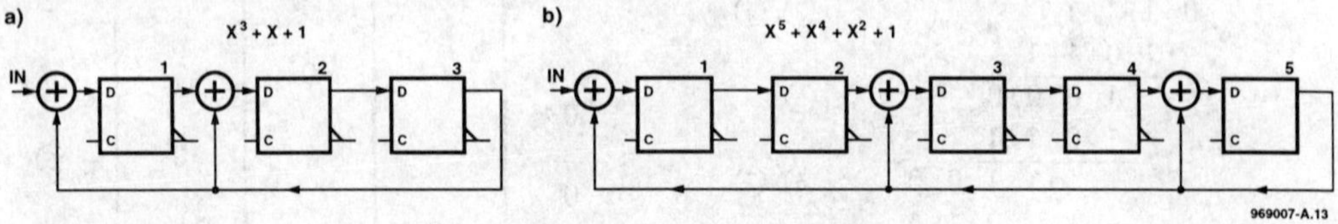

Figure A.13 Two examples of shift registers for generating and checking CRC bit patterns. The characteristic polynomial is shown for each shift register.

The objective is to calculate (polynomial A) ÷ (polynomial B), where polynomial B is the *generator polynomial* of the code or the *characteristic polynomial* of the shift register. A CRC code sequence that is n bits long requires an n-bit shift register that is described by an nth-degree generator polynomial. The corresponding bit sequence (that is used as the divisor) is $n + 1$ bits long.

Preparation: if the remainder of the division is to be generated (CRC generation), then n '0' bits must first be added to the end of the bit chain that corresponds to polynomial A, where n is the degree of the generator polynomial (the length of the CRC control bit sequence). If the remainder of the division is to be tested to see if it is all zeros (CRC checking), then the bit chain already consists of the data and CRC bits and it is not necessary to add any zero bits.

Hardware: the shift register consists of a number of flip-flops that are connected either directly or via XOR gates. The total number of flip-flops is equal to the degree n of the generator polynomial. If a term x^i is present in the generator polynomial, then an XOR gate is connected between the output of flip-flop i and the input of flip-flop $i + 1$, with the output of the most-significant flip-flop connected to the second input of the XOR gate. If the term

Data bit sequence: 1 1 0 0 1 0 1 1

Generator polynomial: $x^3 + x + 1$

CRC bit sequence: 1 0 1 1

CRC check bits: 3

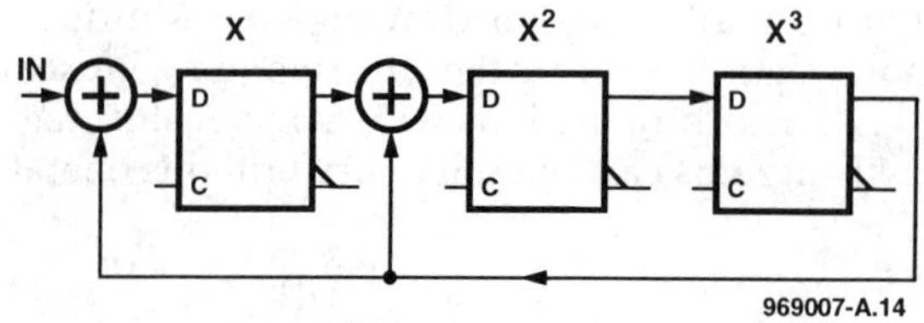

CRC generation:

```
1 1 0 0 1 0 1 1 0 0 0 * ÷ 1 0 1 1
1 0 1 1
  1 1 1 1                ①
  1 0 1 1
    1 0 0 0              ②
    1 0 1 1
      0 1 1 1 1          ③, ④
        1 0 1 1
          1 0 0 0        ⑤
          1 0 1 1
              1 1 0 0    ⑥, ⑦
              1 0 1 1
                1 1 1    ⑧
        remainder = CRC bits
```

Input	X	X²	X³	Step
	0	1	1	
0	1	1	1	①
1	0	0	1	②
0	1	1	0	③
1	1	1	1	④
1	0	0	1	⑤
0	1	1	0	⑥
0	0	1	1	⑦
0	1	1	1	⑧

CRC checking:

```
1 1 0 0 1 0 1 1 1 1 1 ** ÷ 1 0 1 1
1 0 1 1
  1 1 1 1                ①
  1 0 1 1
    1 0 0 0              ②
    1 0 1 1
      0 1 1 1 1          ③, ④
        1 0 1 1
          1 0 0 1        ⑤
          1 0 1 1
              1 0 1 1    ⑥, ⑦
              1 0 1 1
                0 0 0    ⑧
        remainder = 0 ⇒ OK
```

Input	X	X²	X³	Step
	0	1	1	
0	1	1	1	①
1	0	0	1	②
0	1	1	0	③
1	1	1	1	④
1	0	0	1	⑤
1	0	1	0	⑥
1	1	0	1	⑦
1	0	0	0	⑧

Figure A.14 Sample calculations for the characteristic polynomial of Figure A.13a.

x^n is present, then the output of the most-significant flip-flop must be XORed with the shift register input signal to produce the input to the first flip-flop.

Operation: The shift register is first loaded with all zeros. The bits of polynomial *A* are then shifted in, starting with the most significant bit. As long as the output of the most-significant flip-flop stays at zero, the XOR gates are all driven by '0' values and thus simply pass their input values to their outputs ($0 \oplus x = x$). When the most-significant '1' of polynomial *A* reaches the output of the most-significant flip-flop, the XOR gates become effective. Successive shift operations cause polynomial *A* to be 'broken down' in exactly the same manner as when it is divided using pencil and paper, so that the content of the shift register after the last bit of the polynomial has been shifted in is the desired remainder.

A disadvantage (for CRC generation) is that the remainder of the division is not immediately available after the last data bit has been shifted in (the least-significant bit of the original bit chain corresponding to polynomial *A*), due to the extra zero bits. One remedy is to multiply polynomial *A* by a factor x^i instead of adding on zeros, where *i* corresponds to the degree of the generator polynomial *B*. This multiplication has the same effect (adding on zeros), but it can be combined with the division in a single circuit by means of a very simple modification, as illustrated in Figure A.15 (compare this to Figure A.9g).

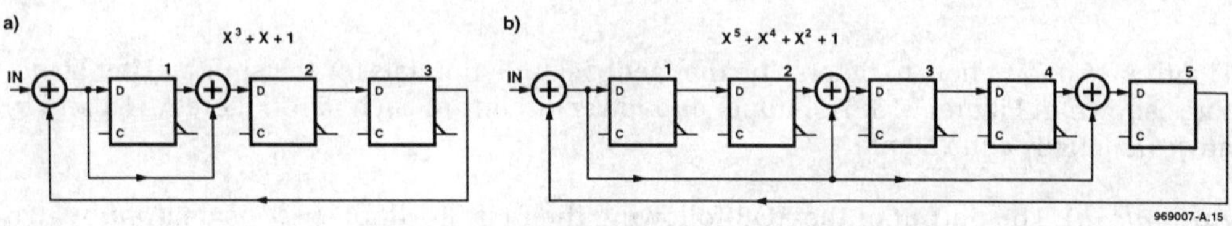

Figure A.15 *Shift registers for generating CRC codes.*
(a) basic circuit as shown in Figure A.13, (b) with 'built-in' multiplication.

Explanation of Figure 4.14:

- *The middle section of Figure 4.14 illustrates CRC generation and the bottom section illustrates CRC checking. (Note: * = zeros added on, ** = CRC bits added on.)*
- *The individual division steps are shown in the sample calculations together with the contents of the shift register. The quotient is not shown, since only the remainder is of interest.*
- *The representation of the content of the shift register always starts with the entry of the most-significant '1' into the first flip-flop. Each row shows the content of the shift register plus the next-following input bit (to the lower left). The immediately following row shows the shift register content that results from the shift operation:*

  $(x_{new} = x^3 \oplus IN, x^2_{new} = x \oplus x^3,\ x^3_{new} = x^2).$
- *The most significant bit is on the left for the written division, but in the shift register it is on the right!*

Various types of shift register structures (see Figures A.9d–f, A.11 and A.13). The shift register shown in Figure A.11 is described by a feedback polynomial, while those shown in Figure A.13 are described by characteristic polynomials. However, these two structures are mutually convertible!. A shift register with a given set of characteristics can thus be built as shown in Figure A.11 or as shown in Figure A.13.

Shift registers as shown in Figure A.11 are thus suitable for CRC applications, and shift registers as shown in Figure A.13 are suitable for generating pseudo-random bit sequences. (The advantage of the latter type is that it can run at higher clock rates, since at most a single 2-input XOR gate is connected between any two flip-flops.)

Conversion: the feedback polynomial is always the reverse of the characteristic polynomial, and vice versa. The reverse polynomial X_{-1} of the polynomial X can be obtained by reversing the bit sequence corresponding to polynomial X (the least significant bit becomes the most significant bit, etc.) and then representing the resulting bit sequence as a polynomial.

For example:

$$X = x^4 + x^3 + x^2 + 1 \Rightarrow 1\,1\,1\,0\,1$$

$$\text{reversal: } 1\,0\,1\,1\,1 \Rightarrow x^4 + x^2 + x + 1 = X_{-1}$$

Figure A.16 shows how to determine the feedback polynomials for the shift register structures shown in Figure A.13. The upper and lower circuits of each pair (Figure A.16a and b) are functionally equivalent.

Example (a): The output of the XOR following the first flip-flop $(x + 1)$ is delayed by two subsequent flip-flops (x^2). The polynomial is:

$$1 + (x + 1) \cdot x^2 = x^3 + x^2 + 1.$$

The same result can be obtained by taking the reverse polynomial of the characteristic polynomial (1 0 1 1 → 1 1 0 1).

Example (b): The output of the XOR following the second flip-flop $(x^2 + 1)$ is delayed by two subsequent flip-flops (x^2) and again XORed (thus $(x^2 + 1) \cdot x^2 + 1$). This signal is delayed by one flip-flop, so the last expression must be multiplied by x. The polynomial is:

$$1 + ((x^2 + 1) \cdot x^2 + 1) \cdot x = x^5 + x^3 + x + 1.$$

The same result can be obtained by taking the reverse polynomial of the characteristic polynomial (1 1 0 1 0 1 → 1 0 1 0 1 1).

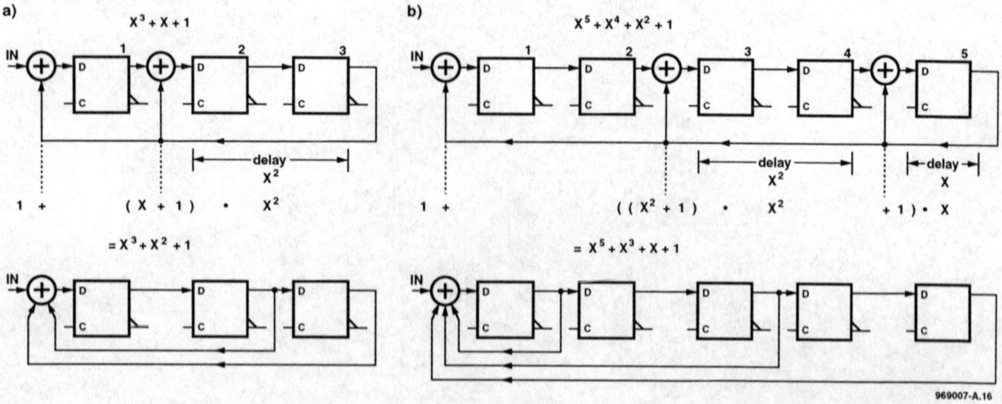

Figure A.16 Converting from the characteristic polynomial to the feedback polynomial.

Notes:

1 A shift register that is constructed according to its characteristic polynomial has a feedback equation with the form $\mathbf{1} + x^n$ (where **1** represents the input value and the x^n terms represent the feedback).

2 Every XOR operation can be represented in the form **(current flip-flop output) + 1**.

3 The x^n terms result from additional XOR operations in the shift path (delayed by any intervening flip-flops).

4 A one-flip-flop delay is equivalent to multiplication by x, a two-flip-flop delay is equivalent to multiplication by x^2 and so on.

Appendix 7
Obtaining on-line information

Information systems. The largest world-wide information system is the Internet. In addition, there are a several commercial providers, such as CompuServe and America Online (AOL), and numerous smaller information systems that are operated by individual companies or hobbyists. The trend is to have a 'presence' on the Internet, so that more and more firms are closing down their own bulletin-board systems and Compuserve forums in favour of the Internet.

Practical tip: have a look at what's on offer! There are more than enough 'starter kits' for new users. You can access the Internet via CompuServe, AOL and other commercial services, via OS/2 and Windows 95, and so on. 'Package deals' consisting of access software, a modem and a discount for your first session(s) are also available.

Accessing the Internet in practice. What is the point of all of these thick books about the Internet? After all, anyone can use the telephone without having to first digest the contents of massive 'how-to' tomes. You can regard the Internet – especially the World Wide Web (WWW) – as simply a sort of international telephone network with automatic call transferring and a graphical display.

What do you need?

- a PC with a suitably modern operating system (Windows 3.1 or better, OS/2 Version 2 or better); a reasonable video subsystem (at least 800 × 600 pixels resolution) and a CD-ROM drive (for installing the access software),
- network access software (a Web browser) – these days the suppliers practically beg to be allowed to give you a browser; you need only install it, and that is usually pleasantly simple;
- a telephone line and a modem;
- enough free space on the hard disk (and preferably also a removable-medium drive – Zip, Jaz, LS-120, Syquest or similar – so that you can make some room on the hard disk from time to time),
- a reasonable-quality printer (preferably a robust laser printer with a large paper tray and at least 300 dpi resolution).

'Paper-free' information? In theory yes, in practice no. While you can easily read short reports (located on Web pages) directly on the screen, perusing a DRAM 'data sheet' (for example) is something else. There are typically umpteen pages that usually have simply been scanned in from a printed book (with small type, multiple columns, vertically-oriented illustrations and so on) and converted into a PDF document.

File formats. Information files are typically available in the following formats:

- HTML (the 'house format' of the World Wide Web),
- PDF (the format used by the Acrobat Reader program),
- DOC (Word for Windows 6 document format),
- PostScript,
- Rich Text Format (RTF),
- Windows help file format (HLP).

Viewers (reader software). You can download the viewer programs 'Acrobat Reader' (Adobe) and 'Word Viewer' for Winword 6 (Microsoft) free of charge (see below).

Printing. You will have to experiment: sometimes 300 dpi is OK and sometimes you have to select 600 dpi; sometimes it is better to configure a given printer as an 'industry standard' (LaserJet) device for Acrobat Reader and as the type that it actually is for Netscape (but using only 'medium' printing quality), and so on.

Double-sided (duplex) printing. In principle, the most economical solution is to print on both sides of the paper. However, the necessary accessory hardware (duplex mechanism) itself costs money, and you have to use more expensive paper. (Certain types of paper are are expressly specified for double-sided laser printing. There is an increased risk of paper jams if ordinary paper is used for double-sided printing)

Paper. You will find that you consume a large amount of paper, once you have developed a taste for it.

Selected network addresses (web sites)

Semiconductor manufacturers:

- http://www.intel.com and http://developer.intel.com
- http://www.ti.com (Texas Instruments)
- http://www.micron.com (memory ICs and memory modules)
- http://www.cyrix.com
- http://www.amd.com
- http://www.mot.com (Motorola)

Note: nearly all Japanese vendors are represented on the WWW under their company names (*for example:* NEC.COM, HITACHI.COM and so on)

General computer technology and software:

- http://www.microsoft.com
- http://www.ibm.com
- http://www.compaq.com
- http://www.digital.com

Manufacturers of test and measurement equipment:

- http://www.tek.com (Tektronix)
- http://www.hp.com (Hewlett-Packard)
- http://www.fluke.com
- http://www.niti.com (Nicolet Technologies)
- http://www.lecroy.com
- http://www.keithley.com (look under *Instruments Division*)

Downloading viewer software:

- http://www.adobe.com (Acrobat Reader)
- http://www.microsoft.com

  Select 'Free Downloads' and then 'Web Browsers and Viewers'. The Winword 6 viewer (Wordview) can also be found here.

Search engines:

- http://www.yahoo.com
- http://www.altavista.digital.com

Appendix 8
Bibliography

References

- Fink and Beaty, *Standard Handbook for Electrical Engineers.*
  New York: The MacGraw-Hill Book Company.

  Includes tutorial articles on electrical engineering topics.

- Fink and Christiansen, *Electronics Engineers' Handbook.*
  New York: The MacGraw-Hill Book Company.

  Encyclopaedic.

- Scadden, *IEE Wiring Recommendations.*
  Stoneham, Massachusetts: Butterworth-Heinemann.

- Scadden, *Electric Wiring (Domestic).*
  Stoneham, Massachusetts: Butterworth-Heinemann.

- *Fuses for Electronics Applications.*
  Utrecht: Littlefuse Tracor bv.

Standard texts

- Barna and Poret, *Introduction to Digital Techniques.*
  New York: John Wiley and Sons.

  A good basic text.

- Higgins, *Electronics with Digital and Analog Integrated Circuits.*
  Englewood Cliffs, New Jersey: Prentice-Hall.

  An alternative to Horowitz and Hill.

- Horowitz and Hill, *The Art of Electronics.*
  Cambridge, England: Cambridge University Press.

  A standard text with many practical circuits and suggestions.

- Millman & Grabel, *Microelectronics.*
  New York: The MacGraw-Hill Book Company.

  An all-around text and reference.

- Millman and Taub, *Pulse, Switching and Digital Waveforms.*
  New York: The MacGraw-Hill Book Company.

  The standard work on the subject.

Techniques for testing digital circuits

- Abramovici, Breuer and Friedman, *Digital Systems — Testing and Testable Design.*
  Computer Science Press (W. H. Freeman & Co.).

  Thorough, academically-oriented treatment of contemporary techniques, with the emphasis on built-in test provisions.

- Lala, *Digital Circuit Testing and Testability.*
  London: Academic Press.

  Survey of contemporary techniques; more practically oriented than Breuer and Friedman.

Measurement practice

- Carr, *Elements of Electronic Instrumentation and Measurement.*
  New York: John Wiley and Sons.

  An introduction to standard workshop and laboratory instruments.

- Gilles, *Instrumentation and Measurement for Electronics Technicians.*
  New York: John Wiley and Sons.

  A popular introduction to measurement techniques; covers the fundamentals and includes seventeen lab experiments.

- Iddings, *Using the Oscilloscope.*
  New York: John Wiley and Sons.

  Designed to teach how to use the oscilloscope, including many advanced features.

- Witte, *Electronic Test Instruments: Theory and Application.*
  New York: John Wiley and Sons.

  Explains basic measurement theory for students with a basic knowledge of electronics, with practical applications.

Practical testing and fault-finding

- Cannon, *Understanding Digital Troubleshooting.*
  New York: John Wiley and Sons.

  A guide to the repair and maintenance of digital systems for the technician and non-engineer.

- Davidson, *Electronic Troubleshooting and Repair Handbook.*
  New York: The MacGraw-Hill Book Company.

  A detailed and entirely practically-oriented presentation of fault-finding in 'conventional' electronic circuitry. Many examples.

- Johnson and Graham, *High-Speed Digital Design — A Handbook of Black Magic.*
  Englewood Cliffs, New Jersey: Prentice-Hall.

  A highly readable, practically-oriented presentation of the subtleties and critical aspects of high-speed digital circuits — not just for designers!

- Matthes, *PC Service and Repair.*
  Dorset, England: Elektor Electronics (Publishing).

  Describes systematic fault-finding in standard PCs on the basis of typical fault symptoms and related fault-location and repair procedures. Also includes suggestions for measurement techniques.

- Pease, *Troubleshooting Analog Circuits.*
  Stoneham, Massachusetts: Butterworth-Heinemann.

  Almost indispensable for anyone who wishes to become familiar with the subtleties of modern analogue technology.

Index

I

J

K

L

M

Q

R

S

T